Probability and Statistical Inference

Probability and Statistical Inference

ROBERT BARTOSZYŃSKI

The Ohio State University

MAGDALENA NIEWIADOMSKA-BUGAJ

West Virginia University

A Wiley-Interscience Publication

JOHN WILEY & SONS, INC.

New York • Chichester • Brisbane • Toronto • Singapore

This book is dedicated to

Ela Pleszczyńska—our friend and teacher

Preface

This book is intended as a textbook for upper-level undergraduate and lower-level graduate students in departments of statistics, as well as in departments where statistics is taught as a tool to be used in research and application: engineering, economics, education, biological sciences, social sciences, business administration, medicine, to mention just a few.

The book, however, is not aimed at any particular specialty. It treats probability theory and statistics as self-contained theories, with their own conceptual structures and methods taken from mathematics. The value of these theories as a means of description and inference about real-life situations lies precisely in their level of abstraction. Indeed, it is a fact—perhaps insufficiently appreciated by those who regard application as more important than theory—that the more abstract a concept, the wider is its applicability.

A novelty of this book lies in the way the material is presented and in the aspects that are emphasized. To put it succinctly, we stress "why" over "how to." This may appear old fashioned, but what we suggest is that readers spend some time and effort on reflection, even at the cost of sacrifying some of the pragmatic values. But we think that today, in an era of undreamed-of computational facilities, reflection in an attempt to understand is not a luxury but a necessity. In our opinion students profit most if they study statistics as the science (and art) of inductive inference and of subsequent decisions based on this inference. In such a study, the methodology of statistics comes out most clearly if it is illustrated by a variety of real-life situations but is taught as an abstract system, not confined to a single domain of applications. Titles such as "Statistics for Social Scientists," "Statistics for Petroleum Engineers," and the like may appear attractive for some audiences, but they involve a degree of deception: Statistical methods are the same regardless of the domain of application.[1] Naturally, real-life examples provide students with some experience with applications, help in

---

[1] This is not to deny that certain statistical methodologies were created largely in response to needs of some applications, such as experimental design theory, developed primarily for best utilization of agricultural experiments (each lasting typically one season!)

developing imagination, and may give an idea of how to apply statistical methods to their own specialties or interests. However, to understand why statistical methods work, it is best to use examples that are to the largest possible extent free of all distracting content, so as to concentrate on the issue in question. The best examples of this kind are provided by coin tossing or throwing dice, and after a while become replaced by "Bernoulli trials with probability p of success."

At some point students must develop enough intuition and experience to understand the meaning and implications of a phrase such as "Let $X_1, \ldots, X_n$ be a random sample from the distribution . . . ." We hope that a few months into the course, to understand the text the student will not need to be given a sequence of numbers representing "pressures in radial tires serviced in the Shell station at the corner of Walnut and College" or "waiting times, in minutes, at a checkout counter in K-Mart." These two examples are taken from currently used textbooks, where they appear in introductory chapters, and students are asked to calculate $\bar{x}$ and s^2.

The numbers given are plausible, and perhaps they do indeed come from actual observations (although it is hard to imagine why anybody should care about that). What is disturbing, though, is that the students are not told whether the observations come from measurements of pressure in all four tires or just one (say, the front left) of each car, and, in the second example, whether the data give waiting times of consecutive customers. The issue is important since any dependence structure in the data affects the inference made on the basis of s^2.

These comments are not intended to suggest that we advocate going into such delicate issues in the beginning chapters of an elementary statistics course. But almost any real-life example poses some subtle questions, too difficult for beginners to handle. Neglecting these questions is a policy which often causes a disrespect for statistics, regarded merely as a skill in crunching numbers, whether it makes sense or not.

Recent tendencies in teaching statistics (as expressed, e.g., in *Amstat News*, December 1991) involve greater use of computers, especially to simulate variability and produce graphics, and also a turning away from mathematics. It is the last part which may in our opinion open up some frightening possibilities. It is, perhaps, justifiable to turn away from mathematics on a lower undergraduate level, where one may concentrate instead on developing some intuition and teaching some skills. But the justification for statistical methods and explanations of why they work is not—and to the maximal extent possible should not be—empirical in character. One must strive ultimately to base these methods on precise formal proofs, with statistics being subject to the same demands of rigor as the rest of mathematics.[2] The level of statistical service supplied by

---

[2] This does not mean that we want to disregard the justification of some properties of statistical procedures through "computer-intensive" methods. Certain problems are so complex that there is no hope for a theoretical solution any time soon; hence the use of "empirical arguments" is unavoidable.

"statistical technicians" (which the undergraduate courses are designed to produce) and the development of statistical theory both depend rather heavily on the level of understanding of the theory, not only on a knowledge of how it is applied.

From kindergarten on, students are taught that "learning is fun." And so it is, but "fun" is not synonymous with "easy." In fact, most joy of learning comes from overcoming difficulties and experiencing moments when some obscure truth becomes clear in a flash of understanding. Such flashes often seem to occur "out of nowhere," but in fact they must be preceded by hard labor. One can only wonder why in popular perception a person who starts his day at 5 a.m. with an hour of strenuous jogging is understood and respected, while someone who spends equally strenuous hours on trying to understand some subtle mathematical reasoning is often perceived as a nerd.

We do hope that advocating an understanding of mathematical subtleties in justification of statistical methods is not an entirely lost cause. Accordingly, we try to provide students with what we feel is a reasonable level of mathematical detail.

A contemporary trend in statistical textbooks is their reliance and frequent reference to computers (e.g., specific statistical packages such as Minitab, SAS or JMP). No doubt, such texts are useful and needed; nevertheless, their objectives are much too "pragmatic" (as opposed to "cognitive") to be reconciled with the objectives of our book. Thus, we do not attempt to teach students how to use computer packages. What we hope is that after a course taught from our book, a good student should be able to carry on various procedures by finding instructions for them in manuals to statistical packages.

The book provides the material for a one-semester course in probability theory, to be followed by a second semester on statistics. Actually, it contains more than that. Depending on the level of the course, the instructor may select the topics and examples, both as to theory and applications. The examples range from simple illustrations of the concepts, to introductions to entire theories (e.g., prediction, extrapolation, and filtration in time series as examples of the use of the concepts of covariance and correlation). Parts of the book that deal with material typically not included in probability courses on the levels specified above are marked with asterisks. These parts range in size from Chapter 5 on Markov chains or Chapter 14 on discrimination, to specific sections (e.g., on the Kolmogorov inequality) or smaller parts. There are also fragments of the text that deal with issues of purely mathematical character; these are meant for students who have some mathematical maturity and are often not satisfied with the level of explanation (or its lack) in most textbooks on statistics. Such parts are distinguished by an asterisk in the margin at the beginning and end of the section. Here we give, for instance, a proof of the extension theorem (Theorem 6.2.4), showing that the cumulative distribution function determines the measure on the line; we outline the construction of Lebesgue, Riemann–Stieltjes, and Lebesgue–Stieltjes integrals, and explain the differ-

other texts: for instance, as a reference or as a source of examples and problems.

The process of writing this book took about six years. The initial drafts were used as texts for courses taught at various universities by us or by our colleagues. In addition, many friends read earlier versions of various parts of the book. As a result, we were lucky to have feedback, which we used to continue to improve the text. We interrupted this process of improvement not because we think that we have reached perfection, but simply because as the text grew longer, the project started taking up too much of our time and energy—not to mention the fact that some of our colleagues became impatient.

It is appropriate that we express our gratitude to all those who have contributed to the present form of the book by pointing out mistakes, suggesting improvements, additions, simplifications, references, and so on. A full list is impossible to give, but the ones who contributed most and to whom we are especially grateful are Włodek Bryc, Doug Critchlow, Jason Hsu, Jacek Jakubowski, Marek Kimmel, Jacek Koronacki, Igo Kotlarski, Bill Notz, Boris Pittel, Ela Pleszczyńska, Bob Sulanke, and Jim Thompson as well as the anonymous reviewers who provided helpful comments and valuable suggestions.

We would also like to thank others who helped to make this book possible: Kate Roach, Kimi Sugeno and Jessica Downey of Wiley for friendly and efficient care during the publishing process; the parents of MNB for help in preparing the figures; and the children of MNB, Kasia, Agata, and Łukasz, for their patient understanding and admirable restraint during the years when this book was being written.

ROBERT BARTOSZYŃSKI

MAGDALENA NIEWIADOMSKA-BUGAJ

March 1996

CHAPTER 1

Experiments, Sample Spaces, and Events

1.1 INTUITIVE BACKGROUND

It may be said that the primary purpose of using statistical methods is to help in making better decisions under conditions of uncertainty. In many cases the consequences of making a decision today depend on what will happen in the future (or at least on that limited part of the world and of the future which is relevant to our decision). Therefore, statistical methods are aimed, to a large extent, at helping to *predict* the future.

Somewhat superficially, judging from the failures of weather forecasts, to more spectacular prediction failures, such as bankruptcies of large companies, stock market crashes, lost wars, and so on, it would appear that statistical methods do not perform too well. This judgment, however, is rather unfair, since—with the possible exception of weather forecasting—the examples above are at best only partially statistical predictions. Moreover, failures tend to be better remembered than successes. Whatever the case, it appears that statistical methods are at present, and are likely to remain indefinitely, our best and most reliable prediction tools.

To analyze a given fragment of reality relevant for the specific purpose at hand, one usually needs to collect some *data*. These may come from past experiences and observations or may result from some controlled processes, such as laboratory or field experiments. The data are then used to hypothesize about the laws (often called '*mechanisms*') which govern the fragment of reality that is of interest. In cases relevant for the present book, we are interested in laws expressed in probabilistic terms: They specify directly, or allow us to compute, the probabilities of some events in the future. Knowledge of these probabilities is, in most cases, the best one can do with regard to prediction and decisions.

Probability theory is a domain of pure mathematics, and as such, it has its own conceptual structure. To enable a variety of applications (typically comprising all areas of human endeavor, ranging from biological, medical,

Example 1.2.5. Let the experiment consist of recording the lifetime of a piece of equipment, say a light bulb. An outcome here is the time until the bulb burns out, so that an outcome typically will be represented by a number $t \geq 0$ ($t = 0$ if the bulb is not working at the start) and therefore S is the nonnegative part of the real axis. In practice, t is measured with some precision (in hours, days, etc.), so that one might take $S = \{0, 1, 2, \ldots\}$. Which of these choices is better depends on the type of subsequent analysis.

Example 1.2.6. Two persons enter a cafeteria and sit at a square table, with one chair on each of its sides. Suppose that we are interested in the event "they sit at a corner" (as opposed to sitting across from one another). To construct the sample space, one may let A and B denote the two persons and take as S the set of outcomes represented by 12 ideograms in Figure 1.1.

One could argue, however, that the sample space above is unnecessarily large. If we are interested only in the event "they sit at a corner," then there is no need to label the persons as A and B. Accordingly, the sample space S may be taken as the set of six outcomes depicted in Figure 1.2.

But even this sample space may be simplified. Indeed, one may use the rotational symmetry of the table and argue that once the first person sits at some chair (it does not matter which), then the sample space consists of the three chairs remaining for the second person (see Figure 1.3).   □

Sample spaces may be classified according to the number of points they contain. *Finite* sample spaces contain finitely many outcomes, elements of *countably infinite* sample spaces may be arranged into an infinite sequence,

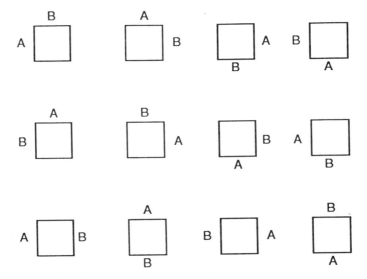

Figure 1.1 Possible seatings of two distinguishable persons at a square table.

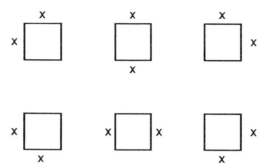

Figure 1.2 Possible seatings of two indistinguishable persons at a square table.

and others are called *uncountable* sample spaces. But even here the choice is not really dictated by the nature of the experiment, as shown by the following example.

Example 1.2.7. Consider a toss of a coin, one of the simplest experiments, most commonly used as an example in probability theory. Typically, the sample space is taken to consist of two outcomes, heads and tails, so that $S = \{H, T\}$. However, one can argue that the result here is determined by a number of parameters, such as the initial orientation, the velocity and angle of toss, the amount of spin, the hardness of the landing surface, and so on, and randomness lies in the selection of values of these parameters. Theoretically at least, all these parameters could be listed, so that it might be possible to describe the experiment of tossing a coin by taking S to be the set of all possible vectors of values of these parameters. Such a space may be awkward and difficult to handle, and we do not recommend its use; the point is, however, that both of the sample spaces above are equally acceptable from the formal viewpoint, and the first is preferred only for practical reasons.

$\square$

The next concept to be defined is that of an *event*. Intuitively, an event is anything about which we can tell whether or not it has occurred, after observing an outcome of the experiment. This leads to the following formal representation of events.

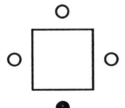

Figure 1.3 Possible seatings of one person if the place of the other person is fixed.

Definition 1.2.2.   An *event* is a subset of the sample space S.

Example 1.2.8.   In Example 1.2.1, concerning two throws of a die, an event such as "the sum equals 7" is a subset of the sample space S consisting of the six outcomes $(1, 6), (2, 5), (3, 4), (4, 3), (5, 2)$, and $(6, 1)$. In Example 1.2.3, the same event consists of one outcome, 7.   □

When an experiment is performed, we observe its outcome. In the interpretation developed in this chapter, this means that we observe a point chosen randomly from the sample space. If this point belongs to the set representing the event A, we say that *the event A occurs*.

We shall let events be denoted either by letters $A, B, C, \ldots$, possibly with identifiers, such as $A_1, B_k, \ldots$, or by more descriptive means, such as $\{X = 1\}, \{a < Z < b\}$, and so on; often we describe events through verbal phrases, such as "two heads in a row occur before the third tail." In these examples X and Z are some numerical attributes of the sample points (formally: random variables, discussed in Chapter 6), and the last example concerns the sample space describing the experiment of tossing a coin. Interpreting a verbal description of an event as a subset of the sample space (and vice versa) is a skill acquired by practice.

In all cases considered thus far, we assumed that an outcome (point of the sample space) can be observed. To put it more precisely: All sample spaces S considered so far were constructed in such a way that their points were observable. Thus, for any event A, we were always able to tell whether or not it occurred.

This approach is satisfactory as long as we are interested only in descriptive aspects of statistics, that is, in statistics as a collection of techniques of summarizing large sets of data so as to stress, or just make visible, some of features of the data. However, the data are usually collected with some purpose in mind, and in such cases it may be worthwhile to define the sample space in a way that will later facilitate solving the problem. This often involves assuming that sample points are only partially observable.

Example 1.2.9 (Selection).   Consider a simple case of selection of candidates on the basis of a test. Each candidate is characterized by his or her level of skills required for the job, say z. The actual value of z is not observable, though; what we observe is the candidate's score x on some test. Thus, the sample point in S is a pair $s = (z, x)$, and only one coordinate of s, namely x, is observable. The statistical analysis of this problem will be continued in Chapter 14.

The objective might be to find selection thresholds z_0 and x_0 such that the rule "accept all candidates whose score x exceeds x_0" would lead to maximizing the (unobservable) number of persons accepted whose true level of skill z exceeds z_0. Naturally, to find such a solution, one needs to know something about the statistical relation between observable x and unobservable z.

Example 1.2.10 (Discrimination). Another situation in which the elements of sample space may be pairs $s = (z, x)$ with unobservable z and observable x may be cases of statistical discrimination. Here z is again some unobservable attribute, except that (unlike the case of selection) it need not be ordered in any way. As an example one may think of archaeological studies in which there is a problem of determining the gender ($z = M$ or F) of a person on the basis of observing some fragments of his or her bones, found in a gravesite. Here x represents results of measurements of bones. Again, the problem lies in determining, with the smallest chance of error, the value of z on the basis of x.

Example 1.2.11 ("Touchy" Questions). Another kind of example when the points in the sample space are only partially observable concerns studies of the incidence of activities about which one may not wish to respond truthfully, or even respond at all. These are typically problems concerned with sexual habits or preferences, abortions, drug use, and so on. Here the idea is as follows. Let Q generally be the activity analyzed, and assume that the researcher is interested in the frequency of persons who (say) ever participated in activity Q (let us call them, for simplicity of formulation, Q-persons). It ought to be stressed that the objective is *not* to identify the Q-persons but only to find the frequency of such persons in the population.

The direct question, reducible to something like "Are you a Q-person?", is not likely to be answered truthfully, if at all. It is therefore necessary to make the respondent safe, guaranteeing that their responses will reveal nothing about them as regards Q. This can be accomplished as follows. The respondent is given a pair of dice, one green and one white. She throws both of them at the same time, in such a way that the experimenter does not know the results of the toss (say, dice are in a box and only the respondent looks into the box after it is shaken). The instruction is: If the green die shows an odd face (1, 3, or 5), then respond to the question "Are you a Q-person?". If the green die shows an even face (2, 4, or 6), then respond to the question "Does the white die show an ace?". The scheme of this response is summarized by the flowchart in Figure 1.4.

The interviewer knows the answer "yes" or "no" but does not know whether it is the answer to the question about Q or the question about the white die. Here a natural sample space consists of points $s = (i, x, y)$, where x and y are outcomes on the green and white die, respectively, while i is 1 or 0 depending on whether or not the respondent is a Q-person. We have $\phi(s) = \phi(i, x, y) =$ "yes" if $i = 1$ and $x = 1$, 3, or 5 for any y, or if $x = 2$, 4, 6 and $y = 1$ for any i. In all other cases $\phi(s) =$ "no."

One can wonder what could be a possible advantage, if any, of not knowing the question asked, and observing only the answer. This seems nonsensical if we need to know the truth about each respondent individually. However, one should remember that we are only after the overall frequency of Q-persons.

Now, we are in effect "contaminating" the question by making the re-

(vi) They arrive less than 15 minutes apart and they do not meet,

(vii) Mary arrives at 5:15 p.m. and meets John, who is already there,

(viii) They almost miss one another.

Problems 1.2.10 and 1.2.11 concern the possibility of expressing some events, depending on the choice of the sample space.

1.2.10 Let $\mathcal{E}$ be the experiment consisting of tossing a coin three times, with H and T standing for head and tail.

    (i) The following set of outcomes is an incomplete list of the points of the sample space S of the experiment $\mathcal{E}$: {HHH, HTT, TTT, HHT, TTH, HTH, THH}. Find the missing outcome.

    (ii) An alternative sample space S' for the same experiment $\mathcal{E}$ consists of the following four outcomes: no heads (0), one head (1), two heads (2), and three heads (3). Which of the following events can be described as subsets of S, but not as subsets of $S' = \{0, 1, 2, 3\}$?

$$A_1 = \text{more than two heads,}$$
$$A_2 = \text{head on the second toss,}$$
$$A_3 = \text{more tails than heads,}$$
$$A_4 = \text{at least one tail, with a head on the last toss,}$$
$$A_5 = \text{at least two faces the same,}$$
$$A_6 = \text{head and tail alternate.}$$

    (iii) Still another sample space S'' for the experiment $\mathcal{E}$ consists of the four outcomes $(0,0), (0,1), (1,0),$ and $(1,1)$, where the first coordinate is 1 if the first two tosses show the same face and zero otherwise, while the second coordinate is 1 if the last two tosses show the same face, and zero otherwise [e.g., if we observe HHT, the outcome is $(1,0)$]. List the outcomes of S that belong to the event $A = \{(1,1), (0,1)\}$ of S''.

    (iv) Which of the following events can be represented as subsets of S but cannot be represented as subsets of S''?

$$B_1 = \text{first and third toss show the same face,}$$
$$B_2 = \text{head on all tosses,}$$
$$B_3 = \text{all faces the same,}$$
$$B_4 = \text{each face appears at least once,}$$
$$B_5 = \text{more heads than tails.}$$

1.2.11 Let $\mathcal{E}$ be the experiment consisting of tossing a die twice. Let S be the sample space with sample points $(i, j), i, j = 1, 2, \ldots 6$, with i and

j being the numbers of dots that appear in the first and second toss, respectively.

(i) Let S' be the sample space for the experiment $\mathcal{E}$ consisting of all possible sums $i + j$, so that $S' = \{2, 3, \ldots, 12\}$. Which of the following events can be defined as subsets of S but not of S'?

$A_1 =$ one face odd, the other even,

$A_2 =$ both faces even,

$A_3 =$ faces different,

$A_4 =$ result on the first toss less than the result on the second,

$A_5 =$ product greater than 10,

$A_6 =$ product greater than 30.

(ii) Let S'' be the sample space for the experiment $\mathcal{E}$ consisting of all possible absolute values of the difference $|i - j|$, so that $S'' = \{0, 1, 2, 3, 4, 5\}$. Which of the following events can be defined as subsets of S but not of S''?

$B_1 =$ one face shows twice as many dots as the other,

$B_2 =$ faces the same,

$B_3 =$ one face shows six times as many dots as the other,

$B_4 =$ one face odd, the other even,

$B_5 =$ the ratio of the numbers of dots on the faces is different from 1.

1.2.12 Referring to Example 1.2.11, suppose that we modify it as follows. The respondent tosses a green die (with the outcome unknown to the interviewer). If the outcome is odd, he responds to the Q-question; otherwise, he responds to the question "Are you born in April?". Again the interviewer observes only the answer "yes" or "no." Apart from the obvious difference in frequency of the answer "yes" to the auxiliary question (one in twelve instead of one in six), are there any essential differences between this scheme and the scheme in Example 1.2.11? Explain your answer.

1.2.13 Consider the set of four elements A, B, C, D and the experiment $\mathcal{E}$ of "drawing two elements without replacement" (so that the element drawn is not returned).

(i) List all elementary outcomes in the sample space, if:

(a) We treat the description literally, so that outcomes are pairs (x, y), with x, y being two distinct letters from among A, B, C, D;

(b) We visualize the selection as a two-step procedure: First the set $\{A, B, C, D\}$ is ordered, and then two initial elements of the ordering constitute our selection (alternatively: all four elements are selected, but we can observe only the initial two choices).

(ii) Describe the event "the second element chosen is D" as a subset of S in cases (a) and (b).

(iii) The sample space in case (b) appears unnecessarily redundant. Can you think of the situation when it would be preferred to that in case (a)?

1.3 ALGEBRA OF EVENTS

Next, we introduce some concepts that will allow us to construct composite events out of simpler ones. We begin with the relations of *inclusion* and *equality*.

Definition 1.3.1. The event A is *contained* in the event B, or B *contains* A, if every sample point of A is also a sample point of B. Whenever this is true, we shall write $A \subset B$, or equivalently, $B \supset A$. □

An alternative terminology often used here is that A *implies* (or *entails*) B.

Definition 1.3.2. Two events A and B are said to be *equal*, $A = B$, if $A \subset B$ and $B \subset A$. □

It follows that two events are equal if they consist of exactly the same sample points.

Example 1.3.1. Consider two tosses of a coin and the corresponding sample space S consisting of four outcomes HH, HT, TH, and TT. Then the event $A =$ "heads on the first toss" $= \{$HH, HT$\}$ is contained in the event $B =$ "at least one head" $= \{$HH, HT, TH$\}$. The events "the results alternate" and "at least one head and one tail" imply one another, and hence are equal.

Definition 1.3.3. The set containing no elements is called the *empty set* and denoted by $\varnothing$. The event corresponding to $\varnothing$ is called *impossible*.

Example 1.3.2*. One may wonder whether it is correct to use the definite article in the definition above and speak of "*the* empty set," since it would appear that there may be many different empty sets. For instance, the set of all kings of the United States and the set of all real numbers x such that $x^2 + 1 = 0$ are both empty, but one consists of people and the other of numbers, so they cannot be equal. This is not so, however, as is shown by the following formal argument (to appreciate this argument, one needs some training in logic). Suppose that $\varnothing_1$ and $\varnothing_2$ are two empty sets. To prove that they are equal, one needs to prove that $\varnothing_1 \subset \varnothing_2$ and $\varnothing_2 \subset \varnothing_1$. Formally, the first inclusion is the implication: "If s belongs to $\varnothing_1$, then s belongs to $\varnothing_2$."

This implication is true, since its premise is false: There is no s that belongs to $\varnothing_1$. The same holds for the second implication, so that $\varnothing_1 = \varnothing_2$. $\square$

We shall now give the definitions of three principal operations on events: *complementation*, *union*, and *intersection*.

Definition 1.3.4. The set that contains all sample points that are not in the event A will be called the *complement* of A and denoted A^c, to be read as "not A."

Definition 1.3.5. The set that contains all sample points belonging either to A or to B (so possibly to both of them) is called the *union* of A and B and denoted $A \cup B$, to be read as "A or B."

Definition 1.3.6. The set that contains all sample points belonging to both A and B is called the *intersection* of A and B, and denoted $A \cap B$, to be read as "A and B." $\square$

An alternative notation for complement is A' or $\overline{A}$, while in case of intersection one often writes AB instead of $A \cap B$.

The operations introduced have the following interpretation in terms of the occurrence of events:

1. Event A^c occurs if event A does not occur.
2. Event $A \cup B$ occurs when either A or B or both events occur.
3. Event $A \cap B$ occurs when both A and B occur.

Example 1.3.3. Consider the experiment of tossing a coin three times, with the sample space consisting of outcomes described as HHH, HHT, and so on. Let A be the event "heads and tails alternate," and let B be "heads on the last toss." The event A^c occurs if either heads or tails occur at least twice in a row, so that $A^c = \{\text{HHH, HHT, THH, HTT, TTT, TTH}\}$, while B^c is "tails on the last toss," hence $B^c = \{\text{HHT, THT, HTT, TTT}\}$. The union $A \cup B$ is the event "either the results alternate, or we have heads on the last toss", hence $A \cup B = \{\text{HTH, THT, HHH, THH, TTH}\}$. Observe that while A has two outcomes and B has four outcomes, their union has only five outcomes, because the outcome HTH appears in both events. This common part is the intersection $A \cap B$. $\square$

Some formulas may be simplified by introducing the operation of the *difference* of two events,

$$A \setminus B = A \cap B^c.$$

The difference $A \setminus B$ contains all sample points that belong to A but not to B. Clearly, $S \setminus A$ is the same as A^c.

Example 1.3.4. In Example 1.3.3, the difference $B^c \setminus A$ has the verbal description "at least two identical outcomes in a row and tails on the last toss," which means the event {HHT, HTT, TTT}.   □

Next, we have the following important concept:

Definition 1.3.7. If $A \cap B = \varnothing$, then the events A and B are called *disjoint*, or *mutually exclusive*.

Example 1.3.5. Referring to Example 1.3.3, the following two events are disjoint: $C = $ "more heads than tails" and the intersection $A \cap B = $ "the results alternate, ending with tails."   □

Example 1.3.5 shows that to determine whether events are disjoint or not, it is not necessary to list the outcomes in both events and check whether there exist common outcomes. Apart from the fact that such a listing is not feasible in almost any case of practical importance (sample spaces are much too large), it is often simpler to employ some logical reasoning, for instance trying to argue that one of the events is contained in the complement of the other (i.e., if one of them occurs, the other does not). In the case above, if the results alternate ending with tails, then the outcome must be THT, so there are more tails than heads, and therefore C does not occur.

The definitions of union and intersection can be extended to the case of a finite and even infinite number of events (discussed in Section 1.4). Thus

$$A_1 \cup A_2 \cup \cdots \cup A_n \left(\text{denoted also } \bigcup_{i=1}^{n} A_i \right) \tag{1.1}$$

is the event that contains the sample points belonging to A_1 or A_2 or $\cdots$ or A_n. Consequently, (1.1) is the event "*at least one A_i (occurs).*"

Similarly,

$$A_1 \cap A_2 \cap \cdots \cap A_n \left(\text{denoted also } \bigcap_{i=1}^{n} A_i \right) \tag{1.2}$$

is the event that contains the sample points belonging to A_1 and A_2 and $\cdots$ and A_n. Consequently, the event (1.2) is "*all A_i's (occur).*"

Example 1.3.6. Suppose that n shots are fired at a target, and let A_i, $i = 1, 2, \ldots, n$, denote the event "the target is hit on the ith shot." Then the union $A_1 \cup \cdots \cup A_n$ is the event "the target is hit" (at least once). Its complement $(A_1 \cup \cdots \cup A_n)^c$ is the event "the target is missed" (on every shot), which is the same as the intersection $A_1^c \cap \cdots \cap A_n^c$.   □

* A perceptive reader may note that the unions $A_1 \cup \cdots \cup A_n$ and intersections $A_1 \cap \cdots \cap A_n$ do not require an extension of the definition of union and intersection for two sets. Indeed, we could consider unions such as

$$A_1 \cup (A_2 \cup (\cdots (A_{n-2} \cup (A_{n-1} \cup A_n)) \cdots)),$$

where only the union of two sets is formed in each set of parentheses. The property of associativity below shows that parentheses can be omitted, so that the expression $A_1 \cup \cdots \cup A_n$ is unambiguous. The same argument applies to intersections.

The operations on events defined in this section obey some laws. The most important ones are listed below.

Idempotence:
$$A \cup A = A, \quad A \cap A = A.$$

Double Complementation:
$$(A^c)^c = A.$$

Absorption:
$$A \cup B = B \text{ iff } A \cap B = A \text{ iff } A \subset B. \tag{1.3}$$

In particular,

$$A \cup \varnothing = A, \quad A \cup S = S, \quad A \cap \varnothing = \varnothing, \quad A \cap S = A,$$

which in view of (1.3) means that $\varnothing \subset A \subset S$.

Commutativity:
$$A \cup B = B \cup A,$$
$$A \cap B = B \cap A.$$

Associativity:
$$A \cup (B \cup C) = (A \cup B) \cup C,$$
$$A \cap (B \cap C) = (A \cap B) \cap C.$$

Distributivity:
$$A \cap (B \cup C) = (A \cap B) \cup (A \cap C),$$
$$A \cup (B \cap C) = (A \cup B) \cap (A \cup C).$$

De Morgan's Laws:
$$(A_1 \cup \cdots \cup A_n)^c = A_1^c \cap \cdots \cap A_n^c,$$
$$(A_1 \cap \cdots \cap A_n)^c = A_1^c \cup \cdots \cup A_n^c. \tag{1.4}$$

1.4 INFINITE OPERATIONS ON EVENTS

As already mentioned, the operations of union and intersection can be extended to infinitely many events. Let $A_1, A_2, \ldots$ be an infinite sequence of events. Then

$$A_1 \cup A_2 \cup \cdots = \bigcup_{i=1}^{\infty} A_i$$

and

$$A_1 \cap A_2 \cap \cdots = \bigcap_{i=1}^{\infty} A_i$$

are events "*at least one A_i occurs*" and "*all A_i's occur.*"

If at least one event A_i occurs, then there is one that occurs first. This remark leads to the following useful decomposition of a union of events into a union of *disjoint* events:

$$\bigcup_{i=1}^{\infty} A_i = A_1 \cup (A_1^c \cap A_2) \cup (A_1^c \cap A_2^c \cap A_3) \cup \cdots, \tag{1.6}$$

where $A_1^c \cap \cdots \cap A_{k-1}^c \cap A_k$ is the event "A_k is the first event in the sequence that occurs."

For an infinite sequence $A_1, A_2, \ldots$ one can define two events:

$$\limsup A_n = \bigcap_{k=1}^{\infty} \bigcup_{i=k}^{\infty} A_i \tag{1.7}$$

and

$$\liminf A_n = \bigcup_{k=1}^{\infty} \bigcap_{i=k}^{\infty} A_i, \tag{1.8}$$

being, respectively, the event that *infinitely many A_i's occur* and the event that *all except finitely many A_i's occur*. Indeed, the inner union in the event (1.7) is the event "at least one event A_i with $i \geq k$ will occur"; call this event B_k. The intersection over k means that the event B_k occurs for every k: No matter how large we take k, there will be at least one event A_i with $i \geq k$ that will occur. But this is possible only if infinitely many A_i's occur.

For the event $\liminf A_n$ the argument is similar. The inner intersection $A_k \cap A_{k+1} \cap \cdots = C_k$, say, occurs if all events A_i with index $i \geq k$ occur. The union $C_1 \cup C_2 \cup \cdots$ means that at least one of the events C_k will occur, and that means that all A_i, except possibly finitely many, will occur.

If all events (except possibly finitely many) occur, then infinitely many of them must occur, so that $\limsup A_n \supset \liminf A_n$. If $\limsup A_n \subset \liminf A_n$, then (see Definition 1.3.2) we say that the sequence $\{A_n\}$ *converges*, and we write $\lim A_n$ for the sets $\limsup A_n = \liminf A_n$.

The most important class of convergent sequences of events consists of *monotone* sequences, when $A_1 \subset A_2 \subset \cdots$ (increasing sequence) or when $A_1 \supset A_2 \supset \cdots$ (decreasing sequence). We have here the following theorem.

Theorem 1.4.1. *If the sequence $A_1, A_2, \ldots$ is increasing, then*

$$\lim A_n = \bigcup_{n=1}^{\infty} A_n,$$

and in case of a decreasing sequence, we have

$$\lim A_n = \bigcap_{n=1}^{\infty} A_n.$$

Proof. If the sequence is increasing, then the inner union in $\lim \sup A_n$ remains the same independently of k, namely $\bigcup_{i=1}^{\infty} A_i$, so that $\lim \sup A_n = \bigcup_{i=1}^{\infty} A_i$. On the other hand, the inner intersection in $\lim \inf A_n$ equals A_k, so that $\lim \inf A_n = \bigcup_{k=1}^{\infty} A_k$, which is the same as $\lim \sup A_n$, as was to be shown. A similar argument holds for decreasing sequences. $\quad\square$

The following two examples illustrate for the concept of convergence of events.

Example 1.4.1. Let $B(r)$ and $C(r)$ be the sets of points on the plane (x, y) satisfying the conditions $x^2 + y^2 < r^2$ and $x^2 + y^2 \leq r^2$, respectively. If $A_n = B(1 + 1/n)$, then $\{A_n\}$ is a decreasing sequence and therefore $\lim A_n = \bigcap_{n=1}^{\infty} B(1 + 1/n)$. Since $x^2 + y^2 < (1 + 1/n)^2$ for all n if and only if $x^2 + y^2 \leq 1$, we have $\bigcap_{n=1}^{\infty} B(1 + 1/n) = C(1)$. On the other hand, if $A_n = C(1 - 1/n)$, then $\{A_n\}$ is an increasing sequence, and $\lim A_n = \bigcup_{n=1}^{\infty} C(1 - 1/n) = B(1)$. We suggest that the reader provide a justification of the last equality.

Example 1.4.2. Let $A_n = B(1 + 1/n)$ for n odd and $A_n = B(\frac{1}{3} - 1/n)$ for n even. The sequence $\{A_n\}$ is now $B(2), B(\frac{1}{6}), B(\frac{3}{2}), B(\frac{1}{4}), \ldots$, so that it is not monotone. We have $\lim \sup A_n = C(1)$, since every point (x, y) with $x^2 + y^2 \leq 1$ belongs to infinitely many A_n. On the other hand, $\lim \inf A_n = B(\frac{1}{3})$. Indeed, if $x^2 + y^2 < \frac{1}{9}$, then $x^2 + y^2 < (\frac{1}{3} - 1/n)^2$ for all n large enough (and also $x^2 + y^2 < 1 + 1/n$ for all n). On the other hand, if $x^2 + y^2 \geq \frac{1}{3}$, then (x, y) does not belong to any A_n with even n.

Thus, $\lim \sup A_n \neq \lim \inf A_n$, and the sequence $\{A_n\}$ does not converge. $\quad\square$

Infinite operations on events will play a very important role in the development of the theory. This role will be connected mostly with determining the values of probabilities of some events through passing to limits.

To prepare the ground for the considerations of the following chapters, we motivate the definitions below as follows. In Chapter 2 we introduce probability as a number assigned to an event. Formally, therefore, we shall be considering numerical functions defined on events, that is, on subsets of the sample space S. As long as S is finite or countably infinite, one can take the class of all subsets of S as the domain of definition of probability. In the case of infinite but *not* countable S (e.g., when S is an interval, the real line, the plane, etc.) it may not be possible to define probability on the class of *all* subsets of S. The reason for this lies beyond the scope of this book. Here we shall merely start outlining the formal construction which will allow us to avoid the difficulty. Roughly, the solution will lie in a suitable restriction of the class of subsets of S, which will be taken as events (hence as the domain of definition of probability, to be given in Chapter 2).

We begin with the concept of *closure* under some operation. Let S be the sample space. We shall consider some operations performed on subsets of S such that the results of these operations will also be subsets of S. Complementation A^c, finite union $A_1 \cup \cdots \cup A_n$, infinite union $A_1 \cup A_2 \cup \cdots$, and limits of sequences $\lim A_n$ are a few examples of such operations.

Definition 1.4.1. We say that the class $\mathcal{A}$ of subsets of S is *closed* under a given operation if the sets resulting from performing this operation on elements of $\mathcal{A}$ are also elements of $\mathcal{A}$.

Example 1.4.3. Let S be the set of all integers $\{0, 1, 2, \ldots\}$ and let $\mathcal{A}$ consist of all subsets of S that are finite. Then $\mathcal{A}$ is closed under finite unions and all intersections, finite or not. Indeed, if $A_1, \cdots, A_n$ are finite sets, then $A = A_1 \cup \ldots \cup A_n$ is also finite. Similarly, if $A_1, A_2, \ldots$ are finite sets, then $\cap_i A_i \subset A_1$, hence $\cap_i A_i$ is also finite. However, $\mathcal{A}$ is *not closed* under complementation: If A is finite (hence $A \in \mathcal{A}$), then A^c is not finite, hence $A^c \notin \mathcal{A}$. On the other hand, if $\mathcal{A}$ is the class of all subsets of S that contain some fixed element, say 0, then $\mathcal{A}$ is closed under all intersections and unions, but is not closed under taking complements.

Example 1.4.4. Let S be the real line and let $\mathcal{A}$ be the class of all intervals closed on the right and open on the left [i.e., intervals of the form $(a, b] = \{x : a < x \leq b\}$]. Assume that we allow $b \leq a$, in which case $(a, b]$ is empty. Then $\mathcal{A}$ is closed under the operation of intersection with $(a, b] \cap (c, d] = (\max(a, c), \min(b, d)]$. $\qquad \square$

The following three concepts play a central role in the construction of probability theory.

Definition 1.4.2. A nonempty class $\mathcal{A}$ of subsets of S which is closed under complementation and all finite operations (i.e., finite union, finite intersection) is called a *field*. If $\mathcal{A}$ is closed under complementation and all countable operations, it is called a *σ-field*. Finally, if $\mathcal{A}$ is closed under mono-tone passage to the limit,[*] it is called a *monotone class*. □

Let us observe that Definition 1.4.2 can be formulated using fewer condi-tions, so in a more efficient way. For $\mathcal{A}$ to be a field, it suffices to require that if $A, B \in \mathcal{A}$, then $A^c \in \mathcal{A}$ and $A \cap B \in \mathcal{A}$ (or $A^c \in \mathcal{A}$ and $A \cup B \in \mathcal{A}$). Any of these two conditions implies (by induction and De Morgan's laws) the closure of $\mathcal{A}$ under all finite operations (including that of taking the differ-ence). Consequently, for $\mathcal{A}$ to be a $σ$-field, it suffices to require that whenever $A_1, A_2, \ldots \in \mathcal{A}$, then $A_i^c \in \mathcal{A}$ and $\bigcap_{i=1}^{\infty} A_i \in \mathcal{A}$ (or $A_i^c \in \mathcal{A}$ and $\bigcup_{n=1}^{\infty} A_i \in \mathcal{A}$); this follows again from De Morgan's laws.[†]

It is important to realize that closure under countable operations is stronger than closure under any finite operations. This means, among other things, that there exist classes of sets that are fields but not $σ$-fields. This is shown by the following example.

Example 1.4.5. Let $S = \{1, 2, 3, \ldots\}$ and let $\mathcal{A}$ be the class of all subsets A of S such that either A or A^c is finite. Then $\mathcal{A}$ is a field but not a $σ$-field. First, if $A \in \mathcal{A}$, then $A^c \in \mathcal{A}$, since the definition of $\mathcal{A}$ is symmetric with respect to complementation. Next, if A and B are both in $\mathcal{A}$, so is their union. Indeed, if A and B are both finite, then $A \cup B$ is finite and hence belongs to $\mathcal{A}$. On the other hand, if either A^c or B^c (or both) are finite, then $A \cup B$ has finite complement: To see this, observe that $(A \cup B)^c = A^c \cap B^c$, which is contained in A^c and also in B^c, and hence must be finite.

Thus, $\mathcal{A}$ is a field. However, $\mathcal{A}$ is not a $σ$-field. Let A_n be the set consisting only of the element n (i.e., $A_n = \{n\}$). Clearly, $A_n \in \mathcal{A}$. Take now $\bigcup_{n=1}^{\infty} A_{2n} = \{2, 4, 6, \ldots\}$. This is a countable union of sets in $\mathcal{A}$ that is not in $\mathcal{A}$ (since the set of all even numbers is not finite, nor does it have a finite complement). □

Typically, it is easy to determine that a class of sets is a field, while direct verification that it is a $σ$-field may be difficult. On the other hand, it may occasionally be easy to verify that a class of sets is a monotone class. Thus, the following theorem may sometimes be useful.

Theorem 1.4.2. *A $σ$-field is a monotone class. Conversely, a monotone field is a $σ$-field.*

Proof. To prove this theorem, assume first that $\mathcal{A}$ is a $σ$-field, and let $A_1, A_2, \ldots$ be a monotone sequence of elements of $\mathcal{A}$. If $A_1 \subset A_2 \subset \cdots$, then

---

[*] In view of the fact proved earlier that all monotone sequences converge, this condition means that (a) if $A_1 \subset A_2 \subset \cdots$ is an increasing sequence of sets in $\mathcal{A}$, then $\cup_i A_i \in \mathcal{A}$, and (b) if $A_1 \supset A_2 \supset \cdots$ is a decreasing sequence of sets in $\mathcal{A}$, then $\cap_i A_i \in \mathcal{A}$.
[†] For various relations between classes of sets defined through closure properties under opera-tions, see, for example, Chow and Teicher (1988) or Chung (1974).

  (iii) Answer the same questions if $A_1 \subset A_2 \subset \cdots \subset A_n$.

  (iv) Answer the same question if $A_1 = \cdots = A_n = \varnothing$.

  (v) Answer questions (i)–(iv) for a σ-field.

1.4.5 Let $S = \{0, 1, 2, \ldots\}$ be the set of all integers. For $A \subset S$, let $f_n(A)$ be the number of elements in the intersection $A \cap \{0, 1, \ldots, n\}$. Let $\mathcal{A}$ be the class of all sets A for which the limit

$$q(A) = \lim_{n \to \infty} \frac{f_n(A)}{n}$$

exists. Show that $\mathcal{A}$ is not a field. (*Hint:* Let $A_1 = \{1, 3, 5, \ldots\}$ and $A_2 = \{$all odd integers between 2^{2n} and 2^{2n+1} and all even integers between 2^{2n+1} and 2^{2n+2} for $n = 0, 1, \ldots\}$. Show that both A_1 and A_2 are in $\mathcal{A}$ but $A_1 \cap A_2 \notin \mathcal{A}$.)

1.4.6 Let $S = (-\infty, +\infty)$. Show that the class of all finite unions of intervals of the form $[a, b]$, (a, b), $[a, b)$, and $(a, b]$, with possibly infinite a or b (i.e., intervals of the form $[a, \infty)$, etc.) forms a field.

REFERENCES

Chow, Y.S. and Teicher, H., 1988. *Probability Theory: Independence, Interchangeability, Martingales.* 2nd ed. Springer Verlag, New York.

Chung, K.L., 1974. *A Course in Probability Theory.* 2nd ed. Academic Press, Orlando.

CHAPTER 2

Probability

2.1 INTUITIVE BACKGROUND

The concept of probability has been an object of debate among philosophers, logicians, mathematicians, statisticians, physicists, and psychologists for the last couple of centuries, and this debate is not likely to be over in the foreseeable future. As advocated by Bertrand Russell in his essay on scepticism, when experts disagree, the layperson would do best by refraining from forming a strong opinion. Accordingly, we shall not enter into the discussion about the nature of probability; rather, we shall start from the issues and principles that are commonly agreed upon.

Probability is a number associated with an event, intended to represent its "likelihood," "chance of occurring," "degree of certainty," and so on. The phrases above have to be explicated so as to obtain workable principles. This can be done in several ways, the most important being (1) the *frequency* (or *objective*) interpretation of probability; (2) the *classical* (sometimes called *logical*) interpretation of probability, and (3) the *subjective* or *personal* interpretation of probability.

2.2 FREQUENCY INTERPRETATION OF PROBABILITY

For the moment it will be enough to formulate the frequency principle underlying the first interpretation. This is the most common interpretation, according to which probability is the "long-run" relative frequency of an event. Before attempting to provide a more formal explication, let us observe that the idea connecting probability and frequency is (and had been for a long time) very well grounded in everyday intuition. For instance, it is known that ancient Romans loaded dice (such dice were occasionally found in their graves). This fact indicates that they were aware of the possibility of making profit by loading dice (i.e., by modifying the long-run frequencies of some outcomes).

In the twentieth century the intuition regarding the relationship between probabilities and frequencies is even more firmly established. For instance, the phrases "there is a 3% chance that an orange picked at random from this shipment will be rotten" and "the fraction of rotten oranges in this shipment is 3%" apppear almost synonymous. But on closer reflection one realizes that the first phrase refers to the probability of an event "a randomly selected orange will be rotten," while the second phrase refers to the population of oranges.

Sometimes this frequency intuition leads to amusing results, as illustrated by the following story.

Example 2.2.1. About 40 years ago, the TV and radio stations introduced probability statements in weather forecasts, such as "about a 30% chance of rain this morning (in such and such an area)." Initially, many viewers and listeners were puzzled and called the stations asking for explanation. Does "a 30% chance of rain" mean that it will rain over 30% of the viewing area and that the remaining 70% will be dry? Or does it mean that one should take the umbrella with probability 0.3, or what? The stations had a hard time explaining (incidentally, try to think what the proper explanation should be!), until someone invented an answer which everybody understood and accepted: "There are 10 meteorologists in our station; 3 of them say that it is going to rain, and 7 say that it is not."   □

The precise nature of relation between probability and frequency is very hard to formulate. But the usual explanation is as follows. Consider an experiment that can be repeated under identical conditions, potentially an infinite number of times. In each of these repetitions, some event, say A, may occur or not. Let $N(A)$ be the number of occurrences of A in the first N repetitions. The frequency principle states that the ratio $N(A)/N$ approximates the probability $P(A)$ of event A with the accuracy of the approximation increasing as N increases.

Let us observe that this principle serves as a basis for estimating probabilities of various events in the real world, especially those probabilities that might not be attainable by any other means (e.g., the probability of heads in tossing a biased coin).

We start this chapter by putting a formal framework (axiom system) on probability regarded as a *function on the class of all events*. That is, we impose some general conditions on a set of individual probabilities. This axiom system, due to Kolmogorov (1933), will be followed by the derivation of some of its immediate consequences. The latter will allow us to compute probabilities of some composite events given the probabilities of some other ("simpler") events.

One could argue that this is a "backward" approach and that one should first establish the techniques for calculating the probabilities of simpler events, and only then develop rules for probabilities of composite events. This is not so: Consequences of general conditions imposed on the sets of probabilities are conceptually simpler and easier to grasp than the rules of calculating individual probabilities. The situation here is somewhat similar to that in formal logic, where one starts from a seemingly more complex task of establishing the truth values of composite propositions given the truth values of constituent propositions.

2.3 AXIOMS OF PROBABILITY

Formally, probability is a function defined on the class of all events,[*] satisfying the following conditions (usually referred to as *axioms*):

Axiom 1 (Nonnegativity)

$$P(A) \geq 0 \text{ for every event } A.$$

Axiom 2 (Norming)
$$P(S) = 1.$$

Axiom 3 (Countable Additivity)

$$P(A_1 \cup A_2 \cup \cdots) = \sum_{i=1}^{\infty} P(A_i)$$

for every sequence of pairwise disjoint events $A_1, A_2, \ldots$ *(i.e., events such that* $A_i \cap A_j = \varnothing$ *for all* i, j *distinct).* □

If the sample space S is finite or countable, one can define a probability function P as follows. Let f be a nonnegative function defined on S, satisfying the condition $\sum_{s \in S} f(s) = 1$. Then P may be defined for all subsets of S by the condition $P(A) = \sum_{s \in A} f(s)$. One can easily check that P so defined satisfies all three axioms.

Indeed, $P(A) \geq 0$ because the function f is nonnegative. Next, $P(S) = \sum_{s \in S} f(s) = 1$, by assumption on f. Finally, let $A_1, A_2, \ldots$ be a sequence of disjoint subsets of S. Then

$$P(A_1) + P(A_2) + \cdots = \sum_{s \in A_1} f(s) + \sum_{s \in A_2} f(s) + \cdots$$

$$= \sum_{s \in A_1 \cup A_2 \cup \cdots} f(s) = P\left(\bigcup_{i=1}^{\infty} A_i\right).$$

[*] The nature of the class of all events will be (to a certain extent) explicated in Section 2.6. See also Section 1.4.

Passing from the first to the second line is allowed because $A_1, A_2, \ldots$ are disjoint, so each term appears only once. The sum in the second line is well defined (i.e., it does not depend on the order of summation, because the terms are nonnegative).

However, if S is not countable, one usually needs to replace summation by integration, $P(A) = \int_A f(s)\,ds$. This imposes some conditions on functions f and on the class of events A for which the definition above holds. We shall not consider those conditions here; the reader is referred to more advanced probability texts [e.g., Chung (1974)].

Example 2.3.1 (Geometric Probability).   One of the first examples of an uncountable sample space is associated with "the random choice of a point from a set." This phrase is usually taken to mean the following: A point is selected at random from a certain set S in a finite-dimensional space (line, plane, etc.), where S has finite measure (length, area, etc.), denoted generally by $|S|$. The choice is such that if A is a subset of S with measure $|A|$, then the probability of the chosen point falling into A is proportional to $|A|$. Identifying S with the sample space, we may then write $P(A) = |A|/|S|$.

For instance, suppose that in shooting at a circular target S, one is certain to score a hit, and that the point where one hits S is assumed to be chosen at random in the sense described above. What is the probability that the distance of the point of hit to the center is more than half the radius of the target? Looking at Figure 2.1, it is clear that the point of hit must lie somewhere in the shaded annulus A. Its area is $|A| = \pi R^2 - \pi(R/2)^2$, so that $P(A) = |A|/\pi R^2 = \frac{3}{4}$. Incidentally, the assumption under which this solution

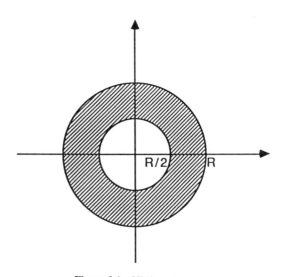

Figure 2.1 Hitting a target.

is obtained is not very realistic: Typically, sets closer to the center are more likely to be hit than sets of the same area located nearer the perimeter.

□

The concept of "random choice" from an uncountable set is sometimes ambiguous. This is illustrated by the following paradox, due to Bertrand.

Example 2.3.2. A chord is chosen at random in a circle. What is the probability that the length of the chord will exceed the length of the side of an equilateral triangle inscribed in the circle?

SOLUTION 1. Choose point A as one of the ends of the chord. The chord is determined uniquely by the angle α (see Figure 2.2). These angles are chosen at random from the interval $(0, \pi)$, and it is clear that the length of the chord exceeds the side of the equilateral triangle if α lies between $\pi/3$ and $2\pi/3$, so that the answer to the question is $\frac{1}{3}$.

SOLUTION 2. Let us draw a diameter QQ' (see Figure 2.3) perpendicular to the chord. Then the length of the chord exceeds the side of the equilateral triangle if it intersects the line QQ' between points B and B'. Elementary calculations give $|BB'| = |QQ'|/2$, which gives the desired probability to be $\frac{1}{2}$.

SOLUTION 3. The location of the chord is uniquely determined by the location of its center (except when the center coincides with the center of the circle, which is an event with probability zero). For the chord to be longer that the side of the equilateral triangle inscribed in the circle, its center must

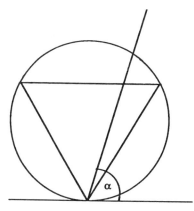

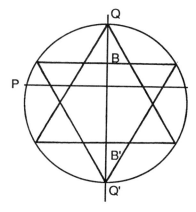

Figure 2.2 First solution of Bertrand's problem.

Figure 2.3 Second solution of Bertrand's problem.

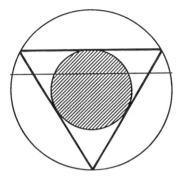

Figure 2.4 Third solution of Bertrand's problem.

fall somewhere inside the shaded circle in Figure 2.4. Again, by elementary calculations we see that the probability of the last event is $\frac{1}{4}$.   □

The discovery of Bertrand's paradox was one of the impulses that made researchers in probability and statistics acutely aware of the need to clarify the foundations of the theory, leading ultimately to the publication of Kolmogorov's book (1933). In the particular instance of the Bertrand "paradoxes," they are explained simply by the fact that each of the solutions refers to a different sampling scheme: (1) choosing a point on the circumference and then choosing the angle between the chord and the tangent at the point selected; (2) choosing a diameter perpendicular to the chord, and then selecting the point of intersection of the chord with this diameter; and (3) choosing a center of the chord. Random choice according to one of these schemes is not equivalent to a random choice according to the other two schemes.

To see why it is so, we shall show that the first and second schemes are not equivalent: The reader is asked to provide analogous arguments for the other two possible pairs of schemes.

Example 2.3.3.   Imagine the persons who constructed devices (e.g., physical mechanisms, computer programs, etc.) for sampling random chords. One scheme chooses a point on the circumference and then the angle α between the chord and the tangent to the circle at the point chosen (Figure 2.2). The second scheme chooses first the direction of the diameter and then the point B on the diameter at which the chord perpendicular to this diameter intersects it (Figure 2.3). From Figure 2.5 it is seen that the angle AOB is α, and therefore $y = |OB| = \cos \alpha$. Thus $dy = (\sin \alpha) \, d\alpha$, which means that equal changes of α do not produce equal changes of y. In fact, those changes are smaller when α is small. Consequently, a device that chooses angles α at random will tend to produce more intersections of the diameter which are farther from the center (i.e., more shorter chords).   □

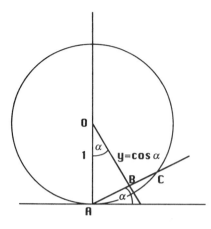

Figure 2.5   Explanation of Bertrand's paradox.

PROBLEMS

2.3.1   Show that the first and third, as well as the second and third schemes of sampling chords (see Bertrand's paradox) are not equivalent.

2.3.2   Use frequency interpretation to label all statements below as true or false.

   (i) If A is more likely to occur than A^c, then $P(A) > 0.5$.

   (ii) If A occurs whenever B does, then $P(B) \leq P(A)$.

   (iii) If $P(A) \leq P(B)$, then whenever B occurs, A does also.

   (iv) If $P(A) = 0.75$, then A must occur every three times out of four.

   (v) If A and B are disjoint, then the sum of their probabilities cannot exceed 1.

   (vi) If A and B are not disjoint, then the sum of their probabilities must exceed 1.

   (vii) If $P(A \cap B), P(A \cap C)$, and $P(B \cap C)$ are all positive, then $P(A \cap B \cap C)$ is also positive.

   (viii) If sample spaces for two experiments are identical, then the probability of the same event A must be the same for both experiments.

2.3.3   Your bathroom floor is covered by square tiles with side length a. You drop a coin with diameter b, where $b < a$, on the bathroom floor. Find:

   (i) The probability that the coin will rest entirely within one tile.

   (ii) The probability that the coin will partially cover four different tiles.

2.3.4 **(i)** A point (a, b) is selected at random from the square $-1 \leq a, b \leq 1$. Find the probability that the equation $ax^2 + bx + 1 = 0$ has two distinct real solutions.

(ii) Answer the same question if the point (a, b) is selected at random from the rectangle $A_1 \leq a \leq A_2, B_1 \leq b \leq B_2$. Discuss all possible choices of A_1, A_2, B_1, B_2.

2.4 SOME CONSEQUENCES OF THE AXIOMS

The simplest consequences of the axioms of probability are as follows.

1. *The probability of the impossible event is zero*:

$$P(\varnothing) = 0. \tag{2.1}$$

This follows from the fact that the sequence $S, \varnothing, \varnothing, \ldots$ satisfies conditions of Axiom 3, so that $1 = P(S) = P(S \cup \varnothing \cup \varnothing \cup \cdots) = P(S) + P(\varnothing) + P(\varnothing) + \cdots$, which is possible only if $P(\varnothing) = 0$. It is important to realize that the converse is not true: The condition $P(A) = 0$ does not imply that $A = \varnothing$. This is shown by the following example.

Example 2.4.1. Consider an experiment consisting of tossing a coin infinitely many times. The outcomes here may be represented as infinite sequences of the form HHTHTTHT..., so that the sample space S contains infinitely many such sequences. The event "heads only," that is, the set consisting of just one sequence HHHH ..., is not empty. However, the chance of such an outcome is, at least intuitively, zero: Tails should come up sooner or later. □

2. *Probability is finitely additive*. That is, for any $n = 1, 2, \ldots$,

$$P(A_1 \cup \cdots \cup A_n) = P(A_1) + \cdots + P(A_n)$$

if the events A_i are pairwise disjoint. This follows by taking an infinite sequence $A_1, \ldots, A_n, \varnothing, \varnothing, \ldots$. Clearly, these events are pairwise disjoint only if $A_1, \ldots, A_n$ are, so Axiom 3 applies. Therefore, using (2.1) we have

$$\begin{aligned} P(A_1 \cup \cdots \cup A_n \cup \varnothing \cup \cdots) &= P(A_1) + \cdots + P(A_n) + P(\varnothing) + \cdots \\ &= P(A_1) + \cdots + P(A_n), \end{aligned}$$

while the left-hand side is $P(A_1 \cup \cdots \cup A_n)$.

3. *Monotonicity*. If $A \subset B$, then $P(A) \leq P(B)$. This follows from the fact that $B = A \cup (B \cap A^c)$. The events on the right-hand side are disjoint, so we

have $P(B) = P(A) + P(B \cap A^c) \geq P(A)$ by Axiom 1. Since $B \cap A^c = B \setminus A$, the equality part gives a useful consequence: If $A \subset B$, then

$$P(B \setminus A) = P(B) - P(A). \tag{2.2}$$

Since $A \subset S$ for every event A, we have, using also Axiom 1, the inequality

$$0 \leq P(A) \leq 1.$$

4. *Probability is countably subadditive.* For every sequence of events $A_1, A_2, \ldots,$

$$P\left(\bigcup_{n=1}^{\infty} A_n\right) \leq P(A_1) + P(A_2) + \cdots. \tag{2.3}$$

This follows from representation (1.6) of a union as a union of disjoint events, and then from monotonocity. Indeed, we have

$$P\left(\bigcup_{n=1}^{\infty} A_n\right) = P(A_1) + P(A_1^c \cap A_2) + P(A_1^c \cap A_2^c \cap A_3) + \cdots$$

$$\leq P(A_1) + P(A_2) + P(A_3) + \cdots.$$

5. *Complementation*

$$P(A^c) = 1 - P(A). \tag{2.4}$$

This follows from Axiom 2 using the fact that A and A^c are disjoint and $A \cup A^c = S$.

6. *Probability of a union of events:*

$$P(A \cup B) = P(A) + P(B) - P(A \cap B). \tag{2.5}$$

Indeed, we may write $A \cup B = A \cup (A^c \cap B)$, so that $P(A \cup B) = P(A) + P(A^c \cap B)$. On the other hand, $B = (A \cap B) \cup (A^c \cap B)$, hence $P(B) =$

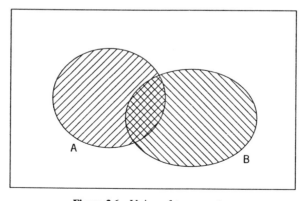

Figure 2.6 Union of two events.

$P(A \cap B) + P(A^c \cap B)$. Solving for $P(A^c \cap B)$ in one equation and substituting into the other, we obtain (2.5).

A more intuitive argument may be made by using Venn diagrams (see Figure 2.6). The sum $P(A) + P(B)$ counts each sample point in the intersection $A \cap B$ twice, so to obtain the probability of the union $A \cup B$ we must subtract the probability $P(A \cap B)$.

Example 2.4.2.   Suppose that $P(A) = 0.4, P(B^c) = 0.6, P(A^c \cap B) = 0.1$. Find $P(A \cup B^c)$.

SOLUTION.   The best strategy for solving this kind of problem is usually to start by finding the probabilities of all the intersections (in this case, $A \cap B, A \cap B^c, A^c \cap B, A^c \cap B^c$). Probability $P(A^c \cap B) = 0.1$ is given. Next, $(A \cap B) \cup (A^c \cap B) = B$, and the events on the left are disjoint, so that $P(A \cap B) + P(A^c \cap B) = P(B) = 1 - P(B^c)$, which means that $P(A \cap B) + 0.1 = 1 - 0.6$, hence $P(A \cap B) = 0.3$. Next, $A = (A \cap B) \cup (A \cap B^c)$, which gives $0.4 = P(A) = P(A \cap B) + P(A \cap B^c) = 0.3 + P(A \cap B^c)$, hence $P(A \cap B^c) = 0.1$. Finally, in the same way, we obtain $P(A^c \cap B^c) = 0.5$. Using formula (2.4), we have $P(A \cup B^c) = P(A) + P(B^c) - P(A \cap B^c) = 0.4 + 0.6 - 0.1 = 0.9$.

□

For the case of three events: A, B, and C, the same argument based on Venn diagrams gives the formula

$$
\begin{aligned}
P(A \cup B \cup C) = {} & P(A) + P(B) + P(C) \\
& - P(A \cap B) - P(A \cap C) - P(B \cap C) \\
& + P(A \cap B \cap C).
\end{aligned} \tag{2.6}
$$

Looking at Figure 2.7, we see that this formula counts every sample point in the union $A \cup B \cup C$ exactly once.

Formula (2.6) may be generalized to the case of the union of any finite number of events. We have the following formula for the probability of union of n events:

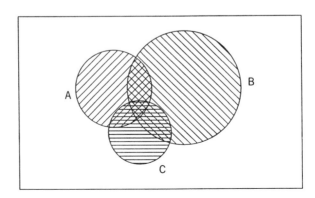

Figure 2.7   Union of three events.

Theorem 2.4.1. *For any events $A_1, \ldots, A_n$,*

$$P(A_1 \cup \cdots \cup A_n) = \sum_i P(A_i) - \sum_{i_1 < i_2} P(A_{i_1} \cap A_{i_2})$$

$$+ \sum_{i_1 < i_2 < i_3} P(A_{i_1} \cap A_{i_2} \cap A_{i_3}) + \cdots + (-1)^{n+1} P(A_1 \cap \cdots \cap A_n). \quad (2.7)$$

Proof. We shall proceed by induction. The theorem is true for $n = 1$. Assume now that formula (2.7) holds and let us write

$$P\left(\bigcup_1^{n+1} A_i\right) = P\left(\bigcup_1^n A_i\right) + P\left(A_{n+1} \setminus \bigcup_1^n A_i\right).$$

We have also

$$P(A_{n+1}) = P\left(A_{n+1} \cap \bigcup_1^n A_i\right) + P\left(A_{n+1} \setminus \bigcup_1^n A_i\right),$$

so that

$$P\left(\bigcup_1^{n+1} A_i\right) = P\left(\bigcup_1^n A_i\right) + P(A_{n+1}) - P\left[\bigcup_1^n (A_i \cap A_{n+1})]\right].$$

Applying formula (2.7) to $P(\bigcup_1^n A_i)$ and to $P\left[\bigcup_1^n (A_i \cap A_{n+1})\right]$, and combining the corresponding terms, we obtain (2.7) with n replaced by $n + 1$.

Example 2.4.3. Suppose that 101 events $A_1, \ldots, A_{101}$ are such that $P(A_1) = \cdots = P(A_{101}) = 0.01, P(A_1 \cap A_2) = P(A_1 \cap A_3) = \cdots = P(A_{100} \cap A_{101}) = r$, while every triple intersection is empty. What is the smallest possible value of the probability of intersection r?

SOLUTION.   Observe that $1 \geq P(A_1 \cup \cdots \cup A_{101}) = P(A_1) + \cdots + P(A_{101}) - P(A_1 \cap A_2) - \cdots - P(A_{100} \cap A_{101}) = 101/100 - r(100 \times 101)/2$, since the number of intersections $A_i \cap A_j$ with $i < j$ is $(100 \times 101)/2$ (to see this, one may arrange all pairs into a square table 101×101. The pairs with $i < j$ lie on one side of the diagonal, so their number is $(101^2 - 101)/2$). There are only two kinds of terms in the sum: Since triple intersections are empty, so are any higher intersections. Thus, we have $1 \geq 101/100 - r(101 \times 100)/2$, hence $r \geq 1/505,000$.   □

It may be seen from all the consequences of the axioms noted above that to handle unions and complements, we need to be able to handle probabilities of intersections. This problem requires introducing a new notion, that of conditional probability, and is discussed in Chapter 4.

30% have (WD), and 20% have all three. What is the probability that a family has at least one of these appliances?

2.4.14 An ordinary die and a die whose faces have 2, 3, 5, 6, 8, and 9 dots are tossed, and the total number of dots is noted. What is the probability that the sum is greater than or equal to 10?

2.4.15 A die is loaded in such a way that the probability of a face with j dots turning up is proportional to j, for $j = 1, 2, \ldots, 6$. What is the probability that in one roll of the die an odd number of dots will turn up?

2.5 CLASSICAL PROBABILITY

In this section we turn to the second of the three interpretations of probability mentioned, the *classical* or *logical* interpretation. Here we make the following assumption: *The sample space S contains a finite number N of outcomes, and all of these outcomes are equally probable.*

Obviously, in this case each of the outcomes has the same probability $1/N$, and for every event A we have

$$P(A) = \frac{\text{number of outcomes in } A}{N},$$

which in a more traditional way is written as

$$P(A) = \frac{\text{number of outcomes favoring } A}{\text{total number of outcomes}}. \tag{2.8}$$

This is not a definition of probability since it is based on the concept of equiprobable events, which has not been defined. However, in many real situations the outcomes in the sample space reveal a certain symmetry, derived from physical laws, logical considerations, or simply from the sampling scheme used. In such cases one may often *assume* that the outcomes are equiprobable and use (2.8) as a rule for computing probabilities. Obviously, the function P in (2.8) satisfies the axioms of probability.

To use some very simple examples, in tossing a die one usually assumes that each outcome has the same probability $\frac{1}{6}$. Then the probability of the event $A=$ "outcome odd" is $P(A) = \frac{3}{6} = \frac{1}{2}$, since there are 6 outcomes, 3 of them being odd.

The case above is rather trivial, but considerations of symmetry can sometimes lead to unexpectedly simple solutions of various problems.

Example 2.5.1. Peter tosses a fair coin n times, and Paul tosses it $n + 1$ times. What is the probability that Paul tosses more heads than Peter?

SOLUTION. Either Paul tosses more heads then Peter (event A) or he tosses more tails then Peter (event B). These two events exclude one another and exhaust all possibilities (since one cannot have ties in number of heads *and* number of tails). Switching the role of heads and tails transforms one of these events into the other. Thus sample space becomes partitioned into two equiprobable events, and we must have $P(A) = \frac{1}{2}$.   □

The solution above is worth a moment of reflection in order to determine where it avoids a common trap of the following "reasoning": Either A occurs or A^c occurs. There are therefore two possibilities, of which only one favors A, so $P(A) = \frac{1}{2}$ [this "argument" shows that $P(A) = \frac{1}{2}$ for *every* event A!].

The use of (2.8) requires the development of techniques for counting the numbers of elements in some sets. These topics, known under the name *combinatorics*, are analyzed in Chapter 3.

PROBLEMS

2.5.1 A number X is chosen at random from the series 4, 9, 14, 19, ... and another number Y is chosen from the series 1, 5, 9, 13, ... Each series has 100 terms. Find $P(X = Y)$.

2.5.2 A coin is tossed 7 times. Assume that each of the $2^7 = 128$ possible outcomes (sequences such as HTTHHTH of length 7) is equally likely. Relate each outcome to a binary number by replacing H by 1 and T by 0 (e.g., THHHTTH is $0111001 = 57$). Find the probability that a number generated in this way lies between 64 and 95 (inclusive on both sides).

2.5.3 A die is tossed three times, with outcomes X_1, X_2, X_3. Assuming that all 216 possible outcomes (x_1, x_2, x_3) are equally likely, find:
   (i) $P(X_1 < X_2 < X_3)$.
   (ii) $P(X_1 > X_2 = X_3)$.
   (iii) $P[\max(X_1, X_2, X_3) = 3]$.
   (iv) $P[\min(X_1, X_2, X_3) = 2]$.

2.5.4 Use formula (2.7) to find the number of primes not exceeding 100. [*Hint:* Assume that you sample one of the numbers 1, 2, ...,100. Let A_i be the event "the number sampled is divisible by i." Determine $p = P(A_2 \cup A_3 \cup A_5 \cup A_7)$. Then the answer to the problem is $100(1 - p) + 3$ (why?).]

2.6 NECESSITY OF THE AXIOMS*

Looking at Axiom 3, one may wonder why we need it for the case of countable (and not just finite) sequences of events. Indeed, the necessity of Axioms 1–3, with only finite additivity in Axiom 3, can easily be justified simply by using probability to represent the limiting relative frequency of occurrences of events. Recalling the symbol $N(A)$ from Section 2.1 for the number of occurrences of the event A in the first N experiments, the nonnegativity axiom is simply a reflection of the fact that the count $N(A)$ cannot be negative. The norming axiom reflects the fact that event S is certain and must occur in every experiment, so that $N(S) = N$, hence $N(S)/N = 1$. Finally (taking the case of two disjoint events A and B), we have $N(A \cup B) = N(A) + N(B)$, since whenever A occurs, B does not, and conversely. Thus, if probability is to reflect the limiting relative frequency, then $P(A \cup B)$ should be equal to $P(A) + P(B)$, since the frequencies satisfy the analogous condition $N(A \cup B)/N = N(A)/N + N(B)/N$.

The need for countable addivity, however, cannot be explained so simply. This need is related to the fact that to build a sufficiently powerful theory, one needs to take limits. If $A_1, A_2, \ldots$ is a monotone sequence of events (increasing or decreasing, that is, such that $A_1 \subset A_2 \subset \cdots$ or $A_1 \supset A_2 \cdots$), then $\lim P(A_n) = P(\lim A_n)$, where the event $\lim A_n$ has been defined in Section 1.4. Upon a little reflection, one can see that such continuity is a very natural requirement. In fact, the same requirement has been taken for granted for over 2000 years in a somewhat different context: In computing the area of a circle, one uses a sequence of polygons with an increasing number of sides, all inscribed in the circle. This leads to an increasing sequence of sets "converging" to the circle, and therefore the area of the circle is taken to be the limit of the areas of approximating polygons. The validity of this idea [i.e., the assumption of the continuity of the function $f(A) = $ area of A] was not really questioned until the beginning of the twentieth century. Research on the subject culminated with the results of Lebesgue.

To quote the relevant theorem, let us say that a function P, defined on a class of sets (events), is *continuous from below at the set A* if the conditions $A_1 \subset A_2 \subset \cdots$ and $A = \cup A_n$ imply that $\lim P(A_n) = P(A)$. Similarly, P is continuous *from above* at the set A if the conditions $A_1 \supset A_2 \supset \cdots$ and $A = \cap A_n$ imply that $\lim P(A_n) = P(A)$. A function that is continuous at every set from above or from below is simply called *continuous* (above or below). Continuity from below *and* from above is simply referred to as *continuity*.

We may characterize countable addivity as follows:

Theorem 2.6.1. *If the probability P satisfies Axiom 3 of countable addivity, then P is continuous from above and from below. Conversely, if a function P satisfies Axioms 1 and 2, is finitely additive, and is either continuous from below, or continuous from above at the empty set $\varnothing$, then P is countably additive.*

Proof. Assume that P satisfies Axiom 3 and let $A_1 \subset A_2 \subset \cdots$ be a monotone increasing sequence. We have

$$\bigcup_{n=1}^{\infty} A_n = A_1 \cup (A_2 \cap A_1^c) \cup (A_3 \cap A_2^c) \cup \cdots$$

$$= A_1 \cup (A_2 \setminus A_1) \cup (A_3 \setminus A_2) \cup \cdots, \qquad (2.9)$$

the events on the right-hand side being disjoint. Since $\cup A_n = \lim A_n$ (see Section 1.5), using (2.9) and the assumption of countable addivity,

$$P(\lim A_n) = P(\cup A_n) = P(A_1) + \sum_{i=2}^{\infty} [P(A_i) - P(A_{i-1})]$$

$$= P(A_1) + \lim_{n \to \infty} \sum_{i=2}^{n} [P(A_i) - P(A_{i-1})]$$

$$= P(A_1) + \lim_{n \to \infty} [P(A_n) - P(A_1)] = \lim_{n \to \infty} P(A_n)$$

(passing from the first to the second line we used the fact that the infinite series is defined as the limit of its partial sums). This proves continuity of P from below. To prove continuity from above, we pass to the complements, and proceed as above.

Let us now assume that P is finitely additive and continuous from below, and let $A_1, A_2, \ldots$ be a sequence of mutually disjoint events. Put $B_n = A_1 \cup \cdots \cup A_n$, so that $B_1 \subset B_2 \subset \cdots$ is a monotone increasing sequence with $\cup A_n = \cup B_n$. We have then, using continuity from below and finite addivity:

$$P(\cup A_n) = P(\cup B_n) = P(\lim B_n) = \lim P(B_n)$$

$$= \lim[P(A_1) + \cdots + P(A_n)] = \sum_{n=1}^{\infty} P(A_n),$$

again by definition of a numerical series being the limit of its partial sums. This shows that P is countably additive.

Finally, let us assume that P is finitely additive and continuous from above at the empty set $\varnothing$ (impossible event). Taking again a sequence of disjoint events $A_1, A_2, \ldots$, let $C_n = A_{n+1} \cup A_{n+2} \cup \cdots$. We have $C_1 \supset C_2 \supset \cdots$ and $\lim C_n = \bigcap_n C_n = \varnothing$. By finite additivity we obtain

$$P\left(\bigcup_{i=1}^{\infty} A_i\right) = P\left(\bigcup_{i=1}^{n} A_i \cup C_n\right) = P(A_1) + \cdots + P(A_n) + P(C_n). \qquad (2.10)$$

Since (2.10) holds for every n, we may write

$$P(\cup A_i) = \lim[P(A_1) + \cdots + P(A_n) + P(C_n)]$$

$$= \lim[P(A_1) + \cdots + P(A_n)] + \lim P(C_n)$$

$$= \sum_{n=1}^{\infty} P(A_i),$$

1. P^* satisfies the probability axioms.
2. $P^*(A) = P(A)$ if $A \in \mathcal{A}$.

A comment that is necessary here concerns the question: What does it mean that a function P defined on a field $\mathcal{A}$ satisfies the axioms of probability? Specifically, the problem concerns the axiom of countable additivity, which asserts that if events $A_1, A_2, \ldots$ are disjoint, then

$$P\left(\bigcup_{n=1}^{\infty} A_n\right) = \sum_{n=1}^{\infty} P(A_n). \tag{2.12}$$

However, if P is defined on a field, then there is no guarantee that the left-hand side of formula (2.12) makes sense, since $\bigcup_{n=1}^{\infty} A_n$ need not belong to the field of events on which P is defined. The meaning of the assumption of Theorem 2.6.3 is that formula (2.12) is true whenever the union $\bigcup_{n=1}^{\infty} A_n$ belongs to the field on which P is defined.

The role of Theorem 2.6.3 is simply that it "allows a statistician to sleep peacefully." He may need to find the probability of some complicated event A. The way of solving this problem is to represent A as a limit of some sequence of events whose probabilities can be computed, and then pass to the limit. Theorem 2.6.3 asserts that this procedure will give the same result, regardless of the choice of sequence of events approximating the event A.

Example 2.6.1 (Densities).   A very common situation in probability theory occurs when $S = (-\infty, +\infty)$. A probability measure P on S can be defined as follows. Let f be a function such that $f(x) \geq 0$ for all x and $\int_{-\infty}^{+\infty} f(x) \, dx = 1$. We shall assume in addition that f is continuous and bounded, although those conditions can be greatly relaxed in general theory.

We now *define* probability on S by putting

$$P(A) = \int_A f(x) \, dx \tag{2.13}$$

(in this case f is referred to as *density* of P). The full justification of this construction lies beyond the scope of this book, but the main points are as follows. First, the definition (2.13) is applicable for all intervals A of the form $(a, b), [a, b], (-\infty, b), (a, \infty), [a, \infty)$, and so on. Then we can extend P to finite unions of disjoint intervals by additivity (the class of all such finite unions forms a field). One may check easily that such an extension is unique; that is,

$$P[(a, b)] = \int_a^b f(x) \, dx = \sum_j \int_{I_j} f(x) \, dx$$

does not depend on the way interval (a, b) is partitioned into the finite union of nonoverlapping intervals I_j. This provides an extension of P to the smallest field of sets containing all intervals. If we show that P defined this way is continuous on the empty set, then we can claim that there exists an extension of P to the smallest σ-field of sets containing all intervals.

Now, the decreasing sequences of intervals converging to the empty set are built of two kinds of sequences: "shrinking open sets" and "escaping sets," exemplified by the following two sequences:

$$I_1 \supset I_2 \supset \cdots \text{ with } I_n = (a, a + \epsilon_n), \epsilon_1 > \epsilon_2 > \cdots \to 0$$

and

$$J_1 \supset J_2 \supset \cdots \text{ with } J_n = (a_n, \infty), a_1 < a_2, \ldots \to \infty.$$

We have $\lim I_n = \bigcap I_n = \varnothing$ and $\lim J_n = \bigcap J_n = \varnothing$. In the first case $P(I_n) = \int_a^{a+\epsilon_n} f(x)\, dx \leq \epsilon_n M \to 0$, where M is a bound for function f. In the second case, $P(I_n) = \int_{a_n}^{\infty} f(x)\, dx = 1 - \int_{-\infty}^{a_n} f(x)\, dx \to 1 - \int_{-\infty}^{+\infty} f(x)\, dx = 0$. $\qquad\square$

Standard problems involving material covered in this section are mostly of a theoretical character, so they will not be given here since they may be too difficult for students from departments other than mathematics and statistics.

2.7 SUBJECTIVE PROBABILITY*

Let us finally consider briefly the third interpretation of probability: namely, as a degree of certainty, or belief, about the occurrence of an event. Most often this probability is associated not so much with an event as with the truth of a proposition asserting the occurrence of this event.

The material of this section assumes some degree of familiarity with the concept of expectation, formally defined only in later chapters. For the sake of completeness, in the simple form needed here, this concept is defined below. In the presentation, we follow more or less the historical development, refining gradually the conceptual structures introduced. The basic concept here is that of a *lottery*, defined by an event, say A, and two objects, say a and b. Such a lottery, written simply aAb, will mean that the participant in the lottery receives object a if the event A occurs, and receives object b if the event A^c occurs.

The second concept is that of *expectation* associated with the lottery aAb, defined as

$$u(a)P(A) + u(b)P(A^c), \tag{2.14}$$

where $u(a)$ and $u(b)$ are measures of how much the objects a and b are "worth" to the participant. In the sequel, to simplify the language, we refer to the participant in the lottery as Mr. X or simply X. In the case when a and b are sums of money (or prices of objects a and b), and we put $u(x) = x$, the quantity (2.14) is sometimes called *expected value*, abbreviated *EV*. In cases when $u(a)$ and $u(b)$ are values that Mr. X attaches to a and b (at a given moment), these values do not necessarily coincide with prices. We then refer to $u(a)$ and $u(b)$ as *utilities* of a and b, and the quantity (2.14) is called *expected utility* (*EU*). Finally, when in the latter case the probability $P(A)$ is the subjective assessment of likelihood of the event A by Mr. X, the quantity (2.14) is called *subjective expected utility* (*SEU*).

It should be remarked that in any of the three cases above, one or both of the objects a and b may be a "ticket" to some other lottery, $a'Bb'$, say. In such a case, it is postulated that EV, EU, or SEU is computed according to the following rule: SEU (say) of the lottery $[a'Bb']Ab$ equals

$$[u(a')P(B) + u(b')P(B^c)]P(A) + u(b)P(A^c)$$

and similarly for lotteries such as $aA[a'Bb']$, $[aAb]B\{cC[dDe]\}$, and so on.

First, it has been shown by Ramsay (1926) that the degree of certainty about the occurrence of an event (of a given person) can be measured. Consider an event A and the following choice suggested to Mr. X (whose subjective probability we want to determine). X is given a choice between the following two options:

1. *"Sure"* option: receive some fixed amount $\$u$ [which is the same as lottery $(\$u)B(\$u)$, for any event B].
2. *Lottery* option: receive some fixed amount (say $\$100$) if A occurs, and receive nothing if A does not occur, that is, lottery $(\$100)A(\$0)$.

One should expect that if u is very small, X will probably prefer the lottery. On the other hand, if u is close to $\$100$, X may prefer the sure option.

There should therefore exist an amount u^* such that X will be indifferent between the sure option with u^* and the lottery option. Taking the amount of money as a representation of its value (or utility), the expected return from lottery, equals

$$0(1 - P(A)) + 100P(A) = 100P(A),$$

which, in turn, equals u^*. Consequently, we have $P(A) = u^*/100$. Obviously, under the stated assumption that utility of money is proportional to the dollar amount, the choice of $\$100$ is not relevant here, and the same value for $P(A)$ would be obtained if we chose another "base value" in the lottery option (this can be tested empirically). This scheme of measurement may provide us with an assessment of the values of the (subjective) probabilities of a given person, for a class of events.

It is of considerable interest that the same scheme was suggested in 1944 by von Neumann and Morgenstern (1944) as a tool for measuring utilities. They assumed that probabilities are known (i.e., the person whose utility is being assessed knows the objective probabilities of events, and his subjective probabilities coincide with the objective ones). If a person is now indifferent between the lottery as above, and the sure option of receiving an object, say q, then the utility $u(q)$ of object q must equal the expected value of the lottery, that is, a $100P(A)$. This allows us to measure utilities on the scale with zero set on nothing (status quo) and "unit" being the utility of \$100. The scheme of von Neumann and Morgenstern (1944) was later improved by some authors, culminating with the theorem of Blackwell and Girshick (1954).

Still, the disadvantages of both approaches was due to the fact that to determine utilities, one needed to assume knowledge of probabilities by the subject, while conversely, to determine subjective probabilities, one needed to assume knowledge of utilities. The discovery that one can determine *both* utilities and subjective probabilities of the same person is due to Savage (1954). We present here the basic idea of the experiment rather than formal axioms (which tend to obscure the issue by technicalities).

Let $A, B, C, \ldots$ denote events, and let $a, b, c, \ldots$ denote some objects, whose probabilities $P(A), P(B), \ldots$ and utilities $u(a), u(b), \ldots$ are to be determined (one should keep in mind that both P and u refer to a particular person X, the subject of the experiment). We now accept the main postulate of the theory: that out of two lotteries, X will prefer the one that has higher SEU.

Suppose that we find an event A with subjective probability $\frac{1}{2}$, so that $P(A) = P(A^c) = \frac{1}{2}$. If X prefers lottery aAb to lottery cAd, then

$$u(a)P(A) + u(b)P(A^c) > u(c)P(A) + u(d)P(A^c),$$

which means that

$$u(a) - u(c) > u(d) - u(b).$$

A number of experiments on selected objects will allow us to estimate the utilities, potentially with an arbitrary accuracy (taking two particular objects as zero and unit of the utility scale). In turn, if we know the utilities, we can determine the subjective probability of any event B: If X is indifferent between lotteries aBb and cBd, then we have

$$u(a)P(B) + u(b)(1 - P(B)) = u(c)P(B) + u(d)(1 - P(B)),$$

which gives

$$P(B) = \frac{u(d) - u(b)}{u(a) - u(b) + u(d) - u(c)}.$$

The only problem lies in finding an event A with subjective probability $\frac{1}{2}$. Empirically, an event A has subjective probability $\frac{1}{2}$, if for any objects a and b, the person is indifferent between lotteries aAb and bAa. Such an event was found experimentally (Davidson et al., 1957); it is related to a toss of a die with three of the faces marked with nonsense combination ZOJ and the other three with nonsense combination ZEJ (these were combinations that evoked least number of associations).

Let us remark at this point that the system of Savage involves determining first an event with probability $\frac{1}{2}$, then the utilities, and then the subjective probabilities. Luce and Krantz (1971) suggested an axiom system (leading to an appropriate scheme) that allows *simultaneous* determination of utilities and probabilities. The reader interested in these topics is referred to the monograph by Krantz et al. (1971).

A natural question arises here: Are Axioms 1–3 of probability theory satisfied here (at least in its finite version, without countable addivity)? On the one hand, this is an empirical question: The probabilities of various events may be determined numerically (for a given person), and then one may check whether the axioms hold. On the other hand, looking superficially one could conclude that there is no reason why Mr. X's probabilities should obey any axioms: After all, subjective probabilities that do not satisfy probability axioms are not logically inconsistent.

However, there is a reason why a person's subjective probabilities should satisfy the axioms. The reason lies in the fact that if any axiom is violated by the subjective probability of Mr. X (and Mr. X accepts the principle of SEU), then one can design a bet which appears favorable for Mr. X (hence a bet that he would accept), and yet, the bet being such that Mr. X is sure to lose.

Indeed, suppose first that the probability of some event A is negative. Consider the bet (lottery) $(-c)A(-b)$ (i.e., a lottery in which X pays the sum c if A occurs, and pays the sum b if A does not occur). We have here (identifying, for simplicity, the amounts of money with their utilities)

$$SEU = -cP(A) - bP(A^c),$$

so that SEU is positive for c large enough, if $P(A) < 0$. Thus, following the principle of maximizing SEU, Mr. X should accept this lottery over the status quo (no bet); but he will lose in any case, either the amount c or the amount b.

Suppose now that $P(S) < 1$. Consider the bet $(-c)Sb$, whose SEU is $-cP(S) + bP(S^c)$. Since $P(S^c) > 0$, making b large enough, the bet appears favorable for X, yet he is bound to lose the amount c on every trial.

If $P(S) > 1$ or if the additivity axiom is not satisfied, one can also design bets that will formally be favorable for Mr. X (SEU will be positive), and such that X will be bound to lose. Determination of these bets is left to the reader.

PROBLEMS

2.7.1   Peter and Tom attend the same college. One day Tom bought a ticket for a rock concert. Tickets were already sold out and were in great demand. Peter, who did not have a ticket, agreed to play the following game with Tom. For a fee of $25, Peter would toss a coin three times and receive the ticket if all tosses show up heads. Otherwise, for an additional fee of $50, Peter would toss a coin twice and receive the ticket if both tosses show up heads. Otherwise, for an additional fee of $100, Peter would toss a coin and receive the ticket if the toss shows up heads. Otherwise, all money would be lost to Tom, and he will also keep the ticket. Assuming that the coin is fair, subjective probabilities of various outcomes coincide with objective probabilities, and that Peter's utility is linear in money, show that Peter's utility of the ticket exceeds $200.

2.7.2   Refer to Problem 2.7.1. Tom would agree on the following conditions: Peter pays him $75 and tosses a coin, winning the ticket if it comes up heads, and otherwise losing $75. In such a situation, should they both agree that Peter buys the ticket from Tom for $180?

2.7.3   Suppose that Tom is confronted with the choice between two options: $O1$, which is simply to receive $1,000,000, or $O2$, which is to receive $5,000,000 with probability 0.1, receive $1,000,000 with probability 0.89, and receive $0 with the remaining probability 0.01. After some deliberation Tom decides that $O1$ is better, mostly because the outcome $0, unlikely as it may be, is very unattractive.

   Tom is also confronted with a choice between two other options, $O3$ and $O4$. In $O3$ he would receive $5,000,000 with probability 0.1 and $0 with probability 0.9. In $O4$ he would receive $1,000,000 with probability 0.11 and $0 with probability 0.89. Here Tom prefers $O3$: the "unattractive" option $0 has about the same probability in both $O3$ and $O4$, while the positive outcome, although slightly less probable under $O3$, is much more desirable in $O3$ that in $O4$.

   Show that these preferences of Tom are not compatible with the assumption that he has utilities A, B, and C of $5,000,000, $1,000,000, and $0, such that $A > B > C$. [This is known as Allais' paradox (Allais, 1953).]

REFERENCES

Allais, M., 1953. "Le Comportement de l'homme rationnel devant le risque: critique des postulats de l'école Americaine," *Econometrica*, **21**, 503–546.

Blackwell, D., and M.A. Girshick, 1954. *Theory of Games and Statistical Decisions*, Wiley, New York.

Chung, K. L., 1974. *A Course in Probability Theory*, Academic Press, New York.

Davidson, D., P. Suppes, and S. Siegel, 1957. *Decision Making: An Experimental Approach*, Stanford University Press, Palo Alto, Calif.

Kolmogorov, A. N., 1933. *Grundbegriffe der Wahrscheinlichkeitsrechnung*, Springer, Berlin (English translation by N. Morrison, *Foundations of the Theory of Probability*, Chelsea, New York, 1956).

Krantz, D. H., R. D. Luce, P. Suppes, and A. Tversky, 1971. *Foundations of Measurement*, vol. 1, Academic Press, New York.

Luce, R. D., and D. H. Krantz, 1971. "Conditional Expected Utility," *Econometrica*, **39**, 253–271.

Ramsay, F. P., 1926, "Truth and Probability" [reprinted in *Studies in Subjective Probability*, H. E. Kyberg and H. E. Smokler (eds.), Wiley, New York, 1964, pp. 61–92].

Savage, L. J., 1954. *The Foundations of Statistics*, Wiley, New York.

von Neumann, J., and O. Morgenstern, 1944. *Theory of Games and Economic Behavior*, Princeton University Press, Princeton, N.J.

CHAPTER 3

Counting

3.1 INTRODUCTION

In Chapter 2 we considered the principle of calculating probabilities, according to which (in the classical case) the probability of an event is the ratio of the number of outcomes that favor this event (imply its occurrence) and the total number of all possible outcomes. A practical implementation of this principle requires developing the techniques for counting the numbers of elements of certain sets (e.g., sets of all possible outcomes of an experiment).

As a rule, the sizes of the sets to be counted are such that simple enumeration of all elements is not feasible. This necessitates developing methods of "counting without counting." The branch of mathematics dealing with such methods is called *combinatorics*, or *combinatorial analysis*. In this chapter we introduce some combinatorial principles and illustrate their use in computing probabilities.

While we give here somewhat more than the typical material covered by textbooks on probability and statistics, a much more complete presentation of combinatorial methods and their applications to probability can be found in Feller (1968), a textbook thus far unsurpassed in its charm, depth, elegance, and diversity of applications.

3.2 PRODUCT SETS, ORDERINGS, AND PERMUTATIONS

We begin with a simple fact concerning counting. Consider two operations of some sort, which may be performed one after another. The notion of "operation" will be interpreted in many ways, so we will leave it vague at the moment.

We assume that:

1. The first operation can be performed in k_1 different ways.
2. For each of the ways of performing the first operation, the second operation can be performed in k_2 ways.

We have the following theorem.

Theorem 3.2.1. *Under assumptions 1 and 2, a two-step procedure consisting of the first operation followed by the second operation can be performed in $k_1 k_2$ distinct ways.*

Proof. Observe that each way of performing the two operations can be represented as a pair (a_i, b_{ij}) with $i = 1, \ldots, k_1$ and $j = 1, \ldots, k_2$, where a_i is the ith way of performing the first operation and b_{ij} is the jth way of performing the second operation if the first operation was performed in ith way. All such pairs can be arranged in a rectangular array with k_1 rows and k_2 columns. $\square$

We now show some applications of Theorem 3.2.1.

Example 3.2.1 (Cartesian Products). One of the most common operations on sets is the Cartesian product. If A and B are two sets, their Cartesian product $A \times B$ is defined as the set of all ordered pairs (a, b) where $a \in A$ and $b \in B$. For instance, if A consists of elements x and y, while B consists of the digits 1, 2, and 3, then the Cartesian product $A \times B = \{x, y\} \times \{1, 2, 3\}$ contains the six pairs

$$\{(x, 1), (x, 2), (x, 3), (y, 1), (y, 2), (y, 3)\}. \tag{3.1}$$

Observe that the Cartesian product $A \times B$ is an operation quite distinct from the usual set-theoretical product $A \cap B$. For instance, in the above case, $A \cap B = \varnothing$, since A and B have no elements in common. Also, while $A \cap B = B \cap A$, for Cartesian products we have in general $A \times B \neq B \times A$. Indeed, $B \times A$ contains, for example, the pair $(1, x)$, which does not belong to the set (3.1). Observe also that we may have $A = B$. The idempotence law $A \cap A = A$ *does not* hold for Cartesian products, since the set $A \times A$ consists of *pairs* (x, y) with $x, y \in A$, hence of elements of a different nature than elements of A, so that $A \times A \neq A$. In cases when there is no danger of confusion, we shall use the term *product* for *Cartesian product*. $\square$

Identifying the first and second operations with "choice of an element from set A" and "choice of an element from set B," we obtain the following consequence of Theorem 3.2.1:

Theorem 3.2.2 (Multiplication Rule). *If A_1 and A_2 are finite sets consisting, respectively, of k_1 and k_2 elements, then the Cartesian product $A_1 \times A_2$ consists of $k_1 k_2$ elements.* $\square$

Theorem 3.2.2 allows for an immediate generalization. We may define Cartesian products of more than two sets. Thus, if $A_1, \ldots, A_n$ are some sets, then their Cartesian product $A_1 \times A_2 \times \cdots \times A_n$ is the set of all n-tuples $(a_1, a_2, \ldots, a_n)$ with $a_i \in A_i, i = 1, \ldots, n$. By easy induction, Theorem 3.2.2 can now be generalized as follows.

Theorem 3.2.3. *If $A_1, \ldots, A_n$ are finite, with A_i consisting of k_i elements $(i = 1, \ldots, n)$, then $A_1 \times \cdots \times A_n$ contains $k_1 \cdots k_n$ elements.*

Example 3.2.2. The total number of possible initials consisting of three letters (name, middle name, family name) is 26^3. Indeed, each three-letter initial is an element of the set $A \times A \times A$, where A is the alphabet, so that $k_1 = k_2 = k_3 = 26$. If we want the total number of possible two- or three-letter initials, we must compute the number of the elements in the union $(A \times A) \cup (A \times A \times A)$, equal to $26^2 + 26^3 = 18{,}252$.

Example 3.2.3 (License Plates). Most states now use the system where a license plate has six symbols. One type (call it A) of such licenses has a prefix of three letters followed by a three-digit number (e.g., CQX 786). Other states (call them B) use a two-letter prefix followed by a four-digit number (e.g., KN 7207). Still other states (call them C) use a mixed prefix, a digit and two letters, followed by a three-digit number (e.g., 2CP 412). In addition, the states try to augment their revenues by allowing (for a special fee) "personalized" plates, such as CATHY3, MY CAR, and the like. Disregarding the personalized plates, which type of states can register most cars?

SOLUTION. Let A and D stand for the alphabet and for the set of 10 digits: $0, 1, \ldots, 9$. Then a license plate from an A-state can be regarded as an element of $A \times A \times A \times D \times D \times D$, while a license plate in a B-state is an element of the set $A \times A \times D \times D \times D \times D$. The numbers of elements in these Cartesian products are $26^3 \times 10^3$ and $26^2 \times 10^4$. The ratio is $26/10 = 2.6$, so that an A-state can register 2.6 times as many cars as a B-state.

As regard states of type C, the answer depends on whether or not 0 is allowed in the prefix. If a plate such as 0HY 314 is not allowed (e.g., because the digit 0 can be confused with the letter O), then the number of possible license plates is only $9 \times 26 \times 26 \times 10 \times 10 \times 10$, which is 10% less than the number of plates possible in states of type B.

If 0 is allowed as the first character, then the numbers of plates of types B and C are the same. □

In the interpretation of Theorem 3.2.1, the set of ways of performing the second operations was always the same regardless of which of the options was selected for the first operation. This need not be true: Theorem 3.2.1 remains valid if the sets of ways of performing the second operation depend on the choice of the first operation. In particular, we can think of the first and second operations as two consecutive choices of an element from *the same set*, without returning the chosen elements. If the set, say A, has n elements,

of such allocations in which birthdays do not repeat. The first number is 365^r by virtue of Theorem 3.2.3, while the second number is P^r_{365} (assuming that $r \le 365$; if $r > 365$, we must have at least one birthday repeating, so $p_r = 0$). Thus, for $r \le 365$ we have

$$p_r = \frac{P^r_{365}}{365^r} = \frac{365 \cdot (365 - 1) \cdots (365 - r + 1)}{365 \cdot 365 \cdots 365}$$

$$= \left(1 - \frac{1}{365}\right)\left(1 - \frac{2}{365}\right)\cdots\left(1 - \frac{r-1}{365}\right).$$

As first approximation, we may take, neglecting all products that have denominators of order 365^2 or higher:

$$p_r \approx 1 - \frac{1 + 2 + \cdots + (r-1)}{365} = 1 - \frac{(r-1)r}{730}. \tag{3.6}$$

This approximation works quite well for small r. To get a better approximation, we may use the formula $\log(1 - x) \approx -x$, so that

$$\log p_r = \log\left(1 - \frac{1}{365}\right) + \cdots + \log\left(1 - \frac{r-1}{365}\right)$$

$$\approx -\frac{1}{365} - \cdots - \frac{r-1}{365} = -\frac{r(r-1)}{730},$$

hence

$$p_r \approx e^{-r(r-1)/730}. \tag{3.7}$$

It is of some interest that people without experience in probability tend to overestimate the values of p_r. For instance, few people realize that for $r = 23$ a repeated birthday is about as likely as no repetition. The smallest r for which p_r is less than 0.01 is 56.

PROBLEMS

3.2.1 The number of all possible permutations of length 2 from a certain set equals 90. What is the number of elements of this set?

3.2.2 The number of permutations of length 3 from a certain set is 10 times larger than the number of permutations of length 2 from this set. What is the number of elements in this set?

3.2.3 A skyscraper is 40 stories high. Five persons enter the elevator on the first floor. Assuming that each person is equally likely to get off at any of the 39 floors $2, 3, \ldots, 40$, what is the probability that all persons will get off at different floors? Find the exact value, and then derive and compute the approximations analogous to (3.6) and (3.7).

3.2.4 Express the product of odd integers $1 \cdot 3 \cdot \ldots \cdot (2n - 1)$ through factorials. This product is sometimes denoted $(2n - 1)!!$ [*Hint:* Start with an easier task of expressing the product of even integers $2 \cdot 4 \cdot \ldots \cdot (2n)$ through factorials.]

3.2.5 The English alphabet has 26 letters. In Polish there are additional 9 letters, namely ą, ę, ż, ć, ń, ó, ś, ź, and ł, but q, v, and x are not used, so that altogether Polish has 32 letters.

  (i) Which is larger: the number of all five-letter words that can potentially be formed in English (starting from a "word" like aaaaa), or all three-letter words that one can potentially form in Polish?

  (ii) Answer the same question if the letters cannot repeat.

  (iii) Answer the same question if the letters can repeat but two neighboring letters must be distinct.

3.2.6 A two-letter code is to be formed by selecting (without replacement) the letters from a given word. Find the number of possible codes if the word is:

  (i) CHART.

  (ii) ALOHA.

  (iii) STREET.

3.2.7 Determine the number of 0's at the end of 16! and at the end of 27!

3.2.8 A telephone number in the United States has 10 digits: area code (three digits), prefix (three digits), and a four-digit number. An area code cannot start with 0 or 1, and it must have 0 or 1 as its second digit. In the prefix, neither the first nor the second digit can be 0 or 1, while the last four digits may be arbitrary. If those were the only restrictions, how may different telephone numbers could there be? [*Comment:* In fact, there are some other restrictions imposed by telephone companies; for example, no area code is 411 or 911; if the prefix is 555, it must be followed by some numbers only, such as 1212, and so on (so that there are fewer than 10,000 telephone numbers with prefix 555).]

3.2.9 Twelve girls and 17 boys went to a dance.

  (i) How many possible dancing pairs (boy–girl) may be formed?

  (ii) The dance floor can accommodate at most 11 pairs at a time. If

$$\pi(x) = \frac{\text{number of permutations in which } x \text{ is a pivot}}{\text{number of all permutations}}. \qquad (3.8)$$

As an example, suppose that the committee consists of four people, A, B, C, and D, with A having two votes and others having one vote each. A simple majority (at least three votes) is needed to pass the issue. In this case the pivot is the person who (in a given permutation) casts the third vote. Thus, A will be the pivot in all permutations in which he is in second or third place, such as BACD or DBAC. There are 12 such permutations, so that $\pi(A) = 12/4! = \frac{1}{2}$. On the other hand, B will be the pivot if he appears on the third place in a permutation, provided that A appears in a later place (i.e., in fourth place), or if he appears in second place following A. There are four such permutations: CDBA, DCBA, ABCD, and ABDC, so that $\pi(B) = 4/24 = \frac{1}{6}$. By symmetry, $\pi(C) = \pi(D) = \frac{1}{6}$, and we see that (in this case) having twice as many votes as the others gives three times as much power in making decisions.

3.2.17 Generalize the preceding example. Find the power of A in a committee of n persons with A having two votes and every other member having one vote, with a simple majority needed to carry the issue.

3.2.18 Generalize the situation of Problem 3.2.17 still further, assuming that A has k votes (other conditions being the same). What is the smallest k for which A is a dictator (i.e., every other member has zero power)?

3.2.19 Determine the voting powers in a committee consisting of six persons, with A having three votes, B having two votes, all others having one vote, and again, a simple majority being required to pass the issue.

3.2.20 **(i)** Determine the voting power in a committee of five persons, each having one vote, with a simple majority needed to pass the issue, under the additional condition that A has veto power (i.e., the positive vote of A is necessary to pass the issue).

(ii) Assume, in addition, that B has two votes. Is it better to have one vote and veto power or two votes without it?

3.2.21 Find the voting powers in a voting body of $2n$ members, among them the chairman. Everybody (including the chairman) has one vote, and the majority carries the issue. In the case of a tie, the chairman's vote prevails.

3.2.22 The United Nations Security Council consists of five permanent members (China, France, the United Kingdom, the United States, and Russia) and 10 nonpermanent members. For the issue to pass it must receive the unanimous votes of all five permanent members (so that each has veto power) and of at least four nonpermanent members. Disregard the possibility of abstaining from a vote, and determine the voting power of each member of the UN Security Council.

3.2.23 Assume that in a corporation the number of votes of a shareholder equals the number of his or her shares. If a simple majority is required, determine the voting powers in a corporation in which there are three shareholders, A with 500,000 shares, B with 499,999 shares, and C with 1 share.

3.3 BINOMIAL COEFFICIENTS

The permutations considered in Section 3.2 concerned the ordered choices from a certain set. Often, the order in which the elements are selected is not relevant, and we are interested only in the total number of possible choices, regardless of the particular order in which they are obtained. Such choices are referred to as *combinations*. We have the following definition.

Definition 3.3.1. A subset of size k selected from a set of size n (regardless of the order in which this subset was selected) is called a *combination* of k out of n. The total number of distinct combinations of k out of n is denoted by C_n^k.

We have

Theorem 3.3.1. *The number of combinations of k out of n is given by*

$$C_n^k = \frac{P_n^k}{k!}. \tag{3.9}$$

Proof. By Theorem 3.2.5 we have P_n^k different permutations of k out of n elements. Each permutation determines the set of k elements selected. Clearly, two permutations determine the same set only if they differ by the order in which the elements have been chosen. Consequently, $k!$ permutations lead to the same set, which proves (3.9). □

The ratio $P_n^k/k!$ appears often in various contexts, and it is convenient to have a special symbol for it.

Definition 3.3.2. The ratio

$$\frac{P_n^k}{k!} = \frac{n(n-1)\cdots(n-k+1)}{k!} \tag{3.10}$$

is called the *binomial coefficient* and is denoted by $\binom{n}{k}$, to be read as "n choose k." $\square$

Using (3.3), we have

$$\binom{n}{k} = \frac{n!}{k!(n-k)!}. \tag{3.11}$$

Observe, however, that (3.11) requires n to be an integer, while in definition (3.10) n can be any real number (k has to be an integer in both cases). We shall make use of this distinction later; in this section we shall tacitly assume that n is an integer with $n \geq k$.

Observe also that the symbol $\binom{n}{k}$ makes sense for $k = 0$ and $k = n$, in view of convention that $0! = 1$. Thus, we have

$$C_n^k = \binom{n}{k} \tag{3.12}$$

for all integers k, n such that $n \geq k$. For $k = 0$ we have $C_n^0 = 1$, since there is only one empty set, and $C_n^n = 1$ since only one set of size n can be selected out of a set of size n. Formula (3.12) gives correct values, namely

$$\binom{n}{n} = \binom{n}{0} = 1. \tag{3.13}$$

We shall now study some properties of the binomial coefficients $\binom{n}{k}$. First, note that

$$\binom{n}{k} = \binom{n}{n-k}, \tag{3.14}$$

which follows at once from the symmetry in formula (3.11). One can also prove (3.14) by observing that choosing a set of size k is equivalent to "leaving out" a set of size $n - k$, so the number of different sets of size k chosen must be equal to the number of different sets of size $n - k$ that are "chosen" by leaving them out.

We shall now prove the following formula.

Theorem 3.3.2 (Pascal's Triangle). *The binomial coefficients satisfy the relation*

$$\binom{n}{k} + \binom{n}{k-1} = \binom{n+1}{k}. \tag{3.15}$$

Proof. The formula can easily be proved by "brute force," expressing the left-hand side using (3.9) and reducing it algebraically to get the right-hand side. It is, however, much more instructive to use the following argument, relying on the principal interpretation of the coefficients $\binom{n}{k}$ as C_n^k. The right-hand side of (3.15) counts the number of sets of size k that can be chosen out of a set of size $n + 1$. Let us take one element of the latter set and label it somehow. We then have a set of n unlabeled and 1 labeled element. Each subset of size k is of one of the following two categories: (1) subsets that contain only k unlabeled elements, or (2) subsets that contain $k - 1$ unlabeled elements and the labeled element. Clearly, the two terms on the left-hand side of (3.15) count the numbers of subsets of the first category and of the second category. $\square$

The name *Pascal's triangle* is connected with the following way of computing the coefficients $\binom{n}{k}$, useful for small values of n. We build a triangle as follows (see Figure 3.1). We start with the top row (counted as the zeroth row), consisting of the single number 1. Each number in any of the subsequent rows is the sum of two numbers directly above it in the preceding row (as marked by arrows in the fifth row). The consecutive numbers in the nth row are, reading from the left, the values of

$$\binom{n}{0}, \binom{n}{1}, \binom{n}{2}, \ldots$$

[so that, for example, $\binom{6}{3} = 20$, as marked on the triangle below].

The name *binomial coefficient* is connected with the following formula.

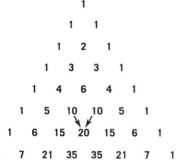

Figure 3.1 Pascal's triangle.

We have

Theorem 3.3.3 (Newton's Binomial Formula).   *For any positive integer* n
and real x, y *we have*

$$(x + y)^n = \sum_{k=0}^{n} \binom{n}{k} x^{n-k} y^k. \tag{3.16}$$

Proof. We shall prove the theorem by induction. For $n = 1$ the right-hand
side equals $\binom{1}{0} x + \binom{1}{1} y = x + y$. Assume now that the assertion holds for
some n and multiply both sides of (3.16) by $(x + y)$. We have

$$(x + y)^{n+1} = (x + y) \sum_{k=0}^{n} \binom{n}{k} x^{n-k} y^k$$

$$= \sum_{k=0}^{n} \binom{n}{k} x^{n+1-k} y^k + \sum_{k=0}^{n} \binom{n}{k} x^{n-k} y^{k+1}$$

$$= \sum_{k=0}^{n} \binom{n}{k} x^{n+1-k} y^k + \sum_{k=1}^{n+1} \binom{n}{k-1} x^{n-(k-1)} y^k.$$

Separating the term for $k = 0$ in the first sum and the term for $k = n + 1$ in
the last sum, we may write

$$(x + y)^{n+1} = x^{n+1} + \sum_{k=1}^{n} \left[\binom{n}{k} + \binom{n}{k-1} \right] x^{(n+1)-k} y^k + y^{n+1}$$

$$= \sum_{k=0}^{n+1} \binom{n+1}{k} x^{(n+1)-k} y^k,$$

where the last equality is due to Theorem 3.3.2.   □

We shall now prove

Theorem 3.3.4.   *The binomial coefficients satisfy the identities*

$$\binom{n}{0} + \binom{n}{1} + \cdots + \binom{n}{n} = 2^n \tag{3.17}$$

and

$$\binom{n}{0} - \binom{n}{1} + \binom{n}{2} - \cdots \pm \binom{n}{n} = 0. \tag{3.18}$$

Proof. It suffices to consider the expansion (3.16) of $(1+1)^n$ and $(1-1)^n$, leading directly to (3.17) and (3.18). Observe, however, that (3.17) can also be shown in the following way, instructive because of its reliance on the interpretation of the binomial coefficient $\binom{n}{k}$ as a number C_n^k of distinct subsets of size k that can be chosen out of a set of n elements. According to this interpretation, the left-hand side of (3.17) equals the total number of all subsets that can be chosen out of a set of size n, including the empty subset and the whole set. Now, the number of all subsets can also be computed differently: We may visualize the process of forming a subset as a process of deciding about each of the elements of the set whether or not to include it in the subset being formed. Each decision here can be made in 2 ways, and there are n decisions altogether, so that the total number of distinct ways of making the string of n decisions is 2^n, by Theorem 3.2.3.   □

We also have the following theorem

Theorem 3.3.5.   *For every $n = 1, 2, \ldots$ and every $k = 0, 1, \ldots, n$ the binomial coefficients satisfy the relation*

$$\binom{n}{0}\binom{n}{k} + \binom{n}{1}\binom{n}{k-1} + \cdots + \binom{n}{k}\binom{n}{0} = \binom{2n}{k}. \qquad (3.19)$$

Proof. Consider the product $(1+x)^n(1+x)^n = (1+x)^{2n}$. Expanding the right-hand side, we obtain

$$\sum_{k=0}^{2n} \binom{2n}{k} x^k. \qquad (3.20)$$

On the other hand, the left-hand side equals the product

$$\left[\sum_{i=0}^{n} \binom{n}{i} x^i\right]\left[\sum_{j=0}^{n} \binom{n}{j} x^j\right]. \qquad (3.21)$$

For $k \leq n$, comparison of the coefficients of x^k in (3.20) and (3.21) gives (3.19).   □

As a consequence of (3.19) we obtain

Corollary 3.3.6.

$$\sum_{j=0}^{n} \binom{n}{j}^2 = \binom{2n}{n}.$$

Proof. Take $k = n$ in (3.19) and use the fact that

$$\binom{n}{n-i} = \binom{n}{i}.$$ □

We shall now present some examples of the use of binomial coefficients in solving various probability problems, some with a long history.

Example 3.3.1.   In probability theory one often considers a choice without replacement from a finite set containing two categories of objects. A typical problem is: An urn contains r red and b blue balls. We choose at random n balls without replacement. What is the probability that there will be exactly k red balls chosen?

SOLUTION.   We apply here the "classical" definition of probability. The choice of n objects without replacement is the same as choosing a subset of n objects from the set of total of $r + b$ objects. This can be done in $\binom{r+b}{n}$ different ways. Now, we must have k red balls; this choice can be made in $\binom{r}{k}$ ways. We must also have $n - k$ blue balls, and this can be accomplished in $\binom{b}{n-k}$ ways. Clearly, each choice of k red balls can be combined with each of the $\binom{b}{n-k}$ choices of blue balls, so that by Theorem 3.2.2 the total number of choices is the product

$$\binom{r}{k}\binom{b}{n-k}.$$

Consequently, the probability in question is

$$P(\text{exactly } k \text{ red balls}) = \frac{\binom{r}{k}\binom{b}{n-k}}{\binom{r+b}{n}}. \tag{3.22}$$

As we shall see later, formula (3.22) is a special case of hypergeometric distribution.

Example 3.3.2.   As an illustration of the application of formula (3.22), consider the problem of estimating the number of fish in a lake (actually, the method described below is more often used to estimate the sizes of bird populations). The lake contains an unknown number N of fish. To estimate

N we first catch c fish, label them, and release them back into the lake. We assume here that labeling does not harm fish in any way, that the labeled fish mix with unlabeled ones in a random manner, and that N remains constant (in practice, these assumptions may be debatable). We now catch k fish, and observe the number, say x, of labeled ones among them. The values c and k are, at least partially, under the control of the experimenter, and therefore are the *design* variables. The unknown parameter is N, while x is the value occurring at random and providing us with the key to estimating N. Let us compute the probability $P_N = P_N(x)$ of observing x labeled fish in the second catch if there are N fish in the lake. We may interpret fish as balls in the urn, labeled and unlabeled ones playing the role of red and blue. Formula (3.22) gives

$$P_N = \frac{\binom{c}{x}\binom{N-c}{k-x}}{\binom{N}{k}}. \tag{3.23}$$

To estimate N we shall use here the principle of *maximum likelihood*, explored in detail in Chapter 12. At present it suffices to say that this principle suggests using as an estimator of N that value of N which maximizes (3.23). Let us call this value $\hat{N}$. It depends on the observed value x, hence is itself random. Thus, $\hat{N}$ is defined by the condition

$$P_{\hat{N}} \geq P_N \qquad \text{for all } N,$$

and our objective is to find the maximizer of P_N. Since N is a discrete variable, we cannot use methods of finding maxima based on derivatives. Instead, the method that works in this case is based on the observation that if the function P_N has a maximum (possibly local) at N^*, then $P_{N^*}/P_{N^*-1} > 1$ and $P_{N^*+1}/P_{N^*} < 1$. If at two neighboring arguments the values are equal, the ratio equals 1. Consequently, we should study the ratio P_N/P_{N-1} and find all arguments at which this ratio crosses the threshold 1. We have here, after some reduction,

$$P_N/P_{N-1} = \frac{(N-c)(N-k)}{N(N-c-k+x)}.$$

This ratio always exceeds 1 if $x = 0$, which means that in this case the maximum is not attained. Assume now that $x > 0$. The inequality

$$P_N/P_{N-1} \geq 1$$

is equivalent to

$$N \leq \frac{kc}{x}. \tag{3.24}$$

with the equality occurring if and only if $P_N/P_{N-1} = 1$. Thus, the maximum is attained at[*]

$$\hat{N} = \left[\frac{kc}{x}\right],$$

and also at $kc/x - 1$ if the latter value is an integer. Let us observe that the result above is consistent with common intuition: The proportion of labeled fish in the whole lake is c/N, and it should be close to the proportion x/k of labeled fish in the second catch. This gives the approximate equation $c/N \cong x/k$, with the solution $N \approx kc/x$.

Example 3.3.3.   To supplement their revenues, many states are sponsoring number games, or lotteries. The details vary somewhat from state to state, but generally, a player who buys a lottery ticket chooses several numbers from a specified set of numbers. We shall carry the calculations for the choice of 6 out of 50 numbers 1,2, ..., 50, which is quite typical. After the sales of tickets close, six winning numbers are chosen at random from the set 1,2, ..., 50. All those (if any) who chose six winning numbers share the Big Prize; if there are no such winners, the Big Prize is added to the next week's Big Prize. Those who have five winning numbers share a smaller prize, and so on. Let $P(x)$ be the probability that a player has exactly x winning numbers. We shall compute $P(x)$ for $x = 6, 5, 4$, and 3. From the point of view of calculations, the situation is the same as if the winning numbers were chosen in advance but remained secret to the players. We can now represent the situation in a familiar scheme of an urn with 6 winning numbers and 44 losing numbers, and the choice of 6 numbers from the urn (without replacement). This is the same problem as that of labeled fish. The total number of choices that can be made is $\binom{50}{6}$, while $\binom{6}{x}\binom{44}{6-x}$ is the number of choices with exactly x winning numbers. Thus,

$$P(x) = \frac{\binom{6}{x}\binom{44}{6-x}}{\binom{50}{6}}.$$

For $x = 6$ we have

$$P(6) = \frac{1}{\binom{50}{6}} = \frac{6!}{50 \cdot 49 \cdot 48 \cdot 47 \cdot 46 \cdot 45} = \frac{1}{15,890,700} = 6.29 \cdot 10^{-8}.$$

---

[*]Here $[a]$, the integer part of a, is the largest integer not exceeding a. For instance, $[3.21] = 3, [-1.71] = -2, [5] = 5$, and so on.

Similarly, $P(5) = 1.66 \cdot 10^{-5}$, $P(4) = 8.93 \cdot 10^{-4}$, and $P(3) = 0.016669$.

Thus, the chances of winning a share in the Big Prize is about 1 in 16 million. It would therefore appear that there should be, on the average, one Big winner in every 16 million tickets sold. The weekly numbers of tickets sold are well known, and it turns out that the weekly numbers of winners (of the Big Prize) vary much more than one could expect. For example, in weeks where the number of tickets sold is about 16 million, one could expect no winner, one winner, or two winners; three winners is rather unlikely. In reality, it is not at all uncommon to have five or more winning tickets in a week with 16 million tickets sold. These observations made some people suspicious as regards the honesty of the process of drawing the numbers, to the extent that there have been attempts to bring suit against the lottery (e.g., accusing the organizers of biasing the lottery balls with certain numbers, so as to decrease slightly their chance of being selected, thus favoring some other numbers).

Actually, the big variability of weekly numbers of winners is to be expected if one realizes that these numbers depend on *two* chance processes: the choice of winning numbers from the urn (which may be, and probably is, quite fair), and the choice of numbers by the players. This choice is definitely not uniform: It favors certain combinations, which seem more "random" to the naive persons than other choices. For instance, the combination 1,2,3,4,5,6 appears less likely than (say) 5, 7, 19, 20, 31, and 45. As a consequence, some combinations are selected more often by the players than others. Each combination has the same chance of being the winning one; but some may have higher numbers of winners associated with them. This point can be illustrated by the following analogy: Imagine that each week a name (first name, such as Mary, Susan, George, etc.) is chosen at random from the set of all names used, and all persons in the state with the selected name share the prize. Then chances of being chosen are the same for John as for Sebastian (as they depend on the process of sampling names of winners). But if the name Sebastian is chosen, each Sebastian will share the prize with many fewer other winners than if the name John were selected (here the numbers of winners to share the prize depend on another process, namely that of selecting names for their children by parents).

Example 3.3.4. We have k urns, labeled $1, \ldots, k$, and n identical (indistinguishable) balls. In how many ways can these balls be distributed in k urns?

SOLUTION. There are no restrictions here on the number of balls in an urn, or the number of empty urns. To get the answer, let us identify each possible allocation with a string of $k + 1$ bars and n circles, of the form

$$| \, \text{OO} \, | \, | \, \text{O} \, \text{O} \, \text{OO} \, | \, \text{O} \, | \cdots \text{O} \, |$$

the only condition being that the string should start and end with a bar.

The spaces between bars represent urns; thus, in the arrangement above, the first urn contains 2 balls, the second none, the third 4 balls, and so on. Clearly, the number of distinct arrangements equals the number of distinct arrangements of $k - 1$ bars and n circles, hence equals $\binom{n + k - 1}{n}$. Indeed, we have here a string of $n + k - 1$ symbols (not counting the two extreme bars), and by specifying n places for the symbol $\bigcirc$ we obtain a possible allocation. $\square$

Example 3.3.4 shows that the binomial coefficient can be interpreted in two ways. On the one hand, $\binom{a + b}{a}$ is the number C_{a+b}^a of distinct sets of size a that can be chosen out of a set of size $a + b$. On the other hand, $\binom{a + b}{a}$ is also the number of distinct strings of a indistinguishable elements of one kind and b indistinguishable elements of another kind. To see this, it suffices to think of choice of such a string as being determined by the choice of "a out of total of $a + b$" slots into which we put elements of the first kind.

Example 3.3.5 (The Matching Problem).   The secretary typed n letters and addressed n envelopes. For some reason the letters were put into envelopes at random. What is the probability of at least one match, that is, of at least one letter being put into the correct envelope?

SOLUTION.   This problem appears in almost every textbook on probability under various formulations (e.g., of guests receiving their hats at random, etc.). One could expect that the probability of at least one match varies greatly with n. In fact, however, the contrary is true: This probability is almost independent of n. Let A_i be the event that the ith letter is placed in the correct envelope. We have here, using formula (2.7),

$$P(\text{at least one } A_i) = P(A_1 \cup \cdots \cup A_n)$$
$$= \sum P(A_i) - \sum_{i<j} P(A_i \cap A_j)$$
$$+ \sum_{i<j<k} P(A_i \cap A_j \cap A_k) - \cdots \pm P(A_1 \cap \cdots \cap A_n).$$

By symmetry, the probability of each intersection depends only on the number of events in the intersection,* so that we may let p_r denote the probability of the intersection of r events, $p_r = P(A_{i_1} \cap A_{i_2} \cap \cdots \cap A_{i_r})$. Clearly, the numbers of terms in the consecutive sums are

$$\binom{n}{1}, \binom{n}{2}, \ldots,$$

---

*This property is called exchangeability of events, to be discussed in more detail in section 4.6.

so that we have

$$P(\text{at least one } A_i) = \binom{n}{1}p_1 - \binom{n}{2}p_2 + \binom{n}{3}p_3 - \cdots \pm \binom{n}{n}p_n. \quad (3.25)$$

To evaluate p_r, we may argue as follows. Assume that envelopes are ordered in some way. The total number of ways that one can order n letters is $n!$. If specific r events, say $A_{i_1}, \ldots, A_{i_r}$, are to occur (perhaps in conjunction with other events), then the letters number $i_1, \ldots, i_r$ must be at their appropriate places in the ordering (to match their envelopes). The remaining $n - r$ letters can appear in any of the $(n - r)!$ orders. Thus

$$p_r = \frac{(n - r)!}{n!}.$$

Consequently, the rth term in the sum (3.25) equals (up to the sign)

$$\binom{n}{r}\frac{(n - r)!}{n!} = \frac{1}{r!},$$

and we obtain

$$P(\text{at least one match}) = \frac{1}{1!} - \frac{1}{2!} + \frac{1}{3!} - \cdots \pm \frac{1}{n!}.$$

Since

$$e^{-1} = \sum_{k=0}^{\infty} \frac{(-1)^k}{k!},$$

we have

$$P(\text{at least one match}) \sim 1 - \frac{1}{e},$$

with the accuracy increasing as $n \to \infty$. The approximation is actually quite good for small n: The limiting value is $0.632121\ldots$, while the exact values of the probability π_n of at least one match for selected values of n are

$$\pi_1 = 1$$

$$\pi_2 = 1 - \frac{1}{2} = 0.5$$

$$\pi_3 = 1 - \frac{1}{2} + \frac{1}{6} = 0.6666667$$

$$\pi_4 = 1 - \frac{1}{2} + \frac{1}{6} - \frac{1}{24} = 0.625$$

$$\pi_5 = 1 - \frac{1}{2} + \frac{1}{6} - \frac{1}{24} + \frac{1}{120} = 0.6333333$$

$$\pi_6 = 1 - \frac{1}{2} + \frac{1}{6} - \frac{1}{24} + \frac{1}{120} - \frac{1}{720} = 0.6319444$$

$$\pi_7 = 1 - \frac{1}{2} + \frac{1}{6} - \frac{1}{24} + \frac{1}{120} - \frac{1}{720} + \frac{1}{5040} = 0.6321429$$

$$\pi_8 = 1 - \frac{1}{2} + \frac{1}{6} - \frac{1}{24} + \frac{1}{120} - \frac{1}{720} + \frac{1}{5040} - \frac{1}{40320} = 0.6321181.$$

Example 3.3.6 (The Ballot Problem).   Suppose that in an election, candidate A receives a votes, while candidate B receives b votes, where $a > b$. Assuming that votes are counted in random order, what is the probability that during the whole process of counting, A will be ahead of B?

SOLUTION.   Note that other votes, if any, do not matter, and we may assume that $a + b$ is the total number of votes. The process of counting votes is determined by the arrangement of the votes, that is, the arrangement of a symbols A, and b symbols B. Clearly, such an arrangement is uniquely determined by specifying the locations of the A's (or, equivalently, B's). It will be helpful for the intuition to use the following graphical representation: Define the function $c(k)$ as the net count for candidate A after inspection of k votes. Thus, if in the first k votes we had r votes for A and $k - r$ votes for B, then $c(k) = r - (k - r) = 2r - k$. We may then represent the process of counting as a polygonal line that starts at the origin and has vertices $(k, c(k)), k = 1, \ldots, a + b$ (see Figure 3.2).

In Figure 3.2 we have the beginning of counting, when the first five votes inspected are AABAB. The problem can now be formulated as finding the probability that the counting function $c(x)$ lies above the x-axis for all $x = 1, 2, \ldots, a + b$. Observe that the first vote counted must be for A (as in Figure 3.2); this occurs with probability $a/(a + b)$.

The remaining votes will give a polygonal line leading from $(1, 1)$ to $(a + b, a - b)$, and we must find the number of such lines that will never touch or cross the x-axis. The number of such lines is equal to the total number of lines from $(1, 1)$ to $(a + b, a - b)$ minus the number of lines from $(1, 1)$ to $(a + b, a - b)$ which touch or cross the x-axis. The total number of lines

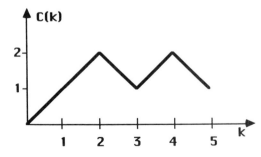

Figure 3.2 Process of counting votes.

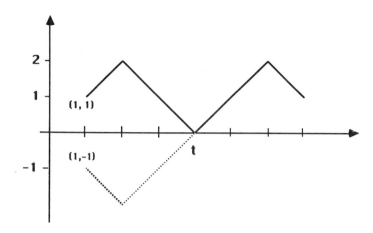

Figure 3.3 Reflection principle.

leading from $(1,1)$ to $(a+b, a-b)$ is $\binom{a+b-1}{a-1}$, since each such line has $a-1$ steps "up" and b steps "down," which can be ordered in any manner. Thus, it remains to count the number of lines from $(1,1)$ to $(a+b, a-b)$ which touch or cross the x-axis. Let V be the set of all such lines. Each line in V must touch the x-axis for the first time at some point, say t (see Figure 3.3). If we reflect the part of this line that lies to the left of t with respect to x-axis, we obtain a line leading from $(1, -1)$ to $(a+b, a-b)$. Moreover, different lines in V will correspond to different lines leading from $(1, -1)$ to $(a+b, a-b)$ and each line in the latter set will be obtained from some line in V. This means that the set V has the same number of lines as the set of lines leading from $(1, -1)$ to $(a+b, a-b)$. But the latter set contains $\binom{a+b-1}{a}$ lines, since each such line must have a steps "up" and $b-1$ steps "down." Consequently, the required probability equals

$$p = \frac{a}{a+b} \times \frac{\binom{a+b-1}{a-1} - \binom{a+b-1}{a}}{\binom{a+b-1}{a-1}},$$

which reduces to

$$p = \frac{a-b}{a+b}.$$

Example 3.3.7 (Poker). We shall now consider the probabilities of several poker hands (some students will probably say that *finally*, the book gives useful information).

In poker, five cards are dealt to a player from a standard deck of 52 cards.

The number of possible hands is therefore $\binom{52}{5} = 2,598,960$. The lowest type of hand is that containing one pair (two cards of the same denomination, plus three unmatched cards). To find the number of possible hands containing one pair, one can think in terms of consecutive choices leading to such hand:

(a) The denomination of the cards that form a pair can be chosen in $\binom{13}{1}$ ways.

(b) The suits of the pair can be chosen in $\binom{4}{2}$ ways.

(c) The choice of denominations of the remaining three cards can be made in $\binom{12}{3}$ ways.

(d) The suits of those three cards may be chosen in 4^3 ways. Thus, combining (a)–(d), we have

$$P(\text{one pair}) = \frac{\binom{13}{1}\binom{4}{2}\binom{12}{3}4^3}{\binom{52}{5}}.$$

The next kind of hand is the one containing two pairs. Here the argument is as follows:

(a) The denominations of the two pairs can be selected in $\binom{13}{2}$ ways.

(b) The suits of cards in these two pairs can be selected in $\binom{4}{2} \times \binom{4}{2}$ ways.

(c) the remaining card may be chosen in 44 ways (two denominations are eliminated).

Combining (a)–(c), we have

$$P(\text{two pairs}) = \frac{\binom{13}{2}\binom{4}{2}^2 \times 44}{\binom{52}{5}}.$$

Finally, we shall calculate one more probability, namely that of a straight (the remaining hands appear in exercises). Straight is defined as a hand containing five cards in consecutive denominations but not of the same suit (e.g., 9, 10, jack, queen, and king). Ace can appear at either end, so that we may have a straight of the form ace, 2, 3, 4, 5, as well as 10, jack, queen, king, ace). The number of hands with a straight can be computed as follows. The

denominations (uniquely determined by, say, the lowest denomination), may
be chosen in 10 ways (starting from ace, $2, 3, \ldots, 10$). Then the suits of the
five cards may be chosen in $4^5 - 4$ ways: 4^5 is the total number of selections
of suits, and we must subtract 4 selections in which all cards are of the same
suit. Thus,

$$P(\text{straight}) = \frac{10(4^5 - 4)}{\binom{52}{5}}.$$

The numerical values of the probabilities of hands in poker are:

Hand	Probability
One pair	0.4226
Two pairs	0.0475
Three-of-a-kind	0.0211
Straight	3.92×10^{-3}
Flush	1.96×10^{-3}
Full house	1.44×10^{-3}
Four-of-a-kind	2.40×10^{-4}
Straight flush	1.39×10^{-5}
Royal flush	1.54×10^{-6}.

PROBLEMS

3.3.1 Which is larger:

(i) $\binom{1000}{50}$ or $\binom{1000}{51}$?

(ii) $\binom{1000}{949}$ or $\binom{1000}{950}$?

(iii) $\binom{1000}{50}$ or $\binom{1001}{50}$?

(iv) $\binom{1000}{50}$ or $\binom{1001}{51}$?

3.3.2 A committee of size 10 is to be formed out of the U.S. Senate. What
is the probability that:

(i) Both senators from Ohio will be represented?

(ii) Exactly one senator from Ohio will be represented?

(iii) No senator from Ohio will be represented?

3.3.3 A committee of size k is to be formed out of the U.S. Senate. How large must k be in order that the probability of at least one senator from Ohio being included exceeds the probability of no senator from Ohio being included?

3.3.4 What is the probability that a randomly chosen committee of 50 senators contains one senator from each of the 50 states?

3.3.5 Points A, B, C, D, and E are such that no three of them are collinear. How many straight lines can be drawn that contain exactly two of these points?

3.3.6 In how many ways can one order the deck of 52 cards so that all four kings are next to each other?

3.3.7 Peter lives at the corner of 2nd Avenue and 72nd Street. His office is in the building at a corner of 7th Avenue and 78th Street. The streets and avenues in the city form a perpendicular grid, with no streets or passages in the middle of the blocks. Peter walks to work along either street or avenue, always in the direction that gets him closer to his office. He always returns home by subway, so that he walks across town only once a day.

A terrorist organization wants to assassinate Peter. They parked permanently a car loaded with explosives near the "center" of Peter's walk, namely on 4th Avenue, in the middle of the block between 75th and 76th Streets. The bomb will be exploded by a remote device when Peter walks in this block.

  (i) How many different paths can Peter choose to go to work?

  (ii) If Peter makes a list of all possible paths and chooses one of them randomly every morning, what is the probability that he will survive a week (5 working days; assume no other dangers).

  (iii) Where would the terrorists park the car if they consulted a statistician? (Assume no parking restriction within blocks; even a terrorist, however, cannot park at a corner.)

3.3.8 (Poker Hands) Find the probability of each of the following hands in poker:

  (a) Royal flush (ace, king, queen, jack, and 10 in one suit).

  (b) Straight flush (five cards of one suit in a sequence, but not a royal flush).

  (c) Flush (five cards in one suit, but not a straight flush or a royal flush).

  (d) Four-of-a-kind (four cards of the same denomination).

  (e) A full house (one pair and one triple of the same denomination).

  (f) Three-of-a-kind (three cards of the same denomination, plus two cards unmatched).

3.3.9   Find the probability that a poker hand will contain no pair, and be neither a straight, nor a flush of any kind.

3.3.10   A poker player has $3\heartsuit, 7\diamondsuit, 8\spadesuit, 9\diamondsuit, Q\clubsuit$. He discards $3\heartsuit$ and $Q\clubsuit$ and obtains 2 cards.*

   (i) What is the probability that he will have a straight?

   (ii) Answer the same question if $Q\clubsuit$ is replaced by $J\clubsuit$ (i.e., he discards $3\heartsuit$ and $J\clubsuit$).

3.3.11   A poker player has $3\heartsuit, 7\diamondsuit, 8\diamondsuit, 9\diamondsuit$, and $Q\clubsuit$. She discards $3\heartsuit$ and $Q\clubsuit$ and obtains 2 cards. What is the probability that she will have:

   (i) A straight flush.

   (ii) A flush, but not a straight flush.

   (iii) A straight, but not a straight flush.

3.3.12   A poker player has three-of-a-kind. He discards the two unmatched cards and obtains two new cards. Find the probability that he will have:

   (i) Three-of-a-kind.

   (ii) Four-of-a-kind.

   (iii) A full house.

3.3.13   Suppose that we distribute n balls in k boxes ($n \geq k$), labeled $1, \ldots, k$. Show that the number of different ways this can be done so that there is at least one ball in each box is $\binom{n-1}{k-1}$.

3.3.14   Suppose that n balls are distributed in k boxes ($n \geq k$). Find the probability that there are no empty boxes. [*Hint:* Let $A(n, k)$ be the number of allocations of balls such that no box is empty. Show that $A(n, k) = \sum_{i=1}^{n-k+1} \binom{n}{i} A(n - i, k - 1)$. Use boundary condition $A(n, 1) = 1$ to find $A(n, k)$.]

3.3.15   If n balls are put at random into n boxes, what is the probability of exactly one box remaining empty?

3.3.16   If n people are seated randomly in a row of $2n$ seats, what is the probability that no two persons will sit on adjacent seats?

3.3.17   Compute probabilities $P(x)$ of winning x numbers in lotteries, where the player chooses:

---

* The cards discarded are not mixed with the deck. Assume that the player receives replacements of the discarded cards from the unused remainder of the deck.

(a) 5 out of 44 numbers.

(b) 6 out of 55 numbers.

3.3.18 Find the number of polygonal lines with vertices $(x, c(x))$, where $c(x)$ is as in Example 3.3.6 and with possible edges leading from $(x, c(x))$ to $(x + 1, c(x) + 1)$ or $(x + 1, c(x) - 1)$, connecting the points:

  (i) $(0, 0)$ and $(10, 0)$.

  (ii) $(0, 0)$ and $(10, 5)$.

  (iii) $(3, -2)$ and $(8, 1)$.

3.3.19 Find the number of polygonal lines (as in Problem 3.3.18), that join the points $(2, 3)$ and $(16, 5)$ and:

  (i) Never touch the x-axis.

  (ii) Never touch the line $y = 7$.

3.3.20 Suppose that in an election, candidate A received a votes, while candidate B received b votes $(a > b)$. Find the probability $p(k, r)$ that in the process of counting votes there will be a time when A will lead by k votes, and also there will be a time when B will lead by r votes, where $k + r < a$. (*Hint:* Use the reflection principle twice.)

3.4 EXTENSION OF NEWTON'S FORMULA

As already remarked, in Definition 3.2.1, in the formula

$$\binom{n}{k} = \frac{n(n-1)\cdots(n-k+1)}{k!}, \tag{3.26}$$

the upper number n need not be an integer. We shall therefore use formula (3.26) to compute binomial coefficients for some specific noninteger n and integer k.

Example 3.4.1. As an illustration, let us evaluate $\binom{-1/2}{k}$, which we shall use later. We have

$$\binom{-1/2}{k} = \frac{(-\frac{1}{2})(-\frac{1}{2} - 1)(-\frac{1}{2} - 2)\cdots(-\frac{1}{2} - k + 1)}{k!}$$

$$= \frac{(-\frac{1}{2})(-\frac{3}{2})(-\frac{5}{2})\cdots(-\frac{2k-1}{2})}{k!} = \frac{(-1)^k 1 \cdot 3 \cdot 5 \cdots (2k-1)}{2^k k!}.$$

Multiplying numerator and denominator by $2 \cdot 4 \cdots (2k) = 2^k k!$, we get

$$\binom{-1/2}{k} = \frac{(-1)^k}{2^{2k}} \binom{2k}{k}. \tag{3.27}$$

□

Newton's binomial formula extends to the following theorem, which can be found in most calculus texts.

Theorem 3.4.1. *For any a we have*

$$(1 + x)^a = \sum_{k=0}^{\infty} \binom{a}{k} x^k. \tag{3.28}$$

Proof. It suffices to observe that the right-hand side of (3.28) is the Taylor expansion of the left-hand side [i.e., $\binom{a}{k}$ equals $\frac{1}{k!} \frac{d^{(k)}}{dx^k}(1 + x)^a$ for $x = 0$].

Example 3.4.2. For $a = -\frac{1}{2}$ and $|x| < 1$ we have

$$\frac{1}{\sqrt{1 - x}} = \sum_{k=0}^{\infty} \binom{-\frac{1}{2}}{k}(-x)^k.$$

Using (3.27), we obtain, after some algebra,

$$\frac{1}{\sqrt{1 - x}} = \sum_{k=0}^{\infty} \binom{2k}{k} \left(\frac{x}{4}\right)^k. \qquad\qquad □$$

Example 3.4.3. Several well-known formulas for sums of geometric series can be obtained as special cases of (3.28). For instance, upon noting that $\binom{-1}{k} = (-1)^k$, we obtain, for $|x| < 1$,

$$\frac{1}{1 - x} = 1 + x + x^2 + \cdots$$

and

$$\frac{1}{1 + x} = 1 - x + x^2 - \cdots.$$

PROBLEMS

3.4.1 Show that for every $x > 0$ (integer or not)

$$\binom{-x}{k} = (-1)^k \binom{x + k - 1}{k}.$$

3.4.2   Show that

$$\binom{k}{k} + \binom{k+1}{k} + \cdots + \binom{k+r}{k} = \binom{k+r+1}{k+1}.$$

3.4.3   Show that

$$\binom{1/2}{n} = (-1)^{n-1}\frac{1}{n}\binom{2n-2}{n-1}2^{-2n+1}.$$

3.4.4   By integrating the series in Example 3.4.3, show that

$$\log(1+x) = x - \frac{x^2}{2} + \frac{x^3}{3} - \frac{x^4}{4} + \cdots. \tag{3.29}$$

3.4.5   Use formula (3.29) to show that

$$\frac{1}{2}\log\frac{1+x}{1-x} = x + \frac{x^3}{3} + \frac{x^5}{5} + \cdots.$$

3.4.6   Show that

$$\binom{-2}{k} = (-1)^k(k+1).$$

3.5   MULTINOMIAL COEFFICIENTS

Choosing a subset of size k out of a set of size n is logically equivalent to partitioning the set of size n into two subsets, one of size k and the other of size $n - k$. The number of such partitions is, by definition,

$$\binom{n}{k} = \frac{n!}{k!(n-k)!}.$$

Generalizing this scheme, we have

   Theorem 3.5.1.   *Let $k_1, \cdots, k_r$ be integers satysfying the relation $k_1 + \cdots + k_r = n$. Then the number of ways a set of n elements can be partitioned into r classes, of sizes $k_1, k_2, \cdots, k_r$, equals*

$$\frac{n!}{k_1!k_2!\cdots k_r!}. \tag{3.30}$$

Proof. We can think of a partition above being accomplished in steps: First we choose k_1 out of n elements, to form the first class of the partition. Next we choose k_2 elements out of the remaining $n - k_1$ elements, and so on; in the last step we have $n - k_1 - k_2 - \cdots - k_{r-2} = k_{r-1} + k_r$ elements, from which we choose k_{r-1} to form the next-to-last class. The remaining k_r elements then form the last class. This can be accomplished, in view of Theorem 3.2.2, in

$$\binom{n}{k_1}\binom{n - k_1}{k_2}\binom{n - k_1 - k_2}{k_3} \cdots \binom{n - k_1 - \cdots - k_{r-2}}{k_{r-1}} \tag{3.31}$$

ways. Simple algebra now shows that formula (3.31) is the same as formula (3.30). $\qquad\square$

As a generalization of Newton's binomial formula, we have

Theorem 3.5.2. *For every integer n,*

$$(x_1 + \cdots + x_r)^n = \sum \frac{n!}{k_1! \cdots k_r!} x_1^{k_1} \cdots x_r^{k_r}, \tag{3.32}$$

where the summation is extended over all m-tuples $(k_1, \ldots, k_r)$ of nonnegative integers with $k_1 + \cdots + k_r = n$.

Proof. In the product $(x_1 + \cdots + x_r)^n$, one term is taken from each factor, so that the general term of the sum is of the form $x_1^{k_1} \cdots x_r^{k_r}$ with $k_1 + \cdots + k_r = n$. From Theorem 3.5.1 it follows that the number of times the product $x_1^{k_1} \cdots x_r^{k_r}$ appears equals (3.30). $\qquad\square$

In analogy with formula (3.17), the sum of all multinomial coefficients equals r^n, which follows by substituting $x_1 = \cdots = x_r = 1$ in (3.32).

To illustrate the theorem and reasoning leading to it, consider the following example.

Example 3.5.1. Suppose that one needs the value of the coefficient of $x^2 y^3 z^4 w$ in the expression $(x + y + z + w)^{10}$. One can argue here as follows. In the multiplication $(x + y + z + w) \cdots (x + y + z + w)$ there are 10 factors, and each term will contain one component from each set of parentheses. Thus we choose x from 2 out of 10 pairs of parentheses, y from 3 out of 10, and so on. This amounts to partitioning 10 pairs of parentheses into four classes, with sizes $k_1 = 2, k_2 = 3, k_3 = 4, k_4 = 1$. The total number of ways such a partition can be accomplished is the coefficient of $x^2 y^3 z^4 w$, and equals $\dfrac{10!}{2!3!4!1!} = 12{,}600$. $\qquad\square$

PROBLEMS

3.5.1 Show that if $a \leq b \leq c \leq n$, then

$$\binom{n}{c}\binom{c}{b}\binom{b}{a} = \binom{n}{a}\binom{n-a}{b-a}\binom{n-b}{c-b}.$$

(i) Use the definition of binomial coefficients as ratios of factorials.

(ii) Use the interpretation of the binomial coefficients directly as the number of subsets of a given size.

3.5.2 Generalize the formula in Problem 3.5.1 to answer the question: In how many ways can one choose an a-element subset from a b-element subset from a c-element subset from a d-element subset from an n-element set (where $a \leq b \leq c \leq d \leq n$)?

3.5.3 The letters B,C,E,E,N,R,S,S,Y,Z,Z,Z,Z are arranged at random. Determine the probability that these letters will spell the word SZCZEBRZESZYN. (Szczebrzeszyn is a town in eastern Poland. If you would like to try, the pronounciation is something like: Shchebzheshin.)

3.5.4 Find the coefficient of the term $x^3 y^4 z^5$ in the expansion of $(x - 2y + 3z)^{12}$.

3.5.5 Use an argument analogous to that in Theorem 3.3.2 to show that if $i \geq 1, j \geq 1, k \geq 1$, and $i + j + k = n + 1$, then

$$\frac{(n+1)!}{i!j!k!} = \frac{n!}{(i-1)!j!k!} + \frac{n!}{i!(j-1)!k!} + \frac{n!}{i!j!(k-1)!}.$$

3.6 STIRLING'S FORMULA

In many cases it is convenient to have an approximation to $n!$. This is given by Stirling's formula.

Theorem 3.6.1 (Stirling's Formula). *We have*

$$n! \sim \sqrt{2\pi} n^{n+1/2} e^{-n}, \tag{3.33}$$

where the sign $\sim$ means that the ratio of the two sides tends to 1 as $n \to \infty$.

We shall not give the proof here, as it lies beyond the scope of this book. However, it is worthwhile to give an argument for the order of the terms. Indeed, $\log(n!) = \log 1 + \log(2) + \cdots + \log n$. Now the logarithm can be approximated from above and from below as

$$\int_{k-1}^{k} (\log x)\, dx \le \log k \le \int_{k}^{k+1} (\log x)\, dx,$$

hence

$$\int_{0}^{n} (\log x)\, dx \le \log 1 + \log 2 + \cdots + \log n \le \int_{1}^{n+1} (\log x)\, dx,$$

which gives

$$n \log n - n \le \log(n!) \le (n+1) \log(n+1) - n.$$

The actual order, namely

$$(n + \tfrac{1}{2}) \log n - n,$$

is the average of the two extreme sides, which gives the terms in (3.33). The determination of the constant $\sqrt{2\pi}$ and the actual proof will be omitted.

Example 3.6.1.   A group of $2n$ boys and $2n$ girls is divided at random into two equal parts. What is the probability p_n, for large n, that boys and girls are divided evenly between two groups?

SOLUTION.   Clearly, the number of ways a group of $4n$ children can be divided into two equal parts is $\binom{4n}{2n}$. The number of ways $2n$ boys can be divided evenly is $\binom{2n}{n}$ and the same holds for girls. Thus,

$$p_n = \frac{\binom{2n}{n}\binom{2n}{n}}{\binom{4n}{2n}}, \tag{3.34}$$

which can be written, using (3.9) and (3.33), as

$$p_n = \frac{[(2n)!]^2}{(n!)^4} \Big/ \frac{(4n)!}{[(2n)!]^2} = \frac{[(2n)!]^4}{(n!)^4 (4n)!}$$

$$\sim \frac{\left[\sqrt{2\pi}(2n)^{2n+1/2}e^{-2n}\right]^4}{\left[\sqrt{2\pi}n^{n+1/2}e^{-n}\right]^4 \left[\sqrt{2\pi}(4n)^{4n+1/2}e^{-4n}\right]} = \sqrt{\frac{2}{n\pi}}.$$

The first of these questions requires using the weighted average, usually referred to as the formula for total probability. To answer the second question, one has to use Bayes' formula.   ☐

The examples and exercises of this chapter are designed to provide practice in recognizing, from the description of the problem, which probabilities are conditional and which are not. The key phrases here for indicating conditional probabilities would be "the proportion (frequency, percentage) of ... *among* ...," or "*if ... occurs*, then the probability of ... is," and so on.

PROBLEMS

4.1.1 Imagine that we have a computer data file concerning households (in a certain city, say). An entry here contains various kinds of data on a a household, concerning income, social conditions, ages, number of children, and so on. The computer is programmed so that it selects one entry at random, each entry having the same probability of being selected. Consequently, probabilities of various events are interpretable as fractions of entries in the data file with the corresponding property.

Let X, Y, and Z be, respectively, the numbers of boys, girls, and cars in the households sampled; let A be the event that a household has a TV set, and let B be the event that it has a swimming pool.

  (i) Interpret the probabilities below as relative frequencies of occurrence of some attributes in certain subpopulations.

   (a) $P(A)$.

   (b) $P(A|Z > 0)$.

   (c) $P(Z > 0|A)$.

   (d) $P(X = 0|X + Y = 3)$.

   (e) $P(B|A^c)$.

   (f) $P[(X + Y = 0)^c|A \cap B]$.

   (g) $P(XY = 0|X + Y > 1)$.

  (ii) Write in symbols the probabilities corresponding to the following relative frequencies.

   (a) Relative frequency of households with two cars.

   (b) Relative frequency of households with no children among households with at least one car.

   (c) Relative frequency of households that have both a swimming pool and a TV set among those which have either a swimming pool or a TV set and have at least one car.

4.2 DEFINITION OF CONDITIONAL PROBABILITY

The conditional probability of event A given event B will be defined as

$$P(A|B) = P(A \cap B)/P(B), \text{ provided that } P(B) > 0. \tag{4.1}$$

The definition of conditional probability in cases when the conditioning event B has probability zero is discussed in later chapters. *In this chapter it is always assumed, even if not mentioned explicitly, that the event appearing in the condition has a positive probability.*

A motivation of definition (4.1) based on the frequential interpretation of probability is as follows. As in Chapter 2, let $N(\cdot)$ denote the number of occurrences of the event in parentheses in the first N repetitions of the experiment. Then $P(A|B)$ is to be approximated by the frequency of occurrence of A among those cases when B occurred. Now, B occurred in $N(B)$ cases, and the number of occurrences of A among them is $N(A \cap B)$. Consequently, $P(A|B)$ should be the limit of $N(A \cap B)/N(B)$, and the latter quantity may be written as $[N(A \cap B)/N]/[N(B)/N]$, which converges to $P(A \cap B)/P(B)$.

For a fixed event B, suppose that we regard the conditional probabilities $P(A|B)$ as a function of event A. Observe that these conditional probabilities satisfy the axioms of probability from Chapter 2. Indeed, nonnegativity holds because $P(A|B)$ is a ratio of two nonnegative numbers, $P(A \cap B)$ and $P(B)$. Next, $P(\mathcal{S}|B) = P(\mathcal{S} \cap B)/P(B) = P(B)/P(B) = 1$. Finally, if the events $A_1, A_2, \ldots$ are mutually exclusive, the same is true for the events $A_1 \cap B, A_2 \cap B, \ldots$, so that

$$P\left(\bigcup_k A_k \Big| B\right) = P\left(\bigcup_k A_k \cap B\right)/P(B)$$
$$= \sum_k P(A_k \cap B)/P(B) = \sum_k P(A_k|B).$$

As we shall see later, it is fruitful to treat conditional probabilities $P(A|B)$ as a function of the event B. More precisely, we shall consider $P(A|B_1)$, $P(A|B_2)$, $P(A|B_3), \ldots$ for fixed A and the class of events B_i which are disjoint and such that one of them must occur. In this case $P(A|B_k)$ is a *random quantity* whose value depends on which of the events B_k occurs. We have encountered such random quantities (examples of *random variables*, discussed in Chapter 6) whenever we used numerical attributes in the description of various experiments, such as the number of children in the sampled family, the age of the sampled person, and so forth.

A simple consequence of (4.1) is the formula

$$P(A \cap B) = P(A|B)P(B) = P(B|A)P(A). \tag{4.2}$$

Indeed, the first equality is equivalent to (4.1), while the second follows from the observation that the left-hand side of (4.2) remains the same when we interchange the roles of A and B. Such an interchange applied to the middle term gives the right-hand-side term.

Formula (4.1) shows how to find conditional probability if we have the corresponding unconditional probabilities. Formula (4.2), on the other hand, shows how one may find the probability of an intersection of two events: It equals the product of the conditional probability of one event given the second, and the unconditional probability of the second event. These two ways of using conditional probability will now be illustrated by some simple examples.

Example 4.2.1.   Consider families with two children, and assume that each of the four combinations of sexes, BB, BG, GB, and GG, is equally likely. What is the probability that a family has two boys given that at least one child is a boy?

SOLUTION.   We have

$$P(\text{two boys}|\text{at least one boy})$$
$$= P([\text{two boys}] \cap [\text{at least one boy}])/P(\text{at least one boy}). \quad (4.3)$$

Since the event "two boys" implies (is contained in) the event "at least one boy," their intersection is the event "two boys," hence the probability in the numerator equals $\frac{1}{4}$. In the denominator we have the event {BB, BG, GB}, and its probability is $\frac{3}{4}$. Thus, the answer is $\frac{1}{3}$.

It is of some interest to observe that

$$P(\text{two boys}| \text{ older child is a boy}) \quad\quad\quad\quad (4.4)$$

reduces to $P(\text{BB})/P(\text{BG or BB}) = \frac{1}{2}$.

At first it might appear that the answers in (4.3) and (4.4) should be the same. To grasp the reason why they are not, observe that the two probabilities refer to the frequencies of families with two boys in different subpopulations. In the first case we look at all families that have at least one boy, so we eliminate families with two girls. In the second case we look at all families whose older child is a boy, so that we eliminate families with two girls, and also those with the combination GB.

Example 4.2.2.   Assume that in some population the ratio of the number of men to the number of women is r. Assume also that color blindness among men occurs with frequency p, while color blindness among women occurs with frequency p^2. If you choose a person at random, what is the probability of selecting a woman who is not color blind? (Incidentally, color blindness is sex-linked attribute, and this is why the frequency of its occurrence among females is the square of the frequency of its occurrence among males; see Example 4.3.4.)

SOLUTION.   Observe first that the answer is not $1 - p^2$, the latter being the conditional probability of a person selected not being color blind given that this person is a woman. Let M, W, and D denote the events "man selected," "woman selected," "person selected is color blind." Our objective is then to find the probability $P(W \cap D^c)$. Using (4.2), we may write $P(W \cap D^c) = P(D^c|W)P(W) = P(W|D^c)P(D^c)$. The third term is useless, since the data do not provide us directly with $P(D^c)$ or $P(W|D^c)$. Now, using the middle term we have $P(D^c|W) = 1 - p^2$. To determine $P(W)$ we note that $P(M)/P(W) = [1 - P(W)]/P(W) = r$, which gives the solution $P(W) = 1/(1+r)$. Consequently, the answer to the question is $P(W \cap D^c) = (1 - p^2)/(1 + r)$. $\qquad\qquad\qquad\qquad\qquad\qquad\qquad\qquad\qquad\qquad\qquad\qquad$ $\square$

As an immediate consequence of the definition of conditional probability, we have the following theorem, often called the chain rule:

Theorem 4.2.1 (Chain Rule).   *For any events $A_1, A_2, \ldots, A_n$ we have*

$$P(A_1 \cap A_2 \cap \cdots \cap A_n)$$
$$= P(A_1)P(A_2|A_1)P(A_3|A_1 \cap A_2) \cdots P(A_n|A_1 \cap A_2 \cap \cdots \cap A_{n-1}),$$
(4.5)

provided $P(A_1 \cap \cdots \cap A_{n-1}) > 0$ (which implies that all other conditional probabilities appearing in (4.5) are well defined).

To prove (4.5) it suffices to write each of the conditional probabilities on the right-hand side as the corresponding ratio of unconditional probabilities according to (4.1). The product cancels then to the probability on the left-hand side. $\qquad\qquad\qquad\qquad\qquad\qquad\qquad\qquad\qquad\qquad\qquad\qquad$ $\square$

It might be helpful to observe that the chain rule is closely related to the counting formula in sampling without replacement in Chapter 3.

Example 4.2.3.   A man has N keys of which only one opens the door. For some reason he tries them at random (eliminating the keys that have already been tried). What is the probability that he will succeed in opening the door on the kth attempt?

SOLUTION.   Let A_i be the event "ith attempt unsuccessful." The problem is then to find $P(A_1 \cap A_2 \cap \cdots \cap A_{k-1} \cap A_k^c)$. Applying the chain formula, we obtain

4.2.2   Assume that A and B are disjoint. Find:

(i) $P(A^c|B)$.

(ii) $P(A^c|A \cup B)$.

4.2.3   Find which, if either, of the following two equalities is true for any A, B such that $0 < P(B) < 1$.

(i) $P(A|B) + P(A^c|B) = 1$.

(ii) $P(A|B) + P(A|B^c) = 1$.

4.2.4   Find $P(A)/[P(A) + P(B)]$ if $P(A \cap B) > 0$ and $P(B|A) = aP(A|B)$.

4.2.5 $P(A) = \frac{3}{4}, P(B) = \frac{1}{2}, P(A|B) - P(B|A) = \frac{1}{9}$. Find $P(A \cap B)$.

4.2.6 $P(A) = P(B) = \frac{1}{2}, P(A|B) = \frac{1}{3}$. Find $P(A \cap B^c)$.

4.2.7   Find $P(A|B)$, if $P(A^c) = 2P(A)$ and $P(B|A^c) = 2P(B|A) = \frac{1}{5}$.

4.2.8   The following probabilities are given: $P(A) = 0.8, P(B) = 0.4$, $P(C) = 0.4, P(A \cup B) = 1, P(A \cup C) = 0.9, P(B \cup C) = 0.6$. Find $P[A \cap B \cap C|(A \cap B) \cup (A \cap C)]$.

4.2.9   Assume that A, B, C are events such that at least one of them occurs and $P(A|B) = P(B|C) = P(C|A) = \frac{1}{2}, P(B \cap C) = 2P(A \cap B) = 4P(C \cap A)/3 = 2P(A \cap B \cap C)$. Find:

(i) The probability that exactly one of these three events occurs.

(ii) The probability that only B occurs.

4.2.10   Three distinct integers are chosen at random from the set of the first 15 positive integers. Find the probability that:

(i) Their sum is odd.

(ii) Their product is odd.

(iii) The sum is odd if it is known that product is odd.

(iv) The product is odd if it is known that the sum is odd.

(v) Answer questions (i)–(iv) if the sampling is from the set of the first 20 integers.

(vi) Suppose now that the sampling is from the set of the first n integers. What general facts can be deduced from the answers to parts (i)–(iv)?

(vii) Answer questions (i)–(iv) by drawing from integers 1 through n, and pass to the limit with $n \to \infty$.

4.2.11   A deck of eight cards contains four jacks and four queens. A pair of cards is drawn at random. What is the probability that both of the cards are jacks if:

(i) At least one of the cards is a jack?

(ii) At least one of the cards is a red jack?

(iii) One of the cards is a jack of hearts?

4.2.12 A fair coin is tossed until a head appears. Given that the first head appeared on an even-numbered toss, what is the probability that the head appeared on the $2n$th toss?

4.2.13 A tennis player has the right to two attempts at a serve: If he misses his first serve, he can try again. A serve can be played "fast" or "slow." If a serve is played fast, the probability that it is good (the ball hits opponent's court) is A; the same probability for a slow serve is B. Assume that $0 < A < B$, that is, the fast serve is more difficult (but not impossible) to make.

If a serve is good, the ball is played and eventually one of the players wins a point. Let a be the probability that the server wins the point if the serve is fast (and good), and let b be the same probability for a slow serve. Assume that $0 < b < a$, that is, a fast serve gives a certain advantage to the server (the ball is harder to return, etc.).

Now, the server has four possible strategies: FF, FS, SF, and SS, where FF is "play first serve fast; if missed, play second serve fast," FS is "play first, serve fast; if missed, play second, serve slow," and so on.

   (i) Determine the probabilities P_{FF}, P_{FS}, P_{SF}, and P_{SS} of winning the point by the server under the four strategies.

   (ii) Show that the strategy SF is always inferior to the strategy FS. (*Hint:* Consider the difference $P_{FS} - P_{SF}$.)

   (iii) Show that if $Aa > Bb$, then strategy FF is better than FS.

   (iv) Show, by choosing appropriate numbers A, B, a, and b (with $A < B$ and $a > b$), that each of the strategies FF and SS may be best among the four. Explain why such cases do not occur among top players, for whom the best strategy is FS.

4.2.14 Three cards are drawn without replacement from an ordinary deck of cards. Find the probability that:

   (i) The first heart occurs on the third draw.

   (ii) There will be more red than black cards drawn.

   (iii) No two consecutive cards will be of the same value.

4.2.15 An urn contains three red and two green balls. If a red ball is drawn, it is not returned, and one green ball is added to the urn. If a green ball is drawn, it is returned, and two blue balls are added. If a blue ball is drawn, it is simply returned to the urn. Find the probability that in three consecutive draws from the urn, there will be a total of one blue ball drawn.

Example 4.3.3. Referring to Example 4.2.2, in a certain group of people the ratio of the number of men to the number of women is r. It is known that the incidence of color blindness among men is p, and the incidence of color blindness among women is p^2. What is the probability that a person chosen randomly from this group is color blind?

SOLUTION. To answer, we use the formula for total probability, taking the events M and W (choice of man and choice of woman) as a partition, and we shall find the probability of the event D (the person selected is color blind). We have here $P(D) = P(D|M)P(M) + P(D|W)P(W)$. As in Example 4.2.2, we have $P(M) = r/(1+r)$ and $P(W) = 1/(1+r)$, so that the answer is $P(D) = pr/(1+r) + p^2/(1+r) = p(p+r)/(1+r)$.

Example 4.3.4. The reason why the frequency of color blindness among females is the square of the corresponding frequency among males is that the gene responsible for color blindness (as well as the gene responsible for other sex-linked attributes, such as hemophilia) are located on the X chromosome. The explanation here is as follows. There are two sex chromosomes, X and Y, and every person has a pair of such chromosomes, with individuals of type XX being females and XY being males.

The color-blindness gene has two forms (alleles), say C and c, with form c being recessive and causing color blindness. Now, every man has one X chromosome, and therefore is either of category C or c, the latter being color blind. Women, having a pair of X chromosomes, are of one of three categories: (1) CC, which we shall call "normal"; (2) Cc, the so-called carriers (i.e., persons who are not color blind but are capable of transmitting the color-blindness gene c to their offspring); and (3) cc, women who are color blind.

Let p be the relative frequency of form c in the population of genes in question. Then p is the relative frequency of color-blind men (since each man carries one color-blindness gene, either C or c). Now, let u and v denote the relative frequency of carriers and of color-blind women, respectively. To find u and v we may proceed as follows. First, a woman is color blind if her father is color blind and either (a) her mother is color blind, or (b) her mother is a carrier and transmits the gene c to the daughter. This gives, using the formula for total probability, the relation

$$v = p\left(v + \frac{1}{2}u\right). \tag{4.11}$$

On the other hand, a woman is a carrier if either (a) her father is color blind and her mother is "normal", (b) her father is color blind and her mother is a carrier and transmits to her the gene C; (c) her father is "normal" and her mother is a carrier and transmits gene c, or (d) her father is "normal" and her mother is color blind.

This gives the relation

$$u = p\left[(1-u-v)+\frac{1}{2}u\right] + (1-p)\left[\frac{1}{2}u+v\right]. \qquad (4.12)$$

The solution of the pair of equations (4.11) and (4.12) gives $v = p^2, u = 2p(1-p)$. The relative frequency of women who are "normal" (i.e., neither color blind nor a carrier) is $1 - u - v = (1-p)^2$, while the relative frequency of color-blind women is p^2, as asserted.

Example 4.3.5. Assume that we have data on the frequencies of various combinations of sexes (BB, GG, BG) among twin births. Now, twins are of two kinds: identical and fraternal. Identical twins develop from a single egg and may essentially be regarded as "one person existing in two copies." One of the consequences here is that the identical twins are *always* of the same sex. On the other hand, fraternal twins develop from separate eggs and may be regarded as any other siblings, which just happened to be born at the same time. Assuming that sex is determined for each person separately as something that could be modeled as a flip of a coin, what is the overall frequency of identical twins?

SOLUTION.   Let I and F be the events "twins identical" and "twins fraternal." Then I and F form a partition: One of these events must occur and they exclude one another. Let us take any combination of sexes, say GG, and try to determine the probability of this combination. We may write here

$$P(GG) = P(GG|I)P(I) + P(GG|F)P(F). \qquad (4.13)$$

Letting $P(I) = x$, we have $P(F) = 1 - x$. We also have $P(GG|I) = \frac{1}{2}$, $P(GG|F) = \frac{1}{4}$: Identical twins have chance $\frac{1}{2}$ of being both girls and $\frac{1}{2}$ chance of being both boys, while fraternal twins have $\frac{1}{4}$ chance of being both girls (like the chance of two heads in two tosses of a fair coin). Consequently, (4.13) becomes $P(GG) = x/2 + (1-x)/4 = 1/4 + x/4$, so that $x = 4P(GG) - 1$. The numerical value of $P(GG)$ is obtainable from hospital data on the frequencies of combinations of sexes in twin births, and this gives access to the value of the frequency x of identical twins. It is a known fact that in twin births, the frequency of two boys and the frequency of two girls are both over 0.25, so that the value of x is positive, as it should be.

Observe that if we took the event GB (instead of GG), formula (4.13) would still simplify, namely $P(GB) = (1-x)/2$, so that $x = 1 - 2P(GB)$.

Example 4.3.6 (Secretary Problem). The following problem, also known

the chances of winning are $1/(t-1)$. Adding over all $t \geq 2$ (for $T = 1$ nobody will be employed!), we obtain

$$p_r = \sum_{t=2}^{n-r+2} \frac{\binom{n-t}{r-2}}{\binom{n}{r-1}} \cdot \frac{1}{t-1}.$$

The maximum in this formula occurs at the same r as in the formula (4.14), but the calculation is now much harder. □

The formula for total probability may be extended to conditional probabilities as follows.

Theorem 4.3.2. *Let $\mathcal{H} = \{H_1, H_2, \ldots\}$ be a positive partition of the sample space and let A, B be two events with $P(B) > 0$. Then*

$$P(A|B) = \sum_j P(A|B \cap H_j)P(H_j|B). \tag{4.15}$$

Proof. The right-hand side can be written as*

$$\sum_j \frac{P(A \cap B \cap H_j)}{P(B \cap H_j)} \times \frac{P(B \cap H_j)}{P(B)} = \frac{1}{P(B)} \sum_j P(A \cap B \cap H_j)$$

$$= \frac{1}{P(B)} P\left((A \cap B) \cap \bigcup_j H_j\right) = \frac{P(A \cap B)}{P(B)} = P(A|B).$$ □

As mentioned at the beginning of this chapter, it is fruitful to consider the conditional probability of an event as a function defined on a partition [i.e., the function, call it $P_{\mathcal{H}}(A|\cdot)$, where the arguments are sets of the partitions]. It was also mentioned that partitions may be used to represent information, with event $H_i \in \mathcal{H}$ signifying the situation when we know that the outcome is in H_i, but we do not have any more specific information.

It is natural to consider now the situation of "coarsening" a partition $\mathcal{H}'$ by grouping its sets, thus forming a new partition $\mathcal{H}$. This corresponds to less precise information, represented by partition $\mathcal{H}$.

We have

Theorem 4.3.3. *Let $\mathcal{H}, \mathcal{H}'$ be two partitions, with $\mathcal{H}$ being a coarsening of $\mathcal{H}'$. Then for any event A, the conditional probabilities with respect to partitions*

---

* Formally, the summation in (4.15) extends only over those j for which $P(B \cap H_j) > 0$, since otherwise the conditional probability is not defined. This does not matter, however, since $P(B \cap H_j) = 0$ implies that $P(H_j|B) = 0$.

*$\mathcal{H}$ and $\mathcal{H}'$ are related as follows: If $H_i = H'_{i_1} \cup \cdots \cup H'_{i_k}$, with $H_i \in \mathcal{H}, H'_{i_j} \in \mathcal{H}'$,
then*

$$P_{\mathcal{H}}(A|H_i) = \sum_{j=1}^{k} P_{\mathcal{H}'}(A|H'_{i_j}) P_{\mathcal{H}}(H'_{i_j}|H_i).$$

Thus the conditional probability with respect to a coarser partition is the average of probabilities with respect to the finer partition.

Proof. The proof is obtained by simple algebra: We have, using that fact that $H'_{i_j}, j = 1, 2, \ldots, k$ form a partition of H_i:

$$P(A|H_i) = P(A \cap H_i|H_i) = P\left(A \cap \bigcup_{j=1}^{k} H'_{i_j}\middle|H_i\right)$$

$$= \sum_{j=1}^{k} P(A \cap H'_{i_j}|H_i) = \sum_{j=1}^{k} \frac{P(A \cap H'_{i_j} \cap H_i)}{P(H_i)}$$

$$= \sum_{j=1}^{k} \frac{P(A \cap H'_{i_j})}{P(H_i)} = \sum_{j=1}^{k} \frac{P(A|H'_{i_j})P(H'_{i_j})}{P(H_i)}$$

$$= \sum_{j=1}^{k} P(A|H'_{i_j}) \frac{P(H'_{i_j} \cap H_i)}{P(H_i)}$$

$$= \sum_{j=1}^{k} P(A|H'_{i_j}) P(H'_{i_j}|H_i).$$

PROBLEMS

4.3.1 An event W occurs with probability 0.4. If A occurs, then the probability of W is 0.6; if A does not occur but B occurs, the probability of W is 0.1, while if neither A nor B occur, the probability of W is 0.5. Finally, if A does not occur, the probability of B is 0.3. Find $P(A)$.

4.3.2 Suppose that initially the urn contains one red and two green balls. We draw a ball and return it to the urn, adding three red, one green, and two blue balls if a red ball was drawn, and three green and one blue ball if a green ball was drawn. Then a ball is drawn from the urn. Find the probability that it is

   (i) Blue.

   (ii) Red.

   (iii) Green.

4.3.3 Suppose that in Problem 4.3.2 we return the second ball to the urn, and add new balls as described, with the condition that if the second ball is blue, we add one ball of each color. Then we draw the third ball. What is the probability that:

   (i) The third ball is blue?

   (ii) The third ball is blue if the first ball is red?

   (iii) The third ball was blue if the second ball was red?

   (iv) The third ball is blue if no blue ball was drawn on any of the preceding draws?

4.3.4 Let A and B be two events with $P(B) > 0$, and let $C_1, C_2, \ldots$ be a possible partition of sample space. Prove or disprove the following formulas.

$$P(A|B) = \sum_i P(A|B \cap C_i)P(C_i),$$

$$P(A|B) = \sum_i P(A|B \cap C_i)P(B|C_i)P(C_i).$$

4.3.5 (Tom Sawyer Problem)   You are given a task, say painting a fence. The probability that the task will be completed if k friends are helping you is p_k $(k = 0, 1, 2, \ldots)$. If j friends already helped you, the probability that the $(j+1)$st will also help is π_j $(j = 0, 1, 2, \ldots)$. On the other hand, if the jth friend did not help, then the $(j+1)$st will not help either.

   (i) Find the probability that the task will be completed.

   (ii) Find the probability that the task will be completed if $p_k = 1 - \lambda^k$, $\pi_j = \mu$ for all j.

Problems 4.3.6–4.3.8 assume an understanding of Example 4.3.4.

4.3.6 If both father and son are color blind and the mother is not color blind, what is the probability that she is a carrier?

4.3.7 Answer the question in Problem 4.3.6 if it is known only that the son is color blind.

4.3.8 Answer the question in Problem 4.3.6 if the son is color blind but the parents are not.

4.4 BAYES' FORMULA

Let us now consider a question opposite to that considered in the formula for total probability, namely: Given that the event A occurred, what is the probability of the event H_k of the partition? The answer is contained in the following time-honored theorem, dating back to the eighteenth century.

Theorem 4.4.1 (Bayes' Formula). *Let* $\mathcal{H} = \{H_1, H_2, \ldots\}$ *be a positive partition of* $\mathcal{S}$, *and* A *be an event with* $P(A) > 0$. *Then for any event* H_k *of the partition* $\mathcal{H}$, *we have*

$$P(H_k|A) = \frac{P(A|H_k)P(H_k)}{P(A|H_1)P(H_1) + \cdots + P(A|H_n)P(H_n)}$$

in the case of a finite partition $\mathcal{H}$, *into n events, and*

$$P(H_k|A) = \frac{P(A|H_k)P(H_k)}{P(A|H_1)P(H_1) + P(A|H_2)P(H_2) + \cdots}$$

in the case of a countably infinite partition $\mathcal{H}$.

Proof. Using formula (4.1) twice we may write

$$P(H_k|A) = \frac{P(H_k \cap A)}{P(A)} = \frac{P(A|H_k)P(H_k)}{P(A)}.$$

The theorem follows if one replaces the denominator on the right-hand side by the formulas given in Theorem 4.3.1.   □

In the special case of a partition into two events, we obtain: For any two events A, B satisfying the assumptions $P(A) > 0$ and $0 < P(B) < 1$, we have

$$P(B|A) = \frac{P(A|B)P(B)}{P(A|B)P(B) + P(A|B^c)P(B^c)}.$$

One of the interpretations of Bayes' theorem is when the partition $\mathcal{H}$ represents all possible mutually exclusive conditions (states of nature, hypotheses, etc.) which are logically possible and the probabilities $P(H_1), P(H_2), \ldots$ represent the prior knowledge, experience, or belief about the likelihood of $H_1, H_2, \ldots$. We then observe an event A, and this fact usually modifies the probabilities of H_i's. Accordingly, $P(H_i)$ and $P(H_i|A)$ are often called *prior* and *posterior* probabilities of H_i.

Example 4.4.1. Returning to Example 4.3.3, suppose that we sampled a person and found this person to be color blind. Intuitively, this should increase the chance that this person is a man. Indeed, we have here

$$P(M|D) = \frac{P(D|M)P(M)}{P(D|M)P(M) + P(D|W)P(W)} = \frac{\dfrac{pr}{1+r}}{\dfrac{pr}{1+r} + \dfrac{p^2}{1+r}} = \frac{r}{r+p},$$

a quantity close to 1 if p is small.   □

A natural problem arises in connection with the above-mentioned interpretation of Bayes' theorem as a means of using the evidence to reassess
the probabilities of "states of nature," or "hypotheses": How should one reassess the probabilities when the evidence comes sequentially? Specifically,
if the evidence comes in two portions, say A' followed by A'', should one
modify the prior probabilities given $A' \cap A''$, or should one first obtain posterior probabilities given A' and then use these posteriors as new priors, to
be modified given A''?

The answer, as might be expected, is that it does not matter. We have the
following theorem.

Theorem 4.4.2 (Updating the Evidence).   *Let* $\mathcal{H} = \{H_1, H_2, \ldots\}$ *be a par-*
tition, and let A' and A'' be two events. If $P(A' \cap A'') > 0$, then for every event
H_k in partition $\mathcal{H}$, we have

$$P(H_k | A' \cap A'') = \frac{P(A' \cap A'' | H_k)P(H_k)}{\sum P(A' \cap A'' | H_j)P(H_j)}$$
$$= \frac{P(A'' | H_k \cap A')P(H_k | A')}{\sum P(A'' | H_j \cap A')P(H_j | A')}. \tag{4.16}$$

Proof. Clearly, the middle term is Bayes' formula applied to the left-hand
side. For the middle term we may write

$$P(A' \cap A'' | H_i)P(H_i) = P(A' \cap A'' \cap H_i)$$
$$= P(A'' | A' \cap H_i)P(A' \cap H_i)$$
$$= P(A'' | A' \cap H_i)P(H_i | A')P(A'),$$

which shows the equality of the middle and right-hand-side terms.

Example 4.4.2.   An urn contains two coins: One is a normal coin, with
heads and tails, while the other is a coin with heads on both sides. A coin is
chosen at random from the urn and tossed n times. The results are all heads.
What is the probability that the coin tossed is a two-headed one?

SOLUTION.   Intuitively, for large n we expect the probability that the coin
selected is a two-headed one to be close to 1, since it is increasingly unlikely
to toss n heads in a row with a normal coin. Let H_1 and H_2 be the events
"normal coin was chosen" and " two-headed coin was chosen." Clearly, H_1
and H_2 form a partition. Let the prior probabilities be $P(H_1) = P(H_2) = \frac{1}{2}$,
and let A_n be the event "n heads in a row"; our objective is to find $P(H_2 | A_n)$.
Since $P(A_n | H_2) = 1$ for all n (a two-headed coin will give only heads), and
$P(A_n | H_1) = 1/2^n$, by Bayes' theorem we have

$$P(H_2|A_n) = \frac{P(A_n|H_2)P(H_2)}{P(A_n|H_1)P(H_1) + P(A_n|H_2)P(H_2)} = \frac{2^n}{2^n + 1}. \qquad (4.17)$$

The probability (4.17) indeed tends to 1 as n increases, as expected.

Suppose now that after A_n was observed, an additional m tosses again produced only heads (event B_m, say). Clearly, $A_n \cap B_m$ is the same as A_{n+m}, so that the posterior probability of the two-headed coin (H_2) given A_{n+m} is $2^{n+m}/(2^{n+m} + 1)$ by replacing n by $n + m$ in (4.17). Using the second part of formula (4.16), we obtain

$$
\begin{aligned}
P(H_2|A_n \cap B_m) &= \frac{P(B_m|H_2)P(H_2|A_n)}{P(B_m|H_1)P(H_1|A_n) + P(B_m|H_2)P(H_2|A_n)} \\
&= \frac{1 \cdot \dfrac{2^n}{2^n + 1}}{\dfrac{1}{2^m} \cdot \dfrac{1}{2^n + 1} + 1 \cdot \dfrac{2^n}{2^n + 1}} \\
&= \frac{2^{n+m}}{2^{n+m} + 1},
\end{aligned}
$$

which agrees with the result of updating "all at once."

It might seem at first that this problem is purely artificial, invented for the sole purpose of providing practice for students. This is not so: The problem is of importance in breeding and selection. Consider a gene with two forms (alleles) A and a. Assume that A is dominant and a is recessive, so that we have two phenotypes: Individuals aa can be distinguished from the others, but individuals AA and Aa cannot be told apart. Moreover, assume that allele a is undesirable and we want to eliminate it, ultimately producing a pure strain of individuals of type AA only. The problem then lies in eliminating individuals Aa. One of the ways to do it is to allow crossbreeding between a tested individual (of type AA or Aa) with an individual of type aa (you may think here of plants which can easily be cross-pollinated). If the tested individual is of the type AA ("two-headed coin"), all offspring will be of type Aa. However, if the tested individual is of the type Aa ("normal coin"), about half of the offspring will be of type Aa and half of type aa (corresponding to results of tosses being heads or tails with probability $\frac{1}{2}$). Now, if n offspring are of type Aa, the posterior probability that the tested individual is AA can be computed as above, except that the prior probability need not be $\frac{1}{2}$ (it may be assessed from genetic theory and other information about the tested individual). Usually, the breeder would accept an individual as pure strain AA if the posterior probability that it is indeed AA exceeds a threshold such as 0.99 or so, hence if sufficiently many offspring of type Aa have been observed.

Example 4.4.3. In a well-known TV game, there are three curtains, say $A, B,$ and C, of which two hide nothing while behind the third there is a Big

Prize. The Big Prize is won if it is guessed correctly which curtain hides it. You choose one of the curtains, say A. Before curtain A is pulled to reveal what is behind it, the game host pulls one of the two other curtains, say B, and shows that there is nothing behind it. He then offers you the option to change your decision (from curtain A to curtain C). Should you stick to your original choice (A) or change to C (or does it matter)? The answer to this question is counterintuitive and stirred a lot of controversy in the past. The common conviction is that it does not matter: There are two closed curtains, and the chances of the Big Prize being behind either of them must be fifty–fifty, so it is irrelevant whether or not you change your choice. Actually, the answer is that *you should* switch to C; in this way you double your chance of winning.

We assume that (1) the game host pulls this one of the two curtains (B or C) that does not hide the Big Prize (this implies that the host knows where the Big Prize is); and (2) if both curtains B and C have nothing behind them, the host selects one at random.

Let A, B, and C be the events "Big Prize is behind curtain A" (respectively, B and C); we assume the original choice was curtain A, and let B^* be the event "Host shows that there is nothing behind curtain B." We want $P(A|B^*)$. By Bayes' formula, taking A, B, C as a partition and assuming that $P(A) = P(B) = P(C) = \frac{1}{3}$,

$$P(A|B^*) = \frac{P(B^*|A)P(A)}{P(B^*|A)P(A) + P(B^*|B)P(B) + P(B^*|C)P(C)}$$

$$= \frac{\frac{1}{2} \cdot \frac{1}{3}}{\frac{1}{2} \cdot \frac{1}{3} + 0 \cdot \frac{1}{3} + 1 \cdot \frac{1}{3}} = \frac{1}{3}.$$

Since $P(B|B^*) = 0$, we must have $P(C|B^*) = \frac{2}{3}$, so indeed, chances of finding the Big Prize behind curtain C (i.e., winning if one changes the original choice) is $\frac{2}{3}$ and *not* $\frac{1}{2}$.

People who claimed originally that the chances are $\frac{1}{2}$ are seldom convinced by a formal argument, which they tend to dismiss as irrelevant. Here is an argument that might help convince them. Imagine that John and Peter are going to play the game a large number of times, say 300 times each. Each originally chooses A, and after being given a choice to switch, John never switches, whereas Peter always switches.

John wins only if his original choice was correct, which happens about 33% of the time. He may expect to win about 100 times out of 300. On the other hand, Peter wins whenever his original choice was *wrong* (because then switching is a correct choice). Consequently, he wins in about 200 cases out of 300.

Example 4.4.4. Let us consider one more example of the use of Bayes' formula, this time applied to the analysis of paternity trials. The results as such are of little importance nowadays, as explained below. However, they were of importance some time ago and are perhaps worth showing, both for historical reasons and as an example of interesting application. The main ideas here are due to Steinhaus (1954), developed later by Łukaszewicz (1956). For a detailed analysis of the problem as a statistical discrimination problem, see Niewiadomska-Bugaj et al. (1991).

Paternity trials are court cases in which a woman claims that a specific man, the defendant in the trial, is the father of her child. Typically, such cases concern alimony, inheritance, and so on; however, the primary issue that the court has to decide is whether or not the defendant is the biological father of the child.

In countries that base their legal system on Napoleonic code, the law is quite specific as to how the trial should proceed; we shall not go into details here, only mentioning that typically, at some time during the trial, a decision is made to perform a blood test. In such a test, the blood types of the mother (M), child (C), and defendant (D) are determined. We shall illustrate the situation for the special case of the most common and best known blood type classification, ABO, into types O, A, B, and AB. (Other known classifications concern the Rh factor, groups M and N, and so on. Altogether, over 15 classifications of blood types have been discovered to date). In the ABO classification, A and B are antigens, which may or may not be present in the blood. It has been established that these antigens may be present in the child's blood only if they are present in the blood of either the mother or the father. This fact leads to the possibility of exclusion of D being the father of C: If C has some antigen, say A, which is not present in the blood of M (say that M is of type B), and D is of type O or B, then he *cannot* be the father of C, since the antigen A has to be inherited from one of the parents, and it is not inherited from the mother. Observe that exclusion of the paternity of D is possible only if the child has some antigen that is absent in the blood of the mother.

If D is not excluded, the natural question arises: Can one determine the probability that he is the biological father of the child C? To answer this question, let us consider one special case (the reasoning in other cases is analogous), say when M is of blood type O and C is of blood type B. Let F denote the event that D is the biological father of C, and let E denote the event that D is excluded by the blood test. Thus we need now to determine $P(F|E^c)$. Since F and F^c form a partition, we have

$$P(F|E^c) = \frac{P(E^c|F)P(F)}{P(E^c|F)P(F) + P(E^c|F^c)P(F^c)}.$$

Clearly, a true father can never be excluded, so that $P(E^c|F) = 1$. Next, $P(E^c|F^c)$—in the case under consideration—is the probability that D has blood type O or A. Letting f_O, f_A, f_B, and f_{AB} be the frequencies of various blood types in the population, we have

$$P(E^c|F^c) = f_B + f_{AB}. \tag{4.18}$$

This formula presupposes that given that the defendant is not the father of C, his blood type is as if he were chosen at random from the general population of men. This, in turn, relies on two assumptions. The first asserts that blood type has nothing to do with physical or psychological characteristics that would make a man more attractive to women. This is, as far as we know, true. The second assumption is that the fact of choosing a man as a defendant by a particular woman has nothing to do with his blood type. This is rather a large assumption, presupposing that a woman is ignorant about either the very fact that blood types may lead to the possibility of exclusion, or that she is ignorant of the blood type of at least the defendant (if not of herself and her child). It is obvious that such an assumption is untenable nowadays. However, knowledge of blood types and especially of their consequences for paternity exclusions was far from widespread, especially in rural areas, up to the end of World War II.

Anyway, if (4.18) may be assumed, putting $p = P(F)$, we get

$$P(F|E^c) = \frac{p}{p + (f_B + f_{AB})(1 - p)}.$$

The problem therefore reduces to estimating the probability p, that is, the prior probability that D is the father of the child. In other words, p is the probability that the mother points to the right man as the father of her child. Steinhaus called this probability the "coefficient of truthfulness of women," a term perhaps not quite well chosen.

An interesting point here is that the value of p may be estimated quite precisely [again accepting the assumptions underlying (4.18)]. To see it, imagine that we have access to a large set of court records in which the blood test was performed. Again for simplicity, let us restrict our attention to the cases with M being of type O and C being of type B. Suppose that this situation occurred N times and that we had N_e exclusions. Let us now find how many exclusions we may expect in N such court cases. First of all, among N defendants, about Np will be true fathers, and none of them will be excluded. Among $N(1 - p)$ defendants who are not true fathers, exclusion occurs whenever D has blood type O or A, so that we have an approximate equality $N(1 - p)(f_B + f_{AB}) = N_e$, in which only p is unknown.

This method of estimating p may be improved greatly if we consider all possible combinations of types of blood of mother and a child, and numbers of exclusions in these categories, compared with expected numbers of exclusions.

Studies on estimation of the value p have been carried out in several countries [the details can be found in Łukaszewicz (1956)], leading to an estimate of p equal to about 0.73. This, in turn, makes it possible to test sex biases in courts. Imagine a judge who in his career has decided (say) 200 cases of paternity. We know that about 146 of the defendants were the true fathers, and this should be reflected in the judge's decisions. If he decided, say, 100 cases against the man and 100 against the woman, we may be quite certain that a large number of these cases were decided wrongly against the woman—even if we cannot point to any specific case as involving a wrong decision.

Finally, let us remark that within recent years it became possible to establish or exclude paternity without any margin of error, by analyzing the whole strands of DNA, thus making the paternity analysis through blood types obsolete.

PROBLEMS

4.4.1 Two different suppliers, A and B, provide the manufacturer with the same part. All supplies of this part are kept in a large bin. In the past, 2% of all parts supplied by A and 4% of parts supplied by B have been defective. Moreover, A supplies three times as many parts as B.

  (i) Suppose that you reach into the bin and select a part. Find the probability that this part is defective.

  (ii) Suppose that you reach into the bin, select a part, and find it nondefective. What is the probability that it was supplied by B?

4.4.2 An urn contains originally 3 blue and 2 green balls. A ball is chosen at random from the urn, returned, and 4 balls of the opposite color are added to the urn. Then a second ball is drawn.

  (i) Find the probability that the second ball is blue.

  (ii) Find the probability that both balls are of the same color.

  (iii) Given that the second ball is blue, find the probability that the first ball was green.

4.4.3 One box contains 6 red and 3 green balls. The second box has 6 red and 4 green balls. A box is chosen at random. From this box 2 balls are selected and found to be green. What is the probability that the pair was drawn from the first box, if the draws are

  (i) Without replacement.

  (ii) With replacement.

4.4.4 Suppose that box A contains 4 red and 5 green chips and box B contains 6 red and 3 green chips. A chip is chosen at random from box A and placed in box B. Finally, a chip is chosen at random from those now in box B. What is the probability that a green chip was transferred given that a red chip was drawn from box B?

4.4.5 We have three dice, each with numbers $x = 1, \ldots, 6$, and with probabilities as follows: Die 1: $p(x) = 1/6$, die 2: $p(x) = (7 - x)/21$, die 3: $p(x) = x^2/91$. A die is selected, tossed, and the number 4 appears. What is the probability that it is die 2 that was tossed?

4.4.6 Players A and B draw balls in turn, without replacement, from an urn containing 3 red and 4 green balls. A draws first. The winner is the person who draws the first red ball. Given that A won, what is the probability that A drew a red ball on the first draw?

4.4.7 Suppose that medical science has developed a test for cancer that is 95% accurate both on those who do and those who do not have cancer. Assume also that 5% of people have cancer. Find the probability that a person has cancer if the test says so.

4.4.8 One of three prisoners, A, B, and C, is to be executed the next morning. They all know about it, but they do not know who is going to die. The warden knows, but he is not allowed to tell them until just before the execution.

In the evening, one of the prisoners, say A, goes to the warden and asks him: "Please, tell me the name of one of the two prisoners, B and C, who is not going to die. If both are not to die, tell me one of their names at random. Since I know anyway that one of them is not going to die, you will not be giving me any information."

The warden thought about it for a while, and replied: "I cannot tell you who is not going to die. The reason is that now you think you have only $\frac{1}{3}$ chance of dying. Suppose I told you that B is not to be executed. You would then think that you have a $\frac{1}{2}$ chance of dying, so that, in effect, I would give you some information."

Was the warden right or was the prisoner right?

4.4.9 A prisoner is sentenced to life in prison. One day the warden comes to him and offers to toss a fair coin for either getting free or being put to death. After some deliberation, the prisoner refuses, on the ground that it is too much risk: He argues that he may escape, be pardoned, and so on. The warden asks him if he would agree to play such a game if the odds for death were 1:9 or less. The prisoner agrees.

Here is the game: An urns contains coins, labeled with digits on both sides. There is one coin labeled 1 and 2. There are nine coins labeled 2 and 3, and there are 81 coins labeled 3 and 4. A coin is to be selected at random by the warden and tossed. The prisoner can see the upper face and has to guess correctly the other face of the coin.

The prisoner decides to use the following guessing strategy: If 1 shows up, he would guess that the other side is 2 (and win). Similarly, if 4 shows up, he would guess that the other side is 3, and win. If 2 shows up, he would guess that the other side is 3, since there is one coin with 1 against nine coins with 3 on the other side. Similarly, if 3 shows up, he would guess 4, since there are nine coins with 2, against 81 coins with 4.

A coin was chosen, tossed, and 2 appeared. The prisoner was about to say "3" when the following doubts occurred to him: "Suppose that this coin has indeed 3 on the other side. It is therefore a 2–3 coin. But this coin, before it was tossed, had equal chances of showing 2 and showing 3. But if it had shown 3, I would have guessed 4, and be put to death. So I played a game with 50–50 chances for death, which I decided not to play at the start."

So, as the story goes, the prisoner decided not to play and spent the rest of his life in prison contemplating whether or not he did the right thing. Find:

(i) The probability that prisoner would go free before the coin is selected

(ii) The probability that prisoner would go free after a coin of a given kind is selected

(iii) The same probability after a given face shows up.

4.5 INDEPENDENCE

The notion of the conditional probability $P(A|B)$ introduced in Section 4.2 concerned the modification of the probability of an event A in light of the information that some other event B has occurred. Obviously, an important special case here is when such information is irrelevant: Whether or not B has occurred, the chances of A remain the same. In such a case we say that the event A is *independent* of the event B. As we shall see, the relation of independence defined in this way is symmetric: When A is independent of B, then B is also independent of A.

To proceed formally, the essence of the idea above is that A is independent of B whenever $P(A|B) = P(A)$. Using the definition of conditional probability, this equality means that $P(A \cap B)/P(B) = P(A)$, and multiplying by $P(B)$—which we may do since $P(B) > 0$ by assumption—we obtain the relation $P(A \cap B) = P(A)P(B)$. This relation is symmetric in A and in B, as asserted; moreover, it holds also if one or both events have probability zero.

Consequently, we introduce the following definition:

Definition 4.5.1. We say that two events, A and B, are *independent* if their probabilities satisfy the *multiplication rule*:

$$P(A \cap B) = P(A)P(B). \tag{4.19}$$

□

In practice, (4.19) is used in two ways. First, we may compute both sides separately and compare them to check whether or not two events are independent. More often, however, we *assume* that A and B are independent and use (4.19) for determining the probability of their intersection (joint occurrence). Typically, the assumption of independence is justified on intuitive grounds (e.g., when the events A and B do not influence each other).

Example 4.5.1.   A box contains r red and g green balls. We draw a ball at random from the box, and then we draw another ball. Let R_1 and R_2 be the events "red ball on the first (second) draw." The question now is whether the events R_1 and R_2 are independent.

Solution.   The answer depends crucially on what happens after the first draw: Is the first ball returned to the box before the second draw or not? More generally: Is the content of the box modified in any way by the first draw?

Suppose first that the ball drawn is returned to the box (the scheme known under the name *sampling with replacement*). Then $P(R_1) = P(R_2) = r/(r + g)$, hence $P(R_1)P(R_2) = [r/(r + g)]^2$. To check the independence, we have to compute $P(R_1 \cap R_2)$. Now, the two draws with replacement may produce $(r + g)^2$ distinct results, by the product rule of the preceding chapter. The number of ways one can draw two red balls is—again by the product rule—equal to r^2. Consequently. $P(R_1 \cap R_2) = r^2/(r + g)^2$, which is the same as $P(R_1)P(R_2)$. This shows that these events are independent.

If we do not return the ball to the box (*sampling without replacement*), we may proceed as follows. We write $P(R_1 \cap R_2) = P(R_1)P(R_2|R_1)$ by (4.2), and we know that $P(R_2|R_1) = (r - 1)/(r + g - 1)$, since if R_1 occurs, there remain $r + g - 1$ balls in the box, of which $r - 1$ are red. Consequently, $P(R_1 \cap R_2) = r(r - 1)/(r + g)(r + g - 1)$. To verify whether or not this last quantity equals $P(R_1)P(R_2)$, we need to evaluate $P(R_2)$. Observe that probability of red ball on the second draw is a random quantity, depending on the outcome of the first draw. Taking events R_1 and G_1 as a partition, and using the total probability formula, we may write $P(R_2) = P(R_2|R_1)P(R_1) + P(R_2|G_1)P(G_1)$, which equals $r/(r + g)$, so that $P(R_1 \cap R_2)$ is not the same as $P(R_1)P(R_2)$, and therefore the events in question are dependent.   □

We now prove an important property of independent events, which will facilitate many calculations.

Theorem 4.5.1.   *If the events A and B are independent, so are the events A and B^c, A^c and B, as well as A^c and B^c.*

Proof. It is sufficient to prove only the independence of A and B^c. Indeed, the independence of the second pair will follow by symmetry, and the independence of the third pair will follow by successive application of the two already proven statements.

Thus, assume that $P(A \cap B) = P(A)P(B)$, and consider the event $A \cap B^c$. We have here $A \cap B^c = A \setminus (A \cap B)$, so that

$$P(A \cap B^c) = P(A) - P(A \cap B)$$
$$= P(A) - P(A)P(B) = P(A)(1 - P(B)) = P(A)P(B^c),$$

which was to be proved.   □

As shown by Example 4.5.1, the difference between sampling with and without replacement lies in the fact that the first leads to independent outcomes, while the second does not. However, one feels that when the population is large, this difference in negligible, in the sense that we should be able to consider one kind of sampling interchangeably with the other. This means that we may choose the kind of sampling that happens to be easier to implement in a given situation, and use the theoretical results pertaining to that sampling which happens to be easier to treat.

Specifically, we may regard two events A and B as "almost independent" if the difference between $P(A \cap B)$ and $P(A)P(B)$ is small. To check whether our intuition about sampling with and without replacement for large population is valid, let us compute the difference

$$Q = |P(R_1 \cap R_2) - P(R_1)P(R_2)|$$

in the case of sampling without replacement (obviously, for sampling with replacement this difference is zero). We have here

$$Q = \left| \frac{r}{r+g} \left(\frac{r-1}{r+g-1} - \frac{r}{r+g} \right) \right| = \frac{rg}{(r+g)^2(r+g-1)},$$

a quantity that is small when either r or g is large.

It appears that we might take $|P(A \cap B) - P(A)P(B)|$ as a measure of degree of dependence between events A and B. This is indeed the case, except that we have to take into account the fact that this difference may be small simply because one of the events A or B has small probability. Consequently, we need to "standardize" the difference. This leads to the following definition.

Definition 4.5.2. Assume that $0 < P(A) < 1$ and $0 < P(B) < 1$. The quantity

$$r(A, B) = \frac{P(A \cap B) - P(A)P(B)}{\sqrt{P(A)(1 - P(A)P(B)(1 - P(B))}} \qquad (4.20)$$

will be called the *coefficient of correlation* between events A and B.

The coefficient of correlation is a measure of dependence in the sense specified by the following theorem.

Theorem 4.5.2.   *The coefficient of correlation $r(A, B)$ is zero if and only if A and B are independent. Moreover, we have $|r(A, B)| \leq 1$, with $r(A, B) = 1$ if and only if $A = B$, and $r(A, B) = -1$ if and only if $A = B^c$.*

Proof. The proof follows by elementary checking, except the "only if" parts of the last two statements. These parts will follow from the general theorem on the properties of correlation coefficients for random variables, to be given in Chapter 7.   □

Next, we consider the generalization of the concept of independence to more than two events.

Definition 4.5.3.   We say that the events $A_1, A_2, \ldots, A_n$ are *independent* if for any set of indices $i_1, i_2, \ldots, i_k$ with $1 \leq i_1 < i_2 < \cdots < i_k \leq n$ we have the following multiplication rule:

$$P(A_{i_1} \cap A_{i_2} \cap \cdots \cap A_{i_k}) = P(A_{i_1})P(A_{i_2}) \cdots P(A_{i_k}). \qquad (4.21)$$

Furthermore, the events in an infinite sequence $A_1, A_2, \ldots$ are called independent if for every n, the first n events of the sequence are independent.
                                                                                    □

Thus, in the case of n events, we have one condition for every subset of the size $k \geq 2$ [clearly, for $k = 1$ condition (4.21) is trivially satisfied]. Since the number of all subsets of a set with n elements is 2^n, the number of conditions in (4.21) is $2^n - n - 1$.

For instance, in the case of three events, A, B, C, definition (4.21) represents in fact four conditions, namely $P(A \cap B \cap C) = P(A)P(B)P(C)$, and three conditions for pairs, $P(A \cap B) = P(A)P(B)$, $P(A \cap C) = P(A)P(C)$, and $P(B \cap C) = P(B)P(C)$. In case of four events A, B, C, and D, the multiplication rule (4.21) must hold for the quadruplet (A, B, C, D), for four triplets $(A, B, C), (A, B, D), (A, C, D)$, and (B, C, D), and for six pairs (A, B), $(A, C), (A, D), (B, C), (B, D)$, and (C, D).

The question arises: Do we really need so many conditions? Taking the simplest case of $n = 3$ events, it might seem that independence of all three possible pairs [i.e., $(A, B), (A, C)$, and (B, C)] should imply the condition $P(A \cap B \cap C) = P(A)P(B)P(C)$, a direct analogue of the defining condition for pairs of events.

Conversely, one might expect that the multiplication rule $P(A \cap B \cap C) = P(A)P(B)P(C)$ implies independence of pairs $(A, B), (A, C)$, and (B, C), as well as conditions of the kind $P(A \cap B^c \cap C^c) = P(A)P(B^c)P(C^c)$, and so on. In fact, however, none of these implications is true, as will be shown by examples.

Example 4.5.2 (Independence in Pairs Does Not Imply the Multiplication Rule for More Events). We shall use the case of three events, A, B, and C, and show that the condition $P(A \cap B \cap C) = P(A)P(B)P(C)$ need not hold, even if all three pairs of events are independent.

Suppose that we toss a die twice, and let $A =$ "odd outcome on the first toss," $B =$ "odd outcome on the second toss," and $C =$"sum odd." We have here $P(A) = P(B) = \frac{1}{2}$. Taking the 6 by 6 square matrix with rows and columns labeled with the results of the first and second tosses as the sample space, a simple count will show that $P(C) = \frac{1}{2}$ also. Now, again, by simple counting, we obtain $P(A \cap B) = P(A \cap C) = P(B \cap C) = \frac{1}{4}$, which shows that each of the possible pairs is independent. However, $P(A)P(B)P(C) = \frac{1}{8}$, while $P(A \cap B \cap C) = 0$, since the events A, B, and C are mutually exclusive: If A and B hold, then the sum of outcomes must be even, so that C does not occur.

Example 4.5.3 (Multiplication Rule Does Not Imply Independence in Pairs). We shall now show that the condition $P(A \cap B \cap C) = P(A)P(B)P(C)$ does not imply independence in pairs. The simplest example is obtained if we take two events A and B which are dependent, and then take C with $P(C) = 0$. An example where all events have positive probability may be obtained as follows. Let the sample space be $S = \{a, b, c, d, \ldots\}$, and let $A = \{a, d\}, B = \{b, d\}$, and $C = \{c, d\}$. Moreover, let $P(a) = P(b) = P(c) = p - p^3$, and $P(d) = p^3$, where $p > 0$ satisfies the inequality $3(p - p^3) + p^3 \leq 1$. The remaining points (if any) have arbitrary probabilities, subject to the usual constrain that the sum of probabilities is 1. We have here $P(A) = P(B) = P(C) = (p - p^3) + p^3 = p$, hence $P(A \cap B \cap C) = P(A)P(B)P(C) = p^3$. However, since $A \cap B = A \cap C = B \cap C = A \cap B \cap C$, we have $P(A \cap B) = p^3$, while $P(A)P(B) = p^2$, and similarly for other pairs. □

The possibility of events being pairwise independent, but not totally independent, is mainly of theoretical interest. In the sequel, unless explicitly stated otherwise, whenever we speak of independent events, we shall always mean total independence in the sense of Definition 4.5.3.

Let us state here the following analogue of Theorem 4.5.1 for the case of n events.

Theorem 4.5.3. *If the events $A_1, A_2, \ldots, A_n$ are independent, the same is true for events $A_1', A_2', \ldots, A_n'$ where for each k, the event A_k' stands for either A_k or its complement A_k^c.*

We present now a formula that allows us to compute the probability of union of independent events. We have the following theorem, giving one of the most commonly used "tricks of the trade" in probability theory.

Theorem 4.5.4. *If the events $A_1, A_2, \ldots, A_n$ are independent, then*

$$P(A_1 \cup \cdots \cup A_n) = 1 - [1 - P(A_1)][1 - P(A_2)] \cdots [1 - P(A_n)].$$

Proof. The proof uses the fact that can be summarized as P(at least one) $= 1 - P$(none). Indeed, by De Morgan's law, $(A_1 \cup A_2 \cup \ldots \cup A_n)^c = A_1^c \cap A_2^c \cap \cdots A_n^c$. Using Theorem 4.5.3 we may write $1 - P(A_1 \cup \cdots \cup A_n) = P(A_1^c \cap A_2^c \cap \cdots A_n^c) = P(A_1^c)P(A_2^c) \cdots P(A_n^c) = [1 - P(A_1)][1 - P(A_2)] \cdots [1 - P(A_n)]$, which was to be shown.

Example 4.5.4. Consider a sequence of independent experiments, such that in each of them an event A (usually labeled "success") may occur or not. Let $P(A) = p$ for every experiment, and let A_k be the event "success at the kth trial." Then $A_1 \cup \cdots \cup A_n$ is the event "at least one success in n trials," and consequently,

$$P(\text{at least one success in } n \text{ trials}) = 1 - (1 - p)^n. \qquad \square$$

At the end, we shall prove the theorem complementing Theorem 2.6.2.

Theorem 4.5.5 (Second Borel–Cantelli Lemma). *If $A_1, A_2, \ldots$ is a sequence of independent events such that*

$$\sum_{n=1}^{\infty} P(A_n) = \infty,$$

then

$$P(\limsup A_n) = 1.$$

Proof. We have $\limsup A_n = \bigcap_{n=1}^{\infty} \bigcup_{k=n}^{\infty} A_k = \{$infinitely many A_i's will occur$\}$. By De Morgan's law, $[\limsup A_n]^c = \bigcup_{n=1}^{\infty} \bigcap_{k=n}^{\infty} A_k^c$. Since the intersections $\bigcap_{k=n}^{\infty} A_k^c$ increase with n, we have

$$P([\limsup A_n]^c) = \lim_{n \to \infty} P\left(\bigcap_{k=n}^{\infty} A_k^c \right)$$

$$= \lim_{n \to \infty} \lim_{N \to \infty} P\left(\bigcap_{k=n}^{N} A_k^c \right).$$

If A_k are independent, so are A_k^c, and we can write for the inner limit, using the inequality $1 - x \le e^{-x}$:

$$\lim_{N \to \infty} P\left(\bigcap_{k=n}^{N} A_k^c \right) = \lim_{N \to \infty} [1 - P(A_n)] \cdots [1 - P(A_N)]$$

$$\le \lim_{N \to \infty} e^{-[P(A_n)+\ldots+P(A_N)]} = 0$$

in view of the assumption that the series $\sum P(A_n)$ diverges. Consequently, $P([\limsup A_n]^c) = 0$, which was to be shown.

PROBLEMS

4.5.1 Label the statements true or false.

    (i) Your goal is to hit the target at least once. If you take three independent shots at the target (instead of one shot), you triple the chances of attaining the goal (assume each shot has the same positive chance of hitting the target).

    (ii) If A and B are independent, then $P(A^c|B^c)$ is $1 - P(A)$.

    (iii) If A, B are independent, then they must be disjoint.

4.5.2 Events A and B are independent, $P(A) = P(B)$, and $P(A \cup B) = \frac{1}{2}$. Find $P(A)$.

4.5.3 Suppose that A and B are independent events such that $P(A \cap B^c) = \frac{1}{3}$ and $P(A^c \cap B) = \frac{1}{6}$. Find $P(A \cap B)$.

4.5.4 Events A and B are independent, $P(A) = kP(B)$, and at least one of them must occur. Find $P(A^c \cap B)$.

4.5.5 Events A and B are independent, A and C are mutually exclusive, B and C are independent. Find $P(A \cup B \cup C)$ if $P(A) = 0.5, P(B) = 0.25, P(C) = 0.125$.

4.5.6 If A and B are events having positive probabilities, and such that $P(A \cap B) = 0$, find all independent pairs among the following pairs: $\varnothing$ and A, A and B, A and S, A and $A \cap B$, $\varnothing$ and A^c.

4.5.7 A ball is drawn at random from a box containing 12 balls labeled 1 through 12. Let X be the number on the ball selected. Let A be the event that X is even, let B be the event that $X \geq 7$, and let C be the event that $X < 4$. Check which of the pairs $(A, B), (A, C), (B, C)$ are independent.

4.5.8 Find the possible values of $P(A)$ if it is known that the probability that A occurs at least once in three independent trials exceeds the probability that A will occur twice in two independent trials.

4.5.9 Suppose that a point is picked at random from the unit square $0 \leq x, y \leq 1$. Let A be the event that it falls in the triangle bounded by the lines $y = 0, x = 1$, and $x = y$, and let B be the event that it falls into the rectangle with vertices $(0,0), (1,0), (1,\frac{1}{2})$, and $(0,\frac{1}{2})$. Find all statements that are true:

(i) $P(A \cup B) = P(A \cap B)$.

(ii) $P(A|B) = \frac{1}{8}$.

(iii) A and B are independent.

(iv) $P(A) = P(B)$.

(v) $P(A \cap B) = \frac{3}{8}$.

4.5.10 Suppose that events A and B are such that $P(A) = p, P(B) = 3p$, where $0 < p < \frac{1}{3}$. Find the correct answers in parts (i) and (ii).

(i) The relation $P(B|A) = 3P(A|B)$ is:

(a) True.

(b) True only if A and B are disjoint.

(c) True only if A and B are independent.

(d) False.

(ii) The relation $P(A \cap B^c) \leq \min(p, 1 - 3p)$ is:

(a) True.

(b) False.

4.5.11 A coin is tossed six times. Find the probability that the number of heads in the first three trials is the same as the number of heads in the last three trials.

4.5.12 Two people take turns rolling a die. Peter rolls first, then Paul, then Peter again, and so on. The winner is the first to roll a six. What is the probability that Peter wins?

4.5.13 A coin with probability p of turning up heads is tossed until it comes up tails. Let X be the number of tosses required. You bet that X will be odd, your opponent bets that X will be even. For what p is the bet advantageous to you? Is there a p such that the bet is fair?

4.5.14 Find the probability that in repeated tossing of a pair of dice, a sum of 7 will occur before a sum of 8.

4.5.15 Three people, A, B, and C, throw a die in turn. The first one to roll a number exceeding 4 wins, and the game is ended. Find the probability that A will win.

4.5.16 Consider a die in which the probability of a face is proportional to the number of dots on this face. What is the probability that in six independent throws of this die each face appears exactly once?

4.5.17 The machine has three independent components, two of which fail with probability p, and one that fails with probability 0.5. The machine operates as long as at least two parts work. Find the probability that the machine will fail.

4.5.18 Two bombers are to be sent to attack a certain target. Each plane carries one bomb. To destroy the target it is enough to score one hit. The crew of each plane consists of a pilot and a bombardier. The planes fly independently. There are two pilots, Jim and Carl, and two bombardiers, Nancy and Lucy. Jim has probability a of finding the target, while the same probability for Carl is b $(b < a)$. If the pilot finds the target, Nancy will hit it with probability c, the same probability for Lucy is d $(d < c)$.

To maximize the probability of success of the mission it is best to form the crews as follows (find a correct answer):

(a) Jim + Nancy and Carl + Lucy (better pilot with better bombardier).

(b) Carl + Nancy and Jim + Lucy (better pilot with weaker bombardier).

(c) Does not matter.

4.5.19 An athlete in a high jump competition has the right of three attempts at each height. Suppose that his chance of clearing the bar at height h is equal $p(h)$, independent of the results of previous attempts. The heights to be attempted are set by the judges to be $h_1 < h_2 < \cdots$. An athlete has the right to skip trying a given height. Let Y be his final result, that is, the highest h that he actually cleared. Let A stand for the strategy "try all heights," and let B stand for the strategy "skip the first height and then try every second height." Find the probability $P(Y = h_{2k})$ under strategies A and B.

4.5.20 The French mathematician D'Alembert claimed that in tossing a coin twice, we have only three possible outcomes: "two heads," "one head," and "no heads." This is a legitimate sample space, of course; however, D'Alembert also claimed that each outcome in this space has the same probability $\frac{1}{3}$.

(i) Is it possible to have a coin biased in such a way so as to make D'Alembert's claim true?

(ii) Is it possible to make two coins, with different probabilities of heads, so as to make D'Alembert's claim true?

4.5.21 Is it possible to bias a die in such a way that in tossing the die twice, each sum 2, 3, ...,12 has the same probability?

4.6 EXCHANGEABILITY; CONDITIONAL INDEPENDENCE

At the end of this chapter we introduce an important type of dependence between events. We start from the following definition.

Definition 4.6.1. Let A, B, and H be three events, and let $P(H) > 0$. We say that events A and B are *conditionally independent given H* if

$$P(A \cap B|H) = P(A|H)P(B|H).$$

Definition 4.6.2. Let $\mathcal{H} = \{H_1, H_2, \ldots\}$ be a positive partition (finite or countably infinite). We say that events A and B are *conditionally independent given $\mathcal{H}$* if A and B are conditionally independent given any set H_i in partition $\mathcal{H}$.

Example 4.6.1. To understand why the events that are conditionally independent with respect to every set of the partition may be dependent, consider the following simple situation. Let the partition consist of two events, H and H^c, with $P(H) = 0.6$. Suppose now that if H occurs, then both A and B are quite likely, say $P(A|H) = 0.8, P(B|H) = 0.9$, conditional independence requiring that $P(A \cap B|H) = 0.8 \cdot 0.9 = 0.72$. On the other hand, if H^c occurs, then both A and B are rather unlikely, say $P(A|H^c) = P(B|H^c) = 0.1$. Again, conditional independence requires that $P(A \cap B|H^c) = (0.1)^2 = 0.01$.

It is easy to check that the events A and B are dependent. Indeed, $P(A) = P(A|H)P(H) + P(A|H^c)P(H^c) = 0.8 \cdot 0.6 + 0.1 \cdot 0.4 = 0.52, P(B) = P(B|H)P(H) + P(B|H^c)P(H^c) = 0.9 \cdot 0.6 + 0.1 \cdot 0.4 = 0.58$. On the other hand, $P(A \cap B) = P(A \cap B|H)P(H) + P(A \cap B|H^c)P(H^c) = 0.72 \cdot 0.6 + 0.01 \cdot 0.4 = 0.436$, which is not equal to $P(A)P(B) = 0.3016$.

What happens here is that the occurrence or nonoccurrence of one of the events allows us to make rather reliable predictions about the other event. In the case under consideration, the events A and B are positively correlated: If one occurs, the other is likely to occur also, and vice versa. Indeed, we have (see Definition 4.5.2)

$$r(A, B) = \frac{0.436 - 0.3016}{\sqrt{0.52 \cdot 0.48 \cdot 0.58 \cdot 0.42}} = 0.545.$$

A real-life situation might help to visualize better the nature of dependence that we have here. Imagine some animals, such as birds, which raise their young every year. If the conditions in a given year are hard (a draught, severe winter, etc.) often all young die, and typically very few survive. If the conditions are favorable, the number of young that make it to next year is typically higher. The fates of different families *in any given year* are independent, in the sense that survival of offspring in one family does not depend on the survival of offspring in the other. If A and B are events occurring in two different families, and H_i's are events describing possible types of conditions in any given year, then A and B are conditionally independent with respect to any given events H_i.

Unconditionally, however, events A and B are dependent, the dependence arising from the fact that all families are subject to the same conditions. Thus, if one family fares badly, then we may expect that others will also fare badly, simply because the conditions are likely to be hard.   □

The definition of conditional independence given an event, and therefore also the definition of conditional independence given a partition, extends naturally to the case of n events. The extension consists of defining first the conditional independence of events $A_1, A_2, \ldots, A_n$ given an event H_i. We shall not give this definition here: It is a repetition of Definition 4.5.3, the only difference being that the unconditional probabilities in formula (4.21) are replaced by conditional probabilities given H_i. Definition 4.6.2 remains unchanged.

Now let $A_1, A_2, \ldots, A_n$ be a set of events, and let $N = \{1, 2, \ldots n\}$ be the set of their indices. We introduce the following definition.

Definition 4.6.3.   The events $A_1, A_2, \ldots, A_n$ are called *exchangeable* if for any subset K of $\{1, 2, \ldots, n\}$, the probability

$$P\left(\bigcap_{i \in K} A_i \cap \bigcap_{j \notin K} A_j^c\right)$$

of joint occurrence of all events with indices in K and nonoccurrence of all events with indices not in K depends only on the size of the set K.   □

This means that for exchangeable events, the probability of occurrence of exactly m of the events does not depend on which events are to occur and which are not. For instance, in the case of three events A, B, and C, their exchangeability means that $P(A \cap B^c \cap C^c) = P(A^c \cap B \cap C^c) = P(A^c \cap B^c \cap C)$ and also $P(A \cap B \cap C^c) = P(A \cap B^c \cap C) = P(A^c \cap B \cap C)$.

We have the following theorem.

Theorem 4.6.1.   *If the events $A_1, A_2, \ldots, A_n$ are conditionally independent given a partition $\mathcal{H} = \{H_1, H_2, \ldots\}$, and $P(A_i|H_k) = P(A_j|H_k)$ for all i, j, k, then they are exchangeable.*

Proof. Let $P(A_i|H_k) = w_k$ (by assumption, this probability does not depend on i). The probability that, say, the first r events will occur and the remaining ones will not, equals (by the law of total probability and conditional independence)

$$P(A_1 \cap A_2 \cap \cdots A_r \cap A_{r+1}^c \cap \cdots \cap A_n^c)$$
$$= \sum_k P(A_1 \cap A_2 \cap \cdots \cap A_r \cap A_{r+1}^c \cap \cdots \cap A_n^c | H_k) P(H_k)$$
$$= \sum_k w_k^r (1 - w_k)^{n-r} P(H_k). \tag{4.22}$$

But it is clear that (4.22) is also the probability of occurrence of *any r* among the events $A_1, \ldots, A_n$ and nonoccurrence of the remaining ones.   □

As may be expected, in case of conditional independence of A' and A'' with respect to the partition $\mathcal{H}$, Theorem 4.4.2 on updating the evidence will take a simpler form. We have

Theorem 4.6.2. *Let $\mathcal{H} = \{H_1, H_2, \ldots\}$ be a partition, and let A', A'' be two events conditionally independent with respect to $\mathcal{H}$. If $P(A' \cap A'') > 0$, then for every event H_k in partition $\mathcal{H}$ we have*

$$P(H_k | A' \cap A'') = \frac{P(A'|H_k) P(A''|H_k) P(H_k)}{\sum_j P(A'|H_j) P(A''|H_j) P(H_j)}$$
$$= \frac{P(A''|H_k) P(H_k|A')}{\sum_j P(A''|H_j) P(H_j|A')}. \tag{4.23}$$

Proof. The middle term shows how the conditional probability of H_k is computed if the "evidence" $A' \cap A''$ is taken jointly. The extreme right sum shows how the updating is done successively if first one updates probabilities using A' and then uses A'' for further modification of posterior probabilities.

For the first formula, observe that the middle term of (4.23) is just the middle term in (4.16), using the assumption of conditional independence. To prove the second equation, observe that the factor $P(A''|H_k \cap A')$ in the right-hand side of (4.16) can be written as

$$P(A''|H_k \cap A') = \frac{P(A' \cap A'' \cap H_k)}{P(A' \cap H_k)}$$
$$= \frac{P(A' \cap A''|H_k) P(H_k)}{P(A'|H_k) P(H_k)}$$
$$= \frac{P(A'|H_k) P(A''|H_k)}{P(A'|H_k)} = P(A''|H_k).$$

PROBLEMS

4.6.1   Subset S of size t, $1 \leq t \leq N$, is selected at random from the set $\{1, \ldots, N\}$ and event A_i $(i = 1, \ldots, n)$ is defined as: "Element i was one of the elements selected." If $S = \{i_1, \ldots, i_t\}$ is chosen, we say that events $A_{i_1}, \ldots, A_{i_t}$ occur, while the remaining events do not.

  (i) Show that events $A_1, \ldots, A_N$ are exchangeable.

  (ii) Find correlation coefficient $r = r(A_i, A_j), i \neq j$ between two events. Give an intuitive explanation why $r < 0$.

4.6.2 Generalizing the scheme of Problem 4.6.1, let p_t be the probability of choosing size t for subsets $(t = 0, 1, \ldots, N)$. After choosing t, a subset $S \subset \{1, \ldots, N\}$ of size t is chosen at random, and all events with indices in S occur, while other events do not.

  (i) Argue that events $A_1, \ldots, A_N$ are exchangeable.

  (ii) Find probabilities $P(A_1^c)$ and $P(A_1 \cap A_2)$.

REFERENCES

Bartoszyński, R., 1974. "On Certain Combinatorial Identities," *Colloquium Mathematicum* **XXX**, no. 2, 289–293.

Chow, Y. S, H. Robbins, and D. Siegmund, 1972. *Great Expectations: The Theory of Optimal Stopping,* Houghton Mifflin, Boston.

Łukaszewicz, J., 1956. "On Paternity Proving" (in Polish), *Zastosowania Matematyki*, **II**, no. 4, 349–379.

Niewiadomska-Bugaj, M., H. Szczotka, and Z. Szczotkowa, 1991. "Paternity Proving" in: T. Bromek and E. Pleszczyńska (eds.), *Statistical Inference, Theory and Practice*, Kluwer Academic Publishers, Dordrecht.

Steinhaus, H., 1954. "On Establishing Paternity" (in Polish), *Zastosowania Matematyki*, **I**, no. 2, 67–82.

CHAPTER 5

Markov Chains*

5.1 INTRODUCTION AND BASIC DEFINITIONS

Thus far we have learned only a mere beginning of the conceptual structure of probability theory: sample spaces in Chapter 1, axioms and the simplest laws that can be deduced from them in Chapter 2, and the concept of conditional probability with accompanying law of total probability and Bayes' rule in Chapter 4. Additionally, in Chapter 3 we show how to compute or assign probabilities in discrete setup. It may therefore come as a surprise that even with such limited tools, it is already possible to develop an extensive and powerful theory with numerous practical applications and cognitive consequences, namely the theory of Markov chains.

Generally, the term *chain* will denote a sequence of events, typically dependent one on another in some way. It will be convenient to use the terminology referring to time: We shall think of events as occurring one after another, so that the event whose occurrence is actually observed is the "present" event, while the events following it belong to the "future" and the remaining ones to the "past." In this way we obtain a description of a process of dynamic changes (of something, which we shall generally call a *system*). Quite naturally, the main problem will be to develop tools for making some inference (prediction, etc.) of the future on the basis of the knowledge of the past.

Specifically, consider a system that evolves in a random manner. The observations are taken at some fixed times $t_0 < t_1 < t_2 < \cdots$, so that in effect we record a sequence of states at which we find the system at the times of the observations. Without loss of generality we may let $t_n = n$, so that the observations are taken every unit of time starting at $t = 0$.

The notions of "system" and "state" in the description above are left unspecified, in order to have flexibility in applying the theory to various situations. Before proceeding any further, it is worthwhile to consider some examples, which will later guide our intuition.

Example 5.1.1. Imagine a gambler who plays a sequence of games. In each game she may win or lose some amount of money. Let w_n denote her

130

net winnings in nth game (loss, if w_n is negative). If we are interested only in the financial aspects of the situation, we can regard the gambler as a "system," with the state of the system identified with the gambler's fortune. Thus, the state in this case is a number. Letting $s(n)$ be the state at the time immediately after nth game, and letting $s(0)$ be the initial fortune, we have $s(n) = s(0) + w_1 + w_2 + \cdots + w_n$.

This scheme has to be made specific by adding appropriate assumptions. First, we need assumptions about the nature of the winnings w_k. In the simplest case, the gambler may play for the same fixed amount, say $1, so that $s(n)$ either increases or decreases by 1 with each game, and the outcomes of consecutive games are independent. Another kind of assumption is that of "boundary" conditions, which specify what happens when the gambler becomes ruined, that is, when $s(n)$ reaches the value 0 [e.g., the game may end, with $s(n)$ remaining 0 forever, or it may continue from the state $s(n) = 1$, if the gambler borrows a dollar to continue playing, etc.].

This scheme is often called a *random walk*, because one may interpret the state $s(n)$ as a position of a pawn, which moves randomly along the x-axis, being shifted at the nth move by the amount w_n. By decreasing appropriately the units of the time scale and of the scale of location of the pawn, one may obtain an approximation to the process of *diffusion* in the following sense. Consider a molecule, or some small object (e.g., a speck of dust) in a liquid medium. The molecule is subject to random hits by particles of liquid, each hit causing a small displacement in random direction. This process, known as *Brownian motion*, is one of the simplest examples of processes of diffusion. Let the location at t of a molecule subject to Brownian motion be $(x(t), y(t), z(t))$. Then the random motion of a "pawn" moved along the x-axis is now an approximation of the coordinate $x(t)$ of Brownian motion. To be somewhat more specific, we may regard $s(t)$ as the location of the pawn at time t. For limiting passage we assume that the moves occur at times $\Delta t, 2\Delta t, \ldots$, so that at time t we have $n = [t/\Delta t]$ moves (where $[x]$ is the integer part of x). The displacement between 0 and t is $w_1 + \cdots + w_n$, and we must make appropriate assumptions about w_i's when we change Δt (and consequently n). One of the possible assumptions is that w_i equals $+\Delta x$ or $-\Delta x$ with equal probabilities $\frac{1}{2}$. The square of the displacement occurring in Δt is now $(\Delta x)^2$, and this suggests passage to the limit as $\Delta t \to 0, \Delta x \to 0$, such that $(\Delta x)^2/\Delta t = D > 0$. The constant $D = (\Delta x)^2/\Delta t$ in the limiting passage, called the *diffusion constant*, is a property of the medium, and it depends, among other things, on the temperature (average kinetic energy of the particles of the medium). Meaningful questions of interest are now: What is the probability that at time t the distance $x(t)$ from the initial location will satisfy the inequality $a < x(t) < b$? We shall return to this problem in later chapters (especially Chapter 10) when we develop tools for analyzing the properties of random sums of the form $w_1 + \cdots + w_n$, representing total displacement after n hits.

Example 5.1.2. Consider the following scheme, aimed at describing the evolution of an epidemic. Imagine a group of subjects, of whom some may be infected with the disease and therefore capable of spreading it to others (even though they may not yet be aware of being infectious), and some other persons may be susceptible. The spread of the disease is generated by the process of contacts, where the contact between an infective and a susceptible could lead to an infection. The state of the system may now be described by a pair of numbers $s(n) = [x(n), y(n)]$, where $x(n)$ is the number of infectives at time $t = n$, and $y(n)$ is the number of susceptibles at time $t = n$.

By taking the time between observations sufficiently small, one may assume that only the following changes from $[x(n), y(n)]$ to $[x(n+1), y(n+1)]$ are possible:

(a) $x(n+1) = x(n) + k, y(n+1) = y(n) - k$ (infection of a group of k susceptibles, $k = 1, 2, ...$).

(b) $x(n+1) = x(n) - 1, y(n+1) = y(n)$ ("removal" of an infective, which may correspond to death, isolation in the hospital, recovery with immunity, etc.).

(c) "Status quo," that is, $x(n+1) = x(n), y(n+1) = y(n)$.

The probabilities of these three types of transitions depend on the nature of the contact process and the mode of transmission of the given disease (in particular, the process simplifies somewhat if we assume that k is always 1). By specifying these probabilities, one may obtain models of spread of various infectious diseases. □

In the sequel, we shall analyze other examples. The two above, however, should help in developing some intuition. They show that the notions of the *system* and *state* are rather flexible, and can be interpreted in a number of ways. Basically, *system* here means any fragment of reality that we want to model and analyze, while *state* is the information about the system that is relevant for the purpose of the study.

In this chapter we regard time as discrete. In general theory, one considers also systems evolving in continuous time, when changes can occur and be observed at any time. The theory designed to describe and analyze randomly evolving systems is called the *theory of stochastic processes*. While it lies beyond the scope of the present book, we shall occasionally introduce some elements of it. Apart from Markov chains in this chapter, we discuss some continuous-time stochastic processes (e.g., Poisson process and it simplest generalizations in Chapter 9), and we also consider certain statistical problems arising in inference about stochastic processes in later chapters of the book, devoted to statistics.

PROBLEMS

5.1.1 After some time spent in a bar, Peter starts to walk home. Suppose that the streets form a rectangular grid. Peter always walks to the nearest corner and then decides on the direction of the next segment of his walk (so that he never changes direction in the middle of the block). Define the "state of the system" in such a way as to accommodate conveniently the assumption that Peter never goes directly back to the corner he just left.

5.1.2 Customers arrive at a service station (e.g., a taxi stand, a cable car lift at a skiing resort, etc.), and form a queue. At times $t = 1, 2, \ldots$, the first m customers ($m \geq 1$) in the queue are served (if there are that many). Let $Y_1, Y_2, \ldots$ denote the numbers of customers arriving during the time intervals between services, and let X_n denote the number of customers in the system at the time immediately preceding the nth service. Argue that $X_{n+1} = (X_n - m)^+ + Y_n$, where $U^+ = \max(U, 0)$. Assuming that the events $\{Y_{n_1} = k_1\}, \ldots, \{Y_{n_j} = k_j\}$ are independent for any $j, k_1, \ldots, k_j$ and any distinct $n_1, \ldots, n_j$ and that $P\{Y_n = k\} = p_k$, independent of n, find $P(X_{n+1} = j | X_n = i)$ for all possible $i, j = 0, 1, 2, \ldots$.

5.2 DEFINITION OF A MARKOV CHAIN

It ought to be clear from the examples in Section 5.1 that a natural sample space that may be useful for describing the evolution of a system consists of all possible sequences of states at the observation times. Each such sequence represents a possible history of the process. Formally, let $\mathcal{E}$ be the set of all possible states of the system, with elements of $\mathcal{E}$ denoted by letter e, possibly with identifying subscripts or superscripts, such as e_1, e', and so on. The sample space $\mathcal{S}_N$, which will represent the history of the system at times $t = 0, 1, \ldots, N$, will consist of all possible sequences $s_N = [s(0), \ldots, s(N)]$, where $s(i)$ is the element of $\mathcal{E}$ describing the state of the system at time $t = i$.

In this chapter we assume that the set $\mathcal{E}$ is either finite or countably infinite. We may then label the states by integers, so that $\mathcal{E} = \{e_1, e_2, \ldots, e_n\}$ or $\mathcal{E} = \{e_1, e_2, \ldots\}$. Under this assumption, each of the sample spaces $\mathcal{S}_N$ is also either finite or countably infinite and we shall define the probabilities on individual sample points of $\mathcal{S}_N$ by explicit formulas. At the same time, however, we shall consider certain limiting passages, with the lengths of the sequences of states increasing to infinity. The space of all infinite sequences of states in $\mathcal{E}$ is uncountable, regardless of whether $\mathcal{E}$ is finite or not. The details of the construction of probability of such spaces (although sketched in Chapter

2) lie beyond the scope of this book. However, we shall use simple limiting passages with the lengths of sequences increasing to infinity, basically treating each space S_N as a space of all beginnings of length N of infinite sequences.

The main types of events that we shall consider will be of the form "state e at time $t = n$." In Chapter 1, events were identified with subsets of the sample space. Accordingly, the event above will be defined as

"state e at time $t = n$ " $= \{s(n) = e\}$
$$= \text{set of all sequences } [s(0), \dots, s(N)] \text{ with } s(n)$$
$$\text{equal } e_n.$$

Typically, the probability of the the next state depends in some way on the preceding states, so that by the chain formula (4.5), for all $n \le N$ we have

$$P[s(0) = e_{i_0}, s(1) = e_{i_1}, \dots, s(n) = e_{i_n}] = P[s(0) = e_{i_0}]$$
$$\times \ P[s(1) = e_{i_1}|s(0) = e_{i_0}] \cdots P[s(n) = e_{i_n}|s(n-1) = e_{i_{n-1}}, \dots, s(0) = e_{i_0}]. \quad (5.1)$$

By tradition, the commas in formula (5.1) signify the operation of intersection of events, thus replacing a rather clumsy notation such as $P[s(0) = e_{i_0} \cap s(1) = e_{i_1} \cap \cdots \cap s(n) = e_{i_n}]$. For reasons of convenience, the events in the conditions are written in the order that starts from the most recent one. Formula (5.1) expresses the probability of a sequence of consecutive states through the probability of the "initial" state $s(0)$, and conditional probabilities involving all preceding states, thus involving longer and longer fragments of the past. The definition below postulates that for Markov chains such conditional probabilities simplify by reaching only to the most recent state.

Definition 5.2.1. The evolution of a system is said to form a *Markov chain* if for every n and every vector $(e_{i_0}, e_{i_1}, \dots, e_{i_{n-1}}, e_{i_n})$ of states, the conditional probabilities satisfy the following relation, called the *Markow property:*

$$P[s(n) = e_{i_n}|s(n-1) = e_{i_{n-1}}, s(n-2) = e_{i_{n-2}}, \dots, s(0) = e_{i_0}]$$
$$= P[s(n) = e_{i_n}|s(n-1) = e_{i_{n-1}}]. \quad (5.2)$$

We therefore have the following theorem.

Theorem 5.2.1. *If the sequence of transitions between states constitutes a Markov chain, then*

$$P[s(0) = e_{i_0}, \dots, s(n) = e_{i_n}]$$
$$= P[s(0) = e_{i_0}] \prod_{j=1}^{n} P[s(j) = e_{i_j}|s(j-1) = e_{i_{j-1}}]. \quad (5.3)$$

Generally, the transition probability $P[s(j) = e_{i_j}|s(j-1) = e_{i_{j-1}}]$ from state $e_{i_{j-1}}$ to state e_{i_j} at time $t = j$ depends on j. If this is *not* the case, the situation greatly simplifies, and this seemingly very special case is the starting point of · a theory. This theory is practically and cognitively useful, leads to rather deep mathematical results and fruitful generalizations, and is pleasantly elegant as well. Accordingly, we introduce the following definition.

Definition 5.2.2.   A Markov chain will be said to have *stationary transition probabilities*, or be *time homogeneous*, if for all states e_i, e_j the probability

$$P[s(t) = e_j | s(t-1) = e_i]$$

does not depend on t.   □

From now on, *all Markov chains under consideration in this chapter will have stationary transition probabilities*. Writing

$$p_{ij} = P[s(t) = e_j | s(t-1) = e_i]$$

and

$$r_i = P[s(0) = e_i],$$

we may rewrite (5.3) as

$$P[s(0) = e_{i_0}, s(1) = e_{i_1}, \ldots, s(n) = e_{i_n}] = r(e_{i_0}) p_{i_0 i_1} \cdots p_{i_{n-1} i_n}.$$

Clearly, $\{r_i, i = 1, 2, \}$, called the *initial probability distribution*, and the set of probabilities p_{ij}, called the *transition probability* matrix, must satisfy the following conditions:

$$\sum_i r_i = 1 \tag{5.4}$$

and

$$\sum_j p_{ij} = 1 \text{ for every } i. \tag{5.5}$$

The last condition means simply that if the system is in some state e_i, it must pass to one of the states e_j (possibly remaining in the same state e_i) in the next step.

Since square matrices (finite or not) with nonnegative entries and row sums equal to 1 are called *stochastic matrices*, we see—in view of (5.5)—that every transition probability matrix of a Markov chain is a stochastic matrix.

Example 5.2.1 (Gambler's Ruin).   Continuing Example 5.1.1 of the gambler, assume that each game is played for a unit stake, with the gambler's probability of winning a game being p. Moreover, assume that the games are independent. In this case, the sequence $s(0), s(1), \ldots$ of the gambler's fortunes after consecutive games forms a Markov chain. Indeed, letting $P[s(n+1) = j | s(n) = i] = p_{ij}$, we have for all $k > 0$,

$$p_{k,k+1} = p, \qquad p_{k,k-1} = 1 - p$$

and $p_{kj} = 0$ for all j other than $k+1$ or $k-1$, regardless of the gambler's fortunes $s(n-1), s(n-2), \ldots$ at the times preceding the nth game. The transition probabilities for $k = 0$ depend on the assumption about what happens when the gambler becomes ruined. For instance, if upon reaching $k = 0$ the game stops, we may put p_{0j} equal to 1 for $j = 0$ and equal to 0 for all other j. If the process starts with the gambler having initial capital M, then r_j equals 1 for $j = M$ and equals 0 for all other j. It seems plausible (and in fact, it is true) that if $p < \frac{1}{2}$ (the games are unfavorable for the gambler), then regardless of the initial state, the state 0 (gambler's ruin) will sooner or later be reached. Less obvious, but still true, is that the same holds for the fair games $(p = \frac{1}{2})$ no matter what the initial fortune is. An interesting question is to determine the probability of the gambler ever reaching 0 (i.e., of becoming eventually ruined) if she plays favorable games, that is, if $p > \frac{1}{2}$. This probability depends on the initial state M and may be shown to be $[(1-p)/p]^M$.

Example 5.2.2 (Division of Stake in Gambler's Ruin). A problem related to the gambler's ruin above is as follows. Two players, A and B, play a sequence of games, each game for a unit stake, until one of them loses all his or her money. Games are independent; in each the probability of A winning is p (and the probability of A losing $1 - p$, which means that there are no ties). For some reason the contest is interrupted (and is not going to be resumed) when A has x dollars and B has $M - x$ dollars (so that the total stake is M). The question is: How can the stake M be divided between the players in a fair way? One of the ways (arguably just) is to divide the stake in proportion to the chances of eventually winning the contest if it was to continue. Consequently, the problem lies in determining the probabilities of eventually winning the contest for each of the two players. We can interpret the situation in terms of the gambler's problem as follows. Let $i = 0, 1, \ldots, M$ represent the state of the system, defined as the capital of A (say). Then the changes of states constitute a Markov chain, as in Example 5.2.1, the only difference being that now $p_{00} = p_{MM} = 1$ (i.e., one cannot leave states 0 and M). The objective is to determine $u(x)$, equal to the probability of A ultimately winning the contest, if the starting state is x. We shall study this kind of problem in more generality in next sections. To illustrate the technique, observe that for $0 < x < M$ the probabilities $u(x)$ must satisfy the equations

$$u(x) = pu(x+1) + (1-p)u(x-1), \qquad (5.6)$$

with boundary conditions $u(0) = 0, u(M) = 1$. Indeed, (5.6) is a special case of the total probability formula (4.9) for the event "ultimate winning by A" and the partition into the two events "next game won by A" and "next game lost by A." This technique of using the partition obtained by considering the possible results of the next transition will be used again and again in the analysis of Markov chains. In the present case, a linear combination of solutions of (5.6) is also a solution of (5.6). Consequently, we need two linearly independent solutions to determine the constants in their linear combinations using the two boundary conditions.

Obviously, $u(x) = 1$ is always a solution of (5.6). To find another solution, let $u(x) = s^x$. This gives $s = ps^2 + q$, where we let $q = 1 - p$. The solutions of the quadratic equation are $[1 \pm (1 - 4pq)^{1/2}]/2p = [1 \pm |p - q|]/2p$, which equal 1 and q/p. Thus, if $p \neq q$ (hence $p \neq \frac{1}{2}$), a general solution of (5.6) is $u(x) = A + B(q/p)^x$, and using the boundary conditions we obtain

$$u(x) = [(q/p)^x - 1]/[(q/p)^M - 1].$$

If $p = q = \frac{1}{2}$, the solutions $u(x) = (q/p)^x$ and $u(x) = 1$ coincide. In this case a solution not identically 1 is, for instance, $u(x) = x$, and the general solution of (5.6) is $u(x) = A + Bx$. Using again the boundary conditions, we obtain

$$u(x) = x/M$$

as the probability of winning in the case of fair games.

Example 5.2.3. Let us now return to Example 5.1.2 of an epidemic. The state of the system is represented by a pair of integers (x, y), with x being the number of infectives and y being the number of susceptibles. As was argued, by taking the time between transitions sufficiently small, we may assume that apart from remaining in the same state, the only transitions possible from (x, y) are to $(x - 1, y)$, representing removal (death, isolation, recovery with immunity, etc.) of one infective, and to $(x + 1, y - 1)$, representing infection of one susceptible. These transitions are depicted in Figure 5.1 as a random walk over the integer lattice on the plane, with the two types of transitions being steps in the western and southeastern directions. We now have to define

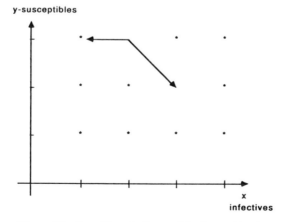

Figure 5.1 Transitions in the model of an epidemic.

the transition probabilities for the two kinds of transitions above so as to obtain a meaningful approximation to the real conditions of an epidemic. Here one could argue that it is reasonable to assume that the probability of the first transition (removal) is proportional to x, while the probability of the second transition (infection) is proportional to both x and y. Thus, the transition probabilities between states will be of the form $p[(x - 1, y)|(x, y)] = ax, p[(x + 1, y - 1)|(x, y)] = bxy$ for suitably chosen a and b.

The nonlinear term (proportional to the product xy) is largely responsible for the fact that the analysis of an epidemic process is extremely difficult. From the point of view of applications, two characteristics of epidemics are of major interest: first, the time until the termination of the epidemic, and second, its total size, that is, the total number of persons who will become infected (and subsequently removed).

The first characteristic may be defined formally as

$$N = \text{the first } n \text{ such that } x(n) = 0$$

[so that the event $N = k$ corresponds to $x(0) > 0, x(1) > 0, \ldots, x(k - 1) > 0$, and $x(k) = 0$]. Graphically, N is the first time when the walk depicted on Figure 5.1 reaches the vertical axis.

The second characteristic cannot be expressed in terms of the variables introduced thus far. The handling of the case is an illustration of the point made repeatedly in Chapter 1 about the nonuniqueness of the choice of the sample space. One may simply enrich the state space by inclusion of another count which keeps track of the removals. Thus, one may consider the states as being described by three variables, $x(n)$, $y(n)$, and $z(n)$, where x and y are the numbers of infectives and susceptibles as before, while z is the number of removals. The two kinds of transitions considered so far are replaced by transitions from (x, y, z) to $(x - 1, y, z + 1)$, representing the removal of an infective, and from (x, y, z) to $(x + 1, y - 1, z)$, representing a new infection. Assuming that the epidemic starts with $z = 0$, the total size Z of epidemic is defined as $Z = z(N)$, where $N =$ the first n with $x(n) = 0$.

As mentioned above, finding the probabilities $P\{N = n\}$ and $P\{Z = k\}$ for given initial conditions of the Markov chain is extremely difficult, and we shall not deal with this problem here. ☐

The examples above illustrate one kind of problem that we analyze in Section 5.5, that of *absorption*. Prior to that, however, we analyze in Section 5.4 another kind of problem, concerning the limiting behavior of the probability of finding the system in a given state after a large number of transitions.

PROBLEMS

5.2.1 Modify Example 5.2.1 by assuming that in each game the player may win with probability p, lose with probability q, or draw (so that his fortune does not change) with probability r, where $p + q + r = 1$. Find the transition probability matrix in this case.

5.2.2 Modify Example 5.2.1 in the same way as in Problem 5.2.1. Write the analogue to equation 5.2.1 for the probability of ruin and solve it.

5.2.3 Suppose it has been found that the results of an election in a certain city depend only on the results of the last two elections. Specifically, letting R and D denote Republican and Democratic victories, the state before any election may be RR, RD, DR, DD, the letters signifying, respectively, the outcome of the next-to-last and last elections. Generally, assume that a, b, c, and d are the probabilities of a Republican victory in each of the four states listed above. Find the transition probability matrix of the resulting Markov chain.

5.2.4 A college professor teaches a certain course year after year. He has three favorite questions and he always uses one of them in the final exam. He never uses the same question twice in a row. If he uses question A in some year, then next year he tosses a coin to choose between questions B and C. If he uses question B, he tosses a pair of coins and chooses question A if both coins show tails. Finally, if he uses question C, then he tosses three coins and uses question A if all of them show heads. Find the transition probability matrix for the resulting Markov chain.

5.2.5 **(Dog Fleas, or the Ehrenfest Model of Diffusion)** Consider two urns (or dogs), and N balls (or fleas), labeled $1, \ldots, N$, allocated between the urns. At times $t = 1, 2, \ldots$ a number 1 through N is chosen at random, and the ball with the selected number is transferred to the other urn. Let $s(n)$ be the number of balls in the first urn at time n. Find the transition probability matrix for the Markov chain $s(n)$.

5.2.6 Consider a specific kind of part needed for the operation of a certain machine (e.g., the water pump of a car). When the part breaks down, it is replaced by a new one. The probability that a new part will last for exactly n days is $r_n, n = 1, 2, \ldots$. Let the state of the system be defined as the age of the part currently in the machine. Find the transition probability matrix of the resulting chain.

5.3 *n*-STEP TRANSITION PROBABILITIES

We shall now find the probability

$$p_{ij}^{(n)} = P\{s(t + n) = e_j | s(t) = e_i\} \tag{5.7}$$

of passing from e_i at time t to e_j at time $t + n$, called the n-step transition probability. Obviously, for Markov chains with stationary transition probabilities, the quantity (5.7) does not depend on t. We have the following theorem.

Theorem 5.3.1. *The n-step transition probabilities satisfy the relations*

$$p_{ij}^{(0)} = 1 \text{ or } 0, \text{ depending on whether or not } j = i \qquad (5.8)$$

and for all $m, n \geq 0$,

$$p_{ij}^{(m+n)} = \sum_k p_{ik}^{(m)} p_{kj}^{(n)}. \qquad (5.9)$$

Proof. We proceed by induction. Formula (5.8) covers the case $m = n = 0$, and is true, since in zero steps the system cannot change (so it must remain, with probability 1, in state e_i). We shall now prove formula (5.9) for $n = 1$. Indeed, if the system passes from e_i to e_j in $m + 1$ steps, then in step m it must be in some state e_k. The events "state e_k after m steps" form a partition, and the total probability formula (4.9) gives in this case $p_{ij}^{(m+1)} = \sum_k p_{ik}^{(m)} p_{kj}$, which was to be shown, since $p_{ij} = p_{ij}^{(1)}$ by definition. The extension to the case of arbitrary n is immediate. □

In matrix notation, Theorem 5.3.1 states simply that $\mathbf{P}^{(m+n)} = \mathbf{P}^{m+n}$, with $\mathbf{P}^0 = \mathbf{I}$, the identity matrix.

One may expect that as the number of transitions increases, the system "forgets" the initial state, that is, the effect of the initial state on the probability of a state at time n gradually wears out. Formally, we may expect that the probabilities $p_{ij}^{(n)}$ converge, as n increases, to some limits independent of the starting state e_i. In Section 5.4 we explore conditions under which this is true, and we also find the limits of the probabilities $p_{ij}^{(n)}$. At present, we motivate the consideration by the following example.

Example 5.3.1. The weather on Markov Island is governed by the following laws. There are only three types of days: sunny, rainy, and cloudy, and the weather on a given day depends only on the weather on the preceding day. There are never two rainy days in row, and after a rainy day, a cloudy day is twice as likely as a sunny day. Fifty percent of days following a cloudy day are also cloudy, while 25% are rainy. Finally, after a sunny day each type of weather is equally likely. How often on the average is it cloudy on Markov Island?

SOLUTION. The fact that the tomorrow's weather depends only on the weather today makes the process of the weather change a Markov process. Using the obvious notation R, S, and C for the three types of weather, we have $p_{RR} = 0$, $p_{RC} = \frac{2}{3}$, $p_{RS} = \frac{1}{3}$; $p_{CC} = \frac{1}{2}$, $p_{CR} = p_{CS} = \frac{1}{4}$; finally, $p_{SS} = p_{SC} = p_{SR} = \frac{1}{3}$, so that the transition matrix is

$$\mathbf{P} = \begin{bmatrix} 0 & \frac{2}{3} & \frac{1}{3} \\ \frac{1}{4} & \frac{1}{2} & \frac{1}{4} \\ \frac{1}{3} & \frac{1}{3} & \frac{1}{3} \end{bmatrix}. \tag{5.10}$$

Now, assume that the limits of the *n*-step transition probabilities exist and do not depend on the starting state. Thus, we have three limits, u_R, u_S, and u_C, where

$$u_R = \lim p_{RR}^{(n)} = \lim p_{CR}^{(n)} = \lim p_{SR}^{(n)},$$

and similarly for u_S and u_C. We may then write, using (5.9) for $n = 1$,

$$\begin{aligned} p_{RR}^{(m+1)} &= p_{RR}^{(m)} p_{RR} + p_{RC}^{(m)} p_{CR} + p_{RS}^{(m)} p_{SR} \\ p_{CC}^{(m+1)} &= p_{CR}^{(m)} p_{RC} + p_{CC}^{(m)} p_{CC} + p_{CS}^{(m)} p_{SC} \\ p_{SS}^{(m+1)} &= p_{SR}^{(m)} p_{RS} + p_{SC}^{(m)} p_{CS} + p_{SS}^{(m)} p_{SS}. \end{aligned} \tag{5.11}$$

Passing to the limit in (5.11) we obtain a system of three linear equations:

$$\begin{aligned} u_R &= u_R p_{RR} + u_C p_{CR} + u_S p_{SR} \\ u_C &= u_R p_{RC} + u_C p_{CC} + u_S p_{SC} \\ u_S &= u_R p_{RS} + u_C p_{CS} + u_S p_{SS}. \end{aligned}$$

Substituting the probabilities from the matrix (5.10), this system reduces to

$$\begin{aligned} -u_R + u_C/4 + u_S/3 &= 0 \\ 2u_R/3 - u_C/2 + u_S/3 &= 0 \\ u_R/3 + u_C/4 - 2u_S/3 &= 0. \end{aligned} \tag{5.12}$$

This is a homogeneous system, so that we need an additional equation to determine the solution. We have here the identity

$$p_{RR}^{(n)} + p_{RC}^{(n)} + p_{RS}^{(n)} = 1, \tag{5.13}$$

and passing to the limit with *n* we get

$$u_R + u_C + u_S = 1. \tag{5.14}$$

The solution of (5.12) and (5.14) is $u_R = 9/41, u_C = 20/41$, and $u_S = 12/41$, which means that almost 50% of days on Markov Island are cloudy.   □

The equations of this example are derived under the assumption that (a) the limits $p_{ij}^{(n)}$ exist and are independent of the starting state i, and (b) the number of states is finite. The latter assumption allowed us to pass to the limit in the sums in (5.11) and (5.13), obtaining (5.12) and (5.14). As regards assumption (a), it is true under some conditions. Basically, we have to exclude two obvious cases when the probabilities $p_{ij}^{(n)}$ either cannot converge, or if they do, the limits depend on the starting state i.

Example 5.3.2. Consider a Markov chain with two states, e_1 and e_2, such that $p_{12} = p_{21} = 1$ (so that $p_{11} = p_{22} = 0$). Clearly, the system must alternate between states 1 and 2 in a deterministic way. Consequently, $p_{11}^{(n)} = 0$ or 1, depending whether n is even or odd, so that the sequence $p_{11}^{(n)}$ does not converge.

Example 5.3.3. Consider again a Markov chain with two states e_1 and e_2 and assume now that the transition probabilities are $p_{11} = p_{22} = 1$ (so that $p_{12} = p_{21} = 0$). In this case the system must forever remain in the initial state, so that we have $p_{11}^{(n)} = 1$ and $p_{21}^{(n)} = 0$ for all n. The limits $p_{ij}^{(n)}$ exist but depend on the initial state i.   □

These two examples, in essence, exhaust all possibilities that may prevent the existence of the limits $p_{ij}^{(n)}$ and their independence of the initial state i. One may summarize them as *periodicity* (in Example 5.3.2) and the *impossibility of reaching some state from some other state or states* (in Example 5.3.3). In the next section we introduce definitions pertaining to the classification of states in a Markov chain. These definitions will allow us to formulate the conditions under which the n-step transition probabilities converge to certain limits, regardless of the starting state.

PROBLEMS

5.3.1 Find all two-step transition probabilities for the Markov chain described in Problem 5.2.3.

5.3.2 Find all two-step transition probabilities for the dog flea model of Problem 5.2.5.

5.3.3 Assume that a man's occupation can be classified as professional, skilled laborer, or unskilled laborer. Assume that of the sons of professional men, A percent are professional, the rest being equally likely to be skilled laborers as unskilled laborers. In the case of sons of skilled laborers, B percent are skilled laborers, the rest being equally likely to be professional men or unskilled laborers. Finally, in the case of unskilled laborers, C percent of the sons are unskilled laborers, the rest again divided evenly between the other two categories. Assume that every man has one son. Form a Markov chain by following a given family through several generations. Set up the matrix of transition probabilities, and find:

(i) The probability that the grandson of an unskilled laborer is a professional man.

(ii) The probability that a grandson of a professional man is an unskilled laborer.

5.3.4   In Problem 5.3.3 it was assumed that every man has one son. Assume now that the probability that a man has a son is r. Define a Markov chain with four states, where the first three states are as in Problem 5.3.3 and the fourth state is entered when a man has no son. This state cannot be left (it corresponds to a male line of the family dying out). Find the probability that an unskilled laborer has a grandson who is a professional man.

5.4   THE ERGODIC THEOREM

The main objective of this section is to formulate and sketch the proof of the *ergodic theorem*, which asserts that the nth-step transition probabilities converge in the manner described in Section 5.3. We begin with some definitions.

Definition 5.4.1.   We say that a set C of states is *closed* if for every state e_i in C we have

$$\sum_{e_j \in C} p_{ij} = 1.$$

□

The condition means that a one-step transition from a state in C leads always to a state in C. By induction, the same must be true for any number of steps, so that if C is closed, then for every state e_i in C and every $n = 1, 2, \ldots$ we have $\sum_{e_j \in C} p_{ij}^{(n)} = 1$ (it is not possible to leave a closed set of states).

Example 5.4.1.   Consider a Markov chain with the transition probability matrix given below (in this matrix "+" stands for a strictly positive probability):

$$
\begin{array}{c c c c c c c}
 & e_1 & e_2 & e_3 & e_4 & e_5 & e_6 \\
\begin{array}{c} e_1 \\ e_2 \\ e_3 \\ e_4 \\ e_5 \\ e_6 \end{array}
\left[
\begin{array}{c c c c c c}
+ & 0 & + & 0 & 0 & 0 \\
0 & + & 0 & 0 & + & 0 \\
+ & 0 & + & 0 & 0 & 0 \\
0 & + & 0 & 0 & + & 0 \\
0 & + & 0 & 0 & 0 & 0 \\
0 & 0 & + & + & 0 & +
\end{array}
\right]
\end{array}
$$

The scheme of possible one-step transitions is shown in Figure 5.2. As we may see, in addition to the whole set of states [which is always closed in view of (5.5)], the closed sets here are $\{e_1, e_3\}, \{e_2, e_5\}, \{e_2, e_4, e_5\}$.

Definition 5.4.2. A Markov chain in which the only closed set of states is the set of all states is called *irreducible*. Next, we say that state e_j is *accessible* from state e_i if $p_{ij}^{(n)} > 0$ for some n (i.e., one can reach e_j in some number of steps starting from e_i). Moreover, we say that states e_i and e_j *communicate* if each of them is accessible from the other (not necessarily in the same number of steps). Finally, we say that state e_i is *transient* if there exists a state e_j such that e_j is accessible from e_i but e_i is not accessible from e_j.   □

Let us remark here that a more general definition of transient states, suitable for chains with infinitely many states, is: A state is transient, if the probability of never returning to it is strictly positive.

We may now characterize irreducibility as follows:

Theorem 5.4.1. *A chain is irreducible if and only if all states communicate.*

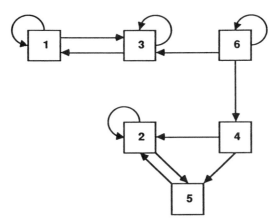

Figure 5.2 Scheme of transitions.

Proof. If two states communicate and C is a closed class of states, then either both states belong to C or neither of them does. It follows that if all states communicate, then they form the unique closed set, and hence the chain is irreducible. Conversely, suppose that the states e_i and e_j do not communicate. This means that at least one of them is not accessible from the other. Assume, for instance, that e_j is not accessible from e_i. For every k, if state k is accessible from e_i, then e_j cannot be accessible from e_k. Indeed, if it were not so, we would have $p_{ik}^{(n)} > 0$ for some n, and also $p_{kj}^{(m)} > 0$ for some m. But then $p_{ij}^{(n+m)} \geq p_{ik}^{(n)} p_{kj}^{(m)} > 0$, hence e_j would be accessible from e_i, contrary to the assumption. It follows that the class C of states accessible from e_i (closed by definition) does not contain state e_j, and the chain is not irreducible.   □

We shall say that state e_i is *periodic* if there exists $d > 1$ such that whenever $p_{ii}^{(n)} > 0$, then n is divisible by d. This means that a return to state e_i is possible only in a number of steps that is a multiple of d. The smallest d with this property is called the *period* of state e_i. If no such d exists, then state e_i is called *aperiodic*. One can show that in an irreducible chain all states must have the same period, or all be aperiodic. In this case we simply say that the chain is aperiodic.

We may now formulate the following theorem.

Theorem 5.4.2.   *Let* $\mathbf{P} = [p_{ij}], i, j = 1, \ldots, M$ *be the transition probability matrix of an aperiodic irreducible Markov chain with a finite number M of states. Then the limits*

$$\lim_{n \to \infty} p_{ij}^{(n)} = u_j \tag{5.15}$$

exist for every $j = 1, \ldots, M$, and are independent of the initial state e_i. Moreover, the u_j satisfy the system of linear equations

$$u_k = \sum_j u_j p_{jk}, \qquad k = 1, \ldots, M, \tag{5.16}$$

$$\sum_k u_k = 1. \tag{5.17}$$

Proof. Assume first that the limits (5.15) exist. As in Example 5.3.1, we have

$$p_{ij}^{(n+1)} = \sum_k p_{ik}^{(n)} p_{kj}.$$

Passing to the limit on both sides, we obtain

$$u_j = \lim_{n \to \infty} \sum_k p_{ik}^{(n)} p_{kj} = \sum_k \lim_{n \to \infty} p_{ik}^{(n)} p_{kj} = \sum_k u_k p_{kj},$$

which proves (5.16). Next, since $\sum_j p_{ij}^{(n)} = 1$ for every n, we may again pass to the limit on both sides, obtaining equation (5.17). In both cases, the interchange of order of summation and limiting passage is allowed in view of the finiteness of the number of terms in the sum. It remains to prove the existence of the limits. While the original proof was purely algebraic (based on an analysis of the powers of a stochastic matrix), we shall outline here a proof based on the concept of *coupling*, a surprisingly powerful probabilistic technique for analyzing Markov processes of various kinds. Let $u_k, k = 1, \ldots, n$ be a solution of the system (5.16)–(5.17). If a Markov chain starts with initial probabilities $r_i = u_i, i = 1, \ldots, M$, then the chain is *stationary* in the sense that $P[s(n) = e_j] = u_j$ for every $j = 1, \ldots, M$ and for every n. Indeed, the statement is true for $n = 0$ by definition. If it holds for some n, then (taking the possible states at the nth step as a partition) we obtain, using (5.16),

$$P\{s(n+1) = e_j\} = \sum_i P\{s(n) = e_i\}p_{ij} = \sum_i u_i p_{ij} = u_j.$$

Consider now two chains running independently in parallel, both with transition probability matrix **P**. One chain starts from state i and the other starts from the initial state chosen at random according to the distribution $\{u_j\}$. Let $s(n)$ and $s'(n)$ be the states at the nth transition of the two chains. We let T denote the first time when the two chains are in the same state, that is, $s(T) = s'(T)$. We say that T is the time of first *coupling* of the two chains.

The following argument is a good example of making use of conditioning with respect to events from a partition. We take as the partition the class of events $T = m$ for $0 \leq m \leq n$, together with the event $T > n$. These events are clearly disjoint and cover all possibilities. We have

$$p_{ij}^{(n)} = P\{s(n) = e_j | s(0) = e_i\}$$

$$= \sum_{m=0}^{n} P\{s(n) = e_j | s(0) = e_i, T = m\}P\{T = m | s(0) = e_i\}$$

$$+ P\{s(n) = e_j | s(0) = e_i, T > n\}P\{T > n | s(0) = e_i\}. \qquad (5.18)$$

But

$$P\{s(n) = e_j | s(0) = e_i, T = m\}$$
$$= P\{s(n) = e_j | s(0) = e_i, s'(k) \neq s(k), k \leq m - 1, s'(m) = s(m)\}$$
$$= P\{s'(n) = e_j | s(0) = e_i, s'(k) \neq s(k), k \leq m - 1, s'(m) = s(m)\}$$
$$= P\{s'(n) = e_j | s'(m) = s(m)\}$$
$$= P\{s'(n) = e_j\} = u_j \qquad (5.19)$$

(by stationarity of the chain s'). Substituting (5.19) to (5.18), we obtain

$$p_{ij}^{(n)} = \sum_{m=0}^{n} u_j P\{T = m | s(0) = e_i\}$$
$$+ P\{s(n) = e_j | T > n, s(0) = e_i\} P\{T > n | s(0) = e_i\}$$
$$= u_j P\{T \le n | s(0) = e_i\}$$
$$+ P\{s(n) = e_j | T > n, s(0) = e_i\} P\{T > n | s(0) = e_i\}.$$

To show that the sum of the last two terms converges to u_j as asserted, it suffices to show that for every i, the term $P\{T > n | s(0) = e_i\}$ converges to zero as n increases. We shall not give a formal proof here, but intuitively it ought to be clear that the probability that T exceeds n, that is, the probability of no coupling of the two processes during the first n transitions, tends to zero. Indeed, the only two ways in which the coupling could be avoided indefinitely is (a) if the two chains are periodic and "out of phase," or (b) if the chain is reducible, with two disjoint closed sets of states. Both possibilities are excluded by the assumption.

Example 5.4.2. Consider a Markov chain with two states, e_1 and e_2. Let the transition matrix be

$$\begin{bmatrix} a & 1-a \\ 1-b & b \end{bmatrix},$$

where $0 < a, b < 1$. Then the chain is irreducible (all states communicate) and aperiodic. The probabilities u_1, u_2 satisfy the system of equations

$$u_1 = au_1 + (1-b)u_2$$
$$u_2 = (1-a)u_1 + bu_2$$
$$u_1 + u_2 = 1.$$

The first two equations are the same. The solution is easily found to be

$$u_1 = (1-b)/(2-a-b), \quad u_2 = (1-a)/(2-a-b). \tag{5.20}$$

If both a and b are close to zero, then the probability of remaining in either of the states is small; hence a typical sequence of states will look like 1212121221212121211212121211..., with the states alternating most of the time. The probability of finding the system in state 1 after a large number of steps is close to $\frac{1}{2}$, so that we may expect u_i close to $\frac{1}{2}$ for $i = 1, 2$, as follows from (5.20).

Next, if a is close to 0 and b is close to 1, then the system tends to leave state 1 fast, and tends to remain in state 2. Thus, a typical sequence of states would be like 122222122222221122222221.... Consequently, u_1 should be close to 0 and u_2 should be close to 1, which agrees with (5.20).

Finally, if a and b are both close to 1, the system tends to remain in its present state, and consequently a typical sequence of states would be like 1112222111222... with the average durations of the series of identical terms depending on the ratio of the probabilities of changing $1 - a$ and $1 - b$.

$\square$

Example 5.4.3 (Quality Inspection Scheme). Consider a machine (production line, etc.) which produces some items. Each of these items may be good or defective, the latter event occurring with probability p, independent of the quality of other items.

The items are inspected according to the scheme (see Taylor and Karlin, 1984) described below, and every item found defective is replaced by a good one. The inspection scheme is described by two positive integers, say A and B. For brevity, we will be using the term (A, B) scheme. Inspection in the (A, B) scheme is of two kinds: *full inspection* (where every item is inspected), and *random inspection*, where one item is chosen for inspection at random from a group of B items. The rules are as follows:

1. The process starts with a full inspection.
2. Each full inspection period continues until one encounters a sequence of A good items. Then a period of random inspection starts.
3. Each period of random inspection lasts until one finds a defective item. In this case one starts the next period of full inspection.

Thus, the periods of full inspection alternate with the periods of random inspection. The first type of periods are costly but result in full elimination of defective items. The second type of periods are less expensive but leave some defective items.

To apply the theory of Markov chains to the analysis of the inspection scheme described above, let us define state i $(i = 0, 1, \ldots, A - 1)$ as the state when the system is in a period of full inspection and the last item inspected was the ith in a string of consecutive good items. Thus, we are in state 0 if the last item was defective, in state 1 if the last item was good but the preceding was defective, and so on. Moreover, let A be the state when the system is in a period of random inspection.

While in practical implementation of the (A, B) inspection scheme may indeed involve collecting B items and sampling one of them at random for inspection, such a procedure is not very convenient to analyze analytically. We shall therefore approximate the random sampling scheme used in real situations by another scheme, which leads to a Markov chain. We will carry out the analysis assuming that while in state A, every item is inspected, but the probability of it being defective is p/B, not p. Thus, the probability of finding no defective item among B items inspected is $(1 - p/B)^B \approx 1 - p$, which is the same as the probability of finding a good item if the choice is random from a set of B items.

Thus, we obtain the transition matrix: for $i = 0, 1, \ldots, A - 1$,

$$p_{i,j} = \begin{cases} 1 - p & \text{for } j = i + 1 \\ p & \text{for } j = 0 \\ 0 & \text{for all other } j, \end{cases}$$

while for $i = A$

$$p_{A,j} = \begin{cases} 1 - p/B & \text{for } j = A \\ p/B & \text{for } j = 0 \\ 0 & \text{for all other } j. \end{cases}$$

Clearly, all states communicate, so that the chain is irreducible. Also, we have $p_{A,A} > 0$, which means that (see the remarks preceding Theorem 5.4.2) the chain is aperiodic. Thus, the limits u_i exist and satisfy the system of equations

$$\begin{aligned} u_0 &= pu_0 + pu_1 + \cdots + pu_{A-1} + (p/B)u_A \\ u_i &= (1 - p)u_{i-1}, i = 1, 2, \ldots, A - 1 \\ u_A &= (1 - p)u_{A-1} + (1 - p/B)u_A \\ u_0 &+ u_1 + \cdots + u_A = 1. \end{aligned} \tag{5.21}$$

The second of equations (5.21) gives, by induction, the formula

$$u_i = (1 - p)^i u_0, \qquad i = 1, \ldots, A - 1. \tag{5.22}$$

The third equation of (5.21) gives $u_A = (B/p)(1 - p)u_{A-1}$, hence

$$u_A = \frac{B}{p}(1 - p)^A u_0. \tag{5.23}$$

Substitution of equation (5.23) in (5.21) gives

$$u_0 \left\{ 1 + (1 - p) + (1 - p)^2 + \cdots + (1 - p)^{A-1} + \frac{B}{p}(1 - p)^A \right\} = 1,$$

hence $u_0 = p\{1 + (B - 1)(1 - p)^A\}^{-1}$, and generally,

$$u_i = \frac{p(1 - p)^i}{1 + (B - 1)(1 - p)^A}, \qquad i = 0, 1, \ldots, A - 1$$

$$u_A = \frac{B(1 - p)^A}{1 + (B - 1)(1 - p)^A}.$$

Since we inspect all items as long as the system is in states $E_0, \ldots, E_{A-1}$ and only one item out of B is inspected in state E_A, the average fraction of items inspected is

$$f = u_0 + \cdots + u_{A-1} + \frac{1}{B}u_A$$

$$= \frac{1}{1 + (B-1)(1-p)^A}.$$

This fraction is the major component in the cost (per item produced) of inspection. If $p = 1$, the cost is 1 (all items inspected). If $p = 0$, the cost is $1/B$, since only one in each B items is then inspected.

On the other hand, the fraction of noninspected items is $1 - f$, and out of those, the fraction p are defective. Since all other defective items are detected and replaced by good ones, the average quality of product (measured by fraction of defectives) resulting from the (A, B) scheme is $(1 - f)p$, equal to

$$Q = \frac{(B-1)(1-p)^A p}{1 + (B-1)(1-p)^A}.$$

In practice, p may be not known (or may be subject to change). Then finding (A, B) optimal against some specific p (with optimality suitably defined through costs and the final quality) is not practicable. One can, however, determine the maximum of the quality Q (i.e., $Q^* = \max\{Q : 0 \le p \le 1\}$). This value depends on the constants A and B of the inspection plan. Similarly, one can find the maximal inspection cost $f^* = \max\{f : 0 \le p \le 1\}$. Knowledge of Q^* and f^* [as functions of (A, B)] allows finding the optimal sampling scheme, as an acceptable compromise between cost of inspection and loss due to the resulting quality of the product. The values Q^* and f^* are conservative in the sense that they provide guaranteed quality and guaranteed cost, not to be exceeded (on the average) in real situations characterized by unknown p. □

Since u_j is the limiting frequency of visits in state e_j, one may expect that the reciprocal $1/u_j$ should be equal to the average number of steps between two consecutive visits in state e_j. We shall return to this problem in subsequent chapters.

Theorem 5.4.2 covers the case of an irreducible chain. We also have the following theorem:

Theorem 5.4.3. *If the chain has a finite number of states, and there exist m and j such that $p_{ij}^{(m)} > 0$ for all i, then the assertions of Theorem 5.4.2 hold: The limits (5.15) exist and satisfy the system of equations (5.16)–(5.17).*

We omit the proof. Observe that in this case the chain need not be irreducible, in the sense that some states may be transient (if e_j is transient, then $\lim p_{ij}^{(n)} = 0$ for all i).

Example 5.4.4. Suppose that the transition matrix is

$$\begin{bmatrix} 1 & 0 \\ p & 1-p \end{bmatrix}.$$

where $0 < p < 1$. Clearly, state 1 is absorbing and state 2 is transient. The chain has two closed sets of states, $\{1\}$ and $\{1,2\}$. The assumptions of Theorem 5.4.3 (there must be a positive column in some power of the transition matrix) hold for $j = 1$ and $m = 1$. The system of equations is $u_1 = u_1 + pu_2, u_2 = (1-p)u_2, u_1 + u_2 = 1$, and the solution is $u_1 = 1, u_2 = 0$, as expected: State 1 will eventually be reached.

PROBLEMS

5.4.1 Determine the period of the dog flea model of Problem 5.2.5.

5.4.2 Argue that if the number of states is M, and state e_j is accessible from state e_i, then it is accessible in no more than $M - 1$ steps.

5.4.3 A stochastic matrix $\mathbf{P} = [p_{ij}]$ is called *doubly* stochastic if the sums of its columns are 1. Show that if an irreducible and aperiodic Markov chain with M states has a doubly stochastic transition matrix, then $u_j = 1/M$ for all j.

5.4.4 By a long-established tradition (or by a tsar's order), each man in a certain remote Siberian village has exactly one son and the only names used are Ivan, Peter, or Boris (one may also assume that these restrictions apply only to the eldest sons, with all younger sons leaving the village). Assume that Ivan's son has a 50% chance of being named Ivan, and a 25% chance each of being named Peter or Boris. Peter's son has a $\frac{7}{12}$ chance of being named Peter, a $\frac{1}{3}$ chance of being named Ivan, and a $\frac{1}{12}$ chance of being named Boris. Finally, $\frac{2}{3}$ of sons of Borises are named Boris, with the remainder split evenly between the remaining two names. The choice of the name for a son is not affected by the name of his grandfather.
  (i) Which of the combinations: Ivan Ivanovich ($=$ Ivan, son of Ivan), Peter Petrovich, or Boris Borisovich is most frequent?
  (ii) Which is more frequent: Ivan Borisovich (Ivan, son of Boris) or Boris Ivanovich?
  (iii) On the average, what proportion of men have the name Peter?

5.4.5 Show that the probabilities u_j for the dog flea model of diffusion are given by the formula

$$u_j = \binom{N}{j} 2^{-N}, \qquad j = 0, 1, \ldots, N.$$

5.4.6 Another model of diffusion, intended to represent the diffusion of noncompressible substances (e.g., liquids) is as follows. There are N red and N green balls, distributed evenly between urns A and B, so that each urn contains exactly N balls. At each step one ball is selected from each urn at random, and the balls are interchanged. The state of the system is defined as (say) the number of red balls in urn A. Find the transition probability matrix, show that the limiting probability distribution u_j exists, and find it.

5.4.7 Let $s(n)$ be a stationary Markov chain with transition probability matrix $\mathbf{P} = [p_{ij}]$ and $P[s(n) = j] = u_j$. Let $q_{ij} = P[s(n-1) = j | s(n) = i]$. Clearly, $\mathbf{Q} = [q_{ij}]$ is also a transition probability matrix (why?). We say that the chain with matrix $\mathbf{Q}$ is obtained from the chain with matrix $\mathbf{P}$ by *reversing time*. If $\mathbf{P} = \mathbf{Q}$, we say that the chain is *time reversible*.
  (i) Find the matrix $\mathbf{Q}$ in terms of $\mathbf{P}$.
  (ii) Check which (if any) of the two diffusion models (from Problems 5.2.5 and 5.4.6) are time reversible.

5.5 ABSORPTION PROBABILITIES

This section is devoted to problems similar to that of the gambler's ruin, discussed earlier. Let T be the set of all transient states, and assume that the remaining states may be partitioned into disjoint closed classes $C_1, C_2, \ldots, C_r$ with $r > 1$.

For $e_j \in T$, and let $q_j(C_k)$ be the probability of eventual absorption by the class C_k, that is,

$$q_j(C_k) = P\{s(n) \in C_k \text{ for some } n | s(0) = e_j\}.$$

Once the system reaches one of the states in C_k, it will remain there forever, and consequently,

$$q_j(C_1) + q_j(C_2) + \cdots + q_j(C_r) \le 1.$$

Here $1 - [q_j(C_1) + \cdots + q_j(C_r)]$ is the probability of the system remaining forever in transient states (this probability is zero if the number of states is finite; however, in case of an infinite number of states, it may be possible that the system will never leave the class T of transient states). We are interested in determining the probabilities $q_j(C_k)$ for any e_j in T and $k = 1, \ldots, r$. As it turns out, one can find these probabilities for each class separately. To simplify the notation, we may therefore omit the index k and look for generic values $q_j(C)$ for some closed class C. We have here the following theorem.

Theorem 5.5.1. *The probabilities $q_j(C)$ satisfy the system of linear equations:*

$$q_i(C) = \sum_{e_j \in T} p_{ij} q_j(C) + \sum_{e_j \in C} p_{ij}, \ e_i \in T. \qquad (5.24)$$

Proof. The proof is immediate: This equation states that from a state e_i in T one can pass in one step to another state e_j in T and then the absorption probability becomes $q_j(C)$, or one can pass to C, the probability of this being the second sum in (5.24). All other transitions lead to closed sets different than C, and then transition to C becomes impossible. $\qquad \square$

Example 5.5.1. Let us consider the basic unit of a tennis match, namely a game. During the game the same player always serves, and the winner of the game is the first player to win four points and to be at least two points ahead of the other player. Suppose that the players are A and B, and that A (the server, say) has probability p of winning each point (so that the probability of B winning a point is $1 - p$). Moreover, assume that the points are played independently. By tradition, the first two points are worth 15 each, and all points thereafter are worth 10 each, so that the tennis game may be regarded as a process of transitions over the partial scores, as depicted in Figure 5.3. The score corresponding to each of the players having won two points is $(30 - 30)$, and it is equivalent to the score corresponding to three points won by each $(40 - 40)$, or to any number $k > 3$ of balls won by each (called *deuce*). Equivalence here means that from this score, it is necessary to win two points more than the opponent in order to win the game. One

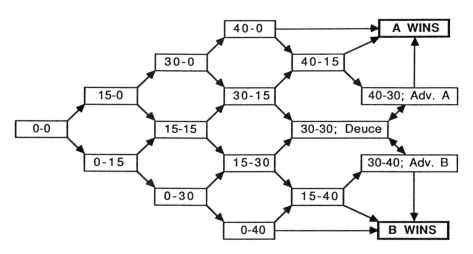

Figure 5.3 Scheme of a game in tennis.

may regard the game of tennis as a random walk on the set of states listed in Figure 5.3. The possible transitions and their probabilities are as marked, with the probability of going "up" being p, and the probability of going "down" being $1 - p$. There are two closed sets, one consisting of the state A labeled "A wins" and the other consisting of the state labeled "B wins." All the remaining states are transient. We want to find the probabilities $q_j(A)$ of player A winning the game if the score is j. In particular, we are interested in the value $q_{0-0}(A)$. For C consisting of the single state A, the system of equations (5.24) becomes the following. For states (scores) from which an immediate victory or loss is not yet possible, we have

$$q_{0-0}(A) = pq_{15-0}(A) + (1 - p)q_{0-15}(A),$$
$$q_{15-0}(A) = pq_{30-0}(A) + (1 - p)q_{15-15}(A),$$
$$\cdots$$
$$q_{30-30}(A) = pq_{40-30}(A) + (1 - p)q_{30-40}(A). \tag{5.25}$$

Next, for states from which an immediate win by A is possible,

$$q_{40-0}(A) = p + (1 - p)q_{40-15}(A),$$
$$q_{40-15}(A) = p + (1 - p)q_{40-30}(A), \tag{5.26}$$
$$q_{40-30}(A) = p + (1 - p)q_{30-30}(A).$$

Finally, for states from which an immediate loss by A is possible, we have

$$q_{0-40}(A) = pq_{15-40}(A),$$
$$q_{15-40}(A) = pq_{30-40}(A), \tag{5.27}$$
$$q_{30-40}(A) = pq_{30-30}(A).$$

To solve this system, one solves first the set of three equations labeled (5.25), (5.26), and (5.27), and then one proceeds recursively backwards, finally arriving at the formula

$$p_{0-0}(A) = (15p^4 - 34p^5 + 28p^6 - 8p^7)/(1 - 2p + 2p^2).$$

(see Bartoszyński and Puri, 1981). The graph of this probability, regarded as a function of the probability p of player A winning a single point, is given in Figure 5.4. Several observations here may be of some interest. First, the graph is fairly close to being linear in the range of p between 0.35 and 0.65, with values $q_{0-0}(A)$ close to 0 or 1 outside this range. This means that tennis is really interesting if it is played between evenly matched opponents. If one of them is much stronger than the other (say, the probability p of winning a point is above 70%), then the opponent has very little chance of winning a game, let alone a set or a match. Second, the derivative of the

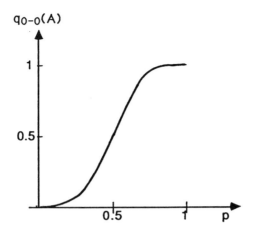

Figure 5.4 Probability $q_{0-0}(A)$ of winning a game by A as function of p.

function $q_{0-0}(A)$ at $p = \frac{1}{2}$ is $\frac{5}{2}$, so that the slope of $q_{0-0}(A)$ in the nearly linear (middle) part of the graph is about 2.5. This shows the role of practice training: for evenly matched players, every 1% increase in the probability of winning a point pays off in about a 2.5% increase in probability of winning a game. This is a substantial increase, if one considers that a match (between men) has at least 18 games.

PROBLEMS

5.5.1 Consider the following simple model of evolution. On a small island there is a room for 1000 members of a certain species. One year a favorable mutant appears. We assume that in each subsequent generation either the mutants take one place from the regular members of the species, with probability 0.6, or the opposite happens. Thus, for example, the mutation disappears in the very first generation with probability 0.4. What is the probability that the mutants eventually take over?

5.5.2 Find the probability of winning a game in tennis directly, using the fact that when the deuce (or 30–30) is attained, the probability of winning is $\sum_{n=0}^{\infty} (2pq)^n p^2$ (why?).

REFERENCES

Taylor, H. M., and S. Karlin, 1984. *An Introduction to Stochastic Modeling*, Academic Press, New York, pp. 132–134.

Bartoszyński, R., and M. L. Puri, 1981. "Some Remarks on Strategy in Playing Tennis." *Behavioral Science*, **26**.

CHAPTER 6

Random Variables: Univariate Case

6.1 INTRODUCTION

In Chapter 1, probability theory was generally characterized as a collection of techniques to describe, analyze, and predict random phenomena. We then introduced the concept of sample space, identified events with subsets of this space, and developed some techniques of evaluating probabilities of events. The latter were generally described through some phrases, and in each case it was necessary to identify the subset of the sample space corresponding to a given phrase.

Random variables, to be introduced now, will initially be regarded merely as useful tools for describing events. A random variable will be defined as a numerical function on the sample space S. Then, if X stands for random variable, inequality such as $X \leq t$ determines a set of all outcomes s in S satisfying the condition $X(s) \leq t$. We shall postulate* that $\{X \leq t\}$ is an event for each t. In this way, we gain a powerful tool for describing events, in addition to techniques used thus far, such as specifying subsets of sample space S by listing their elements or by providing a verbal description.

However, the following question arises. In Chapter 1 it was repeatedly stressed that the concept of sample space is to a large extent a mathematical construction and that one can have several sample spaces, all of them equally acceptable for the same phenomenon. The problem then is how to reconcile

---

*The reader may wonder why we use the term *postulate* here: The set $\{s \in S : X(s) \leq t\}$ is a subset of S, and consequently, is an event. However, as explained in Section 2.4, in the case of nondenumerable sample spaces S, we may be unable to define probability P on the class of *all* subsets of S, and we must therefore restrict considerations to some class of subsets of S, forming a σ-field, say $\mathcal{A}$, of events. In defining random variables, we postulate that $\{s \in S : X(s) \leq t\} \in \mathcal{A}$ for every t. In accepted terminology, we say that random variables X are functions on S which are *measurable* with respect to the σ-field $\mathcal{A}$. In the sequel we shall always tacitly assume that each set of the form $\{X \leq t\}$ is an event, and therefore it is permissible to speak of its probability.

the inherent lack of uniqueness of the sample space used to describe a given phenomenon with the idea of a random variable being a function defined on the sample space. At first sight it would appear that the concept of random variable, being based on another concept (sample space S) which allows subjective freedom of choice, must itself be tainted by subjectivity, and therefore be of limited use.

Specifically, suppose that there are several sample spaces, say $S, S', \ldots$, with the associated probability measures $P, P', \ldots$, that are equally acceptable for describing a given phenomenon of interest, and let $X, X', \ldots$ be random variables defined, respectively, on $S, S', \ldots$.

As we shall explain, those random variables that are useful for analyzing a given phenomenon must satisfy some "invariance" principle that make them largely independent on the choice of the sample space. The general idea here is as follows. Intuitively, a random variable is exactly what the name suggests: *a number depending on chance*. As a matter of fact, phrases of this sort were commonly used as "definitions" of the concept of random variable before probability theory was built on the notion of sample space. What we want to achieve is to define random variables in a way that makes them associated with the phenomenon studied rather than with a specific sample space, hence invariant under the choice of sample space. This is accomplished by introducing the following principle.

Invariance Principle for Random Variables.   Suppose that X' and X'' are two random variables defined on two sample spaces S' and S'' used to describe the same phenomenon. We say that these two random variables are *equivalent* if for every t the event $\{X' \leq t\}$ occurs if and only if the event $\{X'' \leq t\}$ occurs, and moreover, these events have the same probability.

Clearly, the equivalence of random variables defined above satisfies the logical requirements for relation of equivalence: namely, reflexivity, symmetry, and transitivity. We may therefore consider equivalence classes of random variables and speak of representatives of these classes. As a consequence, random variables are indeed useful in describing and analyzing random phenomena: They do not depend on a particular choice of a sample space, hence are free of the subjectivity and arbitrariness involved in the selection of S.

6.2   DISTRIBUTIONS OF RANDOM VARIABLES

Although formally a random variable is a numerical function on sample space S, the time-honored tradition in probability theory is to suppress the dependence on elements of S in notation and use capital letters at the end of the alphabet, X, Y, Z, possibly with subscripts, as symbols for random variables. This tradition stems in part from the fact that random variables had long been used before the formal foundations of probability theory were laid by

Kolmogorov (1933). In the sequel, when considering random variables, we shall assume that they are defined on some sample space S, but we shall refer to S only when needed.

With every random variable we shall associate its *probability distribution*, or simply *distribution*. The latter is defined as follows:

Definition 6.2.1. By a *distribution* of the random variable X we mean the assignment of probabilities to all events defined in terms of this random variable, that is, events of the form $\{X \in A\}$, where A is a set of real numbers.

□

Formally, the event above is a subset of the sample space S: We have

$$\{X \in A\} = \{s \in S : X(s) \in A\} \subset S.$$

Let X be a random variable. The basic type of event that we shall consider will be those when A is an interval, that is,

$$\{a < X < b\}, \{a \le X < b\}, \{a \le X \le b\}, \{a < X \le b\}, \qquad (6.1)$$

where $-\infty \le a \le b \le \infty$.

If we can compute the probability of each of the events (6.1) for all $a \le b$, then we can compute probabilities of more complicated events, by using the rules from Chapter 2. For example,

$$\begin{aligned}
P[\{a < X \le b\}^c] &= P\{X \le a \text{ or } X > b\} \\
&= P\{X \le a\} + P\{X > b\} \\
&= P\{X \le a\} + 1 - P\{X \le b\}.
\end{aligned}$$

Similarly, for the probability of the intersection $\{a < X \le b\} \cap \{c < X \le d\}$, the answer depends on the mutual relationship between a, b, c, and d (in addition to assumptions $a \le b$ and $c \le d$). If, for instance, $a < c < b < d$, then the event in question reduces to $\{c < X \le b\}$, and so on.

Actually, it turns out that it is sufficient to know the probabilities of only one type of event in (6.1) in order to determine the probabilities of other types of events in (6.1). We shall prove one such relation; the proofs of others are similar.

Theorem 6.2.1. *The probabilities of events of the form $\{a < X \le b\}$ for all $-\infty \le a \le b \le \infty$ uniquely determine the probabilities of events of the form $\{a < X < b\}, \{a \le X \le b\}, and \{a \le X < b\}.$*

Proof. We have

$$\{a < X < b\} = \bigcup_n \left\{ a < X \le b - \frac{1}{n} \right\}$$

$$\{a \le X \le b\} = \bigcap_n \left\{ a - \frac{1}{n} < X \le b \right\} \qquad (6.2)$$

$$\{a \le X < b\} = \bigcap_n \bigcup_m \left\{ a - \frac{1}{n} < X \le b - \frac{1}{m} \right\}.$$

We shall prove the last of these identities. Let s be a sample point belonging to the left hand side, so that $a \le X(s) < b$. Then for every n we have $a - 1/n < X(s)$. Also, if $X(s) < b$, then $X(s) \le b - 1/m$ for some m (actually, for all m sufficiently large). Thus, s belongs to the set $\bigcap_n \bigcup_m \{a - 1/n < X \le b - 1/m\}$. On the other hand, if s belongs to the right-hand side of the last equality in (6.2), then we have

$$a - \frac{1}{n} < X(s) \le b - \frac{1}{m}$$

for all n and for some m. Passing to the limit with $n \to \infty$, we obtain $a \le X(s) \le b - 1/m < b$, which means that s belongs to the set $\{a \le X < b\}$.

$\square$

Thus, each of the sets $\{a < X < b\}, \{a \le X \le b\}$ and $\{a \le X < b\}$ can be represented as the union or intersection (or both) of a sequence of events of the form $\{a_n < X \le b_n\}$.*

Let us observe that

$$\{a < X \le b\} = \{X \le b\} \setminus \{X \le a\}.$$

Consequently,

$$P\{a < X \le b\} = P\{X \le b\} - P\{X \le a\}$$

and the probabilities of all events of the form $\{a < X \le b\}$ are determined by the probabilities of events of the form $\{X \le t\}$ for $-\infty < t < \infty$. This justifies the following important definition:

Definition 6.2.2. For any random variable X, the function of the real variable t defined as

$$F_X(t) = P\{X \le t\} \qquad (6.3)$$

is called the *cumulative probability distribution function,*[†] or simply cumulative distribution function (cdf) of X.

---

* In terminology introduced in Chapter 1, we may say that each of the three types of events above belongs to the smallest σ-field generated by the class of all events of the form $\{a < X \le b\}$. Equivalently, we may say that the intervals on the real line of the kinds $(a, b), [a, b),$ and $[a, b]$ belong to the smallest σ-field of sets of real numbers that contains all intervals of the form $(a, b]$.
[†] According to the footnote at the beginning of this chapter, the right-hand side of (6.3) is well defined for each t (i.e., $\{X \le t\}$ is an event).

The following two examples illustrate the concept of cdf.

Example 6.2.1. The experiment consists of shooting once at a circular target T of radius R. Assume that it is certain that the target will be hit and that the probability of hitting a particular set A contained in T is given by $|A|/|T|$, where $|\cdot|$ is the area of the set.

SOLUTION. Let X be the random variable defined as the distance of the point of hitting from the center of the target. We shall determine the cdf of X. A natural sample space in this case is just T: Without loss of generality we may put the center of the coordinate system in the center of T, so that sample points $s = (x, y)$ satisfy $x^2 + y^2 \leq R^2$. If the target is hit at the point $s = (x, y)$, then $X = X(s) = \sqrt{x^2 + y^2}$ is the distance of the point of hit from the origin. Clearly, we have $0 \leq X \leq R$.

Now let $F_X(t) = P\{X \leq t\}$. Obviously, if $t > R$, then $F_X(t) = 1$, and if $t < 0$, then $F_X(t) = 0$. For t with $0 \leq t \leq R$, we have

$$F_X(t) = P\{X \leq t\}$$
$$= P\{\text{point } s \text{ falls in the circle of radius } t \text{ centered at the origin}\}$$
$$= \frac{\pi t^2}{\pi R^2} = \left(\frac{t}{R}\right)^2.$$

Thus,

$$F_X(t) = \begin{cases} 0, & t < 0 \\ \left(\dfrac{t}{R}\right)^2, & 0 \leq t \leq R \\ 1, & t > R. \end{cases} \tag{6.4}$$

The graph of $F_X(t)$ is given in Figure 6.1.   □

Remark. A simple fact, useful in determining the cdf's of random variables, is that if the random variable X is bounded from above, then $F_X(t) = 1$ for all $t > M$, where M is the upper bound for X. Indeed, if $t > M$, then

$$F_X(t) = P\{X \leq t\}$$
$$= P\{X \leq M\} + P\{M < X \leq t\} = P\{X \leq M\} = 1.$$

Similarly, if X is bounded from below, then $F_X(t) = 0$ for $t < m$, where m is the lower bound for X.

Example 6.2.2. Let the experiment consist of three tosses of a coin, and let X be the total number of heads obtained. The natural sample space now

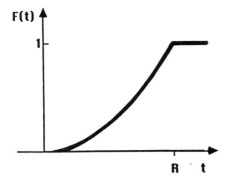

Figure 6.1 Cdf of the distance from the center of the target.

has eight points, listed below together with associated values of X:

S	X
HHH	3
HHT	2
HTH	2
THH	2
TTH	1
THT	1
HTT	1
TTT	0

The random variable X satisfies the condition $0 \le X \le 3$, so that $F_X(t) = 0$ if $t < 0$ and $F_X(t) = 1$ if $t \ge 3$. Moreover, since X can take on only values $0, 1, 2$, and 3, we have $P\{X \in A\} = 0$ for every set A that does not contain any of the possible values of X. Finally, simple counting gives $P\{X = 0\} = P\{X = 3\} = \frac{1}{8}$ and $P\{X = 1\} = P\{X = 2\} = \frac{3}{8}$. Thus, for $0 \le t < 1$ we may write

$$
\begin{aligned}
F_X(t) &= P\{X \le t\} \\
&= P\{X < 0\} + P\{X = 0\} + P\{0 < X \le t\} \\
&= P\{X = 0\} = \tfrac{1}{8}.
\end{aligned}
$$

Similarly, if $1 \le t < 2$, then

$$
\begin{aligned}
F_X(t) &= P\{X \le t\} \\
&= P\{X < 0\} + P\{X = 0\} + P\{0 < X < 1\} \\
&\quad + P\{X = 1\} + P\{1 < X \le t\} \\
&= P\{X = 0\} + P\{X = 1\} \\
&= \tfrac{1}{8} + \tfrac{3}{8} = \tfrac{1}{2}.
\end{aligned}
$$

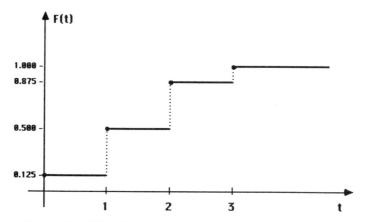

Figure 6.2   Cdf of the number of heads in three tosses of a coin.

Proceeding in a similar way with t in the remaining interval $2 \leq t < 3$, we get

$$
F_X(t) = \begin{cases}
0 & \text{for } t < 0 \\
\frac{1}{8} & \text{for } 0 \leq t < 1 \\
\frac{1}{2} & \text{for } 1 \leq t < 2 \\
\frac{7}{8} & \text{for } 2 \leq t < 3 \\
1 & \text{for } 3 \leq t.
\end{cases}
$$

The graph of $F_X(t)$ is given in Figure 6.2.   □

We shall now investigate in some detail the general properties of cdf's.

Theorem 6.2.2.   *Every cumulative probability distribution function F has the following properties:*

(a) *F is nondecreasing.*
(b) *$\lim_{t \to -\infty} F(t) = 0$, $\lim_{t \to \infty} F(t) = 1$.*
(c) *F is continuous on the right.*

Proof. Let F be the cdf of random variable X. To prove (a), let $t_1 < t_2$. Then

$$
F(t_2) - F(t_1) = P\{X \leq t_2\} - P\{X \leq t_1\} = P\{t_1 < X \leq t_2\} \geq 0.
$$

Consequently, $F(t_1) \leq F(t_2)$, which was to be shown.

To prove the second relation in (b), we have to show that $\lim_{n \to \infty} F(t_n) = 1$ for any sequence $\{t_n\}$ such that $t_n \to \infty$. Without loss of generality we may assume that $t_1 < t_2 < \cdots \to \infty$. The events $\{X \le t_1\}, \{t_1 < X \le t_2\}, \{t_2 < X \le t_3\}, \ldots$ are disjoint and their union is the whole sample space S. By the second and third axioms of probability, we have

$$1 = P(S) = P\{X \le t_1\} + \sum_{k=2}^{\infty} P\{t_{k-1} < X \le t_k\}$$

$$= F(t_1) + \sum_{k=2}^{\infty} [F(t_k) - F(t_{k-1})]$$

$$= F(t_1) + \lim_{n \to \infty} \sum_{k=2}^{n} [F(t_k) - F(t_{k-1})]$$

$$= \lim_{n \to \infty} \{F(t_1) + [F(t_2) - F(t_1)] + \cdots + [F(t_n) - F(t_{n-1})]\}$$

$$= \lim_{n \to \infty} F(t_n),$$

which was to be shown. The proof that $\lim_{t \to -\infty} F(t) = 0$ is analogous.

To prove (c), let $\{t_n\}$ be a sequence such that $t_1 > t_2 > \cdots \to t^*$. We have to show that $\lim_{n \to \infty} F(t_n) = F(t^*)$.

The events $\{X \le t_n\}$ form a monotonically decreasing sequence, with $\bigcap_{n=1}^{\infty} \{X \le t_n\} = \{X \le t^*\}$, so that by Theorem 2.6.1 we have

$$F(t^*) = P\{X \le t^*\} = P\left[\bigcap_{n=1}^{\infty} \{X \le t_n\}\right]$$

$$= P[\lim_{n \to \infty} \{X \le t_n\}] = \lim_{n \to \infty} P\{X \le t_n\} = \lim_{n \to \infty} F(t_n). \qquad \square$$

One of the consequences of the foregoing properties of cdf's is the following fact: For every x

$$P\{X = x\} = F(x) - F(x - 0), \tag{6.5}$$

where $F(x - 0) = \lim_{h \downarrow 0} F(x - h)$ is the left-hand-side limit of F at x. Indeed, we have

$$\{X = x\} = \bigcap_{n=1}^{\infty} \{x - h_n < X \le x\} \tag{6.6}$$

for every decreasing sequence of positive numbers h_n with $h_n \to 0$. Thus, by continuity property of probability [since the sets in the product (6.6) are decreasing],

$$P\{X = x\} = \lim_{n \to \infty} P\{x - h_n < X \le x\}$$

$$= \lim_{n \to \infty} [F(x) - F(x - h_n)] = F(x) - \lim_{n \to \infty} F(x - h_n)$$

$$= F(x) - F(x - 0).$$

Very often it is necessary to identify values of random variable corresponding to a given value of cdf. These values are called *quantiles*.

Definition 6.2.3. Let X be a random variable with cdf F, and let $0 < p < 1$. The *p*th *quantile* ξ_p of X (or of F) is defined as any solution of the simultaneous inequalities

$$P\{X \le x\} \ge p, \qquad P\{X \ge x\} \ge 1 - p. \qquad (6.7)$$

The inequalities (6.7) are equivalent to

$$F(x) \ge p, \qquad F(x - 0) \le p. \qquad \square$$

To illustrate this definition, assume now that X is a random variable with continuous cdf $F(x)$. For any p with $0 < p < 1$ there exists* a point ξ_p satisfying the relation $F(\xi_p) = p$. This means that $P\{X \le \xi\} = p$. If F is continuous, then $P\{X \ge \xi_p\} = P\{X > \xi_p\} = 1 - F(\xi_p) = 1 - p$, so that ξ_p satisfies (6.7). The point ξ_p need not be unique, though; F may satisfy the relation $F(x) = p$ for an interval of values x, and each of them can serve as ξ_p. So for the case of random variables with continuous cdf, the condition (6.7) reduces to

$$P\{X \le \xi_p\} = p, \qquad P\{X \ge \xi_p\} = 1 - p, \qquad (6.8)$$

with each of the relations (6.8) implying the other.

Example 6.2.3. For random variable X in Example 6.2.1, the quantiles can be detemined from the relation $(\xi_p/R)^2 = p$, hence $\xi_p = R\sqrt{p}$. For instance, if $p = 0.25$, then $\xi_{0.25} = R/2$: The chance of hitting the target at a distance less than $\frac{1}{2}$ of the radius is 0.25. $\square$

The reason for using inequalities in condition (6.7) instead of the simpler condition (6.8) lies in the fact that the equation $F(x) = p$ may have no solution. Such a situation will be illustrated by the following example.

Example 6.2.4. Let X assume values 1, 2, and 3 with probabilities 0.2, 0.6, and 0.2, respectively. Then $F(x) = 0.8$ for $2 \le x < 3$ and $F(x) = 1$ for $x \ge 3$. Thus, the equation $F(x) = p$ is solvable only for $p = 0.2$ and $p = 0.8$. To find, for example, a $\xi_{0.5}$, we need to use (6.7), looking for points such that $P\{X \le x\} \ge 0.5$ and $P\{X \ge x\} \ge 0.5$. The first inequality gives $x \ge 2$, while the second gives $1 - F(x - 0) \ge 0.5$, hence $x \le 2$, so both inequalities are satisfied only for $x = 2$ and hence $\xi_{0.5} = 2$. $\square$

---

*The existence of such a point follows from continuity—hence also the Darboux property—of function F: It must assume every value intermediate between its lower bound 0 and upper bound 1.

Certain quantiles have special names. For example, $\xi_{0.5}$ is called the *median*, $\xi_{0.25}$ and $\xi_{0.75}$ are called *lower* and *upper* quantiles. Quantiles may also be called *percentiles*, *p*th quantile being the same as $100p$th percentile.

Note that we have the following theorem.

Theorem 6.2.3. *If X is a random variable with continuous cdf and $0 < \alpha < \beta < 1$, then*

$$P\{\xi_\alpha \le X \le \xi_\beta\} = \beta - \alpha, \tag{6.9}$$

where ξ_α and ξ_β are any quantiles of order α and β respectively.

Proof: The left hand side of (6.9) equals $F(\xi_\beta) - F(\xi_\alpha - 0)$, which equals $F(\xi_\beta) - F(\xi_\alpha)$ by continuity of F, and hence equals $\beta - \alpha$ by the definition of quantiles. □

Example 6.2.5. For example, if F is continuous, there is always 50% probability that a random variable with cdf F will assume a value between its upper and lower quartile. □

Some comments about the concept of cdf's are now in order. First, we defined the cdf of a random variable X using nonstrict inequality, that is, $F(t) = P\{X \le t\}$. It is equally permissible to define cdf by the formula $F^*(t) = P\{X < t\}$. The only difference between F and F^* is that the latter is left continuous. Actually, such a definition of cdf appears prevalent in the Russian and Eastern European statistical literature.

The second comment concerns the sufficiency of conditions (a)–(c) in Theorem 6.2.2. To put it differently: If a function F satisfies conditions (a)–(c), does there exist a random variable X such that F is the cdf of X, that is, $F(t) = P\{X \le t\}$?

We shall not give the proof here, but mention only that the answer is positive: We have the following:

Theorem 6.2.4. *Any function F that satisfies conditions (a)–(c) of Theorem 6.2.2 is a cdf of some random variable.*

The importance of this theorem (as in the case of Theorem 2.6.3) is that it allows a statistician peace of mind by guaranteeing that some phrases that he or she uses make sense. In this case, it guarantees that the phrase "let F be a cdf," with no other assumptions about F than (a)–(c) of Theorem 6.2.2, is not without content.

The next question, quite naturally, is: Do cdf's determine uniquely the random variables associated with them? The answer here is negative: There are many random variables (associated with different phenomena, or even associated with the same phenomenon) which have the same cdf. Thus, it is a random variable that determines its cdf, but not conversely. To see why it is so, let us consider the following example.

Example 6.2.6. Continuing Example 6.2.2, let us consider an experiment consisting of three tosses of a coin and two random variables: $X =$ total number of heads and $Y =$ total number of tails. We have here, as before:

S	X	Y	S	X	Y
HHH	3	0	TTH	1	2
HHT	2	1	THT	1	2
HTH	2	1	HTT	1	2
THH	2	1	TTT	0	3

A simple count shows that $P\{X = k\} = P\{Y = k\}$ for all k, which implies that the cdf's $F_X(t)$ and $F_Y(t)$ are identically equal.

This example illustrates the fact that there exist two distinct variables X and Y defined on *the same* sample space S, which have the same distribution function. Obviously, if X and Y are defined as above, but refer to two distinct triplets of tosses, their distributions would still be the same.   □

*   Finally, the last comment concerning the cumulative distribution function of a random variable X is as follows. In Definition 6.2.1 we defined the distribution of random variable X as an assigment of probabilities $P\{X \in A\}$, where A is a set of real numbers. Now, the cumulative distribution function F_X provides us directly with probabilities of intervals open on the left and closed on the right:

$$P\{X \in (a, b]\} = P\{a < X \le b\} = F_X(b) - F_X(a). \qquad (6.10)$$

The question then is: Does F_X determine also the probabilities $P\{X \in A\}$ for sets A other then intervals $(a, b]$, and if so, what is the class of these sets? Before answering, let us observe the following. The leftmost member of (6.10) can be written as $P\{s \in S : X(s) \in (a, b]\}$, so that the symbol P refers to subsets of sample space S, as it should according to the definition of probability. On the other hand, the rightmost member depends only on a and b, and assigns a number to an interval $(a, b]$. We may therefore say that we deal here with probability on sets of real numbers (at least on intervals). According to the footnote at the beginning of this chapter, the left-hand side is well defined: We know that events (sets on which P is defined) form a σ-field, and we also know that sets $\{X \le t\}$ are events, so that

$$\{a < X \le b\} = \{X \le b\} \setminus \{X \le a\}$$

is also an event.

We can now formulate and prove the following extension theorem.

Theorem 6.2.5. *If F is a cumulative distribution function, then the function m_F defined on the class $\mathcal{B}_0$ of all intervals $(a, b]$ with $-\infty \leq a \leq b \leq +\infty$ by the formula*

$$m_F(a, b] = F(b) - F(a) \tag{6.11}$$

can be extended uniquely to a probability measure m_F on the smallest σ-field $\mathcal{B}$ containing $\mathcal{B}_0$.

Proof. According to Theorem 2.6.3 any σ-additive function on a field $\mathcal{G}$ has a unique extension to a σ-additive function on the smallest σ-field containing $\mathcal{G}$. To prove our theorem it therefore suffices to show (a) that m_F can be extended uniquely to the smallest *field* containing all intervals $(a, b]$, (i.e., to the smallest field, say $\mathcal{B}_1$, containing $\mathcal{B}_0$), and (b) that the function m_F extended to $\mathcal{B}_1$ is σ-additive on $\mathcal{B}_1$.

To prove (a), observe that the complement of $(a, b]$ is the union of $(-\infty, a]$ and $(b, \infty]$, while the intersection $(a, b] \cap (c, d]$ is empty if $b \leq c$ and otherwise equals $(\max(a, c), \min(b, d)]$. Consequently, each set in $\mathcal{B}_1$ is representable as a finite union of disjoint intervals $(a, b]$ from $\mathcal{B}_0$. Using generally the symbol I for sets in $\mathcal{B}_0$, each set $A \in \mathcal{B}_1$ can be written as

$$A = \bigcup_{j=1}^{m} I_j, \qquad \text{where } I_j \cap I_k = \varnothing, \text{ if } j \neq k. \tag{6.12}$$

We then define

$$m_F(A) = \sum_{j=1}^{m} m_F(I_j), \tag{6.13}$$

with $m_F(I_j)$ being defined in (6.11). We only need to show that definition (6.13) is unambiguous, that is, it does not depend on the choice of representation (6.12). Indeed, suppose that

$$A = \bigcup_{j=1}^{m} I_j = \bigcup_{k=1}^{n} I_k^*, \tag{6.14}$$

where $I_j \cap I_r = I_k^* \cap I_s^* = \varnothing$ if $j \neq r, k \neq s$. We have to show that

$$\sum_{j=1}^{m} m_F(I_j) = \sum_{k=1}^{n} m_F(I_k^*). \tag{6.15}$$

We may write, using (6.14),

$$I_j = I_j \cap A = I_j \cap \bigcup_{k=1}^{n} I_k^* = \bigcup_{k=1}^{n} (I_j \cap I_k^*)$$

and similarly,

$$I_k^* = A \cap I_k^* = \left(\bigcup_{j=1}^{m} I_j\right) \cap I_k^* = \bigcup_{j=1}^{m}(I_j \cap I_k^*).$$

Consequently,

$$\sum_{j=1}^{m} m_F(I_j) = \sum_{j=1}^{m} m_F\left(\bigcup_{k=1}^{n}(I_j \cap I_k^*)\right)$$

$$= \sum_{j=1}^{m}\sum_{k=1}^{n} m_F(I_j \cap I_k^*) = \sum_{k=1}^{n}\sum_{j=1}^{m} m_F(I_j \cap I_k^*)$$

$$= \sum_{k=1}^{n} m_F\left(\bigcup_{j=1}^{m}(I_j \cap I_k^*)\right) = \sum_{k=1}^{n} m_F(I_k^*), \qquad (6.16)$$

which shows (6.15).

It remains now to show that the function m_F is σ-additive on $\mathcal{B}_1$. In fact, the argument in (6.15) remains valid for infinite unions, provided we show that m_F is σ-additive on $\mathcal{B}_0$, that is, that for all $a \le b$ the condition

$$(a, b] = \bigcup_{k=1}^{\infty} I_k, \quad I_k \cap I_j = \varnothing \quad \text{for } k \ne j$$

implies that

$$F(b) - F(a) = \sum_{k=1}^{\infty} m_F(I_k).$$

For every fixed n we have

$$\bigcup_{k=1}^{n} I_k \subset (a, b].$$

Rearranging, if necessary, the intervals $I_k = (a_k, b_k]$, we may assume that for every n,

$$a \le a_1 \le b_1 \le a_2 \le b_2 \le \cdots \le b_{n-1} \le a_n \le b_n \le b. \qquad (6.17)$$

Consequently

$$m_F\left(\bigcup_{k=1}^{n} I_k\right) = \sum_{k=1}^{n} m_F(I_k) = \sum_{k=1}^{n}[F(b_k) - F(a_k)]$$

$$\le \sum_{k=1}^{n}[F(b_k) - F(a_k)] + \sum_{k=1}^{n-1}[F(a_{k+1}) - F(b_k)]$$

$$= F(b_n) - F(a_1) \le F(b) - F(a) = m_F(a, b].$$

Letting $n \to \infty$ we may write

$$\sum_{k=1}^{\infty} m_F(I_k) \le m_F(a, b].$$

It remains to prove the reverse inequality. At this point it is perhaps appropriate to make the following comment, without which the proof below might appear incomprehensible. In fact, a mathematically unsophisticated reader may think that the proof is "obvious:" Since $(a, b]$ is partitioned into a countable number of disjoint intervals, one can add their measures m_F simply by canceling negative and positive terms in expressions for contiguous intervals, according to the formula, valid for $a < b < c$, such as

$$[F(b) - F(a)] + [F(c) - F(b)] = F(c) - F(a).$$

The trouble is that it *need not be true that intervals I_k can be arranged into a sequence of contiguous intervals.*

To visualize such a possibility, consider the partition of interval $(-1, 1]$ by points

$$-1, -\frac{1}{2}, -\frac{1}{3}, -\frac{1}{4}, \ldots, \frac{1}{4}, \frac{1}{3}, \frac{1}{2}, 1,$$

so that

$$(-1, 1] = \bigcup_{n=1}^{\infty} \left(\frac{-1}{n}, \frac{-1}{n+1} \right] \cup \bigcup_{n=1}^{\infty} \left(\frac{1}{n+1}, \frac{1}{n} \right].$$

There is no rightmost term in the first union, and no leftmost term in the second union, so no cancellation occurs in passing from one sum to the other. The situation can be much more complicated, since there may be infinitely many such accumulation points as 0 in the example above.

To continue with this proof, let us exclude the trivial case $a = b$ and choose ϵ such that $0 < \epsilon < b - a$. Let the sets I_k in (6.16) be $I_k = (a_k, b_k]$ (observe that no monotonicity of sequences $\{a_k\}$ and $\{b_k\}$ is assumed).

Since function F is continuous on the right, for every n there exists $\beta_n > 0$ such that

$$F(b_n + \beta_n) - F(b_n) < \frac{\epsilon}{2^n}. \tag{6.18}$$

Let $I_n^{\epsilon} = (a_n, b_n + \beta_n)$. We then have

$$[a + \epsilon, b] \subset \bigcup_{n=1}^{\infty} I_n^{\epsilon}.$$

By the Heine–Borel lemma,* there exists a finite N such that

---

*The Heine–Borel lemma asserts that from any covering of a compact set by open sets one can choose a finite covering. In the present case, the closed interval $[a + \epsilon, b]$ is compact, and covering is by open sets $(a_n, b_n + \beta_n)$.

$$[a + \epsilon, b] \subset \bigcup_{n=1}^{N} I_n^{\epsilon}. \tag{6.19}$$

Let $n_1 \leq N$ be such that $b \in I_{n_1}^{\epsilon}$. If $a + \epsilon < a_{n_1}$, choose $n_2 \leq N$ such that $a_{n_1} \in I_{n_2}^{\epsilon}$. We continue in this way until some k such that $a_{n_k} \leq a + \epsilon$. Such a k must exist in view of (6.19). Renumbering, if necessary, the chosen intervals, we have $[a + \epsilon, b] \subset \bigcup_{n=1}^{k} I_n^{\epsilon}$ and

$$a_1 < b < b_1 + \beta_1$$
$$a_{i+1} < a_i < b_{i+1} + \beta_{i+1}, \qquad i = 1, 2, \ldots, k - 1$$
$$a_k \leq a + \epsilon < b_k + \beta_k.$$

Consequently, using (6.18),

$$m_F[a + \epsilon, b] \leq m_F(a_1, b_1 + \beta_1]$$

$$= m_F(a_1, b_1 + \beta_1] + \sum_{j=2}^{k} m_F(a_j, a_{j-1}]$$

$$\leq \sum_{j=1}^{k} m_F(a_j, b_j + \beta_j]$$

$$\leq \sum_{j=1}^{\infty} m_F(a_j, b_j] + \epsilon.$$

Letting $\epsilon \to 0$, we obtain

$$m_F(a, b] \leq \sum_{j=1}^{\infty} m_F(a_j, b_j] = \sum_{j=1}^{\infty} m_F(I_j),$$

which completes the proof.

PROBLEMS

6.2.1 In the statements below F and G stand for cdf's of random variables X and Y, respectively. Classify each of the statements below as true or false,

(i) If X is always strictly positive, then $F(t)$ is strictly positive for all t.

(ii) If $F(37) = F(45)$, then $P(40 < X < 42) = P(43 < X < 44)$.

(iii) If $Y = X + 3$, then $G(t) \leq F(t)$ for all t.

(iv) If $G(17) - F(17) = 1$, then both X and Y are always less than 17.

    (v) If $G(17) - F(17) = 1$, then Y cannot ever exceed X.
    (vi) If $G(17) \cdot F(17) = 1$, then $\max(X, Y) \leq 17$.
    (vii) If $G(17) \cdot F(17) = 0$, then $\min(X, Y) \leq 17$.
    (viii) If $|F(t) - G(t)| < \epsilon$ for all t, then $|X - Y| < \epsilon$.

6.2.2 Show that if $P(X \leq Y) = 1$, then $F(t) \geq G(t)$ for all t.

6.2.3 Figure 6.3 shows the graph of the cdf of random variable X. Find:
    (i) $P(X = 0)$.
    (ii) $P(X \leq 3)$.
    (iii) $P(X = -2)$.
    (iv) $P(X > 2)$.
    (v) $P(X > 2.79)$.
    (vi) $P(-1 < X \leq 0.7)$.
    (vii) $P(X < 3)$.
    (viii) $P(X < 0.13)$.
    (ix) $P(1 \leq |X| \leq 2)$.
    (x) $P(1.5 < |X| \leq 2)$.

6.2.4 A die is biased in such a way that probability of obtaining the result k ($k = 1, 2, \ldots 6$) is proportional to k. Let $X =$ outcome of a single toss of this die. Find $F_X(t)$.

6.2.5 A die from Problem 6.2.4 is tossed twice. Let the outcomes be X_1 and X_2 and let $V = \min(X_1, X_2)$. Determine the cdf of V.

6.2.6 Solve Problem 6.2.5 for $W = \max(X_1, X_2)$.

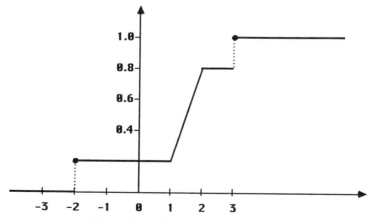

Figure 6.3 Cdf of random variable X.

6.2.7 A die from Problem 6.2.4 is tossed n times. Let $X_1, \ldots, X_n$ be the outcomes of consecutive tosses, and let $V_n = \min(X_1, \ldots, X_n)$, $W_n = \max(X_1, \ldots, X_n)$. Find the cdf's of V_n and W_n.

6.2.8 A point is chosen at random from a square with side a. Let X denote the distance from the selected point to the nearest corner of the square. Find and graph $F_X(x)$.

6.2.9 A coin of diameter d is tossed on a bathroom floor covered with square tiles with side of length $D > d$. Let X be the number of tiles that intersect with the coin. Find the distribution of X.

6.2.10 Let X be a random variable with cdf given by $F_X(x) = 0$ for $x < 0$ and $F_X(x) = 1 - 0.3e^{-\lambda x}$ for $x \geq 0$.
  (i) Determine $P(X = 0)$.
  (ii) Determine λ if $P(X \leq 3) = \frac{3}{4}$.
  (iii) Using the results of part **(ii)**, determine $P(|X| \leq 5)$.

6.2.11 Determine the medians and lower and upper quartiles for random variables with the following cdf's:

$$\textbf{(i)} \ \ F_X(x) = \begin{cases} 0 & \text{for } x < 0 \\ kx^3 & \text{for } 0 \leq x \leq 1/\sqrt[3]{k}, \ k > 0. \\ 1 & \text{for } x > 1/\sqrt[3]{k}, \end{cases}$$

$$\textbf{(ii)} \ \ F_X(x) = \begin{cases} 0 & \text{for } x < 0 \\ 1 - \alpha e^{-x} & \text{for } 0 \leq x < 1 \ \text{(consider all possible } \alpha). \\ 1 & \text{for } x \geq 1, \end{cases}$$

6.2.12 Determine the median of the number of tiles intersecting with the coin (see Problem 6.2.9), as a function of D and d.

6.2.13 Prove the first part of assertion (b) of Theorem 6.2.2.

6.3 DISCRETE AND CONTINUOUS RANDOM VARIABLES

Although the cdf of a random variable X provides all information necessary to determine probabilities $P\{X \in A\}$ for a large class of sets A, there exist wide and practically important classes of random variables whose distributions may be described in simpler and more economical ways. Two such classes are discussed in this section. Accordingly, we introduce the following definition.

Definition 6.3.1. A random variable X will be called *discrete* if there exists a finite or countably infinite set of real numbers $U = \{x_1, x_2, \ldots\}$ such that

$$P\{X \in U\} = \sum_n P\{X = x_n\} = 1. \tag{6.20}$$

Example 6.3.1. Let $U = \{1, 2, \ldots, n\}$ for some n and let $P\{X = j\} = 1/n$ for $j \in U$. Condition (6.20) clearly holds, so that the values of X are restricted to U. This example describes the selection (assumed fair) of the number of winning lottery ticket, where n is the total number of tickets. □

A discrete random variable with a finite set U of values, all of them having the same probability of occurrence, is called *uniform*, or with *uniform distribution* on the set of its possible values U. We shall later analyze this distribution in more detail.

Example 6.3.2 (Binomial Distribution). This distribution plays a central role in probability theory, and will be analyzed from various points of view throughout this book. Here we just give a definition and basic formulas. We consider n independent repetitions of the same experiment. In each repetition an event, called a *success*, may occur with probability p. □

Definition 6.3.2. The *binomial random variable* is defined as a total number of successes in n independent experiments, each of them leading to success with the same probability p. □

We have encountered special cases of binomial random variable in the preceding chapters, in examples analyzing the distribution of the "total number of heads in 3 tosses of a coin" and the like. The set of possible values of X is $\{0, 1, \ldots, n\}$, since the number of successes is an integer, at best equal to n (the number of trials) and at worst equal to 0 (if all repetitions lead to failure). The probability of k successes and $n - k$ failures in any specific order $SFFS \ldots S$ equals $p(1 - p)(1 - p)p \cdots p = p^k(1 - p)^{n-k}$. Since the probability of this string does not depend on its order, we obtain $P\{S_n = k\}$ by taking $p^k(1 - p)^{n-k}$ as many times as there are different orders of k letters S and $n - k$ letters F. This number is $\binom{n}{k}$, since each such order is completely specified by selecting the set of locations for letter S among n slots. Thus

$$P\{X = k\} = \binom{n}{k}p^k(1 - p)^{n-k}, \qquad k = 0, 1, \ldots, n \tag{6.21}$$

and the Newton binomial formula (3.16) shows that

$$\sum_{k=0}^{n} P\{X = k\} = \sum_{k=0}^{n} \binom{n}{k}p^k(1 - p)^{n-k} = [p + (1 - p)]^n = 1,$$

as it should be.

In the sequel we shall use the symbol $\text{BIN}(n,p)$ to denote a binomial distribution with parameters n and p, so that we can say that X has distribution $\text{BIN}(n,p)$ or simply $X \sim \text{BIN}(n,p)$. We also let the individual probabilities in distribution $\text{BIN}(n,p)$ be denoted by $b(k;n,p)$, so that

$$b(k;n,p) = \binom{n}{k} p^k (1-p)^{n-k}, \qquad k = 0,1,\ldots,n. \tag{6.22}$$

In the examples above the set U was finite, equal to $U = \{1,\ldots,n\}$ in Example 6.3.1 and equal to $U = \{0,1,\ldots,n\}$ in Example 6.3.2. In the example below the set U is infinite.

Example 6.3.3. Consider a sequence of independent tosses of a die. Let X be the number of tosses until the first ace. Clearly, X may assume values $1,2,3,\ldots$ and the event $\{X = k\}$ occurs if the kth toss gives an ace (chances are $\frac{1}{6}$) and the first $k-1$ tosses are all different from an ace [chances are $(\frac{5}{6})^{k-1}$]. Thus

$$P\{X = k\} = \frac{1}{6}\left(\frac{5}{6}\right)^{k-1}, \qquad k = 1,2,\ldots \tag{6.23}$$

We check easily that

$$\sum_{k=1}^{\infty} P\{X = k\} = \frac{1}{6}\sum_{k=1}^{\infty}\left(\frac{5}{6}\right)^{k-1} = \frac{1}{6}\cdot\frac{1}{1-\frac{5}{6}} = 1,$$

so that condition (6.20) is satisfied.

This random variable may generally be described as "waiting time for first success" (in this case, success being an ace). Distribution (6.23) is an example of *geometric* distribution, analyzed in detail in Chapter 9.

Example 6.3.4 (Poisson Distribution). We say that random variable X has *Poisson distribution* with parameter $\lambda > 0$ if the possible values of X are nonnegative integers $0,1,2,\ldots$ and

$$P\{X = k\} = \frac{\lambda^k}{k!}e^{-\lambda}, \qquad k = 0,1,2\ldots \tag{6.24}$$

Poisson distribution occurs often in practice and we shall study its properties in some detail in the following sections. At present, observe that (6.24) is correctly defined, since

$$\sum_{k=0}^{\infty} P\{X = k\} = \sum_{k=0}^{\infty}\frac{\lambda^k}{k!}e^{-\lambda} = e^{-\lambda}\sum_{k=1}^{\infty}\frac{\lambda^k}{k!} = e^{-\lambda}\cdot e^{\lambda} = 1.$$

$\square$

In the examples above the set U of possible values of X consisted of integers. This is the most frequent case, since typically discrete random variables represent some counts involved in observing random phenomena. In other words, discrete random variables are typically obtained as "the number of ...." One should remember, however, that the definition allows any values in the set U, not necessarily integers.

Let X be a discrete random variable. Its distribution is determined by the set $U = \{x_1, x_2, \ldots\}$ of possible values of X, and the assignment of probabilities $P\{X = x_i\}$ to all $x_i \in U$, the only condition being (6.20) (i.e., the sum of all probabilities must be 1). The function p_X, defined by the formula

$$p_X(x) = \begin{cases} P\{X = x\} & \text{if } x \in U \\ 0 & \text{if } x \notin U, \end{cases}$$

is often called the *probability mass function*, or *probability function* of random variable X, or (in more engineering-oriented texts) *probability density function*. We shall often use the first two of these terms. The term *density* appears confusing in this context, and we shall reserve it for continuous random variables, discussed later in this section.

The cdf of a discrete random variable may now be determined easily. We have

$$F(t) = P\{X \le t\} = \sum_{x_i \le t} P\{X = x_i\}. \tag{6.25}$$

Example 6.3.5. In the simplest cases, it may be most convenient to represent the distribution of a discrete random variable X as a double array of elements of U with the corresponding probabilities, such as

Values	-3	$\frac{1}{2}$	5
Probability	$\frac{1}{2}$	$\frac{1}{3}$	$\frac{1}{6}$

We have here $U = \{-3, \frac{1}{2}, 5\}$ and $P\{X = -3\} = \frac{1}{2}, P\{X = \frac{1}{2}\} = \frac{1}{3}, P\{X = 5\} = \frac{1}{6}$.

According to (6.25), we have $F(t) = 0$ for all $t < -3$. For $-3 \le t < \frac{1}{2}$ we have

$$F(t) = P\{X \le t\} = P\{X = -3\} = \frac{1}{2}.$$

For $\frac{1}{2} \le t < 5$ we have

$$P\{X \le t\} = P\{X = -3 \text{ or } X = \frac{1}{2}\}$$
$$= P\{X = -3\} + P\{X = \frac{1}{2}\} = \frac{1}{2} + \frac{1}{3} = \frac{5}{6}.$$

Finally, for $t \ge 5$ we have $P\{X \le t\} = P\{X \le 5\} = 1$. The graph of $F(t)$ is shown in Figure 6.4. $\square$

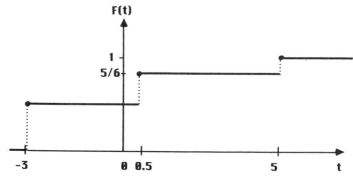

Figure 6.4 Cdf of random variable X.

At this point it is worthwhile to stress that cdf is defined for *all* real arguments, not only for the possible values of the random variable. For some reason, grasping this fact or remembering it is often hard for students, who become confused by a simple question like: "If X is the result of a toss of a die and F is the cdf of X, what is $F(2.27)$?" The point is to remember that the fact that X is an integer does not prevent us from evaluating F at *any* argument, integer or not.

Example 6.3.6.   We shall consider the random variable X from Example 6.3.3, namely $X =$ number of tosses of a die up to and including the first ace. Here the cdf is as follows. First, since $X \geq 1$, we have $F(t) = 0$ for all $t < 1$. If $k \leq t < k + 1$, where $k = 1, 2, \ldots$, then we have

$$F(t) = P\{X \leq t\} = \sum_{j=1}^{k} P\{X = j\} = \sum_{j=1}^{k} \frac{1}{6}\left(\frac{5}{6}\right)^{j-1} = 1 - \left(\frac{5}{6}\right)^{k}.$$

$\square$

In both examples above the cumulative distribution function was a step function, with cdf increasing at every point of the set U, and constant between jumps. One might think that this is a property characterizing discrete random variables. This is, however, not true: There are discrete random variables such that their cdf is not constant on *any* interval.

Example 6.3.7.   Let U be the set of all rational numbers. It is well known that U is countable; that is, all elements of U can be arranged in a sequence $x_1, x_2, \ldots$ (but U *cannot* be arranged in a sequence $x_1, x_2, \ldots$ with $x_n < x_{n+1}$ for all n). Let X be a random variable such that $P\{X = x_n\} = 1/2^n, n = 1, 2, \ldots$. Then cdf of X is not constant on any interval. Indeed, for $t_1 < t_2$ we have

$$F(t_2) - F(t_1) = \sum_{t_1 < x_n \leq t_2} P\{X = x_n\}, \tag{6.26}$$

and the right-hand side is positive, since there exists a rational number x_n between any two distinct real numbers t_1 and t_2. Consequently, F increases between any two points. ☐

Formula (6.26) shows how to calculate the probability of any interval $(t_1, t_2]$, and generally, for any set A we have

$$P\{X \in A\} = \sum_{x_n \in A} P\{X = x_n\}.$$

Example 6.3.8. A die is biased in such a way that the probability of obtaining k dots ($k = 1, \ldots, 6$) is proportional to k^2. Which number of dots is more likely: odd or even?

SOLUTION. Letting X be the result of a toss of the die in question, we have $U = \{1, 2, 3, 4, 5, 6\}$. The corresponding probabilities are

$$P\{X = 1\} = C,$$
$$P\{X = 2\} = 4C,$$
$$P\{X = 3\} = 9C,$$
$$P\{X = 4\} = 16C,$$
$$P\{X = 5\} = 25C,$$
$$P\{X = 6\} = 36C.$$

To find C we use (6.20), according to which we must have

$$C + 4C + 9C + 16C + 25C + 36C = 1,$$

which yields $C = 1/91$. Thus,

$$P(\text{result odd}) = P\{X = 1\} + P\{X = 3\} + P\{X = 5\} = \frac{1 + 9 + 25}{91} = \frac{5}{13},$$

$$P(\text{result even}) = 1 - \frac{5}{13} = \frac{8}{13}.$$ ☐

Example 6.3.9. The random variable X has the distribution given by the following array:

Value	−3	−2	0	2	3
Probability	$\frac{2}{9}$	$\frac{1}{9}$	p	p^2	$\frac{2}{9}$

and we want to find $P(|X + 1| > 1)$. Again, we have to use (6.20) to find the value of p. We have here

$$\frac{2}{9} + \frac{1}{9} + p + p^2 + \frac{2}{9} = 1,$$

which gives the quadratic equation $p^2 + p - \frac{4}{9} = 0$. The solutions are $p = \frac{1}{3}$ and $p = -\frac{4}{3}$, of which only the first is admissible as a probability. Consequently, $P\{X = 0\} = \frac{1}{3}, P\{X = 2\} = \frac{1}{9}$ and

$$P\{|X + 1| > 1\} = P\{X = -3\} + P\{X = 2\} + P\{X = 3\}$$
$$= \frac{2}{9} + \frac{1}{9} + \frac{2}{9} = \frac{5}{9}.$$ □

Let us now introduce another large class of random variables, whose distribution may be described in a convenient way other than by specifying the cdf. We start with the following definition.

Definition 6.3.3. The random variable X will be called *continuous* if there exists a function f, called the *density* of X, such that for all t we have

$$F(t) = \int_{-\infty}^{t} f(x)\, dx, \tag{6.27}$$

where F is the cdf of X. □

Example 6.3.10 (Uniform Distribution). Let $A < B$ and let the density $f(x)$ be constant on the interval $[A, B]$ and 0 outside it:

$$f(x) = \begin{cases} 0 & \text{if } x < A \\ c & \text{if } A \le x \le B \\ 0 & \text{if } x > B. \end{cases} \tag{6.28}$$

If $t < A$, then the integrand in (6.27) is identically 0, so that $F(t) = 0$. For $A \le t \le B$ we may write

$$F(t) = \int_{-\infty}^{A} f(x)\, dx + \int_{A}^{t} f(x)\, dx = \int_{-\infty}^{A} 0\, dx + \int_{A}^{t} c\, dx = c(t - A).$$

Finally, if $t > B$, then

$$F(t) = \int_{-\infty}^{A} f(x)\, dx + \int_{A}^{B} f(x)\, dx + \int_{B}^{t} f(x)\, dx = \int_{A}^{B} c\, dx = c(B - A).$$

This means that $F(t)$ is of the form presented on Figure 6.5. Clearly, since $\lim_{t\to\infty} F(t) = 1$ by Theorem 6.2.2, we must have $c(B - A) = 1$, hence $c = 1/(B - A)$. In other words, if function (6.28) is to be a density of a random variable, then c cannot be arbitrary and is, in fact, uniquely determined by

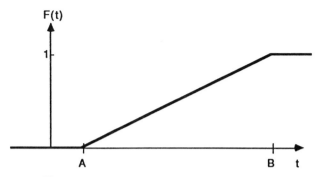

Figure 6.5 Cdf of uniform distribution on $[A, B]$.

the length of the interval $[A, B]$. *Thus, the density of the uniform distribution on $[A, B]$ is*

$$f(x) = \begin{cases} 0 & \text{if } x < A \\ \dfrac{1}{B-A} & \text{if } A \leq x \leq B \\ 0 & \text{if } x > B. \end{cases} \qquad (6.29)$$

□

In general, letting $t \to \infty$ in (6.27), we see that every density function f must satisfy the relation

$$\int_{-\infty}^{+\infty} f(x)\, dx = 1. \qquad (6.30)$$

Also, it follows from (6.27) that $F(t)$ is a continuous function, and consequently, for every x,

$$P(X = x) = F(x) - F(x - 0) = 0. \qquad (6.31)$$

Formula (6.27) combined with (6.31) leads to the following theorem:

Theorem 6.3.1. *If random variable X has density f, then for all $a < b$ we have*

$$P\{a \leq X \leq b\} = P\{a < X \leq b\} = P\{a < X < b\}$$

$$= P\{a \leq X < b\} = \int_{a}^{b} f(x)\, dx = F(b) - F(a).$$

This formula implies that in dealing with continuous random variables one can afford the luxury of being sloppy in handling inequalities. In particular, one can treat an event such as $\{X \leq a\}$ as (equivalent to) a complement of the event $\{X \geq a\}$, and so on. This is in sharp contrast with discrete random variables, where the events $\{X \leq a\}$ and $\{X < a\}$ may have different probabilities.

A question arises as to the extent to which the cdf of a continuous random variable determines its density. Formula (6.27) suggests that we have

$$F'(t) = f(t) \qquad (6.32)$$

and consequently, since F is a nondecreasing function (by Theorem 6.2.2), we must have

$$f(t) \geq 0. \qquad (6.33)$$

In fact, however, formulas (6.32) and (6.33) need not be valid for all points t. The reason is that the density f determines the probabilities of intervals through integration. This means that f is not defined uniquely. Indeed, two densities f_1 and f_2 that differ only at a single point or on a finite set of points will satisfy the condition

$$\int_a^b f_1(x)\,dx = \int_a^b f_2(x)\,dx$$

for all a and b. The same will be true if f_1 and f_2 differ on some set of measure zero.

Consequently, we may claim only that formulas (6.32) and (6.33) are valid almost everywhere, that is, outside a set of measure zero.

Example 6.3.11. The random variable X with density given in Example 6.3.10 is called *uniform* on interval $[A, B]$. It is clear, however, that if we modify the definition of f at boundaries A and B [e.g., by putting $f(A) = f(B) = 0$], the cdf will remain unchanged. A particular consequence of this observation is as follows. In real situations, we shall often deal with discontinuous densities defined by "broken formulas," that is, functions given by different formulas in different domains. In such cases, it does not matter how the density is defined at the boundaries between domains. □

The last remark is true as regards the calculation of probabilities. One should, however, be careful with the following interpretation of density. Since for $\Delta x > 0$, we have

$$P\{x \leq X \leq x + \Delta x\} = \int_x^{x+\Delta x} f(t)\,dt,$$

we may approximate the last integral, for *continuous densities* f, by $f(x)\,\Delta x$. This leads to an "engineer's" interpretation, according to which $f(x)\,dx$ is the "probability that random variable X will assume the value in the infinitesimal interval $(x, x + dx)$." One has to be careful with such an interpretation, however, at points where f is not continuous. For instance, in Example 6.3.10 we had $f(B) = 1/(B - A)$; hence the probability of X assuming a value in the interval $[B, B + h]$ for some small $h > 0$ may be taken as $h/(B - A)$. But this value is positive, while in fact the probability of X exceeding B by *any* amount is zero.

Let us now introduce an important type of continuous distributions called an exponential distribution.

Definition 6.3.4 (Exponential Distribution).   The distribution with the density

$$f(x) = \begin{cases} \lambda e^{-\lambda x} & \text{for } x > 0 \\ 0 & \text{for } x \le 0 \end{cases} \qquad (6.34)$$

where $\lambda > 0$ will be called *exponential* with parameter λ, denoted EXP(λ).   □

To check that (6.34) is indeed a density, we may write

$$\int_{-\infty}^{+\infty} f(x)\, dx = \int_{-\infty}^{0} + \int_{0}^{+\infty} \lambda e^{-\lambda x} dx = -e^{-\lambda x}\Big|_{0}^{\infty} = 1.$$

According to the remark made in Example 6.3.11, the value of the density at $x = 0$ plays no role. Thus, we could have defined $f(x)$ in Definition 6.3.4 as $\lambda e^{-\lambda x}$ for $x \ge 0$ and 0 for $x < 0$.

Example 6.3.12.   Let X be a random variable with density (6.34) for $\lambda = 2$ and let us compute $P\{|X - \frac{1}{2}| > \frac{1}{4}\}$. We have here, from the fact that density determines probabilities through integration,

$$P\left\{|X - \frac{1}{2}| > \frac{1}{4}\right\} = \int_{|x-1/2|>1/4} f(x)\, dx = \int_{x<1/4} f(x)\, dx + \int_{x>3/4} f(x)\, dx$$

$$= \int_{-\infty}^{0} 0\, dx + \int_{0}^{1/4} 2e^{-2x}\, dx + \int_{3/4}^{\infty} 2e^{-2x}\, dx$$

$$= 1 - e^{-1/2} + e^{-3/2} = 0.6166. \qquad □$$

Example 6.3.12 should serve as a warning to some students. It happens often that density is defined by a formula consisting of several parts, like (6.34). In calculating probabilities one has to integrate f over some set, and special care should be taken to use the proper part of the formula for f on appropriate parts of the domain of integration.

In the sequel we shall make the following convention: We shall always choose a version of density that is "most regular." In particular, this means choosing a continuous or piecewise continuous version if possible. When the density is discontinuous, the choice of values at breakpoints is irrelevant, and we shall always choose a version continuous either from the left or from the right, without taking particular care which version we take.

Example 6.3.13 (Normal Distribution).   One of the most common distributions encountered in both probability theory and statistics, as well as in nature, is the *normal distribution*, with density defined by

$$f(x) = f(x; \mu, \sigma) = \frac{1}{\sigma\sqrt{2\pi}} e^{-(x-\mu)^2/2\sigma^2} \tag{6.35}$$

where μ and $\sigma > 0$ are two parameters whose interpretation will be given later. The distribution (6.35) is often denoted $N(\mu, \sigma^2)$. The distribution $N(0,1)$, with density denoted by $\varphi(x)$, namely

$$\varphi(x) = \frac{1}{\sqrt{2\pi}} e^{-x^2/2}, \tag{6.36}$$

is called *standard normal*.

We shall show that the function (6.35) is a density; that is, it integrates to 1. Indeed, letting $z = (x - \mu)/\sigma$, so that $\sigma \, dz = dx$, we have

$$\frac{1}{\sigma\sqrt{2\pi}} \int_{-\infty}^{+\infty} e^{-(x-\mu)^2/2\sigma^2} \, dx = \frac{1}{\sqrt{2\pi}} \int_{-\infty}^{+\infty} e^{-z^2/2} \, dz$$

and it is sufficient to prove that standard normal density (6.36) integrates to 1.

The function $e^{-x^2/2}$ does not have an antiderivative that can be written in a closed form. Thus, to compute the integral

$$I = \frac{1}{\sqrt{2\pi}} \int_{-\infty}^{+\infty} e^{-x^2/2} \, dx$$

we need to use some trick. In this case, the trick is to compute I^2:

$$I^2 = \left(\frac{1}{\sqrt{2\pi}} \int_{-\infty}^{+\infty} e^{-x^2/2} \, dx \right) \cdot \left(\frac{1}{\sqrt{2\pi}} \int_{-\infty}^{+\infty} e^{-y^2/2} \, dy \right)$$
$$= \frac{1}{2\pi} \int_{-\infty}^{\infty} \int_{-\infty}^{+\infty} e^{-x^2+y^2/2} \, dx \, dy. \tag{6.37}$$

Let us pass to polar coordinates (r, θ), where $x = r \cos \theta$, $y = r \sin \theta$. The Jacobian of this transformation is

$$J = \begin{vmatrix} \dfrac{\partial x}{\partial r} & \dfrac{\partial x}{\partial \theta} \\[2mm] \dfrac{\partial y}{\partial r} & \dfrac{\partial y}{\partial \theta} \end{vmatrix} = \begin{vmatrix} \cos \theta & -r \sin \theta \\ \sin \theta & r \cos \theta \end{vmatrix} = r.$$

Changing the variables in (6.37), we obtain

$$I^2 = \frac{1}{2\pi} \int_0^{2\pi} d\theta \int_0^\infty re^{-r^2/2} \, dr = 1;$$

hence $I = 1$, which was to be shown.   □

As already mentioned, the density function $\varphi(x)$ of standard normal distribution (as well as any other normal density function $f(x; \mu, \sigma)$) does not have antiderivative expressible in closed form. The consequence of this fact is that to find probabilities for normal random variables we need suitable tables (or, nowadays, special computer packages).

Suppose that we have the random variable $X \sim N(\mu, \sigma^2)$ and we need to find the numerical value of the probability

$$P\{a \le X \le b\} = \frac{1}{\sigma\sqrt{2\pi}} \int_a^b e^{-(x-\mu)^2/2\sigma^2}\, dx. \tag{6.38}$$

As we shall see, in many situations when the distribution of the random variable depends on a parameter, we need extensive set of tables, especially when the parameter has two components as in the case of binomial distribution. Fortunately, in the case of the normal distribution, one table, that of standard normal distribution, is sufficient. To see why, consider the probability (6.38) and use the change of variable $z = (x - \mu)/\sigma$ in the integral. We have $dx = \sigma\, dz$; hence after substitution to (6.38) we obtain

$$P\{a \le X \le b\} = \frac{1}{\sqrt{2\pi}} \int_{z_1}^{z_2} e^{-z^2/2}\, dz$$
$$= \Phi(z_2) - \Phi(z_1),$$

where $z_1 = (a - \mu)/\sigma$, $z_2 = (b - \mu)/\sigma$, and Φ is the cdf of the standard normal distribution (Table A.2). This table is prepared for $z > 0$, but for $z < 0$ one may use the fact that $\varphi(z)$ is symmetric about 0, so that

$$\Phi(-z) = 1 - \Phi(z). \tag{6.39}$$

The procedure is illustrated by the following example.

Example 6.3.14. Suppose that $X \sim N(-0.7, 4)$ and that U is defined as the integer nearest to X. Find $P(U = -1)$.

SOLUTION. In this case $\mu = -0.7$ and $\sigma^2 = 4$, hence $\sigma = 2$. Consequently,

$$P(U = -1) = P(-1.5 < X < -0.5)$$
$$= \Phi\left(\frac{-0.5 - (-0.7)}{2}\right) - \Phi\left(\frac{-1.5 - (-0.7)}{2}\right)$$
$$= \Phi(0.1) - \Phi(-0.4).$$

Using (6.39) to reduce to positive arguments of Φ, we have

$$P(U = -1) = \Phi(0.1) - (1 - \Phi(0.4))$$
$$= \Phi(0.1) + \Phi(0.4) - 1$$
$$= 0.5398 + 0.6554 - 1 = 0.1952,$$

where the numerical values are given in Table A.2. $\qquad\square$

At the end of this section it is necessary to point out that discrete and continuous random variables do not exhaust all possibilities. First, we may have practical situations of random variables of "mixed" type, partially discrete and partially continuous. Second, we may also have random variables which are neither continuous nor discrete. This second possibility may appear at first as a mathematical pathology of some sort; nevertheless, there are random variables occurring in practice which have such pathological distributions.

Example 6.3.15.   An example of a random variable of "mixed" type may be as follows. We purchase a piece of equipment (e.g., a light bulb, etc.). It has lifetime T. This lifetime is typically a continuous random variable in the sense that T may assume any value from some interval; thus any particular value $T = t$ from this interval has probability zero. In addition, there is one value, namely $T = 0$, that can be assumed with positive probability. In other words, the bulb may either be broken at the time of purchase (in which case $T = 0$), or it may break at some future time $t > 0$, but then the event $\{T = t\}$ has probability zero. Consequently, the cdf of T is a function that is identically 0 for all negative t, and continuously increasing to 1 for positive t. At $t = 0$ the cdf is discontinuous, with $F(0)$ being the probability of purchasing a broken light bulb.   □

Such mixtures of continuous and discrete distributions still do not exhaust all possibilities. This is illustrated by the following example, which is of theoretical interest.

*   **Example 6.3.16.**   We shall now construct a cdf $F(t)$ that is continuous, increases from 0 to 1, and such that $F'(t) = 0$ except on a set of measure 0. The latter condition excludes the existence of density: Indeed, if the density exists, then it equals $F'(t)$ almost everywhere. Thus, we have $f(t) = 0$ almost everywhere, hence $\int_{-\infty}^{+\infty} f(t) \, dt = 0$, which means that F' is not a density.

The construction is based on the *Cantor set*. We let $F(t) \equiv 0$ for $t \leq 0$ and $F(t) \equiv 1$ for $t \geq 1$. Next, we let $F(t) \equiv \frac{1}{2}$ for $\frac{1}{3} \leq t < \frac{2}{3}$. On middle parts of intervals $[0, \frac{1}{3}]$ and $[\frac{2}{3}, 1]$, that is, for $\frac{1}{9} \leq t < \frac{2}{9}$ and for $\frac{7}{9} \leq t < \frac{8}{9}$, we let $F(t) \equiv \frac{1}{4}$ and $F(t) \equiv \frac{3}{4}$, respectively. This process is continued, and at each step, $F(t)$ is the average of values on neighboring intervals in the middle one-third of the "gap." The total length of intervals in $[0, 1]$ where F is constant (hence where $F' = 0$) is

$$ L = \frac{1}{3} + 2 \cdot \frac{1}{9} + 4 \cdot \frac{1}{27} + \ldots = \sum_{n=0}^{\infty} \frac{2^n}{3^{n+1}} = \frac{1}{3} \cdot \frac{1}{1 - \frac{2}{3}} = 1. $$

Moreover, one can show easily that F is continuous at each point. Thus, we have constructed a cdf of a random variable which is neither discrete nor continuous (it is called *singular*).

PROBLEMS

6.3.1 Let X have a continuous distribution with density e^{-x} for $x > 0$ and 0 for $x \leq 0$. Let $Y = [X]$ be the integer part of X, and let Z be the integer nearest to X. Find:

  (i) The distributions of Y and Z.

  (ii) $P(Y = Z)$.

  (iii) $P(Y = 3|Z = 4)$.

  (iv) $P(Z = 4|Y = 3)$.

  (v) $P(Y = 4|Z = 3)$.

  (vi) $P(Z = 3|Y = 4)$.

6.3.2 Let X have the density $\lambda e^{-\lambda x}$ for $x > 0$ and 0 for $x \leq 0$, and let Y and Z be defined as in Problem 6.3.1.

  (i) Find $P(Y = Z)$.

  (ii) Show that $P(Y = k|Z = k+1) = P(Y = Z)$ for all $k = 0, 1, \ldots$.

  (iii) Find $P(Z = k+1|Y = k)$ for $k = 0, 1, \ldots$.

6.3.3 Let X have the exponential distribution with density given by $f(x) = \lambda e^{-\lambda x}$ for $x > 0$ and $f(x) = 0$ for $x \leq 0$. Show that for $s, t > 0$ the following *memoryless* property of exponential distribution holds:

$$P\{X > s+t|X > s\} = P\{X > t\}.$$

6.3.4 Let X be a continuous random variable with cdf F. Assume that F is strictly increasing. After observing the value of X we compute $Y = F(X)$. Find the cdf of the random variable Y.

6.3.5 A computer can generate values of the random variable with distribution uniform on $[0, 1]$. Use the results of Problem 6.3.4 to design a method of using these values to generate observations of a random variable with cumulative distribution F.

6.3.6 Let X have the density $f(x) = Cx$ for $0 \leq x \leq 1, f(x) = C(2-x)/2$ for $1 < x \leq 2$ and $f(x) = 0$ otherwise. Find C and $F(x)$ and compute the following probabilities using the density and the cdf. Interpret the probabilities obtained on the graphs of $f(x)$ and $F(x)$.

  (i) $P(X \geq 3/2)$.

  (ii) $P(|X - 1| \leq \frac{1}{2})$.

  (*Hint:* The problem can be solved using simple geometry, without integration.)

6.3.7 Let X have the density $f(x) = Ce^{-0.4|x|}$, $-\infty < x < +\infty$. Find C and:

  (i) $P(X > -2)$.

  (ii) $P(|X + 0.5| < 1)$.

6.3.8 Let X_n be the difference (possibly negative) between the number of heads and the number of tails in n tosses of a coin. Find:

    (i) The distribution of X_4.

    (ii) The cdf of X_4 at point $x = -0.6$.

    (iii) The probability that X_n is positive given that it is nonnegative for (a) $n = 4$ and (b) $n = 5$.

6.3.9 You have 5 coins in your pocket: 2 pennies, 2 nickels, and a dime. Three coins are drawn at random. Let X be the total amount drawn (in cents). Find:

    (i) The distribution of X.

    (ii) $P(X \leq 10 | X \leq 15)$.

    (iii) The probabilities that two pennies are drawn, if it is known that $X \leq 11$.

6.3.10 Let the random variable X have a uniform distribution on the union of intervals $(0, a)$ and $(a + 2, b)$. Let F be the cdf of X. Assume that $F(4) = 0.2, F(a + 1) = 0.25$. Find:

    (i) a and b.

    (ii) $F(8.39)$.

    (iii) $P(3.01 \leq X \leq 9.14)$.

6.3.11 Let p be a fixed number with $0 < p < 1$, and let ξ denote a random variable with uniform distribution on $[0, 1]$, sampled independently on each run through the block of the flowchart on Figure 6.6 labeled "sample ξ." Find the distribution of X at the time when the program stops.

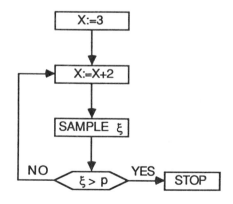

Figure 6.6 Flowchart for Problem 6.3.11.

6.3.12 An oscillator sends the wave $X(t) = A\cos(2\pi t)$, where $A = 1$ or 2 with equal probabilities. We observe the value of $X(t)$ at the point chosen at random, with uniform distribution, on the interval $[n, n+1)$ for some n. Find:

    (i) $P(X(t) \leq 1)$.

    (ii) $P(|X(t)| > 3/2)$.

    (iii) $P(X(t) > 0)$.

6.4 FUNCTIONS OF RANDOM VARIABLES

In this section we discuss the problems of transformations of random variables. The motivation for study of such problems is easy to explain and is very convincing. First, the motivation lies simply in the fact that in practical situations we deal with functions (often referred to more picturesquely as *transformations*) of random variables and we generally face the problem of determining the distributions of transformed random variables, given the distribution of untransformed ones. The examples abound, starting from the simplest cases, such as a change of unit of measurement, or representations on the logarithmic scale. More complicated cases, discussed in Chapter 7, involve pairs of random variables (e.g., conversion of a pair of Cartesian coordinates of a random point on the plane to polar coordinates of this point, or ratios of coordinates). The latter case occurs, for instance, in determining the distribution of velocity, calculated as the ratio of distance to time, both subject to variability (either in the actual measured velocity or due to errors of measurements).

Another compelling reason to study the transformation of random variables is connected with computer simulation. It happens often that we need to generate random numbers that follow a specific distribution. A typical method here involves transformations. A computer (and often even pocket calculators) can generate random numbers from a uniform distribution on $[0,1]$. Then the desired random variables are obtained as suitable transformations of uniform random variables. Obviously, to apply such a method it is necessary to develop techniques of determining the distributions of functions of random variables, at least of those uniformly distributed.

Finally, the third reason is related to statistics. Observations, such as experimental results, values recorded in the sample, and so on, are regarded as values of random variables. These are often summarized into global indices; a typical case in point is the average. Each such index (generally referred to as a *statistic*) is a function of the sample values, hence a function of values of random variables, and for statistical inference it is vital to know the distributions of such indices.

We begin with the conceptually and technically simplest case of transformations of one discrete random variable. Let X assume values in the set $U = \{x_1, x_2, \ldots\}$, with

$$p_i = P\{X = x_i\},$$

where $\sum p_i = 1$. Let φ be a real-valued function of real variable, and let

$$Y = \varphi(X).$$

Obviously, Y has a discrete distribution, and our objective is to find it.

Example 6.4.1. Instead of starting with formulas, we first explain the situation using a simple numerical example. Suppose that X has the following distribution:

Value	-2	-1	0	1	2	3	4
Probability	$\frac{1}{10}$	$\frac{2}{10}$	$\frac{1}{10}$	$\frac{1}{10}$	$\frac{1}{10}$	$\frac{2}{10}$	$\frac{2}{10}$

Suppose that $\varphi(x) = x^2$ so that $Y = X^2$. Then the distribution of Y is

Value	0	1	4	9	16
Probability	$\frac{1}{10}$	$\frac{1}{10} + \frac{2}{10}$	$\frac{1}{10} + \frac{1}{10}$	$\frac{2}{10}$	$\frac{2}{10}$

The sums in the second row are connected to the fact that φ is not one to one: Indeed,

$$P\{Y = 4\} = P\{X^2 = 4\} = P\{X = 2\} + P\{X = -2\} = \frac{1}{10} + \frac{1}{10},$$

and similarly for $P\{Y = 1\}$. $\qquad\qquad\qquad\qquad\qquad\qquad\qquad\qquad\square$

In general, we have

$$P\{Y = y\} = P\{\varphi(X) = y\}$$
$$= P\{x : \varphi(x) = y\} = \sum_{x_i : \varphi(x_i) = y} P\{X = x_i\}.$$

An analogous formula holds for functions of two variables, that is, random variables of the form $Z = \varphi(X, Y)$, where the distribution of Z is expressed in terms of distributions of X and Y. The principle here is in fact extremely simple, so that it is probably always best to start "from scratch" and use common sense. We illustrate this by examples.

Example 6.4.2. Let the experiment consist of two independent throws of a die, where we let X and Y be the result on the first and second throws respectively. We want to find the distribution of $Z = \max(X, Y)$. Thus, the value of the function $\varphi(x, y)$ equals x or y depending on whether $x \geq y$ or $y \geq x$, and the random variable Z may be described as "the best out of two throws."

Clearly, Z may be $1, 2, \ldots, 6$ and it is simplest to find the distribution of Z by direct listing of the outcomes associated with specific values of Z. Thus $Z = 1$ only if $(X, Y) = (1, 1)$, hence $P\{Z = 1\} = \frac{1}{36}$. Next, $Z = 2$ if the outcome is $(2, 1), (1, 2)$, or $(2, 2)$, so that $P\{Z = 2\} = \frac{3}{36}$. Proceeding in this way, we get

Z	1	2	3	4	5	6
Probability	$\frac{1}{36}$	$\frac{3}{36}$	$\frac{5}{36}$	$\frac{7}{36}$	$\frac{9}{36}$	$\frac{11}{36}$

From this table it can be seen how the repetition, or "giving the person second chance," improves the score: Larger values are much more likely than smaller values.

Example 6.4.3. Let us continue Example 6.4.2, generalizing it to the case of n random variables (to be considered formally in Chapter 7). Let us imagine that an adult plays some game with a child, and in order to give the child some advantage the adult allows the child to use the "best of n tosses" of a die. How large should n be to give the child a 99% or more chance to score at least 4?

SOLUTION. We let $X_1, X_2, \ldots, X_n$ denote the result of consecutive tosses, and we let $Z = \max(X_1, \ldots, X_n)$. We want the smallest n with $P\{Z \geq 4\} \geq 0.99$. Let us therefore determine the distribution of Z, exhibiting its dependence of n. Direct enumeration was feasible for $n = 2$, as in the preceding example, but it is cumbersome for larger n. However, it is easy to find $P\{Z \leq k\}$ for $k = 1, \ldots, 6$. Indeed, $P\{Z \leq 1\} = P\{Z = 1\} = 1/6^n$, since only one outcome among the total of 6^n outcomes gives the maximum score 1. Next,

$$\{Z \leq 2\} = \{\max(X_1, \ldots, X_n) \leq 2\} = \{X_1 \leq 2, X_2 \leq 2, \ldots, X_n \leq 2\}$$

and we have

$$P\{Z \leq 2\} = \frac{2^n}{6^n}.$$

In general,

$$P\{Z \leq k\} = \frac{k^n}{6^n} \qquad \text{for } k = 1, \ldots, 6,$$

hence

$$P\{Z = k\} = P\{Z \le k\} - P\{Z \le k - 1\} = \frac{k^n}{6^n} - \frac{(k-1)^n}{6^n}.$$

We want now $P\{Z \ge 4\} \ge 0.99$, that is,

$$P\{Z \le 3\} = \left(\frac{3}{6}\right)^n \le 0.01.$$

The inequality $(\frac{1}{2})^n \le 0.01$ gives $n \log(\frac{1}{2}) \le \log 0.01$, which means that

$$n \ge \frac{\log 0.01}{\log(1/2)} = \frac{\log 100}{\log 2} = 6.64.$$

Thus, $n = 7$ tosses "practically guarantees" that at least one toss will lead to $4, 5$, or 6. □

We shall now consider the case of the transformation of a single continuous random variable. We shall let F and f denote the cdf and the density of X and let $Y = \varphi(X)$, where φ is assumed to be at least piecewise differentiable. The objective is to find the distribution (cdf and/or density) of Y.

Regardless of whether we want cdf or density, the best strategy is to start by finding the cdf of Y. This method, occasionally referred to as the *cdf technique*, will be illustrated by some examples.

We begin with the simplest case, when φ is a strictly monotone function. If φ is increasing, we may write for the cdf of Y:

$$\begin{aligned} F_Y(y) &= P\{Y \le y\} = P\{\varphi(X) \le y\} \\ &= P\{X \le \psi(y)\} = F_X(\psi(y)), \end{aligned}$$

where ψ is the function inverse to φ.

The density is obtained by differentiating the cdf, and consequently,

$$f_Y(y) = \frac{d}{dy} F_X(\psi(y)) = f_X(\psi(y)) \cdot \psi'(y). \tag{6.40}$$

If φ is monotonically decreasing, we may write, taking into account that ψ is now a decreasing function,

$$F_Y(y) = P\{\varphi(X) \le y\} = P\{X \ge \psi(y)\} = 1 - F_X(\psi(y)),$$

and therefore

$$f_Y(y) = -f_X(\psi(y))\psi'(y), \tag{6.41}$$

a quantity that is positive, since now $\psi'(y) < 0$. We may combine (6.40) and (6.41) into the following theorem.

Theorem 6.4.1. *If φ is a continuous differentiable function with inverse ψ and X is a continuous random variable with density f_X, then the density of $Y = \varphi(X)$ is*

$$f_Y(y) = f_X(\psi(y))|\psi'(y)|. \tag{6.42}$$

Example 6.4.4 (Linear Transformations). Let $Y = aX + b$. If $a > 0$, we can write

$$F_Y(y) = P\{aX + b \le y\} = P\left\{X \le \frac{y-b}{a}\right\} = F_X\left(\frac{y-b}{a}\right)$$

and

$$f_Y(y) = \frac{1}{a}f_X\left(\frac{y-b}{a}\right).$$

If $a < 0$, we have

$$F_Y(y) = P\{aX + b \le y\} = P\left\{X \ge \frac{y-b}{a}\right\} = 1 - F_X\left(\frac{y-b}{a}\right)$$

and

$$f_Y(y) = -\frac{1}{a}f_X\left(\frac{y-b}{a}\right),$$

so that we always have

$$f_Y(y) = \frac{1}{|a|}f_X\left(\frac{y-b}{a}\right). \qquad\Box$$

The following transformation is important both theoretically and practically. Let X have cdf F and density f, and let F be strictly increasing, so that F^{-1} exists. We consider the transformation $Y = F(X)$. Obviously, $0 < Y < 1$, so that distribution of Y is concentrated on $[0, 1]$. We therefore have $F_Y(y) = 0$ for $y \le 0$ and $F_Y(y) = 1$ for $y \ge 1$, while for $0 < y < 1$ we may write

$$F_Y(y) = P\{F(X) \le y\} = P\{X \le F^{-1}(y)\}$$
$$= F(F^{-1}(y)) = y.$$

Consequently, $f_Y(y) = \dfrac{d}{dy}F_Y(y) = 1$ on $[0, 1]$, and we have proved the following theorem.

Theorem 6.4.2. *If X has a continuous strictly increasing cdf F, then the distribution of Y = F(X) is uniform on* [0, 1].

This theorem may be formulated as follows: *If Y is uniform on* [0, 1], *then* $X = F^{-1}(Y)$ *has distribution with cdf F.* Thus, we have obtained a method of generating random variables with given continuous invertible cdf F, using random variables uniform on [0, 1].

Example 6.4.5. The assumption of monotonicity of φ (and hence also ψ) is crucial for the validity of (6.42), except for the obvious remark that what is really needed is monotonicity "in the domain where it really matters." Specifically, if C is the set of all point x at which $f(x) > 0$ (called the *support* of X), then it suffices that φ is monotone on C only.

Consider the distribution with density

$$f_X(x) = \begin{cases} \alpha e^{-\alpha x} & \text{for } x \geq 0 \\ 0 & \text{for } x < 0, \end{cases}$$

and let $\varphi(x) = x^2$. Then φ is not monotone for all x, but it is monotone on the support of X, namely on $C = [0, \infty)$. Thus, $\psi(x) = \sqrt{x}, \psi'(x) = 1/2\sqrt{x}$ and the density of $Y = X^2$ is

$$f_Y(y) = \begin{cases} \dfrac{\alpha}{2\sqrt{y}} e^{-\alpha\sqrt{y}} & \text{for } y \geq 0 \\ 0 & \text{for } y < 0. \end{cases} \qquad \square$$

In the case when φ is not monotone, the procedure is as follows. Again, we start from the cdf of Y, writing

$$F_Y(y) = P\{Y \leq y\} = P\{\varphi(X) \leq y\}.$$

This time, however, φ has no inverse, and the inequality $\varphi(X) \leq y$ is not, in general, equivalent to a single inequality for X. Still, the right-hand side can usually be represented through the cdf F_X evaluated at some points dependent on y. Differentiating, we will recover the density f_Y of Y.
This principle will now be illustrated by some examples.

Example 6.4.6. Let X be a random variable with the density

$$f_X(x) = \begin{cases} \frac{5}{2}x^4 & \text{for } |x| \leq 1 \\ 0 & \text{otherwise.} \end{cases} \qquad (6.43)$$

Let $\varphi(x) = x^2$, so that now φ is not monotone, and consider $Y = \varphi(X)$. Since $Y \geq 0$, we have $F_Y(y) = 0$ for $y \leq 0$, while for $y > 0$ we may write

$$F_Y(y) = P\{X^2 \leq y\} = P\{-\sqrt{y} \leq X \leq \sqrt{y}\}$$
$$= F_X(\sqrt{y}) - F_X(-\sqrt{y}).$$

Differentiating, we get for $y > 0$,

$$f_Y(y) = F_Y'(y) = [f_X(\sqrt{y}) + f_X(-\sqrt{y})] \cdot \frac{1}{2\sqrt{y}},$$

so that (remembering that f_X vanishes if the argument exceeds 1 in absolute value)

$$f_Y(y) = \begin{cases} \frac{5}{2}y^{3/2} & \text{for } 0 < y < 1 \\ 0 & \text{otherwise.} \end{cases}$$

Example 6.4.7 (Square of a Normal Random Variable). Let X have a normal distribution with the density

$$f_X(x) = \frac{1}{\sqrt{2\pi}}e^{-x^2/2}$$

and let $\varphi(x) = x^2$, so that now φ is not monotone. We have $F_Y(y) = 0$ for $y \leq 0$, while for $y > 0$ we may write

$$F_Y(y) = P\{X^2 \leq y\} = P\{-\sqrt{y} \leq X \leq \sqrt{y}\}$$
$$= F_X(\sqrt{y}) - F_X(-\sqrt{y}).$$

Differentiating, we get for $y > 0$,

$$f_Y(y) = F_Y'(y) = [f_X(\sqrt{y}) + f_X(-\sqrt{y})] \cdot \frac{1}{2\sqrt{y}} = \frac{1}{\sqrt{2\pi}}y^{-1/2}e^{-y/2}. \qquad (6.44)$$

This is a special case of the *gamma density*, given by the formula $Cy^{a-1}e^{-by}$ ($y > 0$), where $a > 0, b > 0$ and C is the normalizing constant. Density (6.44) is a gamma density for $a = \frac{1}{2}, b = \frac{1}{2}$.

Example 6.4.8 (Folded Normal). Let X be as in Example 6.4.7, and let now $\varphi(x) = |x|$. Then for $y > 0$,

$$F_Y(y) = P\{|X| \leq y\} = P\{-y \leq X \leq y\} = F_X(y) - F_X(-y).$$

Differentiating, we get

$$f_Y(y) = f_X(y) + f_X(-y) = \sqrt{\frac{2}{\pi}} e^{-y^2/2} \text{ for } y > 0, \qquad (6.45)$$

and $f_Y(y) = 0$ for $y < 0$. Formula (6.45) is self-explanatory as regards the adjective "folded" in the name of the distribution.

PROBLEMS

6.4.1 If X is the result of tossing a balanced die, find the distribution of:
  (i) $Y = (X - 1)^2$.
  (ii) $Z = |X - 2.5|$.

6.4.2 Let X have the Poisson distribution with parameter λ, and let $Y = 3X$. Find the distribution of Y.

6.4.3 Find the density of $Y = (X - 1)^3$ if X has the standard normal distribution.

6.4.4 Find the density of $Y = X(1 + X)$ if X has the uniform distribution on $[0, 1]$.

6.4.5 Following Example 6.4.6, find the density of $Y = X^2$ if X has the density
$$f_X(x) = \begin{cases} Cx^4 & \text{for } -2 \le x \le 1 \\ 0 & \text{otherwise.} \end{cases}$$

6.4.6 Let U have uniform distribution on $[0, 1]$.
  (i) Find a function of U that has the distribution of X given in Problem 6.4.5.
  (ii) Find $P(X < 0)$ using the obtained representation of X as a function of U.

6.4.7 Suppose that the measured radius R is a random variable with density
$$f_R(x) = \begin{cases} 12x^2(1 - x) & \text{for } 0 \le x \le 1 \\ 0 & \text{otherwise.} \end{cases}$$

Find the cdf of:
  (i) The diameter of a circle with radius R.
  (ii) The circumference of a circle with radius R.
  (iii) The area of a circle with radius R.
  (iv) A ball with radius R.

6.4.8 Let X have the distribution uniform on $[-1, 1]$. Find the distribution of:

(i) $Y = |X|$.

(ii) $Z = 2|X| - 1$.

6.4.9 Let X have a continuous distribution with cdf F_X and density f_X, such that $F_X(0) = 0$. Find the cdf and density of random variables:

(i) $\sqrt{X}$.

(ii) $\log X$.

(iii) $1/X$.

(iv) e^X.

6.4.10 Let X have density $f(x) = 2(1 - x)$ for $0 \le x \le 1$. Find:

(i) The density of $Y = X(1 - X)$.

(ii) The density of $\max(X, 1 - X)$.

6.4.11 Assume that X is standard normal. Find the density of:

(i) $Y = X^3$.

(ii) $Y = e^X$.

6.4.12 Let X have the exponential distribution with density given by $f(x) = \lambda e^{-\lambda x}$, for $x > 0$. Find:

(i) The cdf and density of $Y = \sqrt{X}$.

(ii) The lower quartile of Y.

6.4.13 Let X be uniform on $[0, 1]$. Find φ such that $Y = \varphi(X)$ is exponential with parameter λ.

6.4.14 The speed of a molecule of gas at equilibrium is a random variable X with density

$$f(x) = \begin{cases} kx^2 e^{-bx^2}, & x > 0 \\ 0, & x \le 0, \end{cases}$$

where k is a normalizing constant and b depends on the temperature of the gas and the mass of the molecule. Find the probability density of the kinetic energy $E = mX^2/2$ of the molecule.

6.4.15 The duration (in days) of the hospital stay following a certain treatment is a random variable $Y = 4 + X$, where X has the density $f(x) = 32/(x + 4)^3$, for $x > 0$. Find:

(i) The density of random variable Y.

(ii) The probability that the hospital period of a patient following the treatment will exceed 10 days.

6.5 SURVIVAL AND HAZARD FUNCTIONS

In this section we consider a special class of random variables, namely continuous, nonnegative random variables. These random variables can serve as possible models for *lifetimes* of living organisms or of some equipment. We shall let the letter T denote such random variables.

In the case of a living organism, T may denote its actual lifetime (i.e., time until death), or time until some event, e.g., recovery from a disease. In the case of equipment, T will usually mean time until some event (e.g., failure).

In the sequel, we let F and f denote the cdf and density of T, so that for $t \geq 0$,

$$F_T(t) = P\{T \leq t\} = \int_0^t f(u) \, du \tag{6.46}$$

[the integration starts at 0, since nonnegativity of T implies that $P\{T \leq 0\} = 0$, and hence $f(t) = 0$ for $t < 0$].

In addition, we let

$$\tilde{F}(t) = 1 - F(t) = P\{T > t\} = \int_t^\infty f(u) \, du, \tag{6.47}$$

and we shall call $\tilde{F}$ the *survival function* of the random variable T (or: the *tail* of the cdf of T).

We introduce the following definition.

Definition 6.5.1.   The function

$$h(t) = -\frac{d \log \tilde{F}(t)}{dt} \tag{6.48}$$

defined for $t > 0$ with $\tilde{F}(t) > 0$ is called the *hazard function* of random variable T.   □

By differentiating $\log[1 - F(t)]$ it follows from (6.48) that at almost all points t we have

$$h(t) = \frac{f(t)}{\tilde{F}(t)}. \tag{6.49}$$

Consequently, we have the following theorem expressing the cdf through the hazard function:

Theorem 6.5.1.   *If $h(t)$ is a hazard function of a random variable T, then its cdf equals*

$$F(t) = 1 - e^{-\int_0^t h(u)du}.$$

Proof: The proof follows immediately from integrating (6.49) between 0 and t. $\square$

The interpretation of hazard functions is as follows: $h(t)\delta t$ can be approximated by

$$\frac{f(t)\delta t}{\tilde{F}(t)} \approx \frac{P\{t < T \leq t + \delta t\}}{P\{T > t\}},$$

hence, using definition of conditional probability,

$$h(t)\delta t \sim P\{t < T \leq t + \delta t | T > t\}$$
$$= P\{\text{ lifetime ends before } t + \delta t | \text{ lifetime longer than } t\}.$$

In other words, $h(t)$ is the death rate at t of those who survived until t (are "at risk" at t). Thus, hazard function describes the process of aging, in terms of changes of risk of death with current age.

Example 6.5.1. Let the density of X be

$$f(t) = \begin{cases} \alpha e^{-\alpha t} & \text{for } t > 0 \\ 0 & \text{for } t \leq 0. \end{cases}$$

Then $F(t) = 1 - e^{-\alpha t}$, so that $\tilde{F}(t) = e^{-\alpha t}$ and

$$h(t) = \frac{\alpha e^{-\alpha t}}{e^{-\alpha t}} = \alpha.$$

Thus, exponential distribution describes lifetime distribution in case of lack of aging, when the risk of death is constant with age.

Example 6.5.2. Suppose you buy a new car, and let T be the time of first breakdown. Typically $h(t)$ is initially high, then declines to a constant and remains on this level for several years, eventually beginning to increase.

The reason is that early failures are typically caused by hidden faults of material, undetected in factory control. If they do not show up immediately, then there are probably no faults and risk of failure remains constant for some time. The later increase is due to wearing out of various parts. $\square$

In engineering contexts, especially in problems of reliability of equipment, the hazard function is often called *failure rate*.

Let us assume that $F(t) < 1$ for all t (which is the most important case of interest). We can then write

$$F(t) = 1 - e^{-H(t)} \tag{6.50}$$

where, by definition,

$$H(t) = -\log[1 - F(t)].$$

On the other hand, in view of Theorem 6.4.1, we also have $H(t) = \int_0^t h(u)\,du$, which gives (6.32). Since $F(t) \to 1$ as $t \to \infty$, we must have

$$\int_0^\infty h(u)\,du = \infty. \tag{6.51}$$

Example 6.5.1 shows that the exponential distribution has a constant hazard, interpreted as lack of aging. If hazard is increasing, we have the phenomenon of aging ("new better than old"), while the opposite is true in the case of decreasing hazard ("old better than new"). A flexible model of both situations is given by the following definition.

Definition 6.5.2.   A nonnegative random variable T with the hazard rate

$$h(t) = Kt^\gamma, \qquad t > 0, K > 0, \gamma > -1 \tag{6.52}$$

is said to have a *Weibull distribution.*

Observe first that the condition $\gamma > -1$ ensures that the relation (6.51) holds, so that T is a genuine random variable in the sense that

$$\lim_{t\to\infty} P\{T \le t\} = 1$$

(i.e., $T = \infty$ is excluded). For reasons of tradition and convenience, one usually puts $\alpha = K/(\gamma + 1), \beta = \gamma + 1$, so that the hazard rate of Weibull distribution takes the form

$$h(t) = \alpha\beta t^{\beta-1}, \qquad \alpha > 0, \beta > 0, t > 0. \tag{6.53}$$

PROBLEMS

6.5.1   Find the hazard function of the distribution uniform on $[0, 1]$. Explain why $h(t)$ is unbounded.

6.5.2   Let X be a random variable with density

$$f(x) = \begin{cases} 0, & x < 0 \\ 0.5, & 0 \le x \le 1 \\ qe^{-\alpha x}, & x > 1. \end{cases}$$

   (i) Find q, assuming that α is known.
   (ii) Find the hazard $h(t)$.

6.5.3   Find the cdf and density of the Weibull distribution.

6.5.4 Assume that the fuel pumps in a certain make of cars have lifetimes with Weibull hazard rate $2/\sqrt[3]{t}$, with t measured in years. Find the probability that a fuel pump is still working after 5 months.

6.5.5 A cancer specialist claims that the hazard function of the random variable $T_c =$ "age at death due to cancer" is a bounded function that for large t has the form $h_c(t) = k/t^2$ (where t is the age in years and k is a constant).

Let $h(t)$ be the hazard of the time of death due to reasons other than cancer (accidents, other diseases, old age, etc.) and assume that the actual time of death T^* has hazard function $h_c(t) + h(t)$. If T_c and T are times of death with hazards $h_c(t)$ and $h(t)$, show that the observed time of death is $T^* = \min(T_c, T)$, so that $\max(T_c, T)$ is unobservable. To get an insight into the feasibility of the assumption of the cancer specialist in question, imagine that *all* reasons of death other than cancer have been eliminated [i.e., $H(t) \equiv 0$] and find the probability that a person will live forever.

REFERENCES

Kolmogorov, A. N., 1933. *Grundbegriffe der Wahrscheinlichkeitsrechnung*, Springer, Berlin (English translation by N. Morrison, *Foundations of the Theory of Probability*, Chelsea, New York, 1956).

CHAPTER 7

Random Variables: Multivariate Case

7.1 BIVARIATE DISTRIBUTIONS

The considerations of Section 6.2 can be extended to the case of several random variables analyzed at once, or equivalently, to the analysis of vector-valued random variables. We begin with the simplest case of a pair of random variables (X, Y), that is, a pair of functions on the sample space S. We may think here of description of some random phenomenon in terms of two attributes, X and Y. Alternatively, one can think of a random choice of a point from the plane.

Example 7.1.1. Let the experiment consist of three tosses of a coin, and let $X =$ number of heads in all three tosses and $Y =$ number of tails in the last two tosses. The sample space S and corresponding values of X and Y are then

S	X	Y
$s_1 = $ HHH	3	0
$s_2 = $ HHT	2	1
$s_3 = $ HTH	2	1
$s_4 = $ THH	2	0
$s_5 = $ HTT	1	2
$s_6 = $ THT	1	1
$s_7 = $ TTH	1	1
$s_8 = $ TTT	0	2

We can summarize all possibilities and their probabilities in the following table:

200

X	Y		
	0	1	2
0	0	0	$\frac{1}{8}$
1	0	$\frac{1}{4}$	$\frac{1}{8}$
2	$\frac{1}{8}$	$\frac{1}{4}$	0
3	$\frac{1}{8}$	0	0

The entries in the table represent the corresponding probabilities. For instance,

$$P\{X = 2, Y = 1\} = \frac{1}{4},$$

which was obtained by counting the number of points s of the sample space where $X(s) = 2$ and $Y(s) = 1$ (there are two such points: $s_2 = $ HHT and $s_3 = $ HTH). Here, as well as in the sequel, we follow the long-established tradition of using commas to denote the intersection of events. Thus we write $P\{X = x, Y = y\}$ instead of the rather clumsy $P[\{X = x\} \cap \{Y = y\}]$.

□

In a natural way, this example leads to a definition of a *discrete* bivariate random variable.

Definition 7.1.1. We say that the pair (X, Y) of random variables has a *discrete distribution* if there exist finite or countable sets A and B such that

$$P\{(X, Y) \in A \times B\} = \sum_{x \in A, y \in B} P\{X = x, Y = y\} = 1.$$

□

In Example 7.1.1 we have $A = \{0, 1, 2, 3\}$ and $B = \{0, 1, 2\}$. It should be remarked that while $P\{X = x, Y = y\} = 0$ if any of the values x or y lies outside the set A (respectively, B), it is also possible that $P\{X = x, Y = y\} = 0$ for some $x \in A$ and $y \in B$; for instance, in Example 7.1.1 we have $P\{X = 0, Y = 1\} = 0$. In other words, the values (x, y) that have positive probability may form a proper subset of $A \times B$.

The continuous bivariate distributions, as may be expected, are defined through their densities.

Definition 7.1.2. Random variables (X, Y) are said to be *jointly continuous* (or: have continuous joint distribution) if there exists a function $f(x, y)$ such that for every rectangle

$$C = \{(x, y) : a \le x \le b, c \le y \le d\}$$

with $-\infty \le a < b \le \infty, -\infty \le c < d \le \infty$, we have

$$P\{(X, Y) \in C\} = \int_C \int f(x, y) \, dx \, dy. \tag{7.1}$$

The function f is called the *joint or bivariate* density of (X, Y). When no confusion is possible, we shall speak simply of density f. ☐

Some comments are in order here. First, as in the one-dimensional (univariate) case, the density is defined only up to sets of measure zero (such as single points or arcs on the plane).

Second, the obvious consequences of Definition 7.1.2 are

$$\int_{R^2} \int f(x, y) \, dx \, dy = 1, \tag{7.2}$$

where R^2 is the plane, and

$$f(x, y) \geq 0 \text{ almost everywhere.} \tag{7.3}$$

These relations are analogues of the corresponding relations (6.30) and (6.33) for univariate densities.

*   The third commentis as follows. Formula (7.1) concerns only rectangles. In analogy with Theorem 6.2.5 , one can show that probability P defined on the class of rectangles by (7.1) determines its unique extension to all sets in the plane that can be approximated through countable operations on rectangles, in particular to all figures that can be triangulated, as well as to circles, ellipses, and so on. In other words, formula (7.1) holds for a much wider class of sets C on the plane.

Conditions (7.2) and (7.3) allow us to use the Fubini theorem and replace the double integral (7.1) by the iterated integral, thus making actual integration possible. In particular, if C is the rectangle specified in Definition 7.1.2, then

$$P\{(X, Y) \in C\} = \int_C \int f(x, y) \, dx \, dy$$

$$= \int_a^b \left[\int_c^d f(x, y) \, dy \right] dx = \int_c^d \left[\int_a^b f(x, y) \, dx \right] dy.$$

In general, we have

$$\int_C \int f(x, y) \, dx \, dy = \int_{C_2} \varphi(y) \, dy = \int_{C_1} \psi(x) \, dx.$$

Here C_1 and C_2 are "shadows" of C on the x-axis and y-axis, that is,

$$C_1 = \{x : (x, y) \in C \text{ for some } y\}$$

$$C_2 = \{y : (x, y) \in C \text{ for some } x\}$$

and

$$\psi(x) = \int_{C_x} f(x, y) \, dy, \qquad \varphi(y) = \int_{C_y} f(x, y) \, dx$$

with C_x and C_y being sections of C at points x and y (see Figure 7.1):

$$C_x = \{y : (x, y) \in C\}, C_y = \{x : (x, y) \in C\}.$$

Some less mathematically experienced readers may be puzzled by our stressing the difference between the double integral

$$\int_{R^2} \int f(x, y) \, dx \, dy$$

and iterated integrals

$$\int_{-\infty}^{+\infty} \int_{-\infty}^{+\infty} f(x, y) \, dx \, dy,$$

interpreted as

$$\int_{-\infty}^{+\infty} \left[\int_{-\infty}^{+\infty} f(x, y) \, dx \right] dy \quad \text{and} \quad \int_{-\infty}^{+\infty} \left[\int_{-\infty}^{+\infty} f(x, y) \, dy \right] dx.$$

The readers are advised to consult a good calculus text for the respective def-

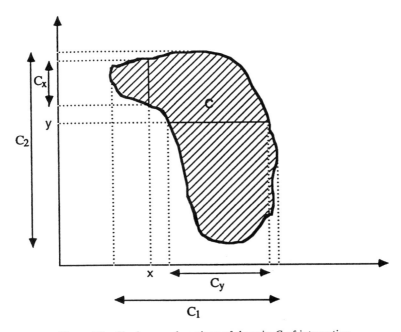

Figure 7.1 Shadows and sections of domain C of integration.

initions, and a statement of the Fubini theorem, which gives conditions under which all three integrals are equal (we discuss these topics in Chapter 8). It is both easy and worthwhile, however, to explain the issues involved here, using a simplified situation of a series (instead of integrals). Thus, $\sum_{k=1}^{\infty} a_k$ is *defined* as $\lim_{n \to \infty} \sum_{k=1}^{n} a_k$, provided that this limit (of a well-defined numerical sequence!) exists. By the same argument, the double sum $\sum_{m=1}^{\infty} \sum_{n=1}^{\infty} a_{mn}$ is defined unambiguously (this sum is an analogue of iterated integral). However, the symbol $\sum_{m,n=1}^{\infty} a_{mn}$ (the analogue of double integral) is *not* well defined, since it does not specify the order in which the two-dimensional array $\{a_{mn}\}$ is to be added. The point here is that the sum of an infinite sequence of numbers may depend on the order of summation. Thus, *some* assumptions about $\{a_{mn}\}$ are needed to make the last sum independent of the order of summation. Under these assumptions (e.g., nonnegativity) the double sum and both iterated sums are either all infinite or all equal to the same finite number.

A similar kind of difficulty appears in the case of double and iterated integrals and is resolved by the Fubini theorem.

We shall now illustrate the calculation of probabilities by some examples.

Example 7.1.2. The density $f(x, y)$ is given by the formula

$$f(x, y) = \begin{cases} Cx(x + y) & \text{if } x \geq 0, y \geq 0, x + y \leq 1 \\ 0 & \text{otherwise.} \end{cases} \tag{7.4}$$

Find $P\{X \leq \frac{1}{2}\}$ (see Figure 7.2).

SOLUTION. The first objective is to determine the constant C in the formula (7.4). As usual, the normalizing condition (7.2) provides the key here:

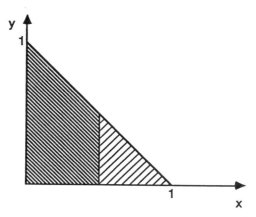

Figure 7.2 Support of density f, and the set $\{X \leq \frac{1}{2}\}$.

$$1 = \iint Cx(x+y)\,dx\,dy = \int_0^1 \left[\int_0^{1-y} C(x^2+xy)\,dx \right] dy$$

$$= C \int_0^1 \left(\frac{x^3}{3} + \frac{x^2}{2}y \Big|_0^{1-y} \right) dy = C \int_0^1 \left[\frac{(1-y)^3}{3} + \frac{(1-y)^2 y}{2} \right] dy$$

$$= \frac{C}{8},$$

so that $C = 8$.

Now, $P\{X \le \frac{1}{2}\}$ is the integral of the density $f(x,y)$ over the doubly shaded area (D) in Figure 7.2. This time it is simpler to integrate over y first, and then over x, namely

$$P\{X \le 1/2\} = \iint_D 8x(x+y)\,dx\,dy$$

$$= 8 \int_0^{1/2} \left[\int_0^{1-x} (x^2+xy)\,dy \right] dx$$

$$= 8 \int_0^{1/2} \left(x^2 y + \frac{xy}{2} \Big|_0^{1-x} \right) dx$$

$$= 8 \int_0^{1/2} \left[x^2(1-x) + \frac{1}{2}x(1-x)^2 \right] dx$$

$$= \frac{7}{16}. \qquad\qquad\qquad \square$$

Although used much less often than in the univariate case, the cumulative distribution function (cdf) is still the most general way of specifying probabilities in the bivariate case, regardless of whether the distribution is discrete, continuous, or neither. Consequently, we introduce the following definition.

Definition 7.1.3. The function of two real variables, defined as

$$F(x,y) = P\{X \le x, Y \le y\},$$

is called the *cumulative distribution function* (cdf) of the pair (X,Y) of random variables. $\qquad\qquad \square$

The relation between the cdf and probability mass function in the discrete case, or the density function in the continuous case, are similar to those for univariate random variables, namely

$$F(x,y) = \sum_{x_i \le x, y_j \le y} P\{X = x_i, Y = y_j\}$$

in the discrete case, and

$$F(x,y) = \int_{-\infty}^{x} \int_{-\infty}^{y} f(u,v)\, dv\, du \tag{7.5}$$

in the continuous case. In particular, from (7.5) it follows that for almost all (x, y) we have

$$f(x,y) = \frac{\partial^2 F(x,y)}{\partial x\, \partial y}.$$

As in the univariate case, one can prove the following analogue of Theorem 6.2.2:

Theorem 7.1.1. *Every bivariate cumulative distribution function $F(x,y)$ has the following properties:*

(a) $\lim_{x,y\to+\infty} F(x,y) = 1$.
(b) *For every y, the function $F(x,y)$ is nondecreasing and continuous from the right in x.*
(c) *For every x, the function $F(x,y)$ is nondecreasing and continuous from the right in y.*
(d) $\lim_{x\to-\infty} F(x,y) = 0$ *for every y, and* $\lim_{y\to-\infty} F(x,y) = 0$ *for every x.*
(e) *For all $x_1 < x_2$ and $y_1 < y_2$,*

$$F(x_2, y_2) - F(x_2, y_1) - F(x_1, y_2) + F(x_1, y_1) \geq 0. \tag{7.6}$$

We omit the proof here, leaving proofs of some of the properties as exercises. The following comments, however, are important. While (a)–(d) are direct analogues of the properties of univariate cdf's, condition (e) has no counterpart. The question therefore arises whether conditions (a)–(d) alone characterize bivariate cdf's. In other words: Is every function satisfying (a)–(d) a cdf of some pair (X, Y) of random variables? The answer is negative, as shown by the following example.

Example 7.1.3. Let $F(x,y)$ be defined by the formula

$$F(x,y) = \begin{cases} 1 & \text{if } x + y \geq 0 \\ 0 & \text{if } x + y < 0. \end{cases} \tag{7.7}$$

It is easy to check that the function (7.7) satisfies conditions (a)–(d) given above. Suppose now that $F(x,y)$ is the cdf of a pair of random variables, and let us compute the probability (see Figure 7.3) $P\{-1 < X \leq 2, -1 < Y \leq 2\}$. Since $F(x,y)$ is the probability of the pair (X, Y) taking a value "southwest" of (x,y) [i.e., to the left and below (x,y)], we have

$$P\{-1 < X \leq 2, -1 < Y \leq 2\}$$
$$= F(2,2) - F(-1,2) - F(2,-1) + F(-1,-1) = 1 - 1 - 1 + 0 = -1,$$

so that $F(x,y)$ cannot be a cdf.   □

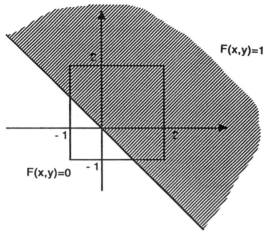

Figure 7.3 Function satisfying (a)–(d) which is not a cdf.

Clearly, condition (e) is necessary, since the left-hand side of (7.6) equals $P\{x_1 < X \le x_2, y_1 < Y \le y_2\}$ and hence must be nonnegative.

It turns out (we omit the proof here) that (a)–(e) characterize a bivariate cdf; that is, any function satisfying (a)–(e) is a cdf of some pair of random variables.

PROBLEMS

7.1.1 Let the joint density of random variables X, Y be $f(x, y) = cx^3 y^2$ for $0 \le x \le 1, x^2 \le y \le 1$ and $f(x, y) = 0$ otherwise. Find $P(X < Y)$.

7.1.2 Let the joint cdf of random variables X, Y be $F(x, y) = \frac{1}{30}xy(x + y)$ for $0 \le x \le 2, 0 \le y \le 3$. Find the density $f(x, y)$.

7.1.3 Let X, Y have the joint distribution given by the following table:

	Y		
X	2	3	4
0	1/48	0	b
1	0	5/48	8/48
2	a	0	11/48
3	5/48	12/48	0

(i) Find a and b if it is known that $P(X = Y) = \frac{1}{3}$.

(ii) Find $P(XY = 0)$.

(iii) If F is a cdf of (X, Y), find $F(-1.5, 3), F(0.7, 2.11)$, and $F(1.5, 18)$.

7.1.4   We toss a fair die twice.
  (i) Find the joint distribution of random variables $X =$ sum of outcomes and $Y =$ best of the two outcomes. Find:
  (ii) $P(X \leq 8, Y \leq 5)$; $P(X = 9, Y \leq 2)$, $P(4 \leq X \leq 7, 1 \leq Y \leq 3)$.
  (iii) $P(Y = 3 | X = 4)$, $P(Y < 6 | X = 7)$, $P(4 < Y \leq 6 | X \leq 8)$, $P(X = 8 | 3 < Y < 6)$.

7.1.5   We toss a fair die twice. Let X be the number of times that 1 came up, and Y be the number of times that 2 came up. Find:
  (i) The joint distribution of X and Y.
  (ii) The correlation coefficient between events $\{X = 1\}$ and $\{Y = 2\}$. [See formula (4.20) in Definition 4.5.2.]

7.1.6   The joint density of X and Y is given by $f(x, y) = y^2(xy^3 + 1)$ on the rectangle $0 \leq x \leq k, 0 \leq y \leq 1$. Find:
  (i) k.
  (ii) $P(X \leq Y)$.

7.1.7   Assume that X, Y have density $f(x, y) = x + y$ for $0 \leq x, y \leq 1$ and $f(x, y) = 0$ otherwise. Find $P\{Y \leq \sqrt[3]{X}\}$.

7.1.8   Assume that (X, Y) have the joint density $f(x, y) = cxy^2$ for $0 \leq x, y \leq 1$ and $f(x, y) = 0$ otherwise. Find:
  (i) c.
  (ii) $P\{X^2 \leq Y \leq X\}$.
  (iii) cdf of (X, Y).

7.1.9   Let (X, R) have the joint density function $f(x, y) = 1/y$ for $0 < x < y < 1$ and $f(x, y) = 0$ otherwise. Find:
  (i) $P(X + Y > \frac{1}{2})$.
  (ii) $P(Y > 2X)$.

7.2 MARGINAL DISTRIBUTIONS; INDEPENDENCE

One can naturally expect that the bivariate distribution (in the form of cdf, joint density, or probability mass function, as the case might be) contains more information than the univariate distributions of X and Y separately. Consequently, we may expect that given a bivariate distribution, one should be able to recover both univariate distributions, of X and of Y (but not conversely).

We begin with the case of discrete bivariate distributions. Let $A = \{x_1, x_2, \ldots\}$ and $B = \{y_1, y_2, \ldots\}$ be the sets of possible values of X and Y, respectively, and let us put

$$p_{ij} = P\{X = x_i, Y = y_j\}. \tag{7.8}$$

Our objective is to express the distributions of X and of Y through p_{ij}. Since the events $\{Y = y_1\}, \{Y = y_2\}, \ldots$ form a partition (in the sense defined in Chapter 4, that is, these events are mutually exclusive and one of them has to occur), we may write for every i

$$\{X = x_i\} = \bigcup_j [\{X = x_i\} \cap \{Y = y_j\}]$$

which, using the convention of replacing intersection by commas, is

$$\{X = x_i\} = \bigcup_j \{X = x_i, Y = y_j\}.$$

Since the events on the right-hand side are disjoint, we have

$$P\{X = x_i\} = \sum_j P\{X = x_i, Y = y_j\} = \sum_j p_{ij}.$$

In a similar way,

$$P\{Y = y_j\} = \sum_i p_{ij}.$$

We now introduce:

Definition 7.2.1. Given the joint distribution (7.8), the distributions of X and Y calculated using the formulas

$$P\{X = x_i\} = p_{i+}, \quad P\{Y = y_j\} = p_{+j},$$

where

$$\sum_j p_{ij} = p_{i+}, \quad \sum_i p_{ij} = p_{+j},$$

will be referred to as *marginal* distributions. □

If we think of the numbers p_{ij} as arranged into a matrix, then $P\{X = x_i\}$ and $P\{Y = y_j\}$ are sums of the corresponding rows (or columns) of the matrix $[p_{ij}]$.

Since the sum of all p_{ij} equals 1, both marginal distributions satisfy the condition that the sum of their probabilities equals 1.

Example 7.2.1. In Example 7.1.1 we considered three tosses of a coin, and we let X be the number of heads in all three tosses, while Y was the number of tails in the last two tosses. Then the joint distibution of X and Y

is summarized by the following table:

		Y		
X	0	1	2	Marginal of X
0	0	0	$\frac{1}{8}$	$\frac{1}{8}$
1	0	$\frac{1}{4}$	$\frac{1}{8}$	$\frac{3}{8}$
2	$\frac{1}{8}$	$\frac{1}{4}$	0	$\frac{3}{8}$
3	$\frac{1}{8}$	0	0	$\frac{1}{8}$
Marginal of Y	$\frac{1}{4}$	$\frac{1}{2}$	$\frac{1}{4}$	1

On the margins here we give the row sums and column sums. The distribution of X is

$$P\{X = 0\} = P\{X = 3\} = \tfrac{1}{8}, \quad P\{X = 1\} = P\{X = 2\} = \tfrac{3}{8},$$

while the distribution of Y is

$$P\{Y = 0\} = P\{Y = 2\} = \tfrac{1}{4}, \quad P\{Y = 1\} = \tfrac{1}{2}. \qquad \square$$

It is worthwhile to remark that the adjective *marginal* refers to the way in which the distribution was obtained and implies nothing about the properties of the distribution.

The definition of marginal distribution for continuous random variables is analogous, with summation replaced by integration:

Definition 7.2.2. If (X, Y) is a pair of continuous random variables with bivariate density $f(x,y)$, then the functions

$$f_1(x) = \int_{-\infty}^{+\infty} f(x,y)\, dy$$

and

$$f_2(y) = \int_{-\infty}^{+\infty} f(x,y)\, dx$$

are called, respectively, the *marginal densities* of random variables X and Y. $\qquad \square$

The justification of Definition 7.2.2 consists of two parts. First, we need to show that f_1 and f_2 are densities, that is, are nonnegative for almost all arguments (i.e., for all arguments, except a set of arguments of measure zero) and that f_1 and f_2 integrate to 1. These properties are immediate consequences of the fact that $f(x,y) \geq 0$ except possibly on a set of measure zero.

Consequently, using the Fubini theorem, we can write

$$1 = \int \int f(x,y)\, dx\, dy = \int_{-\infty}^{+\infty} \left[\int_{-\infty}^{+\infty} f(x,y)\, dx \right] dy$$

$$= \int_{-\infty}^{+\infty} f_2(y)\, dy$$

and similarly for f_1.

The second part of the justification of Definition 7.2.2 consists in showing that, for instance,

$$P\{a \le X \le b\} = \int_a^b f_1(x)\, dx.$$

Here we may write

$$\int_a^b f_1(x)\, dx = \int_a^b \left[\int_{-\infty}^{+\infty} f(x,y)\, dy \right] dx$$

$$= \iint_{a \le x \le b,\, -\infty < y < +\infty} f(x,y)\, dx\, dy$$

$$= P\{a \le X \le b,\, -\infty < Y < +\infty\} = P\{a \le X \le b\}.$$

Example 7.2.2. A man shoots at a circular target. Assume that his skills are such that he is certain to hit the target. However, he is unable to aim with any more precision, so that the probability of hitting a particular part of the target is proportional to the area of this part. Let X and Y be the horizontal and vertical distance from the center of the target to the point of impact. Find the distribution of X.

SOLUTION. Without loss of generality we may introduce the coordinate system with origin in the center of the target, and assume its radius to be R. Then the target is described by the relation $x^2 + y^2 \le R^2$. From the conditions of the problem it is seen that the density of the point of impact is constant on the target, so that

$$f(x,y) = \begin{cases} c & \text{if } x^2 + y^2 \le R^2 \\ 0 & \text{otherwise.} \end{cases}$$

Next, the condition $\iint_{R^2} f(x,y)\, dx\, dy = 1$ implies that c must be the reciprocal of the area of the target, hence $c = 1/\pi R^2$. If $|x| > R$, then $f(x,y) = 0$ for all y (see Figure 7.4). For $|x| \le R$, we have $f(x,y) = 0$ if $x^2 + y^2 > R^2$ (hence if $|y| > \sqrt{R^2 - x^2}$), and $f(x,y) = 1/\pi R^2$ when $x^2 + y^2 \le R^2$ (so $|y| \le \sqrt{R^2 - x^2}$).

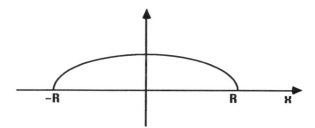

Figure 7.4 Marginal density.

Consequently, for $|x| \leq R$,

$$f_1(x) = \int_{-\sqrt{R^2-x^2}}^{\sqrt{R^2-x^2}} \frac{1}{\pi R^2} dy = \frac{2\sqrt{R^2-x^2}}{\pi R^2},$$

so that the density $f_1(x)$ has the shape given in Figure 7.4.   □

At this point it may be worthwhile to make the following remark on notation. First, the symbols $f_1(x)$ and $f_2(y)$ for marginal densities of X and Y are used in the literature in parallel with more readable symbols such as $f_X(x)$ and $f_Y(y)$, with the name of the variable appearing as an index. This latter system becomes cumbersome when the variables are labeled X_1 and X_2 instead of X and Y, since logically one should then use $f_{X_1}(x)$ and $f_{X_2}(y)$. We shall deliberately avoid keeping rigidly to any fixed system of notation, and use whichever notation appears most natural in a given instance. Another problem concerns the use of an argument in the density or cdf. It appears natural to label the argument x in the density or cdf of X, label it y in the density of Y, and so on. However, it is *not possible* to use such notation consistently: Indeed, if $F(x)$ and $f(x)$ are to be used as the cdf and density of X, then we have $F(x) = \int_{-\infty}^{x} f(u)\,du$, and in the integrand we can use almost any symbol *except* x, since the symbol $\int_{-\infty}^{x} f(x)\,dx$ is unacceptably ambiguous.

We shall now define the concept of independence of two random variables. We recall from Chapter 4 that the independence of two events A and B was defined by the product rule $P(A \cap B) = P(A)P(B)$. A random variable X allows us to define events of the form $\{X \in A\}$, where A is some set of real numbers, and the same is true for the random variable Y. It seems natural to require that these events are independent for independent random variables X and Y. Thus, we have

Definition 7.2.3 (Intention of the Concept). We say that random variables X and Y are *independent* if the events $\{X \in A\}$ and $\{Y \in B\}$ are independent for all sets A, B, that is, if

$$P\{X \in A, Y \in B\} = P\{X \in A\}P\{Y \in B\}. \tag{7.9}$$

□

This definition, as spelled out in its label, concerns the "final effect" of the concept: independence of *every pair* of events in a very large class of such pairs. Clearly, checking independence using this definition would be very difficult if we were to verify the product rule (7.9) for every pair A, B. Consequently it becomes necessary to find a condition that would be verifiable, and strong enough to imply independence in the sense of Definition 7.2.3. It turns out that it is sufficient to require only a seemingly weaker condition.

Definition 7.2.4 (Verifiable Definition of Independence). The random variables X and Y are said to be *independent* if the joint and marginal cdf's satisfy the following condition: For every x, y,

$$F(x, y) = F_1(x)F_2(y), \tag{7.10}$$

where $F_1(x) = P\{X \leq x\}$ and $F_2(y) = P\{Y \leq y\}$ are marginal cdf's of X and Y. □

In the case of discrete random variables with p_{ij} being the probability $P\{X = x_i, Y = y_j\}$, condition (7.10) is implied by:

$$\text{for every } i, j \text{ we have } p_{ij} = p_{i+}p_{+j}. \tag{7.11}$$

For continuous random variables, the condition implying (7.10) is: The marginal densities f_1 and f_2 are such that

$$f(x, y) = f_1(x)f_2(y) \tag{7.12}$$

for all x, y except possibly for a set of points (x, y) of measure zero.

The concept of independence of random variables covers intuition (analogous to the intuition behind the concept of independence of events), according to which if two random variables are independent, then information about the value of one of them provides no information about the value of the other.

Example 7.2.3. Let us return to Example 7.1.1, where we had the experiment of three tosses of a coin, X = number of heads in all tosses, and Y = number of tails in the last two tosses. An inspection of the table of joint and marginal distribution in Example 7.1.1 shows that, for instance, $P\{X = 3, Y = 2\} = 0$ while $P\{X = 3\}P\{Y = 2\} = \frac{1}{8} \cdot \frac{1}{4} \neq 0$. This means that X and Y are not independent.

Example 7.2.4. Let us consider again the case of three tosses of a coin, and let U = number of heads in the first two tosses, and V = number of tails in the last toss. Listing the sample space we have here:

S	U	V
HHH	2	0
HHT	2	1
HTH	1	0
THH	1	0
HTT	1	1
THT	1	1
TTH	0	0
TTT	0	1

The joint probability distribution and the marginals are now

U	V 0	1	
0	$\frac{1}{8}$	$\frac{1}{8}$	$\frac{1}{4}$
1	$\frac{1}{4}$	$\frac{1}{4}$	$\frac{1}{2}$
2	$\frac{1}{8}$	$\frac{1}{8}$	$\frac{1}{4}$
	$\frac{1}{2}$	$\frac{1}{2}$	1

and direct check shows that we have

$$P\{U = x, V = y\} = P\{U = x\}P\{V = y\}$$

for every cell (x, y) of the table above. This shows that U and V are independent. □

Before we illustrate the concept of independence of random variables in the continuous case, it is worthwhile to make some remarks. Verification that two random variables are in fact independent requires checking the multiplicative property for *all* x and y. In the last example, this required checking six cells of the table (i.e., comparing the table of joint distribution with the "multiplication table" obtained by multiplying the marginals). Such a direct verification is not feasible except for discrete variables with small sets of possible values. To handle more complicated cases, we typically must have some algebraic formula for the joint distribution from which we can calculate the marginal distributions and verify the product rule algebraically.

On the other hand, to show that two variables are *not independent* (i.e., that they are *dependent*), it is enough to find *one* pair (x, y) for which the product rule does not hold.

A practical consequence here is that it is generally worthwhile to *think* before one starts calculations, trying to determine on the ground of meaning of the variables in question whether or not we should expect them to be independent. This determines the strategy: Do we aim at checking that variables are independent, or do we aim at showing that they are dependent? These two goals may require somewhat different types of techniques.

To illustrate the point, in Example 7.2.3 we may have expected the variables to be dependent: The more tails in the last two tosses (Y), the lower one may expect to be the total number of heads (X). In particular, we may be able to find a "pure exclusion," such as the fact that one cannot have $X = 3$ and $Y = 2$ simultaneously, while each of these events is not impossible separately.

On the other hand, in Example 7.2.4 we may have expected U and V to be independent, since the values of U and V were determined by nonoverlapping sets of tosses.

In particular, the following criterion may be useful in showing that two variables are *not independent*.

Theorem 7.2.1. *If the table of joint probabilities for (X, Y) contains a zero entry, then X and Y are dependent.*

Proof. To prove this, suppose that $P\{X = x_0, Y = y_0\} = 0$ is the zero entry. Since x_0 and y_0 are the possible values of X and Y, respectively, the row sum and column sum at (x_0, y_0) must be positive, hence the product $P(X = x_0)P(Y = y_0)$ is positive. □

This criterion provides a quick "visual" test for lack of independence. Of course, it works in only one direction: If there are no zeros in the table of joint distribution, the variables may be dependent or not, and further checking is necessary.

We shall now consider some examples of applications of Definition 7.2.4 for the case of continuous random variables.

Example 7.2.5. The joint density of random variables X and Y is

$$f(x, y) = \begin{cases} Cx^n e^{-\alpha x - \beta y} & \text{for } x > 0, y > 0 \\ 0 & \text{otherwise.} \end{cases}$$

The question is whether these random variables are independent. It is interesting that the answer here does not require any calculations. In particular, we do not need to calculate C. Indeed, we may simply represent the joint density on the positive quadrant as

$$f(x,y) = C_1 x^n e^{-\alpha x} \cdot C_2 e^{-\beta y} \qquad (7.13)$$

with C_1 and C_2 such that

$$C_1 \int_0^\infty x^n e^{-\alpha x} dx = 1, \quad C_2 \int_0^\infty e^{-\beta y} dy = 1$$

(then automatically $C = C_1 C_2$). The two factors on the right of (7.13) must be marginal densities (why?) and we showed independence of X and Y. The answer to the question "why?" is not simply "because $f(x,y)$ factors into the product of two functions, each depending on one variable only," since the marginal density depends on the shape of the support of the joint density [i.e., the set of points (x,y) where the density is positive]. To see it, consider the following example.   □

Example 7.2.6. Let the joint density be

$$f(x,y) = \begin{cases} Cxy^2 & \text{if } x \geq 0, y \geq 0, x + y \leq 1 \\ 0 & \text{otherwise.} \end{cases}$$

At first sight the situation here is similar to that in Example 7.2.5. However, this is not the case: The marginal distributions are *not* $C_1 x$ and $C_2 y^2$ for any C_1 and C_2. Indeed (see Example 7.1.2 and Figure 7.2 for limits of integration),

$$f_1(x) = \int_{-\infty}^{+\infty} f(x,y)\, dy$$

$$= \int_{-\infty}^0 0\, dy + \int_0^{1-x} Cxy^2\, dy + \int_{1-x}^\infty 0\, dy$$

$$= Cx \frac{(1-x)^3}{3} \qquad \text{for } 0 < x < 1$$

and $f_1(x) = 0$ for $x < 0$ and $x > 1$. Similarly, $f_2(y) = 0$ if $y < 0$ or $y > 1$, and for $0 < y < 1$,

$$f_2(y) = \int_{-\infty}^{+\infty} f(x,y)\, dx = \frac{Cy^2(1-y)^2}{2}.$$

Since $f(x,y) \neq f_1(x) f_2(y)$, the random variables are not independent.   □

We can now formulate

Theorem 7.2.2. *Assume that the joint density $f(x,y)$ is continuous, and let A be the set on which $f(x,y)$ is strictly positive. If A is not a Cartesian product $A = A_X \times A_Y$, then X and Y are dependent.*

Proof. The argument here is as follows. The marginal densities f_1 and

f_2 are strictly positive in the "shadows" A_X and A_Y of the set A on the x-axis and y-axis (see Figure 7.5). If A is not equal to the Cartesian product $A_X \times A_Y$, there exists a point (x_0, y_0) with $x_0 \in A_X, y_0 \in A_Y$, and $(x_0, y_0) \notin A$. If $x_0 \in A_X$, then there exists y^* such that $(x_0, y^*) \in A$, and therefore $f(x_0, y^*) > 0$. Since f is continuous, it must be positive in some neighborhood of (x_0, y^*), and therefore $f_1(x_0) = \int f(x_0, y)\, dy > 0$. Similarly, $f_2(y_0) > 0$. Consequently, $0 = f(x_0, y_0) \neq f_1(x_0)f_2(y_0) > 0$. The continuity of $f(x, y)$, and hence of $f_1(x)f_2(y)$, implies now that $f(x, y) \neq f_1(x)f_2(y)$ on a set of a positive measure. □

While the continuity of $f(x, y)$ is sufficient for the criterion given by Theorem 7.2.2, it is not necessary. However, it ought to be mentioned here that this criterion should be applied with some caution if f is not continuous, since $f(x, y)$ is defined only up to sets of measure zero. We omit here the precise formulation of this criterion which takes such effects into account.

Example 7.2.7. In Example 7.2.6 the set A is the triangle with vertices $(0, 0), (1, 0)$, and $(0, 1)$, so we could infer without any calculations that X and Y are dependent, even though $f(x, y)$ is not continuous on the line $x + y = 1$. □

At the end of this section let us make the following important remark. Formulas (7.9)–(7.12) from Definitions 7.2.3 and 7.2.4 can be used in two ways. The first is as illustrated so far: to determine whether or not two variables are independent. However, a more frequent use of these formulas is in the reverse direction. Often we know the distributions of X and Y and we know (or assume) that X and Y are independent. Then formulas (7.9)–(7.12) are used to find the joint distribution of X and Y (i.e., joint density, joint cdf, etc.) through the product rule.

Example 7.2.8. A man makes two attempts at some goal. His performance X_1, at the first attempt, measured on the scale from 0 to 1, is a ran-

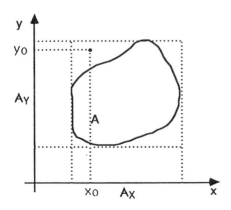

Figure 7.5 Condition for dependence.

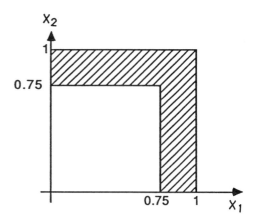

Figure 7.6 Probability of better of two attempts exceeding 0.75.

dom variable with density $f_1(x_1) = 12x_1^2(1 - x_1)$. His performance X_2 at the second attempt is independent on X_1, and tends generally to be lower; its density is $f_2(x_2) = 6x_2(1 - x_2), 0 \le x_2 \le 1$. What is the probability that the man exceeds level 0.75 in the better of the two attempts?

SOLUTION.   The joint density of (X_1, X_2) is

$$f(x_1, x_2) = \begin{cases} 72x_1^2(1 - x_1)x_2(1 - x_2) & \text{on } 0 \le x_1, x_2 \le 1 \\ 0 & \text{otherwise.} \end{cases}$$

The required probability is the integral of $f(x_1, x_2)$ over the shaded area in Figure 7.6, hence equals

$$1 - \int_0^{0.75} \int_0^{0.75} 72x_1^2(1 - x_1)x_2(1 - x_2) \, dx_1 \, dx_2 = 0.3771.$$

$\square$

PROBLEMS

7.2.1   Two cards are drawn at random from the ordinary deck of cards. Let X be the number of aces and let Y be the number of hearts obtained.
   (i) Find the joint probability function of (X, Y).
   (ii) Find the marginal distribution of X and Y.
   (iii) Are X and Y independent?

7.2.2   Let X and Y have the joint distribution $P\{X = x, Y = y\} = C(\lambda^{x+y})/(x!y!), x, y = 0, 1, \ldots$, where $\lambda > 0$.
   (i) Find C,
   (ii) Find the marginal distribution of X.
   (iii) Are X and Y independent?

(iv) Find an expression for $P(X < Y)$ involving a double series (do not try to add this series).

(v) Find an expression for $P(X < Y)$ involving a single series.

7.2.3 An urn contains five balls, two of them red and three green. Three balls are drawn without replacement. Let X and Y denote the number of red (X) and green (Y) balls drawn.

(i) Find the joint distribution of (X, Y).

(ii) Find the marginal distributions of X and Y.

(iii) Are X and Y independent?

(iv) Find the joint distribution of X and $Z = Y - X$.

7.2.4 Random variables X and Y have joint distribution given by the following table:

		Y	
X	1	2	3
1	$\frac{1}{3}$	a	$\frac{1}{6}$
2	b	$\frac{1}{4}$	c

Show that X and Y are dependent variables, regardless of the values a, b, and c.

7.2.5 Random variables have joint distribution given by the table

		Y	
X	1	2	3
1	a	$2a$	$3a$
2	b	c	d

Find a, b, c, d given that X, Y are independent, and $P(X = 2) = 2P(X = 1)$.

7.2.6 Random variables X, Y have distribution uniform on the shape of letter Y (see Figure 7.7). Identify the shapes of the marginal densities of X and Y on Figure 7.8.

7.2.7 Consider a system consisting of three components connected as in Figure 7.9. Let Y_1, Y_2, Y_3 be the lifetimes of components 1, 2, and 3. Assume that $Y_i, i = 1, 2, 3$, are independent and each has exponential distribution with parameter α. Let T be the lifetime of the system.

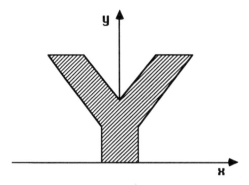

Figure 7.7   Joint distribution of X and Y.

Find:
   (i) The cdf of T.
   (ii) The hazard function of T.

7.2.8   Let X assume values 0, 1, 2, 3 and let Y assume values 0, 1, 2. The joint probability function is of the form $f(x, y) = c|x - y|$.

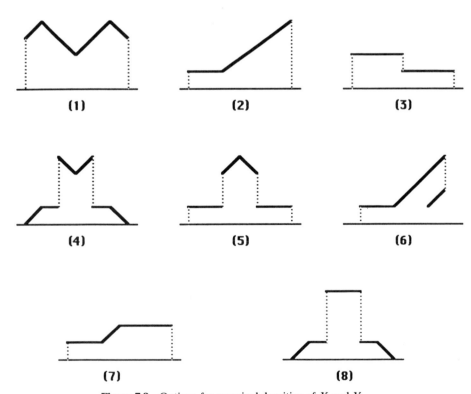

Figure 7.8   Options for marginal densities of X and Y.

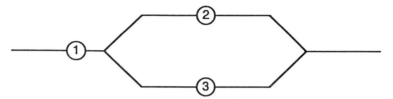

Figure 7.9   Three-component system.

(i) Find c.
(ii) Find $P(X = Y)$.
(iii) Find $P(|X - 1| \leq 1)$.
(iv) Find $P(X + Y \leq 3)$.
(v) Are X and Y independent?

7.2.9   Let (X, Y) have the distribution given by the table

			Y		
X	3	4	5	6
1	1/24	1/2	1/24	1/24
2	1/24	0	1/12	1/12
3	0	1/12	1/12	0

Find a probability distribution of random variables (X', Y') such that X' and Y' have the same marginals as X and Y, but X' and Y' are independent.

7.2.10   Two components of a machine have the following joint density of their lifetimes X and Y:

$$f(x, y) = \begin{cases} xe^{-x(1+y)} & \text{for } x \geq 0, y \geq 0 \\ 0 & \text{otherwise,} \end{cases}$$

(i) Find $P(X \geq 5)$.
(ii) Find the marginal densities of X and Y. Are these variables independent?
(iii) Find the probability that $\max(X, Y) > 2$.

7.2.11   Random variables (X, Y) have a joint density given by

$$f(x, y) = \begin{cases} k(ax + by) & \text{if } 0 < x < 1, 0 < y < 2 \\ 0 & \text{otherwise,} \end{cases}$$

where $a > 0, b > 0$ are given constants. Find:

(i) The value of k (as a function of a and b).

(ii) The marginal distributions of X and Y.

(iii) The cdf of (X, Y).

7.2.12 Assume that (X, Y) are independent random variables, with exponential densities $f(x) = ae^{-ax}, g(y) = be^{-by}, x > 0, y > 0$. Assume that it is not possible to observe both X and Y but that we can observe

$$U = \min(X, Y) \quad \text{and} \quad Z = \begin{cases} 1 & \text{if } X < Y \\ 0 & \text{otherwise.} \end{cases}$$

This kind of situation, called *censoring*, occurs often in medicine and engineering. For instance, X and Y may be the times of death due to cancer (X) and death due to other causes (Y). Then U is the observed lifetime, and Z is the indicator of the observable cause of death.

(i) Find the joint distribution of (U, Z).

(ii) Prove that U and Z are independent

(*Hint:* Note that U is continuous, Z is discrete. It suffices to show that $P\{U \leq u | Z = i\}$ is independent of i.)

7.2.13 Let $h_1(t)$ and $h_2(t)$ be hazard functions of random variables T_1 and T_2. Show that the random variable with the hazard function $h(t) = h_1(t) + h_2(t)$ has the same distribution as $\min(T_1, T_2)$, with T_1, T_2 being independent.

7.2.14 Let T_1 and T_2 be independent random variables with hazard functions $h_1(t)$ and $h_2(t)$. Write a formula for $P(T_1 < T_2)$ expressed through h_1 and h_2.

7.2.15 A box contains 3 coconut candies, 5 hazelnut chocolates and 2 peanut butter chocolates. A sample of 4 sweets is chosen from the box. Let X and Y denote the number of coconut candies and hazelnut chocolates in the sample, respectively. Find:

(i) The joint distribution of (X, Y)

(ii) The marginal distributions of X and of Y

(iii) The coefficient of correlation between the events $\{X = i\}$ and $\{Y = j\}$ for $i = 0, j = 2; i = 1, j = 1; i = 3, j = 4$ (see Definition 4.5.2).

(iv) Are X and Y independent?

(v) Are X and $Z =$ the number of peanut butter chocolates in the sample independent?

7.2.16 Assume that at shooting at a target, the coordinates (X, Y) of the point of impact are independent random variables, each with normal distribution $N(0, \sigma^2)$. Find the density of D, the distance of the point of impact from the center of the target.

7.2.17 An ecologist wishes to select a point inside a circular sampling region of radius R according to a uniform distribution. She decides to do it by sampling first the direction from the center of the region from a uniform distribution on $[0, 360]°$, and then the distance from the center, again from a uniform distribution on $[0, R]$. Find:

   (i) The density $f(x, y)$ of the chosen point in (x, y) coordinates.

  (ii) The marginal distribution of X.

 (iii) The density on $[0, R]$ which would give ecologist the resulting uniform distribution on the region.

7.3 CONDITIONAL DISTRIBUTIONS

A natural question concerning two random variables is how to handle situations when we know the value assumed by one of them. If the variables are dependent, then the information about the value of one of them affects the distribution of the other.

In symbols, the objective now is to determine probabilities of the form

$$P(X \in Q_1 | Y = y) \quad \text{or} \quad P(Y \in Q_2 | X = x) \tag{7.14}$$

for various $Q_1, Q_2, x,$ and y.

We begin with the case of discrete random variables.

Let $V = \{y_1, y_2, \ldots\}$ be the set of all possible values of Y. Then the event $\{Y = y\}$ appearing in the condition of (7.14) has positive probability only if $y \in V$, otherwise $P\{Y = y\} = 0$. We can use here the theory developed in Chapter 4, where we defined the conditional probability $P(A|B)$ of event A given that the event B occurred as $P(A|B) = P(A \cap B)/P(B)$, provided that $P(B) > 0$. If $P(B) = 0$, the probability $P(A|B)$ was left undefined. It is clear that to evaluate (7.14) in the case of discrete random variables we do not need any new concepts. If the condition has probability zero (i.e., if $P\{Y = y\} = 0$), we leave $P\{X \in Q | Y = y\}$ undefined. For $y_j \in V$, we have $P\{Y = y_j\} > 0$, and

$$P(X \in Q | Y = y_j) = \frac{P(X \in Q, Y = y_j)}{P(Y = y_j)}. \tag{7.15}$$

Let $U = \{x_1, x_2, \ldots\}$ be the set of possible values of X. Writing $p(x_i, y_j)$ or p_{ij} for $P\{X = x_i, Y = y_j\}$, the denominator in (7.15) is simply the marginal probability $P\{Y = y_j\} = \sum_i p_{ij} = p_{+j}$, and therefore

$$P\{X \in Q_1 | Y = y_j\} = \frac{\sum_{i: x_i \in Q_1} p_{ij}}{p_{+j}}.$$

An analogous formula with the role of X and Y interchanged is

$$P\{Y \in Q_2 | X = x_i\} = \frac{\sum_{j: y_j \in Q_2} p_{ij}}{p_{i+}}.$$

Example 7.3.1. Let the experiment consist of two throws of a die. We shall find conditional probabilities of the result of the first throw given the sum of both throws, and conversely, conditional probabilities for the sum, given the results of one throw. Accordingly, we shall let X_1 and X_2 denote the results of the first and second throws, and put $Z = X_1 + X_2$.

The natural sample space S is the set of all cells of the following table (where the values of Z are written in the cells).

			X_2			
X_1	1	2	3	4	5	6
1	2	3	4	5	6	7
2	3	4	5	6	7	8
3	4	5	6	7	8	9
4	5	6	7	8	9	10
5	6	7	8	9	10	11
6	7	8	9	10	11	12

The joint distribution of X_1 and Z is obtained by simple counting:

						Z						
X_1	2	3	4	5	6	7	8	9	10	11	12	
1	$\frac{1}{36}$	$\frac{1}{36}$	$\frac{1}{36}$	$\frac{1}{36}$	$\frac{1}{36}$	$\frac{1}{36}$	0	0	0	0	0	$\frac{1}{6}$
2	0	$\frac{1}{36}$	$\frac{1}{36}$	$\frac{1}{36}$	$\frac{1}{36}$	$\frac{1}{36}$	$\frac{1}{36}$	0	0	0	0	$\frac{1}{6}$
3	0	0	$\frac{1}{36}$	$\frac{1}{36}$	$\frac{1}{36}$	$\frac{1}{36}$	$\frac{1}{36}$	$\frac{1}{36}$	0	0	0	$\frac{1}{6}$
4	0	0	0	$\frac{1}{36}$	$\frac{1}{36}$	$\frac{1}{36}$	$\frac{1}{36}$	$\frac{1}{36}$	$\frac{1}{36}$	0	0	$\frac{1}{6}$
5	0	0	0	0	$\frac{1}{36}$	$\frac{1}{36}$	$\frac{1}{36}$	$\frac{1}{36}$	$\frac{1}{36}$	$\frac{1}{36}$	0	$\frac{1}{6}$
6	0	0	0	0	0	$\frac{1}{36}$	$\frac{1}{36}$	$\frac{1}{36}$	$\frac{1}{36}$	$\frac{1}{36}$	$\frac{1}{36}$	$\frac{1}{6}$
	$\frac{1}{36}$	$\frac{2}{36}$	$\frac{3}{36}$	$\frac{4}{36}$	$\frac{5}{36}$	$\frac{6}{36}$	$\frac{5}{36}$	$\frac{4}{36}$	$\frac{3}{36}$	$\frac{2}{36}$	$\frac{1}{36}$	1

We can now answer various questions concerning conditional probabilities. For instance,

$$P(X_1 \text{ odd } | Z = 2) = \frac{P\{X_1 = 1, Z = 2\}}{P\{Z = 2\}} = \frac{\frac{1}{36}}{\frac{1}{36}} = 1$$

while

$$P(Z \text{ odd } |X_1 = 2) = \frac{P\{Z = 3, X_1 = 2\}}{P\{X_1 = 2\}} + \frac{P\{Z = 5, X_1 = 2\}}{P\{X_1 = 2\}}$$
$$+ \frac{P\{Z = 7, X_1 = 2\}}{P\{X_1 = 2\}} = \frac{1}{2}. \qquad \square$$

Let us now consider continuous random variables. In this case, determining a quantity such as $P\{X \in A | Y = y\}$ cannot rely upon the concepts of Chapter 4, since the conditioning event $\{Y = y\}$ has probability zero. We shall therefore proceed directly: We *define* the conditional density and then verify that the defined function is indeed a density and that it meets some conditions that agree with our intuition and common sense.

We let $f(x, y)$ denote the joint density of (X, Y). As in Section 7.2, we let f_1 and f_2 denote the marginal densities of X and Y; finally, we let $g_{12}(x|y)$ and $g_{21}(y|x)$ denote, respectively, the conditional density of X given $Y = y$ and the conditional density of Y given $X = x$.

Definition 7.3.1. The *conditional densities* g_{12} and g_{21} are defined by

$$g_{12}(x|y) = \frac{f(x, y)}{f_2(y)} \qquad (7.16)$$

provided that $f_2(y) > 0$, and

$$g_{21}(y|x) = \frac{f(x, y)}{f_1(x)} \qquad (7.17)$$

provided that $f_1(x) > 0$. $\qquad \square$

If the denominators in (7.16) or (7.17) are zero, the left-hand sides remain undefined. To check that formulas (7.16) and (7.17) define densities, observe first that both functions [regarded as functions of the *first* variable, i.e., x in $g_{12}(x|y)$ and y in $g_{21}(y|x)$], are nonnegative. Moreover, we have

$$\int_{-\infty}^{+\infty} g_{12}(x|y) \, dx = \int_{-\infty}^{+\infty} \frac{f(x, y)}{f_2(y)} dx = \frac{1}{f_2(y)} \int_{-\infty}^{+\infty} f(x, y) \, dx = \frac{f_2(y)}{f_2(y)} = 1$$

and similarly for g_{21}. To justify the definition on "semantic" grounds, assume for simplicity that $f(x, y)$, and hence also $f_2(y)$, are continuous, and let us consider, for some $h > 0$, the probability $P\{X \in A | y \le Y \le y + h\}$. We have here, for small h,

$$P\{X \in A | y \le Y \le y + h\} = \frac{P\{X \in A, y \le Y \le y + h\}}{P\{y \le Y \le y + h\}}$$
$$= \frac{\int_A \int_y^{y+h} f(x, u) \, du \, dx}{\int_y^{y+h} f_2(u) \, du} \approx \frac{\int_A f(x, y) h \, dx}{f_2(y) h} = \int_A g_{12}(x|y) \, dx.$$

This shows that $g_{12}(x|y)$ can indeed be sensibly taken as the density of X given that $Y = y$.

Example 7.3.2. Let the joint density of (X, Y) be given by

$$f(x, y) = \begin{cases} Cxy^2 & \text{if } 0 \leq x \leq 1, 0 \leq y \leq 1, x + y \geq 1 \\ 0 & \text{otherwise.} \end{cases}$$

Thus, the density is positive on the triangle with the vertices $(1, 1), (1, 0)$, and $(0, 1)$, hence X and Y are dependent (by the criterion for dependence of continuous random variables from Section 7.2). The marginal densities are as follows (we do not need to determine the numerical value of C; it will cancel in densities g_{12} and g_{21}). If $0 < x < 1$, then

$$f_1(x) = \int_{-\infty}^{+\infty} f(x, y)\, dy = \int_{1-x}^{1} Cxy^2\, dy$$

$$= Cx \frac{y^3}{3}\bigg|_{1-x}^{1} = \frac{C}{3}(3x^2 - 3x^3 + x^4).$$

Similarly, for $0 < y < 1$,

$$f_2(y) = \int_{1-y}^{1} Cxy^2\, dx = Cy^2 \cdot \frac{x^2}{2}\bigg|_{1-y}^{1} = \frac{C}{2}(2y^3 - y^4). \tag{7.18}$$

Thus

$$g_{12}(x|y) = \frac{f(x, y)}{f_2(y)} = \frac{Cxy^2}{(C/2)(2y^3 - y^4)} = \frac{2xy^2}{2y^3 - y^4},$$

hence

$$g_{12}(x|y) = \begin{cases} \dfrac{2x}{2y - y^2} & \text{for } 0 < y < 1, 1 - y < x < 1 \\ 0 & \text{otherwise.} \end{cases}$$

For instance, if $Y = 0.5$, then

$$g_{12}(x|0.5) = g_{21}(x|Y = 0.5) = \frac{8}{3}x \quad \text{for } 0.5 < x < 1$$

(see Figure 7.10). Similarly,

$$g_{21}(y|x) = \frac{f(x, y)}{f_1(x)} = \frac{Cxy^2}{(C/3)(3x^2 - 3x^3 + x^4)} = \frac{3y^2}{3x - 3x^2 + x^3}$$

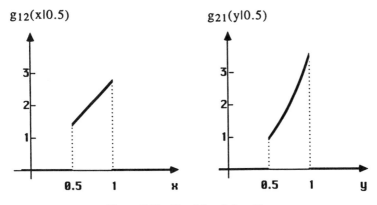

Figure 7.10 Conditional densities.

for $1 - x < y < 1$ and 0 otherwise. Thus, if $X = 0.5$, the density $g_{21} (y|0.5) = 24y^2/7$ for $0.5 < y < 1$ (see Figure 7.10).

Suppose we know that $X = x_0$ and we want to find a point y^* such that Y will have the same chance of being above as below y^* (i.e., the *median* of Y). Thus, we must have

$$\frac{1}{2} = \int_{-\infty}^{y^*} g_{21}(y|x_0)\, dy = \int_{1-x_0}^{y^*} \frac{3y^2}{3x_0 - 3x_0^2 + x_0^3}\, dy$$

$$= \frac{1}{3x_0 - 3x_0^2 + x_0^3}[(y^*)^3 - (1 - x_0)^3],$$

hence

$$y^* = \sqrt[3]{\frac{3x_0 - 3x_0^2 + x_0^3}{2} + (1 - x_0)^3}.$$

Consequently, if $X = 0.75$, then we know that Y will lie in the interval $(0.25, 1)$. Its values, however, are more likely to lie close to 1 than close to 0.25. In fact, one can bet even money that Y will exceed 0.8058.   □

In Chapter 4 we repeatedly stressed the fact that the formula $P(A|B) = P(A \cap B)/P(B)$ can be used in two ways: to determine the conditional probability $P(A \cap B)$ of two events occurring jointly, given the probability of one of those events and the appropriate conditional probability [e.g., $P(B)$ and $P(A|B)$]. The same situation is true in case of conditional distributions in a continuous, discrete, or mixed case: Given one marginal distribution and one conditional distribution one can determine the joint distribution and hence also the marginal distribution of the other variable. We shall illustrate this technique by three examples.

Example 7.3.3. An animal lays a certain number X of eggs, where X is random and has the Poisson distribution

$$P\{X = n\} = \frac{\lambda^n}{n!} e^{-\lambda}, \qquad n = 0, 1, 2, \ldots$$

(see Example 6.3.4). Each egg hatches with probability p independent of hatching of other eggs. Determine the distribution of $Y =$ the number of eggs that hatch.

SOLUTION.   Here the randomness of Y has two sources: first, the number X of eggs laid is random (varies from animal to animal), and second, even for animals that laid the same number of eggs, the randomness in the process of hatching may make the number of offspring different.

Our solution strategy is as follows. We are given the distribution of X. The assumptions of the problem will allow us to determine the conditional distribution of Y given X. These two allow us to determine the joint distribution of (X, Y), and the distribution of Y will be obtained as marginal from the joint distribution of (X, Y).

Thus, we now need to determine the conditional distribution of Y given X, that is, $P\{Y = j | X = n\}$. First, it is clear that $0 \le Y \le X$ (the number of eggs that hatch is nonnegative and cannot exceed the number of eggs laid). Now, the assumption is that eggs hatch with the same probability and independently. Thus Y must have binomial distribution if we regard the process of incubation as an "experiment over an egg" with "success" identified with hatching. Therefore,

$$P\{Y = j | X = n\} = \binom{n}{j} p^j (1 - p)^{n-j}, \qquad j = 0, 1, \ldots, n.$$

By formula (7.15) the joint distribution of (X, Y) is given by

$$\begin{aligned} P\{X = n, Y = j\} &= P\{X = n\} \times P\{Y = j | X = n\} \\ &= \frac{\lambda^n}{n!} e^{-\lambda} \times \binom{n}{j} p^j (1 - p)^{n-j}, \end{aligned}$$

where $n = 0, 1, 2, \ldots$ and $j = 0, 1, \ldots, n$ (for all other combinations n, j the joint probability is zero).

We shall now find the marginal probability $P\{Y = j\}$ for $j = 0, 1, 2, \ldots$. Clearly, we must have $X \ge Y$, hence $n \ge j$, and therefore

$$\begin{aligned} P\{Y = j\} &= \sum_{n=j}^{\infty} P\{X = n, Y = j\} \\ &= \sum_{n=j}^{\infty} \frac{\lambda^n}{n!} e^{-\lambda} \binom{n}{j} p^j (1 - p)^{n-j} \end{aligned}$$

$$= \sum_{n=j}^{\infty} \frac{\lambda^n}{n!} e^{-\lambda} \frac{n!}{j!(n-j)!} p^j (1-p)^{n-j}$$

$$= \frac{(\lambda p)^j}{j!} e^{-\lambda} \sum_{n=j}^{\infty} \frac{[\lambda(1-p)]^{n-j}}{(n-j)!}$$

$$= \frac{(\lambda p)^j}{j!} e^{-\lambda} \sum_{v=0}^{\infty} \frac{[\lambda(1-p)]^v}{v!} = \frac{(\lambda p)^j}{j!} e^{-\lambda} e^{\lambda(1-p)}$$

$$= \frac{(\lambda p)^j}{j!} e^{-\lambda p}.$$

Thus, the marginal distribution of Y is again Poisson, except that the parameter has changed from λ to λp.

The process that leads from X to Y in this example is sometimes called *binomial thinning*. One can visualize it in general as follows. A random process gives the value of X (by assumption, an integer), with some distribution. We can think of X as the number of objects of some kind that are produced, number of elements sampled, number of events of some kind that occur, and so on. Then the process of thinning causes some of the X objects (events, etc.) to disappear (not be counted), due to a certain random process of selection. For instance, some of the X objects produced are defective; some of the X elements sampled are not acceptable or have certain other characteristics; some of the events are unobservable; and so on. Such process of elimination of some X's and the selection of others is called *thinning*. We say that the process of thinning is binomial if the inclusions of X-elements as Y-elements are independent and occur with the same probability. The present example shows that binomial thinning of Poisson random variables leads again to Poisson random variables, with approximately modified parameters.

We shall now consider a situation analogous to that of Example 7.3.3, but for continuous distributions.

Example 7.3.4. A point X is chosen at random from the interval $[A, B]$ according to the uniform distribution (see Example 6.3.10), and then a point Y is chosen at random, again with uniform distribution, from the interval $[X, B]$. Find the marginal distribution of Y.

SOLUTION. We have here

$$f_1(x) = \begin{cases} \dfrac{1}{B-A} & \text{for } A \le x \le B \\ 0 & \text{otherwise,} \end{cases}$$

$$g_{21}(y|x) = \begin{cases} \dfrac{1}{B-x} & \text{for } x \le y \le B \\ 0 & \text{otherwise.} \end{cases}$$

Thus

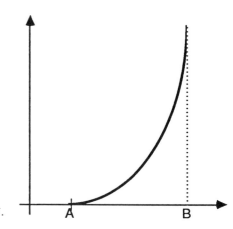

Figure 7.11 Marginal density of Y.

$$f(x,y) = f_1(x)g_{21}(y|x) = \begin{cases} \dfrac{1}{B-A} \cdot \dfrac{1}{B-x} & \text{for } A \le x \le y \le B \\ 0 & \text{otherwise.} \end{cases}$$

Consequently, for fixed y with $A \le y \le B$ we have

$$f_2(y) = \int_A^y \frac{1}{B-A} \cdot \frac{1}{B-x} dx = \frac{1}{B-A} \log \frac{B-A}{B-y}.$$

The graph of density of Y is shown on Figure 7.11. As may have been expected, the values close to B are more likely than those close to A; the density of Y is in fact unbounded as we approach the upper boundary B of the range of Y.

Example 7.3.5.   Finally, we shall consider the case of a mixed distribution, with X being continuous and Y being discrete.

Assume that Y may take one of $n \ge 2$ integer values with

$$P\{Y = i\} = p_i$$

and

$$p_1 + p_2 + \cdots + p_n = 1.$$

In the simplest case $n = 2$ we have

$$P\{Y = 1\} = p, \quad P\{Y = 2\} = 1 - p$$

(i.e., $p_1 = p, p_2 = 1 - p$).

Next, we assume that for given $Y = i$, the random variable X has a continuous distribution with density $\varphi_i(x)$. Thus, we have

$$P\{a \le X \le b, Y = i\} = p_i \int_a^b \varphi_i(x)\, dx.$$

To find the marginal distribution of X, we may use the formula for total probability (4.9) from Chapter 4, using events $\{Y = i\}$ as the partition. Then

$$P\{a \le X \le b\} = \sum_{i=1}^{n} P\{a \le X \le b | Y = i\} \cdot P\{Y = i\}$$

$$= \sum_{i=1}^{n} p_i \int_{a}^{b} \varphi_i(x)\,dx = \int_{a}^{b} \sum_{i=1}^{n} p_i \varphi_i(x)\,dx. \qquad (7.19)$$

It follows that X is a continuous random variable with the density

$$f_X(x) = \sum_{i=1}^{n} p_i \varphi_i(x),$$

called the *mixture* of densities φ_i with *mixing coefficients* p_i.
    If we have only two values of Y, and therefore only two densities φ_1 and φ_2 of X, then

$$f_X(x) = p\varphi_1(x) + (1-p)\varphi_2(x).$$

Let us remark that the above formulas remain valid if Y assumes one of infinitely many values $1, 2, \ldots$, with probabilities $p_1, p_2, \ldots$ such that we have $\sum p_i = 1$. The only potential source of trouble is the interchange of integration and summation in the last step in (7.19). But this interchange is permissible because the terms of the sum are all nonnegative.

PROBLEMS

7.3.1 Let X and Y have the joint density $f(x, y) = \lambda^2 e^{-\lambda y}$ for $0 \le x \le y$ and $f(x, y) = 0$ otherwise. Find:

    (i) The marginal densities of X and Y.
    (ii) The cdf of (X, Y).
    (iii) Conditional density of Y given X.

7.3.2 Let X and Y have a density of the form

$$f(x, y) = \begin{cases} A(y - x)^\alpha & \text{for } 0 \le x < y \le 1 \\ 0 & \text{otherwise.} \end{cases}$$

    (i) What are the values of α under which f can be a density function?
    (ii) Find the value of A for α specified in part (i).
    (iii) Find the marginal densities of X and Y.
    (iv) Find conditional densities of X given Y and of Y given X.

7.3.3 Let X, Y be independent, each with a standard normal distribution. Find the distribution of $V = X/Y$. [*Hint:* Start with the cdf $F_V(v) = P(V \leq v)$.]

7.3.4 Two parts of a paper are typed by two typists. Let X and Y be the numbers of typing errors in the two parts of the paper. Suppose that X and Y have Poisson distributions with parameters λ_1 and λ_2, respectively, and that X and Y are independent. Find:

  (i) The probability that the paper (i.e., combined two parts) has at least one typing error.

  (ii) The probability that the total number of typing errors is m.

  (iii) The probability that the first part of the paper has k typing errors given that there are n typing errors altogether.

7.3.5 The phrase "A stick is broken at random into three pieces" can be interpreted in several ways. Let us identify the stick with interval $[0, 1]$ and let $0 < X < Y < 1$ be the breaking points. Some of the possible mechanisms generating X, Y are as follows:

  (i) A point (U, V) is chosen from the unit square with uniform distribution, and we let

$$X = \min(U, V), \qquad Y = \max(U, V). \qquad (7.20)$$

  (ii) The point U is chosen from $[0, 1]$ with uniform distribution. If $U < \frac{1}{2}$, then V is chosen with uniform distribution from $[U, 1]$, while if $U \geq \frac{1}{2}$, then V is chosen with uniform distribution on $[0, U]$. Then (X, Y) are defined by (7.20).

  (iii) X is chosen from $[0, 1]$ according to the uniform distribution, and then Y is chosen with uniform distribution on $[X, 1]$.

  (iv) U is chosen with uniform distribution on $[0, 1]$. Next, one of the intervals $[0, U]$ or $[U, 1]$ is chosen at random, with probability U and $1 - U$, respectively. Then V is chosen with uniform distribution from the chosen interval, and again, we let (7.20) determine the values (X, Y).

In each of the cases (i)–(iv) find the joint density of (X, Y) and the marginal densities of (X, Y). Which of the mechanisms (i)–(iv) are equivalent?

7.3.6 A fast food restaurant has a dining room and a drive-in window. Let X and Y denote the fraction of time (during a working day) when the dining room (X) and drive-in window (Y) are busy. Suppose that the joint density of (X, Y) is $f(x, y) = k(2x^2 + y^2)$ for $0 \leq x, y \leq 1$ and $f(x, y) = 0$ otherwise. Find:

(i) k.

(ii) Marginal densities of X and of Y.

(iii) The probability that the drive-in window will be busy more than 75% of the time on a day when the dining room is empty less than 10% of the time.

(iv) The density of the fraction of time when dining room is empty on a day when the drive-in window is busy half of the time.

(v) Do you find anything disturbing in this problem? If so, explain.

7.3.7 Let X and Y be independent random variables, each distributed uniformly on $[0,1]$. Find:

(i) $P(X + Y \leq \frac{1}{2} \mid X = \frac{1}{4})$.

(ii) $P(X + Y \leq \frac{1}{2} \mid X \geq \frac{1}{4})$.

(iii) $P(X \geq Y \mid Y \geq \frac{1}{2})$.

7.4 FUNCTIONS OF TWO RANDOM VARIABLES

In Chapter 6 we considered the transformation of random variables. The discrete case was covered completely, and the continuous case was covered for transformation of a single random variable. We shall now consider the case of functions of two continuous random variables X and Y. We let $f(x,y)$ denote the joint density of (X, Y) with support C; this means that we have $f(x,y) > 0$ on C and $f(x,y) = 0$ outside C. The objective is to find the density of $Z = \varphi(X, Y)$, where φ is a differentiable function of two real arguments. By far the simplest here is the cdf technique, introduced in Section 6.4 for the case of a single variable. It may also be applied in multivariate case if we can calculate the explicit form of

$$P\{Z \leq z\} = P\{\varphi(X,Y) \leq z\} = \iint\limits_{\{(x,y):\varphi(x,y)\leq z\}} f(x,y)\, dx\, dy$$

as a function of z. Density can then be obtained by differentiation. This method works especially well for $\varphi(X, Y) = \max(X, Y)$ and $\varphi(X, Y) = \min(X, Y)$, when X and Y are independent. We can avoid the step involving integral completely, writing $P\{\max(X, Y) \leq z\} = P\{X \leq z, Y \leq z\} = F_X(z)F_Y(z)$, hence the density of $Z = \max(X, Y)$ is $f_X(z)F_Y(z) + F_X(z)f_Y(z)$. Now we will present a technique that may be applied for a wider class of cases. It will be given in the form of an algorithm and its use will be illustrated by a number of examples. A formal proof will be omitted; we shall only present some propaganda to convince the reader that the algorithm may be expected to lead to the correct result (the algorithm is, in fact, based on a change of variables in two-dimensional integrals, which may be found in many advanced-level calculus texts).

Determination of densities of bivariate (and multivariate) transformations is typically regarded by students as one of the most incomprehensible, difficult, and therefore terrifying exam topics in probability. We do hope that by presenting it as a purely mechanical procedure—which it largely is—we shall alleviate, or perhaps eliminate, the terror. It is true that the procedure requires constant attention and some level of algebraic skills, but the difficulties are closer to those of proofreading a telephone directory (say) than to those of playing a game of chess.

The algorithm is as follows.

1. Choose a "companion" function, say $w = \eta(x, y)$, such that the pair of equations

$$z = \varphi(x, y)$$
$$w = \eta(x, y) \tag{7.21}$$

can be solved, leading to

$$x = \alpha(z, w) \tag{7.22}$$
$$y = \beta(z, w).$$

2. Determine the image of the support C of density $f(x, y)$ in the (z, w) plane under transformation (7.21). Let this image be D.

3. Find the Jacobian of transformation (7.21), that is, the determinant

$$J = \begin{vmatrix} \dfrac{\partial \alpha}{\partial z} & \dfrac{\partial \alpha}{\partial w} \\[2mm] \dfrac{\partial \beta}{\partial z} & \dfrac{\partial \beta}{\partial w} \end{vmatrix}. \tag{7.23}$$

4. Determine the joint density $g(z, w)$, of random variables $Z = \varphi(X, Y), W = \eta(X, Y)$, given by the formula

$$g(z, w) = \begin{cases} f(\alpha(z, w), \beta(z, w))|J| & \text{for } (z, w) \in D \\ 0 & \text{for } (z, w) \notin D. \end{cases}$$

5. Compute the density of $Z = \varphi(X, Y)$ as the marginal density of the joint density $g(z, w)$:

$$g_Z(z) = \int_{-\infty}^{+\infty} g(z, w)\, dw = \int_{D_z} f(\alpha(z, w), \beta(z, w))|J|\, dw,$$

where D_z is defined as

$$D_z = \{w : (z, w) \in D\}.$$

Out of these five steps, it is only step 1 which requires some moderate amount of thinking (or, at least some experience). The reason is that the choice of the companion transformation η affects all subsequent steps, making the calculations easy, difficult, or perhaps even impossible.

Example 7.4.1 (Sum of Random Variables). The operation of addition of random variables appears so often that it is worthwhile to derive general formulas here. Thus, we have $z = \varphi(x, y) = x + y$. We may take a companion transformation $w = \eta(x, y) = x$ (say), hence the inverse transformation (7.22) is

$$x = w$$
$$y = z - w,$$

so that $\alpha(z, w) = w, \beta(z, w) = z - w$. Thus

$$J = \begin{vmatrix} 0 & 1 \\ 1 & -1 \end{vmatrix}$$

and $|J| = 1$. If (X, Y) have joint density $f(x, y)$, then the joint density of (Z, W) is $f(w, z - w)$. Consequently, the density of Z is

$$g_Z(z) = \int_{-\infty}^{+\infty} f(w, z - w)\, dw. \tag{7.24}$$

In this formula one has only to determine the effective limits of integration, that is, the set of values w (for given z) at which the integrand is positive.
$\square$

Example 7.4.2 (Sum of Exponential Random Variables). Let us apply formula (7.24) to the case when X and Y are independent, with densities

$$f_X(x) = \begin{cases} \alpha e^{-\alpha x} & \text{for } x \geq 0 \\ 0 & \text{for } x < 0 \end{cases} \quad \text{and} \quad f_Y(y) = \begin{cases} \alpha e^{-\alpha y} & \text{for } y \geq 0 \\ 0 & \text{for } y < 0. \end{cases}$$

Thus,

$$f(x, y) = \begin{cases} \alpha^2 e^{-\alpha(x+y)} & \text{if } x, y \geq 0 \\ 0 & \text{otherwise.} \end{cases} \tag{7.25}$$

Consequently, the density $g_Z(z)$, as given by (7.24), is

$$g_Z(z) = \int_{-\infty}^{+\infty} f(w, z - w)\, dw = \int_{D_z} \alpha^2 e^{-\alpha z}\, dw,$$

where D_z is the set $\{w : w \geq 0, z - w \geq 0\}$, since both arguments must be positive for f to be given by the upper part of (7.25). Consequently, we must have $z > 0$ and $0 \leq w \leq z$, hence

$$g_Z(z) = \alpha^2 e^{-\alpha z} \int_0^z dw = \alpha^2 z e^{-\alpha z}, \qquad z \geq 0 \qquad (7.26)$$

and $g_Z(z) = 0$ for $z < 0$. We recognize the density (7.26) as a gamma density with parameters 2 and α (see Example 6.4.7).

Example 7.4.3 (Sum of Two Uniform Random Variables). Suppose now that

$$f(x,y) = \begin{cases} 1 & \text{if } 0 \leq x \leq 1, 0 \leq y \leq 1 \\ 0 & \text{otherwise.} \end{cases}$$

This corresponds to the product of densities $f_X(x)$ and $f_Y(y)$ uniform on $[0,1]$. Formula (7.24) gives now for the sum $Z = X + Y$ of two independent uniform random variables

$$g_Z(z) = \int_{D_z} 1 \, dw, \qquad (7.27)$$

where

$$D_z = \{w : 0 \leq w \leq 1, 0 \leq z - w \leq 1\}.$$

The pair of inequalities defining D_z, namely $0 \leq w \leq 1$ and $z - 1 \leq w \leq z$, can be written as

$$\max(0, z - 1) \leq w \leq \min(z, 1).$$

It is now clear that we must have $0 \leq z \leq 2$, since otherwise D_z is empty. For $0 \leq z \leq 1$, the set D_z is the interval $0 \leq w \leq z$, while for $1 \leq z \leq 2$ we have the interval $z - 1 \leq w \leq 1$. Consequently, (7.27) gives (see Figure 7.12)

$$g_Z(z) = \begin{cases} z & \text{for } 0 \leq z \leq 1 \\ 2 - z & \text{for } 1 \leq z \leq 2 \\ 0 & \text{otherwise.} \end{cases}$$

This distribution, not surprisingly, is called *triangular*.   □

Example 7.4.4. Formula (7.24) was obtained by taking a specific companion transformation $w = \eta(x, y) = x$. There is no compelling reason for this choice. We could, of course, have chosen $w = \eta(x, y) = y$, getting a very similar formula. However, we could also have chosen something more fancy, say $w = \eta(x, y) = x/(x + 5y)$.

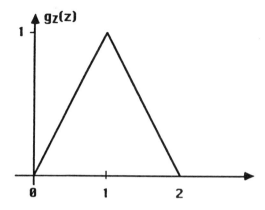

Figure 7.12 Triangular density.

The system of equations

$$z = x + y$$
$$w = \frac{x}{x + 5y}$$

has the solution

$$x = \frac{5wz}{1 + 4w}$$
$$y = z\frac{1 - w}{1 + 4w}.$$

Consequently, the Jacobian equals

$$J = \begin{vmatrix} \dfrac{5w}{1 + 4w} & \dfrac{5z}{(1 + 4w)^2} \\[2mm] \dfrac{1 - w}{1 + 4w} & \dfrac{-5z}{(1 + 4w)^2} \end{vmatrix} = -\frac{5z}{(1 + 4w)^2}.$$

Thus, the density of the sum $Z = X + Y$ can be obtained as the integral

$$g_Z(z) = \int_{-\infty}^{+\infty} f\left(\frac{5wz}{1 + 4w}, z\frac{1 - w}{1 + 4w}\right)\left|\frac{5z}{(1 + 4w)^2}\right| dw, \qquad (7.28)$$

where, again, the effective range of integration depends on the support C of the joint density f.

This example is given here to show that there is no such thing as "the" formula for density of sum of random variables: In fact, (7.24) and (7.28) will both lead to the same final form of $g_Z(z)$ once the integration is carried out. We cannot even say that (7.24) is simpler than (7.28), since the simplicity of integration depends here on the form of f and of its support C.

Example 7.4.5 (Product of Two Random Variables). In this case we have $\varphi(x, y) = xy$, so that we look for the density of the random variable $Z = XY$.

Let us again choose the companion function $\eta(x, y) = x$, so that the system of equations (7.21) is $z = xy, w = x$, and its solution (7.22) is

$$x = \alpha(z, w) = w, \quad y = \beta(z, w) = \frac{z}{w}.$$

The Jacobian of this transformation is

$$J = \begin{vmatrix} 0 & 1 \\ \dfrac{1}{w} & -\dfrac{z}{w^2} \end{vmatrix} = -1/w,$$

hence $|J| = \dfrac{1}{|w|}$. The joint density of (Z, W) is now

$$f\left(w, \frac{z}{w}\right) \frac{1}{|w|}$$

and the density of Z is given by

$$g_Z(z) = \int_{-\infty}^{+\infty} f\left(w, \frac{z}{w}\right) \frac{1}{|w|} \, dw.$$

Again, the effective limits of integration depend on the support C of density $f(x, y)$ and consequently, the sets D_z. $\qquad\qquad\square$

We now give an example that provides us with an algorithm of generating random variables with normal distribution.

Theorem 7.4.1. *If X and Y are independent, uniformly distributed on $[0, 1]$, then the random variables Z and W given by*

$$Z = \sqrt{-2 \log X} \sin(2\pi Y)$$

and

$$W = \sqrt{-2 \log X} \cos(2\pi Y)$$

are independent and each has the standard normal distribution.

Proof. Let X and Y be independent, each with the distribution uniform on $[0, 1]$, so that

$$f(x, y) = \begin{cases} 1 & \text{if } 0 \le x \le 1, 0 \le y \le 1 \\ 0 & \text{otherwise.} \end{cases}$$

We have here

$$z = \sqrt{-2\log x}\,\sin(2\pi y) \quad \text{and} \quad w = \sqrt{-2\log x}\,\cos(2\pi y).$$

By the Pythagorean theorem, $z^2 + w^2 = -2\log x$, hence

$$x = e^{-(z^2+w^2)/2}.$$

On the other hand, $z/w = \tan(2\pi y)$, which gives

$$y = \frac{1}{2\pi}\arctan(z/w).$$

The Jacobian equals

$$J = \begin{vmatrix} -ze^{-(z^2+w^2)/2} & -we^{-(z^2+w^2)/2} \\ \dfrac{1}{2\pi}\dfrac{1/w}{1+z^2/w^2} & \dfrac{1}{2\pi}\dfrac{-z/w^2}{1+z^2/w^2} \end{vmatrix}, \tag{7.29}$$

which, after some algebra, reduces to

$$\frac{1}{\sqrt{2\pi}}e^{-z^2/2} \cdot \frac{1}{\sqrt{2\pi}}e^{-w^2/2}.$$

The unit square, equal to the support of $f(x, y)$, is mapped into the whole plane (z, w). It follows that the joint density of (Z, W), equal in this case to the Jacobian (7.29), is the product of two standard normal densities. □

As already mentioned, the conditional densities given a specific value of a random variable cannot be calculated according to the principles of Chapter 4, since the conditioning event has probability zero. As a warning we present an example, seemingly paradoxical, where by following the rules as explained in this chapter, one obtain two different answers to the same question.

Example 7.4.6 (Borel–Kolmogorov Paradox). Assume that X and Y have joint density

$$f(x, y) = 4xy, \qquad 0 \le x \le 1, 0 \le y \le 1,$$

so that X and Y are independent, with the same marginal densities

$$f_X(t) = f_Y(t) = 2t, \qquad 0 \le t \le 1.$$

We shall try to determine the conditional density of X given the event $X =$

Y. As we shall see, the answer will depend on the representation of event $X = Y$, as $Y - X = 0$ or as $Y/X = 1$.

SOLUTION 1. We introduce new variables, $U = X, V = Y - X$; then our problem becomes equivalent to finding the density of U given $V = 0$. Thus, we have to find the joint density $h(u, v)$ of (U, V), determine the marginal density $h_V(v)$, and our answer will be

$$\varphi_X(x|X = Y) = \frac{h(u, v)}{h_V(v)}\bigg|_{v=0} = \frac{h(u, 0)}{h_V(0)}.$$

The transformation $u = x, y = u + v$, has the inverse $x = u, y = u + v$, so the Jacobian is

$$J = \begin{vmatrix} \dfrac{\partial x}{\partial u} & \dfrac{\partial x}{\partial v} \\ \dfrac{\partial y}{\partial u} & \dfrac{\partial y}{\partial v} \end{vmatrix} = \begin{vmatrix} 1 & 0 \\ 1 & 1 \end{vmatrix} = 1.$$

Consequently, the joint density of (U, V) is

$$h(u, v) = 4u(u + v)$$

on the set $0 \le u \le 1$, $-u \le v \le 1 - u$ (see Figure 7.13). The marginal density for $0 \le v \le 1$ is

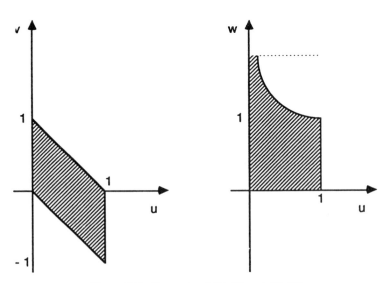

Figure 7.13 Supports of (U, V) and (U, W).

$$h_V(v) = \int_0^{1-v} 4u(u+v)\,du = 4\left[\frac{(1-v)^3}{3} + v\frac{(1-v)^2}{2}\right],$$

while for $-1 \le v \le 0$ we have

$$h_V(v) = \int_{-v}^{1} 4u(u+v)\,du = 4\left[\frac{1}{3} + \frac{v}{2} - \frac{v^3}{6}\right].$$

Both formulas give $h_V(0) = \frac{4}{3}$. Thus, we obtain for the density of U at $V = 0$ the formula

$$\frac{h(u,0)}{h_V(0)} = \frac{4u^2}{\frac{4}{3}} = 3u^2, \qquad 0 \le u \le 1.$$

SOLUTION 2. We introduce the transformation $U = X, W = Y/X$. Then the conditional density of X given $X = Y$ is the same as the conditional density of U given $W = 1$.

The transformation $u = x, w = y/x$ has inverse $x = u, y = uw$, and the Jacobian equals

$$J = \begin{vmatrix} 1 & 0 \\ w & u \end{vmatrix} = u.$$

Consequently, the joint density of (U, W) is

$$g(u, w) = 4u^3 w \qquad \text{for } 0 \le u \le 1, 0 \le w \le \frac{1}{u}$$

(see Figure 7.13). The marginal density $g_W(w)$ is

$$g_W(w) = \int_0^1 4u^3 w\,du = w \qquad \text{for } 0 \le w \le 1,$$

while for $w \ge 1$ we have

$$g_W(w) = \int_0^{1/w} 4u^3 w\,du = \frac{1}{w^3}.$$

For $w = 1$ we have $g_W(1) = 1$, and therefore the conditional density equals

$$\frac{g(u, 1)}{g_W(1)} = 4u^3, \qquad 0 \le u \le 1.$$

Thus, we obtained two different solutions, and we may ask: Which of them— if either—is correct?

The formal answer is that both solutions are correct, and we can choose for the density of X given $X = Y$ any other function as well, simply because

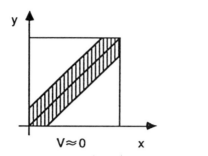
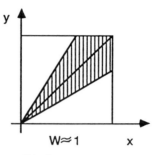

Figure 7.14 Approximations of two conditioning events.

conditional densities of a variable, given conditions with probability zero, can be modified arbitrarily on any set of conditions whose total probability is zero (in particular, on a single condition, or on finitely many conditions). This answer is, however, not quite satisfactory: We also want to understand why the techniques of transformation lead to different answers. To explain this phenomenon, let us assume that we need the density of X given $X = Y$, but we cannot observe exactly whether the condition holds. Specifically, consider two situations: (a) we can observe the difference $X - Y$, within some small error ϵ; (b) we can observe the ratio Y/X, again within a small error ϵ. In such cases, the "natural" approach is to use a limiting passage, with $\epsilon \to 0$. For any fixed ϵ the conditioning events $|X - Y| < \epsilon$ and $|Y/X - 1| < \epsilon$ are different (see Figure 7.14), and it is not surprising that the ratio Y/X favors larger values of u.

PROBLEMS

7.4.1 Suppose that 3 cards are drawn without replacement from an ordinary deck. Let X be the number of aces and let Y be the number of red cards among the cards drawn.

    (i) Find the joint distribution of (X, Y).

    (ii) Suppose that you pay \$10 for drawing the cards, and you win \$8 for each ace and \$4 for each red card. Find the distribution of your net reward $Z = 8X + 4Y - 10$.

7.4.2 A red die and a green die are tossed, each until a 6 shows up. Let X be number of tosses required for the red die and let Y be the number of tosses required for the green die. You win the amount $Y - X$ if $Y > X$, otherwise you pay the amount $X - Y$. Find the distribution of the net amount you win or lose in one game.

7.4.3 Find the cdf and density of $Y = X_1 - X_2$, where the X_i's are independent and have identical exponential distributions with the density e^{-x} for $x > 0$.

7.4.4 Let $X_1, X_2, \ldots, X_n$ be independent, each with distribution with density $f(x)$. Find the cdf and density of

$$Z_1 = \min(X_1, \ldots, X_n) \quad \text{and} \quad Z_n = \max(X_1, \ldots, X_n).$$

[*Hint*: $Z_n \leq t$ is equivalent to $X_i \leq t$ for all i, so $P(Z_n \leq t) = (F(t))^n$, where F is the cdf of X_i. For Z_1 use the fact that $P(Z_1 \leq t) = 1 - P(Z_1 > t)$.]

7.4.5 Let $X_1, \ldots, X_n$ be independent with the same density $f(x)$. Let $U = \min(X_1, \ldots, X_n)$, $V = \max(X_1, \ldots, X_n)$. Find:
  (i) The joint density of (U, V).
  (ii) The density of range defined as $R = V - U$.
      For the questions below, assume that f is uniform on $[0, 1]$. Find:
  (iii) The cdf of R.
  (iv) The probability that the interval $[U, V]$ does not contain point x_0.
  (v) The point x_0 for which the probability in part (iv) is minimum.

7.4.6 Let X and Y be independent random variables, each with standard normal distribution. Find the density of $U = (X - Y)^2/2$.

7.4.7 Let X and Y be independent random variables, with the densities

$$f_X(x) = C_1 x^{\alpha-1} e^{-x}, \quad f_Y(y) = C_2 y^{\beta-1} e^{-y}$$

for $x > 0, y > 0$, where C_1 and C_2 are the normalizing constants. Find the density of $X/(X + Y)$.

7.4.8 A current of I amperes flowing through a resistance of R ohms varies according to the probability distribution with density

$$f(i) = \begin{cases} bi(1 - i) & 0 < i < 1 \\ 0 & \text{otherwise.} \end{cases}$$

If the resistance R varies independently of the current according to a probability distribution with the density

$$g(r) = \begin{cases} 2r & 0 < r < 1 \\ 0 & \text{otherwise,} \end{cases}$$

find the density of the power $W = I^2 R$ watts.

7.4.9 Let X, Y be independent, each with the exponential distribution with density $f(x) = \lambda e^{-\lambda x}$. Show that $Z_1 = X + Y$ and $Z_2 = X/(X + Y)$ are independent.

7.4.10 Let f be the joint density of a pair (X, Y) of random variables. Let $U = aX + b, V = cY + d$, where $ac \neq 0$. Find the joint density of (U, V).

7.4.11 Let R be a nonnegative random variable with density $f(r)$. Let (X, Y) be a bivariate distribution obtained as follows. We first sample a value of random variable R. Then we sample a value of random variable U from the distribution uniform on $[0, 1]$. We now put

$$X = R\cos(2\pi U), \qquad Y = R\sin(2\pi U). \qquad (7.30)$$

Find:
  (i) The joint density of (X, Y).
  (ii) $P\{X^2 + Y^2 \leq t\}$.
  (iii) $P\{XY > 0\}$.
  (iv) $P\{X > 0\}$.
  [*Hint*: For part (i) find the joint density of (R, U) first and then use transformation (7.30).]

7.4.12 Random variables (X, Y) have a joint density given by

$$f(x, y) = \begin{cases} k(ax + by) & \text{if } 0 < x < 2, 0 < y < 2 \\ 0 & \text{otherwise,} \end{cases}$$

where $a > 0, b > 0$ are given constants. Find:
  (i) The value of k (as a function of a and b).
  (ii) The density of random variable $Z = 1/(Y + 1)^2$.
  [*Hint*: Express $F(t) = P\{Z \leq t\}$ in terms of the cdf of Y.]

7.4.13 Let X, Y be independent, each having a density symmetric about 0. Show that Y/X and $Y/|X|$ have the same distribution.

7.4.14 Let X_1, X_2 be independent random variables, each with normal distribution $N(0, \sigma^2)$. Find $P(X_1^2 + X_2^2 < C)$.

7.4.15 Let X, Y be continuous random variables with joint density $f(x, y) = \alpha^2 e^{-\alpha y}$ for $0 \leq x \leq y$ and $f(x, y) = 0$ otherwise. Find:
  (i) The marginal densities of X and of Y.
  (ii) cdf of (X, Y).
  (iii) $P(X/Y \leq \frac{1}{2})$.

7.4.16 Let (X, Y) have the joint density $f(x, y)$. Find densities of
    (i) $aX + bY$.
    (ii) XY.
    (iii) X/Y.

7.4.17 Let X and Y be independent, each with standard normal distribution. Find the density of $X/|Y|$.

7.4.18 Find the cdf and density of the ratio of two independent random variables with the same exponential distribution.

7.5 MULTIDIMENSIONAL DISTRIBUTIONS

We shall now present the generalizations of the concepts introduced thus far from the case of bivariate distributions to the case of multivariate (or: multidimensional) distributions. The motivation for these concepts lies in the frequency of practical situations when the analysis concerns many random variables simultaneously.

The examples here are easy to find. First, often a description of the phenomenon studied (e.g., sampled objects, some process, effects of treatment, etc.) uses several attributes at once. Formally, we have here a number of random variables, $X_1, \ldots, X_n$ (say), defined on a sample space, with X_i being the value of the ith attribute recorded for a given object. In such situations, a natural choice of sample space is to take the population of the objects under consideration, with probability P being generated by a specific scheme of sampling elements from the population.

In another context, we may think of repeated measurements of the same attribute, so that X_i is the result of the ith measurement. Now the probability P describes possible dependence (or lack of it) in the measurement process.

Whatever the interpretation, formal analysis will concern a vector of random variables

$$\mathbf{X} = \mathbf{X}^{(n)} = (X_1, \ldots, X_n).$$

In the discrete case, the joint distribution of $\mathbf{X}$ consists of specifying all probabilities of the form

$$P_{\mathbf{X}}(\mathbf{x}) = P\{\mathbf{X} = \mathbf{x}\} = P\{X_1 = x_1, \ldots, X_n = x_n\},$$

where $\mathbf{x} = (x_1, \ldots, x_n)$. Clearly, we must have

$$\sum_{\mathbf{x}} P_{\mathbf{X}}(\mathbf{x}) = 1,$$

the summation extended over all possible values $\mathbf{x} = (x_1, \ldots, x_n)$ of vector $\mathbf{X}$.

In the continuous case, we have the joint density of the vector **X**, in the form of a nonnegative function $f(\mathbf{x}) = f(x_1, \cdots, x_n)$ of n variables such that

$$P\{\mathbf{X} \in Q\} = \underbrace{\int \cdots \int}_{\mathbf{x} \in Q} f(x_1, \cdots, x_n)\, dx_1 \cdots dx_n$$

with

$$\underbrace{\int \cdots \int}_{R^n} f(x_1, \ldots, x_n)\, dx_1 \cdots dx_n = 1$$

The notions of marginal and conditional distributions and densities remain very much the same as in the bivariate case, except that the marginal distributions may now be themselves multivariate, and the same applies to conditional distributions, with the additional feature that the conditioning event may involve several random variables.

The simplicity of concepts can easily be obscured by confusing notation. While in any special case there is seldom any danger of confusion (we know the meaning of the variables and it is usually clear what is needed and what has to be done), the formulas covering the general cases may be confusing. We shall use the following notation only presenting the theory; in examples we shall try to use the "natural" notation, only identifying the meaning of various symbols from general theory.

First, in case of marginal distribution, we need to specify the variable of interest. They form a subset of the variables $X_1, \ldots, X_n$. Thus, we let $\mathbf{X} = (\mathbf{Y}, \mathbf{Z})$, where $\mathbf{Y}$ are the variables of interest and $\mathbf{Z}$ are the remaining variables. The question of ordering is irrelevant: For instance, if $\mathbf{X} = (X_1, X_2, X_3, X_4, X_5)$ and we are interested in the joint distribution of X_2 and X_5, then $\mathbf{Y} = (X_2, X_5)$ and $\mathbf{Z} = (X_1, X_3, X_4)$. We let $\mathbf{y}$ and $\mathbf{z}$ denote the values of vectors $\mathbf{Y}$ and $\mathbf{Z}$, so that $\mathbf{x} = (\mathbf{y}, \mathbf{z})$ is the partitioning of $\mathbf{x}$ into the corresponding two subsets of coordinates. We now introduce the following definition.

Definition 7.5.1. In the discrete case the *marginal distribution* of $\mathbf{Y}$ is given by

$$p_{\mathbf{Y}}(\mathbf{y}) = P\{\mathbf{Y} = \mathbf{y}\} = \sum_{\mathbf{z}} P\{\mathbf{Y} = \mathbf{y}, \mathbf{Z} = \mathbf{z}\}.$$

In the continuous case, the *marginal density* of $\mathbf{Y}$ is given by

$$f_{\mathbf{Y}}(\mathbf{y}) = \int \cdots \int_{\mathbf{z}} f(\mathbf{y}, \mathbf{z})\, d\mathbf{z},$$

where the integrals represent the multiple integration over all variables X_i that are in vector $\mathbf{Z}$. $\square$

Before presenting some examples, let us give the corresponding definitions for the case of conditional distributions. As in bivariate case, the discrete distributions present no difficulty, and we shall simply state that all formulas can be deduced starting from the definition of conditional probability for events, namely $P(A|B) = P(A \cap B)/P(B)$ if $P(B) > 0$. In the continuous case, we have to partition $\mathbf{X}$ into three components, writing

$$\mathbf{X} = (\mathbf{Y}, \mathbf{Z}, \mathbf{W})$$

where $\mathbf{Y}$ is the set of variables of interest, $\mathbf{Z}$ is the set of variables that will appear in the condition (whose values are assumed known), and $\mathbf{W}$ is the set (possibly empty) of variables that are neither in the condition nor of interest in the given instance. We need to define the conditional density of $\mathbf{Y}$ given $\mathbf{Z} = \mathbf{z}$.

Definition 7.5.2. The *conditional density of* $\mathbf{Y}$ *given* $\mathbf{Z} = \mathbf{z}$ is defined as

$$g_{\mathbf{Y}|\mathbf{Z}}(\mathbf{y}|\mathbf{z}) = \frac{f_{\mathbf{Y},\mathbf{Z}}(\mathbf{y},\mathbf{z})}{f_{\mathbf{Z}}(\mathbf{z})} = \frac{\int f(\mathbf{y},\mathbf{z},\mathbf{w})d\mathbf{w}}{\int \int f(\mathbf{y},\mathbf{z},\mathbf{w})d\mathbf{y}d\mathbf{w}},$$

where the integral symbols represent multiple integration over the variables in vectors $\mathbf{w}$ and $\mathbf{y}$.   □

We illustrate the concepts above by several examples.

Example 7.5.1 (Trinomial Distribution). We consider an experiment in which the outcome can be classified into one of three exclusive and exhaustive categories. We let these categories be A, B, and C, and let α, β, and γ be the probabilities of outcomes A, B, and C, respectively. From the description above it follows that A, B, and C form a partition of all outcomes, so that $\alpha + \beta + \gamma = 1$.   □

In specific situations we may think of A, B, and C as "improvement," "relapse," and "no change" in the case of experiment of testing some treatment method; in cases of quality testing the categories may be "acceptable," "unacceptable and beyond repair," and "repairable," and so on.

Now, we perform the experiment n times, and we assume that the repetitions are independent. Let X_1, X_2 stand for the counts of A and B among n repetitions (the count of C is $n - X_1 - X_2$). We then have

Theorem 7.5.1. *The joint distribution of* $\mathbf{X} = (X_1, X_2)$ *defined above is given by*

$$P\{X_1 = x_1, X_2 = x_2\} = \frac{n!}{x_1!x_2!(n - x_1 - x_2)!}\alpha^{x_1}\beta^{x_2}\gamma^{n-x_1-x_2}, \qquad (7.31)$$

where x_1, x_2 *are nonnegative integers such that* $x_1 + x_2 \leq n$.

Proof. The probability of x_1 outcomes A, x_2 outcomes B, and $n - x_1 - x_2$ outcomes C in a specific order equals

$$P(A)^{x_1} P(B)^{x_2} P(C)^{n-x_1-x_2} = \alpha^{x_1} \beta^{x_2} \gamma^{n-x_1-x_2},$$

by assumption of independence. Since this quantity does not depend on the order of outcomes, the left-hand side of (7.31) equals $\alpha^{x_1} \beta^{x_2} \gamma^{n-x_1-x_2}$ multiplied by the number of different ways of ordering x_1 events A, x_2 events B, and $n - x_1 - x_2$ events C. This number is

$$\frac{n!}{x_1! x_2! (n - x_1 - x_2)!}$$

by Theorem 3.5.1. □

Example 7.5.2. Continuing Example 7.5.1, let us find the marginal distribution of X_2. Formally, we have here for $x_2 = 0, 1, \ldots, n$,

$$p_{X_2}(x_2) = P\{X_2 = x_2\} = \sum_{x_1, x_3} P\{X_1 = x_1, X_2 = x_2, X_3 = x_3\}.$$

However, $x_3 = n - x_1 - x_2$, so that

$$P\{X_1 = x_1, X_2 = x_2, X_3 = n - x_1 - x_2\} = P\{X_1 = x_1, X_2 = x_2\}.$$

Therefore, using (7.31), we may write

$$p_{X_2}(x_2) = \sum_{x_1=0}^{n-x_2} P\{X_1 = x_1, X_2 = x_2\}$$

$$= \sum_{x_1=0}^{n-x_2} \frac{n!}{x_1! x_2! (n - x_1 - x_2)!} \alpha^{x_1} \beta^{x_2} (1 - \alpha - \beta)^{n-x_1-x_2}$$

$$= \frac{n!}{x_2! (n - x_2)!} \beta^{x_2} (1 - \beta)^{n-x_2}$$

$$\times \sum_{x_1=0}^{n-x_2} \frac{(n - x_2)!}{x_1! [(n - x_2) - x_1]!} \frac{\alpha^{x_1}}{(1 - \beta)^{x_1}} \left(1 - \frac{\alpha}{1 - \beta}\right)^{(n-x_2)-x_1}$$

$$= \binom{n}{x_2} \beta^{x_2} (1 - \beta)^{n-x_2},$$

since the last sum above reduces to

$$\sum_{x_1=0}^{n-x_2} \binom{n - x_2}{x_1} \pi^{x_1} (1 - \pi)^{n-x_2-x_1} = 1,$$

where $\pi = \alpha/(1 - \beta)$. □

Thus, the marginal distribution of a variable in a trinomial distribution is binomial, as could have been expected: It was enough to interpret the occurrence of B as a success and the occurrence of anything else (A or C) as failure. Then the probability of success and failure are β and $\alpha + \gamma = 1 - \beta$, and the number of successes X_2 has a binomial distribution, as it should. In the general case, suppose that the outcomes of each experiment are categorized into $m + 1$ classes, their probabilities being $\pi_1, \pi_2, \ldots, \pi_{m+1}$, with $\pi_1 + \pi_2 + \cdots + \pi_{m+1} = 1$. If the experiment is repeated independently n times, and X_i ($i = 1, \ldots, m$) is the number of outcomes of category i among n outcomes, then for nonnegative integers $x_1, \ldots, x_m$ with $x_1 + \cdots + x_m \le n$ we have

$$P\{X_1 = x_1, \ldots, X_m = x_m\} = \frac{n!}{x_1! \cdots x_m! x_{m+1}!} \pi_1^{x_1} \cdots \pi_m^{x_m} \pi_{m+1}^{x_{m+1}},$$

where $x_{m+1} = n - (x_1 + \cdots + x_m)$ and $\pi_{m+1} = 1 - (\pi_1 + \cdots + \pi_m)$. This is the *multinomial distribution*.

Now let $\mathbf{Y} = (X_{i_1}, \ldots, X_{i_k})$ be a subset of variables $\mathbf{X} = (X_1, \ldots, X_m)$ with a multinomial distribution. Proceeding in the same way as in Example 7.5.2 we can show that for $\mathbf{y} = (y_{i_1}, \ldots, y_{i_k})$ with coordinates y_{i_j} being nonnegative integers with $y_{i_1} + \cdots + y_{i_k} \le n$, we have

$$P\{Y_{i_1} = y_{i_1}, \ldots, Y_{i_k} = y_{i_k}\} = \frac{n!}{y_{i_1}! \cdots y_{i_k}! y_{i_{k+1}}!} \pi_{i_1}^{y_{i_1}} \cdots \pi_{i_k}^{y_{i_k}} \pi_{i_{k+1}}^{y_{i_{k+1}}},$$

where $y_{i_{k+1}} = n - y_{i_1} - \cdots - y_{i_k}$ and $\pi_{i_{k+1}} = 1 - \pi_{i_1} - \cdots - \pi_{i_k}$. Thus, a marginal distribution in a multinomial distribution is again multinomial with lower dimensionality.

Example 7.5.3 (Model of Grinding). The following model has been suggested as a description of the process of grinding. A piece of rock (say) is randomly divided into two parts; each of these parts is again divided randomly, and so on, so that in the nth "generation" we have 2^n pieces of rock. Assume that this process continues for a large number of generations, each consisting of independently dividing at random each of the existing parts into two new parts.

The purpose of the model is to obtain the distribution, as $n \to \infty$, of the sizes of the parts. Naturally, the actual sizes will tend to zero, so that the sizes of 2^n pieces in the nth generation will have to be multiplied by the approximate scaling factor. The results allow us to predict what fraction of the initial mass of rock will be ground into a gravel with sizes contained between specific limits a and b, and so on.

We shall not study this process here. Instead, we study only the first two divisions and the resulting sizes. For simplicity we assume that the initial size is 1, represented as the interval $[0, 1]$. In the first division, a point X with uniform distribution partitions the unit interval into two pieces, of length

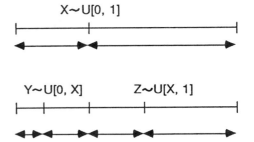

Figure 7.15 First two generations in the process of grinding.

X and $1 - X$. In the next partition, X is divided by Y (with Y distributed uniformly between 0 and X), and the remainder is divided by Z, again distributed uniformly on $[X, 1]$. We therefore have four fragments (see Figure 7.15), of length

$$X_1 = Y, \; X_2 = X - Y, \; X_3 = Z - X, \; X_4 = 1 - Z.$$

We shall start from deriving the joint trivariate density of (X, Y, Z). By assumption, X has density

$$f_X(x) = \begin{cases} 1 & \text{for } 0 \leq x \leq 1 \\ 0 & \text{otherwise.} \end{cases}$$

Given $X = x$, the densities of Y and Z are

$$g_{Y|X}(y|x) = \begin{cases} \dfrac{1}{x} & \text{for } 0 \leq y \leq x \\ 0 & \text{otherwise.} \end{cases}$$

$$g_{Z|X}(z|x) = \begin{cases} \dfrac{1}{1-x} & \text{for } x \leq z \leq 1 \\ 0 & \text{otherwise.} \end{cases}$$

Thus, the joint density is

$$f(x, y, z) = \begin{cases} \dfrac{1}{x} \cdot \dfrac{1}{1-x} & \text{for } 0 \leq y \leq x \leq z \leq 1 \\ 0 & \text{otherwise.} \end{cases} \tag{7.32}$$

Let us find the marginal joint density of Y and Z. This is obtained by integrating x; hence for $0 \leq y \leq z \leq 1$ we have

$$f_{Y,Z}(y, z) = \int_y^z \frac{dx}{x(1-x)} = \int_y^z \left(\frac{1}{x} + \frac{1}{1-x} \right) dx$$

$$= \log \frac{z}{1-z} + \log \frac{1-y}{y}.$$

The marginal density of Y is therefore

$$f_Y(y) = \int_y^1 \left(\log \frac{z}{1-z} + \log \frac{1-y}{y} \right) dz = -\log y$$

for $0 < y < 1$. Finally, let us determine the conditional density of X given $Y = y$ and $Z = z$. We have

$$g_{X|Y,Z}(x|y,z) = \frac{f(x,y,z)}{f_{YZ}(y,z)} = \frac{\dfrac{1}{x(1-x)}}{\log \dfrac{z}{1-z} + \log \dfrac{1-y}{y}}$$

for $0 < y < x < z < 1$. This density is presented on Figure 7.16. As may be seen, given $Y = y$ and $Z = z$, the value of X is more likely to lie closer to y if y is closer to 0 than z to 1 (as in Figure 7.16), while in the opposite case (when z is closer to 1 than y to 0), X is more likely to be closer to z.

Now we have $X_1 = Y, X_2 = X - Y, X_3 = Z - X$, and $X_4 = 1 - Z$. Clearly, $X_1 + X_2 + X_3 + X_4 = 1$, so that we shall derive the density of (X_1, X_2, X_3) only. This is obtained from the trivariate density (7.32) by the transformation

$$x_1 = y$$
$$x_2 = x - y \qquad\qquad (7.33)$$
$$x_3 = z - x.$$

The inverse transformation to (7.33) is

$$y = x_1$$
$$x = x_1 + x_2 \qquad\qquad (7.34)$$
$$z = x_1 + x_2 + x_3.$$

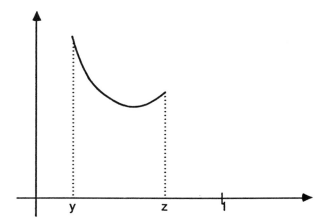

Figure 7.16 Conditional density of X given Y and Z.

whose Jacobian is

$$
J = \begin{vmatrix} \dfrac{\partial y}{\partial x_1} & \dfrac{\partial y}{\partial x_2} & \dfrac{\partial y}{\partial x_3} \\[2mm] \dfrac{\partial x}{\partial x_1} & \dfrac{\partial x}{\partial x_2} & \dfrac{\partial x}{\partial x_3} \\[2mm] \dfrac{\partial z}{\partial x_1} & \dfrac{\partial z}{\partial x_2} & \dfrac{\partial z}{\partial x_3} \end{vmatrix} = \begin{vmatrix} 1 & 0 & 0 \\ 1 & 1 & 0 \\ 1 & 1 & 1 \end{vmatrix} = 1.
$$

Thus, the joint density $g(x_1, x_2, x_3)$ of (X_1, X_2, X_3) is obtained from (7.32) by substituting (7.34), that is,

$$
g(x_1, x_2, x_3) = \begin{cases} \dfrac{1}{(x_1 + x_2)(1 - x_1 - x_2)} & \text{for } x_1, x_2, x_3 > 0, x_1 + x_2 + x_3 < 1. \\[3mm] 0 & \text{otherwise.} \end{cases} \qquad \square
$$

PROBLEMS

7.5.1   Two cards are drawn without replacement from an ordinary deck. Let
X be the number of aces, Y be the number of red cards, and Z be
the number of hearts. Find:

   (i) The joint distribution of (X, Y, Z).
   (ii) The conditional distribution of Z given $Y = y$.
   (iii) $P\{Y = Z\}$.

7.5.2   Let X_1, X_2, X_3 have a continuous joint distribution with a density

$$
f(x_1, x_2, x_3) = \begin{cases} q(ax_1 + bx_2 + cx_3) & \text{for } 0 \le x_i \le 1, i = 1, 2, 3 \\ 0 & \text{otherwise,} \end{cases}
$$

where a, b, c are some positive constants. Find:

   (i) q (as a function of a, b, c).
   (ii) The marginal density of (X_1, X_2).
   (iii) The conditional density of (X_1, X_2) given $X_3 = x_3$.
   (iv) The conditional density of X_3 given $X_1 = x_1, X_2 = x_2$.
   (v) $P\{0.25 < X_3 < 0.75 | X_1 = X_2 = 0.5\}$.

7.5.3   Suppose that X_1, X_2, X_3 have joint density

$$
f(x_1, x_2, x_3) = \begin{cases} x_1 x_2 x_3 & \text{for } 0 < x_i < c, i = 1, 2, 3 \\ 0 & \text{otherwise.} \end{cases}
$$

(i) Find c.

(ii) Let $Y_1 = X_1, Y_2 = X_1X_2, Y_3 = X_1X_2X_3$. Find the joint density of (Y_1, Y_2, Y_3).

7.5.4 A can of Three-Bean-Salad contains beans of varieties A, B and C (plus other ingredients which are of no concern for the problem). Let X, Y and Z denote the relative weights of varieties A, B and C in a randomly selected can (so that $X + Y + Z = 1$).

Let the joint distribution of (X, Y) have the density $f(x, y) = kxy$ for $x > 0, y > 0, x + y < 1$ and $f(x, y) = 0$ otherwise. Find:

(i) k.

(ii) Probability $P(X > \frac{1}{2})$ that beans A take over half of the weight.

(iii) $P(Z > \frac{1}{2})$.

(iv) Probability that none of the three varieties of beans will take more than half of the weight.

(v) The marginal density of (X, Z) and of Z.

7.6 ORDER STATISTICS

Next, we consider a very important case of multidimensional distribution (and also transformation of a multidimensional random variable), arising often in statistics. Let $X_1, X_2, \ldots, X_n$ be independent and identically distributed random variables. In statistical terminology, they represent a random sample of size n from some population. We let F denote the common cdf of X_i's, and we assume that F is a continuous distribution with density f. We consider the vector $X_{1:n}, X_{2:n}, \ldots, X_{n:n}$, where $X_{i:n}$ is the ith in magnitude among $X_1, X_2, \ldots, X_n$. Thus, we have

$$X_{1:n} \leq X_{2:n} \leq \cdots \leq X_{n:n}. \tag{7.35}$$

Since the X_i's are continuous, the probability that two observations will be exactly equal is zero, and we can assume that all inequalities in (7.35) are strict.

Definition 7.6.1. The random variables (7.35) are called *order statistics* of the sample $X_1, X_2, \ldots, X_n$, and $X_{k:n}$ will be referred to as the kth order statistic. □

In particular, the first order statistic is the minimum of the sample, while the nth order statistic is the maximum of the sample. Our objective will be to find the joint distribution of order statistics, as well as their marginal distributions. We shall start from the latter. Let $k \leq n$ and let G_k be the cdf of order statistic $X_{k:n}$. Then $X_{k:n} \leq t$ if k or more elements in the sample $X_1, X_2, \ldots, X_n$ satisfy the inequality $X_i \leq t$. Since the number of elements in the sample that satisfy $X_i \le t$ has binomial distribution with probability of success $p = F(t) = P(X_i \le t)$, we may write

$$G_k(t) = P\{X_{k:n} \le t\}$$
$$= \sum_{r=k}^{n} \binom{n}{r} [F(t)]^r [1 - F(t)]^{n-r}. \tag{7.36}$$

Two particular cases deserve attention here. If $k = 1$, we obtain

$$G_1(t) = P\{\min(X_1, X_2, \ldots, X_n) \le t\} = P\{X_{1:n} \le t\}$$
$$= 1 - [1 - F(t)]^n,$$

and if $k = n$, then

$$G_n(t) = P\{\max(X_1, X_2, \ldots, X_n) \le t\} = P\{X_{n:n} \le t\}$$
$$= [F(t)]^n.$$

Differentiating with respect to t, we obtain the densities

$$g_1(t) = n[1 - F(t)]^{n-1} f(t) \tag{7.37}$$

$$g_n(t) = n[F(t)]^{n-1} f(t). \tag{7.38}$$

To find the density $g_k(t)$ we may differentiate (7.36), obtaining

$$g_k(t) = \sum_{r=k}^{n} \binom{n}{r} r[F(t)]^{r-1} f(t) [1 - F(t)]^{n-r}$$
$$- \sum_{r=k}^{n-1} \binom{n}{r} [F(t)]^r (n-r)[1 - F(t)]^{n-r-1} f(t)$$
$$= \sum_{r=k}^{n} \frac{n!}{(r-1)!(n-r)!} [F(t)]^{r-1} [1 - F(t)]^{n-r} f(t)$$
$$- \sum_{r=k+1}^{n} \frac{n!}{(r-1)!(n-r)!} [F(t)]^{r-1} [1 - F(t)]^{n-r} f(t).$$

Canceling the terms in the last two sums, we obtain

Theorem 7.6.1. *The density of the kth order statistic from a distribution with cdf F and density f is given by*

$$g_k(t) = \frac{n!}{(k-1)!(n-k)!} [F(t)]^{k-1} [1 - F(t)]^{n-k} f(t). \tag{7.39}$$

$\square$

The following argument, although somewhat heuristic, better explains the derivation of (7.39) and also allows some easy extensions. The differential $g_k\,\delta t$ is the probability that the kth order statistic falls into the interval $(t, t + \delta t)$. For this to happen, one observation must fall into this interval, $k - 1$ must fall in the interval $(-\infty, t]$, and the remaining $n - k$ observations must fall in the interval $(t + \delta t, \infty)$. The probability of such an event is multinomial, with probabilities, respectively, $f(t)\delta t, F(t)$, and $1 - F(t + \delta t)$, hence equals

$$\frac{n!}{(k-1)!(n-k)!}[F(t)]^{k-1}f(t)\delta t[1 - F(t + \delta t)]^{n-k}.$$

Passing to the limit with $\delta t \to 0$, we obtain the density (7.39).

In a similar way, one can show that the joint density of $X_{k:n}, X_{l:n}, k < l$ is given by the formula, valid for $s < t$,

$$g_{k,l}(s, t) = \frac{n!}{(k-1)!(l-k-1)!(n-l)!}[F(s)]^{k-1}[F(t) - F(s)]^{l-k-1}$$
$$\times [1 - F(t)]^{n-l}f(s)f(t) \qquad\qquad (7.40)$$

and $g_{k,l}(s, t) = 0$ otherwise.

Example 7.6.1 (Distribution of the Range).   We define $R = X_{n:n} - X_{1:n} = \max(X_1, \dots, X_n) - \min(X_1, \dots, X_n)$. Find the density of R. We present two solutions, each illustrative of some technique.

SOLUTION 1.   This solution consists of finding the joint distribution of $X_{1:n}$ and $X_{n:n}$, and then using the techniques for finding the distribution of transformations of random variables to determine the distribution of $X_{n:n} - X_{1:n}$.

Direct application of (7.40) gives

$$g_{1,n}(s, t) = \frac{n!}{(0)!(n-2)!(0)!}[F(t) - F(s)]^{n-2}f(s)f(t)$$
$$= n(n-1)[F(t) - F(s)]^{n-2}f(s)f(t)$$

for $s < t$.

The same formula could have been obtained by observing that $X_{1:n} > s$ and $X_{n:n} \le t$ if and only if all n observations fall between s and t, so that

$$P\{X_{1:n} > s, X_{n:n} \le t\} = [F(t) - F(s)]^n$$

and the joint density is obtained by taking the mixed derivative

$$-\frac{\partial^2}{\partial s \partial t}P\{X_{1:n} > s, X_{n:n} \le t\}.$$

Since $R = X_{n:n} - X_{1:n}$, we may take the transformation

$$r = t - s$$

and "companion," $w = t$.

Solving for t and s, we obtain $s = w - r, t = w$ with $|J| = 1$. Thus, the density of R is the marginal density, for $r > 0$,

$$
h_R(r) = \int_{-\infty}^{\infty} g_{1,n}(w - r, w) \, dw
$$
$$
= \int_{-\infty}^{+\infty} n(n - 1)[F(w) - F(w - r)]^{n-2} f(w - r) f(w) \, dw. \quad (7.41)
$$

SOLUTION 2. A more direct solution is to condition on one of the variables, say $X_{1:n}$. If $X_{1:n} = s$, then the probability that the range is less than r is $[F(s + r) - F(s)]^{n-1}$, since all remaining observations must fall between s and $s + r$. The density of $X_{1:n}$ is given by (7.37), and therefore

$$
P\{R \le r\} = \int_{-\infty}^{+\infty} [F(s + r) - F(s)]^{n-1} \cdot n(1 - F(s))^{n-1} f(s) \, ds.
$$

Differentiation with respect to r now gives the density of R, in a form somewhat different than (7.41).

Example 7.6.2 (Theory of Outliers). As another illustration of the use of order statistics (and of the use of the foregoing techniques), we shall present basic definitions and results of the theory of outliers suggested by Neyman and Scott (1971).

Intuitively, an outlier is an observation that is sufficiently far away from the remaining observations to justify the suspicion that it results from an observational or recording error, or perhaps a sudden undetected change of conditions that made this particular observation obey other probability distributions than that obeyed by the remaining observations. What we present here is a theory of right outliers. The results for left outliers may be obtained by an obvious modification. Thus, we have a sample, say $x_1, \ldots, x_n$, with values of order statistics $x_{1:n}, \ldots, x_{n:n}$, and $x_{n:n}$ is, in some sense, "too far away" from the rest of observations. The intuition of "being too far away" has been formalized in a number of ways in various approaches to the detection of outliers, typically in terms of the relation of $x_{n:n}$ to the average and standard deviation of the sample.

Neyman and Scott proposed the following definition, based on order statistics. In what follows we shall always assume that $n \ge 3$ and $r > 0$.

Definition 7.6.2. We say that the sample $x_1, \ldots, x_n$ contains an (r, n) *right outlier* if

$$x_{n:n} - x_{n-1:n} > r(x_{n-1:n} - x_{1:n}). \tag{7.42}$$

In the sequel, we simply use the term *outlier* for the right outlier.   □

It is clear that if only $x_{n:n} \neq x_{n-1:n}$, the sample contains an (r, n) outlier for all r satisfying the condition

$$r < \frac{x_{n:n} - x_{n-1:n}}{x_{n-1:n} - x_{1:n}}.$$

Intuitively, the outlier in the usual sense is an (r, n) outlier for r large enough.

Theorem 7.6.2. *A random sample* $X_1, \ldots, X_n$ *from the distribution with cdf F and density f will contain an* (r, n) *outlier with probability*

$$\pi(r, n; F) = n(n-1) \int_{-\infty}^{\infty} \int_{x}^{\infty} \left[F\left(\frac{y + rx}{r+1}\right) - F(x) \right]^{n-2} f(y)f(x) \, dy \, dx \tag{7.43}$$

Proof. Indeed, if we condition on $x_{1:n} = x$ and $x_{n:n} = y > x$, then the sample will contain an (r, n) outlier, if (7.42) holds, hence if

$$x < X_{n-1:n} < \frac{y + rx}{r+1}.$$

This event, in turn, occurs if $n-2$ remaining elements of the sample lie between x and $\frac{y + rx}{r+1}$, which occurs with probability $\left[F\left(\frac{y + rx}{r+1}\right) - F(x) \right]^{n-2}$. Multiplying by joint density of $(X_{1:n}, X_{n:n})$ and integrating, we obtain (7.43).
□

Formula (7.43) allows us to compute the probability of the appearance of an (r, n) outlier for given F and n. As will be explained in later parts of the book, one of the essential features that characterizes situations in statistics (as opposed to those in probability) is that the distribution F that governs the selection of the sample is not known. Typically, we know only that the true F belongs to a certain family $\mathcal{F}$ of distributions. To use some examples, the experimenter may know that the sample she observes has a normal distribution, but she does not know the value(s) of the parameter(s).

Accordingly, it appears natural to introduce the following definitions.

Definition 7.6.3. We say that the family $\mathcal{F}$ of distributions is (r, n)-*outlier resistant* if

$$\sup_{F \in \mathcal{F}} \pi(r, n; F) < 1 \tag{7.44}$$

and is (r, n)-*outlier prone* if

$$\sup_{F \in \mathcal{F}} \pi(r, n; F) = 1. \tag{7.45}$$

Moreover, a family $\mathcal{F}$ of distributions is *totally outlier resistant* if condition (7.44) holds for all $r > 0$ and $n \geq 3$; it is called *totally outlier prone* if condition (7.45) holds for all $r > 0$ and $n \geq 3$.   □

Among other results proved by Neyman and Scott, we mention the following two theorems. One of them asserts that *the family of all normal distributions is totally outlier resistant*. The other theorem asserts that *the family of all gamma distributions is totally outlier prone*.[*] The practical consequences of these two theorems are as follows. If we have a sample $x_1, \ldots, x_n$ from a normal distribution, we can find the largest r for which this sample contains an (r, n) outlier. This r is equal to the ratio $r^* = \dfrac{x_{n:n} - x_{n-1:n}}{x_{n-1:n} - x_{1:n}}$. We may then find the quantity (7.44) for $\mathcal{F}$ being the family of normal distribution and $r = r^*$. If this quantity is sufficiently small, we have good reason to suspect that the element $x_{n:n}$ in the sample is a genuine outlier, in the sense of representing an observation from a distribution other than the rest of the sample. This gives a practical procedure for the rejection of outliers.

On the other hand, if we have a sample $x_1, \ldots, x_n$ from a gamma distribution, we can *never* reject the largest element as an outlier (basing the decision solely on the observed values). Indeed, suppose that we have data such as $x_{1:4} = 0.5, x_{2:4} = 0.55, x_{3:4} = 1, x_{4:4} = 1,000,001$, or something equally peculiar. We cannot reject the observation $1,000,001$ as an outlier, since the probability of a configuration with the ratio

$$\frac{x_{4:4} - x_{3:4}}{x_{3:4} - x_{1:4}} = 2,000,000$$

or more has probability arbitrarily close to 1 for *some* gamma distribution. More precisely, for every $\epsilon > 0$ there is a gamma distribution such that the probability of the configuration above exceeds $1 - \epsilon$.

The conclusion above, asserting that in practice one should never reject any sample element as an outlier if the sample is known to come from some gamma distribution is highly surprising and counterintuitive.

Actually, the conclusion is impeccable, and the answer lies in a quite unexpected place, providing a rather powerful argument for the Bayesian approach to statistics. The point is that it seldom, if ever, happens that the statistician knows *only* that the sample comes from a gamma distribution. He

---

[*] See Example 6.4.7 for the definition of gamma distribution. More detailed information is given in Chapter 9. For further development of the suggested theory of outliers see Green (1974, 1976).

usually has some idea about the parameters, based on his experience, imagination, understanding of the situation, and so on. Such knowledge, vague as it may be, allows the statistician to regard some gamma distributions in the family $\mathcal{F}$ as "more plausible" than others, and perhaps even eliminate some members of $\mathcal{F}$ as impossible in the given situation. Now, if we restrict the family $\mathcal{F}$ to *some* gamma distributions only (by putting a bound on parameter α), then $\mathcal{F}$ is no longer totally outlier prone, and rejection of outliers becomes justifiable.   □

Finally, we give the joint density of all order statistics:

Theorem 7.6.3. *The joint density of the vector $X_{1:n}, X_{2:n}, \ldots, X_{n:n}$ is given by*

$$\phi(x_1, \ldots, x_n) = \begin{cases} n! f(x_1) \cdots f(x_n) & \text{if } x_1 < \cdots < x_n \\ 0 & \text{otherwise.} \end{cases}$$

□

The proof is very simple and will be omitted. Readers are urged, however, to consider the case $n = 3$, draw a picture, and determine the partition of the three-dimensional space which accounts for the appearance of the factor $3! = 6$ in the density. If $n = 3$ is too hard, we suggest starting with the case $n = 2$.

PROBLEMS

7.6.1 Find the density of the median of a sample of size $n = 2k + 1$ from the distribution uniform on $[0, 1]$.

7.6.2 Let $X_{1:n}, \ldots, X_{n:n}$ be order statistics from a distribution with density f, and let $Y_{1:n}, \ldots, Y_{n:n}$ be order statistics from another sample of n observations taken from the same distribution. Assume that the vector $(X_{1:n}, \ldots, X_{n:n})$ is independent of the vector $(Y_{1:n}, \ldots, Y_{n:n})$. Show that $P\{X_{i:n} \leq t\} \geq P\{Y_{j:n} \leq t\}$ for every t if and only if $i \leq j$.

7.6.3 Find $\pi(r, n; F_\theta)$ when F_θ is the uniform distribution on $(0, \theta]$. Find $\pi(r, n; \mathcal{F})$ for the family $\mathcal{F}$ of all uniform distributions on $[0, \theta], 0 < \theta < \infty$.

7.6.4 Eight athletes compete to qualify to the Olympic finals in 100 meter dash, with the best two qualifying. Let T_i be the time of the ith athlete in the qualifying dash, assumed to be a continuous random variable with density f_i $(i = 1, \ldots, 8)$. Moreover, assume that $T_1, \ldots, T_8$ are independent.

    (i) Argue that if $f_1 = \ldots = f_8$, then the probability that athlete # 8 will qualify to the Olympic finals is $\frac{1}{4}$.

   (ii) Without assuming anything about densities $f_1, \ldots, f_8$, derive the formula for the probability that athlete # 8 will qualify to the finals.

   (iii) Show that the formula in part (ii) gives the answer in part (i) under the assumptions $f_1 = \cdots = f_8$.

7.6.5   Let $X_1, \ldots, X_n$ be independent with the same distributon. Find:

   (i) The probability that at least k of the variables will assume values between a and b.

   (ii) The probability that at most j of the variables will assume some fixed value x_0 (consider separately the continuous and discrete cases).

REFERENCES

Green, R. F., 1974. "A Note on Outlier-prone Families of Distributions". *Annals of Statistics*. 2: 1293–1295.

Green, R. F., 1976. "Outlier-prone and Outlier-resistant Families of Distributions." *Journal of the American Statistical Association*. 71: 502–505.

Neyman, J., and Scott, E. L., 1971. "Outlier Proneness of Phenomena and of Related Distributions" in: J. S. Rustagi (ed.), *Optimizing Methods in Statistics*, Academic Press, New York, pp. 413–430.

CHAPTER 8

Expectation

8.1 MOTIVATION AND INTERPRETATION

The concept of random variable, to the extent introduced and developed in Chapters 6 and 7, could be regarded as merely a convenient tool for describing large classes of events. Indeed, once we consider events of the form $\{a < X \le b\}$, it is quite natural to reduce the analysis to even simpler events $\{X \le t\}$. Then the probability of such an event, regarded as a function of the argument t (i.e., the cdf of X) turns out to carry all information about the distribution of X. Once the notion of cdf is introduced, it is only natural to look for classes of cdf's that allow especially simple description (hence the definition of discrete and continuous random variables). Finally, extension from one to many dimensions is one of the first, almost automatic directions of pursuit in mathematics. One may say, therefore, that the analysis of Chapters 6 and 7 did not cover any probabilistic concepts that could not have been developed (admittedly, in a clumsy and awkward way) without the tools of random variables.

As opposed to that, in this chapter we introduce a notion that cannot be formulated without the concept of random variables, namely that of *expectation*, or the *expected value* of a random variable.

The intuitive content of this new notion is as follows. Consider a random variable X, defined on some sample space S. An experiment consists of a random selection of a point s of S, that is, of "letting s occur." We now interpret $X(s)$ as the gain [loss, if $X(s) < 0$] of a hypothetical gambler.

If the experiment is repeated n times, the sample space becomes $S^{(n)} = S \times \cdots \times S$ (n times), and an outcome is $s^{(n)} = (s_1, \ldots, s_n)$, with $s_i \in S, i = 1, \ldots, n$. The gambler's accumulated gain becomes $X(s_1) + X(s_2) + \cdots + X(s_n)$, a quantity that is traditionally written as $X_1 + X_2 + \cdots + X_n$, with $X_i = X(s_i)$ being the outcome of the ith trial.

The average gain per gamble now becomes $\dfrac{X_1 + \cdots + X_n}{n}$, and individual and collective human experience tells us that as n grows larger, this average fluctuates less and less, tending to stabilize at some value, which we shall call expectation of X and denote by $E(X)$. It is not difficult to calculate this value in case of a discrete random variable. Indeed, if X may assume values $x_1, x_2, \ldots, x_m$, with probabilities $p_i = P\{X = x_i\}, i = 1, 2, \ldots, m$, then—regardless of the structure of the sample space—the only events of interest in studying X are $\{X = x_i\}$. Let the experiment be repeated n times, and let the event $\{X = x_i\}$ occur n_i times $(i = 1, \ldots, m)$, where we must have $n_1 + n_2 + \cdots + n_m = n$. According to the frequency interpretation of probability, the ratio n_i/n tends to $p_i = P\{X = x_i\}$ as the number n of experiments increases. The total gain $X_1 + \cdots + X_n$ must be equal $x_1 n_1 + x_2 n_2 + \cdots + x_m n_m$, and therefore the average gain per experiment is

$$x_1 \frac{n_1}{n} + x_2 \frac{n_2}{n} + \cdots + x_m \frac{n_m}{n},$$

which tends to

$$x_1 p_1 + x_2 p_2 + \cdots + x_m p_m.$$

We shall use the last quantity, that is,

$$E(X) = x_1 P\{X = x_1\} + \cdots + x_m P\{X = x_m\} \tag{8.1}$$

as a formal definition of expected value $E(X)$ of a discrete random variable X. We shall later extend the definition to cover random variables that are not discrete.

Formula (8.1) suggests another interpretation of expected value, which will tie it closely to existing powerful mathematical theories. One can regard the right-hand side of (8.1) as an integral of the function X. As we shall see, this opens up the use of both conceptual and computational aspects of theory of integration in probability theory.

8.2 EXPECTED VALUE

We shall start from the definition of expected value as given in most textbooks on probability. This definition is formulated separately for discrete and separately for continuous random variables, the discrete case being an extension of formula (8.1).

Definition 8.2.1. If X is a discrete random variable that assumes values $x_1, x_2, \ldots$ with probabilities $P\{X = x_i\}, i = 1, 2, \ldots$, then the *expected value* of X is defined as

$$E(X) = \sum_i x_i P\{X = x_i\} \tag{8.2}$$

provided that

$$\sum_i |x_i| P\{X = x_i\} < \infty. \tag{8.3}$$

If X is a continuous random variable with the density $f(x)$, then the *expected value* of X is defined as

$$E(X) = \int_{-\infty}^{+\infty} x f(x)\, dx \tag{8.4}$$

provided that

$$\int_{-\infty}^{+\infty} |x| f(x)\, dx < \infty. \tag{8.5}$$

□

This definition covers formula (8.1) as a special case, since if X assumes only finitely many values, condition (8.3) is always satisfied.

The reason for introducing conditions (8.3) and (8.5) is connected with the possibility that the value of an infinite sum may depend on the order of summation (and a similar phenomenon may occur for an integral). The absolute convergence (respectively absolute integrability) guarantees that the expected value is unambiguously defined.

Remark on notation and terminology. The symbols used for the expected value $E(X)$ in probabilistic and statistical literature are also EX, μ_X, m_X, or simply μ, m, if it is clear which random variable is being studied. In physics, the symbol used is $< X >$, while in the Soviet mathematical literature it is $M(X)$. As regards terminology, expected value is also called expectation, mathematical expectation, mean, or average. Moreover, expected value is often associated not with X but with the distribution of X. Consequently, one sometimes uses symbols such as μ_F or m_F, where F is a cumulative distribution function.

We begin with a series of examples of expectations for discrete random variables.

Example 8.2.1. Let A be some event, and let $p = P(A)$. Imagine that you are to receive \$1 if A occurs and nothing if A does not occur (i.e., if A^c occurs). Let X be the random variable defined as "the amount you receive." Then X may assume only two values, and its distribution is given by the array

Value	0	1
Probability	$1 - p$	p

According to formula (8.1), to compute the expectation we multiply the elements of the upper row by corresponding elements of the lower row and add the results. Thus

$$E(X) = 0 \cdot (1 - p) + 1 \cdot p = p.$$

The result agrees with the intuition: If you play such a game over and over, then your average outcome per game will equal the probability of winning p.

Observe that X can also be described as "the number of occurrences of event A in a single experiment." We shall use this interpretation later.

Example 8.2.2. Suppose now that you are to win \$1 if A occurs, but also lose \$1 (i.e., "win" -1) if A does not occur. Let Y be the random variable describing your winning. Then the distribution of Y is

Values	-1	1
Probability	$1 - p$	p

and

$$E(Y) = -1 \cdot (1 - p) + 1 \cdot p = 2p - 1.$$

Thus, $E(Y) > 0$ iff $p > 0.5$, which agrees with our intuition: The game is favorable if probability p of winning exceeds 0.5.

Example 8.2.3 (Expected Value of Binomial Distribution). The binomial random variable was defined in Example 6.3.2 as the total number of successes in n independent trials, where the probability of a success in any trial is p. The possible values of the binomial random variable are $0, 1, \ldots, n$ with probabilities given by (6.21) from Example 6.3.2. Consequently, we may write

$$
\begin{aligned}
E(X) &= \sum_{k=0}^{n} k \binom{n}{k} p^k (1 - p)^{n-k} \\
&= \sum_{k=1}^{n} k \frac{n!}{k!(n - k)!} p^k (1 - p)^{n-k} \\
&= np \sum_{k=1}^{n} \frac{(n - 1)!}{(k - 1)!(n - 1 - (k - 1))!} p^{k-1} (1 - p)^{n-1-(k-1)} \\
&= np \sum_{j=0}^{n-1} \binom{n - 1}{j} p^j (1 - p)^{n-1-j} \\
&= np,
\end{aligned}
$$

where in the last sum we introduced the new index of summation $j = k - 1$. The last sum then turns out to be a Newtonian expansion of $[p + (1 - p)]^{n-1} = 1$.

The fact that the expected value of the binomial random variable X for n

trials and the probability p of success satisfies the relation

$$E(X) = np \tag{8.6}$$

is deeply rooted in our subconscious intuition about probability: For instance, if the chances of failing an exam are 20% (say), and 200 persons take this exam, we expect about 40 of them to fail. This type of intuitive reasoning, when consciously analyzed, appears almost equivalent to the concept of probability.

Example 8.2.4 (Expectation of Geometric Distribution). Let us consider an event A ("success") with probability p. We perform independent experiments until A occurs for the first time, and we let X be the number of trials up to and including first success. As already mentioned (see Example 6.3.3), X is a random variable with geometric distribution. We have here $P\{X = k\} = (1 - p)^{k-1}p$ for $k = 1, 2, \ldots$. Find $E(X)$.

SOLUTION. Letting $q = 1 - p$, we may write

$$E(X) = \sum_{k=1}^{\infty} kP\{X = k\}$$

$$= \sum_{k=1}^{\infty} kq^{k-1}p$$

$$= p\frac{d}{dq}\sum_{k=1}^{\infty} q^k$$

$$= p\frac{d}{dq}\left(\frac{q}{1-q}\right)$$

$$= p \cdot \frac{1}{(1-q)^2} = \frac{1}{p}. \tag{8.7}$$

The answer makes intuitive sense: If (say) probability of success is $p = \frac{1}{5}$, then on the average every fifth trial is a success, and therefore the average number of trials until the next success is 5, which agrees with formula (8.7). ∎

In Example 8.2.4 we used the technique of summing the infinite series by representing it as a derivative of a power series. This technique, if applicable, usually leads quickly to the desired result, and we illustrate the use of this technique in some other calculations. However, for nonnegative integer-valued random variables, the following theorem is also useful.

Theorem 8.2.1. *Let X be a random variable that assumes only nonnegative integer values. Then*

$$E(X) = \sum_{n=0}^{\infty} P\{X > n\}. \tag{8.8}$$

Proof. We have, letting $p_n = P\{X = n\}$,

$$
\begin{aligned}
E(X) &= p_1 + 2p_2 + 3p_3 + \cdots \\
&= p_1 + p_2 + p_3 + \cdots \\
&\quad\ + p_2 + p_3 + \cdots \\
&\qquad\quad\ + p_3 + \cdots
\end{aligned}
$$

$$= P\{X > 0\} + P\{X > 1\} + P\{X > 2\} + \cdots,$$

where a change of order of summation is allowed, since all terms are non-negative. □

Example 8.2.5. Let us use formula (8.8) to find the expectation of the geometric distribution from Example 8.2.4. We have

$$P\{X > n\} = P\{\text{failures on the first } n \text{ trials}\} = q^n,$$

or if this reasoning appears suspiciously simple and one puts more trust in algebra,

$$P\{X > n\} = \sum_{k=n+1}^{\infty} P\{X = k\} = \sum_{k=n+1}^{\infty} q^{k-1}p$$

$$= p\frac{q^n}{1 - q} = q^n.$$

Consequently,

$$E(X) = \sum_{n=0}^{\infty} q^n = \frac{1}{1 - q} = \frac{1}{p}.$$

□

Example 8.2.5 concerned the situation when the expected value was finite. Naturally, such situations are most common in practice, as otherwise the very concept of expected value would make no sense. However, there are situations, of both practical and theoretical significance, when the condition (8.3) or (8.5) is not met.

To be more precise, one should distinguish two cases. We illustrate the possibilities considering condition (8.3) involving a series; the situation with an integral is analogous. Let us consider separately sums $\sum x_i P\{X = x_i\}$ involving positive and negative terms, that is, $U^+ = \sum \max(x_i, 0)P\{X = x_i\}$ and $U^- = -\sum \min(x_i, 0)P\{X = x_i\}$. It is clear that if the series (8.3) diverges, then U^+, U^-, or both must be infinite, for otherwise the series of absolute values (8.3), equal to $U^+ + U^-$, would be finite. If only U^+ is infinite, we may say that $E(X) = +\infty$; similarly, if only U^- is infinite, we may say that $E(X) = -\infty$. It is only when both U^+ and U^- are infinite that we have an expression of the form $\infty - \infty$, to which we cannot assign any value. In the last case we shall say that $E(X)$ *does not exist*.

We start with some examples of random variables with infinite expectations.

Example 8.2.6 (Petersburg Paradox).   Suppose that you have the possibility of participating in the following game. A fair coin is tossed repeatedly until a head occurs. If the first head occurs at the kth toss, you win 2^k dollars. Clearly, this is a game in which you cannot lose, so the question is: What is the amount you should be willing to pay to participate in such game?

SOLUTION.   The most common argument offered in such cases is that one should be willing to pay the fee as long as the expected winnings exceed or equal the fee for participation in the game. Then, "in the long run," one should come even or ahead.

Let X denote the winning in the game (not counting the fee for participation). Then the event $\{X = 2^k\}$ occurs if the first $k - 1$ tosses are tails and the kth toss is heads; chances of this event are $1/2^k$. Consequently, we have

$$E(X) = 2 \cdot \frac{1}{2} + 2^2 \cdot \frac{1}{2^2} + 2^3 \cdot \frac{1}{2^3} + \cdots = \infty.$$

Thus, the expected winnings in this game are infinite, and therefore one should be willing to pay an infinite fee for the right to play. However, in any particular game, the winnings will always be finite, so that one is *certain to lose*.

This phenomenon, discovered in eighteenth century, was named the Petersburg paradox. Attempts to solve it led to the introduction of the concept of utility. Without going into detail, it is postulated that the utility of money—the value a person attaches to a given amount of money—is not constant but depends on how much money the person already has. This appears intuitively acceptable: $20 may seem to be a lot for someone who has nothing but does not seem like very much to a millionaire.

If $u(x)$ is the utility of x, then the postulate above implies that $u(x)$ is not proportional to x. The *expected utility* from participating in the game is

$$E[u(X)] = \sum_{k=1}^{\infty} u(2^k) \cdot \frac{1}{2^k},$$

a quantity that may be finite for some "plausible" choices of utility function $u(x)$. In particular, Bernoulli suggested that $u(x) = c \log x$.

Example 8.2.7 (Bad Luck).   This example is due to Feller (1968). One of the measures of *bad luck* may be as follows. Consider an event to which Peter and some other (perhaps all) people are exposed. As examples, one may think of events such as the onset of a specific illness, breakdown of a newly bought car, etc. We shall let T_0 denote the time of waiting for this event for Peter. Occasionally, Peter learns about some other people who have been exposed to the same situation as he and for whom the event in question occurred. Let their waiting times be $T_1, T_2, \ldots$. We assume symmetry and lack of interference here: Random variables $T_0, T_1, \ldots$ have the same distribution and are independent. Moreover, to avoid the necessity of considering ties, we assume that variables $T_i, i = 0, 1, \ldots$, have a continuous distribution. This means that an equality $T_i = T_j$ has probability zero and hence can be disregarded.

We can measure the "bad luck" of Peter by recording how many persons he has to encounter before finding one "worse off" (i.e., with $T_i < T_0$). Thus, we define the "bad luck" of Peter as measured by random variable N such that $N = n$ if $T_0 < T_1, T_0 < T_2, \ldots, T_0 < T_{n-1}, T_0 > T_n$. Let us find the distribution of N and expected bad luck $E(N)$. We have here $N > n$ if T_0 is the smallest among $T_0, T_1, \ldots, T_n$, that is,

$$P\{N > n\} = P\{T_0 = \min(T_0, \ldots, T_n)\}.$$

But by symmetry T_0 is as likely to be the smallest of $n + 1$ variables $T_0, \ldots T_n$ as any other, hence we must have

$$P\{N > n\} = \frac{1}{n + 1},$$

and consequently,

$$P\{N = n\} = P\{N > n - 1\} - P\{N > n\} = \frac{1}{n} - \frac{1}{n + 1} = \frac{1}{n(n + 1)}.$$

It follows that

$$E(N) = \sum_{n=1}^{\infty} n \cdot P(N = n) = \sum_{n=1}^{\infty} \frac{1}{n + 1} = \infty,$$

and the expected bad luck of Peter (and of everybody else!) is infinite.

Example 8.2.8 (Returns to the Origin in Brownian Motion). Let us consider a physical experiment in which a particle of dust performing Brownian motion in some liquid is observed. Suppose that the particle starts at time $t = 0$ at some location, and we observe the coordinate $x(t)$ of the particle, recording the successive times $T_1 < T_2 < \cdots$ when $x(T_i) = x(0)$ (i.e., when the x-coordinate of the particle returns to the initial value). Now, suppose

that the observations of consecutive times between returns to the initial value
of x-coordinate $T_1, T_2 - T_1, T_3 - T_2, \ldots$ were (say) 10 sec, 15 sec, 7 sec, 2
hours, 3 sec, ... Any experimenter seeing such results would be tempted
to reject the observation of 2 hours as an *outlier* (i.e., observation resulting
from some mistake, observer's falling asleep, etc.). Technically, *outlier* is a
term used in statistics to signify a value in a sequence of observations that
does not follow the same distribution as other values. One of the ways of
modeling outliers consists of assuming that the actual observations follow
the distribution with the density $f(x) = (1 - \epsilon)f_1(x) + \epsilon f_2(x)$ for some small
ϵ (for another approach, see Example 7.6.2). Here f_1 is the density of the
majority of observations, while f_2 is the "contaminating" density and ϵ is the
probability of contamination. Appropriate discrimination rules are then used
to detect observations that follow the distribution f_2 (referred to as "outliers," since typically f_2 is concentrated on some values either much larger
or much smaller than typical values from density f_1). In some cases, outliers
result from mistakes, as mentioned above. In others, they may suggest the
presence of some interesting features of the phenomena analyzed.

In the present case of Brownian motion, however, the situation is different. It turns out that the expected time of return of the x-coordinate to its
initial location is infinite (we shall not give the proof here). The behavior of
observations coming from a distribution with an infinite expectation has been
studied theoretically, and it has been proved that the occurrence of apparent
outliers (of the kind mentioned above) is a rule here. In fact, from time to
time one will observe larger and larger values $T_{n+1} - T_n$, so that one cannot
claim without investigating the underlying distribution theoretically that any
"abnormally" large value is an outlier.   □

Before going further, it may be worthwhile to give the following "mechanical" interpretation of $E(X)$. Consider a discrete random variable, with
possible values x_i and the corresponding probabilities $p_i = P\{X = x_i\}, i =
1, 2, \ldots$.

Let us construct the following mechanical device corresponding to the distribution of X. Consider an infinitely thin and infinitely rigid wire extending
from $-\infty$ to ∞ (this is an abstraction that physicists like to use, such as
ideal void, perfectly black body, etc.). At points with coordinates $x_1, x_2, \ldots$
we attach to this wire weights proportional to probabilities p_i (see Figure 8.1,
where the sizes of the dots symbolize the variability in weight). Then $E(X)$ is
the coordinate of the point at which the whole construction would balance,
its center of gravity.

A similar interpretation holds for continuous random variables. In this case
we have to imagine that an infinitely rigid metal sheet of uniform weight is
cut into the shape of density $f(x)$. Then $E(X)$ is the coordinate of the point
at which the figure will balance if laid flat (see Figure 8.2).

We shall now give some examples of expectations of continuous random
variables.

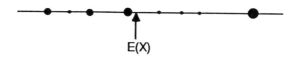

E(X)

Figure 8.1 Interpretation of expected value of a discrete random variable.

Example 8.2.9. Let X have a uniform distribution on interval $[A, B]$, so that (see Example 6.3.10)

$$f(x) = \begin{cases} \dfrac{1}{B - A} & \text{for } A \leq x \leq B \\ 0 & \text{otherwise.} \end{cases}$$

According to the interpretation given above, the metal sheet figure is just the rectangle, hence it will balance if supported in the middle between A and B. Thus, we may expect that $E(X) = (A + B)/2$. Indeed, the computations give

$$E(X) = \int_{-\infty}^{+\infty} x f(x) \, dx = \int_{A}^{B} x \cdot \frac{1}{B - A} \, dx$$

$$= \frac{1}{B - A} \cdot \frac{B^2 - A^2}{2} = \frac{A + B}{2}.$$

Example 8.2.10. We shall now find the expected value of random variable X with exponential distribution with density

$$f(x) = \begin{cases} \lambda e^{-\lambda x} & \text{for } x \geq 0 \\ 0 & \text{for } x < 0. \end{cases}$$

We have here, integrating by parts,

$$E(X) = \int_{0}^{\infty} x \cdot \lambda e^{-\lambda x} \, dx = -\frac{1}{\lambda} e^{-\lambda x} \Big|_{0}^{\infty} = \frac{1}{\lambda}.$$

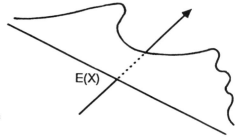

E(X)

Figure 8.2 Interpretation of expected value of a continuous random variable.

Thus, if the lifetime (in hours) of a certain type of electric bulbs has density $0.001e^{-0.001t}$ for $t > 0$, then the expected lifetime equals $(0.001)^{-1} = 1000$ hours.

Example 8.2.11. Find the expectation of the normal density

$$f(x) = \frac{1}{\sigma\sqrt{2\pi}} e^{-(x-\mu)^2/2\sigma^2}.$$

SOLUTION. We substitute $z = (x - \mu)/\sigma, dx = \sigma\, dz$, obtaining

$$
\begin{aligned}
E(X) &= \frac{1}{\sigma\sqrt{2\pi}} \int_{-\infty}^{+\infty} x e^{-(x-\mu)^2/2\sigma^2}\, dx \\
&= \frac{1}{\sqrt{2\pi}} \int_{-\infty}^{+\infty} (\mu + z\sigma) e^{-z^2/2}\, dz \\
&= \mu \cdot \frac{1}{\sqrt{2\pi}} \int_{-\infty}^{+\infty} e^{-z^2/2}\, dz + \frac{\sigma}{\sqrt{2\pi}} \int_{-\infty}^{+\infty} z e^{-z^2/2}\, dz. \\
&= \mu.
\end{aligned}
$$

Indeed, the first integral in the last line equals μ (since the standard normal density integrates to 1) while the second integral is 0 because the integrand is an odd function, so that the integrals over the positive and negative parts of the z-axis (each being finite) cancel each other. Thus, μ is the expected value of the random variable with distribution $N(\mu, \sigma^2)$. □

We shall now derive an analogue of formula (8.8) for an expected value of a nonnegative random variable. We have

Theorem 8.2.2. *Let T be a random variable of continuous type, with density $f(x)$ and cdf $F(x)$. Assume that T is nonnegative, so that $f(t) = F(t) = 0$ for $t < 0$. Then*

$$E(T) = \int_0^\infty [1 - F(t)]\, dt.$$

Proof. Replacing x by $\int_0^x dt$ and changing the order of integration, we have

$$
\begin{aligned}
E(X) &= \int_0^\infty x f(x)\, dx = \int_0^\infty \left(\int_0^x dt \right) f(x)\, dx \\
&= \int_0^\infty \int_t^\infty f(x)\, dx\, dt = \int_0^\infty [1 - F(t)]\, dt.
\end{aligned}
$$

□

Example 8.2.12. Let us again compute the expectation of a random variable with exponential distribution. We have $F(x) = 1 - e^{-\lambda x}$ for $x > 0$; hence using Theorem 8.2.2 we obtain

$$E(X) = \int_0^\infty e^{-\lambda x} \, dx = \frac{1}{\lambda}.$$

$\square$

Example 8.2.13. Finally, let us consider the Cauchy distribution with density given by

$$f(x) = \frac{1}{\pi} \cdot \frac{1}{1 + x^2}.$$

It is easy to check that the function $f(x)$ is indeed a density of a random variable:

$$\frac{1}{\pi} \int_{-\infty}^{+\infty} \frac{dx}{1 + x^2} = \frac{1}{\pi} \arctan x \Big|_{-\infty}^{+\infty} = 1.$$

As regards expectation, the positive part of the defining integral equals

$$\frac{1}{\pi} \int_0^{+\infty} \frac{x \, dx}{1 + x^2} = \frac{1}{\pi} \frac{1}{2} \log(1 + x^2) \Big|_0^{+\infty} = +\infty.$$

In a similar fashion we check that $\int_{-\infty}^0 x f(x) \, dx = -\infty$, hence the expected value of Cauchy distribution does not exist.

PROBLEMS

8.2.1 Random variable X may assume values 0, 1, 2, and 3 with probabilities equal, respectively, to 0.3, a, 0.1, and b. Find a and b if:
  (i) $E(X) = 1.5$.
  (ii) $E(X) = m$. First determine the possible values of m.

8.2.2 Find the expected value $E(|X|)$, where X is a normal random variable with parameters $\mu = 0$ and σ^2 (the distribution of $|X|$ is sometimes called *folded normal*).

8.2.3 Let $X_1, X_2, \ldots, X_n$ be independent, each with exponential density $f(x) = \lambda e^{-\lambda x}, x > 0$. Find $E(\min(X_1, \ldots, X_n))$. [*Hint:* In this and some of the subsequent problems, you should first find the cdf (or density) of the random variable whose expectation is being computed.]

8.2.4 Let X be a nonnegative continuous random variable with hazard rate $h(t) = t$. Find $E(X)$.

8.2.5 The density of the lifetime T of some electronic equipment is $f(t) = \lambda^2 t e^{-\lambda t}, t > 0$. Find $E(T)$.

8.2.6 Suppose that there are $k = 10$ types of toys (e.g., plastic animals, etc.) to be found in boxes of some cereal. Assume that each type occurs with equal frequency. Find $E(X)$, where X is the number of boxes you buy until you collect 3 different types of toys.

8.2.7 Continuing Problem 8.2.6, suppose that your child is collecting the cereal toys and would be happy to get the whole set of 10 different types. Find $E(X)$, where X is the number of boxes you buy until your child has a complete collection.

8.2.8 Show that if the expectation of a continuous random variable X exists, and the density $f(x)$ of X satisfies the condition $f(x) = f(2a - x)$ for all $x \geq 0$, then $E(X) = a$.

8.2.9 Find $E(\min(X_1, \ldots, X_n))$ and $E(\max(X_1, \ldots, X_n))$ under the assumption that $X_1, \ldots, X_n$ are independent, each with the uniform distribution on $[a, b]$.

8.2.10 Let X have the density constant on the union of intervals (a, b) and (c, d), where $a < b < c < d$, and 0 otherwise. Find $E(X)$.

8.2.11 Let X be a continuous random variable with density f. Define Y to be the area under f between $-\infty$ and X. Find $E(Y)$. [*Hint:* Determine first cdf and then density of Y.]

8.3 EXPECTATION AS AN INTEGRAL*

Definition 8.2.1 of expected value in the discrete and continuous cases covers most situations occurring in practice. Many textbooks do not bother to provide the general definition, and some do not even mention the fact that a general definition exists.

One of the consequences of such an omission is the fact that a student capable of some moderate level of reflection should come to the conclusion that the symbol $E(X + Y)$ makes no sense if X is discrete and Y is continuous (since $X + Y$ is neither a continuous nor a discrete random variable). In fact, expectation of a random variable is generally defined as an integral of a random variable, treated as a function on the sample space S. We shall now sketch this definition.

Riemann Integral

Let us start from briefly recalling the definition of an "ordinary" (i.e., Riemann) integral $\int_a^b g(x)\,dx$. For the moment assume that the function g is continuous on $[a, b]$. We first choose a sequence of partitions of the interval $[a, b]$. To simplify the presentation, suppose that the nth partition divides the interval $[a, b]$ into 2^n equal parts, so that the kth point of the nth partition is

$$x_{n,k} = a + \frac{k}{2^n}(b - a), \qquad k = 0, 1, \dots, 2^n.$$

Let $g_{n,k}^{(-)}$ be the minimum of function g in the kth interval, that is,

$$g_{n,k}^{(-)} = \min_{x_{n,k-1} \le x \le x_{n,k}} g(x).$$

We now form the (lower) sum approximating the integral

$$\underline{S}_n = \sum_{k=1}^{2^n} g_{n,k}^{(-)}(x_{n,k} - x_{n,k-1}) = \frac{b-a}{2^n} \sum_{k=1}^{2^n} g_{n,k}^{(-)}. \tag{8.9}$$

The upper sum $\overline{S}_n$ is defined in the same way, with the minimum in (8.9) replaced by the maximum, say $g_{n,k}^{(+)}$.

It is not difficult to show that $\underline{S}_n \le \overline{S}_n$ for all n, and that the sequences $\{\underline{S}_n\}$ and $\{\overline{S}_n\}$ are monotone, the first increasing and the second decreasing. If they converge to the same limit, say S, then this limit is called the Riemann integral of g and is denoted $\int_a^b g(x)\,dx$.

This is the essence of the definition; the details may vary in two respects. First, instead of dividing the interval $[a, b]$ into 2^n equal parts in the nth partition, we can take any points

$$a = x_{n,0} < x_{n,1} < \cdots < x_{n,k_n} = b$$

as long as

$$\lim_{n \to \infty} \max_{1 \le i \le k_n} (x_{n,i} - x_{n,i-1}) = 0$$

(this implies that we must have $k_n \to \infty$).

Second, instead of taking the minimum and maximum of g over the subintervals of partitions (which need not exist if g is not continuous) one may take the value of the function g at a selected point in each subinterval. One then has one partial sum (instead of lower and upper sums), and the requirement is that the limits of these partial sums exist and are the same regardless of the choice of the partitions and the choice of intermediate points.

The main results of the theory built on this definition are very well known:

Continuous functions are integrable on closed finite intervals and if g has antiderivative G (i.e., $G' = g$), then

$$\int_a^b g(x)\, dx = G(b) - G(a).$$

The Riemann integral over an infinite range, $\int_{-\infty}^{+\infty} g(x)\, dx$, is defined in the usual manner through a limiting passage; we omit the details, which can be found in most calculus texts.

We recalled the definition of Riemann integral to better stress the differences and also the close analogy between the principles of definition of Lebesgue and Riemann integrals. The difference between the two definitions seems small and, at first, not essential. Yet the concept of Lebesgue integral is of tremendous consequence, allowing us at once to free the concept of integral of all inessential constraints, and stressing its most crucial and significant features.

Lebesgue Integral

Again, assume at first that the function g, defined on the interval $[a, b]$, is continuous, and let $A < g(x) < B$. Instead of partitioning the interval $[a, b]$ on the x-axis, let us partition the interval $[A, B]$ on the y-axis. For simplicity, let us again take the partitions into 2^n equal parts,

$$A = y_{n,0} < y_{n,1} < \cdots < y_{n,2^n} = B,$$

where

$$y_{n,k} = A + \frac{k}{2^n}(B - A), \qquad k = 0, 1, \ldots, 2^n.$$

The lower sum approximating the integral can now be written as

$$\underline{S}_n^* = \sum_{k=0}^{2^n - 1} y_{n,k} \cdot l(C_{n,k}), \tag{8.10}$$

where l stands for length and $C_{n,k}$ is the set of points x where the function g lies between $y_{n,k}$ and $y_{n,k+1}$; more precisely,

$$C_{n,k} = \{x : y_{n,k} \le g(x) < y_{n,k+1}\}, \qquad k = 0, 1, \ldots, 2^n - 1.$$

The upper sum $\overline{S}_n^*$ is defined in a similar way, with $y_{n,k}$ in (8.10) replaced by $y_{n,k+1}$.

The lower sums $\underline{S}_n$ and $\underline{S}_n^*$ for Riemann and Lebesgue integrals of the same functions are illustrated as parts (R) and (L) in Figure 8.3. The Lebesgue integral is now defined in the same way as the Riemann integral, namely as

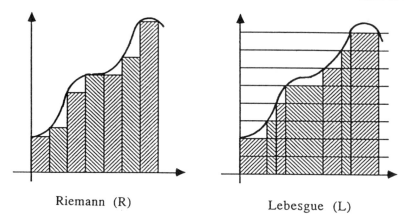

Riemann (R)   Lebesgue (L)

Figure 8.3   Approximating sums for Riemann and Lebesgue integrals.

the common limit (if it exists) of the upper and lower sums. Again, we omit the details of the extension of the definition to an infinite range.

We may capture the difference in the two definitions as follows: In the Riemann integral, when we partition the x-axis, we control the intervals on the x-axis, while the maxima and minima over these intervals depend on the function g. In the Lebesgue integral, when we partition the y-axis, we control the values, but corresponding sets on the x-axis are determined by the function g. The distinction was summarized by Lebesgue as a comparison of two cashiers who, at the end of the day, have to count the cash in their drawers. The "Riemann" cashier takes the money out in the order in which it came in. She counts: "A five-dollar bill, and a penny makes 5.01; then I have a quarter, which makes 5.26; then ...," and so on. On the other hand, the "Lebesgue" cashier proceeds systematically: "I have n_1 twenty-dollar bills, which makes $20n_1$ dollars. Then I have n_2 ten-dollar bills, which makes together $20n_1 + 10n_2$ dollars ...," and so on.

One of the main results in the theory of the Lebesgue integral is that if a function is Riemann integrable, then it is also Lebesgue integrable, and both integrals are equal. Thus, we may use standard calculus to compute Lebesgue integrals, whenever it is applicable. However, it is important to point out that the class of functions that are Lebesgue integrable is substantially wider than the class of functions that are Riemann integrable. To see that it is so, consider the following example.

Example 8.3.1.   Let g be a function continuous on a closed finite interval $[a, b]$, so that g is Riemann integrable. It is clear that if we modify g at a single point x_0 by defining

$$g^*(x) = \begin{cases} g(x) & \text{if } x \neq x_0 \\ c & \text{if } x = x_0, \end{cases}$$

where c is any number such that $c \neq g(x_0)$, then the function $g^*(x)$ is Riemann integrable and $\int_a^b g(x)\, dx = \int_a^b g^*(x)\, dx$. We can modify in this way the function g at any *finite* set of points, and we shall still have a Riemann integrable function, with the same integral as the original function g. However, if we modify g in such a way at a *countable* set of points, then the resulting function may not be Riemann integrable. This is because for a modification at finitely many points, say N, the sums approximating an integral of g and an integral of g^* differ in at most N terms, so that the difference tends to zero, since the sum of lengths of intervals on which the functions g and g^* differ tends to zero. But this is not necessarily the case when g and g^* differ at countably many points. Thus, g^* may be not integrable in the Riemann sense.

However, g^* is integrable in the Lebesgue sense, since two sets that differ by a countable set of points have the same length, so that the sums approximating the Lebesgue integral are the same for g and g^*.   □

We now present two generalizations of the concept of integral introduced thus far. The first generalization concerns both the Riemann and Lebesgue integrals, leading in a natural way to the corresponding concepts of Riemann–Stieltjes and Lebesgue–Stieltjes integrals. The second generalization, more important for our purposes, will specifically concern the Lebesgue integral.

Riemann–Stieltjes Integral

As regards the first generalization, let us observe that in the approximating sums S_n in the Riemann integral given by (8.9), we can write the terms $g_{n,k}^{(-)}(x_{n,k} - x_{n,k-1})$ as $g_{n,k}^{(-)}l([x_{n,k-1}, x_{n,k}])$, where $l(\cdot)$ stands for the length of the set (in this case, of a single interval). This is in analogy with the sums (8.10) approximating the Lebesgue integral.

Now, instead of length, being the difference between the coordinates of endpoints, one can take "length-like" functions, defined as the differences between values of some function, say F, of coordinates. Thus, the terms of the approximating sums are now

$$g_{n,k}^{(-)}[F(x_{n,k}) - F(x_{n,k-1})].$$

Naturally, the function F has to satisfy some conditions if such an extension is to lead to a meaningful concept of integral.

Without striving for a general definition (which can be found in most books on advanced calculus), we shall simply consider the case when F is a cdf [i.e., a nondecreasing function, continuous on the right, satisfying the conditions $\lim_{x \to -\infty} F(x) = 0, \lim_{x \to \infty} F(x) = 1$].

The common limit (if it exists) of the two sequences

$$\underline{S}_n = \sum g_{n,k}^{(-)}[F(x_{n,k}) - F(x_{n,k-1})]$$

and

$$\overline{S}_n = \sum g_{n,k}^{(+)}[F(x_{n,k}) - F(x_{n,k-1})]$$

will be denoted by $\int_a^b g(x)\, dF(x)$, and called the Riemann–Stieltjes (R-S) integral of function g with respect to function F. Again we omit the details of an extension of the concept to the improper integral

$$\int_{-\infty}^{+\infty} g(x)\, dF(x).$$

Example 8.3.2.   Let us consider the special case when F is a cdf of a discrete random variable with the set of possible values $x_1, x_2, \dots$ and the corresponding probabilities $P\{X = x_i\} = p_i$. As we know from Chapter 6, the cdf of X is a step function that is constant between points x_i, and with jumps equal p_i at points x_i.

The approximating (lower) sum equals

$$\underline{S}_n = \sum g_{n,k}^{(-)}[F_X(x_{n,k}) - F_X(x_{n,k-1})].$$

Now, the difference $F_X(x_{n,k}) - F_X(x_{n,k-1})$ of values of F_X at two consecutive points of the nth partition is zero if the interval $(x_{n,k-1}, x_{n,k}]$ does not contain any point x_i of the increase of F_X. When the partition becomes finer as $n \to \infty$, the only terms in $\underline{S}_n$ that remain will be those corresponding to intervals covering the points x_i, and the differences $F_X(x_{n,k}) - F_X(x_{n,k-1})$ for intervals covering point x_i will tend to p_i. Under some regularity assumptions (e.g., continuity of g), the values $g_{n,k}^{(-)}$ corresponding to nonzero terms will converge to corresponding values $g(x_i)$, and the limit will be

$$\int_{-\infty}^{+\infty} g(x)\, dF_X(x) = \sum_i g(x_i) p_i. \tag{8.11}$$

$\square$

From formula (8.11) for the special case $g(x) = x$, we see that

$$\int_{-\infty}^{+\infty} x\, dF_X(x) = \sum x_i P\{X = x_i\},$$

provided that $\sum |x_i| P\{X = x_i\} < \infty$ [which turns out to be the condition for existence of the improper integral $\int_{-\infty}^{+\infty} x\, dF_X(x)$].

We have therefore

Theorem 8.3.1.   *If X is a discrete random variable with cdf F_X and if $E(X)$ exists, then*

$$E(X) = \int_{-\infty}^{+\infty} x \, dF_X(x).$$

Example 8.3.3. Consider the R-S integral in the case when the function F is a cdf of a continuous random variable, so that $F'(x) = f(x)$, where f is the density of random variable X. In this case the lower approximating sum can be written, using the mean value theorem, as

$$\begin{aligned}
\underline{S}_n &= \sum_k g_{n,k}^{(-)}[F(x_{n,k}) - F(x_{n,k-1})] \\
&= \sum_k g_{n,k}^{(-)} f(u_{n,k})(x_{n,k} - x_{n,k-1}),
\end{aligned}$$

where $u_{n,k}$ is a point between $x_{n,k-1}$ and $x_{n,k}$. Again, if g is continuous, the limiting value of $\underline{S}_n$ (and also $\overline{S}_n$) will be the integral of $g(x)f(x)$ between appropriate finite or infinite limits. Thus in this case the R-S integral becomes

$$\int g(x) \, dF(x) = \int g(x) f(x) \, dx. \qquad \square$$

Again, taking $g(x) = x$, we obtain

Theorem 8.3.2. *If X is a continuous random variable with cdf F_X, and if $E(X)$ exists, then*

$$E(X) = \int_{-\infty}^{+\infty} x \, dF_X(x).$$

We therefore have a single expression for the expected value of a random variable, which reduces to formulas (8.2) and (8.4) in the case of discrete and continuous random variables.

Lebesgue–Stieltjes Integral

The Lebesgue–Stieltjes integral is defined in very much the same way as the Lebesgue integral. If g is bounded (say, on interval $[a, b]$) so that $A \le g(x) \le B$ for some A and B, we use the sequence of partitions

$$A = y_{n,0} < y_{n,1} < \cdots < y_{n,2^n} = B$$

and take the (lower) approximating sum as

$$\underline{S}_n = \sum_{k=0}^{2^n} y_{n,k} l_F(C_{n,k}),$$

where

$$\begin{aligned}
C_{n,k} &= \{x : y_{n,k} \le g(x) < y_{n,k+1}\}, k = 0, 1, \ldots, 2^n - 1, \\
C_{n,2^n} &= \{x : g(x) = B\},
\end{aligned}$$

and l_F is now the generalized length of a set. This length is induced by the cdf F, in the sense explained above (see the construction of the Riemann–Stieltjes integral). The upper sum $\overline{S}_n$ is defined similarly, and the common limit (if it exists) of these two sequences of sums is the Lebesgue–Stieltjes integral

$$\int_a^b g(x)\,dF(x).$$

Again, we omit the details of the extension to integrals over an infinite interval.

As before, if the function g is R-S integrable with respect to F, then it is also L-S integrable, and the integrals are equal. If F corresponds to a discrete distribution with masses p_i at points x_i, then

$$\int_{-\infty}^{+\infty} g(x)\,dF(x) = \sum_i g(x_i)p_i,$$

provided that $\sum |g(x_i)|p_i < \infty$. If F is a cdf of continuous distribution with density f, then

$$\int_{-\infty}^{+\infty} g(x)\,dF(x) = \int_{-\infty}^{+\infty} g(x)f(x)\,dx,$$

provided that

$$\int_{-\infty}^{+\infty} |g(x_i)|f(x)\,dx < \infty.$$

Lebesgue Integral. General Case

We now outline the second direction of generalization, applicable only to the case of the Lebesgue (or Lebesgue–Stiltjes) integral.

This generalization, by far the most important and profound extension of the concept of Lebesgue integral, is based on the observation that approximating sums (8.10) make sense also for functions g that are defined on sets other than the real line. In fact, we can consider real-valued functions g defined on an arbitrary set (in particular, the sample space S). The sets $C_{n,k}$ are then subsets of sample space (hence events), and we let probability P play the role of length l.

In more familiar notation, let X be a random variable defined on sample space S. Assume first that $X \geq 0$. In analogy with (8.10), we define the approximating sums as

$$S_n = \sum_{k=0}^{n2^n-1} \frac{k}{2^n} P\left\{ s : \frac{k}{2^n} \leq X(s) < \frac{k+1}{2^n} \right\} \tag{8.12}$$

Observe that as n increases the partitions become finer and their range increases. Observe also that according to the comment made in the footnote on the opening page of Chapter 6, the probabilities in sum (8.12) are well defined (i.e., arguments of probabilities are events if X is a random variable).

We shall omit the details of construction, which can be found in any advanced textbook on probability. Roughly, it is shown that:

(a) For every random variable $X \geq 0$, the sums S_n converge to a finite or infinite limit.
(b) This limit exists and is the same if instead of $y_{n,k} = k/2^n$ we take any other sequence of partitions, provided that these partitions become finer and eventually cover the whole set $[0, \infty)$.

Consequently, we define the integral as the limit of sums (8.12):

$$\int_S X \, dP = \lim S_n.$$

The extension to arbitrary random variables (not necessarily nonnegative) now consists in defining

$$X^+ = \max(X, 0), \ X^- = \max(-X, 0),$$

so that $X = X^+ - X^-$ is represented as a difference of two nonnegative functions. One then puts

$$\int_S X \, dP = \int_S X^+ \, dP - \int_S X^- \, dP,$$

provided that at least one of the integrals on the right-hand side is finite (so that only the indeterminate case $\infty - \infty$ is ruled out).

We now accept the following

Definition 8.3.1.   The *expectation* $E(X)$ of a random variable X is defined as the Lebesgue integral of X:

$$E(X) = \int_S X \, dP,$$

provided that $E|X| = \int_S |X| \, dP < \infty$. If $\int_S X^+ \, dP = \infty$ while we have $\int_S X^- \, dP < \infty$, we define $E(X) = \infty$ [and similarly for $E(X) = -\infty$]. If $\int_S X^+ \, dP = \int_S X^- \, dP = \infty$, we say that expectation of X does not exist.

□

The expectation of a function of two (or more than two) random variables is defined in a similar way. To sketch the construction, consider a nonnegative function $g(X, Y)$ of a pair of random variables (X, Y). The sums approximating the integral $E[g(X, Y)]$ are of the form, in analogy with (8.12):

$$S_n = \sum_{k=0}^{n2^n-1} \frac{k}{2^n} P \left\{ s : \frac{k}{2^n} \leq g(X(s), Y(s)) < \frac{k+1}{2^n} \right\}.$$

The rest of the construction is analogous to the case of a single random variable, leading to definition of the expectation $E[g(X, Y)]$.

8.4 PROPERTIES OF EXPECTATION

The identification of the expected value of a random variable X with Lebesgue integral of X allows us to formulate the properties of the latter as the properties of the expectation. Again, for the proofs we refer the reader to any advanced textbook on probability.

Theorem 8.4.1. *The expectation of random variables has the following properties:*

(a) *Linearity. If $E(X)$ exists, then for all constants α, β:*

$$E(\alpha X + \beta) = \alpha E(X) + \beta. \tag{8.13}$$

(b) *Nonnegativity. If $X \geq 0$, then $E(X) \geq 0$.*
(c) *Modulus Inequality. For any random variable X,*

$$|E(X)| \leq E(|X|).$$

Out of the most important consequences of this theorem, we list here the following. Putting $\beta = 0$ in (8.13), we have $E(\alpha X) = \alpha E(X)$. This means, in particular, that if we change the units of measurement of X, then expectation of X changes in the same way as X. On the other hand, putting $\alpha = 0$, we obtain the property

$$E(\beta) = \beta,$$

so that the expectation of a random variable equal to a constant β equals the same constant β.

In the sequel, we shall often use the following theorem, which gives a sufficient condition for the existence of expected value.

Theorem 8.4.2. *If $|X| \leq Y$, where $E(Y) < \infty$, then $E(X)$ exists and is finite.*

Thus, to prove that the expectation of a random variable exists, it suffices to find another random variable Y that dominates it, whose expectation exists. The domination $|X| \leq Y$ means that $|X(s)| \leq Y(s)$ for every point s in sample space S. Similarly, the symbol $\lim_{n \to \infty} X_n = X$ below is understood as a pointwise limit of random variables as functions on S; that is, $\lim_{n\to\infty} X_n(s) = X(s)$ for every $s \in S$.

Next, we give two principal theorems that connect expectation with convergence, allowing passage to the limit under the integral (expectation) sign.

Theorem 8.4.3 (Monotone Convergence Theorem). *If $X_1, X_2, \ldots$ is a sequence of random variables such that*

$$0 \leq X_1 \leq X_2 \leq \cdots$$

and

$$\lim_{n\to\infty} X_n = X,$$

then

$$\lim_{n\to\infty} E(X_n) = E(X).$$

Theorem 8.4.4 (Dominated Convergence Theorem). *Let $X_1, X_2, \ldots$ be a sequence of random variables satisfying the condition*

$$\lim_{n\to\infty} X_n = X.$$ □

Moreover, assume that there exists a random variable Y with $E(Y) < \infty$ such that $|X| \leq Y, |X_n| \leq Y$ for $n = 1, 2, \ldots$ Then

$$\lim_{n\to\infty} E(X_n) = E(X).$$

In both theorems the assertion is that $\lim E(X_n) = E(\lim X_n)$, which means that we may interchange the order of integration and passage to the limit.

Instead of presenting formal proofs, it may be more instructive to show the necessity of the assumptions.

Example 8.4.1. As regards the monotone convergence theorem, we shall show by an example that the monotonicity alone is not enough for the assertion; nonnegativity is essential. Indeed, let $S = (0, 1)$ and let s be chosen according to the uniform distribution. Now let X_n be defined as

$$X_n(s) = \begin{cases} -1/s & \text{if } 0 < s < 1/n \\ 0 & \text{if } 1/n \leq s < 1. \end{cases}$$

For each fixed s, the sequence $\{X_n(s)\}$ is of the form

$$-1/s, -1/s, \ldots, -1/s, 0, 0, \ldots,$$

hence is monotone, and $\lim X_n \equiv 0$. However, $EX_n = \int_0^{1/n}(-1/s)\, ds = -\infty$, while the integral of the limit random variable $X \equiv 0$ is $EX = 0$.

Example 8.4.2. As regards the dominated convergence theorem, the existence of the bound Y with finite expectation is essential, as shown by the following example. Again let $S = (0, 1)$ as before, with P being uniform on S. Define

$$X_n(s) = \begin{cases} 4n^2 s & \text{for } 0 < s < 1/2n \\ 4n(1 - ns) & \text{for } 1/2n \leq s < 1/n \\ 0 & \text{for } 1/n \leq s < 1. \end{cases}$$

For every fixed s, the sequence of numbers $\{X_n(s)\}$ converges to 0, so again the limiting random variable is $X \equiv 0$, and $E(X) = 0$. But it is easy to check that $E(X_n) = 2$ for every n, so that again the assertion of the theorem does not hold. This time, each random variable X_n is bounded, but there is no integrable common bound for all of them.   □

At the end of this section, it is necessary to connect the two definitions of expectation of a random variable X: as a Lebesgue integral of X, and as a Riemann–Stieltjes integral of function $g(x) \equiv x$ with respect to the cdf of X.

Also, an insightful reader may raise the following doubt: Starting with Chapter 1, we stress the fact that for the same phenomenon (hence for the same random variable), the sample space S can be chosen in a number of different ways. But if we choose different sample spaces to describe the same random variable X, how can we guarantee that the integral of X is the same, regardless of the choice of S?

The answer lies in the following theorem, which provides methods of computing Lebesgue integrals by reducing them to Lebesgue–Stieltjes and Riemann–Stieltjes integrals.

Theorem 8.4.5. *Let X be a random variable defined on some probability space S, and let $F(x) = P\{X \leq x\}$ be its cdf. If $\int_S |X| dP < \infty$, then*

$$\int_S X\, dP = \int_{-\infty}^{+\infty} x\, dF(x). \tag{8.14}$$

More generally, if g is a real function such that $\int_S |g(X)| dP < \infty$, then

$$E[g(X)] = \int_S g(X)\, dP = \int_{-\infty}^{+\infty} g(x)\, dF(x). \tag{8.15}$$

The right-hand sides of formulas (8.14) and (8.15) provide means of computing the expectations. In the case of discrete and continuous random variables, formula (8.14) reduces to (8.2) or (8.4). Similarly, formula (8.15) reduces to

$$E[g(X)] = \sum_i g(x_i)P\{X = x_i\} \qquad (8.16)$$

and

$$E[g(X)] = \int_{-\infty}^{+\infty} g(x)f(x)\, dx. \qquad (8.17)$$

Example 8.4.3. Formulas (8.14) and (8.15) are sometimes referred to as "theorems of the unconscious statistician." The reason for the name is that in calculating expectations of some random variables, a statistician chooses either the left or right side of these formulas, often without being aware that use of the other side may occasionally be simpler. We illustrate the situation by carrying out the calculations for both sides of (8.17).

Imagine that we have a stick of length 1 and we break it at a random point (i.e., the breaking point is chosen according to the uniform distribution from the length of the stick). What is the expected length of the longer of the two parts?

If we take the interval [0, 1] as the sample space, with measure P being uniform (i.e., probabilities are proportional to lengths), then the length of the longer part of the stick, if break occurs at s, is

$$X = \max(s, 1 - s).$$

The graph of X is presented on Figure 8.4(a). It is now clear that $E(X) = \int_S X\, dP = \int_0^1 \max(s, 1-s)ds = \frac{3}{4}$. This can be obtained by actually computing the integral, or by observing that the area under the curve on Figure 8.4(a) is three-fourths of the square of side 1.

We can also find the cdf of X, and finding that X is a continuous random variable—use the right-hand side of (8.17). Clearly, we must have $\frac{1}{2} \leq X \leq 1$. For $\frac{1}{2} \leq x \leq 1$ we may write

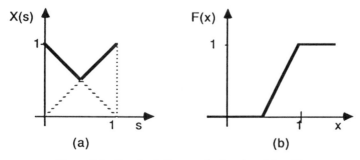

Figure 8.4 Graph of $X = \max(s, 1 - s)$ and its cdf.

$$P\{X \le x\} = P\{\max(s, 1 - s) \le x\}$$
$$= P\{s \le x, 1 - s \le x\}$$
$$= P\{1 - x \le s \le x\} = x - (1 - x) = 2x - 1.$$

Consequently,

$$F(x) = \begin{cases} 0, & x \le \frac{1}{2} \\ 2x - 1, & \frac{1}{2} \le x \le 1 \\ 1, & x \ge 1 \end{cases}$$

[see Figure 8.4(b)]. We now have

$$E(X) = \int_{-\infty}^{+\infty} x \, dF(x) = \int_{1/2}^{1} x \, dF(x) = \int_{1/2}^{1} x \, 2dx = \frac{3}{4}. \qquad \square$$

Let us now consider functions of two random variables. We discuss only the most important two cases, of discrete and continuous bivariate distributions.

Let $g(x, y)$ be a real function such that $E|g(X, Y)| < \infty$. Suppose first that the distribution of (X, Y) is concentrated on $A \times B$, where $A = \{x_1, x_2, \ldots\}$ and $B = \{y_1, y_2, \ldots\}$. Let $p_{ij} = P\{X = x_i, Y = y_j\}$. Then

$$E[g(X, Y)] = \sum_{ij} g(x_i, y_j)p_{ij}. \tag{8.18}$$

Since $E|g(X, Y)| = \sum |g(x_i, y_j)|p_{ij} < \infty$, the sum (8.18) does not depend on the order of summation, so that one can choose whichever order leads to simpler calculations:

$$E[g(X, Y)] = \sum_{i=1}^{\infty} \sum_{j=1}^{\infty} g(x_i, y_j)p_{ij}$$

$$= \sum_{j=1}^{\infty} \sum_{i=1}^{\infty} g(x_i, y_j)p_{ij}$$

$$= \sum_{k=1}^{\infty} \sum_{r=1}^{k} g(x_r, y_{k-r})p_{r,k-r},$$

and so on. The choice depends on properties of the function g and probabilities p_{ij}.

Let (X, Y) have the joint density $f(x, y)$. We then have

Theorem 8.4.6. *If $E|g(X, Y)| < \infty$, then*

$$E[g(X, Y)] = \int_{R^2} \int g(x, y)f(x, y) \, dx \, dy$$

$$= \int_{-\infty}^{+\infty} \left[\int_{-\infty}^{+\infty} g(x, y)f(x, y) \, dx \right] dy$$

$$= \int_{-\infty}^{+\infty} \left[\int_{-\infty}^{+\infty} g(x, y)f(x, y) \, dy \right] dx. \tag{8.19}$$

The last two expressions provide the computational formulas that can be used in practice, since they reduce the computation of a double (two-dimensional) integral to iteration of single (one-dimensional) integrals.

As an illustration of applicability of Theorem 8.4.6, we give the proof (in case of continuous random variables) of the following theorem.

Theorem 8.4.7.   *Assume that $E|X| < \infty$ and $E|Y| < \infty$. Then*

$$E(X + Y) = E(X) + E(Y).$$

Proof. Let $f(x,y)$ be the density of (X, Y), and let f_X and f_Y be the marginal densities of X and Y. We may now write, letting $g(x,y) = x + y$ in (8.19),

$$E(X + Y) = \int_{R^2} \int (x + y) f(x, y)\, dx\, dy$$

$$= \int_{-\infty}^{+\infty} \int_{-\infty}^{+\infty} x f(x, y)\, dx\, dy + \int_{-\infty}^{+\infty} \int_{-\infty}^{+\infty} y f(x, y)\, dx\, dy$$

$$= \int_{-\infty}^{+\infty} x \left[\int_{-\infty}^{+\infty} f(x, y)\, dy \right] dx + \int_{-\infty}^{+\infty} y \left[\int_{-\infty}^{+\infty} f(x, y)\, dx \right] dy$$

$$= \int_{-\infty}^{+\infty} x f_X(x)\, dx + \int_{-\infty}^{+\infty} y f_Y(y)\, dy = E(X) + E(Y).$$

The order of integreation can be changed because of the assumption that $\int_{-\infty}^{+\infty} |x| f_X(x)\, dx < \infty$, $\int_{-\infty}^{+\infty} |y| f_Y(y)\, dy < \infty$. $\qquad\qquad\square$

The property of additivity extends immediately to any finite number of random variables with finite expectations:

$$E(X_1 + \cdots + X_n) = E(X_1) + \cdots + E(X_n),$$

or, combining it with Theorem 8.4.1(a), we have

$$E(\alpha_1 X_1 + \cdots + \alpha_n X_n + \beta) = \sum_{i=1}^{n} \alpha_i E(X_i) + \beta.$$

Finally, nonnegativity property (b) in Theorem 8.4.1 implies that expectation is monotone: If X and Z have finite expectations, then $Y \le Z$ implies that $E(Y) \le E(Z)$. To see this, one may write $Z = Y + (Z - Y)$, where now $Z - Y \ge 0$. By (b) we have

$$E(Z) = E(Y) + E(Z - Y) \ge E(Y).$$

We also have the following important theorem.

Theorem 8.4.8. *If X and Y are independent random variables such that* $E|XY| < \infty$, *then*

$$E(XY) = E(X)E(Y).$$

Proof. We give the proof only for the case of continuous random variables. We have

$$
\begin{aligned}
E(XY) &= \int_{R^2} \int xy f(x, y) \, dx \, dy \\
&= \int_{R^2} \int xy f_X(x) f_Y(y) \, dx \, dy \\
&= \left(\int_{-\infty}^{+\infty} x f_X(x) \, dx \right) \times \left(\int_{-\infty}^{+\infty} y f_Y(y) \, dy \right) \\
&= E(X)E(Y).
\end{aligned}
$$
□

Replacing the double integral by the iterated integrals, as well as the change in the order of integration (or summation), is very often taken for granted, even by persons with a fair degree of mathematical sophistication. Actually, the fact that iterated integrals are equal to one another is not a "law of nature" as some are inclined to believe. It is a fact that is true under specific assumptions, namely under the existence of the double integral.

We shall not provide here the precise statement of the relevant theorem (see the Fubini theorem or Lebesgue integration in any advanced textbook on probability). Instead, we give a simple example which shows that the iterated integrals need not be equal.

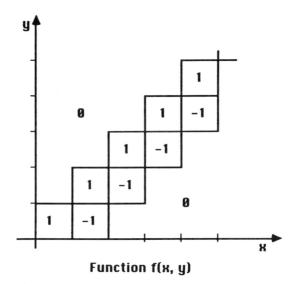

Function f(x, y)

Figure 8.5 Nonintegrable function whose iterated integrals exist and are not equal.

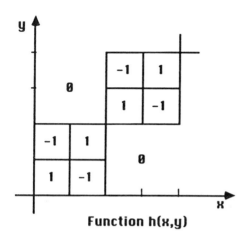

Function h(x,y)

Figure 8.6 Nonintegrable function whose iterated integrals exist and are not equal.

Example 8.4.4. Consider the function $f(x,y)$ described in Figure 8.5. Obviously, $\int_0^\infty f(x,y)\,dx = 0$ for every y, so that $\int_0^\infty \int_0^\infty f(x,y)\,dx\,dy = 0$. On the other hand, $\int_0^\infty f(x,y)\,dy = 1$ for $0 \le x \le 1$ and 0 otherwise, so that $\int_0^\infty \int_0^\infty f(x,y)\,dx\,dy = 1$.

It ought to be clear that the feature responsible for the difference between the values of two iterated integrals is that the double integral (the sum of infinitely many "volumes" + 1 and infinitely many "volumes" −1) does not exist.

Example 8.4.5. Consider now a function $h(x,y)$ similar to that of Example 8.4.4 given in Figure 8.6. In this case we have $\int_0^\infty h(x,y)\,dx = \int_0^\infty h(x,y)\,dy = 0$, so that both iterated integrals exist and are equal zero. Still, the double integral does not exist, for the same reason as in Example 8.4.4.   □

These two examples combined show that if the double integral does not exist, we cannot say anything about the equality of iterated integrals.

PROBLEMS

8.4.1 Show that $E(X - E(X)) = 0$.

8.4.2 Assume that X has the density $f(x) = ax + bx^3$ for $0 \le x \le 2$ and $f(x) = 0$ otherwise. Find a and b if:
   (i) $E(X) = 1$.
   (ii) $E(X^3) = 4$.

8.4.3 Let X be a random variable with the distribution given by

$$P\{X = 2^k\} = \frac{1}{8}, \quad k = 1, 2, 3, 4, 5, 6, 7, 8.$$

Find $E(X), E(1/X)$, and $E(2^X)$.

8.4.4 Let X be a random variable with density $f(x) = \frac{1}{2}$ for $-1 \le x \le 1$ and $f(x) = 0$ otherwise. Find:
   (i) $E(X)$.
   (ii) $E(X^2)$.
   (iii) $E(2X - 3)^2$.
   (iv) $E(X + 1)^3$.

8.4.5 Let X have the density

$$f(x) = \begin{cases} cx & 0 \le x \le 1 \\ c(2 - x) & 1 \le x \le 2 \\ 0 & \text{otherwise.} \end{cases}$$

Find:
   (i) c.
   (ii) $E(X)$.
   (iii) $E(X - 1)^2$.
   (iv) $E(2 - X)^3$.
   (v) $E\left(\dfrac{1}{2 - X}\right)$.

8.4.6 Let $X_1, X_2, \ldots, X_n$ be a random sample from a continuous distribution with density f, and let $X_{1:n}, X_{2:n}, \ldots, X_{n:n}$. Find

$$E\left[\int_{-\infty}^{X_{1:n}} f(t)\, dt\right] \quad \text{and} \quad E\left[\int_{X_{n:n}}^{+\infty} f(t)\, dt\right].$$

8.4.7 An urn contains w white and r red balls. We draw without replacement $n \le r$ balls from the urn, and we let X be the number of red balls drawn. Find $E(X)$, by
   (i) Defining indicator random variables $X_1, \ldots, X_n$ such that $X = X_1 + \ldots + X_n$;
   (ii) Defining indicator variables $Y_1, \ldots, Y_r$ such that $X = Y_1 + \cdots + Y_r$.

8.4.8 A cereal company puts a plastic bear in each box of cereal. Every fifth bear is red. If you have 3 red bears, you get a free box of the cereal. If you decide to keep buying the cereal in question until you get one box free, how many boxes would you expect to buy before getting a free one? [*Hint:* Represent the answer as $E(X_1 + X_2 + X_3)$ where X_i is the number of boxes you buy after getting the $(i-1)$st red bear and until getting the ith red bear.]

8.4.9 We say that X is *stochastically smaller* than Y ($X \leq_{\text{st}} Y$) if $P\{X \leq t\} \geq P\{Y \leq t\}$ for all t. Show that if X and Y have finite expectations and $X \leq_{\text{st}} Y$, then $E(X) \leq E(Y)$. [*Hint:* Start with nonnegative X and Y and use Theorem 8.2.2. Then use the decomposition into positive and negative part]. Show also that the converse assertion is false: there exist random variables X and Y such that $E(X) < E(Y)$ and X is not stochastically smaller than Y.

8.4.10 Show that if X satisfies the condition $P(a \leq X \leq b) = 1$ then $E(X)$ exists and $a \leq E(X) \leq b$.

8.5 MOMENTS

We begin with the following definition.

Definition 8.5.1. Given random variable X, the expectation of X^n, if it exists, will be called nth *moment* (or *moment of the order n*) of X. Letting m_n denote the nth moment, we have

$$m_n = E(X^n),$$

and m_n exists if $E|X|^n < \infty$.   □

Clearly, the moment of the order zero always exists and equals 1, while m_1 is simply the expectation of X. If the nth moment exists, it may be computed from the formula

$$m_n = E(X^n) = \int_{-\infty}^{+\infty} x^n \, dF(x),$$

where F is the cdf of X; this formula is obtained by substituting $g(x) = x^n$ in (8.15). Observe that any bounded random variable X, that is, a random variable such that $P(|X| \leq M) = 1$ for some $M < \infty$, has finite moments of any order. This follows at once from Theorem 8.4.2, and in this case we have $m_n \leq M^n$.

Example 8.5.1. Let X be the random variable with Poisson distribution $P\{X = k\} = \lambda^k e^{-\lambda}/k!, k = 0, 1, 2, \ldots$ Then

$$m_1 = E(X) = \sum_{k=0}^{\infty} k \frac{\lambda^k}{k!} e^{-\lambda}$$

$$= \lambda e^{-\lambda} \sum_{k=1}^{\infty} \frac{\lambda^{k-1}}{(k-1)!} = \lambda.$$

On the other hand,

$$m_2 = E(X^2) = \sum_{k=0}^{\infty} k^2 \frac{\lambda^k}{k!} e^{-\lambda}$$

$$= \sum_{k=1}^{\infty} [k(k-1) + k] \frac{\lambda^k}{k!} e^{-\lambda}$$

$$= \lambda^2 e^{-\lambda} \sum_{k=2}^{\infty} \frac{\lambda^{k-2}}{(k-2)!} + \sum_{k=1}^{\infty} k \frac{\lambda^k}{k!} e^{-\lambda}$$

$$= \lambda^2 + \lambda$$

since the second sum equals $m_1 = \lambda$.

Example 8.5.2. Let X have the distribution uniform on $[a, b]$. Then its density is $f(x) = 1/(b-a)$ for $a \le x \le b$ and $f(x) = 0$ otherwise, so that

$$m_n = \int_a^b x^n \cdot \frac{dx}{b-a} = \frac{1}{n+1} \cdot \frac{b^{n+1} - a^{n+1}}{b-a}.$$

When the distribution is symmetric about 0, then $a = -b$ and we have

$$m_{2k} = \frac{b^{2k}}{2k+1}, \qquad m_{2k+1} = 0. \qquad\qquad \square$$

Before introducing the concepts that will illustrate the usefulness of the notion of moments, let us introduce some definitions pertaining to a variety of types of moments.

Definition 8.5.2. An *absolute* moment of order n is defined as

$$\beta_n = E(|X|^n). \qquad\qquad \square$$

Thus, the existence of absolute moment of a given order implies the existence of "ordinary" moment m_n of the same order.

Note that the order of an absolute moments need not be an integer. The same applies to ordinary moments of positive random variables. Thus, we may speak of moments of order such as $2 + \epsilon$ if $P(X > 0) = 1$, absolute moment of X of order $2 + \epsilon$, and so on.

Next we have

Definition 8.5.3.   Ordinary moments of the random variable $Y = X - E(X)$ are called *central moments* of X, so that

$$\gamma_n = E[(X - m_1)^n],$$

where $m_1 = E(X)$.   □

Clearly, the first central moment γ_1 is always zero. Finally, we have

Definition 8.5.4.   For positive integer n, the value

$$\pi_n = E[X(X - 1) \cdots (X - n + 1)]$$

is called a *factorial* moment of X of order n.   □

We shall now prove the following:

Theorem 8.5.1.   *If an absolute moment of order $\alpha > 0$ exists, then all moments (ordinary, absolute, central, and factorial) of orders $r \le \alpha$ exist.*

Proof. We shall prove first that if $E(|X|^\alpha) < \infty$, then $E(|X|^r) < \infty$ for all $r < \alpha$.

Clearly, if $|X| \ge 1$, then $|X|^r \le |X|^\alpha$, while if $|X| < 1$, then $|X|^r \le 1$. Consequently,

$$|X|^r \le \max(1, |X|^\alpha),   \tag{8.20}$$

and (see Theorem 8.4.2) it remains to prove that the right-hand side of (8.20) has finite expectation.

Now, if $E|X|^\alpha < \infty$, then

$$E(|X|^\alpha) = \int_{-1}^{1} |x|^\alpha \, dF(x) + \int_{|x| \ge 1} |x|^\alpha \, dF(x)$$

and it follows that $\int_{|x| \ge 1} |x|^\alpha \, dF(x) < \infty$ as a difference of two finite quantities. Thus, we can write

$$E(|X|^r) \le E\{\max(1, |X|^\alpha)\}$$
$$= \int_{-1}^{1} dF(x) + \int_{|x| \ge 1} |x|^\alpha \, dF(x) < \infty.$$

It remains to prove that if $E|X|^\alpha < \infty$, then all other types of moments of order α exist. This is true for ordinary moments (by definition). As regards central and factorial moments of order n, they are linear combinations of ordinary moments of order $k \le n$, which proves the theorem.   □

We state here without proof the following theorem:

Theorem 8.5.2 (Liapunov Inequality). *If $0 < \alpha < \beta < \infty$, then*

$$\{E(|X|^{\alpha})\}^{1/\alpha} \le \{E(|X|^{\beta})\}^{1/\beta}. \qquad \Box$$

Note that the first part of the proof of Theorem 8.5.1 is an immediate consequence of the Liapunov inequality.

We now introduce a certain function, which will later prove to be very valuable as a tool in analyzing random variables and limiting behavior of their sums.

Definition 8.5.5. Given a random variable X, the function of real variable t, defined as

$$m_X(t) = E(e^{tX}),$$

is called the *moment generating function* (mgf) of X. $\qquad \Box$

Clearly, for any random variable, its mgf exists for $t = 0$. If X is a positive random variable and mgf of X exists for some t_0, then $m_X(t)$ exists for all $t \le t_0$; this fact follows from Theorem 8.4.2.

Example 8.5.3. If X has the binomial distribution with parameters n and p, then

$$m_X(t) = \sum_{k=0}^{n} e^{tk} \binom{n}{k} p^k q^{n-k}$$
$$= (pe^t + q)^n,$$

so that $m_X(t)$ exists for all t.

Example 8.5.4. If X has exponential distribution with density $f(x) = \lambda e^{-\lambda x}, x > 0$, then

$$m_X(t) = \int_0^{\infty} e^{tx} \cdot \lambda e^{-\lambda x}\, dx = \lambda \int_0^{\infty} e^{-(\lambda - t)x}\, dx = \frac{\lambda}{\lambda - t}, \qquad (8.21)$$

provided that $t < \lambda$. For $t \ge \lambda$ the integral in (8.21) diverges. $\qquad \Box$

We shall now explore some of the properties of moment generating functions. First, observe that the concept of mgf is connected with a distribution rather than with a random variable: Two different random variables with the same distribution will have the same mgf. Consequently, we shall often use the symbol $m_F(t)$, where F is a cdf, to denote an mgf associated with any random variable having cdf $F(x)$.

The name *moment generating function* is related to the following theorem.

Theorem 8.5.3.   *Let X be a random variable with mgf $m_X(t)$, assumed to exist in some neighborhood of $t = 0$.*
   If $E(|X|^n) < \infty$, then for $k = 1, \ldots, n$, the kth moment of X is given by the formula

$$m_k = \frac{d^k}{dt^k} m_X(t)\biggl|_{t=0}.$$

Proof. Differentiating formally the expression for the mgf under the sign of expectation (i.e., under the integral or summation sign), we obtain

$$m_X'(t) = E(Xe^{tX})$$
$$m_X''(t) = E(X^2 e^{tX}),$$

and so on, so that

$$\frac{d^k}{dt^k} m_X(t) = E(X^k e^{tX}), \qquad k = 1, 2, \ldots, n.$$

The substitution $t = 0$ gives the required result.
   The validity of this argument depends crucially on whether or not formal differentiation under the integral sign is allowed. The answer is positive if the corresponding derivatives are absolutely integrable for $t = 0$, which is ensured by the assumption about the existence of the nth moment.

Example 8.5.5.   Since the mgf of random variable with binomial distribution BIN(n, p) is $m_X(t) = (pe^t + q)^n$, we have $m_X'(t) = n(pe^t + q)^{n-1} pe^t$, while $m_X''(t) = n(n-1)(pe^t + q)^{n-2}(pe^t)^2 + n(pe^t + q)^{n-1} pe^t$. Thus, $E(X) = m_X'(0) = np$ and $E(X^2) = m_X''(0) = n(n-1)p^2 + np$.   □

Example 8.5.6.   The moment generating function of the distribution uniform on $[a, b]$ is

$$m_X(t) = \int_a^b e^{tx} \cdot \frac{1}{b-a} \, dx$$
$$= \frac{e^{bt} - e^{at}}{t(b-a)}. \tag{8.22}$$

Determination of the values of derivatives of $m_X(t)$ at $t = 0$ requires repeated use of the de L'Hospital rule. We can, however, expand exponentials into power series, and after some algebra we obtain

$$m_X(t) = 1 + \frac{1}{2!}\frac{b^2 - a^2}{b-a} t + \frac{1}{3!}\frac{b^3 - a^3}{b-a} t^2 \cdots.$$

This is a Taylor expansion of $m_X(t)$ about $t = 0$, so that we must have

$$m_n = m_X^{(n)}(0) = \frac{n!}{(n+1)!} \frac{b^{n+1} - a^{n+1}}{b - a}$$

$$= \frac{1}{n+1}[b^n + b^{n-1}a + b^{n-2}a^2 + \cdots + a^n].$$ $\square$

Next, we shall prove

Theorem 8.5.4. *If X is a random variable with mgf $m_X(t)$, then the random variable $Y = \alpha X + \beta$ has the mgf*

$$m_Y(t) = e^{\beta t} m_X(\alpha t). \tag{8.23}$$

Proof. From properties of the expectation,

$$m_Y(t) = E(e^{tY}) = E(e^{t(\alpha X + \beta)})$$
$$= E(e^{(\alpha t)X} \cdot e^{\beta t}) = e^{\beta t} m_X(\alpha t).$$

Example 8.5.7. Let us find mgf's of random variables U and V, with distributions uniform on $[0, 1]$ and on $[-1, 1]$. Obviously, we could write down the expressions for these mgf's starting from the definition. However, it will be instructive to use Theorem 8.5.4.

If X has a uniform distribution on $[a, b]$, hence with the mgf (8.22), then $U = (X - a)/(b - a)$ has a uniform distribution on $[0, 1]$. Consequently, we may use formulas (8.22) and (8.23) for $\alpha = 1/(b - a)$ and $\beta = -a/(b - a)$. We obtain, after some algebra,

$$m_U(t) = e^{-at/(b-a)} m_X\left(\frac{t}{b-a}\right) = \frac{e^t - 1}{t}. \tag{8.24}$$

Next, if we let $V = 2U - 1$, then V will have uniform distribution on $[-1, 1]$, so taking $\alpha = 2$ and $\beta = -1$ and using (8.24), we get

$$m_V(t) = e^{-t} \cdot \frac{e^{2t} - 1}{2t} = \frac{e^t - e^{-t}}{2t} = \frac{\sinh t}{t}.$$

Example 8.5.8. The moment generating function of a normal distribution can be found as follows. We start by finding the mgf of a standard normal random variable $Z \sim N(0, 1)$. We have

$$m_Z(t) = \frac{1}{\sqrt{2\pi}} \int_{-\infty}^{\infty} e^{tz} e^{-z^2/2} \, dz.$$

Completing the square in the exponent, we obtain

$$m_Z(t) = e^{t^2/2} \frac{1}{\sqrt{2\pi}} \int_{-\infty}^{\infty} e^{-(1/2)(z-t)^2} \, dz = e^{t^2/2}.$$

If $X \sim N(\mu, \sigma^2)$, then $X = \mu + \sigma Z$, and using Theorem 8.5.4, we obtain

$$m_X(t) = e^{\mu t + \sigma^2 t^2/2}. \tag{8.25}$$

$\square$

We now prove

Theorem 8.5.5. *If X and Y are independent random variables with moment generating functions $m_X(t)$ and $m_Y(t)$, then mgf of $X + Y$ is*

$$m_{X+Y}(t) = m_X(t)m_Y(t).$$

Proof. Observe that random variables e^{tX} and e^{tY} are independent for each t, so that by Theorem 8.4.8 we have

$$m_{X+Y}(t) = E(e^{t(X+Y)}) = E(e^{tX})E(e^{tY})$$
$$= m_X(t)m_Y(t).$$

$\square$

Before proceeding with examples, we state one more important theorem. Its proof lies beyond the scope of this book.

Theorem 8.5.6. *If X and Y are two random variables such that their moment generating functions $m_X(t)$ and $m_Y(t)$ coincide in some neighborhood of the point $t = 0$, then X and Y have the same distribution.*

In other words, this theorem asserts that if two mgf's agree in some neighborhood of $t = 0$, then they agree for all t (for which they are defined) and that an mgf determines uniquely the distribution. In still other words, random variables with different distributions must have different mgf's.

Example 8.5.9. Let X and Y be independent and have Poisson distributions with means λ_1 and λ_2. Let us first determine the mgf of X and Y. We have

$$m_X(t) = \sum_{k=0}^{\infty} e^{tk} \frac{\lambda_1^k}{k!} e^{-\lambda_1}$$

$$= e^{-\lambda_1} \sum_{k=0}^{\infty} \frac{(\lambda_1 e^t)^k}{k!} = e^{\lambda_1(e^t-1)}.$$

Similarly, $m_Y(t) = e^{\lambda_2(e^t-1)}$ and by Theorem 8.5.6,

$$m_{X+Y}(t) = m_X(t)m_Y(t) = e^{(\lambda_1 + \lambda_2)(e^t - 1)}. \tag{8.26}$$

We recognize (8.26) as the mgf of Poisson distribution. In view of Theorem 8.5.6, we infer that $X + Y$ has Poisson distribution with mean $\lambda_1 + \lambda_2$.

PROBLEMS

8.5.1 A continuous random variable X has density $f(x) = xe^{-x}$ for $x > 0$. Find its moment generating function.

8.5.2 Find the moment generating function of a discrete random variable X with distribution $P\{X = k\} = 1/n, k = 0, 1, \ldots, n - 1$.

8.5.3 Let X be a nonnegative integer-valued random variable. The function $g_X(s) = Es^X$, defined for $|s| \leq 1$, is called a *probability generating function*, or simply a generating function, of X. Find $g_X(s)$ for random variables with:

    (i) Geometric distribution.

    (ii) Binomial distribution.

    (iii) Poisson distribution.

8.5.4 Find the fourth factorial moment of the random variable X with a Poisson distribution.

8.5.5 Let $X_1, \ldots, X_n$ be independent random variables with the common normal density. Show that for some constants α_n and β_n the density of

$$U = \frac{X_1 + \cdots + X_n - \alpha_n}{\beta_n}$$

is the same as the density of X_1.

8.5.6 A continuous random variable is called symmetric about c if its density satisfies the condition $f(c - x) = f(c + x)$ for all x. Show that:

    (i) If X is symmetric about c and $E(X)$ exists, then $E(X) = c$.

    (ii) If X is symmetric about 0, then all moments of odd order (if they exist) are equal 0.

8.5.7 Let X be a random variable with $E(X) = \mu$, $\text{Var}(X) = \sigma^2$ and such that the third central moment $\gamma_3 = E(X - \mu)^3$ exists. The ratio γ_3/σ^3 is called *coefficient of skewness*, or simply *skewness*. Find skewness of the distributions with densities:

    (i) $f(x) = 1$ for $0 \leq x \leq 1$ and $f(x) = 0$ otherwise.

    (ii) $f(x) = \alpha x^{\alpha - 1}$ for $0 \leq x \leq 1$ and $f(x) = 0$ otherwise ($\alpha > 1$).

(iii) $f(x) = \alpha(1-x)^{\alpha-1}$ for $0 \le x \le 1$ and $f(x) = 0$ otherwise ($\alpha > 1$).

8.5.8 Find skewness (see Problem 8.5.7) of the distribution concentrated at points 0 and 1 with $P(X = 1) = p = 1 - P(X = 0)$. Graph skewness as a function of p.

8.5.9 Let X be a random variable with $E(X) = \mu$, $\text{Var}(X) = \sigma^2$ and such that $\gamma_4 = E(x-\mu)^4$ exists. Then γ_4/σ^4 is called the *coefficient of kurtosis*, or simply *kurtosis*. Find kurtosis of
  (i) Standard normal random variable.
  (ii) Normal $N(\mu, \sigma^2)$ random variable.
  (iii) Random variable from Problem 8.5.8, and graph the kurtosis as a function of p.

8.5.10 Find skewness and kurtosis of Poisson distribution with parameter λ (see Problems 8.5.7 and 8.5.9). Determine the limits of these parameters as $\lambda \to \infty$.

8.5.11 Let X be the number of tosses of a biased coin, up to and including the first head. Assume $P(\text{head}) = p$. Find mgf of X. Use the result to determine the mgf of $Y =$ number of tails preceding the first head.

8.6 CHARACTERISTIC FUNCTIONS*

The main disadvantage of moment generating functions lies in the fact that they may not exist for any $t \ne 0$: There are random variables X such that the random variable e^{tX} has no expectation for any $t \ne 0$. This restricts the usefulness of mgf's as a tool (to be explored in Chapter 10) for obtaining limit theorems. Indeed, for proofs using mgf's to be valid, they have to be confined to classes of random variables for which mgf's exist in some neighborhood of $t = 0$. The corresponding theorems, however, are typically valid without this assumption, and the proofs usually require nothing more than replacing mgf's by the *characteristic functions*. Although we shall not use characteristic functions in this book, it is worthwhile to provide their definition and simplest properties. The concepts below require students to have a rudimentary knowledge of complex numbers.

Definition 8.6.1. Let X be a random variable with the cdf F. The function φ_X of the real argument t ($-\infty < t < +\infty$), defined as

$$\varphi_X(t) = E(e^{itX}) = \int_{-\infty}^{+\infty} e^{itx}\, dF(x), \qquad (8.27)$$

is called the *characteristic function* (ch.f.) of X (or of the cdf F). □

Here i is the imaginary unit (i.e., $i^2 = -1$). From the formula $e^{i\xi} = \cos\xi + i\sin\xi$, we obtain

$$\varphi_X(t) = E\{\cos(tX)\} + iE\{\sin(tX)\}, \tag{8.28}$$

so that $\varphi_X(t)$ is a complex–valued function of a real argument. Since $|e^{itX}| = \cos^2(tX) + \sin^2(tX) = 1$, the expectation in (8.27) exists for all t, so that characteristic functions always exist.

We list below some basic properties of characteristic functions. The proofs are left as exercises. In particular, Theorems 8.5.3, 8.5.4, and 8.5.5 carry over to the case of characteristic functions almost without any change. We have the following theorems, in which $\varphi_X, \varphi_Y, \dots$ are ch.f's of random variables $X, Y, \dots$.

Theorem 8.6.1. *For every random variable X, its ch.f. $\varphi_X(t)$ is uniformly continuous and satisfies the conditions*

$$\varphi_X(0) = 1, \quad |\varphi_X(t)| \le 1 \tag{8.29}$$

for every real t. Moreover, for any real a, b

$$\varphi_{aX+b}(t) = \varphi_X(at)e^{ibt}. \tag{8.30}$$

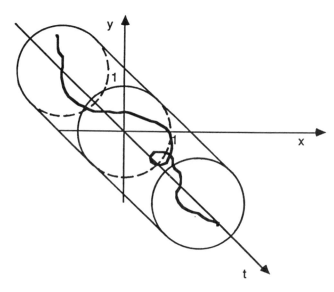

Figure 8.7 Graph of a characteristic function.

In particular, for $a = -1, b = 0$ using (8.28) we obtain

$$\varphi_{-X}(t) = \varphi_X(-t) = \overline{\varphi_X(t)}, \tag{8.31}$$

where $\bar{z}$ stands for the complex conjugate of z. $\qquad\square$

Since $\varphi_X(t)$ is a complex–valued function of a real argument, and since (8.29) holds, it is possible to present graphs of characteristic functions. Indeed, we may choose the axes as labeled in Figure 8.7. The value of the function of the argument t is then a point on the unit disk on the complex plane perpendicular to the argument axis at t. The graph of function $\varphi_X(\cdot)$ is the line (in view of continuity) in the cylinder centered on the argument axis.

Example 8.6.1. If X is binomial $\text{BIN}(n, p)$, then

$$\varphi_X(t) = E(e^{itX}) = \sum_{k=0}^{n} e^{itk} \binom{n}{k} p^k q^n$$
$$= (pe^{it} + q)^n$$
$$= [(p\cos t + q) + ip\sin t]^n.$$

In this case $\varphi_X(t)$ is a periodic function.

Example 8.6.2. If X is uniform on $[0, 1]$, then

$$\varphi_X(t) = \int_0^1 e^{itx}\, dx = \frac{1}{it}(e^{it} - 1).$$

If $Y = 2X - 1$, then Y has a uniform distribution on $[-1, 1]$ and (8.30) gives

$$\varphi_Y(t) = \varphi_{2X-1}(t) = \varphi_X(2t)e^{-it}$$
$$= \frac{1}{2it}(e^{2it} - 1)e^{-it} = \frac{\sin t}{t}.$$

Example 8.6.3. If $X \sim N(0, 1)$, then

$$\varphi_X(t) = \frac{1}{\sqrt{2\pi}} \int_{-\infty}^{+\infty} e^{itx} e^{-x^2/2}\, dx$$
$$= e^{-t^2/2} \frac{1}{\sqrt{2\pi}} \int_{-\infty}^{+\infty} e^{-(1/2)(x-it)^2}\, dx = e^{-t^2/2}. \qquad\square$$

Next, addition of independent random variables corresponds to multiplications of their ch.f.'s. We have

Theorem 8.6.2. *If random variables X and Y are independent, then*

$$\varphi_{X+Y}(t) = \varphi_X(t)\varphi_Y(t). \qquad \square$$

As in the case of mgf's, characteristic functions uniquely determine the distributions, namely:

Theorem 8.6.3. *If two cdfs F and G have the same characteristic function, then $F \equiv G$.* $\square$

Finally, the relationship between moments of X and the behavior of the characteristic function in the neighborhood of $t = 0$ is given by the following theorem.

Theorem 8.6.4. *If X is a random variable such that $E(|X|^k) < \infty$ for some integer $k \geq 1$, then*

$$\varphi_X(t) = \sum_{j=0}^{k} \frac{i^j}{j!} m_j t^j + o(|t|^k)$$

in some neighborhood of the point $t = 0$. Here $m_j = E(X^j)$ denotes the jth moment of X, and $o(x)$ is a function such that $\lim_{x\downarrow 0} o(x)/x = 0$. $\square$

PROBLEMS

8.6.1 Find the characteristic function of:
    (i) Poisson distribution.
    (ii) Geometric distribution.
    (iii) Exponential distribution.

8.6.2 Find the characteristic function of a random variable with normal distribution $N(\mu, \sigma^2)$.

8.6.3 A family $\mathcal{G}$ of distributions is said to be *closed under convolution*, if whenever independent random variables X and Y have distributions in $\mathcal{G}$, the same is true for random variable $X + Y$.
Show that the families of:
    (i) Poisson distributions
    (ii) Normal distributions
are closed under convolutions.

8.6.4 Show that if $\varphi(t)$ is a ch.f., then $|\varphi(t)|^2$ is also a ch.f.

8.6.5 Show that the distribution of X is symmetric about zero if and only if $\varphi_X(t)$ is real.

8.7 VARIANCE

In this section we introduce a concept that will play an extremely important role in almost all statistical analyses. We begin with the following definition.

Definition 8.7.1. If $E(X^2) < \infty$, then the second central moment of X is called the *variance* of X:

$$\text{Var}(X) = E[(X - m_1)^2],$$

where $m_1 = E(X)$. The positive square root of variance is called the *standard deviation* of X. □

The notation for variance differs; the most commonly used symbols are $\text{Var}\,X, D^2(X), V(X)$, and σ_X^2. We begin by listing some of the elementary properties of variance and some preliminary examples. First, observe that we can write

$$\begin{aligned}
\text{Var}(X) &= E(X^2 - 2m_1 X + m_1^2) \\
&= E(X^2) - 2m_1 E(X) + m_1^2 \\
&= E(X^2) - [E(X)]^2.
\end{aligned} \tag{8.32}$$

Formula (8.32) gives an alternative way of calculating the variance. Note that variance is always nonnegative, as the expectation of a nonnegative random variable $(X - m_1)^2$. Consequently, $E(X^2) \geq [E(X)]^2$, that is, $E(X^2)^{1/2} \geq |E(X)|$, which is a special case of the Liapunov inequality given in Theorem 8.5.2.

Next, we have the following theorem.

Theorem 8.7.1. *If* $\text{Var}(X)$ *exists, then*

$$\text{Var}(aX + b) = a^2 \text{Var}(X). \tag{8.33}$$

Proof. Using the fact that $E(aX + b) = aE(X) + b$, we may write

$$\begin{aligned}
\text{Var}(aX + b) &= E[(aX + b)^2] - [aE(X) + b]^2 \\
&= E[a^2 X^2 + 2abX + b^2] - a^2[E(X)]^2 - 2abE(X) - b^2 \\
&= a^2[E(X^2) - (E(X))^2] = a^2 \,\text{Var}(X).
\end{aligned}$$
□

Example 8.7.1. Let X have the Poisson distribution with mean λ. Then the mgf of X is (see Example 8.5.9)

$$m_X(t) = e^{\lambda(e^t - 1)}.$$

After some algebra we obtain

$$m_X''(t) = \lambda(\lambda_{e+1}^t)e^{\lambda(e^t-1)+t},$$

so that

$$E(X^2) = m_X''(0) = \lambda^2 + \lambda.$$

Since $E(X) = \lambda$, we obtain

$$\text{Var}(X) = (\lambda^2 + \lambda) - \lambda^2 = \lambda.$$

Thus, for the Poisson distribution, the mean and variance coincide.

Example 8.7.2. From Example 8.5.5 we know that if X has binomial distribution $\text{BIN}(n, p)$, then $E(X) = np, E(X^2) = n(n-1)p^2 + np$. Consequently, using (8.32), we obtain

$$\text{Var}(X) = n(n-1)p^2 + np - (np)^2 = np(1-p) = npq. \qquad (8.34)$$

Example 8.7.3. Let us find the variance of random variable X with distribution uniform on $[a, b]$. We have here $f(x) = 1/(b-a)$ for $a \le x \le b$, and we easily obtain $E(X) = (a+b)/2$,

$$E(X^2) = \frac{1}{3}\frac{b^3 - a^3}{b - a} = \frac{1}{3}[b^2 + ab + a^2],$$

so that, after some algebra,

$$\text{Var}(X) = \frac{(b-a)^2}{12}.$$

Example 8.7.4. In Example 8.5.4 we found that the mgf of the exponential distribution with density $f(x) = \lambda e^{-\lambda x}, x > 0$ is $m(t) = \lambda(\lambda - t)^{-1}$. Consequently, $m'(t) = \lambda(\lambda - t)^{-2}, m''(t) = 2\lambda(\lambda - t)^{-3}$, so that $E(X) = 1/\lambda, E(X^2) = 2/\lambda^2$, and $\text{Var}(X) = 1/\lambda^2$. Thus, in exponential distribution the mean and standard deviation coincide. □

To interpret the variance, let us consider the case of a discrete random variable. Then the variance equals

$$\text{Var}(X) = \sum_i (x_i - m_1)^2 P(X = x_i). \qquad (8.35)$$

It is clear that for the variance to be small, all terms of the sum (8.35) must be small; hence the values x_i with the large difference $|x_i - m_i|$ must have a low probability. Qualitatively speaking, a small variance means that the values of X are concentrated close to the mean of X, and large deviations are unlikely.

Example 8.7.5.   To look for another interpretation of the variance, consider the function

$$f(\xi) = E(X - \xi)^2.$$

Since $f(\xi) = E(X^2) - 2\xi E(X) + \xi^2$, the function f represents a parabola with branches directed upward. The minimum occurs at the point ξ^* at which $f'(\xi^*) = 0$, so that $\xi^* = E(X)$.   □

This simple fact is important enough to deserve stating it separately as follows.

Theorem 8.7.2.   *The mean square deviation from ξ, namely $E(X - \xi)^2$, is minimized at $\xi = E(X)$, and the minimal value equals $\mathrm{Var}(X)$.*   □

The interpretation of variance (and expectation) is now as follows. Suppose that one plays the game in which one is to predict the value of the random variable X, and if one's prediction is ξ and the actual value is x, then one pays the "penalty" $(x - \xi)^2$. In this game, the best predictor is $\xi = E(X)$, and the average penalty for this prediction is the variance of X.

Example 8.7.6.   A natural question that arises from the problem posed in Example 8.7.5 is to determine the value of ξ that minimizes

$$g(\xi) = E|X - \xi|.$$

The reason that this question can be called "natural" is that cases of predicting a future value of random variable X are very common. In each, we want the prediction to be "as exact as possible," the latter phrase being explicated through the (average) penalty, depending on ξ and X. The function $f(\xi)$ corresponds to the case when, qualitatively speaking, "small errors are almost negligible, large errors are very serious." The function $g(\xi)$ treats the seriousness of error as proportional to its size.   □

The minimization of $g(\xi)$ is not as simple as that of $f(\xi)$. We have the following result.

Theorem 8.7.3.   *The mean absolute deviation from ξ, namely $E|X - \xi|$, is minimized at $\xi = m$, where m is the median of X.*

Proof. We shall present the proof in the case of the continuous random variable X with density f. It suffices to show that

$$E(|X - m|) \leq E(|X - a|)$$

for every a. Assume now that $m < a$. Then

$$E(|X - m|) - E(|X - a|) = \int_{-\infty}^{m} [(m - x) - (a - x)]f(x)\,dx$$

$$+ \int_{m}^{a} [(x - m) - (a - x)]f(x)\,dx$$

$$+ \int_{a}^{\infty} [(x - m) - (x - a)]f(x)\,dx,$$

which simplifies to

$$\int_{-\infty}^{m} (m - a)f(x)\,dx + \int_{m}^{a} (2x - m - a)f(x)\,dx + \int_{a}^{\infty} (a - m)f(x)\,dx.$$

Since $2x - m - a \leq 2a - m - a = a - m$ for $m \leq x \leq a$, combining the second and third interval we obtain the inequality

$$E(|X - m|) - E(|X - a|) \leq (m - a)[P(X \leq m) - P(X \geq m)].$$

It follows that

$$E(|X - m|) - E(|X - a|) \leq 0,$$

since $m - a < 0$ and $P(X \leq m) - P(X \geq m) \geq \frac{1}{2} - \frac{1}{2} = 0$. The proof for the case $a < m$ is analogous. $\square$

We now explore the behavior of the variance under the addition of random variables. This will lead in a natural way to certain new important concepts.

To make the notation more readable, we let $\mu_X = E(X)$ and $\mu_Y = E(Y)$. Then

$$\begin{aligned}
\text{Var}(X + Y) &= E(X + Y)^2 - [E(X + Y)]^2 \\
&= E(X^2 + 2XY + Y^2) - (\mu_X + \mu_Y)^2 \\
&= E(X^2) - \mu_X^2 + E(Y^2) - \mu_Y^2 + 2[E(XY) - \mu_X\mu_Y] \\
&= \text{Var}(X) + \text{Var}(Y) + 2[E(XY) - \mu_X\mu_Y].
\end{aligned}$$

The last quantity appears sufficiently often to deserve a separate name. We therefore introduce the following

Definition 8.7.2. The quantity

$$\text{Cov}(X, Y) = E(XY) - E(X)E(Y) = E[(X - \mu_X)(Y - \mu_Y)] \qquad (8.36)$$

is called the *covariance* between X and Y. $\square$

We shall show that $\text{Cov}(X, Y)$ exists whenever X and Y have finite second moments. Indeed, we have $0 \leq (X - Y)^2 = X^2 - 2XY + Y^2$, which gives $2XY \leq X^2 + Y^2$, and consequently,

$$|XY| \le \frac{X^2}{2} + \frac{Y^2}{2}.$$

Thus, if $E(X^2) < \infty$ and $E(Y^2) < \infty$, then expectations are finite, and $E(|XY|) < \infty$, so that $\text{Cov}(X, Y)$ exists.

We have therefore

Theorem 8.7.4. *If $E(X_1^2)$ and $E(X_2^2)$ exist, then*

$$\text{Var}(X_1 + X_2) = \text{Var}(X_1) + \text{Var}(X_2) + 2\,\text{Cov}(X_1, X_2),$$

and more generally, if $E(X_i) < \infty$ for $i = 1, \ldots, n$, then

$$\text{Var}(X_1 + \cdots + X_n) = \sum_{j=1}^{n} \text{Var}(X_j) + 2 \sum_{i<j} \text{Cov}(X_i, X_j). \qquad \square$$

An important case here is when covariances are zero, so that variances are additive. We therefore introduce the following:

Definition 8.7.3. The random variables X, Y for which $\text{Cov}(X, Y) = 0$ are called *uncorrelated* or *orthogonal*. $\qquad \square$

Observe that in view of Theorem 8.4.8, independent random variables with finite variances are uncorrelated. Consequently, we have

Theorem 8.7.5. *If random variables $X_1, \ldots, X_n$ are pairwise uncorrelated, then*

$$\text{Var}(X_1 + \cdots + X_n) = \text{Var}(X_1) + \cdots + \text{Var}(X_n). \qquad (8.37)$$

In particular, additivity of variance (8.37) holds if $X_1, \ldots, X_n$ are independent.

Somewhat more generally, let us find the variance of a linear combination of random variables, that is, the variance of the sum

$$Y = a_1 X_1 + a_2 X_2 + \cdots + a_n X_n.$$

Using Theorem 8.7.4, we have

$$\text{Var}(Y) = \sum_{i=1}^{n} \text{Var}(a_i X_i) + \sum_{i<j} \text{Cov}(a_i X_i, a_j X_j).$$

By Theorem 8.7.1, $\text{Var}(a_i X_i) = a_i^2\, \text{Var}(X_i)$, while

$$\text{Cov}(a_i X_i, a_j X_j) = E(a_i X_i \cdot a_j X_j) - E(a_i X_i)E(a_j X_j)$$
$$= a_i a_j [E(X_i X_j) - E(X_i)E(X_j)]$$
$$= a_i a_j \text{ Cov}(X_i, X_j).$$

Consequently,

$$\text{Var}(Y) = \sum_{i=1}^{n} a_i^2 \text{ Var}(X_i) + 2 \sum_{i<j} a_i a_j \text{ Cov}(X_i, X_j).$$

In particular, if $X_1, \ldots, X_n$ are uncorrelated, then

$$\text{Var}(Y) = \sum_{i=1}^{n} a_i^2 \text{ Var}(X_i). \tag{8.38}$$

Example 8.7.7. For the variance of a difference of two random variables, we have, taking $a_1 = 1, a_2 = -1$,

$$\text{Var}(X_1 - X_2) = \text{Var}(X_1) + \text{Var}(X_2) - 2 \text{ Cov}(X_1, X_2);$$

hence for uncorrelated random variables

$$\text{Var}(X_1 - X_2) = \text{Var}(X_1) + \text{Var}(X_2).$$

This formula is worth remembering: It is amazing how many students fall into the simple trap of taking the minus sign on the right-hand side and ending up with a negative variance!

Example 8.7.8 (Averaging). Let $X_1, \ldots, X_n$ be independent, with the same distribution with $\text{Var}(X_i) = \sigma^2$, and let

$$\overline{X}_n = \frac{X_1 + \cdots + X_n}{n}.$$

In statistical terminology, the situation above is referred to as a *simple random sample* of size n, and $\overline{X}_n$ is the sample average (sample mean). Here formula (8.38) with $a_i = 1/n$, $i = 1, \ldots, n$, gives

$$\text{Var}(\overline{X}_n) = \frac{\sigma^2}{n}.$$

Thus, averaging decreases the variability (as measured by variance) by the factor $1/n$. The standard deviation of the sample mean is therefore decreased (as compared with the standard deviation of a single observation) by the factor $1/\sqrt{n}$.

Figure 8.8   Length of 16 feet. (Drawing by S. Niewiadomski.)

Example 8.7.9 (A Foot).   The effect of averaging on variability, computed in Example 8.7.8, appears to have been understood long before the beginning of probability and statistics. This is illustrated by the following law from the Middle Ages, which defined the length of one foot. (Before you read any further, think for a while: How could a measure of length be defined in ancient times so as to be—as much as possible—uniform throughout a country?)

The law specified the following procedure (apparently, the standard of one foot was necessary only on market days, in this case Sundays after the mass). The shoes of the first 16 men leaving the church (this was an attempt to get a random sample!) were to be lined up, toe to heel, and $\frac{1}{16}$ of the total length of these 16 shoes was to be used as a measure of 1 foot (see Figure 8.8). The number 16 was clearly chosen because it was easy to divide a string into 16 equal parts, by folding it four times into halves. This procedure cuts down the variability of the obtained standard of one foot, compared with the variability of the length of feet of men (as expressed by the standard deviation) by the factor of 4.   □

Example 8.7.10 (Problem of Design).   Suppose that you have to determine, as precisely as possible, the weights of two objects, A and B. You have at your disposal a scale and a set of weights. The additional condition is that you can use the scale only twice.

If you put an object to be weighted on a scale and balance it with weights, you obtain a measurement of the true weight of the object, say w, with an error. One of the possible assumptions here, met in many practical situations, is that what one observes is a value of random variable, say X, such that $E(X) = w$ and $\text{Var}(X) = \sigma^2$, with different measurements (even of the same object) being independent. In other words, we have $X = w + \epsilon$, where $E(\epsilon) = 0, \text{Var}(\epsilon) = \sigma^2$, with σ being the standard deviation for the error of measurement.

In our situation, it would seem that all one can do is to put A on the scale, balance it with weights, and then do the same with object B, observing two random variables, X and Y, with $E(X) = w_A, E(Y) = w_B$, and $\text{Var}(X) = \text{Var}(Y) = \sigma^2$.

One can, however, proceed differently. Suppose that on the first weighting, one puts both objects A and B on one scale, and then balance the scales, observing the random variable $X_{A+B} = w_A + w_B + \epsilon_1$, where ϵ_1 is the mea-

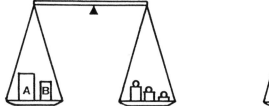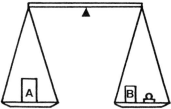

Figure 8.9 Two weightings of A and B.

surement error. On the second weighting, one puts A and B on different scales and add weights as needed for balance (see Figure 8.9).

Thus, on the second measurement, we observe $X_{A-B} = w_A - w_B + \epsilon_2$, where ϵ_2 is independent of ϵ_1, with $E(\epsilon_1) = E(\epsilon_2) = 0$, $\text{Var}(\epsilon_1) = \text{Var}(\epsilon_2) = \sigma^2$. We easily find that

$$\frac{X_{A+B} + X_{A-B}}{2} = w_A + \frac{\epsilon_1 + \epsilon_2}{2}$$

and

$$\frac{X_{A+B} - X_{A-B}}{2} = w_B + \frac{\epsilon_1 - \epsilon_2}{2}.$$

We have here now

$$\text{Var}\left(\frac{\epsilon_1 + \epsilon_2}{2}\right) = \frac{1}{4}\text{Var}(\epsilon_1 + \epsilon_2) = \frac{\sigma^2}{2}, \quad \text{and} \quad \text{Var}\left(\frac{\epsilon_1 - \epsilon_2}{2}\right) = \frac{\sigma^2}{2}.$$

As a result, with two measurements, we obtained the measurements of w_A and w_B, each with standard deviation of the error $\sigma/\sqrt{2} = 0.707\sigma$. This means an error reduction by about 30%, obtained *at no additional cost* (i.e., with the same number of observations). This is one of the simplest examples of the effects of choosing a proper *design of experiment*. □

Let us now investigate more closely the concept of covariance. First, observe that according to formula (8.36),

$$\text{Cov}(X, X) = E[(X - m_X)^2] = \text{Var}(X), \tag{8.39}$$

so that covariance, in some sense, is a generalization of variance.

Next, we have the following identity describing the behavior of covariance for a linear combination of random variables.

Theorem 8.7.6. *If $X_1, \ldots, X_n$ and $Y_1, \ldots, Y_m$ are two sets (not necessarily disjoint) of random variables with finite second moments, then for any constants $a_1, \ldots, a_n, b_1, \ldots, b_m$ and c, d,*

$$\text{Cov}\left(\sum_{i=1}^{n} a_i X_i + c, \sum_{i=1}^{m} b_j Y_j + d\right) = \sum_{i=1}^{n} \sum_{j=1}^{m} a_i b_j \, \text{Cov}(X_i, X_j). \qquad (8.40)$$

The proof involves straightforward checking and will be omitted.

The property that $\text{Cov}(X, Y) = 0$ if X and Y are independent suggests taking the covariance as a measure of dependence. To achieve comparability across various measurements, it is only necessary to standardize covariance. Accordingly, we introduce the following definition.

Definition 8.7.4. Let X, Y be two random variables with finite second moments. Then

$$\rho = \rho_{X,Y} = \frac{\text{Cov}(X, Y)}{\sqrt{\text{Var}(X)\, \text{Var}(Y)}}$$

is called the coefficient of *correlation* between X and Y.

Example 8.7.11. Let U, V, W be independent random variables such that $\text{Var}(U) = \sigma_U^2$, $\text{Var}(V) = \sigma_V^2$, and $\text{Var}(W) = \sigma_W^2$. Compute the coefficient of correlation between $X = U + W$ and $Y = V + W$.

SOLUTION. We have $\sigma_X^2 = \sigma_U^2 + \sigma_W^2$ and $\sigma_Y^2 = \sigma_V^2 + \sigma_W^2$, while $\text{Cov}(X, Y) = \text{Cov}(U + W, V + W) = \text{Cov}(U, V) + \text{Cov}(U, W) + \text{Cov}(W, V) + \text{Cov}(W, W) = \sigma_W^2$ by (8.39). Consequently,

$$\rho_{XY} = \frac{\sigma_W^2}{\sqrt{(\sigma_U^2 + \sigma_W^2)(\sigma_V^2 + \sigma_W^2)}}. \qquad (8.41)$$

Situations such as those described in this example are quite common. We deal often with variables that are related simply because they are influenced by a common component (W). One can imagine here some mental or physical attributes, such as IQ or height of two brothers, being affected by a common genetic factor W, and otherwise independent factors U and V. As another example, one can imagine that the prices X and Y of two items depend on the price W of some components, which are used in the construction of both X and Y. To see the effect of σ_W^2, let us write (8.41) in the form

$$\rho_{XY} = \frac{1}{\sqrt{[(\sigma_U^2/\sigma_W^2) + 1][(\sigma_V^2/\sigma_W^2) + 1]}}.$$

If σ_W^2 is large compared with both σ_U^2 and σ_V^2, then $\rho_{X,Y}$ is close to 1. On the other hand, if σ_W^2 is small compared to one (or both) of σ_U^2 and σ_V^2, then $\rho_{X,Y}$ is close to zero. These results are as expected if $\rho_{X,Y}$ is to be a measure of dependence between random variables X and Y.

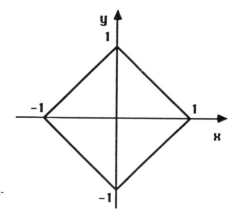

Figure 8.10 Dependent but uncorrelated random variables.

Example 8.7.12. Let X, Y have a joint density uniform on the square with vertices $(-1, 0), (0, 1), (1, 0)$ and $(0, -1)$ (see Figure 8.10). Clearly, $E(X) = E(Y) = 0$ by symmetry of marginal distributions. Also, $E(XY) = 0$, since in the integral $\int \int xyf(x, y) \, dx \, dy$ the contribution arising from domains with $xy > 0$ and $xy < 0$ cancel one another. Thus $\text{Cov}(X, Y) = 0$ and therefore $\rho_{X,Y} = 0$. Yet the random variables are dependent. This is immediately obvious from Theorem 7.2.2, as well as from the "commonsense" consideration that if one knows that X is close to 1 (say), then one can infer that Y must be close to zero.

This example shows that correlation, as a measure of dependence, is not perfect: There are dependent random variables for which $\rho_{X,Y} = 0$. To analyze the properties of the correlation coefficient, we begin by proving an important inequality.

Theorem 8.7.7 (Schwarz Inequality). *For any random variables X and Y,*

$$[E(XY)]^2 \leq E(X^2) \cdot E(Y^2). \tag{8.42}$$

Proof. If $E(X^2)$ or $E(Y^2)$ is infinite, the inequality (8.42) holds. On the other hand, if $E(X^2) = 0$, then we must have $P(X = 0) = 1$, so that $P(XY = 0) = 1$ and $E(XY) = 0$, hence again inequality (8.42) holds. The same argument applies if $E(Y^2) = 0$, so that we may assume that $0 < E(X^2) < \infty$ and $0 < E(Y^2) < \infty$.

Consider now the random variable $Z_t = tX + Y$. We have, for every t,

$$0 \leq E(Z_t^2) = E(t^2 X^2 + 2tXY + Y^2)$$
$$= t^2 E(X^2) + 2t E(XY) + E(Y^2). \tag{8.43}$$

The right-hand side is a quadratic function of t, which is nonnegative for all t. Thus, its discriminant must satisfy the condition

$$4[E(XY)]^2 - 4E(X^2)E(Y^2) \leq 0,$$

which is the same as (8.42).   □

We can now prove the following theorem, asserting the basic properties of the correlation coefficient.

Theorem 8.7.8. *The coefficient of correlation ρ_{XY} satisfies the inequality*

$$-1 \leq \rho_{X,Y} \leq 1 \tag{8.44}$$

with $|\rho_{X,Y}| = 1$ if and only if there exist constants $a \neq 0$ and b such that $P\{Y = aX + b\} = 1$.

Moreover, $|\rho|$ is invariant under linear transformation of random variables; more precisely, for any a, b, c, d with $ac \neq 0$,

$$\rho_{aX+b,cY+d} = \epsilon \rho_{X,Y} \tag{8.45}$$

with $\epsilon = +1$ if $ac > 0$ and -1 if $ac < 0$.

Proof. By formula (8.40), we have $\text{Cov}(aX + b, cY + d) = ac\,\text{Cov}(X,Y)$. On the other hand, $\text{Var}(aX + b) = a^2\,\text{Var}(X), \text{Var}(cY + d) = c^2\,\text{Var}(Y)$, so that

$$\rho_{aX+b,cY+d} = \frac{ac\,\text{Cov}(X,Y)}{\sqrt{a^2\,\text{Var}(X) \cdot c^2\,\text{Var}(Y)}} = \frac{ac}{|ac|} \cdot \rho_{X,Y},$$

which proves (8.45). We can now prove (8.44). By what is already shown in (8.40), we can assume that $EX = EY = 0$, so that

$$\rho_{X,Y} = E(XY)/\sqrt{E(X^2)E(Y^2)}.$$

But now the condition $|\rho_{X,Y}| \leq 1$ is simply equivalent to the Schwarz inequality (8.42).

It remains to prove that $\rho^2 = 1$ is equivalent to the existence of a linear relationship between X and Y. Again, we may assume that $E(X) = E(Y) = 0$. If $Y = aX$ with $a \neq 0$, then $E(XY) = E[X(aX)] = aE(X^2)$ and $E(X)E(Y) = a[E(X)]^2$. Consequently, $\text{Cov}(X,Y) = a\{E(X^2) - [E(X)]^2\} = a\,\text{Var}(X)$, and since $\text{Var}(Y) = a^2\,\text{Var}(X)$, we get

$$\rho_{X,Y} = \frac{a\,\text{Var}(X)}{\sqrt{\text{Var}(X) \cdot a^2\,\text{Var}(X)}} = \frac{a}{|a|} = \pm 1.$$

Conversely, assume that $\rho_{X,Y}^2 = 1$. This means that [assuming that $E(X) = E(Y) = 0$] we have $[E(XY)]^2 = E(X^2)E(Y^2)$ [i.e., X and Y are such that we have equality in (8.43)]. The inspection of the proof of Theorem 8.7.7 shows that this occurs if the discriminant of the right-hand side of (8.43) is 0, hence if there exists t^* such that $E(t^*X + Y)^2 = 0$. But expectation of a nonnegative random variable is 0 if and only if this variable assumes only the value zero, so that $P(t^*X + Y = 0) = 1$. If we had $t^* = 0$, then we would have $P(Y = 0) = 1$, hence $E(Y^2) = 0$, a case we eliminated. This shows that $t^* \neq 0$, hence there is a linear relationship between X and Y. $\qquad\square$

Example 8.7.13.   Consider the following situation: We observe the values of two random variables X and Y, and we want to predict the value of random variable Z. The prediction should be such as to minimize the expected square error.

SOLUTION.   The optimal solution to this problem is well known. Adapting Example 8.7.5 to the case of conditional prediction, we see that given $X = x$ and $Y = y$, one should predict the conditional expectation $\xi = E(Z|X = x, Y = y)$. However, the practical implementation of this solution presupposes knowledge of the joint distribution of (X, Y, Z), and—which may be analytically difficult—the conditional expectation of Z given (X, Y).

Quite often we simply do not have this knowledge. In such cases, one can use the best *linear* prediction, which requires only knowledge of the first two moments of the joint distribution of (X, Y, Z), that is, expectations, variances, and covariances (or equivalently, correlations). The solution is as follows.

First, without loss of generality we may assume that $E(X) = E(Y) = E(Z) = 0$ (since if we know the mean, predicting the random variable is equivalent to predicting its deviation from the mean). We look for a predictor of the form

$$\xi = \alpha X + \beta Y$$

which minimizes

$$E(Z - \xi)^2 = E(Z - \alpha X - \beta Y)^2. \tag{8.46}$$

Condition (8.46) can be expanded to

$$
\begin{aligned}
E(Z - \alpha X - \beta Y)^2 &= E(Z^2) + \alpha^2 E(X^2) + \beta^2 E(Y^2) \\
&\quad - 2\alpha E(XZ) - 2\beta E(YZ) + 2\alpha\beta E(XY) \\
&= \sigma_Z^2 + \alpha^2 \sigma_X^2 + \beta^2 \sigma_Y^2 \\
&\quad - 2\alpha\rho_{XZ}\sigma_Z\sigma_X - 2\beta\rho_{YZ}\sigma_Z\sigma_Y + 2\alpha\beta\rho_{XY}\sigma_X\sigma_Y.
\end{aligned}
$$

Differentiating with respect to α and β and setting the derivatives equal to zero, we obtain the system of linear equations

$$
\begin{aligned}
\alpha\sigma_X + \beta\rho_{XY}\sigma_Y &= \rho_{XZ}\sigma_Z \\
\alpha\rho_{XY}\sigma_X + \beta\sigma_Y &= \rho_{YZ}\sigma_Z.
\end{aligned}
$$

The solution exists if $\rho_{XY}^2 \neq 1$, and equals

$$\alpha = \frac{\rho_{XZ} - \rho_{XY}\rho_{YZ}}{1 - \rho_{XY}^2} \cdot \frac{\sigma_Z}{\sigma_X}$$

$$\beta = \frac{\rho_{YZ} - \rho_{XY}\rho_{XZ}}{1 - \rho_{XY}^2} \cdot \frac{\sigma_Z}{\sigma_Y}$$

(if $\rho_{XY}^2 = 1$, then $X = Y$ or $X = -Y$, so that X and Y provide the same information). This solution is not very important per se. More important is the fact that this method can easily be applied to determine the best linear predictor of one or more random variables. All that is needed are the means and all second order moments of all variables in question—those to be predicted and those serving as predictors. The next two examples will illustrate the class of situations to which such a prediction method can usefully be applied.   □

Example 8.7.14 (Autoregressive Process).   Many phenomena in economics, such as periodic variations of some index (price, demand, etc.), are described by the following model. We observe some index X_t, where $t = \ldots, -1, 0, 1, \ldots$ represents time, measured in some units, and X_t represents the value of the index at time t. For the sake of definitiveness, we shall count time in days, and think of X_t as being the price of some commodity at t. The assumption is that for every t,

$$X_t = \xi_t + \alpha X_{t-1}, \tag{8.47}$$

where $|\alpha| < 1$ and ξ_t's are independent random variables with the same distribution (sometimes referred to as "shocks"). We let $E\xi_t = \mu$, $\text{Var}(\xi_t) = \sigma^2$.

We imagine that the process has been going on for a long time. Equation (8.47) postulates that the price X_t on day t (today) is the yesterday's price X_{t-1}, discounted by some factor α plus the "shock" ξ_t that occurred today. The value ξ_t represents all unpredictable factors that affect the fluctuations of the price, while αX_{t-1} represents the effect of the past, accumulated so far. The problem is to predict, in the best possible way, the future values of X_t, given the observed values in the past up to the present. The basic elements needed for such prediction are, as explained in Example 8.7.13, the means, variances, and correlations between the values of the process at different times.

We have here

$$\begin{aligned}
X_t &= \xi_t + \alpha X_{t-1} \\
&= \xi_t + \alpha[\xi_{t-1} + \alpha X_{t-2}] \\
&= \xi_t + \alpha\xi_{t-1} + \alpha^2[\xi_{t-2} + \alpha X_{t-3}] \\
&= \cdots \\
&= \sum_{j=0}^{\infty} \alpha^j \xi_{t-j}. \tag{8.48}
\end{aligned}$$

This shows that the value of the process is uniquely determined by all the shocks $\xi_t, \xi_{t-1}, \ldots$ that occurred so far. We shall proceed somewhat informally, and assume that the series (8.48) converges, and that we can compute its expectation, variance, and so on. by formal manipulations.

We have, first,

$$E(X_t) = E\left(\sum_{j=0}^{\infty} \alpha^j \xi_{t-j}\right) = \sum_{j=0}^{\infty} \alpha^j E(\xi_{t-j}) = \sum_{j=0}^{\infty} \alpha^j \mu = \frac{\mu}{1-\alpha}. \tag{8.49}$$

Similarly,

$$\text{Var}(X_t) = \text{Var}\left(\sum_{j=0}^{\infty} \alpha^j \xi_{t-j}\right) = \sum_{j=0}^{\infty} (\alpha^j)^2 \text{Var}(\xi_{t-j}) = \sigma^2 \cdot \frac{1}{1-\alpha^2}. \tag{8.50}$$

Finally, for $m > 0$,

$$E(X_t X_{t-m}) = E\left\{\left(\sum_{j=0}^{\infty} \alpha^j \xi_{t-j}\right)\left(\sum_{k=0}^{\infty} \alpha^k \xi_{t-m-k}\right)\right\}$$

$$= \sum_{\substack{j,k=0 \\ j \neq m+k}}^{\infty} \alpha^{j+k} E(\xi_{t-j}\xi_{t-m-k}) + \sum_{k=0}^{\infty} \alpha^{m+2k} E(\xi_{t-m-k}^2)$$

$$= \sum_{\substack{j,k=0 \\ j \neq m+k}}^{\infty} \alpha^{j+k} \mu^2 + \sum_{k=0}^{\infty} \alpha^{m+2k}(\mu^2 + \sigma^2)$$

$$= \sum_{j,k=0}^{\infty} \alpha^{j+k} \mu^2 + \sum_{k=0}^{\infty} \alpha^{m+2k} \sigma^2$$

$$= \left(\frac{\mu}{1-\alpha}\right)^2 + \frac{\alpha^m \sigma^2}{1-\alpha^2}.$$

Consequently, we obtain

$$\text{Cov}(X_t, X_{t-m}) = \frac{\alpha^m \sigma^2}{1-\alpha^2}, \tag{8.51}$$

hence

$$\rho_{X_t,X_{t-m}} = \alpha^m.$$

Observe that the process $\{X_t\}$ is *stationary*, in the sense that its mean and variance do not depend on t, while the correlation between the value of the process at time t and $t - m$ depends only on the temporal distance m that separates them.

We can now build the predictors of the value X_{t+N} of the process at some future time $t + N$, given the past and present values $X_t, X_{t-1}, \ldots$ proceeding along the lines sketched in Example 8.7.14. □

In connection with this example, some coments are in order. First, it is not necessary that the value X_{t+N} that one wants to estimate is in the future of the process. We can consider the problem of estimating the value X_τ, given the data on the process at times X_t for $t > \tau$ and for $t < \tau$. In such a case, one typically speaks on *interpolation* rather than prediction; the techniques, however, remain the same.

Second, an often encountered situation is that we observe the sum of two processes, say $Z_t = X_t + Y_t$ (but not X_t and Y_t separately). Typically, one of those processes, say X_t, is of interest, the other is not. Standard terminology here is "signal" and "noise," and the problem is to estimate the value of X_τ (signal) given the observation of Z_t at some times t (when now we may have $t < \tau, t = \tau$ or $t > \tau$). In this case, the term *filtering* is typically used.

In all three cases, of prediction, interpolation, and filtering, the basic techniques are the same and require knowledge of the first and second moments of the relevant processes and their correlations.

Third, it is worth mentioning that in practical applications, the processes, such as X_t analyzed in this example, represent only the "random" component. In addition, we typically have a trend of some kind, so that the observations concern the process of the form $Y_t = f_t + X_t$, when f_t is some deterministic function, either known or to be estimated.

Clearly, we cannot in this book enter in any depth into the theory that deals with the models and phenomena mentioned above. This theory is called Time Series. We presented some elements of it here, and we continue in the next example to illustrate the importance and applicability of the concept of covariance and correlation.

Fourth, in connectionwith this example, it is necessary to point out some issues of interest for more mathematically sophisticated readers. By formal manipulations we derived the formula (8.48), namely

$$X_t = \sum_{j=0}^{\infty} \alpha^j \xi_{t-j}.$$

Since ξ_t's are random, the natural question arises: How do we know that the series for X_t converges? Surely, $\alpha^j \to 0$ as $j \to \infty$, so that the effect of past shocks ξ_{t-j} diminishes with an increase of j. But what if the past values ξ_{t-j} were so large as to offset the effect of multiplication by α^j? Intuitively, we feel that such an event has probability zero in view of the fact that all ξ_t's have the same finite variance. The answer is positive: The series for X_t converges with probability 1. This can be deduced from the Three Series Theorem of Kolmogorov, which is given in Chapter 10.

A somewhat simpler issue is: Given that X_t is finite, can its expectation (as well as variance, etc.) be calculated in the way presented in formulas (8.49) and (8.50)? We give here the answer in two special cases, for the expectation (8.49).

First, if we assume that the shocks ξ_t are all nonnegative and $0 < \alpha < 1$, then we can argue as follows. We have

$$X_t = \sum_{j=0}^{\infty} \alpha^j \xi_{t-j} = \lim_{N \to \infty} \sum_{j=0}^{N} \alpha^j \xi_{t-j}$$

$$= \lim_{N \to \infty} S_N \text{ (say)}.$$

In this case, the terms being nonnegative, we have $0 \le S_1 \le S_2 \le \cdots$ and we can use the monotone convergence theorem (Theorem 8.4.3), which asserts that $E(X_t) = E(\lim S_N) = \lim E(S_N)$ and the latter quantity, $E(S_N)$, as involving only finite operations can easily be calculated.

On the other hand, if the shocks can be positive or negative, or if $-1 < \alpha < 0$, we cannot use the monotone convergence theorem. But if the distribution of each ξ_t has the same range (i.e., if $P\{|\xi_t| \le M\} = 1$ for some M and all t), then we can use the dominated convergence theorem: We still have $X_t = \lim_{N \to \infty} S_N$, and all S_N's, as well as X_t, can be bounded by the random variable Y equal to the constant $\sum |\alpha|^j M = M/(1 - |\alpha|)$. The general case, again, lies beyond the scope of this book.

Example 8.7.15 (Moving Averages). Consider again the process $\{X_t\}$ of (say) prices of some commodity. In many cases, in addition to randomness, the process $\{X_t\}$ is subject to seasonal variations. For example, if t is measured in months and X_t represents the average monthly price of (say) tomatoes, then X_t varies with the period of 12. One of the methods of detecting trends (or prediction), which takes such periodic seasonal variation into account, is based on taking the averages of X_t over the most recent complete period. Thus, we may define

$$Y_t = \frac{X_t + X_{t-1} + \cdots + X_{t-11}}{12}.$$

Here a possible assumption about X_t may be

$$X_t = f_t + \xi_t,$$

where f_t is some nonrandom function with period 12 (i.e., $f_t = f_{t-12}$ for every t) and ξ_t are, as before, independent "shocks," representing random effects. Under this assumption

$$Y_t = C + \frac{\xi_t + \xi_{t-1} + \cdots + \xi_{t-11}}{12}$$

where $C = \frac{1}{12}[f_t + f_{t-1} + \cdots + f_{t-11}]$, which is a constant independent of t.

Clearly, we may assume without loss of generality that $E(\xi_t) = 0$ for all t. Let $\mathrm{Var}(\xi_t) = E(\xi_t^2) = \sigma^2 > 0$. Then

$$E(Y_t) = C, \quad \mathrm{Var}(Y_t) = \frac{\sigma^2}{12}$$

and

$$\mathrm{Cov}(Y_t, Y_{t-m}) = \mathrm{Cov}\left(\frac{\xi_t + \xi_{t-1} + \cdots \xi_{t-11}}{12}, \frac{\xi_{t-m} + \cdots \xi_{t-m-11}}{12}\right)$$

$$= \frac{1}{144}\mathrm{Cov}(\xi_t + \cdots + \xi_{t-11}, \xi_{t-m} + \cdots + \xi_{t-m-11}).$$

$$= K \cdot \frac{\sigma^2}{144},$$

where K is the number of overlapping terms in the sums $\xi_t + \xi_{t-1} + \cdots + \xi_{t-11}$ and $\xi_{t-m} + \cdots + \xi_{t-m-11}$. Obviously, the number of such terms is 0 if $m \geq 12$ and $12 - m$ otherwise.

Consequently, we have

$$\mathrm{Cov}(Y_t, Y_{t-m}) = \max\left(0, \frac{12 - m}{144}\sigma^2\right)$$

so that

$$\rho_{Y_t, Y_{t-m}} = \begin{cases} 1 - \frac{m}{12} & \text{for } m = 0, 1, \dots, 12 \\ 0 & \text{for } m > 12. \end{cases} \tag{8.52}$$

PROBLEMS

8.7.1 Find the expressions for $E(X)$ and $\mathrm{Var}(X)$ in terms of the moment generating function $m_X(s)$.

8.7.2 The moment generating function of a certain random variable is $m_X(t) = e^{4(e^t - 1)}$. Find $P\{|X - E(X)| \leq \sigma\}$.

8.7.3 Find the variance of X if $E(X) = 3$ and $E[X(X - 1)] = 52$.

8.7.4 Assume that $E(X) = 2, E(Y) = 1, E(X^2) = 10, E(Y^2) = 3$ and $E(XY) = c$.
   (i) Find $\mathrm{Var}(3X - 5Y)$ as a function of c.
   (ii) Find the coefficient of correlation between X and Y.
   (iii) Determine the range of values of c for which the assumptions of the problem are consistent.

8.7.5 Let X be a random variable with cdf given by

$$F(x) = \begin{cases} 0 & \text{for } x < 0 \\ \sqrt{x} & \text{for } 0 \le x \le 1 \\ 1 & \text{for } x > 1. \end{cases}$$

Find $E(X)$ and $\text{Var}(X)$.

8.7.6 A church lottery sells 8000 tickets, each for $2. The prize is a $5000 car.

    (i) What is the expectation of gain X of Mrs. Smith, who buys 1 ticket?

    (ii) What is the expectation of gain Y of Mr. Jones, who buys 20 tickets?

    (iii) Find $\text{Var}(X)$ and $\text{Var}(Y)$.

    (iv) Find the correlation coefficient between X and Y.

8.7.7 The random variable X has binomial distribution with mean 5 and standard deviation 2. Find $P\{X = 6\}$.

8.7.8 Let X and Y have mean 0, variance 1, and correlation ρ. Find the mean and variance of $X - \rho Y$ and the correlation between $X - \rho Y$ and Y.

8.7.9 Let X and Y have joint density f uniform on the triangle with vertices $(0,0), (1,0), (1,2)$. Find means and variances of X and Y and the correlation between X and Y.

8.7.10 Let X and Y have joint density f uniform on the quadrangle with vertices $(0,0), (a,0), (a,1)$, and $(2a,1)$, where $a > 0$. Find the means and variances of X and Y and correlations between X and Y.

8.7.11 Let X, Y be independent, with means μ_X, μ_Y and variances σ_X^2, σ_Y^2. Show that $\text{Var}(XY) = \sigma_X^2 \sigma_Y^2 + \sigma_X^2 \mu_Y^2 + \sigma_Y^2 \mu_X^2$.

8.7.12 Let $X_1, \ldots, X_n$ be independent random variables, having the same distribution with mean μ and variance σ^2. Let $\overline{X} = (X_1 + \cdots + X_n)/n$. Show that $E\{\sum_{i=1}^{n}(X_i - \overline{X})^2\} = (n-1)\sigma^2$. [*Hint:* Since $X_i - \overline{X} = (X_i - \mu) - (\overline{X} - \mu)$, we have $\sum_{i=1}^{n}(X_i - \overline{X})^2 = \sum_{i=1}^{n}(X_i - \mu)^2 - n(\overline{X} - \mu)^2$.]

8.8 CONDITIONAL EXPECTATION

In Chapter 4 we introduced the concept of conditional probability, and in Chapters 6 and 7 we extended this concept to the case of conditional distribution of a random variable. Let (X, Y) be a pair of random variables. We know how to find the conditional distribution of X given $Y = y$. We can therefore compute the expectation of this distribution (provided that it exists). For every particular value $Y = y$, the conditional expectation is a number $E(X|Y = y)$. We may therefore introduce:

Definition 8.8.1. Given two random variables X and Y, the *conditional expectation* $E(X|Y)$ is defined as the random variable that assumes the value $E(X|Y = y)$ when $Y = y$.

Example 8.8.1. Let Y have Poisson distribution with parameter λ, and given $Y = n$, let X be binomial with parameters n and p. We can think here (recall Example 7.3.3) of Y being the number of eggs laid by bird and of X as the number of those eggs that hatch (assuming that eggs hatch independently, each with probability p).

In this case, given Y we have $E(X|Y) = Yp$. This means that the expected number of eggs that hatch, if $Y = n$, is $E(X|Y = n) = np$.

Example 8.8.2. Let (X, Y) have the uniform distribution on the triangle with vertices $(0, 0)$, $(0, 1)$, and $(1, 1)$. Given Y, the distribution of X is uniform on the intersection of the line $Y = y$ and the support of the distribution. Since the expectation of the uniform distribution on an interval is the midpoint, we have $E(X|Y) = Y/2$ and similarly, $E(Y|X) = (1 + X)/2$, where $0 \leq X, Y \leq 1$.

Example 8.8.3. Consider now a "mixed" case, one variable being discrete and the other continuous. Let $P\{Y = 1\} = \alpha$ and $P\{Y = 2\} = 1 - \alpha$, and given Y, let X have the density

$$f(x|Y) = Yx^{Y-1}, \qquad 0 \leq x \leq 1.$$

In this case, if $Y = 1$, then X is uniform on $[0, 1]$, hence $E(X|Y = 1) = \frac{1}{2}$. If $Y = 2$, then X has density $f(x|Y = 2) = 2x$ on $[0, 1]$, hence $E(X|Y = 2) = \frac{2}{3}$.

On the other hand, given $X = x$, we have

$$P\{Y = 1|X = x\} = \frac{\alpha}{\alpha + 2x(1 - \alpha)}, \qquad P\{Y = 2|X = x\} = \frac{2x(1 - \alpha)}{\alpha + 2x(1 - \alpha)}.$$

This follows from Bayes' formula, interpreting the values of the density $f(x|Y)$ as infinitesimal probability $P\{X = x|Y\}dx$. Thus,

$$E(Y|X = x) = 1 \times P(Y = 1|X = x) + 2 \times P(Y = 2|X = x)$$
$$= \frac{\alpha + 4x(1 - \alpha)}{\alpha + 2x(1 - \alpha)}. \qquad \square$$

We shall prove now

Theorem 8.8.1. *For any random variables X, Y, we have*

$$E[E(X|Y)] = E(X) \tag{8.53}$$

provided that $E(X)$ exists.

Proof. We shall present the proof only in the case of continuous random variables. We have, letting $f_1(x)$ and $f_2(y)$ denote the marginal densities of the joint density $f(x, y)$ of (X, Y),

$$E(X|Y = y) = \int xg(x|Y = y)\, dx = \int x\frac{f(x, y)}{f_2(y)}\, dx,$$

hence, taking the expectation with respect to y, we obtain

$$E[E(X|Y)] = \int \left[\int \frac{xf(x, y)}{f_2(y)}\, dx\right] f_2(y)\, dy$$
$$= \int x \left[\int f(x, y)\, dy\right] dx = \int xf_1(x)\, dx$$
$$= E(X).$$

The interchange of the order of integration is legitimate in view of the assumption that $E(X)$ exists.

Example 8.8.4. In Example 8.8.1 we had $E(X|Y) = Yp$, so that remembering that Y has Poisson distribution with mean λ, we obtain $E(X) = E(Y) \cdot p = \lambda p$. This result follows also from the fact (see Example 7.3.3) that X has Poisson distribution with mean λp.

Example 8.8.5. In Example 8.8.2 we have
$$E(X) = E[E(X|Y)] = E(Y/2) \quad \text{and} \quad E(Y) = E[E(Y|X)] = E\left(\frac{1 + X}{2}\right)$$

which gives
$$E(X) = \frac{E(Y)}{2}, \qquad E(Y) = \frac{1}{2} + \frac{E(X)}{2}.$$

This system of two equations can easily be solved, leading to $\mu_X = \frac{1}{3}$ and $\mu_Y = \frac{2}{3}$.

Example 8.8.6. Finally, in Example 8.8.3 we have

$$E(X) = E[E(X|Y)] = \alpha E(X|Y = 1) + (1 - \alpha)E(X|Y = 2) = \frac{1}{2}\alpha + \frac{2}{3}(1 - \alpha).$$

On the other hand, since $f_1(x) = \alpha + 2x(1 - \alpha)$, we have

$$
\begin{aligned}
E(Y) &= E[E(Y|X)] \\
&= \int_0^1 \frac{\alpha + 4x(1 - \alpha)}{\alpha + 2x(1 - \alpha)} f_1(x)\, dx \\
&= \int_0^1 [\alpha + 4x(1 - \alpha)]\, dx = 2 - \alpha.
\end{aligned}
$$

$\square$

Theorem 8.8.1 may be very helpful in determining the expectation of a random variable X, by conditioning with respect to some other variable Y. Clearly, the choice of Y is crucial here if a simplification is to be achieved.

For calculation of variance we have the following analogue of Theorem 8.8.1.

Theorem 8.8.2. *For any random variables X, Y, if $E(X^2)$ exists, then*

$$\text{Var}(X) = E[\text{Var}(X|Y)] + \text{Var}[E(X|Y)]. \tag{8.54}$$

Proof. Let $\mu_X = E(X)$. We have

$$
\begin{aligned}
\text{Var}(X) &= E(X - \mu_X)^2 \\
&= E\left\{[X - E(X|Y)] + [E(X|Y) - \mu_X]\right\}^2 \\
&= E[X - E(X|Y)]^2 + 2E\left\{[X - E(X|Y)] \times [E(X|Y) - \mu_X]\right\} \\
&\quad + E[E(X|Y) - \mu_X]^2 \\
&= A + B + C.
\end{aligned}
$$

Conditioning on Y the terms A and C, we obtain using Theorem 8.8.1

$$
\begin{aligned}
A &= E\left\{E[X - E(X|Y)]^2 | Y\right\} = E[\text{Var}(X|Y)], \\
C &= E[E(X|Y) - \mu_X]^2 = \text{Var}[E(X|Y)],
\end{aligned}
$$

where the last equality follows from the fact that $\mu_X = E[E(X|Y)]$ is the expectation of random variable $E(X|Y)$. It remains to prove that $B = 0$.

We have, as before,

$$
\begin{aligned}
B &= E\left\{[X - E(X|Y)][E(X|Y) - \mu_X]\right\} \\
&= E\left(E\left\{[X - E(X|Y)] \times [E(X|Y) - \mu_X] | Y\right\}\right) \\
&= E\left(E[U(X, Y) \times V(Y)|Y]\right).
\end{aligned}
$$

The random variable $V(Y) = E(X|Y) - \mu_X$ is constant for every fixed value of Y. Consequently, it may be factored out, so that we have

$$B = E\{V(Y)E[U(X, Y)|Y]\}.$$

But $E[U(X,Y)|Y] = E[X - E(X|Y)|Y] = E(X|Y) - E(X|Y) = 0$. Thus,

$$B = E[V(Y) \cdot 0] = 0,$$

which completes the proof.

Example 8.8.7. For an application of Theorem 8.8.2, let us return again to Example 8.8.1. We have there $X \sim \text{BIN}(Y,p)$, so that $E(X|Y) = Yp$, $\text{Var}(X|Y) = Ypq$. Consequently, remembering that Y has Poisson distribution, so that $E(Y) = \text{Var}(Y) = \lambda$, we obtain

$$\begin{aligned}
\text{Var}(X) &= E[\text{Var}(X|Y)] + \text{Var}[E(X|Y)] \\
&= E(Ypq) + \text{Var}(Yp) \\
&= pq\lambda + p^2\lambda = \lambda p,
\end{aligned}$$

again in agreement with our finding from Example 7.3.3 that marginal distribution of X is Poisson with parameter λp. □

Finally, as another example of an application, we shall give the proof of the following (relatively little known) theorem connecting mean, median, and standard deviation of any random variable. The theorem is due to Hotelling and Solomons (1932). The present proof is a slight modification of the proof by O'Cinneide (1990).

Let X be a random variable with $E(X^2) < \infty$, and let μ and σ be its mean and standard deviation. Moreover, let m be any median of X, that is, a number such that $P(X \geq m) \geq \frac{1}{2}$ and $P(X \leq m) \geq \frac{1}{2}$. We then have the following theorem.

Theorem 8.8.3. *For any random variable X,*

$$|\mu - m| \leq \sigma,$$

that is, the mean is within one standard deviation of any median.

Proof. Let $\pi^- = P(X < m)$, $\pi^+ = P(X > m)$, and $\pi^0 = P(X = m)$, so that $\pi^- + \pi^0 \geq \frac{1}{2}$ and $\pi^0 + \pi^+ \geq \frac{1}{2}$. Let Y be the random variable defined as follows:

(a) If $\pi^0 = 0$, then

$$Y = \begin{cases} 1 & \text{if } X < m \\ 2 & \text{if } X > m. \end{cases} \tag{8.55}$$

(b) If $\pi^0 > 0$, then in addition to (8.55), we let

$$P\{Y = 1|X = m\} = \frac{1}{2} - \pi^-$$

$$P\{Y = 2|X = m\} = \frac{1}{2} - \pi^+.$$

Clearly, $P\{Y = 1\} = P\{Y = 2\} = \frac{1}{2}$.

For simplicity, put $\mu_i = E(X|Y = i), i = 1, 2$. Then $\mu = E(X) = E[E(X|Y)] = \frac{1}{2}E(X|Y = 1) + \frac{1}{2}E(X|Y = 2)$, that is,

$$0 \le \mu = \frac{1}{2}\mu_1 + \frac{1}{2}\mu_2. \tag{8.56}$$

Assume that $m \le \mu$ (the argument in case of the opposite inequality is analo-gous). Clearly, we have $\mu_1 \le m$, which gives, in view of (8.56), the inequality

$$\mu - m \le \frac{1}{2}(\mu_2 - \mu_1). \tag{8.57}$$

Using Theorem 8.8.2, we may now write

$$\sigma^2 = \text{Var}(X) = E[\text{Var}(X|Y)] + \text{Var}[E(X|Y)]$$
$$\ge \text{Var}[E(X|Y)] = \left(\frac{\mu_1 - \mu_2}{2}\right)^2 \ge (\mu - m)^2$$

so that $\sigma \ge |\mu - m|$, as asserted.

PROBLEMS

8.8.1 Let X and Y have the joint density uniform on the triangle with vertices $(0, 0), (2, 0)$, and $(3, 1)$. Find:
   (i) $E(X|Y)$ and $E(Y|X)$.
   (ii) $\text{Var}(X|Y)$ and $\text{Var}(Y|X)$.
   (iii) The expectations and variances of X and Y directly, and by using formulas (8.53) and (8.54) from Theorems 8.8.1 and 8.8.2.

8.8.2 Let X, Y be continuous random variables with a joint density f. Let $\text{Var}(Y|x)$ be the variance of the conditional distribution of Y given that $X = x$. Assume that $E(Y|X = x) = \mu$ for all x. Show that

$$\text{Var}(Y) = \int \text{Var}(Y|x)f_X(x)\, dx,$$

where $f_X(x)$ is the marginal density of X.

8.8.3 The number of accidents that occur in a factory in a week is a random variable Y with mean μ and variance σ^2. The numbers of persons injured in an accident are independent random variables, each with mean m and variance k^2. Find the mean and variance of the number of persons injured in a week.

8.9 CHEBYSHEV INEQUALITY

We start by proving the following theorem, which will give us some additional insight into this interpretation of variance and standard deviation.

Theorem 8.9.1. *If V is a random variable such that*

$$E(V) = 0, \qquad \text{Var}(V) = 1$$

then for every $t > 0$,

$$P\{|V| \geq t\} \leq 1/t^2. \tag{8.58}$$

Proof. We shall give the proof in case of continuous random variables, the proof in the discrete case (as well as in the general case) requiring only notational changes. We write, letting f denote the density of V,

$$1 = \text{Var}(V) = E(V^2)$$
$$= \int_{-\infty}^{+\infty} v^2 f(v)\, dv \geq \int_{|v| \geq t} v^2 f(v)\, dv$$
$$\geq t^2 \int_{|v| \geq t} f(v)\, dv = t^2 P\{|V| \geq t\}$$

which was to be shown. □

Clearly, inequality (8.58) is not informative for $t \leq 1$.

If X is a random variable with finite positive variance $\text{Var}(X) = \sigma^2 > 0$, then letting $\mu = E(X)$, we may always *standardize* X, by defining

$$V = \frac{X - \mu}{\sigma}. \tag{8.59}$$

Obviously, $E(V) = 0$ and $\text{Var}(V) = 1$; transformation of X into V amounts to introducing a new scale of expressing values of X, with the origin of the scale located at the expectation of X, and the unit of measurement being the standard deviation σ. Applying formula (8.58) to random variable (8.59), we obtain a traditional form of the Chebyshev inequality:

Theorem 8.9.2 (Chebyshev Inequality). *If $Var(X) = \sigma^2 < \infty$, then for every $t > 0$,*

$$P\{|X - \mu| \geq t\sigma\} \leq \frac{1}{t^2} \tag{8.60}$$

or, putting $\epsilon = t\sigma$,

$$P\{|X - \mu| \geq \epsilon\} \leq \frac{\sigma^2}{\epsilon^2}. \tag{8.61}$$

Both (8.61) and (8.60) show the role of standard deviation and variance: We obtain the bounds on probabilities of random variable X deviating from its mean μ by more than a certain amount [expressed as a multiple of σ in (8.60) or in original units in (8.61)].

For an analogue of Chebyshev inequality obtained without the assumption of finiteness of variance, see the problems for this section connected with the Markov inequality.

Example 8.9.1. For any random variable, the probability that it will deviate from its mean by more than three standard deviations is at most $\frac{1}{9}$. This probability, however, can be much less than $\frac{1}{9}$ for some random variables. For instance [see (9.57)] it equals 0.0026 for normal distribution.

It follows therefore that the *three-sigma rule* (which says that we can practically disregard the probability that a random variable will deviate from its mean by more than three standard deviations) should be used with caution. It may safely be used for random variables with distribution either normal, or close to normal. In the general case, however, the probability $\frac{1}{9}$ can hardly be disregarded. □

Let us now find the bounds given by the Chebyshev inequality in case of some other distributions.

Example 8.9.2. Suppose that we toss a coin 20 times. Let us estimate the probability that the number of heads will deviate from 10 by 3 or more (then we shall have less than 7 or more than 13 heads). Here $\mu = np = 10, \sigma^2 = npq = 5$ (see Examples 8.2.3 and 8.7.2). We have

$$P\{|X - 10| \geq 3\} \leq \frac{\sigma^2}{3^2} = \frac{5}{9} = 0.555.$$

The exact probability (see Table A.1) equals 0.1716. □

Example 8.9.3. Let X have the distribution uniform on the interval $(-a, a)$. Then $E(X) = \mu = 0$, and $\text{Var}(X) = E(X^2) = a^2/3$. The Chebyshev inequality (8.61) gives

$$P\{|X| \geq \epsilon\} \leq \frac{a^2}{3\epsilon^2}.$$

For $\epsilon = a$ the left-hand side is 0, while the right-hand side gives the bound $\frac{1}{3}$. □

Example 8.9.4. Let us find the estimates for exponential distribution with density $f(x) = \lambda e^{-\lambda x}, x > 0$. We have here $\mu = \sigma = 1/\lambda$ (see Example 8.7.4). Consequently,

$$P\left\{\left|X - \frac{1}{\lambda}\right| > \frac{t}{\lambda}\right\} \le \frac{1}{t^2}.$$

For instance, if $t = 1$ we obtain a noninformative bound 1, while the probability that X will deviate from $1/\lambda$ by more than $1/\lambda$ is (remember that exponential random variable can assume only positive values)

$$P\{X \ge 2/\lambda\} = e^{-\lambda(2/\lambda)} = e^{-2} = 0.13534. \qquad \square$$

As may be seen from these examples, the quality of bounds given by the Chebyshev inequality is not impressive. However, what is most important here is that Chebyshev inequality gives a universal bound, valid for *all* random variables with finite variance. In fact, the bound as such cannot be improved, as shown by the following example.

Example 8.9.5. Let $P\{X = -1\} = P\{X = 1\} = \frac{1}{2}$. Then $E(X) = 0$, $\mathrm{Var}(X) = E(X^2) = 1$. The Chebyshev inequality gives $P\{|Y| \ge t\} \le 1/t^2$. If $t = 1$, then both sides are equal 1. $\qquad \square$

The Chebyshev inequality can be used to make some inference about data sets of any nature, not necessarily related to random phenomena. We shall illustrate it by the following example.

Example 8.9.6. A meteorologist at a TV station kept the record of all maximal daily temperatures during a year. Suppose that the data, say $t_1, \ldots, t_{365}$, are such that $\sum t_i = 20075$ and $\sum t_i^2 = 1{,}113{,}250$. What is the minimal possible number of days in the year when the temperature was strictly between 46 and 64 degrees Fahrenheit?

SOLUTION. To answer this question, we may use the Chebyshev inequality as follows. Imagine that the numbers $t_1, \ldots, t_{365}$ are written on slips of paper and put into an urn (if a value t is repeated among $t_1, \ldots, t_{365}$, it is put into the urn on as many tickets as many times it appears, so that there are altogether 365 tickets in the urn).

Assume that one ticket is sampled at random. This leads to a random variable, say T, where $T =$ "the number on the chosen ticket." Obviously, $E(T) = \frac{1}{365}\sum t_i = 20075/365 = 55$, while $\sigma^2 = \mathrm{Var}(T) = \frac{1}{365}\sum t_i^2 - [E(T)]^2 = 25$. If M is the number of days with $46 < t_i < 64$, then

$$\frac{M}{365} = P\{46 < T < 64\} = P\{-9 < T - 55 < 9\}$$

$$= P\{|T - 55| < 9\} = P\left\{|T - E(T)| < \frac{9}{5} \cdot \sigma\right\}.$$

Consequently,

$$1 - \frac{M}{365} = P\left\{|T - E(T)| \geq \frac{9}{5}\sigma\right\} \leq \frac{1}{81/25},$$

which gives $M \geq 365(1 - \frac{25}{81}) = 252.3$. Thus, on at least 253 days of the year, the maximal daily temperature was within the range 46 to 64. This conclusion is valid for any data, real or artificial: in any set of 365 numbers whose sum is 20,075 and sum of squares is 1,113,250, at least 253 of those numbers are between 46 and 64, hence at most 112 can lie outside the range. □

Among the most important consequences of the Chebyshev inequality are the laws of large numbers. We explore this topic in some detail in Chapter 10. Here we illustrate the situation by the following example.

Example 8.9.7 (Binomial Distribution). Consider the binomial random variable $S_n = $ number of successes in n trials. We have then $E(S_n) = np$ and $\mathrm{Var}(S_n) = np(1 - p)$ when p is the probability of success. Consequently, $E(S_n/n) = p$, $\mathrm{Var}(S_n/n) = pq/n$, and the Chebyshev inequality gives

$$P\left\{\left|\frac{S_n}{n} - p\right| \geq \epsilon\right\} \leq \frac{pq}{n\epsilon^2}.$$

□

Letting $n \to \infty$, we obtain the following theorem:

Theorem 8.9.3. *If S_n has binomial distribution with parameter p, then for every $\epsilon > 0$*

$$\lim_{n\to\infty} P\left\{\left|\frac{S_n}{n} - p\right| \geq \epsilon\right\} = 0.$$

□

This theorem appears to explain why the empirical frequency of an event, S_n/n, approaches, as the number of trials n increases, the probability p of the event. It tells us that for any positive number ϵ it becomes increasingly unlikely that the empirical frequency will deviate from theoretical probability by more than ϵ.

Chebyshev Inequality assumes that the random variable has finite second moment. One ocasionally needs a bound for a tail of the distribution in situation when we cannot assume that second moment is finite. In such cases we have the following theorem.

Theorem 8.9.4 (Markov Inequality). *If X is a nonnegative random variable with $E(X) < \infty$, then for every $t > 0$*

$$P\{X \geq t\} \leq \frac{E(X)}{t} \qquad\qquad (8.62)$$

Proof. We give the proof for the discrete case. Let $x_1, x_2, \ldots$ be possible values of X. Then

$$E(X) = \sum_i x_i P(X = x_i) \geq \sum_{x_i \geq t} x_i P(X = x_i)$$

$$\geq t \sum_{x_i \geq t} P(X = x_i) = t P(X \geq t).$$

PROBLEMS

8.9.1 Assume that $P(X \geq 0) = 1$ and $P(X \geq 1) = 0.1$. Show that $E(X) \geq 0.2$.

8.9.2 Assume that $E(X) = 12, P(X \geq 14) = 0.12, P(X \leq 10) = 0.18$. Show that the standard deviation of X is at least 1.2.

8.9.3 Prove the Markov inequality when X is a continuous random variable.

8.9.4 Derive the Chebyshev inequality from the Markov inequality.

8.9.5 Let X be a random variable such that $E(e^{|X|}) = c < \infty$. Then for every $t > 0$,

$$P\{|X| \geq t\} \leq c e^{-t}.$$

8.9.6 Show that if X has an mgf bounded by the mgf of exponential distribution (i.e., $\lambda/(\lambda - t), t < \lambda$), then for $\lambda \epsilon > 1$ we have

$$P\{X \geq \epsilon\} \leq \lambda \epsilon e^{-(\lambda \epsilon - 1)}.$$

(*Hint:* Use the mgf of exponential distribution to obtain a bound for $P\{X \geq \epsilon\}$, then determine minimum of the bound.)

8.9.7 Let X have the Poisson distribution with mean λ. Show that

$$P\{X \leq \lambda/2\} \leq 4/\lambda \quad \text{and} \quad P\{X \geq 2\lambda\} \leq 1/\lambda.$$

8.9.8 Let X be a random variable such that the mgf $m_X(t)$ exists for all t. Use the same argument as in the proof of the Chebyshev inequality to show that $P\{X \geq y\} \leq e^{-ty} m_X(t)$, for any $t \geq 0$.

8.9.9 Show that if X has the Poisson distribution with mean λ, then $P(X \leq \lambda/2) \leq (2/e)^{\lambda/2}$ and $P(X \geq 2\lambda) \leq (e/4)^{\lambda}$. (*Hint:* Use the inequality in Problem 8.9.8, and find the minimum of the right-hand sides for t.)

8.10 KOLMOGOROV INEQUALITY*

In this section we prove an inequality due to Kolmogorov, which will be used in Chapter 10. The underlying assumption are as follows. We consider a sequence of independent random variables $X_1, X_2, \ldots$ with finite variances $\sigma_j^2 = \mathrm{Var}(X_j), j \geq 1$. Without loss of generality we assume that $E(X_j) = 0, j \geq 1$ [in the case of nonzero means we can always consider new variables $X_j' = X_j - E(X_j)$, which have zero mean]. We let $S_0 = 0, S_j = X_1 + X_2 + \cdots + X_j, j \geq 1$. We have the following:

Theorem 8.10.1 (Kolmogorov Inequality).   *Under the assumption above, for every $t > 0$ we have*

$$P\left\{ \max_{1 \leq j \leq n} |S_j| \geq t \right\} \leq \frac{\mathrm{Var}(S_n)}{t^2}. \tag{8.63}$$

Proof. Observe first that the right-hand side equals $(\sigma_1^2 + \ldots + \sigma_n^2)/t^2$ and is the same as the right-hand side in the Chebyshev inequality bound for $P\{|S_n| \geq t\}$. However, the left-hand side describes a considerably more complicated event than $\{|S_n| \geq t\}$. Indeed, the event $B = \{\max_{1 \leq j \leq n} |S_j| \geq t\}$ pertains to the whole "history" of the sequence $X_1, \ldots, X_n$, not only to the final sum S_n. For the proof, let us analyze the nature of this event and introduce some notation.

If the event B occurs, then one or more among the sums $|S_1|, \ldots, |S_n|$ must exceed t . Consequently, there must be the *first* index j when $|S_j| \geq t$. Let

$$B_j = \{|S_1| < t, \ldots, |S_{j-1}| < t, |S_j| \geq t\} \tag{8.64}$$

so that

$$B = \bigcup_{j=1}^{n} B_j. \tag{8.65}$$

Obviously, events B_j exclude one another.

We shall write expected value as an integral over the sample space S (instead of the Riemann, or Riemann–Stieltjes integral). This will simplify the notation considerably, by allowing us to write expressions such as $\int (X_1 + \cdots + X_j)^2 \, dP$ instead of j-fold integral $\int_{-\infty}^{+\infty} \cdots \int_{-\infty}^{+\infty} (x_1 + \cdots + x_j)^2 \, dF_1(x_1) \cdots dF_j(x_j)$, and so on.

To prove (8.63), let us write, as in the case of the Chebyshev inequality,

$$\mathrm{Var}(S_n) = E(S_n^2) = \int_S S_n^2 \, dP$$

$$\geq \int_B S_n^2 \, dP = \sum_{j=1}^{n} \int_{B_j} S_n^2 \, dP. \tag{8.66}$$

where $B \subset S$ is the event (8.65).

Let us now investigate the integrals on the right-hand side of (8.66). We have

$$\int_{B_j} S_n^2 \, dP = \int_{B_j} [S_j + (S_n - S_j)]^2 \, dP$$

$$= \int_{B_j} S_j^2 \, dP + 2 \int_{B_j} S_j(S_n - S_j) \, dP + \int_{B_j} (S_n - S_j)^2 \, dP$$

$$\geq \int_{B_j} S_j^2 \, dP + 2 \int_{B_j} S_j(S_n - S_j) \, dP. \tag{8.67}$$

To evaluate the last integral in (8.67), let $I_{B_j}(\omega)$ be the function equal 1 for $\omega \in B_j$ and 0 otherwise. We have, therefore,

$$\int_{B_j} S_j(S_n - S_j) \, dP = \int_S I_{B_j} S_j(X_{j+1} + \cdots + X_n) \, dP$$

$$= E\{I_{B_j} S_j(X_{j+1} + \cdots + X_n)\}.$$

Observe now that the random variable $I_{B_j} S_j$ depends only on $(X_1, \ldots X_j)$, so that it is independent of $X_{j+1} + \cdots + X_n$.

By Theorem 8.4.8 we have therefore

$$E\{I_{B_j} S_j(X_{j+1} + \cdots + X_n)\} = E\{I_{B_j} S_j\} E(X_{j+1} + \cdots + X_n) = 0$$

in view of the assumption that $EX_i = 0$ for all i. Consequently, (8.67) can be written, remembering that we have $|S_j| \geq t$ on B_j:

$$\int_{B_j} S_n^2 \, dP \geq \int_{B_j} S_j^2 \, dP \geq t^2 \int_{B_j} dP = t^2 P(B_j). \tag{8.68}$$

Combining (8.65), (8.66), and (8.68) we obtain

$$\text{Var}(S_n) \geq t^2 \sum_{j=1}^{n} P(B_j) = t^2 P(B),$$

which was to be shown.   □

REFERENCES

Feller, W., 1968. *An Introduction to Probability Theory and Its Applications*, vol. 2, Wiley, New York.

Hotelling, H., and L. M. Solomons, 1932. "The Limits of a Measure of Skewness," *Annals of Mathematical Statistics*, **3**, 141–142.

O'Cinneide, C. A., 1990. "The Mean Is Within One Standard Deviation of Any Median," *American Statistician*, **44**, 292–293.

CHAPTER 9

Some Probability Models

9.1 INTRODUCTION

In this chapter we review the most commonly used probability models. Almost all of these models were introduced in one form or another in the preceding chapters, mostly as examples. We now present them in a systematic way, adding new information in the process. We also provide references to the examples in preceding chapters, which should help the readers to identify the concepts introduced with situations that have already been discussed, although perhaps without using the appropriate terminology. The models presented now are related to (1) Bernoulli trials, (2) the Poisson process, and (3) normal distribution.

9.2 BERNOULLI TRIALS

Bernoulli trials refer to one of the most common types of situation modeled in the theory of probability, that of independent repetitions of some experiment in which we are interested in an event A that may occur in each trial with the same probability p. We refer to event A as "success," often calling the event A^c a "failure." The decision as to which of the two events of interest, A and A^c, is labeled success is quite arbitrary, and usually implies nothing about the nature of event A in any practical applications.

The Bernoulli random variable, a building block of the theory, is just a count of the number of successes in a single trial. Thus X is 1 or 0, depending on whether A or A^c occurred, and consequently, the distribution of X is given by

$$P\{X = 0\} = 1 - p, \qquad P\{X = 1\} = p. \tag{9.1}$$

It will be convenient to let $q = 1 - p$. The expectation and variance are easily found to be

$$E(X) = p, \qquad \text{Var}(X) = pq = p(1 - p).$$

Moreover, since $0^n = 0$ and $1^n = 1$, the Bernoulli random variable is the only

333

(nondegenerate) random variable X that satisfies the relations $X^n = X$ for all $n \geq 1$. Consequently, the moments of X satisfy for all $n \geq 1$ the condition

$$m_n = E(X^n) = E(X) = p,$$

and, as always, $m_0 = 1$. Consequently, using the fact that $\frac{d^n}{dt^n} m_X(t)|_{t=0} = p$ for all $n \geq 1$, the Taylor expansion of the moment generating function of X is

$$m_X(t) = m_0 + \sum_{n=1}^{\infty} m_n \frac{t^n}{n!} = 1 + p \sum_{n=1}^{\infty} \frac{t^n}{n!}$$
$$= 1 + p(e^t - 1) = pe^t + q.$$

This result, of course, could have been obtained in a much simpler way using the distribution (9.1) of X.

PROBLEMS

9.2.1 Assume that we score $Y = 1$ for a success and $Y = -1$ for a failure. Express Y as a function of the number X of successes in a single Bernoulli trial, and find moments $E(Y^n)$, $n = 1, 2, \ldots$.

9.2.2 Suppose that random variables $X_1, \ldots, X_n$ are independent, each with the same Bernoulli distribution. Find the probability that $X_1 = 1$ given that $\sum_{i=1}^{n} X_i = r$.

9.2.3 Suppose that the random variables $X_1, \ldots, X_n$ are independent, each with the same Bernoulli distribution. For fixed $1 \leq i < j \leq n$, find the covariance between X_i and X_j given that $X_1 + \cdots + X_n = r$.

9.3 BINOMIAL DISTRIBUTION

The binomial distribution plays the central role in probability theory, as a model of one of the most common situations, namely of a count of the total number of successes in n Bernoulli trials. In this chapter we let S_n denote the binomial random variable, and we use the representation

$$S_n = X_1 + \cdots + X_n \tag{9.2}$$

where $X_1, \ldots, X_n$ are the Bernoulli random variables describing the outcomes of successive trials (i.e., $X_i = 1$ or 0, depending whether the ith trial results in success or in failure).

We have encountered binomial random variable in the preceding chapters (e.g., in Examples 6.3.2, 7.3.3, 8.2.3, and 8.7.2). We also know that

$$P\{S_n = k\} = \binom{n}{k} p^k (1-p)^{n-k} = b(k; n, p) \qquad (9.3)$$

and [see (8.6) and (8.34)]

$$E(S_n) = np, \qquad \text{Var}(S_n) = np(1-p).$$

In the sequel we use the symbol $\text{BIN}(n, p)$ to denote binomial distribution with parameters n and p, so that we can say that S_n has distribution $\text{BIN}(n, p)$ or simply $S_n \sim \text{BIN}(n, p)$. Clearly, the Bernoulli random variable has the distribution $\text{BIN}(1, p)$.

To find the most likely number of successes, we proceed as follows. We have

$$\frac{b(k; n, p)}{b(k-1; n, p)} = \frac{\binom{n}{k} p^k (1-p)^{n-k}}{\binom{n}{k-1} p^{k-1}(1-p)^{n-k+1}} = \frac{n-k+1}{k} \cdot \frac{p}{q}.$$

Since $\dfrac{n-k+1}{k} \cdot \dfrac{p}{q} > 1$ if and only if $k < (n+1)p$, the probabilities $b(k; n, p)$ increase initially and then decrease. If $(n+1)p$ is not an integer, the unique maximum of probabilities $b(k; n, p)$ is attained at $k^* = [(n+1)p]$. If $(n+1)p$ is an integer, the maximum is attained at two values, $(n+1)p$ and $(n+1)p - 1$.

The moment generating function of binomial distribution has been found in Example 8.5.3, namely

$$m_{S_n}(t) = E(e^{tS_n}) = \sum_{k=0}^{n} e^{tk} \binom{n}{k} p^k q^{n-k} = (pe^t + q)^n.$$

The same result may be obtained using Theorem 8.5.5 and formula (9.2), by observing that $m_{S_n}(t) = [m_X(t)]^n$, where $m_X(t)$ is the moment generating function of the Bernoulli random variable.

Calculating the numerical values of binomial probabilities is simple in principle but may be cumbersome, especially when—as is often the case—we need to know probabilities of the form $P\{a \le S_n \le b\}$, which require calculating the individual probabilities $P\{S_n = k\}$ for all k between a and b.

For small and moderate n, the situation is somewhat remedied by tables (see Table A.1 in the Appendix). This table, as most of the binomial tables found in the literature, gives values of the cdf of S_n, that is, of probabilities $P\{S_n \le k\}$ for selected n, k, and p. Then $P\{a \le S_n \le b\} =$

$P\{S_n \leq b\} - P\{S_n \leq a - 1\}$, and the required probabilities are obtained by subtraction of the two terms (rather than adding $b - a + 1$ terms).

Since the cdf $P\{S_n \leq x\}$ attains value 1 at $x = n$ for every p, the nth row is omitted for each n. The rows for $x < n$ are also omitted if the values in all columns are equal 1 after rounding-off.

Example 9.3.1. Assume it is known (from past experience, research surveys, etc.) that 40% of buyers of FATCOW butter prefer unsalted butter and the remaining 60% prefer salted butter. The store expects to sell no more than 20 packages of FATCOW butter per day, so they put on the shelf 8 packs of unsalted and 12 packs of salted FATCOW butter. It happened that only 15 persons bought FATCOW butter on a given day, each person buying one package. What is the probability that all buyers found the kind of butter they wanted?

SOLUTION. We assume that the choices made by different people are independent (which may be a reasonable assumption; however, we advise the reader to think of a situation when assumption of independence may not be adequate). If we let S_{15} denote the number of persons (out of 15 buyers) who bought unsalted FATCOW butter, then S_{15} has binomial distribution BIN(15; 0.4). If all customers are to find the kind of butter they want, we must have $S_{15} \leq 8$ and also $15 - S_{15} \leq 12$, that is, $S_{15} \geq 3$. Thus, we need $P\{3 \leq S_{15} \leq 8\}$. Rather than computing the sum $\sum_{k=3}^{8} \binom{15}{k}(0.4)^k(0.6)^{15-k}$, one may find it convenient to observe that the required probability is $P\{S_{15} \leq 8\} - P\{S_{15} \leq 2\}$, and read the two values from Table A.1 for $n = 15$ and $p = 0.4$. □

Table A.1 gives binomial probabilities for selected values of p only up to $p = 0.5$. To use these tables in problems where $p > 0.5$ it suffices to observe the following: If S_n has distribution BIN(n, p), then $S_n' = n - S_n$ has the distribution BIN$(n, 1 - p)$. To put it simply, reverse the role of "successes" and "failures" and you obtain a binomial random variable with parameter $1 - p$.

Example 9.3.2 (A Warning). The probability that a Montesuma rose will blossom during the first year after planting is 80%. Mrs. Smith bought 20 bushes of Montesuma rose and planted them in her garden. What is the probability that less than 75% of her roses will blossom the first year?

SOLUTION. We regard the number S_{20} of rose bushes that will blossom in the first year as a random variable with distribution BIN(20, 0.8). We want probability $P\{S_{20} < 15\} = P\{S_{20} \leq 14\}$, which we can read from the tables of cdf of binomial distribution with $n = 20$ and $p = 0.8$. Most tables do not give probabilities for $p > 0.5$. In such a case we may use the fact that $S_{20}' = 20 - S_{20}$ is the number of roses that do not blossom, hence S_{20}' has binomial distribution BIN(20, 0.2). Now, $P\{S_{20} < 15\} = P\{20 - S'_{20} < 15\} = P\{S'_{20} > 5\} = 1 - P\{S'_{20} \leq 4\} = 1 - 0.6296 = 0.3704$, the latter value obtained from Table A.1.

A moment of reflection should suffice to realize that the *real* answer to this problem is that the required probability simply cannot be computed. Indeed, the probability 0.8 of blossoming in the first year after planting represents presumably some kind of overall success average, obtained from data for various years, soil conditions, gardening techniques, and so on. The roses of Mrs. Smith are likely to be subjected to the same conditions, same type of care, and so on. Consequently, the probability of success for Mrs. Smith's garden need not be 0.8, and—more important—blossoming of her roses are unlikely to be independent of one another.

A more realistic assumption here might be that the probability of blossoming p is random. Keeping the assumption of independence for every p, this will make the blossoming of Mrs. Smith's roses exchangeable (but not independent) events (see Chapter 4). Thus, the solution above is obtained only at the cost of accepting the assumption of independence and $p = 0.8$.

This example is placed here to make the reader aware that application of probability to real situations is a process that requires accepting some assumptions. Justification of these assumptions is a rather delicate issue, discussed in some detail in the part of the book concerned with statistics. □

A difficulty with tabulating the binomial distribution lies in the fact that it is necessary to have a separate table for each pair n and p, for $0 < p \leq 0.5$. Moreover, the tables become increasingly cumbersome when n increases. As we show in subsequent sections, the situation can be remedied substantially for large n. As a preparation for the approximation introduced later in this chapter, let us consider an example.

Example 9.3.3. Assume that about 1 birth in 80 is a twin birth. What is the probability that there will be no twin births among the next 200 births in the maternity ward of a given hospital?

SOLUTION. Clearly, we have here a binomial situation: If S_{200} denotes the number of twin births among the next 200 births, we need $P\{S_{200} = 0\}$, where $S_{200} \sim \text{BIN}(200; 1/80)$. Thus, we may write, remembering that $\lim_{n\to\infty}(1 - c/n)^n = e^{-c}$,

$$P\{S_{200} = 0\} = \binom{200}{0}\left(\frac{1}{80}\right)^0 \left(\frac{79}{80}\right)^{200} = \left(1 - \frac{1}{80}\right)^{200}$$

$$= \left(1 - \frac{200/80}{200}\right)^{200} \approx e^{-200/80} = 0.0821. \tag{9.4}$$

□

The approximation works well if the probability of success p is small and the number of trials n is large. Specifically, the error, equal to $|P\{S_n = 0\} - e^{-np}|$, depends on the value of the product np. As we shall see, (9.4) is an example of the Poisson approximation to the binomial distribution, which we discuss later.

At the end of this section we give an example that shows some rather unexpected possibilities of application of probability. The problem concerns the identification of the remnants of a Polish ruler of Piast dynasty.

Example 9.3.4. Prince Bolesław was born in 1086. He was later called Krzywousty (Wrymouth), because of a jaw deformity acquired either at birth or in childhood; this deformity caused a speech impairment, as recorded in the chronicles of Polish history written by a monk residing at Bolesław's father court. Bolesław ruled from 1102 to 1138. He was buried next to his father in the cathedral of Płock, a Polish city some 100 miles northwest of Warsaw. According to some later written sources, however, Bolesław died in Sochaczew, about 50 miles south of Płock (and presumably was buried there).

As is often the case with conflicting sources, the historians are divided in their opinions as to where Bolesław's tomb is: in Sochaczew (which has no buildings from that period left), or in Płock, where the cathedral is still standing, with two tombs marked as those of Bolesław and his father.

In a history-conscious country such as Poland, the resting place of one of its rulers, especially a famous[*] one, is a matter of some importance, and the obvious possibility that Bolesław died in Sochaczew and that his body was carried to Płock immediately or at a later time cannot be accepted without supporting evidence. In this situation, a decision was made to open the tomb in Płock. Somewhat surprisingly, inside were found remnants of 18 people, all mixed together, so that skulls could not be matched with other bones. One of the 18 skulls was of an adult who had a jaw defect.

Inspection of the cathedral records cleared up part of the mystery: It appeared that at a later time the residing bishop ordered all the tombs in the cathedral to be opened, and for some reason (perhaps an impending enemy attack?) all remnants were put into one tomb. At any rate, the identity of all persons buried in the cathedral is known, and there is no evidence suggesting that any of them (other than Bolesław) had a skull deformation.

Carbon dating was performed on all 18 skeletons and it was found that two of them come from the eleventh or twelfth century (the precision of carbon

---

[*] Bolesław Krzywousty is best known for his consistent policy of resisting—both on the battlefield and through various treaties and alliances—the attempts to impose German domination on Poland. Spending most of his youth evading the attempts on his life by supporters of his half-brother, Bolesław left a will that carefully spelled out the laws of succession among the offspring of his four sons, among whom he divided the country. It is not clear, however, whether the princes would obey the will of Bolesław, since the situation became almost immediately complicated when his fifth son was born posthumously. A period of local squabbles and wars followed, and it was only in 1305 that the country was united again under one ruler.

dating is ± 60 years or so). However, the dating is destructive and could not be performed on skulls.

Intuitively, all these findings together strongly suggest that the deformed skull is the skull of Bolesław Krzywousty. Nobody would question this if it were not for the single fact that a written source claims that he died not in Płock but in Sochaczew. The problem facing a statistician is: Can one assess the likelihood that the deformed skull is that of Bolesław Krzywousty?

At first it might seem that there is nothing here that could be quantified, except in a highly subjective way, and therefore the problem lies outside the applicability of statistical methods. One could, however, argue as follows (this argument is due to David Blackwell, personal communication). The state of medicine was very much the same throughout the Middle Ages. Thus, the probability of surviving damage to the skull (at birth or in childhood) was constant for at least several hundred years. Let this probability be p (actually, we are interested in p not for the general population but for the better-off: specifically, only for those who are candidates for burial in a cathedral).

The value of p can be estimated: By now, all tombs in European cathedrals from the period in question have been opened, inspected, catalogued, and so on, and p can be taken as the frequency of deformed skulls among the skulls found in such tombs (this estimation may not be easy, as the data are scattered among various places; the point is, however, that such estimation is feasible).

Now, in the case of Bolesław Krzywousty, we know that he had a deformed skull. Consequently, if he is buried in Płock, then we can account for one deformed skull, and the probability of finding 17 healthy skulls is $(1 - p)^{17}$. On the other hand, if he is *not* buried in Płock, then the probability of a particular finding (1 skull deformed, 17 healthy) is given by the binomial probability of obtaining exactly one success in 18 trials, hence is equal to $18p(1 - p)^{17}$. The ratio of these two probabilities, $1/18p$, can be taken as the odds for the event that Bolesław Krzywousty is buried in the cathedral in Płock.

This result agrees with common intuition. If p is small, then the odds $1/18p$ are high, as one should expect: The more uncommon it was to have the same kind of deformity as that of Bolesław Krzywousty, the more certain one can be that the skull in question belonged to him.

PROBLEMS

9.3.1 Label the statements below as true or false:

    (i) Suppose that 6% of all cars in a given city are Toyotas. Then the probability that there are 4 Toyotas in a row of 12 cars parked in the municipal parking is $\binom{12}{4}(0.06)^4(0.94)^8$.

    (ii) Suppose that 6% of all Europeans are French. Then the probability that in a random sample of 12 inhabitants of a major European capital there are 4 Frenchmen is $\binom{12}{4}(0.06)^4(0.94)^8$.

9.3.2 An experiment consists of tossing a fair coin 13 times. Such an experiment is repeated 17 times. Find the probability that in a majority of repetitions of the experiment the tails will be in minority.

9.3.3 Show that if S_n is a binomial random variable, then for $k = 1, 2, \ldots, n$,

$$P\{S_n = k\} = \frac{(n - k + 1)p}{k(1 - p)} P\{S_n = k - 1\}.$$

9.3.4 Two players (or two teams) are negotiating the rules for determining the championship. The two possibilities are "best of five" or "best of seven." This means that whoever wins three (respectively, four) games is the champion. Assume that games are independent and that the probability of winning a game by the first player (there are no ties) is p. For what values of p should the first player favor the scheme "best out of five"?

9.3.5 Show that for the binomial distribution we have

$$P\{S_n \leq k\} = (n - k)\binom{n}{k} \int_0^{1-p} x^{n-k-1}(1 - x)^k \, dx.$$

9.3.6 Assume that X_1 and X_2 are independent random variables, with binomial distributions with parameters n_1, p and n_2, p, respectively. Find the correlation coefficient between X_1 and $X_1 + X_2$.

9.4 GEOMETRIC DISTRIBUTION

Another random variable connected with Bernoulli trials is the number X of failures preceding the first success. We have encountered this variable in Examples 6.3.3 and 8.2.4. Clearly,

$$P\{X = k\} = q^k p, \qquad k = 0, 1, 2, \ldots. \tag{9.5}$$

Since the tails of the distribution of X, that is,

$$P\{X \geq k\} = q^k p + q^{k+1} p + \cdots = p \frac{q^k}{1 - q} = q^k, \tag{9.6}$$

are obtained by summing the geometric series, the distribution (9.5) is often called *geometric*.

A distribution closely associated with (9.5) is that of the number Y of trials up to and including the first success, so that

$$P\{Y = n\} = P\{X = n - 1\} = q^{n-1} p, \qquad n = 1, 2, \ldots \tag{9.7}$$

Actually, both distributions, (9.5) and (9.7), are called geometric in the literature. Since $Y = X + 1$, these distributions are closely related, and most results are valid under either definition. In this book we use both definitions, specifying (in cases where it would make a difference), whether (9.5) or (9.7) is used. We use both definitions deliberately, not to confuse readers but to make them flexible in using terminology that is not well established. We shall also use the same symbol for both distributions (9.5) and (9.7), namely we shall use the notation $GEO(p)$ to denote the geometric distribution with probability of success being p.

The expected value of the geometric distribution was found in Example 8.2.4. We have, for distribution (9.7),

$$E(Y) = \frac{1}{p},\qquad\qquad(9.8)$$

so that for distribution (9.5) we obtain

$$E(X) = E(Y - 1) = \frac{1}{p} - 1 = \frac{q}{p}.\qquad\qquad(9.9)$$

The variance of geometric distribution is the same for both X and Y (see Theorem 8.7.1) and is easily found to be

$$\mathrm{Var}(X) = \mathrm{Var}(Y) = \frac{q}{p^2}.\qquad\qquad(9.10)$$

Finally, the mgf of the geometric distribution equals

$$m_X(t) = \frac{p}{1 - qe^t}$$

for $qe^t < 1$, hence for $t < -\log q$. Consequently,

$$m_Y(t) = m_{X+1}(t) = \frac{pe^t}{1 - qe^t},$$

by Theorem 8.5.4.

Let us observe that for all $m, n = 0, 1, 2, \ldots$ we have

$$P\{X \geq m + n | X \geq m\} = P\{X \geq n\}.\qquad\qquad(9.11)$$

Indeed, since $X \geq m + n$ implies that $X \geq m$, we may write

$$P\{X \geq m + n | X \geq m\} = \frac{P\{X \geq m + n\}}{P\{X \geq m\}} = \frac{q^{m+n}}{q^m}$$
$$= q^n = P\{X \geq n\}.$$

Formula (9.11) is said to express the *memoryless property* of geometric distribution: If waiting time for the first success is at least m, then the probability that it will be at least $m + n$ is the same as the probability that the waiting time for the first success will be at least n.

Let us observe that formula (9.11) is valid for random variable X, namely the number of failures preceding the first success but not for the random variable Y (number of trials up to and including the first success). Indeed, for Y we have the corresponding formula

$$P\{Y > m + n | Y > m\} = P\{Y > n\} \tag{9.12}$$

with the strict inequalities.

It is of interest that formula (9.11) characterizes geometric distribution (9.5); equivalently, (9.12) characterizes distribution (9.7). We shall prove

Theorem 9.4.1. *If Y is a discrete random variable that may assume values $1, 2, \ldots$ and satisfies (9.12), then Y has the distribution (9.7) for some p and $q = 1 - p$.*

Proof. Let $\pi_k = P\{Y = k\}$ and $\eta_k = \pi_{k+1} + \pi_{k+2} + \cdots = P\{Y > k\}$. Then (9.12) means that we must have $\eta_{m+n}/\eta_m = \eta_n$ for all $m, n = 1, 2, \ldots$. In particular, $\eta_2 = \eta_1^2$, and by induction, $\eta_k = \eta_1^k$. Consequently, $\pi_k = \eta_k - \eta_{k+1} = \eta_1^k - \eta_1^{k+1} = \eta_1^k(1 - \eta_1)$ for all k, which means that Y has geometric distribution with $q = \eta_1 = P\{Y > 1\}$. $\square$

In fact, it is sufficient to require that (9.12) holds for $m = 1$ and all n or for $n = 1$ and for all m. What is more remarkable, one can replace m by a random variable. We have (see Srivastava, 1981):

Theorem 9.4.2. *Let U be a random variable that assumes only positive integer values. If*

$$P\{Y > U + n | Y > U\} = P\{Y > n\}, \tag{9.13}$$

then Y has geometric distribution (9.7) for some p and $q = 1 - p$.

Example 9.4.1. In Chapter 5 we discussed Markov chains, being processes that describe the evolution of some system which may at any time be in one of a specified set of states and which may change its state at times $t = 1, 2, \ldots$. The main assumption is that if at some time t the system is in state i, then the probability that it will pass to state j at time $t + 1$ is p_{ij}, regardless of the history of transitions prior to time t. This means that given the present (state at time t), the future and the past are independent. Suppose now that at some time t the system enters state i, and let T_i be the duration of the current visit to state i. More precisely, we define $T_i = k$ if the state at times $t, t+1, t+2, \ldots, t+k$ is i but the state at time $t+k+1$ is different from i. Then we have

$$P\{T_i = k\} = p_{ii}^k(1 - p_{ii}), \qquad k = 1, 2, \ldots$$

and we see that the duration of a visit in state i has geometric distribution with $q = p_{ii}$ and $p = 1 - p_{ii}$.

The memoryless property of geometric distribution is in fact the Markov property. Indeed, suppose that the system stayed in state i for m units of time prior to time t. Then the probability of staying there for at least additional n units is the same as of staying in state i for at least $m + n$ units of time. Given the present state, the future (in particular the duration of remaining stay in the present state) is independent of the past (in particular, on how long the system has already stayed in the present state).

Theorem 9.4.2 asserts a seemingly identical property, but a moment of reflection shows that the condition is now much stronger: The "present" is not some fixed time but is a time that may be random. In particular, this randomness may be affected by the process itself.

To give an example, let $X = T_i$, the time of remaining in the state i just entered, and let U be the longest visit in state i recorded thus far (so that U depends on the history of the process up to the present time). Theorem 9.4.2 asserts, in particular, the following: Given that the duration of the present state in state i will break the record $(T_i > U)$, the probability of breaking the record by at least 3 units (i.e., $T_i > U + 3$) is the same as the probability of a visit lasting longer than 3 units $(T_i > 3)$. This property stands in contrast with what one observes in sport records: The consecutive improvements of the world record in any discipline tend to be smaller.

The property asserted in Theorem 9.4.2 is generally called a *strong* Markov property: It means that given the present, the future is independent of the past, even if the "present" is selected at random, as long as this randomness depends only on the past and not on the future. □

Theorems 9.4.1 and 9.4.2 are examples of *characterizations* of probability distributions. This kind of theorem singles out a certain property of a type of distribution and shows that this property is valid only for distributions of this type (in the case above, the memoryless property for geometric distribution).

The characterization theorems, quite apart from their intrinsic interest as mathematical facts, are of great practical value. As we shall see in the chapters on statistics, if one is interested in more than merely summarizing and presenting statistical data, it is useful to regard the data (results of observations, experiments, surveys, etc.) as values of random variables. Roughly speaking, the more we know about the distribution of these random variables, the better is our understanding of the phenomenon studied, as well as the better our possibilities of prediction and control.

To fix the idea, imagine that a statistician's data can be regarded as independent observations of some random variable X. It happens quite often that identification of the distribution of X is accomplished in two steps. The first step consists of identifying a class of distributions that contain the distribution of X. The second step is the identification of the particular distribution in this class.

Now, sometimes the first step is easy: The statistician knows the type of distribution of X because she *makes* it belong to a given type by appropriate sampling. For instance, in independent sampling, with each element sampled being classified as being in one of the two categories, such as defective/nondefective, treatment successful/treatment unsuccessful, and so on, the total number X of elements of a given type is binomial, as long as the sample is really random. Out of two parameters, n is controlled by the experimenter; p (the fraction of elements of a given kind in the population) is unknown and is to be estimated.

In many cases, however, determining the class of distributions that contains distribution of X is not so simple. In such situations one can sometimes use a characterization theorem: If one knows that the distribution has some property that turns out to be characteristic for a given class, then the distribution must be in that class (e.g., if one knows that the distribution of X is discrete and memoryless, then X must be geometric).

We give one more example connected with geometric distribution.

Example 9.4.2 (Family Planning).   Assume that the genders of consecutive children born to the same parents are independent, with the probability of a child being a girl equal to π. We disregard twin births and consider family sizes under various plans.

Plan 1. The couple favors girls (say), and it decides to stop having children as soon as their first girl is born. Let ξ_1 be the number of children according to this plan. Then ξ_1 is a geometric random variable with $P\{\xi_1 = n\} = (1 - \pi)^{n-1}\pi$, so that $E(\xi_1) = 1/\pi$. If the chances of a child being a boy are the same as those of being a girl, we have $\pi = 0.5$ and $E(\xi_1) = 2$.

Plan 2. The couple decides to have children until they have a boy and a girl and then stop. Let ξ_2 be the number of children under this plan. To determine $P(\xi_2 = n)$, let G and B denote the event "first child is a girl (boy)." Then, by formula (4.9) for total probability, for $n \geq 2$ we have

$$P(\xi_2 = n) = \pi P(\xi_2 = n|G) + (1 - \pi)P(\xi_2 = n|B)$$
$$= \pi P\{Y^B = n - 1\} + (1 - \pi)P\{Y^G = n - 1\}.$$

Here Y^B and Y^G are the numbers of children (excluding the first) a family will have until the first boy (girl) is born. Clearly, Y^B and Y^G have the distribution given by (9.6), with probability of success being $1 - \pi$ for Y^B and π for Y^G. Thus, for $n = 2, 3, \ldots,$

$$P\{\xi_2 = n\} = \pi[\pi^{n-2}(1 - \pi)] + (1 - \pi)[(1 - \pi)^{n-2}\pi]$$
$$= \pi^{n-1}(1 - \pi) + (1 - \pi)^{n-1}\pi.$$

To find the expected number of children under plan 2 we can write

$$E(\xi_2) = \pi \sum_{n=2}^{\infty} nP\{Y^B = n - 1\} + (1 - \pi) \sum_{n=2}^{\infty} nP\{Y^G = n - 1\},$$

and putting $k = n - 1$,

$$\sum_{n=2}^{\infty} nP\{Y^B = n - 1\} = \sum_{n=2}^{\infty} [(n - 1) + 1]P\{Y^B = n - 1\}$$
$$= \sum_{k=1}^{\infty} kP\{Y^B = k\} + \sum_{k=1}^{\infty} P\{Y^B = k\}$$
$$= \frac{1}{1 - \pi} + 1.$$

By symmetry, $\sum_{n=2}^{\infty} nP\{Y^G = n - 1\} = 1/\pi + 1$, so that

$$E(\xi_2) = \pi \left(\frac{1}{1 - \pi} + 1 \right) + (1 - \pi) \left(\frac{1}{\pi} + 1 \right) = \frac{\pi}{1 - \pi} + \frac{1 - \pi}{\pi} + 1$$
$$= \frac{1}{\pi(1 - \pi)} - 1.$$

Again, if $\pi = \frac{1}{2}$, then $E(\xi_2) = 3$.

PROBLEMS

9.4.1 Six dice are tossed simultaneously until for the first time all of them show the same face. Find $E(U)$ and $\text{Var}(U)$, where U is the number of tosses until this happens.

9.4.2 Assume that the probability that a birth is a multiple one (twins, triplets, etc.) is π. Given that a birth is a multiple one, probabilities $\alpha_2, \alpha_3, \ldots$ of twins, triplets, $\ldots$, satisfy the condition $\alpha_{k+1} = \gamma \alpha_k$. Find π, α_2, and γ if it is known that the expected number of children born in 100 births is $100 + c$, and the expected number of single births observed before recording a multiple birth is M (assume M and c given).

9.4.3 Assume that the probability of twins being identical is β, and that the genders of children are determined independently, with the probability of a boy being b (possibly $b \neq \frac{1}{2}$). Find the expected number of twin births recorded in the hospital before the first pair of:

   (i) Boys.

   (ii) Girls.

   (iii) Different genders.

   Note that identical twins must be of the same gender. (see Example 4.4.5).

9.4.4 Find an expression for the cdf of a geometric random variable.

9.4.5 Let X be a random variable with a geometric distribution. Find

$$a_m = P\{X \text{ is divisible by } m\}, \qquad m = 2, 3, \ldots.$$

9.5 NEGATIVE BINOMIAL DISTRIBUTION

The geometric distribution discussed in Section 9.4 allows an immediate and natural extension: Rather than to consider the number of Bernoulli trials up to the first success, we may consider the number Y of Bernoulli trials up to and including the rth success.

In analogy with the geometric distribution, we consider also the random variable X defined as the number of failures preceding the rth success. We start by deriving the probability distribution of the random variables X and Y.

Clearly, the possible values of Y are integers $r, r+1, r+2, \ldots$. The event $\{Y = n\}$ occurs if:

   (a) The nth trial results in success.

   (b) The first $n - 1$ trials give exactly $r - 1$ successes (and therefore $n - r$ failures).

Indeed, the conjunction of (a) and (b) ensures that the rth success occurs at trial n. The events (a) and (b) are independent (since their occurrence is determined by disjoint sets of trials), and their probabilities are p and $\binom{n-1}{r-1} p^{r-1} q^{n-r}$, respectively. Consequently,

$$P\{Y = n\} = \binom{n-1}{r-1} p^r q^{n-r}, \qquad n = r, r+1, r+2, \ldots. \tag{9.14}$$

The number X of failures preceding rth success satisfies the relation

$$X + r = Y,$$

so that

$$P\{X = k\} = P\{Y = k + r\} = \binom{k + r - 1}{r - 1} p^r q^k, \qquad k = 0, 1, 2, \ldots. \quad (9.15)$$

Both random variables X and Y are referred to as having a *negative binomial* distribution; in either case, the distribution depends on two parameters, r and p. We shall use the symbol $NB(r, p)$ to denote the negative binomial distribution, whether defined by (9.14) or (9.15).

Let us verify that (9.15) and (9.14) are indeed probability distributions. Thus, we must show that

$$\sum_{k=0}^{\infty} \binom{k + r - 1}{r - 1} p^r q^k = 1. \quad (9.16)$$

Let us observe first that, using formulas (3.14) and (3.26),

$$
\begin{aligned}
\binom{k + r - 1}{r - 1} &= \binom{r - 1 + k}{k} \\
&= \frac{(r - 1 + k)(r - 1 + k - 1) \cdots (r - 1 + k - k + 1)}{k!} \\
&= \frac{r(r + 1) \cdots (r + k - 1)}{k!} \\
&= \frac{(-1)^k (-r)(-r - 1) \cdots (-r - k + 1)}{k!} \\
&= (-1)^k \binom{-r}{k}.
\end{aligned}
$$

Thus, we can write, using Newton's formula (3.28),

$$
\begin{aligned}
\sum_{k=0}^{\infty} \binom{k + r - 1}{r - 1} p^r q^k &= p^r \sum_{k=0}^{\infty} (-1)^k \binom{-r}{k} q^k \\
&= p^r (1 - q)^{-r} = 1,
\end{aligned}
$$

which was to be shown. Obviously, the proof that

$$\sum_{n=r}^{\infty} \binom{n - 1}{r - 1} p^r q^{n-r} = 1$$

is similar, and we can omit the details.

To determine the mean and variance of a random variable with negative binomial distribution, one can proceed in several ways, of which we shall demonstrate two. The first uses direct calculations, while the second uses a representation as a sum of simpler random variables. Observe that since $Y + r = X$, we have $E(Y) + r = E(X)$ and $\text{Var}(Y) = \text{Var}(X)$ (the latter property being a consequence of Theorem 8.7.1). Thus, it suffices to study one of the random variables X and Y, say X.

We begin by finding the moment generating function of X. We have, proceeding as in proof of (9.16),

$$m_X(t) = Ee^{Xt} = \sum_{k=0}^{\infty} e^{kt} \binom{k+r-1}{r-1} p^r q^k = p^r \sum_{k=0}^{\infty} \binom{k+r-1}{r-1} (qe^t)^k$$

$$= p^r \sum_{k=0}^{\infty} \binom{-r}{k} (-qe^t)^k = \frac{p^r}{(1-qe^t)^r},$$

provided that $|-qe^t| < 1$ [i.e., $t < \log(1/q)$].

Consequently,

$$E(X) = m'_X(0) = \frac{rq}{p},$$

and therefore

$$E(Y) = r + E(X) = \frac{r}{p}. \tag{9.17}$$

After some algebra, we also obtain

$$\text{Var}(X) = \text{Var}(Y) = m''_X(0) - [E(X)]^2 = \frac{rq}{p^2}. \tag{9.18}$$

Example 9.5.1. A salesman calls prospective buyers to make a sales pitch. Assume that the outcomes of consecutive calls are independent and that on each call he has a 15% chance of making a sale. His daily goal is to make 3 sales, and he can make only 20 calls in a day. What is the probability that he achieves his goal in 18 trials? What is the probability that he does not achieve his daily goal?

SOLUTION. The "strategy" of solving this kind of problem is to identify the type of distribution that we are analyzing. Assuming that we identify the situation as Bernoulli trials (i.e., repeated independent trials with the same probability p of success), the problem most typically concerns either a binomial or a negative binomial distribution. The crucial question here is: Is the number of trials fixed (and then the number of successes is random) or is the number of successes fixed (and the number of trials is random)? In the first case we have the binomial distribution, in the second case, the negative binomial distribution.

For our salesman, we want the probability that his third sale (success) comes at trial 18. Thus, the number of successes $r = 3$ is fixed (this is the salesman's goal), and it is the number of calls that is random. We have $p =$

0.15 and we ask for $P\{X = 18\}$, where X is the number of trials up to and including the third success. Substitution in formula (9.14) gives

$$P\{X = 18\} = \binom{17}{2}(0.15)^3(0.85)^{15} = 0.0401.$$

The second question is about the probability of the salesman not attaining his daily goal. Here the number of trials is fixed ($n = 20$) and we can treat the problem as involving the binomial distribution. Thus, if S_{20} is the number of successes in 20 trials, the salesman does not achieve his goal if $S_{20} \leq 2$. The answer is, therefore,

$$P\{S_{20} \leq 2\} = P\{S_{20} = 0\} + P\{S_{20} = 1\} + P\{S_{20} = 2\}$$
$$= \sum_{k=0}^{2} \binom{20}{k}(0.15)^k(0.85)^{20-k} = 0.4049.$$

We can also use the negative binomial distribution here. If X is the number of trials up to and including the third success, then the salesman does not attain his goal if $X > 20$, so that the answer is

$$P\{X > 20\} = \sum_{n=21}^{\infty} P\{X = n\} = \sum_{n=21}^{\infty} \binom{n-1}{2}(0.15)^3(0.85)^{n-3},$$

a quantity much harder to evaluate numerically than $P\{S_{20} \leq 2\}$.   □

The example above suggests that we have the following useful identity connecting the probabilities in binomial and negative binomial distributions with the same probability p of success. In what follows S_n stands for a binomial random variable (number of successes in n trials) and $Y^{(r)}$ stands for a negative binomial random variable, namely the number of trials up to and including the rth success.

Theorem 9.5.1. *We have, for every k,*

$$P\{Y^{(r)} > k\} = P\{S_k < r\}. \tag{9.19}$$

Proof. Observe that both sides of (9.19) refer to probability of the same event: The waiting time for the rth success exceeds k if and only if fewer than r successes occur in the first k trials.   □

As a word of warning, it is worth realizing that $Y^{(r)}$ and S_k are both discrete random variables, and it matters whether or not the inequalities are strict. Thus, (9.19) can be written in any of the forms, such as

$$P\{Y^{(r)} > k\} = P\{S_k \leq r - 1\}, \; P\{Y^{(r)} < k + 1\} = P\{S_k \geq r\},$$

and so on.

An inspection of the proof of the formula (9.16) shows that the fact that r is an integer was never used. Formally, therefore, a negative binomial distribution is defined for any $r > 0$ and $0 < p < 1$ (although the interpretation of the probabilities in terms of Bernoulli trials is no longer valid).

Let us also note that the mgf of the negative binomial distribution is the rth power of mgf of the geometric distribution. For integer values of r this fact shows that the random variable $X^{(r)}$ (respectively, $Y^{(r)}$) is a sum of r independent geometric random variables (of the form X or Y, depending on whether we represent $X^{(r)}$ or $Y^{(r)}$). This representation (e.g., in case $X^{(r)}$) means that

$$X^{(r)} = X_1 + \cdots + X_r,$$

where X_i is interpreted as the number of failures falling between the $(i-1)$st and ith successes (e.g., in the case of $r = 3$, if the consecutive trials are FFFSSFFFFS, then $X^{(3)} = 7 =$ number of failures preceding the third success, with $X_1 = 3, X_2 = 0, X_3 = 4$ being the numbers of failures between the three consecutive successes). This representation gives also

$$E(X^{(r)}) = E(X_1) + \cdots + E(X_r)$$

and (because of independence)

$$\text{Var}(X^{(r)}) = \text{Var}(X_1) + \cdots + \text{Var}(X_r),$$

which in view of (9.8) and (9.10) gives formulas (9.17) and (9.18).

The representation of random variables $Y^{(r)}$ is analogous and will be omitted.

PROBLEMS

9.5.1 A hospital needs 20 volunteers to the control group in testing the efficiency of some treatment. The candidates are subject to psychological and medical screening, and on average, only one in 15 candidates is found acceptable for the experiment. The cost of screening, whether a candidate is found acceptable or not, is $50 per person. The granting agency argues that one needs about 50×15 dollars to find one acceptable candidate, and therefore allows $20 \times 50 \times 15 = 15,000$ dollars for the cost of screening.

    (i) Write the formula for the probability that the sum allocated will be enough to find 20 acceptable candidates.

    (ii) Use the Chebyshev inequality to find the sum that would give at least a 90% chance of finding 20 acceptable candidates before the testing money runs out.

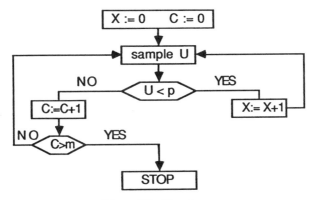

Figure 9.1 Flowchart.

9.5.2 Assume that a given team in the World Series has probability p of winning a game, and that the results of games are independent. Find the probability $p_{k,r}$ that the team will win the World Series given that it already lost k games and won r games $(k, r = 0, 1, 2, 3)$.

9.5.3 Assume that in a tennis match between A and B, the probability of winning a tennis set by player A is p and that the results of sets are independent. Let T be the number of sets played in a match. Find the distribution of T and $E(T)$ as a function of p, assuming that the match is played by (i) men; (ii) women (note that men play "best out of five" while women play "best out of three" sets).

9.5.4 Show that if X has negative binomial distribution $\text{NB}(r, p)$, then
$$E\left(\frac{r-1}{r+X-1}\right) = p.$$

9.5.5 In the flowchart of Figure 9.1, $m > 0$ is an integer and $0 < p < 1$. The block "sample U" means that the computer samples a value of the random variable U with distribution uniform on $(0,1)$, with sampling being independent each time the program executes the instruction "sample U." Find the distribution of $X, E(X)$, and $\text{Var}(X)$.

9.6 HYPERGEOMETRIC DISTRIBUTION

The hypergeometric distribution is an analogue of the binomial distribution in the case of sampling without replacement from a finite population. We have encountered it already in Examples 3.3.1 and 3.3.2.

To proceed systematically, we assume that initially the population has the

N elements of which a are of one kind ("successes") and $b = N - a$ are of another kind ("failures"). We sample n elements without replacement and we let X denote the number of successes in the sample. Let us first determine the range of the random variable X. Clearly, $0 \leq X \leq a$, but we must also have an analogous inequality for the number of failures, namely $0 \leq n - X \leq N - a$. This yields

$$\max(0, n - (N - a)) \leq X \leq \min(n, a). \qquad (9.20)$$

If k satisfies the constraint (9.20), then

$$P\{X = k\} = \frac{\binom{a}{k}\binom{N-a}{n-k}}{\binom{N}{n}}. \qquad (9.21)$$

Let us check that the probabilities in (9.21) add up to 1. Direct calculation is cumbersome, and it is best to use a trick. Let us consider the identity

$$(1 + x)^N = (1 + x)^a (1 + x)^{N-a}$$

and compare the coefficients of x^n on both sides. On the left-hand side the coefficient is $\binom{N}{n}$. On the right-hand side the coefficient equals to the sum of all possible terms of the form $\binom{a}{k}\binom{N-a}{n-k}$, where k must satisfy the constraint (9.20). This shows that the sum of all terms (9.21) is 1, as asserted.

Example 9.6.1. A class consists of 10 boys and 12 girls. The teacher selects 6 children at random for some task. Is it more likely that she chooses 3 boys and 3 girls or that she chooses 2 boys and 4 girls?

SOLUTION.   If X is the number of boys in the set of children selected, then

$$P\{X = 3\} = \frac{\binom{10}{3}\binom{12}{3}}{\binom{22}{6}} = 0.3538;$$

$$P\{X = 2\} = \frac{\binom{10}{2}\binom{12}{4}}{\binom{22}{6}} = 0.2985,$$

so that the first probability is higher than the second.   □

It may be expected that for large populations it does not matter too much whether we sample with or without replacement. We shall now formulate the corresponding approximation theorem.

Theorem 9.6.1. *Let $N \to \infty$ and $a \to \infty$ in such a way that $a/N \to p$, where $0 < p < 1$. Then for every fixed n and $k = 0, 1, \ldots n$ we have*

$$\frac{P\{X = k\}}{b(k; n, p)} \to 1 \qquad (9.22)$$

as $N \to \infty$. Here $P\{X = k\}$ is the probability (9.21) for hypergeometric distribution, while $b(k; n, p)$ is the binomial probability given by (9.3).

Proof. We can write, letting J_N denote the ratio in (9.22),

$$J_N = \frac{\binom{a}{k}\binom{N-a}{n-k}}{\binom{N}{n}\binom{n}{k}p^k(1-p)^{n-k}}$$

$$= \frac{a!(N-a)!(N-n)!}{(a-k)!(N-a-n+k)!N!p^k(1-p)^{n-k}},$$

which cancels to

$$\frac{[a(a-1)\cdots(a-k+1)][(N-a)(N-a-1)\cdots(N-a-n+k+1)]}{N(N-1)\cdots(N-n+1)p^k(1-p)^{n-k}}.$$

Multiplying and dividing by N^n, we can write $J_N = A_N B_N C_N$, where

$$A_N = \frac{a}{N}\left(\frac{a}{N} - \frac{1}{N}\right)\cdots\left(\frac{a}{N} - \frac{k-1}{N}\right),$$

$$B_N = \left(1 - \frac{a}{N}\right)\left(1 - \frac{a}{N} - \frac{1}{N}\right)\cdots\left(1 - \frac{a}{N} - \frac{n-k-1}{N}\right),$$

$$C_N = \left(1 - \frac{1}{N}\right)\left(1 - \frac{2}{N}\right)\cdots\left(1 - \frac{n-1}{N}\right)p^k(1-p)^{n-k}.$$

Since n and k are fixed and $a/N \to p$, the two products in the numerator converge to p^k and $(1-p)^{n-k}$ respectively, while $(1 - 1/N)\cdots[1 - (n-1)/N] \to 1$. This shows that $J_N \to 1$, and therefore proves the theorem. □

The practical use of Theorem 9.6.1 consists mostly of replacing probabilities

$$P\{X = k\} = \frac{\binom{a}{k}\binom{N-a}{n-k}}{\binom{N}{n}}$$

by their approximations $\binom{n}{k} \left(\dfrac{a}{N}\right)^k \left(1 - \dfrac{a}{N}\right)^{n-k}$, which are much easier to compute. The relative error of such approximation depends on all four factors N, n, a, and k.

Let us now find the expectation and variance of the hypergeometric random variable X with distribution (9.21). We shall use the method that we already used in case of binomial and negative binomial random variables, namely representation of X as a sum of simpler random variables. We let

$$X = \xi_1 + \xi_2 + \cdots + \xi_n, \tag{9.23}$$

where ξ_i equals 1 or 0 depending whether the ith draw results in success or failure.

Observe that $\xi_1, \ldots, \xi_n$ are dependent random variables; this fact will affect our calculation of the variance of X. Generally, we shall use the formulas

$$E(X) = E(\xi_1) + \cdots + E(\xi_n),$$
$$\mathrm{Var}(X) = \mathrm{Var}(\xi_1) + \cdots + \mathrm{Var}(\xi_n) + 2 \sum_{i<j} \mathrm{Cov}(\xi_i, \xi_j).$$

Note that ξ_i's are Bernoulli random variables (i.e., they assume only values 0 and 1), so that, letting $p_i = P\{\xi_i = 1\}$,

$$E(\xi_i) = p_i, \qquad \mathrm{Var}(\xi_i) = p_i(1 - p_i). \tag{9.24}$$

Observe that $\xi_i \xi_j$ is also a Bernoulli random variable, so that letting $p_{ij} = P\{\xi_i = 1, \xi_j = 1\}$, we have

$$\mathrm{Cov}(\xi_i, \xi_j) = E(\xi_i \xi_j) - E(\xi_i)E(\xi_j) = P\{\xi_i \xi_j = 1\} - p_i p_j$$
$$= P\{\xi_i = 1, \xi_j = 1\} - p_i p_j = p_{ij} - p_i p_j. \tag{9.25}$$

To determine the probabilities p_{ij} we need the joint distribution of (ξ_i, ξ_j). Generally, one of the ways of visualizing the distribution of the vector $(\xi_1, \ldots, \xi_n)$ is as follows: The elements of the population (of a successes and $N - a$ failures) are ordered at random. This gives the sample space of $N!$ permutations, all of them being equally likely. The random vector $(\xi_1, \ldots, \xi_n)$ is defined at any sampled point (permutation) as the initial n elements of this permutation (the reader should note that this is only one of the possible ways of defining the sample space and random variables for the experiment described in plain language as "choose a random sample of size n, without replacement, from a population of size N").

It is now clear that the joint distribution of (ξ_i, ξ_j) does not depend on (i, j) [i.e., it is the same as, say, joint distribution of (ξ_1, ξ_2)]. Indeed, the probability that $(\xi_i = x, \xi_j = y)$ depends on the number of permutations in the sample space that have specific elements x and y at places i and j. By symmetry, this number is the same as the number of permutations that have elements x and y at the first two places (or on any other designated pair of places).

Consequently, the marginal distributions of ξ_i are the same, and for all i we must have

$$p_i = P\{\xi_i = 1\} = \frac{a}{N}.$$

Similarly,

$$p_{ij} = P\{\xi_1 = 1, \xi_2 = 1\} = P\{\xi_1 = 1\}P\{\xi_2 = 1|\xi_1 = 1\} = \frac{a}{N}\frac{a-1}{N-1}.$$

From (9.24) and (9.25) we get $E(\xi_i) = a/N$, $\text{Var}(\xi_i) = a/N(1 - a/N)$ and therefore

$$\begin{aligned}
\text{Cov}(\xi_i, \xi_j) &= E(\xi_i \xi_j) - E(\xi_i)E(\xi_j) = p_{ij} - E(\xi_i)E(\xi_j) \\
&= \frac{a}{N}\frac{a-1}{N-1} - \left(\frac{a}{N}\right)^2 \\
&= -\frac{1}{N}\frac{a(N-a)}{N(N-1)} = -\frac{1}{N-1}\text{Var}(\xi_i).
\end{aligned}$$

Thus, $E(X) = n(a/N)$. For variance, observe that the number of pairs (i, j) with $i < j$ is $\binom{n}{2} = \frac{n(n-1)}{2}$. Consequently,

$$\begin{aligned}
\text{Var}(X) &= n\frac{a}{N}\frac{N-a}{N} - 2\frac{n(n-1)}{2}\frac{1}{N}\frac{a(N-a)}{N(N-1)} \\
&= n\frac{a}{N}\frac{N-a}{N}\left(1 - \frac{n-1}{N-1}\right).
\end{aligned}$$

Letting $a/N = p$ denote the overall probability of success, we have the following theorem.

Theorem 9.6.2. *If X has the hypergeometric distribution* (9.21), *then*

$$E(X) = n\frac{a}{N} = np$$

$$\text{Var}(X) = n\frac{a}{N}\left(1 - \frac{a}{N}\right)\frac{N-n}{N-1} = np(1-p)\frac{N-n}{N-1}. \qquad \square$$

Thus, the expected number of successes in a sample of size n is np, regardless of whether we sample with or without replacement. The variance in the case of sampling without replacement is smaller than the corresponding binomial variance by the factor $(N - n)/(N - 1)$ (sometimes referred to as *correction factor* for finiteness of population).

It is worthwhile to note the striking difference in the behavior of the variances of binomial and hypergeometric distribution as one increases the sample size n. In a binomial distribution, each new element of the sample contributes the same amount to variance, so that the latter grows linearly with n. In sampling without replacement, variance changes in proportion to the product $n(N - n)$; hence it initially grows to reach a maximum when sampling exhausts half the population, and then declines to zero when $n = N$ (indeed, when $n = N$ we exhaust the entire population, and there is no variability in the sample). Variance for $n = 1$ is the same as for $n = N - 1$; this is clear, since the variability involved in sampling one element is the same as variability involved with leaving just one element unsampled.

At the end of this section we present one more fact connecting the binomial and hypergeometric distributions. It concerns the situation when the successes come from two sources, in each following the binomial distribution, and we are interested in the number of successes coming from one of the sources given the total number of successes. Thus, we shall prove

Theorem 9.6.3. *Let X and Y be independent random variables with binomial distribution $X \sim \mathrm{BIN}(m, p)$ and $Y \sim \mathrm{BIN}(n, p)$. Then*

$$P\{X = k | X + Y = r\} = \frac{\binom{m}{k}\binom{n}{r-k}}{\binom{m+n}{r}}. \tag{9.26}$$

Proof. Since $X + Y \sim \mathrm{BIN}(m + n, p)$, we have

$$P\{X = k | X + Y = r\} = \frac{P\{X = k, X + Y = r\}}{P\{X + Y = r\}}$$
$$= \frac{P\{X = k\}P\{Y = r - k\}}{P\{X + Y = r\}},$$

and substitution of binomial probabilities gives (9.26).   □

The right-hand side of (9.26) is the hypergeometric probability of k successes in sampling r elements from a population with a total of $m + n$ elements, of which m are successes.

Example 9.6.2. To show a possible application of Theorem 9.6.3, we consider a problem of statistical nature, which we shall continue later in the book. Suppose that two music teachers prepare their students for some competition. Suppose further that teacher A trains m contestants, while teacher B trains n contestants. Our objective is to compare the quality of teaching of A and B. Suppose that A and B have access to the same pool of talented students from which they choose their pupils. Moreover, one can argue that *winning* the competition by a pupil is not the best measure of quality of the teacher, as it depends too much on a single performance in the final, small variation of student's talents, and so on. However, *getting to the finals* through several stages of elimination can be taken as a good measure. Assume that

p_A and p_B are probabilities that a student of teacher A (respectively, B) gets to the finals, and let us agree that the levels of teachers are adequately reflected by probabilities p_A and p_B.

Since we are interested in comparing A and B, we may disregard the performance of all other contestants and concentrate on the numbers X_A and X_B of finalists in the competition who were taught by A and by B. To fix our attention, assume that A prepared 5 participants in the competition, of whom 2 reached the finals (i.e., $m = 5, X_A = 2$). Similarly, B prepared 6 finalists, of whom 1 reached the finals (i.e., $n = 6, X_B = 1$). On the level of reasoning accessible to most musicians, 40% of students of A reached the finals, while only about 16.6% of students of B reached the finals, so A is better than B.

Convincing as it may sound, this reasoning misses an important point, namely the effect of randomness on the process of getting to the finals. To decide whether we can convincingly claim that $p_A > p_B$, a statistician may suggest here *assuming* that $p_A = p_B$, then determining how likely it might be, under this assumption, to have the result obtained, or even better for A.

We have $X_A + X_B = 3$ (altogether, 3 students of the teachers A and B reached the finals). The value of X_A was 2, while it could also have been 0, 1, or 3. The problem is now: How much chance is there of having $X_A = 2$ or $X_A = 3$ given that $X_A + X_B = 3$ and given that in fact $p_A = p_B$?

The importance of Theorem 9.6.3 lies in the fact that the probability (9.26) does not depend on p. Consequently, we may eliminate this parameter. We have

$$P\{X_A = 2 \text{ or } 3 | X_A + X_B = 3\} = \frac{\binom{5}{2}\binom{6}{1}}{\binom{11}{3}} + \frac{\binom{5}{3}\binom{6}{0}}{\binom{11}{3}}$$

$$= \frac{70}{165} = 0.424.$$

Thus, if only 3 students of both teachers reach the final, then even if both teachers are equal in their teaching ability, there is over 40% probability that the percentage of finalists of A will exceed the percentage of finalists of B. Consequently, a claim that the results discussed show that A is a better music teacher than B is not justified.   □

The scheme of sampling without replacement can be generalized as follows. Assume that the population consists initially of a elements of one kind (successes) and $b = N - a$ elements of the second kind (failures). Each time an element is sampled, it is returned, and c elements of the same kind as just sampled are added to the urn. This process continues for n samplings. Let X be the number of successes in the sample. This random variable is said to have *Pólya distribution*, and the sampling described above is called the *Pólya scheme*.

The reason for designing this scheme was as follows. Adding elements of the same kind to the population increases the probability of drawing elements of the kind most recently sampled. Such an effect is known to occur when one samples from a population to determine the fraction of persons infected with a disease. Typically, if one finds one person with the disease, then the chances of finding others with the same disease increase. Pólya introduced this scheme of sampling to model such effects.

Observe that we may formally put $c = -1$ (elements are simply not returned). Thus, the special case $c = -1$ yields the random variable X with a hypergeometric distribution. To find $P\{X = k\}$ in the general case, let us first find the probability of sampling the elements in a specific order, say first k successes and then $n - k$ failures. This probability, by the chain rule (4.5), is

$$\frac{a}{N}\frac{a+c}{N+c}\cdots\frac{a+(k-1)c}{N+(k-1)c}\frac{b}{N+kc}\frac{b+c}{N+(k+1)c}\cdots\frac{b+(n-k-1)c}{N+(n-1)c}.$$
(9.27)

Let us observe that the probability (9.27) remains the same regardless of the order of k successes and $n - k$ failures: Each time the number of elements in the population increases by c, therefore the product of all denominators is $N(N+1)\cdots(N+(n-1)c)$. Similarly, the numerators corresponding to sampling successes are $a, a+c, a+2c, \ldots, a+(k-1)c$, regardless of when the successes are to occur, and the same holds for failures. Consequently, we have

Theorem 9.6.4.   *In the Pólya scheme, we have for $k = 0, 1, \ldots, n$,*

$$P\{X = k\} = \frac{n!p(p+\gamma)\cdots(p+(k-1)\gamma)q(q+\gamma)\cdots(q+(n-k-1)\gamma)}{k!(n-k)!(1+\gamma)(1+2\gamma)\cdots(1+(n-1)\gamma)},$$

where $p = a/N, q = b/N = 1 - a/N$, and $\gamma = c/N$.

At the end, it is worth mentioning that the Pólya scheme can be modified in the following way. Again, we have an urn, initially containing a balls of one kind ("successes") and b balls of another kind ("failures"). Balls are drawn successively. After a ball is drawn, it is returned; if it was a success, c_1 balls representing *failure* are added; if it was a failure, c_2 balls representing *success* are added. Thus a success causes an increase in the probability of failure on the next trial, and conversely, a failure causes an increase in the probability of success on the next trial. This scheme has been used to model accidents at work. Here the situation is such that an accident causes an increase in observing safety measures, awareness of danger, and so on. On the other hand, the longer the time without an accident, the more laxity in observing safety regulations, and so on. Under the appropriate choice of a, b, c_1, and c_2, it is possible to model the "aftereffects of accidents" as described above, thus modeling the distribution of times between accidents, number of accidents in a given period, and so on.

Unfortunately, the formulas are much more complicated than those for Pólya scheme, and we shall not pursue this topic here.

PROBLEMS

9.6.1 An urn contains nine balls, five of them red and four blue. Three balls are drawn without replacement. Find the distribution of X = number of red balls drawn.

9.6.2 An urn contains five balls, two red and three green. Three balls are drawn without replacement. Find $E(X)$ and $\text{Var}(X)$, where X = number of red balls in the sample.

9.6.3 A population consists of N individuals, of whom k have a certain attribute (e.g., are carriers of a specific gene, etc). We sample n individuals without replacement, and let X be the number of individuals in the sample who have the attribute in question. For what value of n does the variance of X attain its maximum?

9.6.4 Instead of (9.23), write $X = \eta_1 + \eta_2 + \cdots + \eta_a$, where $\eta_j = 1$ or 0 depending on whether the jth element representing success was selected or not. Use this representation to derive formulas for the mean and variance of X as given in Theorem 9.6.2.

9.7 POISSON DISTRIBUTION

We start from the following definition.

Definition 9.7.1. A random variable X is said to have a *Poisson distribution* if for some $\lambda > 0$,

$$P\{X = n\} = \frac{\lambda^n}{n!} e^{-\lambda}, \qquad n = 0, 1, \dots. \tag{9.28}$$

In this case we shall also use the notation $X \sim \text{POI}(\lambda)$.   □

Thus, a random variable with Poisson distribution is discrete, with the range being all nonnegative integers. Let us check first that the terms in (9.28) add to 1. We have

$$\sum_{n=0}^{\infty} P\{X = n\} = e^{-\lambda} \sum_{n=0}^{\infty} \frac{\lambda^n}{n!} = e^{-\lambda} \cdot e^{\lambda} = 1.$$

To determine the moments of the Poisson distribution, let us compute the mgf. We have

$$m_X(t) = Ee^{tX} = \sum_{n=0}^{\infty} e^{tn} P\{X = n\} = \sum_{n=0}^{\infty} e^{tn} \frac{\lambda^n}{n!} e^{-\lambda}$$

$$= e^{-\lambda} \sum_{n=0}^{\infty} \frac{(\lambda e^t)^n}{n!} = e^{-\lambda} \cdot e^{\lambda e^t} = e^{\lambda(e^t - 1)}.$$

Thus, moment generating function of the Poisson distribution is defined for all t and is differentiable an arbitrary number of times. We have

$$E(X) = m'_X(t)|_{t=0} = \lambda e^t e^{\lambda(e^t - 1)}|_{t=0} = \lambda.$$

An easy differentiation yields $E(X^2) = m''_X(t)|_{t=0} = \lambda^2 + \lambda$, and therefore

$$\text{Var}(X) = \lambda.$$

We shall now prove a theorem asserting the closure property of the family of Poisson distributions under addition of independent random variables. We have

Theorem 9.7.1. *If X and Y are independent, with Poisson distribution with parameters λ_1 and λ_2, respectively, then $X + Y$ has Poisson distribution with parameter $\lambda_1 + \lambda_2$.*

Proof. We present first a direct proof to show the kinds of calculations involved in evaluating the distribution of the sum, and then we present a simple proof (see Example 8.5.9) using moment generating functions. We have

$$P\{X + Y = n\} = \sum_{j=0}^{n} P\{X = j, Y = n - j\}$$

$$= \sum_{j=0}^{n} P\{X = j\} P\{Y = n - j\}$$

$$= \sum_{j=0}^{n} \frac{\lambda_1^j}{j!} e^{-\lambda_1} \frac{\lambda_2^{n-j}}{(n-j)!} e^{-\lambda_2}$$

$$= \frac{1}{n!} e^{-(\lambda_1 + \lambda_2)} \sum_{j=0}^{n} \binom{n}{j} \lambda_1^j \lambda_2^{n-j}$$

$$= \frac{(\lambda_1 + \lambda_2)^n}{n!} e^{-(\lambda_1 + \lambda_2)} \text{ (by Newton's formula)},$$

which completes the proof.

Alternatively, observe that the mgf of $X + Y$ is

$$m_{X+Y}(t) = m_X(t) m_Y(t) = e^{\lambda_1(e^t - 1)} \cdot e^{\lambda_2(e^t - 1)} = e^{(\lambda_1 + \lambda_2)(e^t - 1)},$$

which we recognize as the mgf of Poisson distribution with parameter $\lambda_1 + \lambda_2$.

□

Finally, we prove an analogue of Theorem 9.6.3. We have

Theorem 9.7.2. *If X and Y are independent and $X \sim POI(\lambda_1), Y \sim POI(\lambda_2)$, then for $k = 0, 1, \ldots, n$,*

$$P\{X = k | X + Y = n\} = \binom{n}{k} \left(\frac{\lambda_1}{\lambda_1 + \lambda_2} \right)^k \left(\frac{\lambda_2}{\lambda_1 + \lambda_2} \right)^{n-k}. \qquad (9.29)$$

Thus, the conditional distribution of X given $X + Y$ is binomial, with number of trials $X + Y$ and probability of success $p = \lambda_1/(\lambda_1 + \lambda_2)$.

Proof. We have, using Theorem 9.7.1,

$$P\{X = k | X + Y = n\} = \frac{P\{X = k, X + Y = n\}}{P\{X + Y = n\}}$$

$$= \frac{P\{X = k\} P\{Y = n - k\}}{P\{X + Y = n\}}$$

$$= \frac{\frac{\lambda_1^k}{k!} e^{-\lambda_1} \frac{\lambda_2^{n-k}}{(n-k)!} e^{-\lambda_2}}{\frac{(\lambda_1 + \lambda_2)^k}{n!} e^{-(\lambda_1 + \lambda_2)}}$$

which reduces to the right-hand side of (9.29).   □

Before going any further with exploring the properties of a Poisson distribution, one should perhaps ask when in practical situations one can expect to encounter a random variable with a Poissson distribution. We therefore begin with one such situation, characterized by the following theorem.

Theorem 9.7.3. *If $p \to 0$ and $n \to \infty$ in such a way that $\lim np = \lambda > 0$, then for $k = 0, 1, \ldots$,*

$$\lim_{n \to \infty} \binom{n}{k} p^k (1-p)^{n-k} = \frac{\lambda^k}{k!} e^{-\lambda}. \tag{9.30}$$

Proof. We shall prove (9.30) under the simplifying assumption $np = \lambda$ for all n. The proof in the general case is based on the same idea, but obscured by some technical points. We write, replacing p by λ/n,

$$\binom{n}{k} p^k (1-p)^{n-k} = \frac{n(n-1)\cdots(n-k+1)}{k!} \left(\frac{\lambda}{n}\right)^k \left(1-\frac{\lambda}{n}\right)^{n-k}$$

$$= \frac{\left(1-\dfrac{\lambda}{n}\right)^n \left(1-\dfrac{1}{n}\right)\cdots\left(1-\dfrac{k-1}{n}\right)\lambda^k}{k! \left(1-\dfrac{\lambda}{n}\right)^k}.$$

The factor $(1 - \lambda/n)^n$ converges to $e^{-\lambda}$, while each of the remaining factors involving n converges to 1. Since the number of such factors is constant (does not depend on n), their product also tends to 1, which proves the theorem. □

To see the applicability of Theorem 9.7.3, observe that it may be used as an approximation, valid for small p and large n:

$$\binom{n}{k} p^k (1-p)^{n-k} \approx \frac{(np)^k}{k!} e^{-np}. \tag{9.31}$$

On the left-hand side we recognize the binomial probability of k successes in n trials.

Example 9.7.1. Continuing Example 9.3.3, we assume that on average, one birth in 80 is a multiple birth. What is the probability that among 400 births that occurred in the maternity ward of a given hospital during the first three months of a year there were fewer than 4 multiple births?

SOLUTION. If X stands for the number of multiple births during the period analyzed, then $X \sim \text{BIN}(400, \frac{1}{80})$. We have

$$P\{X < 4\} = \sum_{k=0}^{3} P\{X = k\} = \sum_{k=0}^{3} \binom{400}{k} \left(\frac{1}{80}\right)^k \left(\frac{79}{80}\right)^{400-k}.$$

This sum may be computed directly, and the value is

$$0.00653 + 0.03306 + 0.08348 + 0.14019 = 0.26326.$$

We have here $np = 400 \cdot \frac{1}{80} = 5$, so that the approximation (9.31) by a Poisson distribution gives

$$P\{X < 4\} \approx \sum_{k=0}^{3} \frac{5^k}{k!} e^{-5}$$

$$= 0.00674 + 0.03369 + 0.08422 + 0.14037 = 0.26503.$$

The relative errors of consecutive approximating terms are, respectively, 3.2%, 1.9%, 0.89% and 0.13%, while the relative error of the final answer is 0.67%.

Whether or not this may be regarded as a good approximation depends on the goal of finding the probability in question. For most purposes the relative error below 1% is quite acceptable. One can imagine, however, situations where it need not be so. For instance, an insurance company that is to cover the cost of delivery and hospital care for multiple births might want to know the probability of fewer than 4 multiple births with a precision better than the second decimal in order to decide on the premium.   □

Example 9.7.2. Suppose that on average, one in every 100 passengers does not show up for a flight. An airline sold 250 tickets for a flight serviced by an airplane that has 247 seats. What is the probability that every person who shows up for a flight will get a seat?

SOLUTION.   Let X be the number of passengers who do not show up for the particular flight in question, and let us treat X as a binomial random variable with $n = 250$, $p = 0.01$, so that $np = 2.5$. We want the probability of the event $X \geq 3$, hence we may write, using a Poisson approximation,

$$P\{X \geq 3\} = 1 - P\{X < 3\}$$
$$= 1 - P\{X = 0\} - P\{X = 1\} - P\{X = 2\}$$
$$\approx 1 - e^{-2.5} - \frac{2.5}{1!} e^{-2.5} - \frac{(2.5)^2}{2!} e^{-2.5}$$
$$= 0.45619.$$

The value calculated from the binomial distribution, namely

$$P\{X \geq 3\} = 1 - \sum_{j=0}^{2} \binom{250}{j} (0.01)^j (0.99)^{250-j}, \qquad (9.32)$$

equals 0.45683. This time the approximation by a Poisson distribution to binomial probabilities (9.32) has a relative error of 0.14%.

One should make the following comment here. While in Example 9.7.1 the claim that X has a binomial distribution was fully justified (whether or not a birth is a multiple birth is independent of the multiplicity of other births in the same period), the situation is not as clear in the case of passengers missing an airline flight. The point is that people often fly together (typically, in families or other groups). In these cases, the fact that one person misses the flight may affect the chances of some other persons missing the same flight. Consequently, X is at best *approximately* binomial. Our calculations therefore give a relative error of an approximation to a number that is already an approximation (to the actual probability).

Example 9.7.3. Let us analyze what could be the possible effects of the lack of independence mentioned at the end of Example 9.7.2. To this end, assume a very simple situation when people travel in pairs, with both members of the pair either missing the flight or not. The pairs are assumed to be independent one from another.

As we want to compare this situation with that of Example 9.7.2, let us assume that $N = 2m$ passengers bought the tickets, and let p be the probability of a passenger not showing up. Let X_1 and X_2 be numbers of no-shows under assumptions of Example 9.7.2 (X_1) and under the present assumption (X_2). Thus $X_1 \sim \text{BIN}(N, p)$, hence $E(X_1) = Np$, $\text{Var}(X_1) = Np(1-p)$. As regards X_2, we have $X_2 = 2Y$, where Y is the number of pairs of passengers who do not show up.

Now $Y \sim \text{BIN}(m, p)$, so $E(X_2) = 2E(Y) = 2mp = Np = E(X_1)$, and the expectation is the same in both models. However, remembering formula (8.33), we have here $\text{Var}(X_2) = 2^2 \text{Var}(Y) = 4 \cdot mp(1-p) = 2Np(1-p) = 2\text{Var}(X_1)$. We see therefore that the effect of grouping passengers in pairs doubles the variance of no-shows. One can show that *any* grouping (not necessarily into pairs), with members of the group more likely to behave similarly than to behave differently will tend to increase the variance.

This analysis is based on a conjecture about the probable behavior of families (or other groups) in case of one member being unable to show up for the departure. We can easily collect the data on numbers of passengers missing a given flight on different days, and calculate sample mean and variance. In later chapters we develop techniques of deciding whether the differences observed between sample mean and sample variance can be attributed to chance or whether they are due to the fact that the mean and variance of the random variable analyzed are in fact different. Depending on the outcome of such an experiment, one can design further models of group sizes and the distribution of no-shows in a group.

Example 9.7.4. We shall now consider another class of situations in which we may expect random variables to follow a Poisson distribution. We start from a situation that will call for the use of Theorems 9.7.2 and 9.7.3.

Consider accidents occurring on an intersection controlled by traffic lights.

Suppose a traffic engineer suspected that one of the factors contributing to accidents was a particular sequence of light changes used on this intersection. He suggests that another sequence of light changes should decrease the accident rate. To test the validity of this idea it is necessary to eliminate as many other factors contributing to the accident rate as possible, such as weather conditions, road surface, traffic intensity, pedestrian traffic, and so on. Thus, another intersection is selected, identical in every respect to the one tested. A new traffic light pattern was installed on one intersection and the old one was left on the other intersection. Let X_1 and X_2 be the numbers of accidents observed in the same period on the two intersections.

We first use Theorem 9.7.3 to argue that one can reasonably expect X_1 and X_2 to have Poisson distributions. To see this, let us divide the observation period into a large number n of smaller intervals of the same duration. Let us regard each of these periods as a trial in which we can observe a "success" (accident) and "failure" (no accident). If we choose the time interval sufficiently short, we may disregard the possibility of two or more accidents occurring in the same time interval. Now, if we may assume that the events during one time interval do not affect the chances of events in other time intervals, then we are in the situation of large number of Bernoulli trials with a low probability of success. Consequently, the total number of successes (accidents) has—at least approximately—a Poisson distribution. What we want to investigate is the effect of change in traffic light patterns, that is, we want to find out whether or not $\lambda_1 = \lambda_2$, where λ_i is the expected number of accidents in a given observation period on the ith intersection. For instance, we may want to find if $\lambda_1 < \lambda_2$, which would mean that intersection 1, which has the new lights pattern, has a lower expected number of accidents.

We may proceed here as in Example 9.6.2, but in this case using Theorem 9.7.2. We know that the conditional distribution of X_1 given $X_1 + X_2$ is binomial BIN$(X_1 + X_2, \lambda_1/(\lambda_1 + \lambda_2))$. In particular, if the change of traffic light pattern is irrelevant as regards the accident rate, we have $\lambda_1/(\lambda_1 + \lambda_2) = \frac{1}{2}$. Suppose, for instance, that in the period of observation there were no accidents on the first intersection and five accidents on the second one (i.e., $X_1 = 0, X_2 = 5$). According to Theorem 9.7.2, the probability of such an event is $P\{X_1 = 0 | X_1 + X_2 = 5\} = \binom{5}{0}(0.5)^0(0.5)^5 = 0.03125$, which is about 3%. In future chapters we discuss the problem of whether such a finding may be taken as an indication that $\lambda_1 < \lambda_2$ (since if $\lambda_1 = \lambda_2$, we would have observed an event that happens rather rarely, on the average about 3 times in every 100 observations).

PROBLEMS

9.7.1 Let X have the Poisson distribution with parameter λ. Find the mode of X (i.e., the most likely value of X).

9.7.2   Let X have the Poisson distribution with parameter λ. Find $P\{X$ is even$\}$. (*Hint:* Write the Taylor expansions for e^λ and $e^{-\lambda}$. Any ideas?)

9.7.3   **(Does Nature Prefer Even Numbers?)**   Generalizing Problem 9.7.2, let X be an integer-valued random variable such that $X = X_1 + X_2$, where X_1, X_2 are independent, identically distributed integer-valued random variables. Show that $P\{X$ is even$\} \geq \frac{1}{2}$ (this property has been pointed out to us by Steve MacEachern, personal communication).

9.7.4   Continuing Problem 9.7.3, let $\mathcal{K}^{\text{even}}$ and $\mathcal{K}^{\text{odd}}$ be the classes of integer-valued random variables X such that

$$P\{X \text{ is even}\} \geq \tfrac{1}{2} \text{ for } X \text{ in } \mathcal{K}^{\text{even}}$$

and

$$P\{X \text{ is odd}\} \geq \tfrac{1}{2} \text{ for } X \text{ in } \mathcal{K}^{\text{odd}}.$$

Show that if X', X'' are independent with $X', X'' \in \mathcal{K}^{\text{even}}$ or $X', X'' \in \mathcal{K}^{\text{odd}}$, then $X' + X'' \in \mathcal{K}^{\text{even}}$.

9.7.5   Suppose that the number of eggs X laid by an animal has Poisson distribution. Each egg hatches with probability p, independently of what happens to other eggs. Let V_1 and V_2 denote the numbers of eggs that hatch and the number of eggs that do not hatch (laid by a given animal). Show that V_1 and V_2 are independent.

9.7.6   Suppose that a certain store makes, on average, 2 sales per hour between 9:00 p.m. and 2:00 p.m., and 3 sales per hour between 2:00 p.m. and 9:00 p.m. The numbers of sales in different time periods are independent and have Poisson distribution. Find:

(i) The probability of more than 3 sales between 10:00 a.m. and noon, and similarly between 1:00 p.m. and 3:00 p.m.

(ii) The probability that the number of sales between 10:00 a.m. and 11:00 a.m. will be the same as number of sales between 6:00 p.m. and 7:00 p.m.

9.7.7   A book with 500 pages contains, on average, 3 misprints per 10 pages. What is the probability that there will be more than 1 page containing at least 3 misprints?

9.7.8   The accidents in a given plant occur at a rate of 1.5 per month. The number of accidents in various months are independent and follow Poisson distribution. Find the probability of:

(i) Five accidents in a period of five consecutive months.

(ii) One accident in each of five consecutive months.

9.7.9 Suppose that the daily numbers of ships arriving to a certain port are independent, each with Poisson distribution with mean $\lambda = 3$. Find:

(i) The expected number of days in April when there are no arrivals,

(ii) The expected number and variance of days during the summer months (June, July, August) with the number of arrivals equal to the mean daily arrival rate.

9.7.10 Let X be the number of failures preceding the rth success in a sequence of Bernoulli trials with probability of success p. Show that if $q \to 0, r \to \infty$ in such a way that $rq = \lambda > 0$, then

$$P\{X = k\} \to \frac{\lambda^k}{k!}e^{-\lambda}$$

for every $k = 0, 1, 2, \ldots$. This shows that the negative binomial distribution can be, for large r and small q, approximated by a Poisson distribution. (*Hint:* Use an argument similar to that in the proof of Theorem 9.7.3.)

9.7.11 Weekly numbers of accidents at intersections A, B, and C are independent, each with a Poisson distribution. It is known that, on average, the number of accidents at intersection A is the same as the number of accidents at intersections B and C combined, while the average number of accidents at intersection B is half of that at intersection C.

(i) If there were, in a given week, 16 accidents at intersections A, B, and C, what is the probability that exactly 4 of them were at intersection C?

(ii) What is the probability that there were more accidents at intersection C than at intersection A?

9.7.12 Find the approximate probability that in 1000 randomly chosen persons there are exactly

(i) Two born on New Year and two born on Christmas.

(ii) Four born on either Christmas or New Year.

9.8 POISSON PROCESS

We shall now try to capture those features of the situation analyzed in Example 9.7.4, which are responsible for the fact that the number of occurrences of some event in a given interval of time follows a Poisson distribution.

We consider the class of situations in which a certain event occurs at random points in time. Examples are quite common: arrivals of customers at service stations, twin births in a hospital, earthquakes of specified intensity occurring in a given region, fire alarms in a given town, marriage proposals received by a given lady, and so on. To increase practical applicability, the theory disregards all specific features of the events under consideration, concentrating only on the times of their occurrence. The random variable one needs to analyze here is the number $N_{[t_1,t_2]}$ of events that occur between times t_1 and t_2. The theory built for analyzing such processes in most general cases is called the *theory of point processes*. We shall analyze only a special case of point processes, the *Poisson process*.

The assumptions underlying Poisson processes attempt to capture the intuitive notion of "complete randomness." In particular, in a Poisson process, knowledge of the past provides no clue as regards the future.

To express the properties which will imply that a given stream of events is a Poisson process, it will be convenient to introduce a mathematical notation, which will later be useful also in other contexts. We shall namely introduce the symbol $o(x)$ to denote any function $f(x)$ such that

$$\lim_{x \to 0} \frac{f(x)}{x} = 0.$$

Example 9.8.1. A power function x^a is $o(x)$ if $a > 1$, and so is every function of the form $x^a h(x)$ if $a > 1$ and h is continuous at 0 (hence bounded in the neighborhood of 0). For instance, if S_n is a binomial random variable, then $P\{S_n = k\} = o(p)$ for $k > 1$. Indeed,

$$\frac{P\{S_n = k\}}{p} = \binom{n}{k} p^{k-1} (1 - p)^{n-k},$$

which converges to 0 when $p \to 0$ for $k > 1$.   □

In the sequel, we shall often use the following facts:

   I. *If* $\lim_{x \to 0} h(x)/x = c \neq 0$, *then* $h(x) = cx + o(x)$.
  II. *If the functions* $f_1, f_2, \ldots, f_N$ *are* $o(x)$, *then* $f_1 + \cdots + f_N$ *is also* $o(x)$.

We may now formulate the postulates of the Poisson process.

Postulate 1. *The numbers of events occurring in two nonoverlapping time intervals are independent.*

Postulate 2. *The probability of at least one event occurring in an interval of length* Δt *is* $\lambda \Delta t + o(\Delta t)$ *for some constant* $\lambda > 0$.

Postulate 3. *The probability of two or more events occurring in an interval of length* Δt *is* $o(\Delta t)$.

The first postulate is the one that asserts that knowledge of the past is of no help in predicting the future. The second postulate asserts stationarity, in the sense that the probability of an event occurring in a short time interval is (roughly) proportional to the length of this interval but does not depend on the location of this interval. Finally, the third postulate asserts that events occur one at a time: Chances of two events occurring within an interval of duration Δt become negligible as Δt tends to 0.

Let us now fix the zero on time scale, and let $P_n(t)$ denote the probability of exactly n events prior to t (counting from $t = 0$), so that $P_n(t) = P\{N_{[0,t)} = n\}$. We shall prove

Theorem 9.8.1. *Under Postulates 1–3,*

$$P_n(t) = \frac{(\lambda t)^n}{n!} e^{-\lambda t}, \qquad n = 0, 1, \ldots \tag{9.33}$$

Proof. Observe first that by Postulates 2 and 3, we have for every t and $\Delta t > 0$

$$P\{N_{[t,t+\Delta t)} = 1\} = \lambda \Delta t + o_1(\Delta t)$$
$$P\{N_{[t,t+\Delta t)} = 0\} = 1 - \lambda \Delta t + o_2(\Delta t). \tag{9.34}$$

For $n = 0$ we may write, using Postulate 1 and (9.34):

$$P_0(t + \Delta t) = P\{N_{[0,t)} = 0, N_{[t,t+\Delta t)} = 0\} = P_0(t)[1 - \lambda \Delta t + o_2(\Delta t)],$$

which gives the difference ratio

$$\frac{P_0(t + \Delta t) - P_0(t)}{\Delta t} = -\lambda P_0(t) + \frac{1}{\Delta t} o_2(\Delta t) P_0(t). \tag{9.35}$$

Passing to the limit with $\Delta t \to 0$, we obtain*

$$P_0'(t) = -\lambda P_0(t). \tag{9.36}$$

The initial condition is $P_0(0) = 1$, and the relation of (9.36) gives

$$P_0(t) = e^{-\lambda t}. \tag{9.37}$$

Now, if $n \geq 1$, we may write

---

* The limit of the right-hand side exists, and equals the right-side derivative of P_0 at t. To justify the existence of the left derivative in (9.35) observe that one can replace t by $t - \Delta t$ in (9.35). Since $o_2(\Delta t)$ does not depend on t, we see that P_0 is continuous and that the left derivative is equal to the right derivative.

$$P_n(t + \Delta t) = \sum_{j=0}^{n} P\{N_{[0,t)} = n - j, N_{[t,t+\Delta t)} = j\}$$

$$= \sum_{j=0}^{n} P_{n-j}(t)P\{N_{[t,t+\Delta t)} = j\}$$

$$= P_n(t)[1 - \lambda\Delta t + o_2(\Delta t)] + P_{n-1}(t)[\lambda\Delta t + o_1(t)] + o_3(t),$$

where $o_3(t)$ is the term obtained from combining together all terms involving two or more events occurring between t and $t + \Delta t$.

Forming the difference ratio and passing to the limit with $\Delta t \to 0$ we obtain the equations, valid for $n = 1, 2, \ldots,$

$$P_n'(t) = -\lambda P_n(t) + \lambda P_{n-1}(t), \tag{9.38}$$

which can be solved recursively using (9.37), with the initial conditions being now $P_n(0) = 0, n = 1, 2, \ldots.$ Alternatively, we may use induction to check that probabilities (9.33) satisfy (9.37) and (9.38).   □

We shall now discuss some examples of Poisson processes. Let us start with a continuation of Example 9.7.4.

Example 9.8.2.   Instead of using the argument based on approximation of binomial distribution by Poisson, we can refer to Postulates 1–3 to argue that the number of accidents on an intersection should have Poisson distribution. Some care has to be taken, though. First, Postulate 2 may be considered as a good description of the situation but only during the periods when one can disregard variations of the traffic intensity (e.g., only during rush hours, etc.). Alternatively, one can consider intervals of time long enough to allow the within-a-day variations of traffic rate to "average out," say periods such as a week, and so on. Second, to ensure that Postulate 3 holds, one should count accidents and not, say, numbers of cars involved or number of persons injured (to ensure that events that are counted occur one at a time).

Example 9.8.3.   The maternity ward in a certain hospital has, on average, 30 births per week. Given that there were 6 births on a given day, what is the probability of (a) 3 births on each of the following two days; (b) total of 6 births during the following two days? (c) What is the expected number of days with exactly one birth during the month of May?

SOLUTION.   We assume here that the births in the maternity ward in question form a Poisson process. Consequently, the number of births on a given day does not affect the number of births in future intervals. To answer questions (a)–(c), we first choose the unit of time. This a totally arbitrary choice, but the important point is that once this choice is made, one should express the parameter λ in the chosen units. Then λ is the expected number of events in the unit of time.

In our case, let us choose a time unit equal to one day. Then $\lambda = 30/\text{week} = 4.286/\text{day}$. Consequently, the probability of 3 births in a given day is $(\lambda^3/3!)e^{-\lambda} = 0.1806$, and the probability of this happening on two consecutive days is $\left[(\lambda^3/3!)e^{-2\lambda}\right] = 0.0326$. As regards (b), the probability of 6 births in 2 days is $\left[(2\lambda)^6/6!\right]e^{-2\lambda} = 0.1043$. Finally, in (c) the number of days in May when there is exactly 1 birth is binomial with $n = 31$ and $p = \lambda e^{-\lambda} = 0.0590$, so that the expectation equals $31p = 1.83$. $\qquad\square$

As mentioned, the postulates of the Poisson process attempt to capture the idea of "complete randomness." The theorems below indicate to which extent this attempt was indeed successful.

To facilitate formulation of the theorems, we let $X_t = N_{(0,t)}$ denote the number of events occurring in $(0, t)$, and also let $T_1, T_2, \ldots$ denote the times of occurrence of successive events. Thus,

$$T_k = \inf\{t : X_t \geq k\}.$$

In the analogy with (9.19), we have the following identity:

$$T_k \leq t \quad \text{iff } X_t \geq k$$

and consequently,

$$P\{T_k \leq t\} = P\{X_t \geq k\}.$$

Since X_t has Poisson distribution with parameter λt, we have

$$P\{X_t \geq k\} = \sum_{j=k}^{\infty} \frac{(\lambda t)^j}{j!} e^{-\lambda t} = 1 - \sum_{j=0}^{k-1} \frac{(\lambda t)^j}{j!} e^{-\lambda t}.$$

Consequently, for the cdf of T_k we obtain

$$P\{T_k \leq t\} = P\{X_t \geq k\} = 1 - \sum_{j=0}^{k-1} \frac{(\lambda t)^j}{j!} e^{-\lambda t}. \qquad (9.39)$$

Let now $U_1 = T_1, U_2 = T_2 - T_1, U_3 = T_3 - T_2, \ldots$ be the time to the first event (U_1) and consecutive times between events $(U_2, U_3, \ldots)$. From (9.39) we have for $k = 1$,

$$P\{U_1 \leq t\} = P\{T_1 \leq t\} = 1 - e^{-\lambda t},$$

so that U_1 has exponential distribution with mean $1/\lambda$. Next,

$$P\{U_2 > t | T_1 = \tau\} = P\{U_2 > t | U_1 = \tau\}$$
$$= P\{\text{no events in } (\tau, t + \tau) | T_1 = \tau\}$$
$$= P\{\text{no events in } (\tau, t + \tau)\}$$
$$= e^{-\lambda t},$$

which means that U_2 also has exponential distribution with mean $1/\lambda$ and is independent of U_1. Since the argument can be repeated for all other U_j's, we proved

Theorem 9.8.2. *In a Poisson process, the time U_1 until the first event and the times $U_2, U_3, \ldots$ between subsequent events are independent random variables, each with the same exponential distribution with mean $1/\lambda$.*

Since the origin of the time scale $t = 0$ was chosen arbitrarily, this theorem asserts that if we start observing a Poisson process at an arbitrarily selected time, fixed or randomly chosen,* then the waiting time for the first event has the same distribution as the subsequent interevent times. This property is connected closely with the memoryless property of geometric distribution, specified in Theorem 9.4.2.

Indeed, we have

Theorem 9.8.3. *If X is a random variable with exponential distribution, then for all $s, t > 0$,*

$$P\{X > s + t | X > s\} = P\{X > t\}. \tag{9.40}$$

Conversely, if X may assume only positive values and satisfies (9.40) for all $s, t > 0$, then X has exponential distribution.

Proof. Let X have exponential distribution. The left-hand side of (9.40) is the ratio

$$\frac{P\{X > s + t\}}{P\{X > s\}} = \frac{e^{-\lambda(s+t)}}{e^{-\lambda s}} = e^{-\lambda t} = P\{X > t\}.$$

Conversely, (9.40) implies that the tail of the cdf of X, that is, the function $\phi(x) = P\{X > x\}$, satisfies the equation $\phi(s + t) = \phi(s)\phi(t)$. One can show (see, e.g., Feller, 1957) that any bounded solution of this equation must be of the form $\phi(t) = e^{-\lambda t}$ for some $\lambda > 0$. □

---

*The phrase "randomly chosen" ought to be qualified here. Indeed, suppose that the "random choice" is to start observing the Poisson process 5 minutes *before* the next event. Technically, such a choice gives a random moment of beginning observation (since the time of event is random), and naturally, for such a choice $U_1 = 5$ minutes. The qualification of "random choice" here is that the decision may depend on the past but *not* on the future of the process.

The property of a Poisson process discussed above is one of the arguments for the claim that assumptions of a Poisson process capture "maximal randomness." One may try to contrast Poisson process, where knowledge of the past does not give the clue to the future, with other streams of events. For instance, if one arrives at a bus stop just after a bus has left (knowledge of the past), one may expect a longer wait for the next bus.

The following theorem shows that in Poisson processes, knowledge of the number of events in the past gives us, in a sense, no additional information not only about the future, but also about the past. We have the following:

Theorem 9.8.4. *The events in Poisson process satisfy the following property: For every $0 < u < t$*

$$P\{T_1 < u | X_t = 1\} = \frac{u}{t}. \tag{9.41}$$

Proof. We have

$$P\{T_1 < u | X_t = 1\} = \frac{P\{T_1 < u, X_t = 1\}}{P\{X_t = 1\}}$$

$$= \frac{P\{T_1 < u, X_t = 1\}}{\lambda t e^{-\lambda t}}. \tag{9.42}$$

Now, conditioning on the time of occurrence of the first event and using the fact that T_1 has exponential distribution, we may write for the numerator in (9.42)

$$P\{T_1 < u, X_t = 1\} = \int_0^u P\{X_t = 1 | T_1 = z\}\lambda e^{-\lambda z} dz.$$

If $T_1 = z$, the event $X_t = 1$ occurs if the time U_2 between first and second events exceeds $t - z$, so that

$$P\{X_t = 1 | T_1 = z\} = P\{U_2 > t - z\} = e^{-\lambda(t-z)}.$$

Substituting (9.40), we obtain the formula (9.41). □

This theorem asserts that if we know that only one event occurred between 0 and t, then the conditional distribution of the time of occurrence of this event is uniform on $(0, t)$. In a sense, therefore, we have no information as to when the event occurred. Actually, this theorem can be generalized as follows.

Theorem 9.8.5. *For Poisson process, the conditional joint distribution of $T_1, T_2, \ldots, T_n$ given $X_t = n$ is the same as the joint distribution of n order statistics from the uniform distribution on $(0, t)$.* □

We omit the proof here: The argument is similar as that used in the proof of Theorem 9.8.4. The joint density of $(T_1, \ldots, T_n)$ given $X_t = n$ can be obtained from Theorem 7.6.3.

Finally, let us observe that we can obtain the unconditional density of T_k by differentiating cdf given by formula (9.39). We have, after simple algebra:

Theorem 9.8.6. *The density $f_{T_k}(t)$ of the time T_k of the kth event in Poisson process is*

$$f_{T_k}(t) = \frac{\lambda^k}{(k-1)!} t^{k-1} e^{-\lambda t}, \qquad t > 0. \tag{9.43}$$

□

The distribution of T_k is sometimes called the *Erlang distribution*. As we shall see later, (9.43) is a special case of the gamma density.

Example 9.8.4. Fires in a certain town occur according to a Poisson process. If there were 10 fires in a given week, what is the probability that at least one of them occurred on Friday?

SOLUTION. What is of interest here is that we do not need to know the intensity λ of the Poisson process in question. We know that given that the number of fires was 10, their times of occurrence fall within a week according to the uniform distribution. Probability that a single fire does not fall on Friday is $\frac{6}{7}$; hence the chance of at least one of 10 fires falling on Friday is $1 - (\frac{6}{7})^{10} = 0.7859$.

Example 9.8.5. Suppose that accidents at a given intersection occur according to a Poisson process, with the rate on Saturdays being twice the rate on weekdays and the rate on Sundays being double the rate on Saturdays. The total rate is about 5 accidents per week. What is more likely: two accidents on each of two consecutive weekends (Saturday + Sunday), or a total of four accidents on weekdays in a given week?

SOLUTION. If λ is the average number of accidents on a weekday, then it is 2λ on a Saturday and 4λ on a Sunday. Consequently, we have $5\lambda + 2\lambda + 4\lambda = 5$, which gives $\lambda = \frac{5}{11}$. The number of accidents on a weekend is the sum of the numbers of accidents on Saturday and on Sunday. These are independent Poisson random variables; hence their sum (see Theorem 9.7.1) also has Poisson distribution, with mean $2\lambda + 4\lambda = \frac{30}{11}$. Consequently, the first probability is $P\{X = 2\}^2 = [(\frac{30}{11})^2 e^{-30/11}/2!]^2 = 0.0592$, since $X \sim \text{POI}(\frac{30}{11})$. The second probability is $P\{Y = 4\} = (\frac{25}{11})^4 e^{-25/11}/4! = 0.1145$, where now $Y \sim \text{POI}(5\lambda) = \text{POI}(\frac{25}{11})$,

To continue, suppose that on a Friday (which happened to be Friday the 13th) the number of accidents was as high as the number of accidents on the following weekend. Is such an event unusual?

To determine whether such an event can be attributed to sheer chance, or rather, that some "dark forces" are operating here, let us evaluate the chances of such an occurrence without reference to magic connected with Friday the 13th, that is, probability $P\{X = Y\}$, where X and Y are independent, $X \sim$ POI(λ) and $Y \sim$ POI(6λ) for $\lambda = \frac{5}{11}$.

We have

$$P\{X = Y\} = \sum_{j=0}^{\infty} P\{X = j\}P\{Y = j\}$$

$$= \sum_{j=0}^{\infty} \frac{\lambda^j}{j!} e^{-\lambda} \frac{(6\lambda)^j}{j!} e^{-6\lambda}$$

$$= \sum_{j=0}^{\infty} \frac{(6\lambda^2)^j}{(j!)^2} e^{-7\lambda}$$

$$= 0.0415 + 0.0515 + 0.0159 + 0.0022 + 0.0002 + \cdots$$

$$= 0.1113.$$

As we see, the chances are here slightly over 10%; hence such an event need not necessarily be regarded as highly unusual.

However, as an observant reader may note, almost all of this probability is due to the "unattractive" possibility that $X = Y = 0$ or $X = Y = 1$. Similarly, the probability that there are more accidents on Friday the 13th than on the entire following weekend is still not negligible:

$$P(X > Y) = \sum_{k=1}^{\infty} P\{X = k\} \sum_{j=0}^{k-1} P\{Y = j\} = 0.0406.$$

This time most of the probability is contributed by the terms $P\{X = 1\}$ $P\{Y = 0\}$ and $P\{X = 2\}[P\{Y = 0\} + P\{Y = 1\}]$.

Example 9.8.6. Peter decided to sell his bicycle. He posted ads at various places on the campus. From time to time he receives offers from prospective buyers at a rate, say, of 2 offers per day. The offers occur according to a Poisson process. Assume also that the amounts offered by consecutive buyers, say $X_1, X_2, \ldots$, are random variables, independent of one another and independent of the Poisson process of times of offers. Moreover, assume that $X_1, X_2, \ldots$ all have the same distribution $G(u) = P\{X_i \leq u\}, i = 1, 2, \ldots, u > 0$.

Peter decided not to haggle about price, but simply to sell the bike to the first person who offered the price exceeding some threshold u_0. The questions one may now ask are:

(a) What is the probability distribution of the time when Peter sells his bike?

(b) How many offers can he expect to reject before making the sale?

(c) What is the expected price that he will get?

SOLUTION.   Of the questions above, only the first really concern Poisson processes; that is, it involves the temporal dynamics of receiving the offers. Indeed, as regards (c), the expected price is simply $E(X|X > u_0)$. If we assume that X has density g, then the conditional density of X, given $X > u_0$, is $g(x)/[1 - G(u_0)]$, so that

$$E(X|X > u_0) = \frac{1}{1 - G(u_0)} \int_{u_0}^{\infty} xg(x) \, dx.$$

To answer (b), note that the successive offers are either "unacceptable," which happens with probability $P\{X_i \le u_0\} = G(u_0)$, or "acceptable," with probability $1 - G(u_0)$, and that Peter sells the bike at the first acceptable offer. Thus, the number of offers he turns down is the number of failures preceding the first success in Bernoulli trials, with the probability of success being $1 - G(u_0)$. The expected number of rejected offers is therefore $G(u_0)/[1 - G(u_0)]$ by formula (9.9).

As regards (a), we are in the situation of a Poisson process (of offers) in which the events are classified into two categories: acceptable and unacceptable, this classification being independent from event to event. The probability that Peter will not sell his bike before time t is the probability that all offers (if any) before t are unacceptable. Letting T^* be the time of sale, we have

$$
\begin{aligned}
P\{T^* > t\} &= \sum_{n=0}^{\infty} P\{X_t = n\}[G(u_0)]^n \\
&= \sum_{n=0}^{\infty} \frac{(\lambda t)^n}{n!} e^{-\lambda t}[G(u_0)]^n \\
&= e^{-\lambda t} \sum_{n=0}^{\infty} \frac{[\lambda G(u_0)t]^n}{n!} \\
&= e^{-\lambda[1 - G(u_0)]t}.
\end{aligned}
$$

Thus, the time T^* of sale has exponential distribution with parameter $\lambda[1 - G(u_0)]$ and therefore

$$E(T^*) = \frac{1}{\lambda[1 - G(u_0)]}.$$

It is reasonable to assume that Peter wants to maximize the amount received, and also to minimize the waiting time until the sale. These two are negatively related, so Peter might wish to find u_0 that maximizes (say) the function

$$\phi(u_0) = \frac{A}{1 - G(u_0)} \int_{u_0}^{\infty} xg(x)\,dx - \frac{1}{\lambda[1 - G(u_0)]},$$

where A is a constant measuring how "impatient" he is (i.e., to which extent he may trade the loss due to the decreasing price obtained for the bike and the gain due to decreasing waiting for that price). □

The concept of a Poisson process admits a number of generalizations, of which some are presented below. First, note that the symbol t need not be interpreted as time: It may be some other attribute interpretable as linear dimension. Thus, one can regard faults on a magnetic tape, or misprints in a text (regarded as a continuous string of letters), as Poisson processes, provided that one can reasonably expect that the postulates of a Poisson process hold. We illustrate the situation by the following example.

Example 9.8.7. Consider weak spots occurring on a rope at random places along its length. Suppose that the locations of these weak spots satisfy the postulates of Poisson process. Moreover, assume that the strength of the rope at each weak spot is random. Let $\xi_1, \xi_2, \ldots$ be the strengths at consecutive weak spots (i.e., the rope will break at the ith weak spot if the force applied to it exceeds ξ_i). We assume that ξ_i are independent random variables, all with the same distribution $G(u) = P\{\xi_i \le u\}$. Between weak spots the rope is assumed to be infinitely strong. The rope is of uniform thickness, with the unit length of the rope weighting c. A piece of the rope of length L is hanging loose. What is the probability that the rope will break under its own weight?

SOLUTION. Without loss of generality we may measure the length from the lowest end of the rope. Let $t_1 < t_2 < \cdots < t_N < L$ be the locations of the weak spots on the hanging piece of rope. Here N is random, with

$$P\{N = n\} = \frac{(\lambda L)^n}{n!} e^{-\lambda L}, \qquad n = 0, 1, \ldots$$

where λ is the average number of weak spots in unit length. The rope will break if for some $i \le N$ the weight ct_i of the fragment of the rope below the ith weak spot exceeds the strength ξ_i. Thus, the rope will not break if

$$ct_i < \xi_i, \qquad i = 1, 2, \ldots, N.$$

For fixed N and $t_1 < t_2 < \cdots < t_n$, the probability that the rope will not break is

$$\prod_{i=1}^{n} P\{\xi_i > ct_i\} = \prod_{i=1}^{n} [1 - G(ct_i)].$$

We know from Theorem 9.8.5 that, given $N = n$, the locations $t_1 < \cdots < t_n$ of weak spots have the same joint distribution as the order statistics of n points sampled independently with uniform distribution from $[0, L]$. Thus, the probability that the rope will not break at a weak spot randomly located on $[0,L]$ is

$$\pi = \int_0^L [1 - G(ct)] \frac{dt}{L}.$$

Consequently, the probability that the length L of the rope will not break under its own weight is

$$P_L = \sum_{n=0}^{\infty} \frac{(\lambda L)^n}{n!} e^{-\lambda L} \left\{ \int_0^L [1 - G(ct)] \frac{dt}{L} \right\}^n$$

$$= e^{-\lambda L} \sum_{n=0}^{\infty} \frac{1}{n!} \left\{ \lambda \int_0^L [1 - G(ct)] dt \right\}^n$$

$$= e^{-\lambda \int_0^L G(ct) dt}.$$

Clearly, $\lim_{L \to \infty} P_L = 0$, which agrees with our expectation that a sufficiently long piece of rope is bound to break under its own weight.

Example 9.8.8.   A small town has, on average, 20 fires a year. The fire department has one fire truck, which is quite sufficient for most purposes, except when a new fire starts soon (say, within 4 hours) after the beginning of the current fire.

The decision regarding the purchase of the second fire truck depends on the probability of two fires starting within 4 hours at least once during the year, so the objective is to estimate this probability. Let us assume that fires start according to a Poisson process. Taking a day as a unit of time, we have $\lambda = 20/365 = 0.0548$. If $X = X_{365}$ is the number of fires in a given year, and A is the event "at least one pair of fires within time $\tau = \frac{1}{6}$ (4 hours) from one another," then conditioning on X,

$$P(A) = \sum_{k=0}^{\infty} P\{A|X = k\} P\{X = k\} = \sum_{k=0}^{\infty} P\{A|X = k\} \frac{(365\lambda)^k}{k!} e^{-365\lambda}.$$

$$(9.44)$$

Clearly, $P\{A|X = 0\} = P\{A|X = 1\} = 0$, while for $k \geq 2$ we have

$$P\{A|X = k\} = P\left\{ \min_{2 \leq i \leq k} (\xi_i - \xi_{i-1}) < \frac{1}{6} \right\}, \qquad (9.45)$$

where $0 < \xi_1 < \xi_2 < \cdots < \xi_k < 365$ are order statistics from the sample $(U_1, \ldots, U_k)$, where U_i's are independent, each with distribution uniform on $(0, 365)$.

Since the joint distribution of $(\xi_1, \ldots, \xi_k)$ is known (see Theorem 7.6.3), the probabilities (9.45) can be calculated, at least in principle. However, these calculations are quite cumbersome, and it may be worthwhile to look for some approximations to probability (9.45). We shall present here two such approximate solutions.

Approximation 1. Given that we have a fire, the probability α that the next one will start within 4 hours can be obtained from Theorem 9.8.2. If T and T' are times of beginnings of two consecutive fires, then

$$\alpha = P\left\{T' - T < \frac{1}{6} \middle| T = \tau\right\} = P\left\{T' - T < \frac{1}{6}\right\} = 1 - e^{-(1/6)\lambda}$$
$$= 0.0091.$$

Since this probability α is only about 1%, we can consider the following approximation: Each fire during the year may be "single" or "double," the latter happening with probability α. Then

$$P\{A|X = k\} = P\{\text{at least one double fire } |k \text{ fires}\}$$
$$= 1 - (1 - \alpha)^k.$$

Substitution to (9.44) gives

$$P(A) = \sum_{k=0}^{\infty} [1 - (1 - \alpha)^k] \frac{(365\lambda)^k}{k!} e^{-365\lambda}$$
$$= 1 - \sum_{k=0}^{\infty} (1 - \alpha)^k \frac{(365\lambda)^k}{k!} e^{-365\lambda}$$
$$= 1 - e^{-365\lambda\alpha} = 1 - e^{-0.182}$$
$$= 0.1664.$$

Thus, the chances of at least one multiple fire in a year are about 16%; hence such an event may be expected to occur about once in every six years.

Approximation 2. One can also try to approximate $P\{A|X = k\}$ as follows. If $X = k$, then the times of k fires are distributed uniformly over the period of one year. Thus, for $k \geq 2$, we may write

$$P\{A^c|X = k\} \approx \left(1 - \frac{1}{3 \times 365}\right)\left(1 - \frac{2}{3 \times 365}\right) \cdots \left(1 - \frac{k-1}{3 \times 365}\right). \quad (9.46)$$

The argument for (9.46) is as follows. After choosing the time t_1 of one fire (not necessarily the first), let us "remove" the interval $(t_1 - 4 \text{ hours}, t_1 + 4 \text{ hours})$, sample the time t_2 of the next fire from the remainder of the year, and so on. Each time we should remove the interval of 8 hours $(\frac{1}{3 \times 365}$ of a year), or less if the intervals around fires overlap but their centers are still 4 or more hours apart. Since X has average 20 and variance 20, standard deviation is $\sqrt{20} = 4.47$. Even conservative bounds, such as the Chebyshev inequality (8.60), make it unlikely that X deviates from its mean by more than 3 standard deviations, so X is unlikely to exceed 35 fires (say). For such a small number of fires, the chances that the intervals of length of 8 hours (which surround the times of fires) overlap are negligible, and approximation (9.46) is acceptable. Numerical calculations now yield

$$P(A) = 1 - P(A^c)$$

$$= 1 - [P(X = 0) + P(X = 1) + \sum_{k=2}^{35} \prod_{j=1}^{k-1} \left(1 - \frac{j}{3 \times 365}\right) P(X = k)]$$

$$= 0.1655.$$

Extending the summation above to $k = 40$ gives the answer 0.1651, which means that the possibility of over 35 fires does not contribute significantly to the answer. Both approximations give essentially the same answer.   □

Finally, another generalization of the Poisson process concerns the extension to a higher dimension. Instead of events occurring in time (i.e., random points on a line), one can consider the case of points allocated at random on the plane or in space. The postulates of the Poisson process in such cases are analogous to the postulates in one dimension. The basic random variable is $X(A) = $ number of points falling into A, where A is a set on the plane or in space. The main postulate asserts that the number of points that fall into disjoint sets are independent. The second postulate asserts that the probability that $X(A) = 1$ depends on the size of the set A, not on its location, and equals $\lambda|A| + o(|A|)$, where $|A|$ stands for area or volume of A. Finally, the third postulate asserts that the probability that $X(A) \geq 2$ is of the order $o(|A|)$. Under these postulates, one can show that

$$P\{X(A) = k\} = \frac{(\lambda|A|)^k}{k!} e^{-\lambda|A|}, \qquad k = 0, 1, \ldots, \tag{9.47}$$

where $|A|$ stands for the area or volume of the set A.

As in the one-dimensional case, λ is the expected number of points falling into a region of unit size.

Example 9.8.9.   The data below are taken from Feller (1957), who lists them among examples of phenomena fitting Poisson distribution. The observations concerned points of hits in south London by flying bombs during World War II. The entire area under study was divided into $N = 576$ small areas of 0.25 square kilometer each, and the numbers N_k of areas that were hit k times were counted. The total number of hits is $T = \sum kN_k = 537$, so that the average number of hits per area is $\lambda = T/N = 0.9323$. The fit of the Poisson distribution is excellent, as may be seen from comparison of the actual numbers N_k and expected numbers $NP\{X = k\} = N(\lambda^k/k!)e^{-\lambda}$ for $\lambda = 0.9323$.

k	0	1	2	3	4	5 and over
N_k	229	211	93	35	7	1
$NP\{X = k\}$	226.74	211.39	98.54	30.62	7.14	1.57

The chi-square goodness-of-fit criterion (discussed in subsequent chapters) shows that in about 88% of cases one should expect worse agreement.

This example has become sort of "classic" in the sense of being reproduced in numerous textbooks on statistics, invariably without any comments (other than remarks that the fit is very good). The readers may therefore get the impression that statisticians have somewhat ghoulish interests. In fact, however, the fit to Poisson distribution was a piece of information of considerable value as military intelligence: It showed the state of German technology as regards the precision of their aiming devices. Perfect randomness of hits in a given large area indicated that it was not possible to select a specific target within this area.

We cite here the comment by Feller. He writes "It is interesting to note that most people believed in a tendency of the points of impact to cluster. If this were true, there would be a higher frequency of areas with either many hits or no hit, and a deficiency in the intermediate classes. The data indicates perfect randomness and homogeneity of the area; we have here an instructive illustration of the established fact that to the untrained eye randomness appears as regularity or tendency to cluster." It appears that Feller fully knew the reason for collecting and analyzing the data in question, but could only make a veiled allusion: His book was first published in 1950, just five years after the end of World War II, when many things were still secret.

Example 9.8.10. Assume that trees in a Poisson forest are located in such a way that centers of their trunks follow a Poisson process on the plane. Assume, in addition, that each tree has a trunk of the same diameter, say $d = 2$ feet (we disregard here the possibility of centers of tree trunks being less than 2 feet apart).

You take a walk in the Poisson forest. From time to time you stop at a randomly chosen place, look in a randomly chosen direction, and measure how far you can see in this direction. After a number of such measurements it turns out that on average, you can see at a distance of 800 feet. (a) What is the number of trees growing on a square mile of Poisson forest? (b) What is, on average, the distance from a tree to its nearest neighbor?

SOLUTION.   To answer (a) we must know λ, the average number of centers of tree trunks (hence the average number of trees) per unit area. Let us choose 1 foot as a unit of length. The key to connecting λ with the data is to find the theoretical formula for the average distance of unobstructed vision. Suppose that you are standing at point A, and let V be the distance of unobstructed vision in the horizontal direction. Then the event $V > x$ occurs if the rectangle with height d (diameter of the tree) and base x, as well as the semicircle at the end of the area (see Figure 9.2), are free of the centers of the trunks of trees. The probability of this event is the zero term of the Poisson distribution (9.47), where $|A|$ is the area of the figure in question, equal to $|A| = xd + \pi d^2/2$. Consequently,

$$P\{V > x\} = e^{-\lambda(xd + \pi d^2/4)} \approx e^{-d\lambda x}$$

(the approximation is justified in view of the fact that "typical" such rectangle of vision is several hundred feet long, so the contribution of the area of the semicircle at one end is small). We therefore see that V has exponential distribution with mean $1/\lambda d$. Since $d = 2$, we have the relation $1/2\lambda = 800$, hence $\lambda = 1/1600$. Thus there is, on average, one tree in 1600 square feet. Now a mile has about 5000 feet, so the square mile has 25,000,000 square feet; hence there are about $25 \cdot 10^6/1600 = 15{,}625$ trees per square mile in the Poisson forest.

As regards question (b), let R be the distance to the nearest neighbor from a randomly chosen tree (as measured between centers of trunks). Then $R > x$ if the circle of radius R around the center of the chosen tree contains no other centers of trees. This is given again as the zero term of the Poisson distribution (9.47), with $|A| = \pi x^2$. Thus

$$P\{R > x\} = e^{-\lambda \pi x^2},$$

and differentiating, we obtain the density of R as

$$f_R(x) = 2\lambda \pi x e^{-\lambda \pi x^2}.$$

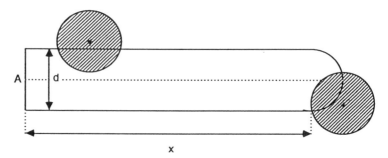

Figure 9.2   Visibility in a Poisson forest.

Consequently,

$$E(R) = \int_0^\infty x f_R(x)\, dx = 2\lambda\pi \int_0^\infty x^2 e^{-\lambda\pi x^2}\, dx.$$

The last integral can easily be evaluated (e.g., integrating by parts or adjusting the constant so as to obtain an integral expressing the variance of normal distribution); it equals $1/2\sqrt{\lambda} = 20$. Thus, on average, the nearest neighbor is 20 feet away.

Example 9.8.11. At the end, we shall give an example involving a Poisson process in three dimensions.

The Poisson Bakery makes a special kind of cookies called Four-Raisin cookies. These are made as follows: 10,000 raisins are added to the dough for 2500 cookies. After thorough mixing, the dough is divided mechanically into 2500 equal parts, of which individual cookies are formed and baked. The first question one may ask: What is the proportion of Four-Raisin cookies that have, in fact, no raisins at all? What proportion have exactly four raisins?

SOLUTION. We have here a spatial Poisson process, with raisins playing the role of points located randomly in space. If we take a cookie as a unit of volume, then $\lambda = 4$, as there are, on average, four raisins per cookie. If we let X denote the number of raisins in a randomly selected cookie, then $P\{X = 0\} = e^{-4} = 0.0183$, so that slightly below 2% of all Four-Raisin cookies are raisin-less. On the other hand, $P\{X = 4\} = (4^4/4!)e^{-4} = 0.1954$.

Now suppose that you buy a box of 200 Four-Raisin cookies. What is the probability that no more than two of them have no raisins? We have here a situation of Poisson approximation to the binomial distribution, with "success" being a cookie with no raisins, so that $np = 200e^{-4} = 3.66$. The number Y of raisinless cookies in the box has binomial distribution BIN(200, 0.0183), which is approximated by a Poisson distribution with $\lambda = np = 3.66$. Thus, $P\{X \leq 2\} = P\{Y = 0\} + P\{Y = 1\} + P\{Y = 2\} \approx (1 + 3.66 + 3.66^2/2)e^{-3.66} = 0.29$; hence $P\{Y \geq 3\}$ is about 0.71, which means that 71% of all boxes will contain three or more Four-Raisin cookies which in fact have no raisins.

PROBLEMS

9.8.1 Accidents on a given intersection occur following a Poisson process. Given that 10 accidents occurred in June, what is the probability that the seventh accident occurred before June 10?

9.8.2 Referring to Problem 9.8.1, given that n accidents occurred in April, what is the expected number of accidents that occurred during the second week of April?

9.8.3   Consider two independent Poisson processes with the same parameter λ. Let $N_i(t)$, $i = 1, 2$, be the number of events in the ith process that occurred up to time t, and let U_T be the total length of all those time-intervals of t with $0 \le t \le T$ at which $N_1(t) = N_2(t)$. Given that $N_1(T) = N_2(T) = 2$, find $E(U_T)$. (*Hint:* Use the fact that the sum of two independent Poisson processes is a Poisson process and Theorem 9.8.5. The answer is $8T/15$.)

9.8.4   Answer Problem 9.8.3 given that $N_1(T) = 2$, $N_2(T) = 3$.

9.8.5   Consider k independent Poisson processes running concurrently. Let λ_j be the intensity of the jth process (you can imagine various kinds of accidents at a given intersection, each kind occurring according to a Poisson process). Find the distribution, mean, and variance of the time to the nearest event (in whichever process), counting from time $t = 0$.

9.8.6   Assume that chocolate chips are distributed within a cake according to a Poisson process with parameter λ. A cake is divided into two parts of equal volume (disregard the possibility of cutting through a chocolate chip). Show that the probability that each part of the cake has the same number of chips is given by

$$e^{-\lambda} \sum_{k=0}^{\infty} \frac{(\lambda/2)^{2k}}{(k!)^2}.$$

9.9   EXPONENTIAL AND GAMMA DISTRIBUTIONS

We have encountered the exponential distribution on a number of occasions, notably as the distribution of times between events in Poisson process in one dimension. To repeat the definition for completeness, a random variable X has exponential distribution EXP(λ) if for some $\lambda > 0$,

$$F(x) = P\{X \le x\} = 1 - e^{-\lambda x}, \qquad x \ge 0$$

so that the density of X is

$$f(x) = \lambda e^{-\lambda x}, \qquad x \ge 0.$$

We know that

$$E(X) = \frac{1}{\lambda}, \qquad \text{Var}(X) = \frac{1}{\lambda^2}$$

and the mgf of X is

$$m_X(s) = \frac{\lambda}{\lambda - s}$$

for $s < \lambda$.

The hazard function (see Section 6.5) of exponential random variable is constant:

$$h(x) = \frac{f(x)}{1 - F(x)} = \frac{\lambda e^{-\lambda x}}{e^{-\lambda x}} = \lambda,$$

a property closely related to the memoryless property of exponential distribution, asserted in Theorem 9.8.3.

As a warning to readers, one should mention that the phrase "exponential distribution with parameter λ" is ambiguous, since many authors tend to write the density of exponential distribution in the form $f(x) = \frac{1}{\lambda}e^{-x/\lambda}, x > 0$. In this notation we have $E(X) = \lambda$ and $\text{Var}(X) = \lambda^2$. Consequently, unless it is clear whether the parameter appears in the numerator or denominator of the exponent, we shall use phrases that convey the information about the mean. Thus, "exponential distribution with mean θ" will have density $(1/\theta)e^{-x/\theta}$, and so on.

Next, the sums of independent and exponentially distributed random variables appeared also in the Poisson process, as times $T_1, T_2, \ldots$ of consecutive events. The cdf's of these random variables were obtained using the identity

$$P\{T_r > t\} = P\{X_t < r\} = \sum_{k=0}^{r-1} \frac{(\lambda t)^k}{k!} e^{-\lambda t}.$$

The density of T_r follows now by the differentiation, namely

$$f_r(t) = \frac{\lambda^r}{(r-1)!} t^{r-1} e^{-\lambda t}, \qquad t > 0. \tag{9.48}$$

Since T_r is the sum of r independent waiting times, each with the same exponential distribution, the mgf of T_r exists for $s < \lambda$ and equals

$$m_{T_r}(s) = \left(\frac{\lambda}{\lambda - s} \right)^r.$$

The definition of distribution of T_r as the sum of waiting times involves using an integer value of r. We have a complete analogy with the binomial distribution (count of successes in a fixed number of trials) and negative binomial distribution (count of number of trials up to a fixed number of successes) on the one hand, and Poisson distribution (number of events until a fixed time) and distribution (9.48) (time until a fixed number of events occurs). *Negative Poisson distribution* could be an appropriate name. When used in the context of servicing systems, the name *Erlang distribution* is used. However, as in the case of negative binomial distribution, r need not be an integer, and we can define the class of distributions comprising densities (9.48) as a special case. To this end let us introduce the following definition.

Definition 9.9.1.   For $t \geq 0$ we define the *gamma function* as

$$\Gamma(t) = \int_0^\infty x^{t-1} e^{-x}\, dx. \tag{9.49}$$

□

One can show that this function is well defined for all $t \geq 0$. Moreover, integration by parts gives

$$\Gamma(t) = (t-1)\Gamma(t-1), \tag{9.50}$$

and since $\Gamma(1) = 1$, we obtain by induction, for any integer $n \geq 1$,

$$\Gamma(n) = (n-1)!$$

We may now introduce the following definition.

Definition 9.9.2.   A random variable with density of the form

$$f(x) = \begin{cases} Cx^{\alpha-1}e^{-\lambda x} & \text{for } x > 0 \\ 0 & \text{for } x < 0 \end{cases}$$

for some $\alpha > 0$ and $\lambda > 0$ is said to have a *gamma distribution* with *shape* parameter α and *scale* parameter λ. Here C is the normalizing constant.

□

We shall use the symbol $X \sim \text{GAMMA}(\alpha, \lambda)$ if X has a gamma distribution with parameters α and λ.

We start by determining the value of C. We must have

$$C \int_0^\infty x^{\alpha-1} e^{-\lambda x}\, dx = 1.$$

Substituting $\lambda x = z$, we obtain easily from (9.49)

$$C = \frac{\lambda^\alpha}{\Gamma(\alpha)}. \tag{9.51}$$

We can now compute the moments of gamma distribution. Indeed,

$$\begin{aligned}
E(X) &= \int_0^\infty x f(x)\, dx \\
&= \frac{\lambda^\alpha}{\Gamma(\alpha)} \int_0^\infty x^\alpha e^{-\lambda x}\, dx \\
&= \frac{\lambda^\alpha}{\Gamma(\alpha)} \cdot \frac{\Gamma(\alpha+1)}{\lambda^{\alpha+1}} = \frac{\alpha}{\lambda}
\end{aligned} \tag{9.52}$$

in view of (9.50). Similarly,

$$E(X^2) = \int_0^\infty x^2 f(x)\, dx$$

$$= \frac{\lambda^\alpha}{\Gamma(\alpha)} \int_0^\infty x^{\alpha+1} e^{-\lambda x}\, dx$$

$$= \frac{\lambda^\alpha}{\Gamma(\alpha)} \frac{\Gamma(\alpha+2)}{\lambda^{(\alpha+2)}} = \frac{\alpha(\alpha+1)}{\lambda^2},$$

so that

$$\text{Var}(X) = E(X^2) - [E(X)]^2 = \frac{\alpha}{\lambda^2}. \tag{9.53}$$

The moment generating function of the gamma distribution can be evaluated as follows:

$$m(t) = \frac{\lambda^\alpha}{\Gamma(\alpha)} \int_0^\infty e^{tx} x^{\alpha-1} e^{-\lambda x}\, dx$$

$$= \frac{\lambda^\alpha}{\Gamma(\alpha)} \int_0^\infty x^{\alpha-1} e^{-(\lambda-t)x}\, dx$$

$$= \frac{\lambda^\alpha}{\Gamma(\alpha)} \frac{\Gamma(\alpha)}{(\lambda-t)^\alpha} = \frac{1}{(1-t/\lambda)^\alpha},$$

provided that $t < \lambda$. It follows at once that we have the following closure property of gamma distributions:

Theorem 9.9.1. *If X and Y are independent with $X \sim \text{GAMMA}(\alpha_1, \lambda)$, $Y \sim \text{GAMMA}(\alpha_2, \lambda)$, then $X + Y \sim \text{GAMMA}(\alpha_1 + \alpha_2, \lambda)$.*

Since the waiting time T_k for the kth event in a Poisson process has distribution $\text{GAMMA}(k, \lambda)$, Theorem 9.9.1 (in the case of integers α_1 and α_2) expresses the simple fact that waiting time $T_{\alpha_1+\alpha_2}$ is the sum of two independent waiting times, T_{α_1} and T_{α_2} (for event number α_1 in the original process and for event number α_2 in the process whose observation starts at time T_{α_1}).

Let us note that exponential distribution with parameter λ is the same as $\text{GAMMA}(1, \lambda)$. Also, let us recall that in Example 6.4.7 we found the density of the square of the standard normal variable [i.e., of $Y = Z^2$, where $Z \sim N(0,1)$]. This density equals $f(y) = (1/\sqrt{2\pi}) y^{-1/2} e^{-y/2}$, which we recognize as $\text{GAMMA}(\frac{1}{2}, \frac{1}{2})$. The sums of squares of independent standard normal variables appear so often in statistics that the distribution of such sums bears its own name. We have the following definition and theorem.

Definition 9.9.3. For integer ν, the distribution $\text{GAMMA}\left(\frac{\nu}{2}, \frac{1}{2}\right)$ is called the *chi-square* distribution with ν *degrees of freedom*. A random variable with such a distribution is typically denoted by χ_ν^2. □

Theorem 9.9.2. *If $Z_1, \ldots, Z_n$ are independent, each with standard normal distribution, then*

$$\chi_n^2 = Z_1^2 + \cdots + Z_n^2$$

has chi-square distribution with n degrees of freedom.

As regards the numerical values of probabilities for gamma distributions, the situation is as follows. If $X \sim \text{GAMMA}(\alpha, \lambda)$, then (substituting $u = \lambda x$)

$$P\{a \le X \le b\} = \frac{\lambda^\alpha}{\Gamma(\alpha)} \int_a^b x^{\alpha-1} e^{-\lambda x}\, dx$$

$$= \frac{1}{\Gamma(\alpha)} \int_{a\lambda}^{b\lambda} u^{\alpha-1} e^{-u}\, du, \tag{9.54}$$

so that the scale factor λ can always be eliminated. Repeated integration by parts can reduce the exponent α to the range $[0, 1]$, when one could begin using special tables of gamma integrals.

Moreover, if α is a multiple of $\frac{1}{2}$, one can use the tables of chi-square distribution. In this case the substitution in (9.54) is $\lambda x = u/2$, so that for $\alpha = \nu/2$ we have

$$P\{a \le X \le b\} = \frac{1}{2^\alpha \Gamma(\alpha)} \int_{2\lambda a}^{2\lambda b} u^{\alpha-1} e^{-u/2}\, du$$

$$= P\{2\lambda a \le \chi_\nu^2 \le 2\lambda b\}. \tag{9.55}$$

The answer, in principle, can now be obtained from tables of chi-square distribution with ν degrees of freedom (see Table A.4).

One should remark here, however, that tables of chi-square distribution that can be found in statistics textbooks give few selected quantiles or upper quantiles for numbers of degrees of freedom between 1 and 30. Such tables are not sufficient to evaluate probabilities (9.55).

Nowadays, there are numerous statistical packages that give values of cdf as well as quantiles of chi-square distributions for various numbers of degrees of freedom. In other words, for given ν and x, one can obtain $P\{\chi_\nu^2 \le x\}$, as well as for given ν and p, one can obtain x such that $P\{\chi_\nu^2 \le x\} = p$.

PROBLEMS

9.9.1 Show that for $n = 1, 2, \ldots,$

$$\frac{1}{\Gamma(n)} \int_0^y t^{n-1} e^{-t}\, dt = 1 - e^{-y} \sum_{j=0}^{n-1} \frac{y^j}{j!}.$$

(*Hint:* Integrate by parts and use induction.)

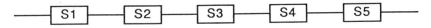

Figure 9.3 Series system.

9.9.2 Let X_1, X_2 be independent, each with exponential distribution EXP(λ). Find the density of $Y = X_1 - X_2$.

9.9.3 Find an expression for the pth quantile of a random variable X with exponential distribution EXP(λ). (The pth quantile of a continuous random variable X is x_p, defined by $P\{X \le x_p\} = p$. The value x_p always exists, but needs not to be unique.)

9.9.4 A system consists of five elements connected in series (see Figure 9.3), so that the system fails as soon as one of its components fails. Suppose that lifetimes of the components are independent, with exponential distributions EXP(λ_1), ..., EXP(λ_5). Find the cdf and density of $T = $ time to failure of the system.

9.9.5 Suppose now that the system of Problem 9.9.4 is connected in parallel (see Figure 9.4) Thus the system works as long as at least one component is operating. Find the cdf and density of $T = $ time to failure of the system.

9.9.6 Find the cdf and density of T if the system of Problem 9.9.4 is connected as in Figure 9.5.

9.9.7 In the flowchart of Figure 9.6, the block denoted "sample U" means that the computer samples a value of random variable U with distribution uniform on $(0, 1)$, the samplings being independent each time

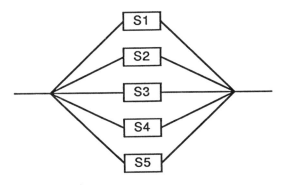

Figure 9.4 Parallel system.

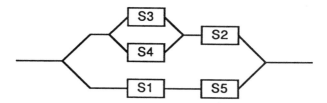

Figure 9.5   Series–parallel system.

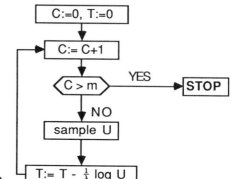

Figure 9.6   Flowchart.

the program executes this instruction. Assume that m is a positive integer and $\lambda > 0$. Find:

(i) $P\{T \leq 2\}$ if $\lambda = 2$ and $m = 3$.

(ii) The cdf and density of T, $E(T)$ and $\text{Var}(T)$ in general case.

9.9.8 Let T be a nonnegative random variable with the hazard rate $h(t) = t^r$. Find $E(T)$.

9.9.9 Find the expectation and variance of Weibull distribution (see Section 6.5) with parameters α and β.

9.9.10 Find $\Gamma(\frac{1}{2})$ and $\Gamma(\frac{7}{2})$.

9.10   NORMAL DISTRIBUTION

We have already encountered the normal distribution in Chapter 6. Let us recall that the univariate normal distribution with parameters μ and σ^2, denoted $N(\mu, \sigma^2)$, has density

$$f(x) = \frac{1}{\sigma\sqrt{2\pi}} e^{-(x-\mu)^2/2\sigma^2}, \qquad -\infty < x < \infty.$$

It has been shown in Example 6.3.13 that f is indeed a density. Moreover, if X has distribution $N(\mu, \sigma^2)$, then

$$E(X) = \mu, \qquad \text{Var}(X) = \sigma^2,$$

which gives a direct interpretation of the parameters.

The moment generating function of normal distribution can be written, completing the square in the exponent, as

$$m(t) = \frac{1}{\sigma\sqrt{2\pi}} \int_{-\infty}^{+\infty} e^{tx} \cdot e^{-(x-\mu)^2/2\sigma^2}\, dx$$

$$= e^{\mu t + \frac{1}{2}\sigma^2 t^2} \cdot \frac{1}{\sigma\sqrt{2\pi}} \int_{-\infty}^{+\infty} e^{-(1/2\sigma^2)[x-(\mu+\sigma^2 t)]^2}\, dx$$

$$= e^{\mu t + \frac{1}{2}\sigma^2 t^2}.$$

We therefore have the following closure properties of the normal distribution.

Theorem 9.10.1.   *A linear transformation of a normally distributed random variable again has a normal distribution. In particular, if random variable X has normal distribution* $N(\mu, \sigma^2)$, *then the random variable*

$$Z = \frac{X - \mu}{\sigma}$$

has the standard normal distribution $N(0,1)$.

Proof. We have, using (8.23),

$$m_Z(t) = m_{(1/\sigma)X - (\mu/\sigma)}(t) = m_X(t/\sigma)e^{-(\mu/\sigma)t} = e^{t^2/2},$$

and the right-hand side is the mgf of a standard normal random variable. $\square$

The property asserted in Theorem 9.10.1 has important practical consequences: It shows that to determine probabilities for *any* normally distributed random variable, it suffices to have access to probabilities for a standard normal random variable.

This property is especially important in view of the fact that the cdf of normal distribution, namely

$$\Phi(x; \mu, \sigma^2) = \frac{1}{\sigma\sqrt{2\pi}} \int_{-\infty}^{x} e^{-(t-\mu)^2/2\sigma^2}\, dt,$$

cannot be integrated in closed form. Thus, it is necessary to use the tables, and

Theorem 9.10.1 implies that one table is sufficient to calculate probabilities for all normal distributions. In the sequel, we let Z denote the standard normal random variable, with density

$$\phi(x) = \frac{1}{\sqrt{2\pi}} e^{-x^2/2}$$

and the cdf

$$\Phi(z) = \frac{1}{\sqrt{2\pi}} \int_{-\infty}^{z} e^{-x^2/2} \, dx. \tag{9.56}$$

The standard normal density is a familiar bell-shaped curve, symmetric about $x = 0$. Since $\phi'(x) = -x\phi(x)$ and $\phi''(x) = (x^2 - 1)\phi(x)$ we have $\phi''(x) = 0$ and $\phi'(x) \neq 0$ at $x = \pm 1$. This means that standard normal density has two inflexion points, at 1 and at -1. It follows therefore that the density of normal distribution $N(\mu, \sigma^2)$ is symmetric about μ and has inflexion points at $x = \mu \pm \sigma$.

Theorem 9.10.1 asserts, in effect, that given the tables of function (9.56), we have, for any $X \sim N(\mu, \sigma^2)$,

$$P\{a \le X \le b\} = P\left\{\frac{a-\mu}{\sigma} \le Z \le \frac{b-\mu}{\sigma}\right\}$$

$$= \Phi\left(\frac{b-\mu}{\sigma}\right) - \Phi\left(\frac{a-\mu}{\sigma}\right).$$

Since $\phi(x)$ is symmetric about 0, we have $P\{Z \le -z\} = P\{Z \ge z\}$; hence the cdf of the standard normal distribution satisfies the relation

$$\Phi(-z) = 1 - \Phi(z).$$

Consequently, many statistical tables give values of $\Phi(z)$ only for $z \ge 0$.

An inspection of Table A.2 shows that $\Phi(3) = 0.9987$. Since for $z > 0$ we have

$$P\{|Z| > z\} = P\{Z > z\} + P\{Z < -z\}$$
$$= 1 - \Phi(z) + \Phi(-z)$$
$$= 2\Phi(-z) = 2[1 - \Phi(z)],$$

we see that $P\{|Z| > 3\} = 0.0026$. Using Theorem 9.10.1, we see that if $X \sim N(\mu, \sigma^2)$, then

$$P\{|X - \mu| > 3\sigma\} = P\{|Z| > 3\} = 0.0026. \tag{9.57}$$

This explains the origin of the *three-sigma rule*, according to which one is allowed to disregard the possibility of a random variable deviating from its mean by more than 3 standard deviations.

While this rule works well for normal random variables, the Chebyshev inequality gives the bound $\frac{1}{9}$ for the left-hand side in (9.57). Example 8.9.5 shows that the bound in the Chebyshev inequality is strict, so that there are random variables for which $P\{|X - E(X)| \geq 3\sigma_X\} = \frac{1}{9}$.

Most tables of standard normal distribution do not give values of $\Phi(z)$ for $z > 3$. If such values are needed, we have

Theorem 9.10.2. *For $z > 0$, the function $\Phi(z)$ satisfies the inequality*

$$\left(\frac{1}{z} - \frac{1}{z^3}\right) \phi(z) \leq 1 - \Phi(z) \leq \frac{1}{z}\phi(z), \tag{9.58}$$

and consequently, as $z \to \infty$,

$$1 - \Phi(z) \sim \frac{1}{z}\phi(z). \tag{9.59}$$

Proof. We have the obvious inequality

$$\left(1 - \frac{3}{x^4}\right)\phi(x) \leq \phi(x) \leq \left(1 + \frac{1}{x^2}\right)\phi(x).$$

Simple check shows that this inequality is equivalent to

$$-\frac{d}{dx}\left[\left(\frac{1}{x} - \frac{1}{x^3}\right)\phi(x)\right] \leq -\frac{d}{dx}[1 - \Phi(x)] \leq -\frac{d}{dx}\left[\frac{1}{x}\phi(x)\right]. \tag{9.60}$$

Integrating (9.60) between z and ∞, we obtain (9.58). $\qquad\square$

Numerically, the relation (9.59) gives $1 - \Phi(4) \cong 3.36 \times 10^{-5}, 1 - \Phi(5) \cong 2.97 \times 10^{-7}, 1 - \Phi(6) \cong 1.01 \times 10^{-9}$. Such small probabilities are of interest in estimating the chances of, say, an accident in a nuclear power plant.

From a practical viewpoint, knowledge of these probabilities allows us to assume the normality of distribution in many cases where "logically" the distribution cannot possibly be normal. To illustrate the point, it is often assumed that such an attribute as, say, height in human population is normally distributed (e.g., among men, we have mean μ about 70 inches and standard deviation σ about 2 inches). Now, normally distributed random variable may *always* assume both positive and negative values. Consequently, one could argue that height—which cannot be negative—cannot have a normal distribution. But the chances of a random variable X with normal distribution N(70, 4) being negative are on the order $\Phi(-35) = 1- \Phi(35) \cong 0.0069 \times e^{-612}$. Events with such a probability can safely be disregarded, and it turns out that it is much easier to work with the assumption of normality than to invent a distribution for height that does not allow negative values.

We shall now prove a very useful and important property of normal distribution, and illustrate its use.

Theorem 9.10.3. *Let X and Y be independent random variables with distributions $X \sim N(\mu_1, \sigma_1^2)$ and $Y \sim N(\mu_2, \sigma_2^2)$. Then the random variable $U = \alpha X + \beta Y$ has distribution $N(\alpha\mu_1 + \beta\mu_2, \alpha^2\sigma_1^2 + \beta^2\sigma_2^2)$.*

Proof. Since $m_X(t) = \exp\{\mu_1 t + \sigma_1^2 \frac{t^2}{2}\}, m_Y(t) = \exp\{\mu_2 t + \sigma_2^2 \frac{t^2}{2}\}$, we have, using the independence of αX and βY,

$$m_U(t) = m_{\alpha X + \beta Y}(t) = m_{\alpha X}(t)m_{\beta Y}(t)$$
$$= m_X(\alpha t)m_Y(\beta t)$$
$$= \exp\left\{\mu_1 \alpha t + \sigma_1^2 \frac{\alpha^2 t^2}{2}\right\} \exp\left\{\mu_2 \beta t + \sigma_2^2 \frac{\beta^2 t^2}{2}\right\}$$
$$= \exp\left\{(\mu_1\alpha + \mu_2\beta)t + (\sigma_1^2\alpha^2 + \sigma_2^2\beta^2)\frac{t^2}{2}\right\}.$$

Since the mgf uniquely determines the distribution, we recognize the right-hand side as an mgf of distribution $N(\mu_1\alpha + \mu_2\beta, \sigma_1^2\alpha^2 + \sigma_2^2\beta^2)$. □

Example 9.10.1. Assume that the height of men in a certain population is normal with mean $\mu_M = 70$ inches and standard deviation $\sigma_M = 2$ inches. The height of women is also normal, with mean $\mu_W = 68$ inches and $\sigma_W = 1.5$ inches. One man and one woman are selected at random. What is the probability that the woman selected is taller than the man selected?

SOLUTION. Let X and Y be the heights of the randomly selected man (X) and woman (Y). We need $P\{Y > X\}$.

Without Theorem 9.10.3 we could proceed as follows (this solution is applicable to *any* distribution of X and Y, and is therefore of some general interest). If F and G are cdf's for X and Y, and f and g are the densities, then conditioning on values of Y, we have

$$P\{Y > X\} = \int P\{Y > X|Y = y\}g(y)\,dy$$
$$= \int P\{X < y\}g(y)dy = \int F(y)g(y)\,dy. \qquad (9.61)$$

We may also condition on values of X, obtaining

$$P\{Y > X\} = \int P\{Y > X|X = x\}f(x)\,dx$$
$$= \int P\{Y > x\}f(x)\,dx$$
$$= \int [1 - G(x)]f(x)\,dx.$$

Using now the assumption of normality of X and Y, we can write for (9.61) the double integral

$$P\{Y > X\} = \int_{-\infty}^{+\infty} \left[\int_{-\infty}^{y} \frac{1}{\sigma_1 \sqrt{2\pi}} e^{-(x-\mu_1)^2/2\sigma_1^2} \, dx \right] \frac{1}{\sigma_2 \sqrt{2\pi}} e^{-(y-\mu_2)^2/2\sigma_2^2} \, dy$$

$$= \frac{1}{2(3/2)2\pi} \int_{-\infty}^{+\infty} \int_{-\infty}^{y} e^{-(x-70)^2/2 \cdot 2^2 - (y-68)^2/2 \cdot (3/2)^2} \, dx \, dy.$$

However, using Theorem 9.10.3, one can proceed as follows: We have

$$P\{Y > X\} = P\{Y - X > 0\} = P\{U > 0\},$$

where $U = Y - X$ has normal distribution with mean $\mu = \mu_W - \mu_M = 68 - 70 = -2$ and $\sigma_U^2 = \sigma_W^2 + \sigma_M^2 = (\frac{3}{2})^2 + 2^2 = \frac{25}{4}$, hence $\sigma_U = 2.5$ inches (we use Theorem 9.10.3 here with $\alpha = -1, \beta = 1$). Consequently, since U is normally distributed,

$$P\{Y > X\} = P\left\{ Z > \frac{0 - (-2)}{2.5} \right\} = P\{Z > 0.8\}$$
$$= 1 - \Phi(0.8) = 1 - 0.7881 = 0.2119. \qquad \square$$

Let us now consider the multivariate normal distribution. We cover in detail the case of two dimensions only. Some situations with more than two dimensions will be considered in Chapter 15, but we omit the general theory, as it would involve too much digression into matrix algebra.

Definition 9.10.1. The pair (X, Y) of random variables is said to have a *bivariate normal distribution* if the joint density is of the form

$$f(x, y) = \frac{1}{2\pi \sigma_1 \sigma_2 \sqrt{1 - \rho^2}}$$

$$\exp\left\{ -\frac{1}{2(1-\rho^2)} \left[\frac{(x-\mu_1)^2}{\sigma_1^2} - 2\rho \frac{x - \mu_1}{\sigma_1} \cdot \frac{y - \mu_2}{\sigma_2} + \frac{(y-\mu_2)^2}{\sigma_2^2} \right] \right\},$$
$$(9.62)$$

where $-\infty < \mu_1, \mu_2 < +\infty$, $\sigma_1 > 0, \sigma_2 > 0$, and $|\rho| < 1$. $\qquad \square$

To find marginal distributions of X and Y and determine the interpretation of parameters, let us change the variables into

$$U = \frac{X - \mu_1}{\sigma_1}, \qquad V = \frac{Y - \mu_2}{\sigma_2}.$$

This amounts to substitution

$$x = \sigma_1 u + \mu_1, \qquad y = \sigma_2 v + \mu_2$$

with the Jacobian $J = \sigma_1 \sigma_2$. Consequently, the joint density of (U, V) is

$$g(u, v) = \frac{1}{2\pi\sqrt{1 - \rho^2}} \exp\left\{-\frac{1}{2(1 - \rho^2)}[u^2 - 2\rho uv + v^2]\right\}.$$

Since $u^2 - 2\rho uv + v^2 = (u - \rho v)^2 + (1 - \rho^2)v^2$, we may write

$$g(u, v) = \frac{1}{\sqrt{2\pi}\sqrt{1 - \rho^2}} \exp\left\{-\frac{1}{2}\left(\frac{u - \rho v}{\sqrt{1 - \rho^2}}\right)^2\right\} \frac{1}{\sqrt{2\pi}} \exp\left\{-\frac{v^2}{2}\right\}.$$

We have

$$\frac{1}{\sqrt{2\pi}\sqrt{1 - \rho^2}} \int_{-\infty}^{+\infty} \exp\left\{-\frac{1}{2}\left(\frac{u - \rho v}{\sqrt{1 - \rho^2}}\right)^2\right\} du = 1$$

for every v, hence the second factor is the marginal density of V, while the first factor is the conditional density of U given $V = v$. Thus, V is standard normal, while $U|V \sim N(\rho V, 1 - \rho^2)$. By symmetry, U must also be standard normal and $V|U \sim N(\rho U, 1 - \rho^2)$. We proved therefore

Theorem 9.10.4. *If (X, Y) have bivariate normal distribution given by (9.62), then both X and Y have normal distributions, with*

$$E(X) = \mu_1, \qquad \text{Var}(X) = \sigma_1^2$$
$$E(Y) = \mu_2, \qquad \text{Var}(Y) = \sigma_2^2.$$

Moreover, the conditional distributions of X given Y and Y given X are also normal; specifically,

$$X|Y = y \sim N\left(\rho\frac{\sigma_1}{\sigma_2}(y - \mu_2) + \mu_1, \sigma_1^2(1 - \rho^2)\right) \tag{9.63}$$

and

$$Y|X = x \sim N\left(\rho\frac{\sigma_2}{\sigma_1}(x - \mu_1) + \mu_2, \sigma_2^2(1 - \rho^2)\right). \tag{9.64}$$

Proof. Only the last two statements require some proof. Normality of the conditional distribution is obvious in view of the normality of conditional distributions of U given V. We have

$$E(X|Y) = E(\sigma_1 U + \mu_1 | \sigma_2 V + \mu_2) = \sigma_1 E(U | \sigma_2 V + \mu_2) + \mu_1$$
$$= \sigma_1 E(U|V) + \mu_1 = \sigma_1 \rho V + \mu_1 = \sigma_1 \rho \frac{Y - \mu_2}{\sigma_2} + \mu_1.$$

Similarly,

$$\text{Var}(X|Y) = \text{Var}(\sigma_1 U + \mu_1 | \sigma_2 V + \mu_2)$$
$$= \sigma_1^2 \text{Var}(U|V) = \sigma_1^2 (1 - \rho^2).$$

which proves (9.63). The proof of (9.64) is analogous.   □

It remains to interpret the parameter ρ. As may be expected, we have

Theorem 9.10.5.   *If X, Y have the bivariate normal distribution (9.62), then ρ is the coefficient of correlation between X and Y.*

Proof. Clearly, $\rho_{X,Y} = \text{Cov}(U, V) = E(UV)$, and we may write

$$E(UV) = \int \int uvg(u,v)\, du\, dv$$

$$= \frac{1}{2\pi\sqrt{1-\rho^2}} \int \int uv \exp\left\{-\frac{1}{2(1-\rho^2)}[u^2 - 2\rho uv + v^2]\right\} du\, dv$$

$$= \frac{1}{\sqrt{2\pi}} \int v e^{-v^2/2} \left[\frac{1}{\sqrt{2\pi}\sqrt{1-\rho^2}} \int u \exp\left\{-\frac{1}{2}\left(\frac{u-\rho v}{\sqrt{1-\rho^2}}\right)^2\right\} du \right] dv$$

$$= \frac{1}{\sqrt{2\pi}} \int v e^{-v^2/2}[\rho v]\, dv = \rho \frac{1}{\sqrt{2\pi}} \int v^2 \exp\left\{-\frac{v^2}{2}\right\} = \rho.   □$$

Example 9.10.2.   Assume that the height X of men in some population is normal with mean 70 inches and standard deviation 2 inches, while the height Y of women in this population is also normal with mean 68 inches and standard deviation 1.5 inches. Assume also that the heights of siblings are correlated with $\rho = 0.6$. If we sample a brother and a sister, what is the probability that she is taller than her brother?

SOLUTION.   We want now $P\{Y > X\}$. Proceeding as in Example 9.10.1, we can write $P\{Y > X\} = P\{Y - X > 0\}$, and we know that $Y - X$ is normal, with mean $E(Y) - E(X) = 68 - 70 = -2$ inches. For the variance we have

$$\text{Var}(Y - X) = \text{Var}(Y) - 2\,\text{Cov}(X, Y) + \text{Var}(X)$$
$$= \sigma_Y^2 - 2\rho\sigma_X\sigma_Y + \sigma_X^2$$
$$= (1.5)^2 - 2 \times 0.6 \times 2 \times 1.5 + 2^2 = 2.65.$$

Consequently,

$$P\{Y > X\} = P\left\{Z > \frac{0 - (-2)}{\sqrt{2.65}}\right\}$$
$$= 1 - \Phi(1.23) = 0.1093. \qquad \square$$

To develop intuitions concerning the bivariate normal distribution and correlation coefficient, let us consider some possible schemes that lead to the appearance of bivariate normal distribution with correlated variables.

Example 9.10.3 (Sequential Formation). It is possible that one of the values, say X, is formed at random, following the normal distribution $N(\mu_1, \sigma_1^2)$; this means that some random process leads to the creation of an element of the population with specific value x of an attribute X. Subsequently, the value of attribute Y is formed by some other random process which generates Y according to the normal distribution with mean $\rho(\sigma_2/\sigma_1)(x - \mu_1) + \mu_2$ and standard deviation $\sigma_2\sqrt{1 - \rho^2}$.

Examples of such "sequential" generation of attributes are quite common: We can think here of X and Y as being the temperatures at noon at a specific place today and tomorrow, water levels on the same river at two specific times, or two places, height of father and son, and so on. In a sense, we have a "natural" ordering, X being the first variable, whose value affects the second variable Y.

To enable clear geometrical representation, let $\sigma_1 = \sigma_2 = 1$ and let μ_1, μ_2 be removed sufficiently far from the origin to have clear plots of marginals, say $\mu_1 = \mu_2 = 5$ (see Figure 9.7). The conditional expectation of Y given X (the *regression* of Y on X, discussed in detail in Chapter 15) is now, by formula (9.64),

$$E(Y|X = x) = \rho(x - 5) + 5,$$

which is the line $y = \rho x + 5(1 - \rho)$ with slope ρ, passing through the point

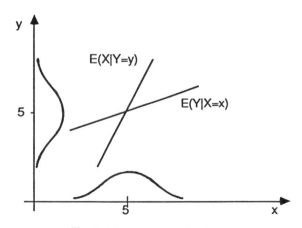

Figure 9.7 Two regression lines.

$(\mu_1, \mu_2) = (5, 5)$. Observe that the regression of X on Y is the line $E(X|Y = y) = \rho(y - 5) + 5$, that is, $y = (1/\rho)x - 5[(1 - \rho)/\rho]$, which also passes through the point $(5, 5)$ but has the slope $1/\rho$.

After the value of $X = x$ is sampled (formed by some physical process, etc.) the value of Y is sampled from the normal distribution centered at the appropriate point on the regression line, and variance $1 - \rho^2$. Remembering that the total variance of Y is 1, we have the decomposition of $\text{Var}(Y)$ into a sum of the form $\rho^2 + (1 - \rho^2)$, the second term being the variance of the deviation $Y - [\rho X - 5(1 - \rho)]$ of Y from the regression line. The first term, ρ^2, is therefore the contribution to the variability of Y coming from the variability of X. In generally accepted terminology, $100\rho^2$ is the "percentage of variance of Y *explained* by the variability of X."

A glance at Figure 9.7 reveals that if ρ were larger, the line would rise more steeply and this would increase the contribution of X to the variability of Y. [Remember that X and Y are standardized, so that their variances remain equal 1. This explains why the regression line $E(Y|X)$—in the case of standardized variables—cannot be steeper than the diagonal, and when ρ increases, to keep $\text{Var}(Y) = 1$ one has to decrease the variance of deviations from the regression line.]

Example 9.10.4. Another interpretation of the correlation coefficient ρ, not related to any temporal or casual ordering of X and Y, is connected with the following situation. Suppose that we have a population of objects of some kind, and that an attribute ξ of objects in this population has a normal distribution $N(\mu, \sigma^2)$. Elements of this population are sampled and their attribute ξ is measured twice. The measurements are subject to error, and the errors ϵ_1 and ϵ_2 of the two measurements are independent of ξ and from one another, with the same normal distribution $\epsilon_1 \sim N(0, \sigma_\epsilon^2)$, $\epsilon_2 \sim N(0, \sigma_\epsilon^2)$. The observed results of measurement are $X = \xi + \epsilon_1$, $Y = \xi + \epsilon_2$. In this case X and Y are normal with the same means and variances $\mu_1 = \mu_2 = \mu$, $\sigma_1^2 = \sigma_2^2 = \sigma^2 + \sigma_\epsilon^2$. The covariance between X and Y is $E(XY) - \mu_\xi^2 = E(\xi + \epsilon_1)(\xi + \epsilon_2) - \mu^2 = \sigma^2$, in view of the assumed independence and $E(\epsilon_i) = 0$. Consequently,

$$\rho = \frac{\text{Cov}(X, Y)}{\sqrt{\sigma_1^2 \sigma_2^2}} = \frac{\sigma^2}{\sigma^2 + \sigma_\epsilon^2}.$$

In this case $\rho > 0$, which means that the results of measurements of the same (random) quantity, subject to independent errors, are *always positively correlated*.

It should be pointed out here that—as opposed to Example 9.10.3—ρ (not ρ^2) now represents the fraction of the variance of X (or Y) "explained" by the variability of ξ. The situation is different from that in Example 9.10.3, since now we "explain" variance of one variable (X), not through the variability of the second variable of the pair (Y) but through the variability of some other variable (ξ) that affects both X and Y.   □

PROBLEMS

In all problems in this section, Z stands for a standard normal variable, and Φ is its cdf.

9.10.1   Use the tables of normal distribution to dermine the probabilities:
   (i) $P(0 \leq Z \leq 1.34)$.
   (ii) $P(0.14 \leq Z \leq 2.01)$.
   (iii) $P(-0.21 \leq Z \leq -0.04)$.
   (iv) $P(-0.87 \leq Z \leq 1.14)$.
   (v) $P(|Z| \geq 1.02)$.
   (vi) $P(Z \geq 1.11)$.
   (vii) $P(Z \leq -1.49)$.

9.10.2   Determine x in the following cases (extrapolate, if necessary):
   (i) $\Phi(x) = 0.72$.
   (ii) $\Phi(x) = 0.35$.
   (iii) $P(|Z| \leq x) = 0.95$.
   (iv) $P(1.4 \leq Z \leq x) = 0.02$.

9.10.3   Find $P(|X - 2| \leq 0.5)$ if $X \sim N(1,4)$.

9.10.4   Find $E(X)$ if X is normal, $\mathrm{Var}(X) = 3$ and $P(X \leq 12) = 0.77$.

9.10.5   Find $\mathrm{Var}(X)$ if X is normal, $E(X) = 2$ and $P(X \geq 5) = 0.33$.

9.10.6   Assume that X_1 and X_2 are independent, normally distributed, $E(X_1) = 3, E(X_2) = -1, \mathrm{Var}(X_1) = 6, \mathrm{Var}(X_2) = 2$. Find:
   (i) $P(3X_1 - 2X_2 \geq 14)$.
   (ii) $P(X_1 < X_2)$.

9.10.7   Assume that X_1 and X_2 have a bivariate normal distribution with $E(X_1) = 3, E(X_2) = 2, \mathrm{Var}(X_1) = 4, \mathrm{Var}(X_2) = 1$, and $\rho = -0.6$. Find:
   (i) $P\{X_1 \leq 4 | X_2 = 3\}$.
   (ii) $P\{|X_2 - 1| \geq 1.5 | X_1 = 2\}$.

9.10.8   A "100-year water," or flood, is the water level that is exceeded once in a hundred years (on average). Suppose that the threatening water levels occur once a year and have a normal distribution. Suppose also that at some location the 100-year water means a level of 30 feet above average. What is the 10,000-year water level?

9.11   BETA DISTRIBUTION

The family of distributions most commonly used to model researchers' uncertainty about the unknown probability p of some event is the family of beta distributions, defined as follows.

Definition 9.11.1.   We say that random variable X has *beta distribution* with parameters $\alpha > 0$ and $\beta > 0$ if the density of X equals

$$f(x) = \frac{\Gamma(\alpha + \beta)}{\Gamma(\alpha)\Gamma(\beta)} x^{\alpha-1}(1-x)^{\beta-1} \tag{9.65}$$

for $0 \le x \le 1$ and $f(x) = 0$ outside interval $[0,1]$. Here Γ is the function defined by (9.49).   □

We shall write $X \sim \text{BETA}(\alpha, \beta)$ if X has beta distribution with density (9.65).

We have first to show that (9.65) is indeed a density. Let us write

$$\Gamma(\alpha)\Gamma(\beta) = \left(\int_0^\infty u^{\alpha-1}e^{-u}\, du \right) \left(\int_0^\infty v^{\beta-1}e^{-v}\, dv \right)$$
$$= \int_0^\infty \int_0^\infty u^{\alpha-1}v^{\beta-1}e^{-(u+v)}\, du\, dv. \tag{9.66}$$

For the change of variables

$$z = u + v, \qquad x = \frac{u}{u+v},$$

we have $0 \le x \le 1, 0 \le z \le \infty$. Moreover,

$$u = xz, \qquad v = z(1-x);$$

hence the Jacobian equals z. Consequently, after substitution in (9.66), the variables separate and we obtain

$$\Gamma(\alpha)\Gamma(\beta) = \int_0^\infty z^{\alpha-1}z^{\beta-1}z\, dz \int_0^1 x^{\alpha-1}(1-x)^{\beta-1}\, dx$$
$$= \Gamma(\alpha + \beta) \int_0^1 x^{\alpha-1}(1-x)^{\beta-1}\, dx,$$

which was to be shown.

Once we know that

$$\int_0^1 x^{\alpha-1}(1-x)^{\beta-1}\,dx = \frac{\Gamma(\alpha)\Gamma(\beta)}{\Gamma(\alpha+\beta)},$$

we may easily compute moments of beta distribution. Indeed, if $X \sim \text{BETA}$ (α, β), then

$$
\begin{aligned}
E(X^k) &= \frac{\Gamma(\alpha+\beta)}{\Gamma(\alpha)\Gamma(\beta)} \int_0^1 x^{k+\alpha-1}(1-x)^{\beta-1}\,dx \\
&= \frac{\Gamma(\alpha+\beta)}{\Gamma(\alpha)\Gamma(\beta)} \cdot \frac{\Gamma(k+\alpha)\Gamma(\beta)}{\Gamma(k+\alpha+\beta)} \\
&= \frac{\Gamma(\alpha+\beta)\Gamma(k+\alpha)}{\Gamma(\alpha)\Gamma(k+\alpha+\beta)}.
\end{aligned}
$$

In particular, for $k = 1$ we obtain, using formula (9.50),

$$E(X) = \frac{\alpha}{\alpha+\beta}.$$

For $k = 2$ we have

$$E(X^2) = \frac{\alpha(\alpha+1)}{(\alpha+\beta)(\alpha+\beta+1)},$$

so that

$$\text{Var}(X) = \frac{\alpha\beta}{(\alpha+\beta)^2(\alpha+\beta+1)}. \tag{9.67}$$

The variety of shapes of the beta distribution is illustrated in Figure 9.8. If $\alpha > 1, \beta > 1$, the distribution is bell-shaped, with the peak becoming narrower for larger α and/or β. For $\alpha = \beta = 1$ the distribution is uniform. If $\alpha < 1$ and $\beta < 1$, the distribution is U-shaped, while if $\alpha < 1, \beta \geq 1$ or $\alpha \geq 1, \beta < 1$, the distribution is J-shaped.

As mentioned, the most typical application of the beta distribution is when we consider the binomial distribution with unknown p and represent knowledge (or uncertainty) about p by assuming that p is random with a beta distribution. The examples below illustrate this approach.

Example 9.11.1. One of the questions posed by Laplace is: What is the probability that the sun will rise tomorrow?

SOLUTION. The question may appear facetious or silly in the phrasing above. But in a slight reformulation it is: Suppose that some event A occurred in all of a large number N of trials. What is the probability that it will occur again on the next trial?

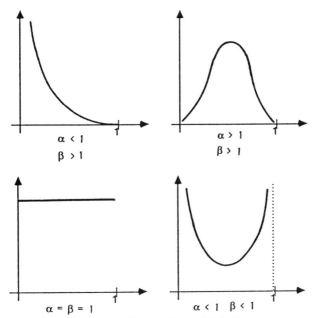

Figure 9.8 Shapes of beta distributions.

We assume the Bernoulli model of independent trials, and let A_n and $A^{(n)}$ stand for the events "A occurs on n consecutive trials" and "A occurs on nth trial." Following Laplace, we take the uniform distribution [i.e., BETA(1,1)] as the distribution of p (presumably, such a choice suggests total impartiality).

We want the expected value of the conditional probability $P(A_{N+1}|A_N)$, the expectation taken with respect to the conditional distribution of p given the data. According to Theorem 4.4.2 on combining evidence, we can proceed in either of two ways:

1. Find the expected probability of the joint occurrence of evidence so far and the event in question (hence $A_N \cap A^{(N+1)} = A_{N+1}$) with respect to the unconditional distribution of p.
2. Find the expected probability of the event in question ($A^{(N+1)}$) with respect to the conditional probability of p given the evidence (A_N).

Using approach 1 the probability of the conjunction $A_N \cap A^{(N+1)}$ is p^{N+1}, so the answer is, assuming that $p \sim \text{BETA}(1,1)$,

$$E(p^{N+1}) = \int_0^1 p^{N+1}\, dp = \frac{N+1}{N+2}.$$

A rather conservative estimate of the time when we *know* that the sun was rising every day is about 7000 years. Much as we may be convinced that it is not the case, it is possible that before the first written records, the sun did not rise but was (say) switched on suddenly in the sky or operated under some other principles (after all, why would Laplace bother to calculate the probability of the sun rising tomorrow if he did not allow the possibility of some other mechanism?). Thus, letting $N = 365 \times 7,000$, we have $(N + 1)/(N + 2) \approx 1 - 4 \times 10^{-7}$.

Example 9.11.2. Continuing the Laplace problem in a somewhat more realistic setup, we may ask the following question: We observe Bernoulli trials where we know that p is random with distribution BETA(α, β). So far we know that there were 3 successes in 4 trials. What is the probability of 2 successes in the next 3 trials?

SOLUTION. Reasoning in exactly the same way as Example 9.11.1, the combined data and event in question is "3 successes in 4 trials and 2 successes in next 3 trials" (which is *not* the same as 5 successes in 7 trials; why?). Conditioning on p, the probability is

$$\binom{4}{3}p^3(1 - p)\binom{3}{2}p^2(1 - p) = 12p^5(1 - p)^2.$$

Taking expectation with respect to $p \sim$ BETA (α, β) we have, for the expected probability $E(2, 3)$ of 2 successes in 3 trials,

$$E(2, 3) = \frac{\Gamma(\alpha + \beta)}{\Gamma(\alpha)\Gamma(\beta)} \int_0^1 12p^5(1 - p)^2 p^{\alpha - 1}(1 - p)^{\beta - 1}\, dp$$

$$= \frac{\Gamma(\alpha + \beta)}{\Gamma(\alpha)\Gamma(\beta)} 12 \frac{\Gamma(5 + \alpha)\Gamma(2 + \beta)}{\Gamma(7 + \alpha + \beta)}.$$

Using formula (9.50), which asserts that $\Gamma(t) = (t - 1)\Gamma(t - 1)$, we obtain, after some algebra,

$$E(2, 3) = 12 \frac{\alpha(\alpha + 1)\cdots(\alpha + 4)\beta(\beta + 1)}{(\alpha + \beta)(\alpha + \beta + 1)\cdots(\alpha + \beta + 6)}.$$

Example 9.11.3. To make the situation still more realistic, consider a client who comes to a statistical consulting center for help. It turns out that she is currently negotiating the purchase of a large shipment of some merchandise. Some of the items conform to the specification, some do not (let us call them defective). Suppose that testing the quality of all items in the lot is not possible because of the prohibitive cost of such an operation or because the process of testing is destructive. Some tests have been made, though, and the data are such that among 25 tested items 3 were found to be defective. What can one reasonably say about the fraction of defective items in the whole lot?

SOLUTION.   Suppose that the lot in question is not the first lot purchased by the client. It turns out that she has been buying the same kinds of merchandise from the same company (or from other companies). In her experience, as judged by tests, data on returns, customer complaints, and so on, she estimates that on the average, the percentages of items below specification (defectives) in various lots purchased were about 12%, with a variability of about 2% in either direction.

The problem of the client of the statistical consulting center may be formulated, for instance, as follows: Determine the chances that the fraction p of defective items in the lot whose purchase is presently being negotiated is below 10%. Thus, we have to evaluate $P\{p < 0.1\}$ in light of the tests of the lot under consideration (3 out of 25 below specification), using the experience accumulated in the past. If we accept the assumption that the fractions in various lots follow a beta distribution, we have for the mean

$$\frac{\alpha}{\alpha + \beta} = 0.12. \tag{9.68}$$

Next, the relatively small variability ($\pm 2\%$) suggests that we have $\alpha > 1, \beta > 1$, and consequently, the density of beta distribution is bell-shaped. Using the three-sigma rule, we may take $3\sigma \approx 0.02$, and using (9.67), we obtain another equation for α and β, namely

$$\frac{\alpha\beta}{(\alpha + \beta)^2(\alpha + \beta + 1)} = \left(\frac{0.02}{3}\right)^2. \tag{9.69}$$

The solution of (9.68) and (9.69) is $\alpha^* = 285, \beta^* = 2090$.

Now, we assume here that the fraction p of defective items is a random variable, whose distribution is BETA(α^*, β^*). If no testing of the currently negotiated lot were made, the answer would be

$$P\{p < 0.1\} = \frac{\Gamma(\alpha^* + \beta^*)}{\Gamma(\alpha^*)\Gamma(\beta^*)} \int_0^{0.1} x^{\alpha^*-1}(1 - x)^{\beta^*-1}\, dx.$$

We have, however, additional information, namely that the event $A =$ "3 defective in sample of 25" occurred. Consequently, we should integrate the conditional density of p given A. We have

$$P(A|p) = \binom{25}{3}p^3(1 - p)^{22}.$$

The conditional density is obtained from Bayes' formula, with summation replaced by integration. Schematically,

$$\varphi_p(x|A) = \frac{P(A|x)\varphi_p(x)}{\int P(A|u)\varphi_p(u)\, du}, \tag{9.70}$$

where $\varphi_p(x)$ and $\varphi_p(x|A)$ are the unconditional and conditional densities of p at x.

Typical reasoning is as follows. The denominator in (9.70) is independent of x, and both sides represent densities in x. Collecting the terms that depend on x and representing all other factors as a constant, the right-hand side of (9.70) is of the form

$$Cx^3(1-x)^{22}x^{\alpha^*-1}(1-x)^{\beta^*-1}, \tag{9.71}$$

so that it is a beta density with parameters $\alpha^* + 3$ and $\beta^* + 22$. This gives

$$C = \frac{\Gamma(\alpha^* + \beta^* + 25)}{\Gamma(\alpha^* + 3)\Gamma(\beta^* + 22)}$$

and one could get the answer to the original question by integrating (9.71) between 0 and 0.1. □

Observe that the analysis above has some wider consequences: We know that if the distribution of p is BETA(α, β) and we observe k successes in n trials (event A), then the conditional distribution of p given A is BETA$(\alpha + k, \beta + n - k)$. One says that the family of beta distributions constitutes the *conjugate priors* for binomial data. We shall return to those problems in Chapter 12.

PROBLEMS

9.11.1 Let X have the distribution BETA(α, β). Find the distribution of $Y = 1 - X$.

9.11.2 Let X have the distribution BETA(α, β). Find $E\{X^k(1 - X)^m\}$.

9.11.3 The joint density of the extreme order statistics $X = X_{1:n}$ and $Y = X_{n:n}$ from the uniform distribution on $[0, 1]$ is $f(x, y) = n(n - 1)(y - x)^{n-2}$ for $0 \leq x \leq y \leq 1$ and $f(x, y) = 0$ otherwise. Find:
   (i) $E(X^m), E(Y^m)$, and $\rho(X, Y)$.
   (ii) $E(Y|X)$.

9.11.4 Let X have a symmetric beta distribution. Find α and β if the coefficient of variation (ratio of standard deviation to the mean) is k. Does a solution exist for all k?

9.11.5 Let $X_1, X_2, \ldots$ be iid with an exponential distribution. Find the distribution of the ratio

$$T_{m,n} = \frac{X_1 + \cdots + X_m}{X_1 + \cdots + X_{m+n}}$$

for any positive integers m and n.

9.12 SOME DISTRIBUTIONS RELATED TO NORMAL

In this section we introduce certain distributions that are important in statistics. Consequently, the applications of these distributions are discussed in later chapters. Here we define them and discuss their most important properties.
We start from the following theorem:

Theorem 9.12.1. *Let $X_1, \ldots, X_n$ be a random sample from the normal distribution $N(\mu, \sigma^2)$ (i.e., X_i's are independent, each with the distribution $N(\mu, \sigma^2)$). Let*

$$\overline{X} = (X_1 + \cdots + X_n)/n \tag{9.72}$$

and

$$U = \sum_{i=1}^{n}(X_i - \overline{X})^2. \tag{9.73}$$

Then $\overline{X}$ and U are independent random variables. Moreover, $\overline{X}$ has the distribution $N(\mu, \sigma^2/n)$ and U/σ^2 has a chi-square distribution with $n - 1$ degrees of freedom.

Proof. We use a multidimensional version of the method of finding the density of a function of random variables. Let us observe that

$$\sum_{i=1}^{n}(x_i - \mu)^2 = \sum_{i=1}^{n}(x_i - \overline{x} + \overline{x} - \mu)^2$$

$$= \sum_{i=1}^{n}(x_i - \overline{x})^2 + n(\overline{x} - \mu)^2,$$

since $2\sum_{i=1}^{n}(x_i - \overline{x})(\overline{x} - \mu) = 2(\overline{x} - \mu)\sum_{i=1}^{n}(x_i - \overline{x}) = 0$. Consequently, the joint density of $(X_1, \ldots, X_n)$ can be written as

$$f(x_1, \ldots, x_n) = \frac{1}{\sigma^n (2\pi)^{n/2}} \exp \left\{ -\frac{1}{2\sigma^2} \sum_{i=1}^{n} (x_i - \mu)^2 \right\}$$

$$= \frac{1}{\sigma^n (2\pi)^{n/2}} \exp \left\{ -\frac{1}{2\sigma^2} \sum_{i=1}^{n} (x_i - \bar{x})^2 - \frac{n}{2\sigma^2} (\bar{x} - \mu)^2 \right\}$$

$$= \frac{1}{\sigma^n (2\pi)^{n/2}} e^{-(u/2\sigma^2)} e^{-(n/2\sigma^2)(\bar{x} - \mu)^2}. \tag{9.74}$$

Let us now make the transformation

$$x_j = \bar{x} + v_j \sqrt{u} \qquad j = 1, 2, \ldots, n. \tag{9.75}$$

Clearly, we have

$$\sum_{j=1}^{n} v_j = 0, \qquad \sum_{j=1}^{n} v_j^2 = 1, \tag{9.76}$$

which means that two of the v_j's are functions of the remaining ones. We therefore solve (9.76) for v_{n-1} and v_n, obtaining two solutions:

$$v_{n-1} = \frac{A - B}{2}, \qquad v_n = \frac{A + B}{2}$$

or

$$v_{n-1} = \frac{A + B}{2}, \qquad v_n = \frac{A - B}{2},$$

where

$$A = -\sum_{k=1}^{n-2} v_k, \quad B = \left\{ 2 - 3 \sum_{k=1}^{n-2} v_k^2 - \sum_{\substack{k,l=1 \\ k \neq j}}^{n-2} v_k v_j \right\}^{1/2}. \tag{9.77}$$

Thus, to each vector $(v_1, \ldots, v_{n-2}, \bar{x}, u)$ with $u > 0$ and $\sum_{k=1}^{n-2} v_k^2 < 1$ there correspond two systems $(x_1, \ldots, x_n)$:

$$x_j = \bar{x} + v_j \sqrt{u}, \qquad j = 1, \ldots, n - 2$$
$$x_{n-1} = \bar{x} + \frac{A - B}{2} \sqrt{u}, \qquad x_n = \bar{x} + \frac{A + B}{2} \sqrt{u} \tag{9.78}$$

and

$$x_j = \bar{x} + v_j \sqrt{u}, \qquad j = 1, \ldots, n - 2$$
$$x_{n-1} = \bar{x} + \frac{A + B}{2} \sqrt{u}, \qquad x_n = \bar{x} + \frac{A - B}{2} \sqrt{u}. \tag{9.79}$$

The Jacobian of transformation (9.78) is

$$
J = \begin{vmatrix}
\dfrac{\partial x_1}{\partial \bar{x}} & \cdots & \dfrac{\partial x_n}{\partial \bar{x}} \\[2mm]
\dfrac{\partial x_1}{\partial \sqrt{u}} & \cdots & \dfrac{\partial x_n}{\partial \sqrt{u}} \\[2mm]
\dfrac{\partial x_1}{\partial v_1} & \cdots & \dfrac{\partial x_n}{\partial v_1} \\[2mm]
\cdots & \cdots & \cdots \\[1mm]
\dfrac{\partial x_1}{\partial v_{n-2}} & \cdots & \dfrac{\partial x_n}{\partial v_{n-2}}
\end{vmatrix}
$$

$$
= \begin{vmatrix}
1 & 1 & \cdots & 1 & 1 & 1 \\[2mm]
v_1 & v_2 & \cdots & v_{n-2} & \dfrac{A-B}{2} & \dfrac{A+B}{2} \\[3mm]
\sqrt{u} & 0 & \cdots & 0 & \dfrac{\sqrt{u}}{2}\left(\dfrac{\partial A}{\partial v_1} - \dfrac{\partial B}{\partial v_1}\right) & \dfrac{\sqrt{u}}{2}\left(\dfrac{\partial A}{\partial v_1} + \dfrac{\partial B}{\partial v_1}\right) \\[3mm]
\cdots & \cdots & \cdots & \cdots & \cdots & \cdots \\[1mm]
0 & 0 & \cdots & \sqrt{u} & \dfrac{\sqrt{u}}{2}\left(\dfrac{\partial A}{\partial v_{n-2}} - \dfrac{\partial B}{\partial v_{n-2}}\right) & \dfrac{\sqrt{u}}{2}\left(\dfrac{\partial A}{\partial v_{n-2}} + \dfrac{\partial B}{\partial v_{n-2}}\right)
\end{vmatrix}
\tag{9.80}
$$

Since for transformation (9.79) the Jacobian is the same, but with the last two columns interchanged, the absolute value of the Jacobian for transformation (9.79) is the same as that for transformation (9.78). We have

$$
|J| = u^{(n-2)/2} h(v_1, \ldots, v_{n-2}),
\tag{9.81}
$$

where h is a function whose exact form is not needed for our purposes.

Substitution of new variables in (9.74) and multiplication by (9.81) gives the same joint density of vector $(\bar{X}, U, V_1, \ldots, V_{n-2})$ for transformations (9.78) and (9.79), namely

$$
g(\bar{x}, u, v_1, \ldots, v_{n-2}) = \frac{1}{\sigma^n (2\pi)^{n/2}} u^{(n-2)/2} e^{-u/2\sigma^2} e^{-(n/2\sigma^2)(\bar{x}-\mu)^2} h(v_1 \ldots, v_{n-2}).
\tag{9.82}
$$

Since the density (9.82) can be written as the product of densities

$$
c_1 u^{(n/2)-1} e^{-(u/2\sigma^2)} \cdot c_2 e^{-(n/2\sigma^2)(\bar{x}-\mu)^2} \cdot c_3 h(v_1, \ldots, v_{n-2}),
$$

it is now clear that the random variables $U, \bar{X}$ and $(V_1, \ldots, V_{n-2})$ are independent. An inspection of terms involving $\bar{x}$ and u shows that after adjusting constants, $\overline{X}$ is normal $N(\mu, \sigma^2/n)$ and U/σ^2 has a chi-square distribution with $n - 1$ degrees of freedom.   □

The most remarkable fact here is that $\overline{X}$ and U are independent, even though $\overline{X}$ appears explicitly in the definition of U. This independence is characteristic for normal distribution: We have the following theorem, which we leave without proof:

Theorem 9.12.2.   *Let $X_1, \ldots, X_n$ be independent, with the same distribution. If $\overline{X}$ and U are independent, then the distribution of X_i is normal.*

We now introduce the following definition:

Definition 9.12.1.   Let Z be a standard normal random variable, and let U be a chi-square distributed random variable independent of Z, with ν degrees of freedom. Then the random variable

$$X = \frac{Z}{\sqrt{U/\nu}} \tag{9.83}$$

is said to have *Student t distribution* with ν degrees of freedom.   □

Let us start from deriving the density of X. We follow the steps outlined in Chapter 7: Starting from the joint density of Z and U, we add a "companion" variable Y to X, find the Jacobian of the transformation $(z, u) \rightarrow (x, y)$, and finally, find the density of X as the marginal in the joint density of (X, Y).

The joint density of (Z, U) is, by the assumption of independence,

$$f(z, u) = \frac{1}{\sqrt{2\pi}} e^{-z^2/2} \cdot \frac{1}{2^{\nu/2}\Gamma(\nu/2)} u^{(\nu/2)-1} e^{-u/2}, \qquad u > 0 \tag{9.84}$$

Let us take as a companion variable to X the variable $Y = U$. We then have

$$z = x\sqrt{y/\nu}, \qquad u = y$$

and the Jacobian equals $J = \sqrt{y/\nu}$; hence the joint density of (X, Y) is

$$\varphi(x, y) = Q y^{(\nu-1)/2} e^{-(1/2)(1+x^2/\nu)y} \tag{9.85}$$

where

$$Q = \frac{1}{\sqrt{\pi\nu} 2^{(\nu+1)/2}\Gamma(\nu/2)}. \tag{9.86}$$

The density (9.85), except for a constant, is a gamma density with parameters $(\nu + 1)/2$ and $\frac{1}{2}(1 + x^2/\nu)$. Thus, the marginal density of X is obtained by integration as

$$f_\nu(x) = \int_0^\infty Q y^{(\nu-1)/2} e^{-(1/2)(1+x^2/\nu)y} \, dy$$

$$= Q \frac{\Gamma\left(\dfrac{\nu+1}{2}\right)}{[(1/2)(1+x^2/\nu)]^{(\nu+1)/2}}$$

$$= \frac{\Gamma\left(\dfrac{\nu+1}{2}\right)}{\sqrt{\pi\nu}\,\Gamma(\nu/2)} \left(1+\frac{x^2}{\nu}\right)^{-(\nu+1)/2}. \tag{9.87}$$

It is evident that the density of the t-distribution is symmetric about $x = 0$. When $\nu = 1$ we have $\varphi(x) = K(1+x^2)^{-1}$, which is the density of the Cauchy distribution (see Example 8.2.13). Integrating $f_\nu(x)$ and comparing coefficients, we find that $\Gamma(\frac{1}{2}) = \sqrt{\pi}$. Observe that for $\nu = 1$ the Student t distribution has no mean (hence no higher moments either).

To see what happens when $\nu \to \infty$, let us write, disregarding the constants

$$\varphi(x) \sim \left[\left(1+\frac{x^2}{\nu}\right)^\nu\right]^{-1/2} \left[\left(1+\frac{x^2}{\nu}\right)\right]^{-1/2}.$$

The first term converges to $(e^{x^2})^{-1/2} = e^{-x^2/2}$, while the second term converges to 1. It follows that the constants must converge to $1/\sqrt{2\pi}$, to preserve integrability to 1. We see therefore that when the number of degrees of freedom increases, the Student t distribution approaches the standard normal distribution.

Let $X_1, \ldots, X_n$ be a random sample from normal distribution $N(\mu, \sigma^2)$. By Theorem 9.12.1, $\overline{X}$ and $\sum_{i=1}^n (X_i - \overline{X})^2$ are independent, with $(\overline{X} - \mu)/\sigma$ being standard normal and $\sum_{i=1}^n (X_i - \overline{X})^2/\sigma^2$ having chi-square distribution with $n-1$ degrees of freedom. Consequently, the ratio

$$\frac{\overline{X} - \mu}{\sigma} \bigg/ \sqrt{\frac{\sum_{i=1}^n (X_i - \overline{X})^2}{(n-1)\sigma^2}} = \frac{\overline{X} - \mu}{\sqrt{\sum_{i=1}^n (X_i - \overline{X})^2}} \sqrt{n-1} \tag{9.88}$$

has Student distribution with $n-1$ degrees of freedom. As we shall see in Chapter 12, the random variable (9.88) will play an important role in building schemes for inference on μ.

Table A.3 gives upper percentiles for Student t distribution for selected number of degrees of freedom, where the upper percentile $t_{\alpha,\nu}$ is defined by the relation $P\{X \geq t_{\alpha,\nu}\} = \alpha$, with X having the Student t distribution with ν degrees of freedom (so that upper α-percentile is $(1-\alpha)$-quantile).

Now let the random variables U and V be independent, each with a chi-square distribution, with numbers of degrees of freedom ν_1 and ν_2, respectively.

Definition 9.12.2. The random variable

$$X = \frac{U/\nu_1}{V/\nu_2} \tag{9.89}$$

is said to have *F-distribution* with numbers of degrees of freedom (ν_1, ν_2).

Let us derive the density of random variable X. The joint density of (U, V) is, in view of the independence of U and V,

$$\varphi(u, v) = \frac{1}{2^{\nu_1/2}\Gamma(\nu_1/2)} u^{(\nu_1/2)-1} e^{-u/2} \cdot \frac{1}{2^{\nu_2/2}\Gamma(\nu_2/2)} v^{(\nu_2/2)-1} e^{-v/2}; \qquad u, v > 0 \tag{9.90}$$

Let us take the companion variable $Y = V$. The inverse transformation $(x, y) \to (u, v)$ is then

$$u = \frac{\nu_1}{\nu_2} xy, \qquad v = y,$$

so that the Jacobian equals $J = (\nu_1/\nu_2)y$. Thus, the joint density of X and Y is, after substitution to (9.90),

$$\psi(x, y) = \frac{\nu_1^{\nu_1/2}}{\nu_2^{\nu_1/2} 2^{(\nu_1+\nu_2)/2} \Gamma(\nu_1/2)\Gamma(\nu_2/2)} x^{(\nu_1/2)-1} y^{[(\nu_2+\nu_1)/2]-1} e^{-(1/2)[(\nu_1 x/\nu_2)+1]y}.$$

The marginal density of X equals, again using the fact that we can separate a gamma-type integral for y,

$$\varphi_{\nu_1, \nu_2}(x) = \int_0^\infty \psi(x, y)\, dy$$

$$= \frac{\nu_1^{\nu_1/2} x^{(\nu_1/2)-1}}{\nu_2^{\nu_1/2} 2^{(\nu_1+\nu_2)/2} \Gamma(\nu_1/2)\Gamma(\nu_2/2)} \int_0^\infty y^{(\nu_2+\nu_1)/2-1} e^{-(1/2)[(\nu_1/\nu_2)x+1]y]}\, dy$$

$$= \frac{\nu_1^{\nu_1/2} x^{(\nu_1/2)-1} \Gamma\left(\frac{\nu_1+\nu_2}{2}\right) 2^{(\nu_1+\nu_2)/2}}{\nu_2^{\nu_1/2} 2^{(\nu_1+\nu_2)/2} \Gamma(\nu_1/2)\Gamma(\nu_2/2)[(\nu_2/\nu_1)x + 1]^{(\nu_2+\nu_1)/2}}$$

$$= \frac{\Gamma\left(\frac{\nu_1+\nu_2}{2}\right)}{\Gamma(\nu_1/2)\Gamma(\nu_2/2)} \nu_1^{\nu_1/2} \nu_2^{\nu_2/2} \frac{x^{(\nu_1/2)-1}}{(\nu_1 x + \nu_2)^{(\nu_1+\nu_2)/2}}.$$

The probabilities of the F distribution have been tabulated (see Tables A.6). Since the distribution depends on a pair (ν_1, ν_2) of degrees of freedom (given on the margins of the table), each entry gives just one quantile with the corresponding numbers of degrees of freedom, namely the 0.95 quantile [hence 0.05 upper quantile; Table A.6(a)] and 0.99 quantile [0.01 upper quantile; Table A.6(b)]. Actually (see Problem 9.12.4), these tables also provide 0.05 and 0.01 quantiles.

PROBLEMS

9.12.1 Show that if X is a random variable with Student distribution with ν_1 degrees of freedom, then $E(X^k)$ exists only for all integers $k < \nu_1$.

9.12.2 Since the density $f_{\nu_1}(x)$ of the Student distribution with ν_1 degrees of freedom is symmetric about $x = 0$, we have $\int_{-A}^{A} x^{2r+1} f_{\nu_1}(x)\, dx = 0$ for every $r = 0, 1, 2, \ldots$ and every A. Thus $\lim_{A \to \infty} \int_{-A}^{A} x^{2r+1} f_{\nu_1}(x)\, dx = 0$. Consequently, all moments of odd order equal zero; hence they exist. Since each even integer is less than some odd integer, it follows that all moments exist. Compare with the result from Problem 9.12.1 and find the error.

9.12.3 Let X be a random variable with Student t distribution with ν_1 degrees of freedom. Show that X^2 has the F distribution with numbers of degrees of freedom $(1, \nu_1)$.

9.12.4 Show that if x_α is the α-quantile of a random variable X with F_{ν_1, ν_2} distribution, then $1/x_\alpha$ is $1 - \alpha$ quantile of a random variable with F_{ν_2, ν_1} distribution.

9.12.5 Let X, Y, Z be independent, with distributions $X \sim N(0,1)$, $Y \sim N(1,1)$, and $W \sim N(2,4)$. Using Table A.6(a), find k such that

$$P\left\{ \frac{X^2 + (Y-1)^2}{X^2 + (Y-1)^2 + (W-2)^2/4} > k \right\} = 0.05.$$

REFERENCES

Feller, W., 1957. *An Introduction to Probability Theory and Its Applications*, vol. 1, Wiley, New York.

Srivastava, R. C., 1981. "On Some Characterization of the Geometric Distribution" in: C. Taillie, G. P. Patil, and B. A. Baldessari (eds.), *Statistical Distribution in Scientific Work*, D. Reidel Publishers, Dordrecht, pp. 349–355.

CHAPTER 10

Limit Theorems

10.1 INTRODUCTION

We have already encountered several limit theorems in past chapters. Before proceeding with a systematic exploration of the topic, it may be worthwhile to recall those theorems briefly and try to find their common features.

First, on several occasions we looked at the differences between sampling with and without replacement, and noted that these two schemes of sampling become closer one to another "as population becomes larger." Second, we proved the Poisson approximation theorem, which asserts that as n becomes larger and p becomes smaller, the binomial and Poisson probabilities of some events become close. Finally, in Chapter 5 we showed that (under some conditions) as the number of transitions of a Markov chain increases, the probability of finding the system in a given state approaches a limiting value.

The common feature of these theorems was that in each case probabilities of certain events—or, more generally, distributions of certain random variables—approached some limits with the appropriate change of one or more parameters. Typically, the parameter is an index of some sequence, such as sample size n, but the case of Poisson approximation shows that it may also be a simultaneous change of two parameters that drives the probabilities to their limiting values.

In addition to the limit theorem mentioned above, we have also encountered a different kind of limit theorem, exemplified by the law of large numbers (Example 8.9.7). We had there convergence not only of the distributions, but of the random variables themselves.

To grasp the difference between those two classes of situations, observe that one can study convergence of a sequence of distributions $\{F_n\}$ without considering any random variables. On the other hand, if $X_1, X_2, \ldots$ is a sequence of independent random variables with the same distribution F, then the sequence $\{F_n\}$ of their distributions clearly converges to F, but we cannot expect any regularity in the behavior of the sequence $\{X_n\}$.

414

10.2   CONVERGENCE OF RANDOM VARIABLES

In what follows we consider a sequence $\xi_1, \xi_2, \ldots$ of random variables defined on the same sample space S with σ-field $\mathcal{F}$ of events, and probability measure P on S. Our first objective will be to distinguish various possible modes in which sequence $\{\xi_n\}$ may converge, and discuss their interrelationships. To connect the analysis with subsequent statistical concepts, it is worthwhile to start with some examples of sequences $\{\xi_n\}$ which may arise in statistical research and practice.

Example 10.2.1.   One of the most common situations in statistics is when we deal with the simple random samples. In terms of probability theory, this means that we observe the beginning of a sequence of independent identically distributed (iid) random variables $X_1, X_2, \ldots$ Depending on the goal of analysis, given the observations $(X_1, \ldots, X_n)$, the statistician computes the value of some function of the observation, $\xi_n = H(X_1, \ldots, X_n)$ and uses ξ_n as means of inference. We say then that ξ_n is a *statistic*, and the behavior of the sequence $\{\xi_n\}$ as n increases tells the statistician to which extent it would be worthwhile to increase the sample size n.

Typical examples of ξ_n may be sample mean $\xi_n = \overline{X}_n = (X_1 + \cdots + X_n)/n$, sample variance $\xi_n = \frac{1}{(n-1)} \sum_{i=1}^{n} (X_i - \overline{X}_n)^2$, maximal element in the sample $\xi_n = \max(X_1, \ldots, X_n)$, and so on.   □

We begin with the definition that captures the type of convergence encountered in the law of large numbers (Theorem 8.9.3).

Definition 10.2.1.   We say that the sequence $\{\xi_n\}$ *converges in probability to a constant c* if for every $\epsilon > 0$,

$$\lim_{n\to\infty} P\{|\xi_n - c| \geq \epsilon\} = 0. \tag{10.1}$$

More generally, we say that $\{\xi_n\}$ converges in probability *to a random variable ξ*, if for every $\epsilon > 0$,

$$\lim_{n\to\infty} P\{|\xi_n - \xi| \geq \epsilon\} = 0. \tag{10.2}$$

Convergence in probability, especially to a constant, is often called *stochastic convergence*.   □

The meaning of (10.1) is that as n increases, it becomes less and less likely that ξ_n will deviate from c by more than ϵ. In (10.2) the interpretation is the same, except that the constant c is replaced by a random quantity, so that, written explicity, (10.2) reads

$$\lim_{n\to\infty} P\{s : |\xi_n(s) - \xi(s)| \geq \epsilon\} = 0.$$

We shall use the symbol $\xrightarrow{P}$ to denote convergence in probability.

Example 10.2.2 (Laws of Large Numbers). In terms of the foregoing concept, the law of large numbers proved in Section 8.9 asserts convergence in probability of empirical frequencies of an event. Such theorems (asserting convergence of probability of averages of random variables) are called *weak laws of large numbers* (WLLNs), to distinguish them from *strong* laws of large numbers, discussed later.   □

For instance, in the Bernoulli case we have the following rephrasing of Theorem 8.9.3.

Theorem 10.2.1. *If S_n has binomial distribution* BIN(n, p), *then*

$$\frac{S_n}{n} \xrightarrow{P} p.$$

A stronger type of convergence of sequences of random variables is given in the following definition.

Definition 10.2.2. Let $\xi_1, \xi_2, \ldots$ be a sequence of random variables defined on some probability space $(\mathcal{S}, \mathcal{F}, \mathcal{P})$. If $\lim \xi_n(s) = \xi(s)$ exists for all points $s \in U$ where $P(U) = 1$, then we say that ξ_n converges to ξ *almost everywhere* (a.e.), *almost surely* (a.s.), or *with probability* 1.   □

We shall use the symbol $P\{\xi_n \to \xi\} = 1$ or $\xi_n \to \xi$ a.s. Other symbols used often are $\xi_n \to_{a.s.} \xi$, $\xi_n \to_{a.e.} \xi$, or $\xi_n \to \xi$ a.e.

*   Before discussing the concept of almost sure convergence any further, let us prove first that the set U on which the sequence ξ_n converges belongs to the σ-field $\mathcal{F}$, so that we are always allowed to assign probability to it. We have to prove therefore that U can be obtained from events (elements of $\mathcal{F}$) by countable operations. This can be accomplished as follows.

If $s \in U$, then $\lim_{n\to\infty} \xi_n(s)$ exists, which happens if and only if the (numerical) sequence $\{\xi_n(s)\}$ satisfies the Cauchy condition: For every $\epsilon > 0$ there exists N such that $|\xi_n(s) - \xi_m(s)| < \epsilon$ whenever $n, m \geq N$.

Let

$$B_{n,m}(\epsilon) = \{s : |\xi_n(s) - \xi_m(s)| < \epsilon\}.$$

Clearly, $B_{n,m}(\epsilon)$ is an event (it belongs to $\mathcal{F}$), since ξ_n and ξ_m are random variables. To show that U is also an event, it is sufficient to show that U can be built out of sets $B_{n,m}(\epsilon)$ through countably many operations.

The key is to replace phrases such as "there exists," "for some" by unions, and phrases such as "for all," "for every" by intersections. For instance,

$$\bigcup_{N=1}^{\infty} \bigcap_{m=N}^{\infty} \bigcap_{n=N}^{\infty} B_{n,m}(\epsilon) \qquad (10.3)$$

is the set of all points s for which there exists N such that for all $m, n \geq N$ we have $|\xi_n(s) - \xi_m(s)| < \epsilon$. Since we require the property asserted above for every $\epsilon > 0$, it would seem that it suffices to take the intersection of all sets (10.3) for $\epsilon > 0$. However, such an intersection is not countable, and we could not claim that the resulting set is in $\mathcal{F}$, hence has a probability assigned to it. But we obtain the same intersection by restricting it to a sequence of numbers ϵ_k which tends to zero, for instance $\epsilon_k = 1/k$. In other words, the convergence set U satisfies

$$U = \bigcap_{\epsilon>0} \bigcup_{N=1}^{\infty} \bigcap_{m=N}^{\infty} \bigcap_{n=N}^{\infty} B_{n,m}(\epsilon)$$

$$= \bigcap_{k=1}^{\infty} \bigcup_{N=1}^{\infty} \bigcap_{m=N}^{\infty} \bigcap_{n=N}^{\infty} B_{n,m}(1/k)$$

and the right-hand term shows that U is indeed in $\mathcal{F}$, as obtained through countable operations on sets in $\mathcal{F}$.

    Observe that the set U was constructed without using knowledge of the limit of the sequence ξ_n. If $\lim_n \xi_n(s) = \xi(s)$ for $s \in U$ is known, then one can represent U differently: letting

$$B_n^*(\epsilon) = \{s : |\xi_n(s) - \xi(s)| < \epsilon\},$$

we have

$$U = \bigcap_{k=1}^{\infty} \bigcup_{N=1}^{\infty} \bigcap_{n=N}^{\infty} B_n^*(1/k). \qquad (10.4)$$

In this case, $\xi_n \to \xi$ a.s. if $P(U) = 1$. Note that

$$P(U) = \lim_{k \to \infty} P\left(\bigcup_{N=1}^{\infty} \bigcap_{n=N}^{\infty} B_n^*(1/k)\right)$$

by continuity of P, and since the sets $\bigcup_{N=1}^{\infty} \bigcap_{n=N}^{\infty} B_n^*(1/k)$ decrease with k, we must have

$$P\left(\bigcup_{N=1}^{\infty} \bigcap_{n=N}^{\infty} B_n^*(1/k)\right) = 1$$

for every k. Next, intersections $\bigcap_{n=N}^{\infty} B_n^*(1/k)$ increase as N increases, so that (10.4) implies $\lim_{N \to \infty} P[\bigcap_{n=N}^{\infty} B_n^*(1/k)] = 1$. This, in turns, implies that for every k

$$\lim_{N \to \infty} P[B_N^*(1/k)] = 1. \qquad (10.5)$$

But the last condition is simply

$$\lim_{N\to\infty} P\left\{|\xi_N - \xi| < \frac{1}{k}\right\} = 1 \quad \text{for } k = 1, 2, \ldots,$$

which means that ξ_N converges to ξ in probability. We proved therefore

Theorem 10.2.2. *If $\xi_n \to \xi$ a.s., then $\xi_n \xrightarrow{P} \xi$.*

The converse to Theorem 10.2.2 is not true, as illustrated by the following example.

Example 10.2.3. Let $\xi_1, \xi_2, \ldots$ be independent random variables, such that $P\{\xi_n = 1\} = 1/n, P\{\xi_n = 0\} = 1 - 1/n$. Thus, if $0 < \epsilon < 1$, then $P\{|\xi_n| \geq \epsilon\} = P\{\xi_n = 1\} = 1/n \to 0$, which shows that the sequence $\{\xi_n\}$ converges to 0 in probability. For any sample point, the sequence $\{\xi_n(s)\}$ is simply a sequence of 0's and 1's, and in order to converge to 0, there must be only a finite number of terms equal 1 (i.e., all terms must be 0, starting from some N). But letting $A_n = \{\xi_n = 1\}$, we have $\sum P(A_n) = \sum \frac{1}{n} = \infty$, and by the second Borel–Cantelli lemma (Theorem 4.5.5) we know that with probability 1 infinitely many events A_n will occur. Thus $P\{\xi_n \text{ converges}\} = 0$, which shows that convergence in probability does not imply a.s. convergence. □

Formula (10.4) allows us to formulate the criterion for a.s. convergence: If $\xi_n \to \xi$ a.s., then $P(U^c) = 0$, so that

$$P\left(\bigcup_{k=1}^{\infty}\bigcap_{N=1}^{\infty}\bigcup_{n=N}^{\infty}[B_n^*(1/k)]^c\right) = 0.$$

Next, probability of a union of events is 0 if all these events have probability 0, so that for every k we must have

$$P\left(\bigcap_{N=1}^{\infty}\bigcup_{n=N}^{\infty}\{|\xi_n - \xi| \geq 1/k\}\right) = 0,$$

and—since inner unions decrease with the increase of N—we must also have

$$\lim_{N\to\infty} P\left(\bigcup_{n=N}^{\infty}\{|\xi_n - \xi| \geq 1/k\}\right) = 0.$$

Remembering that union of events means occurrence of at least one of the events, we may formulate the following theorem.

Theorem 10.2.3.   *The sequence $\{\xi_n\}$ of random variables converges a.s. to a random variable ξ if and only if for every $k = 1, 2, \ldots$,*

$$\lim_{N \to \infty} P\{|\xi_n - \xi| \geq 1/k \text{ for some } n \geq N\} = 0, \qquad (10.6)$$

or equivalently,

$$\lim_{N \to \infty} P\{\sup_{n \geq N} |\xi_n - \xi| \geq 1/k\} = 0. \qquad \square$$

We shall now use (10.6), together with the Kolmogorov inequality (Theorem 8.10.1) to prove the sufficiency part of the Kolmogorov's three series theorem. This theorem deals with the conditions for the a.s. convergence of the series $\sum_{j=1}^{\infty} X_j$ of independent random variables $X_1, X_2, \ldots$.

We shall also introduce the method of truncation, a powerful technique of handling limits of sequences of random variables.

If X is a random variable and $c > 0$, we define the *truncation of X at c* as a random variable $Y = Y^c$, defined by

$$Y = \begin{cases} X & \text{if } |X| \leq c \\ 0 & \text{if } |X| > c. \end{cases}$$

Observe that Y is a bounded random variable, so that $E(Y)$ and $\text{Var}(Y)$ both exist. We have

Theorem 10.2.4 (Kolmogorov Three Series Theorem).   *Let $X_1, X_2, \ldots$ be a sequence of independent random variables, and let Y_n be the truncation of X_n at level $c > 0$. Then $\sum_{n=1}^{\infty} X_n$ converges a.s. if and only if for some $c > 0$:*

(a) $\sum_{n=1}^{\infty} P\{|X_n| > c\} < \infty.$
(b) $\sum_{n=1}^{\infty} E(Y_n) < \infty$
(c) $\sum_{n=1}^{\infty} Var(Y_n) < \infty.$

Proof. As mentioned, only the sufficiency of conditions (a)–(c) will be shown; the proof of necessity is beyond the scope of this book.

Let us fix N, k, and $n > N$, and consider the sums

$$\sum_{j=N}^{r} (Y_j - E(Y_j))$$

for $r = N, \ldots, n$. By Kolmogorov inequality (8.63) we have

$$P\left\{\max_{N \leq r \leq n} \left| \sum_{j=N}^{r} (Y_j - E(Y_j)) \right| \geq \frac{1}{k} \right\} \leq \frac{\sum_{j=N}^{n} \text{Var}(Y_j)}{(1/k)^2}.$$

Letting $n \to \infty$, we have

$$P\left\{\sup_{N \le r}\left|\sum_{j=N}^{r}(Y_j - E(Y_j))\right| \ge \frac{1}{k}\right\} \le k^2 \sum_{n=N}^{\infty} \mathrm{Var}(Y_n).$$

In view of (c), the right-hand side tends to 0 as $N \to \infty$ for every fixed k. By Theorem 10.2.3 the sequence $\sum_{i=1}^{n}(Y_j - E(Y_j)), n = 1, 2, \ldots,$ converges a.s. Since $\sum_{i=1}^{\infty} E(Y_i)$ converges [condition (b)], we infer that $\sum_{i=1}^{\infty} Y_i$ converges a.s. To complete the proof, observe that $P\{|X_n| > c\} = P\{X_n \ne Y_n\}$. In view of condition (a) and the first Borel–Cantelli lemma (Theorem 2.6.2), with probability 1 only finitely many terms Y_n will differ from terms X_n. Consequently, $\sum X_n$ and $\sum Y_n$ will a.s. differ only by a finite quantity, and since $\sum Y_n$ converges, so does $\sum X_n$.   □

Example 10.2.4.   It is well known that for the harmonic series we have $1 + \frac{1}{2} + \frac{1}{3} + \cdots = \infty$. On the other hand, with alternating signs, we have $1 - \frac{1}{2} + \frac{1}{3} - \frac{1}{4} + \cdots = \pi/4$. What if the signs are allocated at random, by a flip of a fair coin? In other words, we ask about convergence of the series

$$X_1 + X_2 + X_3 + \cdots,$$

where X_n assumes values $\pm 1/n$ with probability $\frac{1}{2}$ and X_n's are independent. Taking $c = 2$, say, we have $Y_n = X_n$ for all n, and all terms of the series (a) in Theorem 10.2.4 are zero. Next, we have $E(Y_n) = 0$ and $\mathrm{Var}(Y_n) = E(Y_n^2) = 1/n^2$. Thus, all three series (a)–(c) converge, hence $\sum X_n$ converges a.s. In other words, the probability that random signs + and − in harmonic series will come out so unbalanced as to make the series diverge is zero.

Observe that this is not only the question of "balancing" the numbers of positive and negative signs. Indeed, the series $\sum 1/\sqrt{n}$ diverges, and the series $\sum \epsilon_n/\sqrt{n}$, where ϵ_n's are independent and $\epsilon_n = \pm 1$ with probability $\frac{1}{2}$, also diverges a.s. [since in this case the series (c) of variances diverges].   □

The third concept of convergence that will play an important role is as follows. Let $\xi_0, \xi_1, \xi_2, \ldots$ be a sequence of random variables, and let $F_n(t) = P\{\xi_n \le t\}, n = 0, 1, 2, \ldots$ be their cdf's.

Definition 10.2.3.   We shall say that the sequence $\{\xi_n\}$ *converges in distribution* to ξ_0 if

$$\lim_{n \to \infty} F_n(t) = F_0(t)$$

for every t at which $F_0(t)$ is continuous. In this case we shall write $\xi_n \xrightarrow{d} \xi_0$. Alternatively, we shall also use the symbol $F_n \Longrightarrow F_0$.   □

Before presenting the theorems characterizing convergence in distribution, it is worthwhile to make some comments, which should help in clarifying the motivation and intention of this concept.

First, observe that convergence in distribution does not imply anything about the behavior of random variables. For instance, if the variables $\xi_0, \xi_1, \dots$ are independent and identically distributed (iid), then $F_n(t) \equiv F_0(t)$, so $\xi_n \xrightarrow{d} \xi_0$, but we cannot expect any regularity in behavior of the observed values of random variables. In fact, since we require only convergence of distribution functions, we need not have any specific random variables ξ_n in mind. This explains the dual notation in Definition 10.2.3.

The second question concerns the reasons for requiring the convergence of $F_n(t)$ to $F_0(t)$ only at points of continuity of F_0 (and not at all points). The reason here lies in the special role played by discontinuities of cdf's. Consider a sequence of degenerate random variables (i.e., constants), defined as $\xi_n = 1/n, n = 1, 2, \dots$ and $\xi_0 = 0$. Obviously, in this case $\lim \xi_n = \xi_0$ in the "usual" sense (of convergence of observed values of ξ_n to ξ_0), and we would like to cover this case by the definition of convergence in distribution.

We have

$$F_n(t) = P\{\xi_n \le t\} = \begin{cases} 0 & \text{if } t < 1/n \\ 1 & \text{if } t \ge 1/n \end{cases}$$

and

$$F_0(t) = \begin{cases} 0 & \text{if } t < 0 \\ 1 & \text{if } t \ge 0. \end{cases}$$

For any $t \ne 0$ (where F_0 is continuous), we have $\lim F_n(t) = F_0(t)$. However, at $t = 0$ we have $F_0(0) = 1$, while $F_n(0) = 0$ for all n. Observe that if we put $\xi_n = -1/n$, then $F_n(t)$ will converge to $F_0(t)$ at all points t, including $t = 0$.

We now present some theorems that connect the three types of convergence introduced above, and also some criteria for convergence in distribution.

Theorem 10.2.5 (Slutsky). *If ξ_n and η_n are the sequences of random variables such that $\xi_n - \eta_n \xrightarrow{P} 0$ and $\eta_n \xrightarrow{d} \xi$, then $\xi_n \xrightarrow{d} \xi$.*

Proof. Let F be the cdf of ξ, and let F_n be the cdf of ξ_n. Let t be a continuity point of F, and let $\epsilon > 0$ be such that $t + \epsilon$ and $t - \epsilon$ are also continuity points of F. We can write

$$\begin{aligned} F_n(t) &= P\{\xi_n \le t\} \\ &= P\{\xi_n \le t, |\xi_n - \eta_n| < \epsilon\} + P\{\xi_n \le t, |\xi_n - \eta_n| \ge \epsilon\} \\ &\le P\{\eta_n \le t + \epsilon\} + P\{|\xi_n - \eta_n| \ge \epsilon\}. \end{aligned}$$

By similar reasoning we obtain

$$P\{\eta_n \le t - \epsilon\} \le F_n(t) + P\{|\xi_n - \eta_n| \ge \epsilon\}.$$

As $n \to \infty$, $P\{\eta_n \leq t \pm \epsilon\} \to F(t \pm \epsilon)$, and $P\{|\xi_n - \eta_n| \geq \epsilon\} \to 0$ by the assumption of the theorem. Consequently,

$$F(t - \epsilon) \leq \liminf F_n(t) \leq \limsup F_n(t) \leq F(t + \epsilon).$$

Since $\epsilon > 0$ is arbitrary (subject only to the conditions that F is continuous at $\pm\epsilon$), we see that we must have

$$\lim_{n \to \infty} F_n(t) = F(t),$$

which was to be shown.   □

Taking $\eta_n = \xi$, $n = 1, 2, \ldots$, we obtain the following:

Theorem 10.2.6. *If $\xi_n \xrightarrow{P} \xi$, then $\xi_n \xrightarrow{d} \xi$.*

Since we already know that convergence a.s. implies convergence in probability, we also have

Theorem 10.2.7. *If $\xi_n \to \xi$ a.s., then $\xi_n \xrightarrow{d} \xi$.*

Although, as already explained, convergence in distribution does not imply convergence in probability, such an implication holds in one important case, namely when the convergence is to a constant. We have

Theorem 10.2.8. *If $\xi_n \xrightarrow{d} c$, then $\xi_n \xrightarrow{P} c$.*

Proof. The condition $\xi_n \xrightarrow{d} c$ means that $\{\xi_n\}$ converges in distribution to a random variable ξ such that $P\{\xi = c\} = 1$. Consequently, again letting $F_n(t) = P\{\xi_n \leq t\}$, we have $F_n(x) \to 0$ for all $x < c$ and $F_n(x) \to 1$ for all $x > c$. But this means that $P\{|\xi_n - c| > \epsilon\} = P\{\xi_n < c - \epsilon\} + P\{\xi_n > c + \epsilon\} = F_n(c - \epsilon - 0) + [1 - F_n(c + \epsilon)] \to 0$ for every $\epsilon > 0$, which was to be proved.   □

We shall now state a theorem that shows the extent to which one is allowed to carry the algebraic manipulations on sequences converging in distribution and in probability.

Theorem 10.2.9. *If ξ_n, α_n, and β_n are sequences of random variables such that $\xi_n \xrightarrow{d} \xi$, $\alpha_n \xrightarrow{P} \alpha$, $\beta_n \xrightarrow{P} \beta$, where α and β are constants, then*

$$\alpha_n \xi_n + \beta_n \xrightarrow{d} \alpha \xi + \beta.$$

In particular, we have the following corollary.

Corollary 10.2.10. *If* $\xi_n \overset{d}{\to} \xi$, *and* a, b, a_n, b_n *are constants such that* $a_n \to a, b_n \to b$, *then*

$$a_n \xi_n + b_n \overset{d}{\to} a\xi + b.$$

Finally, we state without proof the following theorem, which will serve us as one of the main tools in proving convergence in distribution to normally distributed random variables.

Theorem 10.2.11. *Let* $\xi_n (n = 1, 2, \ldots)$ *and* ξ *be random variables, such that their moment generating functions* $m_n(t)$ *and* $m(t)$ *exist. We have then* $\xi_n \overset{d}{\to} \xi$ *if and only if* $m_n(t) \to m(t)$ *for every* t *in some interval around the point* $t = 0$. $\qquad \square$

The analogue of Theorem 10.2.11 for characteristic functions may be formulated as follows:

Theorem 10.2.12. *Let* $\xi_1, \xi_2, \ldots$ *be a sequence of random variables, and let* $\varphi_n(t)$ *be the characteristic function of* ξ_n. *Suppose that* $\varphi(t) = \lim_{\to \infty} \varphi_n(t)$ *exists for every* t, *and that* φ *is continuous at* $t = 0$. *Then* $\xi_n \overset{d}{\to} \xi$, *where the random variable* ξ *has characteristic function* φ.

The strength of Theorem 10.2.12, as compared with that of Theorem 10.2.11 lies in two facts. First, the characteristic functions exist for every distribution, so that it is not necessary to restrict theorems to random variables for which mgf exist. Second, the fact that the limit of sequence of distributions of ξ_n is itself a distribution (i.e., that ξ is a random variable) need not be assumed, but is implied by the fact that the limit of characteristic functions is continuous at $t = 0$.

The actual use of Theorem 10.2.11 will be based, in most cases, on the assumption of the existence of moments and the corresponding Taylor expansion of mgf's. We have the following theorem, which is an analogue of Theorem 10.2.12 for characteristic functions.

Theorem 10.2.13. *Let the random variable* X *have moment generating function* $m(t)$. *Assume that* $E(|X|^k) < \infty$ *for some* k, *and let* $m_j = E(X^j)$ *for* $j = 0, 1, \ldots, k$. *Then*

$$m(t) = \sum_{j=0}^{k} \frac{m_j t^j}{j!} + o(|t|^k), \tag{10.7}$$

where $o(|t|^k)$ *is some function such that* $o(|t|^k)/|t|^k \to 0$ *as* $t \to 0$.

Proof. The proof consists of using the well-known theorem from calculus on Taylor expansions, and recalling that (Theorem 8.5.3)

$$m_j = E(X^j) = \frac{d^{(j)} m(t)}{dt^j}\bigg|_{t=0}.$$

                                                                                                          □

We also state here another fact known from calculus:

Theorem 10.2.14. *If $\{c_n\}$ is a sequence of numbers such that $\lim_{n\to\infty} c_n = c$, then*

$$\lim_{n\to\infty} \left(1 + \frac{c_n}{n}\right)^n = e^c.$$

As an illustration of applicability of Theorem 10.2.11 we sketch a proof of Theorem 8.9.3 (law of large numbers in the Bernoulli case). We know [see (8.5.3)] that the mgf of the binomial random variable S_n is $m_{S_n}(t) = E e^{tS_n} = (q + pe^t)^n$. We can write here for every t, using the fact that $m_{S_n/n}(t) = m_{S_n}(t/n)$,

$$m_{S_n/n}(t) = \left[q + p(1 + \frac{t}{n} + \frac{t^2}{2n^2} + \cdots)\right]^n$$

$$= \left(1 + \frac{pt}{n} + \frac{pt^2}{2n^2} + \cdots\right)^n \to e^{pt}.$$

The last step consists of using Theorem 10.2.14 with

$$c_n = pt + \frac{pt^2}{2n} + \cdots \to pt.$$

The rest of the proof consists of observing that we showed that a moment generating function of S_n/n converges for every t (hence also in some neighborhood of $t = 0$) to the moment generating function of random variable equal identically to p. Theorem 10.2.11 allows us to infer that $S_n/n \xrightarrow{d} p$, and by Theorem 10.2.7 we have $S_n/n \xrightarrow{P} p$.

PROBLEMS

10.2.1 The random variable is said to have a Pareto distribution with parametera a, b ($a > 0, b > 0$) if its density is

$$f(x; a, b) = \frac{a}{b(1 + x/b)^{a+1}}, \qquad x > 0.$$

Let $X_1, \ldots, X_n$ be independent and identically distributed (iid) with Pareto distribution $f(x; 1, 1)$. Find the limiting distribution of random variable $U_n = n \min(X_1, \ldots, X_n)$. [*Hint:* Find cdf $F(x)$ of X_1 and determine cdf of U_n in terms of $F(x)$.]

10.2.2 Continuing Problem 10.2.1, let $V_n = \max(X_1, \ldots, X_n)$. Show that V_n does not have a proper limiting distribution; specifically, $\lim_{n \to \infty} P\{V_n \le t\} = 0$ for every t.

10.2.3 For Problem 10.2.2, find the limiting distribution of V_n/n.

10.2.4 Let $X_1, \ldots, X_n$ be a random sample from logistic distribution with cdf $F(x) = 1/(1 + e^{-x})$. Let $V_n = \max(X_1, \ldots, X_n)$. Then $V_n \xrightarrow{P} \infty$, but $V_n - \log n$ converge to a limiting distribution. Find $\lim_{n \to \infty} P\{V_n - \log n \le 0\}$ and $\lim_{n \to \infty} P\{|V_n - \log n| \le 1\}$.

10.2.5 Let $X_1, \ldots, X_n$ be a random sample from Poisson distribution with mean λ. Show that

$$e^{-\overline{X}_n} \xrightarrow{P} P(X_1 = 0).$$

10.2.6 Prove Theorem 9.7.3 about Poisson approximation to binomial distribution using Theorem 10.2.12.

10.2.7 Solve Problem 9.7.10 (concerning Poisson approximation to negative binomial distribution) using Theorem 10.2.12.

10.3 WEAK LAWS OF LARGE NUMBERS

We begin this section with proving some of weak laws of large numbers using the Chebyshev inequality (Theorem 8.9.1) as the main tool.

Theorem 10.3.1. *Let $X_1, X_2, \ldots$ be a sequence of independent identically distributed random variables. Assume that $E(X_i) = \mu$ and $\mathrm{Var}(X_i) = \sigma^2 > 0$. Then for every $\epsilon > 0$*

$$\lim_{n \to \infty} P\left\{\left|\frac{X_1 + \cdots + X_n}{n} - \mu\right| \ge \epsilon\right\} = 0. \tag{10.8}$$

Proof. We have

$$E\left(\frac{X_1 + \cdots + X_n}{n}\right) = \mu \quad \text{and} \quad \mathrm{Var}\left(\frac{X_1 + \cdots + X_n}{n}\right) = \frac{\sigma^2}{n}$$

(see Example 8.7.8). Applying the Chebyshev inequality to the random variable $(X_1 + \cdots + X_n)/n$, we obtain

$$P\left\{\left|\frac{X_1 + \cdots + X_n}{n} - \mu\right| \ge \epsilon\right\} \le \frac{\sigma^2}{n\epsilon^2}, \tag{10.9}$$

which tends to 0 as $n \to \infty$.   □

This theorem confirms, in some sense, our belief in "law of averages": formula (10.9) tells us that if we take the averages $(X_1 + \cdots + X_n)/n$ of larger and larger numbers of observations (of some phenomenon, described by random variable X, whose "copies" $X_1, X_2, \ldots$ are being observed), then it becomes less and less likely that the average $(X_1 + \cdots + X_n)/n$ deviates by more than ϵ from "true average" $\mu = E(X)$.

The assertion of Theorem 10.3.1 constitutes, to a large extent, the basis of the common understanding of why "statistics works," that is, why increasing number of observations pays off in the form of being able to make better inference about some quantities. The latter pertain either to the "population as a whole" or to "theoretical averages" of observations of a given phenomenon (e.g., repeated measurements of a certain quantity, where measurements are subject to errors).

This width and generality of the assertion of Theorem 10.3.1 is in stark contrast with the narrowness of the assumptions. Indeed, the latter specifies that $X_1, X_2, \ldots$ are independent and identically distributed (as mentioned earlier, typically abbreviated "iid" or "i.i.d.") random variables. The importance—practical and cognitive—of assertion (10.8), makes it worthwhile to analyze to which extent the assumption of Theorem 10.3.1 can be relaxed.

The full answer lies beyond the scope of this book; we shall, however, analyze the question of to what extent the assumptions may be relaxed *under the present proof*, based on the Chebyshev inequality.

First, let us observe that we did not fully utilize the assumption that X_i's have the same distribution. In fact, we used only a special consequence of this assumption, namely the fact that $E(X_i) = \mu$ and $\text{Var}(X_i) = \sigma^2$ are the same for all i. Thus, we may drop the requirement that X_i's be identically distributed, as long as we retain the stationarity of mean and variance. To use the example of measurement, we may allow measurements of the same quantity μ using different measuring instruments or methods, provided that $E(X_i) = \mu$ and $\text{Var}(X_i) = \sigma^2$ (such measurements are called unbiased and have the same precision).

But even the requirement that $E(X_i) = \mu$ is not necessary: When $E(X_i) = \mu_i$, relation (10.8) may be replaced by

$$P\left\{\left|\frac{X_1 + \cdots + X_n}{n} - \frac{\mu_1 + \cdots + \mu_n}{n}\right| \geq \epsilon\right\} \to 0, \qquad (10.10)$$

which is the same as

$$P\left\{\left|\frac{U_1 + \cdots + U_n}{n}\right| \geq \epsilon\right\} \to 0,$$

where $U_i = X_i - \mu_i$ is the deviation of the ith random variable from its own mean.

Next, an inspection of the proof of Theorem 10.3.1 shows that it is not necessary that variances are all equal. What is required is that variance of the average $(X_1 + \cdots + X_n)/n$ decreases to zero as n increases. We can therefore formulate the following version of law of large numbers.

Theorem 10.3.2. *Let $X_1, X_2, \ldots$ be a sequence of independent random variables, with $E(X_i) = \mu_i$ and $\mathrm{Var}(X_i) = \sigma_i^2$. If*

$$\lim_{n \to \infty} \frac{1}{n^2} \sum_{i=1}^{n} \sigma_i^2 = 0, \tag{10.11}$$

then relation (10.10) *holds for every $\epsilon > 0$.*

Let us now consider to which extent it is possible to relax the assumption of independence. Again, an inspection of the proof of Theorem 10.3.1 shows that the property which was used is the additivity of variance, specifically the fact that $\mathrm{Var}(X_1 + \cdots + X_n) = \sigma_1^2 + \cdots + \sigma_n^2$. But this property holds under the assumption that the random variables are uncorrelated. We have therefore

Theorem 10.3.3. *If random variables $X_1, X_2, \ldots$ are uncorrelated, then the condition* (10.11) *is sufficient for $\frac{1}{n} \sum_{i=1}^{n}(X_i - \mu_i) \xrightarrow{P} 0$.*

Finally, let us observe that even the latter condition can be relaxed: What we really need to make the proof valid is that $(1/n^2)\mathrm{Var}(X_1 + \cdots + X_n) \to 0$. This in turn, is implied by the assumption (10.11) when all covariances are zero or negative. We therefore have

Theorem 10.3.4. *Let $X_1, X_2, \ldots$ be a sequence of random variables such that $\mathrm{Cov}(X_i, X_j) \le 0$ for $i \ne j$, and $\sigma_j^2 = \mathrm{Var}(X_j)$ satisfying* (10.11) *as $n \to \infty$. Then for every $\epsilon > 0$*

$$P\left\{ \left| \frac{1}{n} \sum_{j=1}^{n} (X_j - E(X_j)) \right| \ge \epsilon \right\} \to 0$$

as $n \to \infty$.

This is as far as we will go using the techniques of proof based on the Chebyshev inequality. It is of some interest that in case of identical distributions one can prove a weak law of large numbers without the assumption of existence of variance. We shall prove the following extension of Theorem 10.3.1, proved by Khintchin in 1929.

Theorem 10.3.5. *Let $X_1, X_2, \ldots$ be a sequence of iid random variables with $E(X_i) = \mu$, and such that their common mgf exists in some neighborhood of $t = 0$. Then the relation* (10.8) *holds.*

Proof. From Theorem 10.2.7 it follows that it is enough to show that $(X_1 + \cdots + X_n)/n \xrightarrow{d} \mu$, and therefore (Theorem 10.2.11) that $m_{S_n/n}(t) \to e^{\mu t}$ in some neighborhood of $t = 0$.

Letting $m(t) = Ee^{tX}$, we have (since X_i's are identically distributed and independent) $m_{S_n/n} = [m(t/n)]^n$. The existence of expectation of X implies the existence of the derivative m' and the relation $m'(0) = \mu$. Then the Taylor expansion of $m(t)$ is (see Theorem 10.2.13)

$$m(t) = 1 + \mu t + o(|t|),$$

so that

$$m_{S_n/n}(t) = \left(1 + \frac{\mu t}{n} + o\left(\frac{|t|}{n}\right)\right)^n.$$

To complete the proof it suffices to use Theorem 10.2.14 for $c_n = \mu t + no(|t|/n) \to \mu t$.

Comment. Observe that the assumption of the existence of an mgf is not necessary for the validity of the theorem. The proof of this strenghtened version requires nothing more than replacing moment generating functions with characteristic functions. All the steps of the proof remain unchanged.

PROBLEMS

10.3.1   Let $X_1, X_2, \ldots$ be a sequence of independent random variables, with the distribution given by

$$P\{X_n = n^\alpha\} = P\{X_n = -n^\alpha\} = 0.5, \qquad n = 1, 2, \ldots.$$

Show that if $\alpha < 0.5$, then the sequence $\{X_n\}$ obeys the weak law of large numbers. (*Hint:* $1^{2\alpha} + 2^{2\alpha} + \cdots + n^{2\alpha} \sim n^{2\alpha+1}/(2\alpha + 1)$ as $n \to \infty$. To see why, observe that the left-hand side is approximated by $\int_0^n x^{2\alpha}\, dx$.)

10.4   STRONG LAWS OF LARGE NUMBERS

The strong laws of large numbers (SLLNs) are theorems that assert almost sure convergence of sequences of random variables obtained by averaging some underlying sequences of random variables. We shall prove here two such theorems, both due to Kolmogorov.

Theorem 10.4.1.   *Let* $X_1, X_2, \ldots$ *be independent, with* $EX_i = \mu_i$, $\mathrm{Var}X_i = \sigma_i^2$. *If*

$$\sum_{n=1}^{\infty} \frac{\sigma_n^2}{n^2} < \infty,$$

then

$$\frac{1}{n} \sum_{i=1}^{n} (X_i - \mu_i) \to 0 \ a.s. \tag{10.12}$$

Proof. To simplify notation, let $S_n = \sum_{j=1}^{n} X_i$ and $m_n = \sum_{i=1}^{n} \mu_i$. For a fixed $\epsilon > 0$ let

$$C_k = \left\{ \max_{2^{k-1} < n \leq 2^k} \frac{1}{n} |S_n - m_n| \geq \epsilon \right\}$$

To prove the theorem, it suffices to show that $\sum P(C_k) < \infty$, since then, by the Borel–Cantelli lemma (Theorem 2.6.2), only finitely many events C_k will occur a.s., which implies (10.12).

If C_k occurs, then at least one of the inequalities

$$|S_n - E(S_n)| \geq \epsilon n \geq \epsilon 2^{k-1}, \qquad n = 2^{k-1} + 1, \ldots, 2^k$$

occurs. By the Kolmogorov inequality (Theorem 8.10.1) we obtain

$$P(C_k) \leq \frac{\text{Var}(S_{2^k})}{\epsilon^2 (2^{2k-2})} = \frac{4}{\epsilon^2} \frac{\sum_{j=1}^{2^k} \sigma_j^2}{2^{2k}}.$$

We can therefore write

$$\sum_{k=1}^{\infty} P(C_k) \leq \frac{4}{\epsilon^2} \sum_{k=1}^{\infty} \frac{1}{2^{2k}} \sum_{j=1}^{2^k} \sigma_j^2$$

$$= \frac{4}{\epsilon^2} \sum_{j=1}^{\infty} \sigma_j^2 \sum_{2^k \geq j} \frac{1}{2^{2k}}$$

$$= \frac{16}{3\epsilon^2} \sum_{j=1}^{\infty} \frac{\sigma_j^2}{j^2} < \infty.$$

The last step is obtained as follows:

$$\sum_{2^k \geq j} \frac{1}{2^{2k}} = \sum_{k \geq \log_2 j} 2^{-2k} = \frac{2^{-2 \log_2 j}}{1 - \frac{1}{4}}.$$ □

We shall now prove another strong law of large numbers, also due to Kolmogorov, which covers the iid case.

Theorem 10.4.2. *Let $X_1, X_2, \ldots$ be iid random variables, and let $S_n = X_1 + \cdots + X_n$. If $\mu = E(X_i)$ exists, then $S_n/n \to \mu$ a.s.*

Proof. Let us truncate the random variables $\{X_n\}$, by letting

$$Y_k = \begin{cases} X_k & \text{if } |X_k| \leq k \\ 0 & \text{if } |X_k| > k. \end{cases}$$

We have

$$\sum_{k=1}^{\infty} P\{Y_k \neq X_k\} = \sum_{k=1}^{\infty} P\{|X_k| > k\}$$

$$= \sum_{k=1}^{\infty} P\{|X_1| > k\} < \infty$$

by the assumption that $E(X_1) < \infty$ and since X_k has the same distribution as X_1. Thus, the inequality $X_k \neq Y_k$ will occur a.s. only a finite number of times, and it suffices to prove that $(1/n) \sum_{k=1}^{n} Y_k \to \mu$ a.s. To this end, we shall use Theorem 10.4.1, proved above.

We let

$$\sigma_k^2 = \text{Var}(Y_k) \leq E(Y_k^2)$$

and therefore we can write

$$\sum_{k=1}^{\infty} \frac{\sigma_k^2}{k^2} \leq \sum_{k=1}^{\infty} \frac{E(Y_k^2)}{k^2} = \sum_{k=1}^{\infty} \frac{1}{k^2} \int_{-k}^{k} x^2 \, dF(x)$$

$$= \sum_{k=1}^{\infty} \frac{1}{k^2} \sum_{j=1}^{k} \int_{j-1 < |x| \leq j} x^2 \, dF(x)$$

$$= \sum_{j=1}^{\infty} \int_{j-1 < |x| \leq j} x^2 \, dF(x) \sum_{k=j}^{\infty} \frac{1}{k^2}$$

$$\leq \sum_{j=1}^{\infty} \int_{j-1 < |x| \leq j} x^2 \, dF(x) \cdot \frac{C}{j}$$

$$\leq C \sum_{j=1}^{\infty} \int_{j-1 < |x| \leq j} |x| \, dF(x) = C \int_{-\infty}^{\infty} |x| \, dF(x) < \infty.$$

We used the estimate $1/j^2 + 1/(j+1)^2 + \cdots < C/j$ valid for some C. It follows that $(1/n) \sum_{j=1}^{n} (Y_j - E(Y_j)) \to 0$ a.s. But $E(Y_j) \to E(X_1) = \mu$ as $j \to \infty$, hence also $(1/n)[E(Y_1) + \cdots + E(Y_n)] \to \mu$, so that we must have

$$\frac{1}{n}\sum_{j=1}^{n} Y_j \to \mu \text{ a.s.,}$$

which completes the proof.

PROBLEMS

10.4.1 Theorem 10.3.5 asserts that (under additional assumption of existence of moment generating functions), a sequence of iid random variables obeys weak law of large numbers if its expectation exists. Theorem 10.4.2 asserts the same (even without any additional requirement) for the strong law of large numbers. On the other hand, the existence of the first moment μ is necessary for S_n/n converging to μ, in probability or almost surely. Thus, in the iid case, the existence of the first moment is necessary and sufficient (hence equivalent) to *both* strong and weak law of large numbers. It follows that these two laws are equivalent. Find an error in this reasoning.

10.5   CENTRAL LIMIT THEOREM

The term *central limit theorem* (CLT) is a generic name used to designate any of the series of theorems which assert that the sums of large numbers of random variables, after standardization (i.e., subtraction of the mean and division by standard deviation) have approximately the standard normal distribution.

As suggested by the adjective *central*, the search for conditions under which sums of large number of components have approximate normal distribution has been (and to a large extent still is) one of the leading research topics in probability theory for about last 200 years. We begin with the simplest case of iid sequences.

Theorem 10.5.1 (Lindeberg and Lèvy). *Let* $X_1, X_2, \ldots$ *be a sequence of iid random variables. Let* $\mu = E(X_i)$ *and* $\sigma^2 = \text{Var}(X_i)$, *where we assume that* $0 < \sigma^2 < \infty$. *Then for every* x *we have, letting* $S_n = X_1 + \cdots + X_n$,

$$\lim_{n\to\infty} P\left\{\frac{S_n - n\mu}{\sigma\sqrt{n}} \le x\right\} = \frac{1}{\sqrt{2\pi}} \int_{-\infty}^{x} e^{-t^2/2}\, dt. \qquad (10.13)$$

Proof. Observe that $E(S_n) = n\mu$ and $\text{Var}(S_n) = n\sigma^2$, so that the left-hand side of (10.13) is simply the limit of cdf's of standardized sums

$$S_n^* = \frac{S_n - E(S_n)}{\sqrt{\text{Var}(S_n)}}.$$

The right-hand side is the cdf of standard normal random variable, denoted $\Phi(x)$. Thus, the theorem asserts that cdf's of S_n^* converge to $\Phi(x)$ for every x, hence at every point of continuity of $\Phi(x)$. Letting Z denote the standard normal random variable, Theorem 10.5.1 asserts that $S_n^* \xrightarrow{d} Z$.

Once the assertion is phrased in this way, it becomes clear that a possible strategy of the proof is to use moment generating functions and Theorem 10.2.11. Let $m_X(t)$ be the common mgf of random variables X_i (assumed to exist). The existence of the first two moments suggests using the Taylor expansion. We have

$$S_n^* = \sum_{i=1}^n \frac{X_i - \mu}{\sigma \sqrt{n}},$$

so that

$$m_{S_n^*}(t) = \left[m_{(X-\mu)/\sigma \sqrt{n}}(t) \right]^n$$
$$= \left[m_{(X-\mu)/\sigma} \left(\frac{t}{\sqrt{n}} \right) \right]^n. \qquad (10.14)$$

Now

$$E\left(\frac{X - \mu}{\sigma} \right) = 0, \qquad E\left(\frac{X - \mu}{\sigma} \right)^2 = 1, \qquad (10.15)$$

so that

$$m_{(X-\mu)/\sigma}(t) = 1 + \frac{t^2}{2} + o(t^2),$$

and consequently, using (10.15),

$$m_{S_n^*} = \left[1 + \frac{t^2}{2n} + o\left(\frac{t^2}{n} \right) \right]^n$$
$$= \left[1 + \frac{(t^2/2) + no(t^2/n)}{n} \right]^n.$$

Using Theorem 10.2.14 for $c_n = t^2/2n + no(t^2/n) \to t^2/2$, we obtain

$$m_{S_n^*}(t) \to e^{t^2/2}$$

which completes the proof in the special case when the underlying random variables X_i have mgf's. For the general case, see the comment following the proof of Theorem 10.3.5.   □

Example 10.5.1. Assume that you buy a supply of household items (e.g., $n = 20$ batteries to be used for some specific purpose, one after another). Assume also that the lifetime of a battery in question is a random variable with mean 2 weeks and standard deviation of 3 days. The batteries are re-placed as soon as they break, and the batteries that are unused are not aging.

What is the (approximate) probability that the supply of batteries will last more than 9 but less than 10 months? (i.e., more than 270 and less than 300 days?).

SOLUTION.   The question here is about the probability that $S_{20} = X_1 + \cdots + X_{20}$ satisfies the inequality $270 < S_{20} < 300$. We have $n\mu = 20 \cdot 14 = 280$ and $\sigma\sqrt{n} = 3 \cdot \sqrt{20} = 13.4$. Thus $270 < S_{20} < 300$ occurs if and only if $(270 - 280)/13.4 < S_{20}^* < (300 - 280)/13.4$, (i.e., if $-0.75 < S_{20}^* < 1.49$).

We may therefore write, letting Z stand for the standard normal random variable,

$$P\{270 < S_{20} < 300\} \approx P\{-0.75 < Z < 1.49\}$$
$$= \Phi(1.49) - \Phi(-0.75) = 0.9319 - 0.2266 = 0.7053.$$

□

Theorem 10.5.1 is a generalization of the oldest central limit theorem, covering binomial distribution, due to Laplace. If S_n is the number of successes in n independent trials, each with probability of success p, then $S_n = X_1 + \cdots + X_n$, where $P\{X_i = 1\} = p = 1 - P\{X_i = 0\}$, $i = 1, 2, \ldots, n$. We have here $ES_n = np$, $\mathrm{Var}(S_n) = npq$, so that

$$S_n^* = \frac{S_n - np}{\sqrt{npq}}.$$

The following theorem is true.

Theorem 10.5.2 (Laplace).   *If S_n has binomial distribution* $\mathrm{BIN}(n, p)$, *then for any $z_1 < z_2$,*

$$\lim_{n\to\infty} \sum_{A_n \le j \le B_n} P\{S_n = j\} = \lim_{n\to\infty} P\{z_1 \le S_n^* \le z_2\}$$
$$= \Phi(z_2) - \Phi(z_1), \tag{10.16}$$

where $A_n = np + z_1\sqrt{npq}$, $B_n = np + z_2\sqrt{npq}$.

Suppose now that we want to find $P\{a \le S_n \le b\}$; according to Theorem 10.5.2, we have, for large n,

$$P\{a \le S_n \le b\} = P\left\{ \frac{a - np}{\sqrt{npq}} \le S_n^* \le \frac{b - np}{\sqrt{npq}} \right\}$$
$$\approx \Phi\left(\frac{b - np}{\sqrt{npq}}\right) - \Phi\left(\frac{a - np}{\sqrt{npq}}\right). \tag{10.17}$$

However, approximation (10.17) can be improved somewhat if we observe that in the present case S_n is an integer-valued random variable, and for *integer a* and *b*, the exact expression is

$$P\{a \le S_n \le b\} = \sum_{j=a}^{b} P\{S_n = j\}. \tag{10.18}$$

Each individual term on the right-hand side of (10.19) can be approximated by an area under the normal curve between $\dfrac{j - \frac{1}{2} - np}{\sqrt{npq}}$ and $\dfrac{j + \frac{1}{2} - np}{\sqrt{npq}}$.

Adding such approximations, the terms for neighboring j cancel, and we obtain the following formula:

$$P\{a \le S_n \le b\} \approx \Phi\left(\frac{b + 0.5 - np}{\sqrt{npq}}\right) - \Phi\left(\frac{a - 0.5 - np}{\sqrt{npq}}\right) \tag{10.19}$$

for any integers $a \le b$.

Example 10.5.2. We toss a fair coin $n = 15$ times. Find the approximate probability that the number S_{15} of heads will satisfy the inequality $8 \le S_{15} < 10$.

SOLUTION. Observe first that the inequality $8 \le S_{15} < 10$ is the same as $8 \le S_{15} \le 9$. We have here $np = 15 \cdot 0.5 = 7.5$, $\sqrt{npq} = \sqrt{15 \cdot 0.5 \cdot 0.5} = 1.94$. The approximation (10.16) gives

$$P\{8 \le S_{15} \le 9\} \approx \Phi\left(\frac{9 + 0.5 - 7.5}{1.94}\right) - \Phi\left(\frac{8 - 0.5 - 7.5}{1.94}\right)$$

$$= \Phi(1.03) - \Phi(0) = 0.8485 - 0.5 = 0.3485.$$

The exact value is

$$P\{S_{15} = 8\} + P\{S_{15} = 9\} = \binom{15}{8}\left(\frac{1}{2}\right)^{15} + \binom{15}{9}\left(\frac{1}{2}\right)^{15}$$

$$= 0.1964 + 0.1527 = 0.3491. \qquad \square$$

Generally, the quality of approximation improves as n increases, and—for the same n—decreases as p moves away from $\frac{1}{2}$ in either direction. Also, observe that use of the *continuity correction* [addition and subtraction of $\frac{1}{2}$ from the limits a and b in formula (10.16)] makes sense only if $0.5/\sqrt{npq}$ exceeds the difference between the consecutive arguments in the tables of normal distribution that are actually being used.

Example 10.5.3 (Decision Problem). Suppose that we design a theater with 1000 seats. The theater has two entrances, A and B, situated in such a way with respect to the parking lot, public transport, and so on, that the patrons have equal chances of choosing any of the entrances.

Suppose also that our theater is located in a city where climate causes the patrons to wear overcoats, which they leave in a checkroom. There are two checkrooms, each located near one of the entrances, and while it is not impossible to enter through one entrance and leave the coat in a checkroom near the other entrance, it is inconvenient to do so. How many places for coats should each checkroom have?

SOLUTION. Clearly, the problem is not precise enough as stated: We have to specify the criterion that we want to attain. One of the objectives is mini-mization of the cost of equipping the coatroom in hangers, racks, and so on. We simply do not want too many of them to remain empty. On the other hand, one does not want to inconvenience the patrons by making too many of them go to the other checkroom. The two extremes, each satisfying one of the foregoing objectives, is to equip each checkroom with 1000 places for coats, and to equip each with exactly 500 places for coats.

A possible objective may be: We want to equip each coatroom with $500 + x$ places for coats, where x is the smallest number such that (say), on 95% of nights where the theater is sold out, everybody will be able to leave his or her coat at the checkroom nearest the entrance used.

To solve the problem, we have to make some assumptions about indepen-dence of choice of entrances A and B by the patrons. As the first approxi-mation, assume that the patrons arrive one at a time and each chooses the entrance independent of other patrons. Let S_{1000} be the number of patrons (among $n = 1000$) who choose entrance A. We want to have $S_{1000} \leq 500 + x$ (then everybody who enters through A is not inconvenienced), and also $1000 - S_{1000} \leq 500 + x$ (which is the analogous condition for those who choose entrance B). Thus, we want the event

$$500 - x \leq S_{1000} \leq 500 + x$$

to occur with probability 0.95 or more.

We have $p = \frac{1}{2}$, so that $np = 1000 \times \frac{1}{2} = 500$, while for standard deviation we have $\sqrt{npq} = \sqrt{1000 \times \frac{1}{2} \times \frac{1}{2}} = 15.8$. Therefore,

$$P\{500 - x \leq S_{1000} \leq 500 + x\} = P\left\{\frac{-x - \frac{1}{2}}{15.8} \leq S^*_{1000} \leq \frac{x + \frac{1}{2}}{15.8}\right\}$$

$$\approx \Phi\left(\frac{x + \frac{1}{2}}{15.8}\right) - \Phi\left(\frac{-x - \frac{1}{2}}{15.8}\right) \geq 0.95.$$

Inspection of Table A.2 shows that we have $\Phi(1.96) - \Phi(-1.96) = 0.95$. We must therefore take x as the smallest integer for which $(x + \frac{1}{2})/15.8 \geq 1.96$, which gives $x = 31$. Thus, to achieve our objective, one should install 531 places for coats in each coatroom.

Example 10.5.4. Continuing Example 10.5.3, a more realistic assumption is that people come to theater in pairs and that both members of a pair enter by the same entrance. We now have $n = 500$ pairs, and letting S_{500} denote the number of pairs who choose entrance A, we must have $2S_{500} \le 500 + x$ and $1000 - 2S_{500} \le 500 + x$. Now $E(S_{500}) = 500 \times 0.5 = 250$ and $\text{Var}(S_{500}) = 500 \times 0.5 \times 0.5 = 125$. The objective therefore becomes $250 - x/2 \le S_{500} \le 250 + x/2$. Using formula (10.16), we get

$$P\left\{250 - \frac{x}{2} \le S_{500} \le 250 + \frac{x}{2}\right\} = P\left\{\frac{-(x/2) - (1/2)}{\sqrt{125}} \le S_{500}^* \le \frac{(x/2) + (1/2)}{\sqrt{125}}\right\}$$

$$= \Phi\left(\frac{x+1}{22.3}\right) - \Phi\left(-\frac{x+1}{22.3}\right).$$

Again, x is the smallest integer for which $(x+1)/22.3 \ge 1.96$, so that $x = 43$. We see that grouping (in this case into pairs, but the effect is the same for other groupings) of the same set of persons, with groups choosing the entrance independently, increases the variability: we now need 543 places for coats in each checkroom to meet the requirement.   □

The central limit theorem proved thus far concerns the rather narrow case of independent and identically distributed components. In this case the sum has asymptotically normal distribution provided only that variance is finite. This theorem was often utilized to explain the frequent appearance of normal distribution in nature. The argument typically goes along the following lines. Consider an attribute such as the height of a man or the error of measurement of some quantity (e.g., the speed of light). In each case the observed value is influenced by a large number of factors, some having a negative and some having a positive effect. Some of these factors are known but their effect cannot be predicted exactly, while some other factors cannot even perhaps be named. What matters is that they operate largely independently of one another and each of them in isolation is small compared with the total effect of all factors (more precisely, factors that are known to have large effects are treated differently and are not included in these considerations). The central limit theorem asserts that the total effect of such "small" factors is random and has approximately a normal distribution. We already know this to be true in the case of factors that are iid. However, independence and identical distribution can hardly be justified in almost any real situation, and can at best be regarded as approximations to reality. Consequently, one may expect that central limit theorems are valid in wider classes of situations, where the iid assumption does not hold. We state below two theorems that provide conditions for asymptotic normality in case of independent random variables that are not identically distributed.

We consider a sequence of random variables $X_1, X_2, \ldots$ and the corresponding sequence of partial sums $S_n = X_1 + X_2 + \cdots + X_n$.

In the theorem below we shall use the following notation and assumption. We let

$$\mu_i = E(X_i), \quad \sigma_i^2 = \text{Var}(X_i) \tag{10.20}$$

and

$$\gamma_i = E|X_i - \mu_i|^3.$$

Moreover, we put

$$m_n = \sum_{j=1}^{n} \mu_j, \quad s_n^2 = \sum_{j=1}^{n} \sigma_j^2, \quad \Gamma_n = \sum_{j=1}^{n} \gamma_j.$$

We now formulate the following theorem:

Theorem 10.5.3 (Liapunov).   *Let $X_1, X_2, \ldots$ be a sequence of independent random variables with finite third moments. Then the condition*

$$\lim_{n \to \infty} \frac{\Gamma_n}{s_n^{3/2}} = 0 \tag{10.21}$$

is sufficient for convergence:

$$\frac{S_n - m_n}{s_n} \xrightarrow{d} N(0, 1).$$

The proof of this theorem lies beyond the scope of this book.

As an illustration of application of the Liapunov theorem, consider a sequence of independent trials, such that in the nth trial the probability of success is p_n. We let S_n denote the number of successes in first n trials, so that $S_n = X_1 + \cdots + X_n$, where $X_i = 1$ or 0 depending on whether or not the ith trial leads to a success. We then have $E(X_i) = p_i$, $\text{Var}(X_i) = p_i q_i$, while the third absolute central moment γ_i is

$$\gamma_i = E|X_i - p_i|^3 = (1 - p_i)^3 P(X_i = 1) + |0 - p_i|^3 P(X_i = 0)$$
$$= q_i^3 p_i + p_i^3 q_i.$$

Thus,

$$\gamma_i = p_i q_i (p_i^2 + q_i^2) \le p_i q_i,$$

$$\frac{\Gamma_n}{s_n^{3/2}} \le \frac{\sum_{i=1}^{n} p_i q_i}{(\sum_{i=1}^{n} p_i q_i)^{3/2}} = \frac{1}{(\sum_{i=1}^{n} p_i q_i)^{1/2}}.$$

Consequently, Liapunov condition (10.21) holds if $\sum_{i=1}^{n} p_i q_i = \infty$, and we have:

Theorem 10.5.4. *Consider a sequence of independent Bernoulli trials, with the probability of success in the ith trial being p_i. If S_n is the number of successes in n first trials, then*

$$\frac{S_n - \sum_{i=1}^{n} p_i}{\sqrt{\sum_{i=1}^{n} p_i q_i}} \xrightarrow{d} N(0, 1)$$

if $\sum_{i=1}^{\infty} p_i q_i = \infty$.

The practical question one may ask about assertions such as the Liapunov theorem is: How can we know whether or not the assumptions are satisfied? The problem is that the conditions are expressed through a limiting passage, hence—at least in theory—they require knowledge of infinitely many parameters (e.g., variances and third central absolute moments). Since such knowledge is clearly unattainable, some statisticians are inclined to regard Theorem 10.5.4 (as well as many other limiting theorems) as irrelevant for statistics, at least in its "applied" part.

This criticism is perhaps somewhat too severe. First, in some cases we need not know the properties of the component random variables, except in some general terms, and we need not know their number, except that we need to know it to be large. An example here may be the diffusion process, where the displacement of a particle suspended in a liquid (say) is due to a large number of random hits by molecules of the liquid. While particular random variables here (displacements due to single hits) are not observable, their general properties may sometimes be inferred from physical laws. Such knowledge may be sufficient to infer that the assumptions of (say) the Liapunov theorem hold, and hence that the displacement in diffusion has a normal distribution.

Another case when one may be able to apply limit theorems of probability theory in statistics occurs in some cases of sampling. It is possible to devise sampling schemes under which the observations have some specific distributions. To use the simplest example, sampling until first success leads to geometric distribution for the number of trials, so that the statistician in fact controls the form of the distribution, reducing the statistical problem to (say) finding a parameter of this distribution.

Finally, there may be cases when one really observes only a fixed number n of random variables, perhaps because there are no other variables (imagine the data on some measurements, one for each of the 50 U.S. states). One could argue easily that limit theorems are irrelevant in this case. However, suppose that one can sensibly evaluate m_i, σ_i^2, and γ_i for each of the variables (e.g., from some previous data, etc.). Then if $\sum_{i=1}^{n} \gamma_i$ is small compared with $(\sum_{i=1}^{n} \sigma_i^2)^{3/2}$, one can expect that the sum of random variables in question will have a distribution close to normal (after all, each distribution is an approximation to any other distribution, except that the approximation may be very poor. In the case in question, it may not be so poor). At any rate, knowing the first three moments, one can use appropriate theorems (e.g., the Berry–Esseen bounds) to estimate the difference between the cdf of $(S_n - m_n)/s_n$ and the cdf of the standard normal distribution.

We close this chapter by stating the theorem that completed the long search for conditions implying limiting normality in the case of independent random variables (the cases of dependent random variables are still the object of intense research, but we shall not discuss this topic). We have the following theorem.

Theorem 10.5.5 (Lindeberg and Feller). *Let $X_1, X_2, \ldots$ be a sequence of independent random variables with finite second moments. Assume that $s_n^2 \to \infty$ and $\max_{1 \leq j \leq n} \sigma_j^2 / s_n^2 \to 0$ as $n \to \infty$. Then*

$$\frac{S_n - m_n}{s_n} \xrightarrow{d} N(0, 1)$$

if and only if for every $\epsilon > 0$,

$$\lim_{n \to \infty} \frac{1}{s_n^2} \sum_{j=1}^{n} \int_{|x - \mu_j| \geq \epsilon s_n} (x - \mu_j)^2 dF_j(x) = 0, \tag{10.22}$$

where F_j is the cdf of X_j.

The "if" part was proved by Lindeberg, and (10.22) is called the *Lindeberg condition*. The "only if" part is due to Feller.

The proof lies beyond the scope of this book. Some comments, however, are in order. First, the Lindeberg condition is expressed only in terms of the first two moments, so it is weaker than the Liapunov condition. At the same time, however, it is much more complicated to verify, since it involves variances not only of original variables, but also of variances of the truncated variables. To get some insight into the Lindeberg criterion, assume that $\mu_j = 0$ for all j (this can always be accomplished without loss of generality, by centering all random variables at their expectations). Then an equivalent form of Lindeberg condition is : For every $\epsilon > 0$ we have

$$\lim_{n \to \infty} \frac{1}{s_n^2} \sum_{j=1}^{n} \int_{|x| \leq \epsilon s_n} x^2 dF_j(x) = 1,$$

which can also be written as

$$\lim_{n \to \infty} \frac{1}{\text{Var}(S_n)} \sum_{j=1}^{n} E(X_{j, \epsilon s_n}^2) = 1,$$

where $X_{j,\epsilon s_n} = X_j$ or 0 depending whether $|X_j| \leq \epsilon s_n$ or $|X_j| > \epsilon s_n$ (i.e., $X_{j,\epsilon s_n}$ is the truncation of X_j at the level ϵs_n). Intuitively, Lindeberg condition means that as $n \to \infty$, less and less "mass" of variables $X_1, \ldots, X_n$ lies outside the range $(-\epsilon s_n, +\epsilon s_n)$.

Finally, it is worthwhile to stress the role of conditions $s_n^2 \to \infty$ and $\max_{1 \leq j \leq n} \sigma_j^2 / s_n^2 \to 0$. Both of them are implied by the Lindeberg condition, so that (10.22) is actually sufficient for asymptotic normality. To see why $s_n^2 \to \infty$ is needed, observe that if $X_2 = X_3 = \cdots = 0$, the limiting distribution of S_n is simply the distribution of X_1, which can be arbitrary.

On the other hand, assume that variances σ_n^2 increase so fast that the variance σ_n^2 is of the order higher than the sum $\sigma_1^2 + \cdots + \sigma_{n-1}^2$. Then $\max_{1 \leq j \leq n} \sigma_j^2 / s_n^2 \geq \sigma_n^2 / (\sigma_1^2 + \cdots + \sigma_{n-1}^2 + \sigma_n^2) \to 1$, and it is (at least intuitively) clear that S_n has a distribution close to X_n (the effect of $X_1 + \cdots + X_{n-1}$ is "squashed" by X_n). Again, the limiting distribution of S_n need not exist, and even if it does, it need not be normal.

PROBLEMS

10.5.1 Assume that 500 students in a certain college will graduate on a given day. From past experience it is known that 50% of students have both parents attending the graduation ceremony, 30% of students have one parent attending, and the remaining 20% have no parent attending. Assuming that only parents will be admitted, how many chairs are needed in order to have at least a 95% chance that all attending parents will have seats?

10.5.2 Passengers on an international flight have a baggage weight limit B. The actual weight W of the passenger's baggage is such that W/B has a beta distribution with parameters a and b such that $a/(a+b) = 0.9$.

Assume that the weights of baggage of various passengers are independent, and let the plane have 220 seats. Find a and b if it is known that if the plane is fully booked, then there is a 5% chance that the total weight of baggage will exceed $200B$.

10.5.3 Let $X_1, \ldots, X_n$ be a random sample from the distribution with the density $f(x) = xe^{-x}, x > 0$. Find c if it is known that $P\{\overline{X}_n > c\} = 0.75$ for $n = 250$.

10.5.4 Let $X_1, \ldots, X_n$ be a random sample from the beta distribution with density $f(x) = Cx^2(1-x), 0 < x < 1$. Let $S_n = X_1 + \cdots + X_n$. Find the smallest n for which $P\{S_n \geq \frac{3}{4}n\} \leq 0.01$.

10.5.5   A regular dodecahedron (12-sided Platonian solid) has 6 red and 6 white faces, faces of each color labeled 1, ..., 6. If you toss a face with label k you pay or win $\$k$, depending on whether the color is red or white. Find the probability that after 50 tosses you are ahead by more than $\$10.00$.

10.5.6   A fair coin is tossed n times. How large must be n if it is known that the probability of number of heads equal to number of tails is less than 0.2?

10.5.7   Let $X_1, \ldots, X_{360}$ represent the outcomes of 360 throws of a fair die. Let S_{360} be the total score $X_1 + \cdots + X_{360}$, and for $j = 1, \ldots, 6$, let Y_j be the total number of throws which gave outcome j. Give the normal approximation to the probabilities:

    (i) $P(55 < Y_3 < 62)$.
    (ii) $P(1200 < S_{360} < 1300)$.
    (iii) $P(1200 < S_{360} < 1300 | Y_1 = 55)$.
    (iv) $P(1200 < S_{360} < 1300 | Y_4 = 55)$.
    (v) $P(1200 < S_{360} < 1300 | X_1 = X_2 = \cdots = X_{55} = 4)$.

10.5.8   A die is unbalanced in such a way that the probability of tossing k ($k = 1, \ldots, 6$) is proportional to k. You pay $\$4.00$ for a toss and win $\$k$ if you toss k. Find approximate probability that you are ahead after 100 tosses.

10.5.9   Referring to Example 8.7.9, assume that a man's shoe has an average length of 1 foot and $\sigma = 0.1$ foot. Find the (approximate) probability that the mean of 16 lengths of men's shoes exceed 1 foot by more than 1 inch.

10.5.10   (**Why Don't We Need Tables for Chi-Square Distribution for Large** n?)   Show that as $n \to \infty$, the chi-square random variable χ_n^2 converges in distribution to normal. Find the normalizing constants and find the approximate 95% and 99% quantiles of χ_{47}^2 and χ_{113}^2.

CHAPTER 11

Outline of Inferential Statistics

11.1 INTRODUCTION

The role of this chapter is to provide an informal bridge between the theory of probability covered in Chapters 1 to 10, and statistics, the object of the rest of the book. The questions asked and answered in statistics complement those asked and answered in probability in the following sense. In probability theory, we tried to develop techniques that could lead to predicting the form of future observation (data): Given certain general information (e.g., about independence of some events, etc.), we deduced the distributions of observed random variables. In practice we answered questions such as: how many future observations will fall into a certain set, what the average of those observations will be, and so forth.

As opposed to that, in statistics—to be more precise, in inferential statistics, which will be our main object of study—the question is: Given the data, what can we say about specific aspects of the stochastic mechanisms that governed the occurrence of those data? The actual data are regarded as a result of a random process, in the sense that if the data collection were to be repeated, the outcome would typically be different. Consequently, whatever inference we make from the actual data, it is subject to error. This error—the central concept of statistics—is not meant to be a "mistake" of any kind (i.e., something that could be avoided). It refers to the unavoidable randomness: If the data were collected again, we might have reached a different conclusion.

At first one could think that this randomness will be eliminated if we increase the precision of measurement. A moment of reflection suffices to realize the surprising fact that the opposite is true. For instance, if we measure length of a table in integer number of feet, the result will be the same under repetition. When we increase the precision, to an inch, then to $\frac{1}{2}$ an inch, and so on, the variability of the results under repetition will become more and more pronounced.

The question arises, then: What inference about the underlying phenomenon can be drawn from premises (data) which may differ from occasion to occasion? Viewed in this way, statistics is a part of theory of inductive inference. But this does not mean that the *theory* of statistics is itself inductive: As a theory, inferential statistics is a fragment of mathematics, in the same way as probability theory. Thus, mathematical statistics has its own structure of specific concepts (motivated mostly by applications) and its own theorems, which are proved under the same requirements of rigor as in other domains of mathematics.

Each theorem in mathematics asserts that some conclusions hold, provided that certain assumptions are satisfied, and theorems in statistics are no exception in this respect. Some of the assumptions refer to the process of data collection, or the properties of underlying random variables. The applicability of statistical methods (i.e., the empirical validity of the consequences of the appropriate theorems) depends crucially on the degree to which assumptions are met in real situations. Sometimes this degree of validity is under the control of the experimenter (see Example 11.1.1); in some other cases, possibly after performing appropriate tests, we simply may have no conclusive evidence that the assumptions are violated; in particular situations (see Example 11.1.2) we may feel justified in disregarding the fact that the assumptions are false.

Example 11.1.1. Suppose that we want to apply a method of statistical inference, for which we need an assumption that a certain random variable X is binomial. Often X represents the number of elements of some kind in the sample (e.g., number of defective items, number of patients who recovered after specific treatment, etc.). Whether or not the assumption that the process of collecting observation is really a sequence of Bernoulli trials depends on various factors, some of which the experimenter can control. Of these, the principal factor is the independence of selections of the sample. In case of defective items, it may require sampling with replacement; in the case of patients, it may require checking that the sample does not have identical twins or other persons whose reaction to the treatment in question may be similar for genetic reasons.

Example 11.1.2. Suppose that the assumption requires some observed variable to have a normal distribution. Imagine that we observe freshmen scores X on an aptitude test. Often it is reasonable to assume that X has normal distribution if such an assumption leads to sufficiently good approximations of relevant probabilities. The fact remains, however, that test scores cannot be normally distributed, since X must be an integer, and it also cannot assume negative values. □

In cases such as above, we typically feel justified in using the consequences of theorems which rely on the assumption that X has normal distribution, even if we know that this assumption is not satisfied. But there may be some "more serious" violations of the assumptions. Checking the validity of an assumption may involve using some elaborate statistical techniques. Quite often, however, it is sufficient to just have a glance on some preliminary graphical presentation of the data, or on the values of some crude statistics. In either case, we deal with some reduction of the data.

The methods of such "initial reduction" of the data belong properly to what is called *descriptive statistics*. Although the main object of the book is inferential (rather than descriptive) statistics, we present some basic ideas of descriptive statistics in the next section. That section is followed by a brief outline of some principal domains of mathematical statistics. It is necessary to stress at once that there is no classification of the area of statistics according to some Aristotelian principles which would satisfy more refined aesthetic tastes. This is due to the fact that most of the development of statistics is connected with specific types of applications such as regression analysis, analysis of variance, sampling methods, and so on. To be sure, there are questions asked regardless of the type of application, and from this viewpoint, one can distinguish two major areas—estimation and hypothesis testing, discussed in Chapters 12 and 13. These do not exhaust the theoretical research—we also have discriminant analysis, methods of statistical optimization, and methods of optimal designs of experiments, to mention just the basic areas.

Finally, we have still another classification, into methods suitable for data measured on a scale of certain type. For instance, methods for data where only ordinal relations matter are generally known as nonparametric statistics. We discuss these topics briefly in this chapter, providing illustrations by typical examples, stressing especially those topics that will be discussed in subsequent parts of the book.

11.2 DESCRIPTIVE STATISTICS

Some simple summary presentation of the data can lead to the discovery of surprising and important consequences. In our overview of descriptive statistics we do not attempt to be complete, since the field grows rapidly as more and more methods are developed. Most of them may be found in statistical packages. Such techniques of descriptive statistics may save labor and allow us to avoid making certain suggestive mistakes.

Let us begin with two examples, both concerning World War II.

Example 11.2.1.   The main route for supplying the Allied armies fighting Nazi Germany in Europe during World War II was through the Atlantic Ocean. The convoys were attacked regularly, mostly by German U-boats. As the war progressed, more and more data accumulated. It turned out that the average number of ships lost in an attack was relatively constant; in particular, it did not depend on the size of the convoy. This observation led to a simple conclusion: *To decrease losses, make convoys as big as possible.* Indeed, two separate convoys might expect to be detected and attacked about twice as many times as a convoy obtained by combining them. Since the average losses per attack are independent on the convoy size, such joining of two convoys cuts losses by half. This simple idea contributed substantially to winning the war.

Example 11.2.2. Bombers were sent on missions over Germany. On route to and from, as well as over the target, they were subject to antiaircraft fire. The direct hits were not very frequent, but the AA shells were set to explode at specific heights, spraying the planes with shrapnels. The planes that returned from the mission were examined for locations of shrapnel hits, and all these locations were recorded on a silhouette of the plane. As more and more data became available, the silhouette was more densely covered with recorded locations of hits. There were, however, some areas that were hit less often than others.

The surprising order was then given: *Strengthen (by putting armor plates, etc.) those areas that were hit SELDOM.* Here the argument is that the locations of hits of bombers by shrapnels must have distribution uniform over the silhouette of the plane. Any "white areas" in the data must therefore indicate the locations of hits for the planes that did not return from missions. $\qquad$ □

We cite these two examples not only because of their combination of simplicity of premises and unexpectedness of conclusion, but also because they are both based on elementary ways of representing the data: plotting average losses per attack against convoy size, or making a scatter diagram of shrapnel hits. Certainly, it is not often that one gets a chance to contribute to victory in war by a visual inspection of some descriptive statistics. Nevertheless, it is worth knowing some simple "tricks of the trade" in preparing the data. The ones briefly mentioned in this section concern univariate data. The goal is to present the data in such a way as to exhibit certain aspects of interest, or making certain patterns visible.

Dot Diagram

First, if the number of data points is small, one may often get a good insight into the structure of the data by drawing a "dot diagram." This is accomplished by marking the data values as dots on the x-axis, with repeated data represented by dots piled up one on another. For instance, the data points

$$28, 36, 37, 52, 36, 45, 39, 45, 38, 35, 36$$

will be represented as a diagram in Figure 11.1. If the data are to be grouped into classes, with class boundaries and class counts replacing individual data values, the dot diagram is a good device to use to help choose class boundaries.

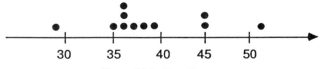

Figure 11.1 Dot diagram.

Stem-and-Leaf Display

This is another method of quick presentation of data. Again, it is best explained by an example. Suppose that the data are

9.5	10.8	8.8	11.2	10.2	10.3	10.2	11.3	10.0	8.8
10.7	9.9	11.4	9.8	10.5	9.8	9.9	10.9	8.1	10.5
10.6	8.2	8.6	9.2	9.9	10.0	11.0	9.2	10.7	10.9

We then choose "stems" and data values are presented as follows:

Stem	Leaves
8	8 8 1 2 6
9	5 9 8 8 9 2 9 2
10	8 2 3 2 0 7 5 9 5 6 0 7 9
11	2 3 4 0

Both the dot-diagram and stem-and-leaf method involve no reduction of information contained in the data. Most of the other methods involve partial reduction of the information; this concerns grouping and representing the data by various kinds of indices.

Box Plot

This is a very simple but quite informative representation of data sets, printed out by many computer routines (see Figure 11.2). The box plot is based on five values and provides a graphical presentation which allows us not only to visualize them, but also to gather information about skewness and outliers.

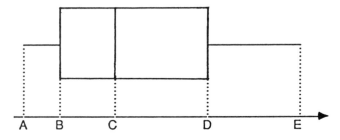

Figure 11.2 Box plot.

Thus, A and E are the smallest and largest data points (so the difference $E - A$ is the range of the data). Next, B and D are the lower and upper quartile of the data, so that the box (whose height is irrelevant) contains the central 50% of data points. Finally, C is the median of the data.

Median as well as upper and lower quartiles are the sample counterparts of the corresponding quantiles of the distribution. Let $y_1, \ldots, y_n$ be the data points, and let $y_{1:n} \le y_{2:n} \cdots \le y_{n:n}$ be the ordered sample. There are several definitions of the sample pth quantile. Some texts define it as $y_{[np]+1:n}$, where $[x]$ stands for the largest integer not exceeding x. The disadvantage of such a definition is that the pth and rth quantiles coincide if p is sufficiently close to r. Another definition of the pth sample quantile is

$$y_{[(n+1)p]:n} + \{(n+1)p - [(n+1)p]\}(y_{[(n+1)p]+1:n} - y_{[(n+1)p]:n}). \tag{11.1}$$

Rather than contemplating this formula, consider a simple example. We have $n = 4$ data points, which (arranged in increasing order) are

$$y_{1:4} = 5, y_{2:4} = 8, y_{3:4} = 15, y_{4:4} = 20. \tag{11.2}$$

Suppose that we want to compute the 37% sample quantile for the data. Here $p = 0.37$, and $[np] + 1 = [4 \times 0.37] + 1 = [1.48] + 1 = 2$. The first of the mentioned methods would give the 37% sample quantile for the data (11.2) as $y_{2:4} = 8$. The same value 8 would be a 38% and even a 39% sample quantile.

The method based on formula (11.1) uses the most obvious linear approximation. The four data points partition the range into five classes, determined by quantiles for 20%, 40%, 60%, and 80%. The 37% quantile lies between the 20% and 40% quantile, hence between $y_{1:4} = 5$ and $y_{2:4} = 8$, at a point whose distance from 5 is $(37 - 20)/(40 - 20) = 0.85$ of the distance between 5 and 8. Thus, the 37% quantile equals $5 + 0.85(8 - 5) = 7.55$.

Observe that formula (11.1) does not apply if p is close to 0 or close to 1. The reason is that if $p < 1/(n+1)$, then $[(n+1)p] = 0$ and $y_{0:n}$ is not defined, while if $p > n/(n+1)$, then $y_{[(n+1)p]+1:n}$ does not exist.

For $p = \frac{1}{2}$, formula (11.1) gives the sample median as

$$y_{[\frac{n+1}{2}]:n} + \left\{ \frac{n+1}{2} - \left[\frac{n+1}{2}\right] \right\} \left(y_{[\frac{n+1}{2}]+1:n} - y_{[\frac{n+1}{2}]:n} \right).$$

Formula for the median when n is odd simplifies to $y_{(n+1)/2:n}$, since in that case $[(n+1)/2] = (n+1)/2$. If n is even, we have $[(n+1)/2] = n/2$ and consequently median is equal to

$$\frac{1}{2}(y_{(n/2):n} + y_{(n/2)+1:n}).$$

One of the examples when sample quantiles are used may be a *normal quantile plot*, known also as a *Q-Q plot*, a graphical method used for check-

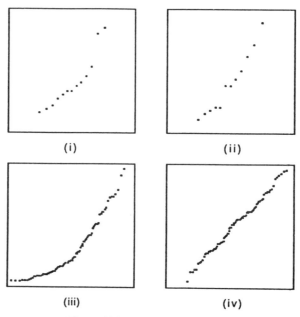

(i)   (ii)

(iii)   (iv)

Figure 11.3   Normal quantile plots.

ing assumption that the data come from a normal population. The graph is obtained by plotting quantiles of standard normal distributions against corresponding (with the same p) sample quantiles obtained by formula (11.1) for the data checked for normality. If the sample comes from a normal population, the Q-Q plot should form a straight line. Figure 11.3(i) and (ii) show Q-Q plots for data from exponential and normal distribution for small samples. As one can see, the random fluctuations obscure the picture, so that it is hard to decide which plot, if any, forms a straight line. Figure 11.3(iii) and (iv) show similar situations, but with large sample sizes ($n = 300$). Now the situation is obvious. A large sample represents the population adequately, and looking at the graphs one can easily conclude which data come from the normal distribution and which ones do not. The final conclusion about normality of the data is based on subjective judgment resulting from a visual inspection. In Chapters 13, 16, and 17 we show other methods of testing normality, for which decision is based on assessment of certain probabilities rather than on subjective judgments.

Grouping

Large sets of data are unmanageable in their original form, and typically, the data points are grouped into classes. One then gives the class boundaries (or equivalently, class midpoints and class widths) and the class counts (also called class frequencies). The only formal requirement here is that the classes must be disjoint and cover all actual, and also potentially possible, data values (so that each data point falls into exactly one class, and consequently, all class frequencies add to the total number of data points).

Histograms

Graphical representation of the grouped data has usually the form of a histogram. These are formed as follows: The horizontal axis is divided into intervals centered at class midpoints and extending in either direction by one half of the class width. Then rectangles are built on these intervals in such a way that the *areas of rectangles are proportional to the class frequencies.*

This definition of histograms sets them apart from some other graphical representations. It is especially important to remember that histograms represent frequencies by areas of rectangles (and not by their heights); this distinction is vital in case of unequal class widths.

The empirical questions here are: What is the "proper" number of classes? Should all classes be of the same width? How does grouping affect the values of summary indices such as mean?

Class Width and Number of Classes
There is no unique answer as to the proper class width (or number of classes). When the data are divided into too many classes (assume equal class widths), the histogram looks jagged. When there are too few classes, one loses too many details in the data. The appropriate number of classes depends not only on sample size but also on the shape of the distribution, and one can formulate the criteria in different ways. For instance, Scott (1979) formulates the problem of the choice of class width as a problem of optimization of certain criteria and shows that the optimal width is on the order of $3.49\hat{\sigma}/\sqrt[3]{N}$, where N is the number of data points and $\hat{\sigma}$ is the estimate of standard deviation. Thus, for $N = 100$ data points with the shape of the histogram close to normal, the range* R is about $4\hat{\sigma}$, and the best class width is about $3.49(R/4)/\sqrt[3]{100} \approx R/5$, so that the histogram should have 5 classes. For $N = 1000$ the range R is about $6\hat{\sigma}$ and the best class width is about $R/16$ (so 16 classes).

Effect on Indices
Indices such as mean are calculated for grouped data by replacing each data value within a class by the class midpoint. Consequently, the values of the same index calculated from the raw data and from the grouped data differ. Under the assumption of equal widths of classes, it is typically a relatively

---

* The distribution of the range $R = X_{N:N} - X_{1:N}$ is given by (7.43). The exact formula for $E(R)$ is not available in case of sampling from normal distribution. Crude approximations are $E(R) \approx 4\sigma$ for $N = 100$ and $E(R) \approx 6\sigma$ for $N = 1000$. For exact values one needs to inspect appropriate statistical tables.

simple task to estimate the upper bound for the error of a given index due to grouping.

For instance, in the case of the mean, the bound is as follows. Each of the N data points is off from its class midpoint by at most $L/2$, where L is the class width. If all these deviations are the largest possible and in the same direction (no cancellation), the difference between the exact and approximate mean is $N(L/2)/N = L/2$. Thus, the error due to grouping is at most $L/2$.

11.3 BASIC MODEL

A typical user of statistical methods is someone who must choose an action in a situation of partial uncertainty regarding the factors that affect the consequences of his action. The uncertainty may be to some extent alleviated by the fact that our decision maker can "spy on nature" through taking some observations that may help him to identify the relevant factors. The formal representation of this rather general description is actually the cornerstone of most of the theory of mathematical statistics. Figure 11.4 presents a general scheme of making decisions in situations of uncertainty. Statistical inference constitutes only a part of this scheme.

First, except in some special cases, the nature of action to be selected is of no concern for statistics and will not be included in the formal structure of the theory. Now, let us imagine ourselves as decision makers. Using the terminology accepted in general decision theory, we say that we are in one of the situations labeled as "world θ_1," ..., "world θ_n," but we do not know

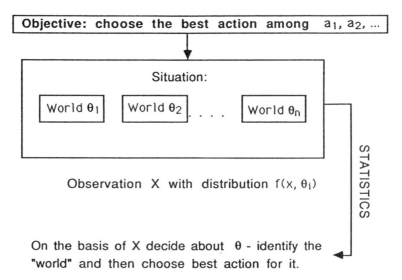

Figure 11.4 Decision scheme.

which one. It is usually assumed that there is one action appropriate for each "world" (although this assumption is not really necessary). Thus, if we knew the "world" we are in (i.e., knew the value θ_j), we would know which is the best action to take.

The only way to identify the actual "world θ_j" is to make some experiments. Let **X** be a generic symbol for the results of such experiments. It may be a single observation of some variable, sequence of observations, a result of some complicated physical experiment, a score on a battery of tests, and so on, depending on the context. Naturally, to make the entire setup meaningful, there must be some relation between **X** and θ_j. This is expressed by assumption that the observation **X**, while being random, comes from a distribution that depends on θ_j. The randomness here is the crucial part of the entire setup. It may be due to sampling variability, or be inherent in the phenomenon, or both.

It may happen, of course, that observation **X** will allow us to identify the value θ_j without ambiguity. This occurs when the sets of possible values of **X** are disjoint for different θ_j's. Such cases, however, are rare in real life, and—what is more important—are trivial from a theoretical point of view. Most often we face a challenging case when the same outcome **X** is possible under several (or even all) θ_j, but occurs with different probabilities, depending on θ_j.

Example 11.3.1.   The examples of situations falling under this scheme abound in science, engineering, management, and in everyday life. A doctor faces a patient who has a headache and high fever. Particular "worlds" are possible illnesses of the patient. The doctor orders some tests, and on the basis of their results **X** makes a diagnosis (identifies the illness, perhaps incorrectly) and chooses the best action in view of the illness diagnosed. As another example, take a pharmaceutical company that has developed a drug against a specific disease, hopefully superior to drugs used thus far. The "worlds" may be numbers θ describing the relative advantage of the new drug versus the best drug available so far, with $\theta > 1$ indicating that the new drug is better, and $\theta \leq 1$ indicating that it is no better (or even inferior) to the drug used thus far. The observation **X** is here a series of studies, specified in great detail by FDA standards. $\square$

Example 11.3.1 concerns medicine, but it is obvious that any new method is subject to the same scheme of tests before it is established whether or not (and to what extent) it is superior to some other method.

As seen from Figure 11.4, statistical theory is concerned with the problem of inference about θ on the basis of **X**. In the case when there is a choice of the variable that we may observe, the statistical theory is also concerned with choosing the "best" variable (these are called problems of design of experiment).

11.4 BAYESIAN STATISTICS

An inspection of Figure 11.4 reveals that one level is missing, representing the possibility that the user may have some knowledge, previous experience, or perhaps other reasons (including prejudice or superstition) which make some "worlds" appear more likely than others before any observations **X** are taken.

There is a general agreement among statisticians that in cases when the "world" is actually chosen randomly from the set of possible worlds, with probabilities having frequential interpretation, those probabilities should be taken into account in the process of deciding about θ_j on the basis of observation **x** (one can then simply compute the conditional probability of θ_j given $\mathbf{X} = \mathbf{x}$). For example, in case of a physician seeing a patient with a headache and high fever, the possible "worlds" (in this case illnesses causing this particular set of symptoms) comprise initial stages of flu as well as initial stages of plague (black death). The incidence of the latter disease is so rare, however, that the doctor may feel perfectly justified in not ordering any test to check the possibility of the disease being black death. Here the physician relies simply on his own and his colleagues' experience about the incidence of various diseases that might start with fever and headache at a given time and geographic location.

The situation is not so straightforward in other cases. The researcher may still be able to assign probabilities to various θ_j's, but these probabilities need not have frequential interpretation. For instance, in case of testing a new drug, the frequency of cases when a new drug proved better than the existing one may be irrelevant (e.g., if the new drug tested is based on an entirely new idea).

Statisticians differ in their opinion whether or not the probabilities of various "worlds" reflecting researcher's experience, intuition, "hunches," and so on, should be used in statistical methodologies. Those who allow such probabilities to be used are called *Bayesians*, and the resulting statistical methodology is called *Bayesian statistics*.

A rather strong argument for the Bayesian approach is provided by the theory of outliers (see Section 7.6). In a non–Bayesian setup, where $\theta = (\alpha, \lambda)$ is the parameter in the family of all gamma distributions, no configuration of data values can be rejected as containing an outlier, even if (say) $n - 1$ observations fall into interval $(0, 1)$ and the nth observation exceeds 10^6. This is because there exists a $\theta = (\alpha, \lambda)$ for which such a configuration of data points is very likely to occur. This counterintuitive example suggests that a statistician should eliminate as "unlikely" some domains of the parameter space, using whatever information or experience he or she has. In the sequel, we shall often provide Bayesian solutions to various problems under considerations. This means that we shall show how the problem is, or may be, solved if the prior probabilities of various θ_j's are available. A systematic exposition of the theory of Bayesian statistics lies, however, beyond the scope of this book.

11.5 SAMPLING

In theory of statistics it is typically assumed that the data are realizations of some random variable whose distribution is modeled in a certain way. The applicability of statistical methods depend therefore on how well the assumption of randomness and the assumption about the distribution is satisfied. In this section we show some of the possible "traps" that one may encounter in implementing statistical methods in practice.

Let us start from an example.

Example 11.5.1.   Suppose that the objective is to estimate an unknown parameter θ, which is the average of some attribute in a population. It may be, for instance, the average yearly income of a family in a given town, the average age of a cancer patient at the time of detection of the disease, and so on. In such cases one typically takes a sample of elements from the population and measures the values of the attribute in question. Thus, $\mathbf{X} = (X_1, \ldots, X_n)$, where $E(X_i) = \theta$.

Example 11.5.2.   The following story illustrates some of the problems that one may encounter in trying to implement in practice the scheme of Example 11.5.1.

Three social science students, Jim, Joe, and Susan, were each assigned a task of estimating the average size of class (number of students) in a given school district. Jim decided to make a card for each class in each school, shuffle the cards, sample one or more of them, and then find the number of children in each class sampled.

Joe found a somewhat simpler scheme; he decided to prepare cards with names of schools and first sample a school (or several schools). Then for each school chosen, he decided to make cards with labels of classes and take a sample of those cards, at the end determining the numbers of children in each class sampled.

Susan applied a still simpler scheme: She decided to take a sample of children from the school district under study and ask each child about the size of the class that he or she attends. The question is: Which of the three students, if any, measured the parameter "average size of the class in a given school district"?

To answer this question, we shall consider only the case when Jim, Joe, and Susan each take a single observation. We may do so because an increase in sample size affects only the precision of the estimator, not the parameter that is being estimated.

Suppose that there are k schools in the district in question, with the ith school having n_i classes, of sizes $C_{ij}, i = 1, \ldots, k, j = 1, \ldots, n_i$. Then the total number of classes is $N = \sum_{i=1}^{k} n_i$, and the average class size is

$$\theta = \frac{1}{N} \sum_{i=1}^{k} \sum_{j=1}^{n_i} C_{ij}.$$

The objective is to estimate θ. If X, Y, and Z denote the random variables constructed, respectively, by Jim, Joe, and Susan, then it is clear that $X = C_{ij}$ with probability $1/N$ (sampling is from the set of all classes). Thus $E(X) = \theta$.

As regards random variable Y, we have $Y = C_{ij}$ if Joe selects the ith school (probability $1/k$) and the jth class in the ith school (probability $1/n_i$). Consequently,

$$E(Y) = \sum_{i=1}^{k} \sum_{j=1}^{n_i} \frac{1}{kn_i} C_{ij} = \frac{1}{k} \sum_{i=1}^{k} \frac{1}{n_i} \sum_{j=1}^{n_i} C_{ij}.$$

We have $E(Y) \neq \theta$ except in a special case when all n_i's are equal (i.e., if each school has the same number of classes).

Finally, for the random variable Z, constructed by Susan, the situation is as follows. Let $C = \sum_{i=1}^{k} \sum_{j=1}^{n_i} C_{ij}$ be the total number of children in all classes. With the probability C_{ij}/C a child from the jth class of the ith school will be selected, and then the value of Z will be C_{ij}. We have, therefore,

$$E(Z) = \sum_{i=1}^{k} \sum_{j=1}^{n_i} C_{ij} P\{Z = C_{ij}\}$$

$$= \sum_{i=1}^{k} \sum_{j=1}^{n_i} C_{ij} \frac{C_{ij}}{C} = \frac{1}{C} \sum_{i=1}^{k} \sum_{j=1}^{n_i} C_{ij}^2.$$

Again, we have $E(Z) \neq \theta$ unless all classes are of the same size.

Thus, it is only Jim, whose method provides an estimate of the parameter θ. For random variables suggested by Joe and Susan we generally have $E(Y) \neq \theta$ and $E(Z) \neq \theta$. As regards the method suggested by Joe, it is known as *stratified sampling*: The population is divided into *strata*, and one first samples strata and then takes a sample from each stratum (in this case the role of strata are played by schools).

Let us mention here that if Joe decided to take a sample of schools and then collect the data about *all* class sizes in selected schools, he would use what is known as *cluster sampling*. With some prior information available (e.g., about relative sizes of the strata), one can adjust easily Joe's estimator (by taking appropriate weighted averages) to build random variable Y with $E(Y) = \theta$ (called an *unbiased* estimator of θ).

As regards the method of Susan, the situation is not so straightforward and cannot easily be remedied. It is connected with *importance sampling*, where the probability of choosing an element with larger value of the attribute is higher than the probability of choosing the element with smaller value of the attribute. □

The bias due to the phenomenon of importance sampling occurs quite often and evades notice. The following example (suggested by R. F. Green, personal communication) provides some surprising insight into the issue.

Example 11.5.3 (Siblings). Suppose that in a certain society, the distribution of the number of children in a family is Poisson with mean 4. What is the average number of siblings of a child in this society?

SOLUTION. An almost automatic response of most persons (including some statisticians) is 3. In fact, however, the answer is 4 (in the special case of Poisson distribution); in general it is *more* than 3, except the special case when all families have exactly 4 children. The situation is very much the same as with sampling by Susan from Example 11.5.2.

Before presenting the solution, one should observe that the distribution given in the problem concerns the population of *families*, while the question concerns the average in the population of *children*. Let $p_0, p_1, \ldots$ be the distribution of number of children in the family, so that in the special case under analysis we have $p_k = (\lambda^k / k!) e^{-\lambda}$ for $\lambda = 4$. Then $\mu = p_1 + 2p_2 + \cdots$ and $\sigma^2 = \sum k^2 p_k - \mu^2$ are the mean and variance of the distribution $\{p_k\}$. Suppose that the population consists of a large number N of families. Clearly, $N p_0$ families have no children, $N p_1$ families have one child, and so on. The total size of population of children is therefore

$$M = N p_1 + 2N p_2 + 3N p_3 + \cdots = N\mu.$$

The probability of chosing a child from family with k children is

$$\frac{kN p_k}{N\mu} = \frac{k p_k}{\mu},$$

for $k = 1, 2, \ldots$ and such a child has $k - 1$ siblings. Then the average number of siblings is

$$Q = \sum_{k=1}^{\infty} (k - 1) \frac{k p_k}{\mu} = \frac{1}{\mu} \left\{ \sum_{k=1}^{\infty} k^2 p_k - \sum_{k=1}^{\infty} k p_k \right\}$$

$$= \frac{1}{\mu} \{ \sigma^2 + \mu^2 - \mu \} = \frac{\sigma^2}{\mu} + \mu - 1.$$

As may be seen, we have $Q \geq \mu - 1$ with $Q = \mu - 1$ in the case when $\sigma^2 = 0$ (i.e., when all families have the same number μ of children). For the Poisson distribution $\sigma^2 = \mu = \lambda$, so that $Q = \mu = 4$. □

A sampling bias closely related to the bias from importance sampling is connected with the following phenomenon, which caused some controversy before it became properly understood. To explain it, we shall again use an anecdotal example.

Example 11.5.4 (Renewal Paradox). The statistical objective is to estimate the average lifetime of electric bulbs of some particular type, all produced by the same company. We shall assume that the distribution of the
lifetime T is exponential with mean $E(T) = 1/\lambda$.

The usual procedure would be to take a random sample of bulbs and
observe the lifetimes $T_1, T_2, \ldots, T_N$ of selected bulbs. Such data could be used
to estimate $E(T)$. If testing were run in parallel it would take the time $T_{N:N} =$
$\max\{T_1, \ldots, T_N\}$ to collect the data, so one might speed up the procedure
by observing only the k shortest lifetimes $T_{1:N} < T_{2:N} < \cdots < T_{k:N}$ and then
interrupt the data collection, recording only that $T_{k+1:N} > T^*$, where T^* is
some threshold (so that $N - k$ lifetimes are not fully observed, and we only
know that they all exceed T^*). This is called *censoring of the data.*

In some cases, one could use another way of collecting the data, which does
not involve waiting. Suppose that there exists a large building (e.g., an office
skyscraper downtown), where the maintenance personnel uses only bulbs of
the type of interest for us. Whenever a bulb fails, it is replaced immediately
by a new one, and this change is recorded. As a consequence we have access
to the record, reaching into the past, of the times of replacement of every
bulb. We may then select a sample in the following way. Fix some time t^*
(preferably in the past) and use as the sample all lifetimes of the bulbs that
were operating at time t^*. Clearly, if t^* is sufficiently far back in the past,
each of the bulbs that operated at t^* has already been replaced, and its
lifetime $T(t^*)$ is known. Otherwise, one may still have to wait for some time
to observe the value of $T(t^*)$ for that bulb. The situation is best explained in
Figure 11.5.

In general, for the ith bulb T_i' is the *spent* lifetime at t^*, and T_i'' is the
residual lifetime at t^*, and it is clear that if $t_p - t^*$ is large enough, all residual
times will be observable. For bulbs 1 and 2, the lifetimes in this method
of sampling are $T_1(t^*) = T_1' + T_1''$ and $T_2(t^*) = T_2' + T_2''$. The value T_3 is not
observable at present, but is at least $T_3' + (t_p - t^*)$. The sample $T_1, T_2, \ldots, T_N$
of all N bulbs that were operating at t^* can be observed and the average
$(1/N) \sum_{i=1}^{N} T_i(t^*)$ can serve as an estimator $E(T)$.

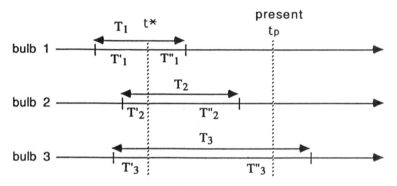

Figure 11.5 Sampling interreplacement times.

Now, the question arises: Is such a method of collecting data correct or not? It should be stressed that we are *not* concerned here with the practical implementation of the scheme. Thus, we assume that the time t^* is chosen without the knowledge of the replacement records (a condition preventing conscious or unconscious bias). We also disregard the fact that in reality, the bulbs are used only part of the time, so that bulbs at some locations are used more often or are subject to different conditions (e.g., outdoor and indoor lights, etc.). Since the problem concerns the theory, we deliberately idealize the situation and assume that the data concern only bulbs that operate constantly and in the same conditions, so that the interreplacement times along each time axis in Figure 11.5 are sampled independently from the same exponential distribution.

To find the answer, consider a single process of changes of bulbs at one place (see Figure 11.6). The process starts at time $t = 0$; the consecutive lifetimes are $\xi_1, \xi_2, \ldots$ and the times of replacements are $S_0 = 0, S_n = S_{n-1} + \xi_n$ for $n \geq 1$. The lifetime recorded in the sample, $T(t^*)$, is the value $\xi_n = S_n - S_{n-1}$ such that $S_{n-1} \leq t^* < S_n$. We assume that $\xi_1, \xi_2, \ldots$ are iid with exponential distribution.

The distribution of the time S_n of nth replacement has been found in Section 9.8. It is a gamma distribution with parameters n and λ, so that the density of S_n is

$$g_n(t) = \frac{\lambda^n}{\Gamma(n)} t^{n-1} e^{-\lambda t} = \frac{\lambda^n}{(n-1)!} t^{n-1} e^{-\lambda t}. \tag{11.3}$$

For further use, observe that we have

$$\sum_{n=1}^{\infty} g_n(t) = \lambda \sum_{n=1}^{\infty} \frac{(\lambda t)^{n-1}}{(n-1)!} e^{-\lambda t} = \lambda. \tag{11.4}$$

To find $E[T(t^*)]$, we shall find first the cdf of $T(t^*)$, namely $F_{T(t^*)}(x) = P\{T(t^*) \leq x\}$. Assume first that $x < t^*$. Then a replacement must have occurred before time t^* since otherwise the original bulb would still be working at t^* and we would have $T(t^*) = \xi_1 > t^* > x$.

Thus, for some $n = 2, 3, \ldots$ we must have $S_{n-1} = z \leq t^* < S_n = S_{n-1} + \xi_n$. In this case $\xi_n > t^* - z$, so that $t^* - z < \xi_n \leq x$. The condition $\xi_n \leq x$ implies that $t^* - z \leq x$, which means that the time z of $(n-1)$st replacement satisfies $t^* - x \leq z \leq t^*$. Partitioning with respect to $n = 2, 3, \ldots$ and conditioning on time z of the time of the $(n-1)$st replacement, we obtain, using (11.4),

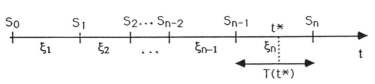

Figure 11.6 Renewal process.

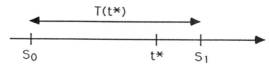

Figure 11.7.

$$F_{T(t^*)}(x) = P\{T(t^*) \le x\} = \sum_{n=2}^{\infty} \int_{t^*-x}^{t^*} P\{t^* - z < \xi_n \le x\} g_{n-1}(z)\, dz$$

$$= \sum_{n=2}^{\infty} \int_{t^*-x}^{t^*} [e^{-\lambda(t^*-z)} - e^{-\lambda x}] g_{n-1}(z)\, dz$$

$$= \int_{t^*-x}^{t^*} [e^{-\lambda(t^*-z)} - e^{-\lambda x}] \sum_{n=2}^{\infty} g_{n-1}(z)\, dz$$

$$= \lambda \int_{t^*-x}^{t^*} [e^{-\lambda(t^*-z)} - e^{-\lambda x}]\, dz$$

$$= 1 - e^{-\lambda x} - \lambda x e^{-\lambda x}.$$

If $x > t^*$, the derivation above has to be modified in two ways. First, we have to add the probability that the first bulb is still working at t^*, and its lifetime ξ_1 satisfies $t^* < \xi_1 \le x$ (see Figure 11.7). Second, if a replacement occurred before t^*, the time $S_{n-1} = z$ of the last replacement before t^* satisfies the inequality $0 < z \le t^*$. Consequently, we have in this case

$$P\{T(t^*) \le x\} = e^{-\lambda t^*} - e^{-\lambda x} + \sum_{n=2}^{\infty} \int_{0}^{t^*} [e^{-\lambda(t^*-z)} - e^{-\lambda x}] g_{n-1}(z)\, dz$$

$$= 1 - e^{-\lambda x} - \lambda t^* e^{-\lambda x}.$$

Differentiating, we obtain the density of $T(t^*)$:

$$f_{T(t^*)}(x) = \begin{cases} \lambda^2 x e^{-\lambda x} & \text{for } x < t^* \\ \lambda(1 + \lambda t^*) e^{-\lambda x} & \text{for } x \ge t^*. \end{cases} \tag{11.5}$$

An easy integration gives the formula for $E[T(t^*)]$, and one can show that

$$\lim_{t^* \to \infty} E[T(t^*)] = \frac{2}{\lambda},$$

so that, on average, the lifetime of the bulb that is in operation at time t^* is about *twice as long* as the average lifetime of other bulbs. Thus, the described method of sampling is biased.

It is of interest to note that the result above (as regards the existence of bias) remains valid as long as the lifetimes $\xi_1, \xi_2, \ldots$ are iid. The specific assumption of exponentiality of distribution implies the form (11.5) of the density, but the fact that $\lim_{t^* \to \infty} E[T(t^*)] > E(\xi)$ is valid for any nondegenerate distribution of ξ's [see Karlin and Taylor (1975)].

Since $T(t^*) = T' + T''$ [i.e., $T(t^*)$ is the spent lifetime at t^* plus the residual lifetime at t^*], we could try to derive our result as follows. The residual time T'' is exponential, in view of memoryless property of exponential distribution, so that $E(T'') = 1/\lambda$. For the spent lifetime T', it cannot exceed t^*, and it exceeds x (where $x < t^*$) if there are no replacements between $t^* - x$ and t^*. Since replacements form a Poisson process, the latter probability is $e^{-\lambda(t^* - x)}$. Consequently,

$$P\{T' > x\} = \begin{cases} e^{-\lambda(t^* - x)} & \text{for } x < t^* \\ 0 & \text{for } x \geq t^*. \end{cases}$$

Thus, by Theorem 8.2.2, $E(T') = \int_0^\infty P\{T' > x\}\, dx = \int_0^{t^*} e^{-\lambda(t^* - x)}\, dx = (1/\lambda)(1 - e^{-\lambda t^*})$, and therefore $E[T(t^*)] = E(T') + E(T'') = 2/\lambda - (1/\lambda)e^{-\lambda t^*}$. This calculation, however, relies on the assumption that memoryless property of exponential distribution is valid for residual waiting time counted from a *randomly selected* moment (and not only from a fixed moment). This property happens to hold (see Problem 11.5.2), provided that one considers "appropriate" randomness.

To see what could do wrong, suppose that we observe waiting time for nearest replacement from the moment *preceding* a given replacement by some fixed constant, say 5 hours. Such a moment is random, and yet, waiting time is at most 5 hours by definition, and therefore does not have exponential distribution. This simple example shows that the notion of "random moment" has to be qualified: We are not allowed to know the future at the time of selecting our moment to start observations.

We shall not go into details here, but a perceptive reader is advised to reflect a little on the meaning of the phrase "choice at t^* is independent on the future" and try to suggest some ways of explicating such a phrase as a formal definition.

PROBLEMS

11.5.1 Find $E[T(t^*)]$ using the density (11.5).

11.5.2 Let $T'' = T''(t^*)$ be the residual lifetime at t^* in a Poisson process [defined formally as $S_N - t^*$, where N is a random index (i.e., a random variable) such that $S_{N-1} < t^* \leq S_N$]. Show that for $x > 0$,

$$P\{T'' \leq x\} = e^{-\lambda t^*} - e^{-\lambda(x + t^*)}$$
$$+ \sum_{n=1}^{\infty} \int_0^{t^*} [e^{-\lambda(t^* - y)} - e^{-\lambda(x + t^* - y)}] g_n(y)\, dy,$$

where $g_n(y)$ is the density of S_n. Then show that T'' has exponential distribution with mean $1/\lambda$ (which means that memoryless property indeed holds also for "random" starting times).

11.6 MEASUREMENT SCALES

To explain the motivation for the topics of the last three chapters of the book, it is necessary to introduce some concepts connected with the level of measurements. Consider a set of objects of some kind. Typically, for the purpose of analysis, statistical description, and so on, each of these objects can be represented by a number $b(x)$ assigned to object x. Often, that number represents the result of measurement in some units, but sometimes it identifies only the class to which the object belongs.

If $b(x) > b(y)$, then x has "more ... " than y (e.g., is heavier, longer, hotter, older, etc.). The question we want to address now is which type of statements expressed through the values of b are meaningful, and which are not. To take an example, if b represents length, and $b(x) = 10, b(y) = 5$, we may say that x is *twice as long* as y, because $b(x)/b(y) = 2$. This statement will remain valid whether we express length in inches, centimeters, or miles: The values $b(x)$ and $b(y)$ will change, but their ratio will remain 2. However, if b represents temperature and $b(x) = 10, b(y) = 5$, it makes no sense to say that x is "twice as hot" as y. Indeed, it suffices to express temperature on a different scale (e.g., change from Fahrenheit to Celsius): The ratio of temperatures will change as the scale changes. The question therefore is: Why are the ratios of scale values meaningful for length, duration, or weight but not meaningful* for temperatures?

The full impact of such questions became apparent only when physical, mathematical, and statistical methods started to be used in the social sciences. Here the attributes that one considers are typically "soft," and the use of certain methods may lead to conclusions that are illegitimate but have the deceptive appearance of being very precise (the statement that a new brand of instant coffee "tastes 11.3% better" may be quite effective in commercials, mainly because of the deceptive use of decimal point, but its meaning, if any, is obscure).

It is clear that the decision which statements expressed through values $b(x)$ are allowed and which are not must lie in the analysis of the nature of the measured attribute. Such an analysis has the following general form. The starting point is an *empirical relational system*, consisting of a set, say A (of objects under consideration), and a number of relations on A, say

---

*Observe that the precision of measurements has nothing to do with the answer: Ratios of lengths make sense even if the lengths are determined imprecisely; the ratios then are simply subject to bigger errors. But the ratios of the temperatures make no sense even if the temperature is measured with the most precise devices available.

$R_1, R_2, \ldots, R_k$. These relations represent some empirically observable relationships between the pairs, triplets, and so on, of objects in A.

The measurement is an assignment of numbers to objects in A (hence it is a function on A). This function must satisfy conditions that "mimic" the empirical relations $R_1, R_2, \ldots$. If such a function exists, we say that measurement exists. The level of freedom of choice of such a function (if it exists) determines the type of the measurement scale.

We shall now illustrate the foregoing, somewhat nebulous description by examples.

Example 11.6.1 (Scale of Hardness). Let the objects under consideration be minerals, and let the empirical relation $\succ$ be defined as $x \succ y$ if mineral x scratches mineral y. One can assign numbers $h(x)$ to objects x in A (minerals) representing their hardness, in the sense that $h(x) > h(y)$ whenever $x \succ y$. One of such assignments gives the value 10 to diamonds and lower values to other minerals. This choice of assignment of numbers to minerals is arbitrary, except that the relation $>$ between values of function h must mimic the relation $\succ$ between arguments of h (i.e., minerals). It is precisely this degree of arbitrariness of h that makes it meaningless to claim that "x is twice as hard as y" if $h(x)/h(y) = 2$.

Example 11.6.2. It is clear that even in such a simple case as above, the possibility of assigning numerical scale values $h(x)$ to minerals x in A results from the fact that the relation $\succ$ (of scratching) is transitive: If $x \succ y$ and $y \succ z$, then $x \succ z$. Still, the existence of a function h that mimics the relation $\succ$ is not obvious: To see the difficulty, imagine that infinitely many new minerals, each of different hardness, are suddenly discovered. Could one still find enough distinct numbers to label those minerals? The answer is positive, but we shall not discuss the proof here.

Example 11.6.3. As an example of relations that order pairs, but are not transitive, consider the set A consisting of versions of state budget, with the relation of preference $\succ$ among them defined as: $x \succ y$ if majority (of some voting body) prefers x to y. We can then have a disturbing possibility that $x \succ y, y \succ z$ but $z \succ x$. To see it, take the simplest case, when there are three voters, a, b, and c, such that a prefers x to y to z, b prefers z to x to y, while c prefers y to z to x. Then a and b (hence a majority) prefer x to y; a and c prefer y to z, while b and c prefer z to x. We therefore have $x \succ y \succ z \succ x$ and there is no numerical assignment of scale values to x, y, and z, which mimics the relation $\succ$.

Example 11.6.4. Consider now an attribute such as length and a possible measurement system for it. We have here the relation of comparison with respect to length, accomplished empirically by putting the object parallel to each other, with one end lined up (like pencils). We obtain then a relation, say $\succeq$, interpreted as "not shorter than." The ultimate goal is to show the existence of a function b (length) defined on A, such that $x \succeq y$ if and only if $b(x) \geq b(y)$. Clearly, we must require that $\succeq$ satisfies some conditions (axioms), such as transitivity:

$$\text{if } x \succeq y \text{ and } y \succeq z, \text{ then } x \succeq z,$$

and so on.

However, one relation $\succeq$ is not sufficient to define length as we know it, since there may be many functions b satisfying the requirement that $\succeq$ agrees with the order induced by the values of b. Clearly, something more is needed to force all functions b to differ one from another only by the unit of measurement (i.e., such that if b and b^* are two assignments of lengths, then $b^* = \alpha b$ for some $\alpha > 0$).

In this case we need a ternary relation describing the operation of concatenation, that is, putting the objects end to end. Letting $\circ$ denote such operation, we obtain new objects, such as $x \circ y$, corresponding to x and y aligned one after another. This operation can be identified with a relation that holds between x, y, and z if $x \circ y \sim z$, where $\sim$ is defined by $a \sim b$ if $a \succeq b$ and $b \succeq a$. Clearly, we want $b(x \circ y) = b(x) + b(y)$, and to achieve that, the relations $\succeq$ and $\circ$ jointly must satisfy a number of conditions. We list here some most obvious ones; for example, for all x, y, z,

$\text{if } x \succeq y, \text{ then } x \circ z \succeq y \circ z \text{ for every } z,$
$(x \circ y) \circ z \sim (z \circ x) \circ y,$

and so on. A less obvious is the requirement of the Archimedean property: *For every x, y there exists n such that $x^{\circ n} \succeq y$, where*

$$x^{\circ n} = x \circ x \circ \cdots \circ x \ (n \text{ times}).$$

A measurement theory of length is then a relational system consisting of set A, relations $\succeq$ and $\circ$, a set of conditions for $\succeq$ and $\circ$ (referred to as *axioms*), and a theorem asserting that if these axioms are satisfied, then:

(i) *There exists a real-valued strictly positive function b on A such that*

$$b(x \circ y) = b(x) + b(y) \text{ and } x \succeq y \text{ iff } b(x) \geq b(y).$$

(ii) *If b^* is any other function satisfying* (i), *then $b^* = \alpha b$ for some $\alpha > 0$.*

The first part of the assertion provides the existence of measurement b, while the second part provides information about uniqueness. In case of length, the measurement is unique up to the choice of unit of measurement, so that the ratios of lengths are invariant. This means that for any objects x, y the ratio $b(x)/b(y)$ does not depend on the choice of function b. ☐

According to "purists," every attribute that one wants to measure (i.e., represent numerically) ought to be analyzed in this way (for a discussion of this topic, see Valleman and Wilkinson, 1993). The systems vary in two respects: (i) the nature of relations in the relational systems and consequently, the axioms that they must satisfy, depend on what is being observed and how; and (ii) the uniqueness part of the conclusion about the measurement scale varies depending on what is being measured.

We give here one more example, again without formulating all the axioms.

Example 11.6.5. Consider psychophysical measurement, where each subject is presented with a pair of stimuli and has to judge which of them is "more ...." For instance, the subjects may be handed two objects and asked which they think is heavier; in another experiment they may be confronted with the shades of gray and asked which is darker; two salt solutions may be judged as to which is more salty; two sounds may be judged regarding their pitch or regarding their intensity; and so on.

In each of these cases, there exists a physical continuum on which the stimuli differ, and the experimenter may present the subject with pairs that are very close one to another. However, one can imagine this experiment carried on stimuli that do not have such a physical continuum (e.g., the subject may be confronted with two descriptions of crimes and asked which in his or her opinion is "more serious").

At any rate, the main issue here is that when the stimuli are "close," the subject may not be sure of the response and may be inconsistent (i.e., judge x more ... than y, y more ... than z, and z more ... than x). The subject may also change this response on subsequent presentation. Thus, for each (ordered) pair (x, y) of stimuli we may (at least in theory) observe the probability $p(x, y)$ that x will be judged as "more ..." than y.

The question now is to what extent the values $p(x, y)$ determine the existence of measurement b of the objects, satisfying the main requirement that $p(x, y) \geq \frac{1}{2}$ if and only if $b(x) \geq b(y)$: A higher measure is assigned to those stimuli which are more often judged as "more ..." than "less ...." Here the axiom systems vary, depending on whether there exist stimuli such that $p(x, y) = 1$, or whether $0 < p(x, y) < 1$ for all x, y. One such system has been suggested by Bartoszyński (1974). The assertion of the theorem is that if $p(x, y)$ satisfies certain axioms, then there exists a function b defined on the set of stimuli such that

(ia) *If $p(x, y) \geq \frac{1}{2}$, then $b(x) \geq b(y)$.*

(ib) *If $0 \leq p(x, y) = p(u, v) < 1$, then $b(x) - b(y) = b(u) - b(v)$.*

(ii) *If b^* is any other function satisfying* (ia) *and* (ib), *then $b^*(x) = \alpha b(x) + \beta$ for some α, β with $\alpha > 0$.*

As in the problem of length, (ia) and (ib) assert the existence of measurement, while (ii) provides information on uniqueness: In this case measurement b is unique up to the choice of zero of the scale and choice of the unit. For such scales the ratios of the values are not invariant, but the ratios of differences are. Thus, for each x, y, u, v, the ratio $[b(x) - b(y)]/[b(u) - b(v)]$ does not depend on the choice of b: Indeed, for any other function b^* such that $b^* = \alpha b + \beta$, we have $b^*(x) - b^*(y) = \alpha[b(x) - b(y)]$, and factor α cancels when the ratio of differences is taken. □

Each scale has its own level of "uniqueness," described by a condition corresponding to condition (ii) in the two examples given above. There may be, of course, infinitely many types of scales. In practical situations, however, one encounters only four major types of scales, as specified by the following definition.

Consider a relational system, where b and b^* are any two assignments of numbers to objects in the set A, which constitute measurement scales as specified by the axioms of the relational system.

Definition 11.6.1

(i) If there exists $\alpha > 0$ such that $b^*(x) = \alpha b(x)$, then we say that measurement is on the *ratio* scale.

(ii) If there exists $\alpha > 0$ and β such that $b^*(x) = \alpha b(x) + \beta$, then measurement is on the *interval* scale.

(iii) If there exists a monotone increasing function u such that $b^*(x) = u[b(x)]$, then measurement is on the *ordinal* scale.

(iv) If there exists a one-to-one function v such that $b^*(x) = v[b(x)]$, then measurement in on the *nominal* scale. □

In other words, for ratio scales (see Example 11.6.4), the measurement is unique up to a choice of unit (examples are length, duration, etc.). For interval scales one can choose not only the unit but also the zero of the scale (e.g., temperature). For ordinal scales only the order matters. For instance, if $b(x) = 10$ and $b(y) = 2$, then all we can say is that x is "more ..." than y, on an attribute designated by .... We cannot meaningfully say that the "difference" between x and y is 8 or that x is "five times ..." as y. The reason is that $b(x) = 10, b(y) = 2$, can be replaced by $b^* = 100, b^*(y) = -3$, or $b^* = 1, b^*(y) = 0.99$, or any other two values, as long as the first exceeds the second.

In physical measurement, the best known such scale is that of hardness (see Example 11.6.1) when the main empirical relation is: $x \succeq y$ if "x scratches y." In the social sciences, the situation is not so clear. The relation between stimuli elicited by asking subjects to evaluate them "on the scale of 1 to 10" or by marking responses such as "strongly agree," "agree," and so on, are of the ordinal nature, but often are treated as if they were expressed on an interval or ratio scale.

The same concerns education, where grades are averaged and compared as if they were measured on an interval scale. Thus, a student who gets two A's and four B's on six exams has average 3.33. Next year he gets three A's, two B's, and a C, and his average increases to 3.5. This is taken as an indicator that he "improved." If the scoring (function b defined on grades $A, B, C, \ldots$) were changed to $b(A) = 4, b(B) = 3, b(C) = 1$, his average would be 3.16 and he would be judged as getting worse than last year.

A moment of reflection suffices to realize that there is nothing sacred in assigning the values $4, 3$, and 2 to grades A, B, and C, considering that the process of averaging often concerns scores on exams in different subjects, done by different teachers, with criteria often formulated rather vaguely. Nevertheless, tradition and the practical needs to assess the progress, or effects of teaching techniques, force one to fix the scoring system for grades and regard them as measurement on an interval scale.

Finally, the nominal scale—the weakest of them—can hardly be regarded as measurement, since the numbers serve only for the purpose of labeling (e.g., numbers on jerseys of football players).

The four scales mentioned above form a linear order: Indeed, any transformation $b^*(x) = \alpha b(x), \alpha > 0$ allowed for ratio scale is a particular case of a transformation $b^*(x) = \alpha b(x) + \beta, \alpha > 0$ that defines an interval scale. This transformation is monotone, hence allowed for ordinal scale, and in turn, each monotone transformation is one to one, allowed for nominal scale. This explains why one can always lower the level of measurement. On the contrary, to use a higher measurement level is incorrect. Monotone transformation does not necessarily have the form $b^*(x) = \alpha b(x) + \beta, \alpha > 0$ and so on.

The theory of measurement scales, as outlined above, has been introduced by Stevens (1946); it rapidly gained popularity in psychology and the social sciences and was developed into a highly sophisticated theory (see, e.g., Krantz et al., 1971, or Roberts, 1979). The four types of scales, as defined above, are most commonly encountered and popularly known. Unfortunately, this knowledge also contributed to the popular belief that *all* possible scale types (as defined by classes of allowed transformations) form a linear order from "stronger" to "weaker." In fact, the order is only partial, with some scales being noncomparable. The noncomparable scales correspond to classes of transformations that are not contained one in another.

Although, as explained above, there exist potentially infinitely many scales of measurement (each described by a group of transformation of scale values), only four are considered in practice: ratio, interval, ordinal, and nominal scales. The types of statistics that one may sensibly use depend on the type of the scale.

There is not much difference between ratio scale and interval scale. While the mean μ and standard deviation σ make sense in both cases, one should not use statistics such as coefficient of variation σ/μ in case of data values measured on an interval (but not ratio) scale. This is because under transformation $y = \alpha x + \beta$ $(\alpha > 0)$ the standard deviation becomes multiplied by α, while the mean becomes multiplied by α and shifted by β. Therefore, the value σ/μ depends on the choice of zero of the scale.

On the other hand, for data measured on the ordinal scale only, the mean and standard deviation are not invariant, and only statistics expressed through ranks (e.g., median and other quantiles) should be used. Finally, for data on nominal scale, one is allowed to use only class frequencies.

In most of what follows (Chapters 12 to 15), it is assumed that the data represent measurements on an interval scale. This allows us, among others, to use such characteristics as mean and standard deviation and consider the data which follow a normal distribution.* Chapter 16 concerns methods of handling the data measured on ordinal scales only (more precisely, only when ordinal relations are taken into account). Finally, Chapter 17 deals with the case of *categorical data*, data expressed on a nominal and/or ordinal scale with data grouped into classes.

REFERENCES

Bartoszyński, R., 1974. "On a Metric Structure Derived from Subjective Judgements: Scaling under Perfect and Imperfect Discrimination," *Econometrica*, **42**, no. 1, 55–71.

Karlin, S. and H. M. Taylor, 1975. *A First Course in Stochastic Processes*, Academic Press, New York.

Krantz, D. H., R. D. Luce, P. Suppes, and A. Tversky, 1971. *Foundations of Measurement*, vol. I, Academic Press. New York.

Roberts, F. S., 1979. *Measurement Theory with Applications to Decisionmaking, Utility and the Social Sciences*, Addison–Wesley, Reading, Mass.

Scott, D. W., 1979. "On Optimal and Data-Based Histograms," *Biometrika*, **66**, 605–610.

Stevens, S. S., 1946. "On the Theory of Scales of Measurement," *Science*, **103**, 677–680.

Valleman, P. P., and L. Wilkinson, 1993. "Nominal, Ordinal, Interval, and Ratio Typologies are Misleading," *American Statistician*, **47**, no. 1, 65–72.

---

* Strictly speaking, if X is measured on a ratio scale, it cannot have normal distribution (since negative values are ruled out for ratio scales). Such a level of adherence to the rules imposed by scale types would drastically impoverish the scope of statistical applications, by absurdly disallowing us to treat attributes such as height, weight, size, duration, and so on, in certain populations as normally distributed, merely because these attributes cannot be represented by a negative number.

CHAPTER 12

Estimation

12.1 INTRODUCTION

Informally, estimation is the process of extracting information about the value of a certain parameter from the data. As a rule, the data will be assumed to result from random sampling from some population, and the parameter to be estimated will be some global characteristic of the population. Let $x_1, x_2, \ldots$ denote the outcomes of successive observations. The crucial issue in applying statistical methods is to regard the data points $x_1, x_2, \ldots$ as values of some random variables $X_1, X_2, \ldots$. This implies making use of the fact that if the same experiment of taking n observations were to be repeated, the new data points would probably be different. The statistical laws that govern the variability of data under hypothetical or actual repetition constitute the base on which one can build the theory of statistical inference. In estimation theory it is assumed that the distribution of the ith observation X_i is known, except for the value of the parameter, say θ. An estimator of θ is then a rule that allows us to calculate an approximation of θ, based on sample $X_1, \ldots, X_n$.

Example 12.1.1. To fix the ideas, let $X_1, X_2, \ldots$ represent the weights of successive trout caught at a certain location. We may be interested in a parameter such as $\theta = P\{X_i > w_0\}$, where w_0 is some fixed weight, so that θ is the fraction of trout whose weight exceeds a given threshold w_0.

Given the observed weights $X_1, \ldots, X_n$, we may estimate θ as

$$T_n = T(X_1, \ldots, X_n) = \frac{\text{number of indices } i \text{ such that } X_i > w_0}{n}.$$

The formula above gives an obvious and rather unsophisticated estimator of θ.

There is another method of estimating the same parameter. Assume that weights of trout follow a normal distribution $N(\mu, \sigma^2)$. Given the sample $X_1, \ldots, X_n$, we may calculate the sample mean $\overline{X} = (1/n)(X_1 + \cdots + X_n)$ and

467

sample variance $S^2 = (1/n)\sum(X_i - \overline{X})^2$. We may then regard $\overline{X}$ and S^2 as approximations of μ and σ^2, and estimate θ as

$$U_n = U(X_1, \ldots, X_n) = 1 - \Phi\left(\frac{w_0 - \overline{X}}{S}\right), \qquad (12.1)$$

where Φ is the cdf of standard normal variable. Here the rationale is that

$$\theta = P\{X \geq w_0\} = P\left\{Z > \frac{w_0 - \mu}{\sigma}\right\} = 1 - \Phi\left(\frac{w_0 - \mu}{\sigma}\right)$$

which is approximated by (12.1).   □

The example above shows that there may be several estimators of the same parameter (i.e., several distinct rules of calculating an approximation of θ given the sample). Once this fact is realized, and as soon as one remembers that every estimator, such as T_n and U_n in Example 12.1.1, is a random variable, there appear a number of questions:

1. How can we judge the performance of estimators (with the idea of choosing the best estimator)?
2. Are there methods of obtaining estimators (other than "ad hoc" methods, seemingly used to construct estimators T_n and U_n in Example 12.1.1)?

A systematic attempt to answer questions 1 and 2 is one of the greatest "success stories" in statistics, leading to a theory with clear and elegant conceptual structure, highlighted by powerful theorems. The empirical situation is such that we observe values of independent random variables $X_1, X_2, \ldots$, sampled from the same distribution $f(x, \theta)$; here f may stand for density or probability function, depending on whether X_i's are continuous or discrete. This distribution depends on a parameter which in an actual situation may assume some value θ (unknown to the observer) in a parameter space Θ. In other words, we deal with a family of distributions, indexed by θ. The examples abound: X_i's may be Bernoulli observations for unknown probability of success θ, or normally distributed observations with unknown mean θ and known standard deviation, or with known mean and unknown standard deviation θ, and so on.

Let $T_n = T_n(X_1, \ldots, X_n)$ be the estimator for sample size n [i.e., some function of observations $X_1, \ldots, X_n$, which we choose as our means to approximate θ]. The first requirement for a "good" estimator is its consistency, defined by the requirement that $T_n \to \theta$. Since T_n's are random variables, we must specify the type of convergence (e.g., in probability or almost surely). The need of consistency is obvious: Estimators that are not consistent do not guarantee that by increasing the sample size one gets closer to the true value of the parameter.

If $E(T_n) = \theta$ for every n, then estimator T_n is called unbiased. This means that if the estimation were to be repeated several times for different samples (but with the use of the same estimator), the results would be, on average, "on target." The "quality" of an unbiased estimator may be defined as its variance. This corresponds to taking $(t_n - \theta)^2$ as the "penalty" or "loss" due to the error of accepting the value t_n of T_n as the approximation of θ. While the squared error is not the only loss function possible, it is realistic in many problems. It is also the only loss function for which estimation theory has been developed to its logical conclusions, in the sense that all major questions have been answered.

To sketch this development: First, one has the powerful Rao–Cramèr inequality, which states that there exists a lower bound for variances of all unbiased estimators of θ (for any fixed sample size n). This bound therefore shows the best that can be achieved in estimating a given parameter, in the sense of providing a yardstick by which one can tell how close a given estimator is to the "ideal" (i.e., to estimator with the smallest possible variance).

At this point it is necessary to turn attention to question 2: How can estimators be constructed? Several methods are proposed, and they may be divided into two groups: One group gives methods (e.g., maximum likelihood, moments, etc.) of finding an estimator. The second group tells how to modify an estimator to improve it (e.g., to make its variance closer to the possible minimum value). One of such improvement methods is based on the concept of the sufficient statistic, that is, a statistic that reduces the data (e.g., to a single number), yet without reducing the information about the parameter. One then has the Rao–Blackwell theorem, which says that if T is an unbiased estimator of θ, and S is a sufficient statistic for θ, then $T^* = E(T|S)$ is also an estimator of θ that is better (or: not worse) than T. One therefore gets a powerful tool of improving estimators: Start with any unbiased estimator T and find its conditional expectation T^* with respect to a sufficient statistic. If this new estimator T^* is not the best, one can continue the process, conditioning with respect to another sufficient statistic, and so on. Instead of such conditioning "one step at a time," one can condition T with respect to the *minimal sufficient statistic* (i.e., maximal reduction of the data, which preserves information about θ). The improvement T^* of T obtained by such conditioning cannot be improved any further. This raises the obvious questions: Is T^* the best possible estimator? Does the estimator T^* depend on the initial starting estimator T? These two questions turn out to be intimately related: If the distributions under consideration form a complete class, then T^* is the best estimator in the sense of attaining the lower bound given by the Rao–Cramèr theorem, and it is also obtainable by conditioning any unbiased estimator of θ with respect to the minimal sufficient statistic.

As already mentioned, the sketch above gives a "success story with a happy ending" in statistical research. Presented in this manner it may appear easy and effortless. But one should remember that bringing the theory to its present form required analysis of countless examples, proving or disproving various conjectures, formulating weaker and weaker sets of assumption under which such or such assertion is true, and this took about half of a century of the efforts of many statisticians.

In this chapter we present some other topics and results in estimation, without which knowledge of the theory would be incomplete: Bayesian estimation, asymptotic properties of maximum likelihood estimators, confidence intervals, other methods of obtaining estimators, including resampling methods (such as jackknifing, bootstrapping, etc.). Some of these are given in more detail, some are merely sketched. But none of them matches in depth, elegance, or completeness the theory of estimation for quadratic loss function.

12.2 OVERVIEW

Let us begin with a simple estimation problem and a number of possible solutions. This will allow us to formulate various questions, as well as to suggest some natural generalizations discussed in this chapter. Recall that many situations where statistical methods apply can be reduced to the framework where one observes a simple random sample $X_1, \ldots, X_n$ from a distribution whose functional form is known, except for the value of a certain parameter θ. This means that the actual observations $x_1, \ldots, x_n$ are the values of iid random variables $X_1, \ldots, X_n$, each with a distribution that will be denoted by the symbol $f(x, \theta)$, where f will stand either for density or for the probability function, depending on whether X_i's are continuous or discrete.

We also assume that θ is an element of the *parameter space* Θ. We begin with the situation when θ is a single number, so that Θ is a subset of the real line: In general, Θ may be a multidimensional space or even a space with more complicated structure. In what follows, we often assume that Θ is an interval of the real line, so that we may employ standard optimization techniques (e.g., involving the differentiation of various quantities with respect to θ).

Example 12.2.1. A politician needs to estimate the proportion of voters who favor a certain issue. In a public opinion poll, n persons are sampled from the population and their responses $X_1, \ldots, X_n$ are noted, where $X_i = 1$ or 0, depending whether or not the ith person polled favors the issue in question. Letting θ denote the proportion of voters in favor of the issue, we have

$$f(x, \theta) = P\{X_i = x | \theta\} = \begin{cases} \theta & \text{for } x = 1 \\ 1 - \theta & \text{for } x = 0. \end{cases}$$

Here $0 \leq \theta \leq 1$ (i.e., $\Theta = [0, 1]$), and we may write a single formula

$$f(x, \theta) = \theta^x (1 - \theta)^{1-x}, \qquad x = 0, 1.$$

Example 12.2.2. Measurement with error can typically be represented as an estimation problem. We are to measure the value of an attribute of an object (e.g., its weight, dimension, temperature, content of some substance, etc). The true value of this attribute to be determined is θ. The measurements are subject to errors. For instance, it may often be assumed that the ith measurement is $X_i = \theta + \epsilon_i$, where $\epsilon_1, \epsilon_2, \ldots$ are iid random variables. If we assume that $\epsilon_i \sim N(0, \sigma^2)$, then $X_i \sim N(\theta, \sigma^2)$, so that

$$f(x, \theta) = \frac{1}{\sigma\sqrt{2\pi}} e^{-(x-\theta)^2/(2\sigma^2)},$$

where σ is the standard deviation of measurements (assumed known). In this case the parameter space Θ is the subset of real line representing the possible values of the measured attribute.

Example 12.2.3. Continuing Example 12.2.2, it is possible that we do not know the standard deviation of measurement σ. If we are measuring the unknown attribute μ, then θ is a two-dimensional parameter $\theta = (\mu, \sigma)$, and we may be interested in estimating one or both components of θ. In this case the parameter space Θ is a subset of the half-plane $\{(\mu, \sigma) : \sigma > 0\}$.

□

As explained in Chapter 11, the objective is to construct a statistic (i.e., a function of the sample), say $T = T(X_1, \ldots, X_n)$, the values of which approximate θ. The following example will show several such statistics for the same parameter. This will illustrate the fact that sometimes we may have an embarrassing variety of choices for an estimator.

Before proceeding further, few comments about notation and terminology are in order. First, we typically use a symbol such as T to denote a statistic, even if it refers to various possible sample sizes. For instance, we may consider a statistic, described verbally as "mean square": If $X_1, \ldots, X_n$ is the sample, this statistic equals $(1/n) \sum_{j=1}^{n} X_j^2$. In fact, however, we are considering here a *sequence* of statistics, each being a function of a different number of arguments. To formulate this properly, it will sometimes be convenient to use the notation involving the sample size. We shall then write $T^{(n)}$ for the estimator based on a sample of size n, so that $T^{(n)} = T^{(n)}(X_1, \ldots, X_n)$, and we shall let T stand for the "generic" element of the sequence $T^{(1)}, T^{(2)}, \ldots$.

Second, we call a statistic an *estimator* (of θ) when this statistic is *used to estimate* θ. Thus, formally, when sample size n is not specified, an estimator is a sequence of statistics, the nth one depending on the observations $X_1, \ldots, X_n$. The value of an estimator obtained for a particular sample will be called an *estimate* of θ. Thus, an estimator is a random variable (or a sequence of random variables), while an estimate is a number obtained by using specific values observed in a sample.

Finally, in presenting the general theory we use the symbol θ for the unknown parameter. In examples we shall often use the symbols conforming to traditions, such as p for probability of success, σ for standard deviation, and so on.

Example 12.2.4.   Suppose that we take a random sample $X_1, \ldots, X_n$ from the distribution uniform on the interval $[0, \theta]$ so that

$$f(x, \theta) = \begin{cases} 1/\theta & \text{if } 0 \leq x \leq \theta \\ 0 & \text{otherwise.} \end{cases}$$

The objective is to estimate the range θ.

SOLUTION.   Using intuition and common sense as a guide, we shall now suggest several estimators of θ. First, we get some information about θ from the largest element of the sample. This suggests taking

$$T_1 = \max(X_1, \ldots, X_n) = X_{n:n} \tag{12.2}$$

as an estimator of θ. We feel (and shall justify it later) that as the sample size increases, the values of T_1 will tend to become closer and closer to θ.

A disadvantage of T_1 is that it always underestimates θ: We have $T_1 \leq \theta$ regardless of the sample $X_1, \ldots, X_n$. We may remedy this in a number of ways.

For instance, we can argue as follows. With the sample of size n, the n observed values $X_1, \ldots, X_n$ partition the interval $[0, \theta]$ into $n + 1$ intervals. Since $X_1, \ldots, X_n$ tend to be "evenly dispersed" over $[0, \theta]$, each of these $n + 1$ intervals will have, on average, the same length, $\theta/(n + 1)$. Thus, we should "push" T_1 to the right by the amount $(1/n)T_1$, which suggests using the estimator

$$T_2 = \left(1 + \frac{1}{n}\right) T_1 = \frac{n + 1}{n} X_{n:n}. \tag{12.3}$$

Another way of "adjusting" T_1 may be based on observing that by symmetry, on average, maximal observation is at the same distance from the upper bound θ as the minimal observation is from the lower bound 0. Thus we may use the estimator

$$T_3 = \min(X_1, \ldots, X_n) + \max(X_1, \ldots, X_n) = X_{1:n} + X_{n:n}. \tag{12.4}$$

We can also argue that the minimum of the sample allows us to estimate θ. In fact, the same argument as above, with partition of the range $[0, \theta]$ into $n + 1$ parts of about equal length, suggests the estimator

$$T_4 = (n + 1) \min(X_1, \ldots, X_n). \tag{12.5}$$

We feel intuitively that this estimator is not very reliable, and therefore is, in some sense, inferior to the others suggested so far. In particular, it may happen that $T_4 < X_{n:n}$, in which case using T_4 makes no sense. Moreover, in general, maximum and minimum of the sample have—by symmetry—the same variance. The estimator T_4 magnifies this variance by factor $(n+1)^2$. We shall show later how this fact affects the precision of the estimators.

Finally, we can argue that the average $\overline{X} = (1/n)(X_1 + \cdots + X_n)$ should be close to the midpoint $\theta/2$. This suggests taking as an estimator of θ the statistic

$$T_5 = 2\overline{X} = \frac{2}{n}(X_1 + \cdots + X_n). \tag{12.6}$$

$\square$

Example 12.2.4 shows that we may have a choice of one of several estimators for the same parameter. This poses a natural question of establishing some criteria for such a choice. Since all estimators are random variables, their performance in one particular instance (i.e., for some particular sample) gives no clue because the situation may be quite different when a new sample is taken. The next few sections give criteria to judge the performance of estimators.

Another question concerns the methods of finding estimators, especially "good" ones: Rather than rely on intuition and common sense, it is desirable to have a scheme that would produce estimators in some more or less "automatic" fashion. We shall also present some modern generalizations and extensions of estimation methods, especially those connected with the availability of fast computing (called resampling methods).

The next problem discussed in this chapter concerns questions of reduction. This issue—in case of estimation—lies in restricting the class of statistics that one might use as estimators of a given parameter. We shall show that good estimators are based on the *sufficient statistics:* Any estimator that is not a sufficient statistic can be improved.

Finally, we discuss the methods of presenting the results of estimation. The point is that it is rare that the observed value of an estimator coincides exactly with the true value of the parameter θ. Most often, the observed value of T is "close" to θ, and it is desirable to give the probability associated with any degree of closeness between T and θ.

PROBLEMS

12.2.1 Consider the problem of estimating the parameters of a beta distribution with density

$$f(x, \alpha, \beta) = \frac{\Gamma(\alpha + \beta)}{\Gamma(\alpha)\Gamma(\beta)} x^{\alpha-1}(1 - x)^{\beta-1}, \qquad 0 \le x \le 1.$$

Specify the parameter space Θ in each of the following cases:
(i) β is known, α is to be estimated.

(ii) Both α and β are to be estimated.

(iii) Both α and β are to be estimated, and the distribution is known to be bimodal.

12.2.2 Referring to Example 12.2.4, suggest an estimator of θ using intuition similar to that used to justify T_5 but based on the fact that $E(X^2) = \theta^2/3$.

12.3 CONSISTENCY

One of the basic properties of a good estimator is that it provides increasingly more precise information about θ with the increase of the sample size n. Accordingly, we introduce the following definition.

Definition 12.3.1. The estimator $T = \{T^{(n)}, n = 1, 2, \ldots\}$ of parameter θ is called *consistent* if $T^{(n)}$ converges to θ in probability; that is, for every $\epsilon > 0$

$$\lim_{n \to \infty} P\{|T^{(n)} - \theta| \leq \epsilon\} = 1. \tag{12.7}$$

The estimator $T^{(n)}$ will be called *strongly consistent* if $T^{(n)}$ converges to θ almost surely, that is,

$$P\{\lim_{n \to \infty} T^{(n)} = \theta\} = 1. \tag{12.8}$$

When both kinds of consistency are considered at the same time, estimators satisfying (12.7) are called *weakly consistent*. □

We shall now analyze the consistency of the five estimators $T_1, \ldots, T_5$ considered in Example 12.2.4.

Example 12.3.1. We can easily obtain the distribution of $T_1 = T_1^{(n)} = \max(X_1, \ldots, X_n)$. Indeed, since X_i's are uniform on $[0, \theta]$, we have

$$P\{X_i \leq t\} = \begin{cases} 0 & \text{if } t < 0 \\ t/\theta & \text{if } 0 \leq t \leq \theta \\ 1 & \text{if } t > 0. \end{cases}$$

Clearly, we are interested only in probabilities for t between 0 and θ, and since $\max(X_1, \ldots, X_n) \leq t$ if and only if $X_i \leq t$ for all i, we have by independence

$$P\{T_1^{(n)} \leq t\} = P\{X_1 \leq t, \ldots, X_n \leq t\} = \left(\frac{t}{\theta}\right)^n. \tag{12.9}$$

If $0 < \epsilon < \theta$, then

$$P\{|T_1^{(n)} - \theta| \leq \epsilon\} = P\{T_1^{(n)} \geq \theta - \epsilon\} = 1 - \left(\frac{\theta - \epsilon}{\theta}\right)^n.$$

Since $[(\theta - \epsilon)/\theta]^n \to 0$ as $n \to \infty$, estimator T_1 is consistent. Using the notation of Chapter 10, we may write

$$T_1^{(n)} \xrightarrow{P} \theta.$$

Example 12.3.2.   Next, $T_2^{(n)} = [(n+1)/n]T_1^{(n)}$ and since $(n+1)/n \to 1$, we have $T_2^{(n)} \xrightarrow{P} \theta$.

Example 12.3.3.   By the law of large numbers, we have

$$\frac{X_1 + \cdots + X_n}{n} \xrightarrow{P} E(X_1) = \frac{\theta}{2},\qquad (12.10)$$

so that

$$T_5^{(n)} = 2 \cdot \frac{X_1 + \cdots + X_n}{n} \xrightarrow{P} 2\frac{\theta}{2} = \theta,\qquad (12.11)$$

which shows that the estimator T_5 is also consistent.

Example 12.3.4.   As regards the consistency of estimator T_3, observe first that for $0 < \epsilon < \theta$

$$P\{\min(X_1, \ldots, X_n) > \epsilon\} = \prod_{i=1}^{n} P\{X_i > \epsilon\} = \left(1 - \frac{\epsilon}{\theta}\right)^n \to 0 \qquad (12.12)$$

which means that

$$\min(X_1, \ldots, X_n) = X_{1:n} \xrightarrow{P} 0.$$

Consequently, we have

$$T_3^{(n)} = T_1^{(n)} + X_{1:n} \xrightarrow{P} \theta,$$

in view of consistency of T_1, so that T_3 is also consistent.

Example 12.3.5.   It remains to analyze the estimator T_4. As remarked in Example 12.2.4, this estimator is, intuitively speaking, inferior to the others. To confirm this feeling, we shall show that the estimator T_4 is not consistent. Let $0 < \epsilon < \theta$. Then

$$P\{|T_4 - \theta| \le \epsilon\} = P\{|(n+1)X_{1:n} - \theta| \le \epsilon\}$$

$$= P\left\{\frac{\theta - \epsilon}{n+1} \le X_{1:n} \le \frac{\theta + \epsilon}{n+1}\right\}$$

$$= F\left(\frac{\theta + \epsilon}{n+1}\right) - F\left(\frac{\theta - \epsilon}{n+1}\right),$$

where F is the cdf of $\min(X_1, \ldots, X_n)$, so that the tail of F is given by (12.12). Consequently, using the fact that $(1 + x/n)^n \to e^x$, we have

$$P\{|T_4 - \theta| \le \epsilon\} = 1 - \left(1 - \frac{\theta + \epsilon}{\theta(n+1)}\right)^n - \left(1 - \left(1 - \frac{\theta - \epsilon}{\theta(n+1)}\right)^n\right)$$

$$= \left(1 - \frac{\theta - \epsilon}{\theta(n+1)}\right)^n - \left(1 - \frac{\theta + \epsilon}{\theta(n+1)}\right)^n$$

$$\to e^{-(\theta-\epsilon)/\theta} - e^{-(\theta+\epsilon)/\theta} < 1.$$

Thus, $\lim_{n \to \infty} P\{|T_4 - \theta| > \epsilon\} > 0$, which shows that T_4 is not a consistent estimator for θ.   □

Let us observe that from the strong law of large numbers in the iid case (Theorem 10.4.2) it follows at once that the estimator T_5 is strongly consistent. Showing strong consistency of estimators T_1, T_2, and T_3 will be left as exercises.

PROBLEMS

12.3.1 Show that the estimators T_1, T_2, and T_3 of Example 12.2.4 are strongly consistent.

12.3.2 Show that the estimator T_4 from Example 12.2.4 is unbounded, in the sense that for every positive A there exists n such that $P\{T_4^{(n)} > A\} > 0$. [Hint: $P\{T_4^{(n)} > A\} = \left(1 - \frac{A}{\theta(n+1)}\right)^n$.]

12.3.3 Suppose that you start observations at some unknown sample size n and record the values $X_{1:n}, X_{1:n+1}, \ldots, X_{1:n+N}$ such that

$$X_{1:n} = X_{1:n+1} = \cdots = X_{1:n+N-1} > X_{1:n+N}.$$

Suggest an estimator of θ based on observing $X_{1:n}, \ldots, X_{1:n+N}$. [Hint: Find the probability distribution of N and express $E(N)$ as a function of θ.]

12.3.4 Let $X_1, \ldots, X_n$ be a random sample from exponential distribution with density $f(x; \lambda) = \lambda e^{-\lambda x}, x > 0$. Does there exist a function g such that $U_n = n \min(X_1, \ldots, X_n)$ is a consistent estimator of $g(\lambda)$?

12.3.5 Assume that the observations are taken from the distribution uniform on $[0, \theta]$, and let U_n be the number of observations (out of first n) which are less than 5. Show that if $\theta > 5$, then $T_n = 5n/U_n$ is a consistent estimator of θ.

12.4   LOSS, RISK, AND ADMISSIBILITY

In this section we continue our review of the criteria of evaluating estimators. This time we attempt to evaluate performance of estimators for a fixed sample size n.

Consider again an experimenter who takes n independent observations $X_1, \ldots, X_n$ of a random variable with distribution* $f(x, \theta)$ in order to estimate θ. This estimate, say θ^*, will subsequently be used for some specific goal, and the degree of success in attaining this goal depends on how close θ^* and θ are.

In the simplest case, this degree of success may be a decreasing function of the error $|\theta^* - \theta|$, so that the smaller the error, the more likely is a high degree of success. One can easily imagine, however, situations that are not so simple, for instance when the error of overestimating θ (i.e., $\theta^* > \theta$) is less serious than the error of underestimating it ($\theta^* < \theta$). To build a general theory, we may assume that the situation can be adequately represented by specifying the *loss function* $\mathcal{L}(\theta^*, \theta)$, describing the negative consequences of proceeding as if the value of the parameter was θ^*, while in reality it is θ (the fact that we consider "loss" and negative consequences is a matter of convention only: by changing the sign, we can convert the considerations to "rewards" and positive consequences).

Suppose now that the experimenter decides to use the estimator $T = T(X_1, \ldots, X_n)$; hence his loss depends on the sample and is equal to $\mathcal{L}(T, \theta)$. The performance of estimator T can now be evaluated as the average loss

$$R_T(\theta) = E_\theta\{\mathcal{L}[T(X_1, \ldots, X_n), \theta]\}, \tag{12.13}$$

where E_θ stands for expected value with respect to the distribution $f_n(x, \theta)$. The function R above is called the *risk function*.

With the help of the concept of a risk function, one can now introduce two important notions.

Definition 12.4.1.   The estimator T_1 is *R-dominating* estimator T_2, or is *R-better* than T_2, if for all $\theta \in \Theta$ we have

$$R_{T_1}(\theta) \le R_{T_2}(\theta),$$

where the inequality is strict for at least one value of θ.

Moreover, an estimator T will be called *R-inadmissible* if there exists an estimator T' which is *R-better* than T. Otherwise, T will be called *R-admissible*.   □

---

* We remind the convention that $f(x, \theta)$ is either the density or—in the discrete case—the probability function.

The constructions based on this definition appear in a number of contexts in statistics, not restricted to comparison of estimators. It is therefore worthwhile to make some general comments here.

The basis for evaluation of a procedure (in this case an estimator) is the risk function R, which depends on the unknown parameter θ. Clearly, if we have two estimators and their risk functions are such that one of them is below the other (or equal to it) *regardless of the value of* θ, we can decide that the corresponding estimator is better. Quite often, however, the two risk functions cross each other (i.e., one is below the other for some θ and above it for some other θ). In such a case the estimators are *not comparable*: Since we do not know θ, we do not know which risk function is smaller at the actual (true) value of θ, so we cannot decide which estimator is better. Thus, we obtain only a *partial order* of estimators; estimators that are not dominated by any other are admissible. Within the class of admissible estimators, by definition none dominates the others and therefore we still need other criteria for choice. But at least the problem becomes reduced, in the sense that all inadmissible estimators are ruled out.

In practice, the search for the class of admissible estimators for a specific loss function may be difficult. Moreover, even if such a class were found, the result would be of limited applicability, restricted only to the situations for which the analyzed loss function gives an adequate representation of preferences among consequences of actions.

To build a general theory, it was therefore necessary to choose a loss function that could be acceptable (at least as a first approximation) to a wide class of users of statistics. One can argue that such a loss function is the *squared error*, that is,

$$\mathcal{L}(\theta^*, \theta) = (\theta^* - \theta)^2. \tag{12.14}$$

The use of this loss function* makes sense only for estimators with finite second moment (i.e., finite variance). We shall tacitly make this assumption. The risk of an estimator T for loss (12.14) is therefore $E_\theta[T(X_1, \ldots, X_n) - \theta]^2$. This risk appears so frequently that it acquired its own name.

Definition 12.4.2. The risk of an estimator T computed for the loss function (12.14) is called the *mean square error* of T and is typically denoted as

---

* Another loss function, as appealing as (12.14), is $\mathcal{L}(\theta^*, \theta) = |\theta^* - \theta|$. This function, suggested by Laplace at the beginning of nineteenth century, leads to a theory that is much less tractable mathematically and has begun to be developed only recently. One could argue, however, that an adequate loss function should reflect consequences of errors made *after* the parameter has been estimated (see Gafrikova and Niewiadomska–Bugaj, 1992). The starting point in the latter approach is the observation that we use the distribution $f(x, \theta)$ for some purpose, such as prediction, calculation of some probabilities, and so on. Thus, the loss $\mathcal{L}(\theta^*, \theta)$ should depend on how much the (estimated) distribution $f(x, \theta^*)$ differs from the true distribution $f(x, \theta)$. Such a "difference" between distributions can be expressed through a measure of difficulty in discriminating between the two distributions.

$$\text{MSE}_\theta(T) = E_\theta[T(X_1, \ldots, X_n) - \theta]^2. \qquad (12.15)$$

Also, when no risk function is specified, admissibility will always mean admissibility with respect to the mean square error.   □

It is clear that when building a theory based on mean square error, the first two moments of T will play a prominent role. Accordingly, we introduce the following definitions.

Definition 12.4.3. An estimator T such that

$$E_\theta(T) = \theta \qquad (12.16)$$

for every θ will be called *unbiased*. Generally, the difference

$$B_\theta(T) = E_\theta(T) - \theta \qquad (12.17)$$

will be called the *bias function*, or simply the *bias* of T.   □

Clearly, $B_\theta(T) = 0$ if and only if T is unbiased. The estimator T will be called positively (respectively, negatively) biased, depending on whether $B_\theta(T) > 0$ or $B_\theta(T) < 0$. More generally, if

$$\lim_{n \to \infty} E_\theta[T(X_1, \ldots, X_n)] = \theta, \qquad (12.18)$$

then T will be called *asymptotically unbiased*. Let us write

$$\begin{aligned}
\text{MSE}_\theta(T) &= E_\theta(T - \theta)^2 \\
&= E_\theta[(T - E_\theta(T)) + (E_\theta(T) - \theta)]^2 \\
&= E_\theta[T - E_\theta(T)]^2 + [E_\theta(T) - \theta]^2 \\
&\quad + 2E_\theta[(T - E_\theta(T))(E_\theta(T) - \theta)] \\
&= \text{Var}_\theta(T) + [B_\theta(T)]^2
\end{aligned}$$

(the expectation of the cross product is easily seen to be zero). We have therefore the following

Theorem 12.4.1. *The mean square error of an estimator is the sum of its variance and square of the bias.*

Consequently, the MSE of an unbiased estimator is equal to its variance.
We shall now illustrate the application of these concepts, by finding the bias and MSE for the five estimators of the range θ in uniform distribution (Example 12.2.4). Observe first that for $n = 1$ we have $T_1 = X_1, T_2 = T_3 = T_4 = T_5 = 2X_1$. Next, $E_\theta(T_1) = \theta/2$, while the remaining estimators are unbiased. Finally, $\text{Var}(T_1) = \text{Var}(X_1) = \theta^2/12$; hence mean square errors of

estimators T_2 to T_5 are the same, $\theta^2/3$. Thus, they are either all admissible or all inadmissible (if there exists a better estimator). In the examples below, we consider only the case $n > 1$.

Example 12.4.1. Let us begin with the estimator $T_1 = X_{n:n}$ from Example 12.2.4. Clearly, since $T_1 < \theta$, the estimator is biased. The cdf of T_1 is given by (12.9), so that the density of T_1 is

$$f_{T_1}(t) = \frac{nt^{n-1}}{\theta^n}, \qquad 0 \le t \le \theta \tag{12.19}$$

and $f_{T_1}(t) = 0$ otherwise. Thus,

$$E_\theta(T_1) = \frac{n}{\theta^n} \int_0^\theta t^n \, dt = \frac{n}{n+1} \theta, \tag{12.20}$$

a result that we derived intuitively from the symmetry of order statistics in Example 12.2.4 and used to justify the need to "correct" T_1 to obtain unbiased estimators T_2 and T_3.

Thus the bias of T_1 is negative:

$$B_\theta(T_1) = \frac{n}{n+1} \theta - \theta = -\frac{\theta}{n+1}. \tag{12.21}$$

To determine the MSE, let us first find the second moment of T_1. We have

$$E_\theta(T_1^2) = \int_0^\theta t^2 f_{T_1}(t) \, dt = \frac{n}{\theta^n} \int_0^\theta t^{n+1} \, dt = \frac{n}{n+2} \theta^2. \tag{12.22}$$

Thus,

$$\text{Var}_\theta(T_1) = \frac{n}{n+2} \theta^2 - \frac{n^2}{(n+1)^2} \theta^2 = \frac{n}{(n+1)^2(n+2)} \theta^2.$$

and consequently,

$$\text{MSE}_\theta(T_1) = \text{Var}_\theta(T_1) + [B_\theta(T_1)]^2 = \frac{2\theta^2}{(n+1)(n+2)}. \tag{12.23}$$

Example 12.4.2. As regards T_2, it was constructed from T_1 so as to remove the bias by appropriate multiplication by a constant: $T_2 = [(n+1)/n]T_1$. Consequently, $B_\theta(T_2) = 0$ and

$$\text{MSE}_\theta(T_2) = \text{Var}_\theta(T_2) = \frac{(n+1)^2}{n^2} \text{Var}_\theta(T_1) = \frac{\theta^2}{n(n+2)}. \tag{12.24}$$

Comparing with $\text{MSE}_{T_1}(\theta)$, we have

$$\frac{\text{MSE}_\theta(T_2)}{\text{MSE}_\theta(T_1)} = \frac{n+1}{2n} < 1,$$

which means that the risk of T_2 is, for large n, about half of the risk of T_1. This also shows that the estimator T_1 is not admissible: There exists an estimator that is better for all θ, namely T_2 (this does *not* imply that T_2 is admissible!).

Example 12.4.3. Let us now consider the estimator

$$T_3 = \min(X_1, \ldots, X_n) + \max(X_1, \ldots, X_n) = X_{1:n} + X_{n:n}.$$

This, again, is a modification of T_1 intended to remove the bias, so that $B_\theta(T_3) = 0$. Indeed, $E(X_{n:n}) = [n/(n+1)]\theta$, and (by symmetry), $E(X_{1:n}) = [1/(n+1)]\theta$. One can easily show (we leave the proof as an exercise; see Problem 12.4.2) that

$$\text{Var}_\theta(X_{1:n}) = \text{Var}_\theta(X_{n:n}) = \text{Var}_\theta(T_1).$$

Therefore,

$$\begin{aligned}
\text{MSE}_\theta(T_3) &= \text{Var}_\theta(T_3) \\
&= \text{Var}_\theta(X_{1:n}) + \text{Var}_\theta(X_{n:n}) + 2\,\text{Cov}_\theta(X_{1:n}, X_{n:n}) \\
&= 2\,\text{Var}_\theta(T_1) + 2\,\text{Cov}_\theta(X_{1:n}, X_{n:n}). \tag{12.25}
\end{aligned}$$

We now need to find the joint density of maximum and minimum of the sample of size n from uniform distribution on $[0, \theta]$. We have, for $0 \le u < t \le \theta$,

$$P\{X_{1:n} \ge u, X_{n:n} \le t\} = \prod_{i=1}^{n} P\{u \le X_i \le t\} = \left(\frac{t-u}{\theta}\right)^n.$$

The joint density is obtained by differentiating with respect to t and u and changing the sign (why?), so that

$$\varphi(t, u) = \frac{n(n-1)}{\theta^n}(t-u)^{n-2}, \qquad 0 \le u < t \le \theta.$$

Since $E_\theta(X_{1:n}) = \theta/(n+1)$ and $E_\theta(X_{n:n}) = n\theta/(n+1)$, we have $\text{Cov}_\theta(X_{1:n}, X_{n:n})$ equal to

$$\frac{n(n-1)}{\theta^n} \int_0^\theta \int_0^t tu(t-u)^{n-2}\,du\,dt - \frac{n\theta^2}{(n+1)^2} = \frac{\theta^2}{n+2} - \frac{n\theta^2}{(n+1)^2}$$

$$= \frac{\theta^2}{(n+1)^2(n+2)}.$$

After some algebra we finally obtain

$$\mathrm{MSE}_\theta(T_3) = \frac{2\theta^2}{(n+1)(n+2)}. \tag{12.26}$$

Thus, $\mathrm{MSE}_\theta(T_3) > \mathrm{MSE}_\theta(T_2)$, so that T_3 is not admissible.

Example 12.4.4. As regards $T_4 = (n+1)\min(X_1, \ldots, X_n)$, it is again unbiased, so that

$$\mathrm{MSE}_\theta(T_4) = \mathrm{Var}_\theta(T_4) = (n+1)^2\,\mathrm{Var}_\theta(X_{1:n})$$
$$= (n+1)^2\,\mathrm{Var}_\theta(X_{n:n}) = (n+1)^2\,\mathrm{Var}_\theta(T_1) = \frac{n}{n+2}\theta^2.$$

Comparing with T_2, we see that T_4 is inadmissible (observe that the variance of T_4 does not even tend to 0).

Example 12.4.5. For the estimator T_5 we may write, using the fact that the variance of X_i equals $\theta^2/12$,

$$\mathrm{Var}(T_5) = \mathrm{Var}\left[\frac{2}{n}(X_1 + \cdots + X_n)\right] = \frac{4}{n^2}\cdot n\,\mathrm{Var}(X_i) = \frac{\theta^2}{3n}.$$

This time $\mathrm{MSE}_\theta(T_2)/\mathrm{MSE}_\theta(T_5) = 3/(n+2)$, which for $n > 1$ is less than 1. Again, therefore, T_5 is not admissible. □

It follows from the examples above that for $n > 1$ the best of the five suggested estimators is T_2: In each case, the risk is a parabola $k\theta^2$, with the smallest coefficient k for estimator T_2. As we concluded, T_1, T_3, T_4, and T_5 are inadmissible. This, however, does not imply automatically that T_2 is admissible: There may exist some estimator with risk smaller than the risk of T_2. To determine the admissibility of T_2 let us simply find the coefficient α such that $T = \alpha X_{n:n} = \alpha T_1$ has the smallest MSE. We have here

$$E_\theta(T) = \alpha E(T_1) = \frac{\alpha n}{n+1}\theta$$

and therefore

$$B_\theta(T) = \frac{\alpha n}{n+1}\theta - \theta = \frac{\alpha n - (n+1)}{n+1}\theta.$$

Next,

$$\mathrm{Var}_\theta(T) = \alpha^2\,\mathrm{Var}_\theta(T_1) = \frac{\alpha^2 n}{(n+1)^2(n+2)}\theta^2.$$

We have, therefore,

$$\mathrm{MSE}_\theta(T) = \left[\frac{\alpha^2 n}{(n+1)^2(n+2)} + \frac{\alpha^2 n^2 - 2\alpha n(n+1) + (n+1)^2}{(n+1)^2}\right]\theta^2$$
$$= \left[\frac{\alpha^2 n + (n+2)(\alpha^2 n^2 - 2\alpha n(n+1) + (n+1)^2)}{(n+1)^2(n+2)}\right]\theta^2.$$

Differentiating the numerator with respect to α, we obtain the value of α that minimizes $\text{MSE}_\theta(T)$ as

$$\alpha = \frac{n+2}{n+1}.$$

For this value of α, the estimator

$$T = \frac{n+2}{n+1} X_{n:n} \tag{12.27}$$

has the risk

$$\text{MSE}_\theta(T) = \frac{\theta^2}{(n+1)^2} < \frac{\theta^2}{n(n+2)} = \text{MSE}_\theta(T_2).$$

Thus, T_2 is also inadmissible, and the best estimator (of those that are proportional to the maximal observation) is the estimator T given by (12.27). Observe that the same conclusion holds also for $n = 1$: The estimator $\frac{3}{2}X_1$ has MSE equal to $\theta^2/4$.

PROBLEMS

12.4.1 Show that the estimator T_3 defined by (12.4) is unbiased. (*Hint:* Find the distribution of $X_{1:n}$ and then determine the expectation.)

12.4.2 Let $X_1, \ldots, X_n$ be a random sample from the distribution uniform on $[0, \theta]$. Show that variances of $X_{1:n} = \min(X_1, \ldots, X_n)$ and $X_{n:n} = \max(X_1, \ldots, X_n)$ are equal. Use two methods: (i) direct evaluation of distributions of $X_{1:n}$ and $X_{n:n}$, and then determining means and variances, and (ii) using the fact that variables X_i and $\theta - X_i$ have the same distribution.

12.4.3 Let T_1 and T_2 be two unbiased estimators of θ. Find values a and b such that estimator $aT_1 + bT_2$ is unbiased.

12.4.4 Let T_1 and T_2 be unbiased independent estimators of θ, with variances σ_1^2 and σ_2^2. For which a and b is the estimator $aT_1 + bT_2$ a minimum variance unbiased estimator (MVUE) of θ?

12.4.5 Solve Problem 12.4.4 replacing the assumption of independence of T_1 and T_2 by the assumption that $\text{Cov}(T_1, T_2) = C$.

12.4.6 Let $X_1, \ldots, X_n$ be a random sample from an unknown distribution with cdf F. Let v be fixed, and let $\theta = F(v)$. Construct an unbiased estimator of θ and find its variance.

12.4.7 Let $X_1, \ldots, X_n$ be a random sample from normal distribution $N(\mu, \sigma^2)$ with σ^2 known and $-\infty < \mu < +\infty$. Consider the estimator T of μ defined as

$$T(X_1, \ldots, X_n) = 3,$$

(i.e., we estimate μ as 3, regardless of the observation). Show that T is admissible.

12.4.8 Let $X_1, \ldots, X_n$ be a random sample from a distribution uniform on the interval $[\theta - 1, \theta + 1]$. Let $U = \min(X_1, \ldots, X_n)$, $V = \max(X_1, \ldots, X_n)$.

(i) Show that $\overline{X}$ and $(U + V)/2$ are both unbiased estimators of θ.

(ii) Determine mean square errors of the estimators in part (i).

12.4.9 Let $X_1, \ldots, X_n$ be a random sample from normal distribution $N(\mu, \sigma^2)$ with both μ and σ^2 unknown. Let

$$S^2 = \frac{1}{n} \sum_{i=1}^{n} (X_i - \overline{X})^2, \quad S_1^2 = \frac{1}{n-1} \sum_{i=1}^{n} (X_i - \overline{X})^2$$

be two estimators of σ^2. Find their mean square error and determine which of these estimators is better according to this criterion.

12.4.10 Continuing Problem 12.4.9, consider estimators of σ^2 of the form

$$S_k^2 = k \sum_{i=1}^{n} (X_i - \overline{X})^2$$

and find k for which S_k^2 has smallest MSE. Explain why in practice the only divisors used are (suboptimal): $k = 1/n$ and $k = 1/(n-1)$.

12.5 EFFICIENCY

To begin let us consider a simple situation of what is probably the most commonly known estimator, the sample average. Thus, we assume that the observations $X_1, X_2, \ldots$ are iid random variables, with $E(X_i) = \theta$ and $\mathrm{Var}(X_i) = \sigma^2 < \infty$, the latter assumed known (so that σ^2 is not the parameter to be estimated). The actual functional form of the distribution of X_i is of no concern for us at the moment.

If we want to estimate the mean θ, the obvious choice of estimator is $T = (X_1 + \cdots + X_n)/n$, the estimator traditionally denoted by $\overline{X}$ or $\overline{X}_n$, if we need to stress the dependence on the sample size. We know that

$$E(\overline{X}_n) = \frac{1}{n}E(X_1 + \cdots + X_n) = \theta, \tag{12.28}$$

so that $\overline{X}_n$ is always an unbiased estimator of the mean θ. Furthermore, by the Chebyshev inequality and the fact that $\mathrm{Var}(\overline{X}_n) = \sigma^2/n$, we have, for every $\epsilon > 0$,

$$P\{|\overline{X}_n - \theta| > \epsilon\} \le \frac{\mathrm{Var}(\overline{X}_n)}{\epsilon^2} = \frac{\sigma^2}{n\epsilon^2} \to 0, \tag{12.29}$$

and therefore $\overline{X}_n$ is a consistent estimator of θ, if only $\mathrm{Var}(X_i) < \infty$. Moreover, by the strong law of large numbers (Theorem 10.4.2), we also know that

$$P\{\lim_{n \to \infty} \overline{X}_n = \theta\} = 1, \tag{12.30}$$

hence $\overline{X}_n$ is strongly consistent. The risk of $\overline{X}_n$ is equal (in view of unbiasedness) to its variance, that is,

$$\mathrm{MSE}_\theta(\overline{X}_n) = \sigma^2/n. \tag{12.31}$$

Let us now compare the situation with that of estimating the range θ of uniform distribution by using estimators such as T_1 or T_2 given by (12.2) and (12.3) (i.e., estimators based on the maximum of the sample). There are two main points of difference here. The first is that in the case of estimating the mean by $\overline{X}_n$, the risk (for fixed n) is constant (does not depend on the estimated parameter θ). In case of estimators T_1 and T_2, the risk (for fixed n) given by (12.23) and (12.24) is a quadratic function of θ: The larger the value estimated, the larger is the variance of the results.

The second difference—much more important—is the rate at which the risk changes with the sample size n. In the case of estimating the mean by the sample average $\overline{X}_n$, the risk changes inversely with n, while in estimating the range of uniform distribution by T_1, or T_2, it changes (approximately) inversely with n^2. Naturally, the latter case is more desirable practically: Increasing the sample size by the factor of 10 (say) reduces the mean square error by a factor of 100, that is, to about 1% of the mean square error for the original sample size. As opposed to that, in estimating the mean by $\overline{X}_n$, an increase in sample size by the factor of 10 decreases the mean square error by a factor of 10 only, that is, to about 10% percent of the mean square error for the original sample size.

The question that comes to mind is: Why do statisticians use the sample average as an estimator if there are other estimators that are so much better?

The answer is that the situation such as with the sample mean—when the mean square error decreases inversely with n—is much more common than the situation when the mean square error decreases with n^2. In fact, in most cases, the best that one can achieve is to build estimators whose mean square error decreases in proportion to $1/n$.

Intuitively, in all cases that could be called "regular," it is possible to gather information about a parameter only at a rate proportional to the sample size. To make this statement precise, we need to introduce some assumptions and new concepts.

We shall consider the usual setup of observing iid random variables $X_1, X_2,$ $\ldots$, each with the distribution $f(x, \theta)$. We shall assume that θ takes values in some open domain and that the set of points x at which $f(x, \theta) > 0$ does not depend on θ. The last assumption rules out the case of observations X_i having uniform distribution on $[0, \theta]$, since then the set of points at which $f(x, \theta)$ is positive depends on θ.

The concept which we shall attempt to formalize is that of the average "amount of information about θ contained in a single observation of X." The reason for distinguishing the "regular" case, when the set of points x where $f(x, \theta)$ is positive (i.e., the domain of "possible" x for a given θ) is the same for all θ, is as follows. If this assumption were violated, a single observation may eliminate some values of θ with certainty (e.g., in the case of uniform distribution on $(0, \theta]$, observation $X = 3$ rules out all $\theta < 3$). Such an elimination is typically unattainable in practical situations: hence we concentrate the theory on "regular" cases.

Let us start with some intuitions. Suppose one gets a message that some event A occurred. The amount of information in this message depends, in an obvious way, on what event A is and who receives the message (think here of the message that TWA flight 503 will arrive 2 hours late). Apart from such semantic and personal information, there is also some information contained in the message (that A occurred) depending only on how likely event A is. If $P(A)$ is close to 1, the amount of information is close to 0, while for $P(A)$ close to 0, the amount of information is high.

This agrees with common intuition, as well as with practice (e.g., in newspaper publishing). The fact that Mr. Smith found a dead roach in the can of beer he drank is of interest and worth reporting precisely because such an event is rare, hence its occurrence carries lots of information. If, on the other hand, dead roaches were commonly found in cans of beer, nobody would care to report it.

It turns out that one obtains an interesting and coherent theory if one assumes that the amount of information in the occurrence of an event with probability p is $\log(1/p) = -\log p$. It is customary in information theory to take as a unit the amount of information in the occurrence of an event with chance $\frac{1}{2}$ (e.g., the result of a toss of a coin). This is equivalent to taking the base of the logarithm as equal to 2. For our purposes, the choice of the unit is of no consequence, so we shall use natural logarithm unless otherwise specified.

Now, the situation in the case of estimation is somewhat more complicated, since we want to define the amount of information *about* θ in the event $X = x$, where X is a random variable with distribution (density or pdf) $f(x, \theta)$. Here $f(x, \theta)$ plays the role of probability p of the event $X = x$. Since we are interested in information about θ, it appears intuitively natural to consider the rate of change of $\log(1/[f(x,\theta)])$ under varying θ at the point x, that is, the derivative

$$\frac{\partial}{\partial\theta}\left[\log\frac{1}{f(x,\theta)}\right] = -\frac{\frac{\partial}{\partial\theta}f(x,\theta)}{f(x,\theta)}. \tag{12.32}$$

We get rid of the effect of the sign by considering the square of quantity (12.32), and we eliminate the restriction to specific value x by taking the average.

These considerations lead to the following definition. Let X be a random variable with distribution $f(x,\theta)$, such that the set of points x at which $f(x,\theta) > 0$ is the same for all θ. We assume that the function $f(x,\theta)$ is twice differentiable with respect to θ for every x, and we consider the random variable

$$J(X,\theta) = \frac{\partial}{\partial\theta}[\log f(X,\theta)]. \tag{12.33}$$

Definition 12.5.1. We define the *Fisher information* about θ in a single observation X as

$$I(\theta) = E\{J(X,\theta)\}^2, \tag{12.34}$$

if the expectation (12.34) exists.   □

Thus, in the case of the continuous random variable X, the quantity (12.34) is the integral

$$I(\theta) = \int [J(x,\theta)]^2 f(x,\theta)\, dx = \int \left[\frac{\partial f(x,\theta)}{\partial\theta}\right]^2 \frac{1}{f(x,\theta)}\, dx.$$

In the case of a discrete random variable, integration is replaced by summation.

Example 12.5.1. Let X have normal distribution $N(\theta,\sigma^2)$. Then

$$\log f(x,\theta) = -\log(\sigma\sqrt{2\pi}) - (x-\theta)^2/2\sigma^2$$

and

$$J(X,\theta) = \frac{\partial}{\partial\theta}[\log f(X,\theta)] = \frac{X-\theta}{\sigma^2}. \tag{12.35}$$

Thus

$$I(\theta) = E\left(\frac{X-\theta}{\sigma^2}\right)^2 = \frac{1}{\sigma^4}E(X-\theta)^2 = \frac{\sigma^2}{\sigma^4} = \frac{1}{\sigma^2}.$$

Example 12.5.2. Consider now a single Bernoulli trial with probability θ. We have $P\{X=1|\theta\} = \theta$ and $P\{X=0|\theta\} = 1-\theta$, and this probability distribution can be written as

$$f(x, \theta) = \theta^x(1 - \theta)^{1-x}, \qquad x = 0, 1. \qquad (12.36)$$

Thus

$$\log f(x, \theta) = x \log \theta + (1 - x) \log(1 - \theta)$$

and

$$\left[\frac{\partial}{\partial \theta} \log f(x, \theta)\right]^2 = \left(\frac{x}{\theta} - \frac{1 - x}{1 - \theta}\right)^2 = \begin{cases} \dfrac{1}{(1 - \theta)^2} & \text{if } x = 0 \\ \dfrac{1}{\theta^2} & \text{if } x = 1. \end{cases} \qquad (12.37)$$

Consequently, taking expectation, we obtain

$$I(\theta) = \frac{1}{(1 - \theta)^2} \cdot (1 - \theta) + \frac{1}{\theta^2} \cdot \theta = \frac{1}{\theta(1 - \theta)}. \qquad (12.38)$$

Example 12.5.3.   Let us modify slightly Example 12.5.2: Suppose now that $P\{X = 1|\theta\} = \theta^2$ and $P\{X = 0|\theta\} = 1 - \theta^2$; thus, we have a single Bernoulli trial, but probability of success is now θ^2. However, we are still interested in the amount of information about θ (not θ^2). A practical example here may be found in genetics, where θ is a frequency of a recessive gene (e.g., a gene causing a person to have blue eyes), so that $X = 1$ corresponds to finding a person with some features, caused by a recessive gene (requiring both parents to transmit this gene). We now have

$$f(x, \theta) = (\theta^2)^x(1 - \theta^2)^{1-x},$$

hence

$$\left[\frac{\partial}{\partial \theta} \log f(x, \theta)\right]^2 = \left\{\frac{\partial}{\partial \theta}[2x \log \theta + (1 - x) \log(1 - \theta^2)]\right\}^2$$

$$= \left\{\frac{2x}{\theta} - 2\theta \frac{(1 - x)}{1 - \theta^2}\right\}^2$$

$$= \begin{cases} \dfrac{4\theta^2}{(1 - \theta^2)^2} & \text{if } x = 0 \\ \dfrac{4}{\theta^2} & \text{if } x = 1. \end{cases}$$

Consequently,

$$I(\theta) = \frac{4\theta^2}{(1 - \theta^2)^2}(1 - \theta^2) + \frac{4}{\theta^2}\theta^2$$

$$= \frac{4\theta^2}{1 - \theta^2} + 4 = \frac{4}{1 - \theta^2}. \qquad (12.39)$$

Comparison of this result with the result in Example 12.5.2 is quite instructive. In a single Bernoulli trial with probability of success θ, the average amount of information in a trial is a function which assumes its minimal value 4 at $\theta = \frac{1}{2}$. As θ moves away from $\frac{1}{2}$ toward either $\theta = 0$ (success very rare) or $\theta \approx 1$ (failure very rare), the average amount of information about θ increases to infinity.

On the other hand, if we can only observe whether or not an event with probability θ^2 has occurred, then the average amount of information about θ is close to 4 if $\theta \approx 0$ and increases to infinity when $\theta \to 1$. The symmetry of the preceding example is lost. Indeed, only when θ is close to 1 is θ^2 also close to 1, so that successes which are sufficiently frequent provide a good estimator of θ^2, hence also of θ. For small θ we have a poor estimate of θ^2, hence also a poor estimate of θ.   □

Before proceeding further, it will be convenient to derive alternative formulas for $I(\theta)$. Observe first that

$$
\begin{aligned}
E_\theta[J(X, \theta)] &= \int \left[\frac{\partial}{\partial \theta} \log f(x, \theta) \right] f(x, \theta)\, dx \\
&= \int \left(\frac{\frac{\partial}{\partial \theta} f(x, \theta)}{f(x, \theta)} \right) f(x, \theta)\, dx \\
&= \int \frac{\partial}{\partial \theta} f(x, \theta)\, dx = \frac{\partial}{\partial \theta} \int f(x, \theta)\, dx \\
&= \frac{\partial}{\partial \theta}(1) = 0.
\end{aligned}
\tag{12.40}
$$

This is valid under conditions that allow interchange differentiation with respect to θ and integration with respect to x in the antepenultimate equality sign. Thus, in view of (12.40),

$$
\operatorname{Var}_\theta[J(X, \theta)] = E_\theta[J(X, \theta)]^2 - \{E[J(X, \theta)]\}^2 = I(\theta).
\tag{12.41}
$$

Still another formula for $I(\theta)$ can be obtained by noting that

$$
\frac{\partial}{\partial \theta} J(x, \theta) = \frac{\partial}{\partial \theta} \left[\frac{\frac{\partial}{\partial \theta} f(x, \theta)}{f(x, \theta)} \right] = \frac{\left(\frac{\partial^2}{\partial \theta^2} f(x, \theta) \right) f(x, \theta) - \left(\frac{\partial}{\partial \theta} f(x, \theta) \right)^2}{[f(x, \theta)]^2}
$$

$$
= \frac{\frac{\partial^2}{\partial \theta^2} f(x, \theta)}{f(x, \theta)} - (J(x, \theta))^2.
$$

Taking expectations of both sides, we get

$$E_\theta \left(\frac{\partial}{\partial \theta} J(x, \theta) \right) = \int \frac{\partial^2}{\partial \theta^2} f(x, \theta) \, dx - E_\theta \, [J(X, \theta)]^2 = -I(\theta),$$

since, assuming that one can interchange integration and differentiation,

$$\int \frac{\partial^2}{\partial \theta^2} f(x, \theta) \, dx = \frac{\partial^2}{\partial \theta^2} \int f(x, \theta) \, dx = \frac{\partial}{\partial \theta^2} (1) = 0.$$

We have therefore the following

Theorem 12.5.1. *If the density $f(x, \theta)$ is twice differentiable in θ and the equality $\int f(x, \theta) \, dx = 1$ can be differentiated twice under the integral sign, we have*

$$I(\theta) = E_\theta [J(X, \theta)]^2 = \text{Var}_\theta [J(X, \theta)]$$
$$= -E_\theta \left(\frac{\partial}{\partial \theta} J(x, \theta) \right). \tag{12.42}$$

□

We can now determine the average information contained in a random sample

$$\mathbf{X} = (X_1, \ldots, X_n).$$

The density of $\mathbf{X}$ is

$$f(\mathbf{X}, \theta) = f(X_1, \theta) \cdots f(X_n, \theta),$$

so that

$$J(\mathbf{X}, \theta) = \frac{\partial}{\partial \theta} \log f(\mathbf{X}, \theta) = \frac{\partial}{\partial \theta} \sum_{i=1}^{n} \log f(X_i, \theta)$$
$$= \sum_{i=1}^{n} J(X_i, \theta).$$

Differentiating both sides, we have

$$\frac{\partial}{\partial \theta} J(\mathbf{X}, \theta) = \sum_{i=1}^{n} \frac{\partial}{\partial \theta} J(X_i, \theta). \tag{12.43}$$

Taking expectation and using the fact that X_i's have the same distribution, we have the following theorem:

Theorem 12.5.2. *The information $I_n(\theta)$, in a random sample of n obser-vations is n times the information of a single observation:*

$$I_n(\theta) = nI(\theta). \tag{12.44}$$

We shall prove now a theorem that connects the amount of information about θ contained in a single observation, with the variance of an estimator of θ. Let $T(\mathbf{X}) = T(X_1, \ldots, X_n)$ be an arbitrary estimator of θ, assumed to have a finite variance. Letting $\mathbf{x} = (x_1, \ldots, x_n)$ and letting $m(\theta)$ denote the expectation of T, we have

$$m(\theta) = E_\theta(T) = \int \cdots \int T(\mathbf{x}) f(\mathbf{x}, \theta) \, dx_1 \cdots dx_n.$$

Consequently, assuming again that we can differentiate under the integral sign,

$$
\begin{aligned}
m'(\theta) &= \int \cdots \int T(\mathbf{x}) \left[\frac{\partial}{\partial \theta} f(\mathbf{x}, \theta) \right] dx_1 \cdots dx_n \\
&= \int \cdots \int T(\mathbf{x}) \left[\frac{\frac{\partial}{\partial \theta} f(\mathbf{x}, \theta)}{f(\mathbf{x}, \theta)} \right] f(\mathbf{x}, \theta) \, dx_1 \cdots dx_n \\
&= \int \cdots \int T(\mathbf{x}) J(\mathbf{x}, \theta) f(\mathbf{x}, \theta) \, dx_1 \cdots dx_n \\
&= E_\theta \{ T(\mathbf{X}) J(\mathbf{X}, \theta) \} = \text{Cov}_\theta \{ T(\mathbf{X}), J(\mathbf{X}, \theta) \}. \tag{12.45}
\end{aligned}
$$

The last equality follows from the fact that $E[J(\mathbf{X}, \theta)] = 0$, which can be established in the same way as for a single random variable X, in view of the relation

$$J(\mathbf{X}, \theta) = \sum_{i=1}^{n} J(X_i, \theta).$$

Now, by the Schwarz inequality (8.7.7) and (12.45) we may write

$$[m'(\theta)]^2 = [\text{Cov}_\theta(T(\mathbf{X}), J(\mathbf{X}, \theta))]^2 \leq \text{Var}_\theta(T(\mathbf{X})) \cdot \text{Var}_\theta[J(\mathbf{X}, \theta)]. \tag{12.46}$$

Consequently, using (12.44), we obtain

Theorem 12.5.3 (Rao–Cramér Inequality). *For any estimator T of θ* (*based on sample of size n*), *we have*

$$\text{Var}_\theta(T) \geq \frac{[m'(\theta)]^2}{nI(\theta)}. \tag{12.47}$$

In particular, if T is unbiased, then $m(\theta) = \theta, m'(\theta) = 1$ and we see that the variance (hence also MSE) of T is bounded from below by the reciprocal of the amount of information about θ contained in the sample.

It is worth mentioning that the estimator T for which $E_\theta(T) = m(\theta)$ is an unbiased estimator of $m(\theta)$. If m is a differentiable function and regularity conditions are satisfied, then the variance of T satisfies the inequality (12.47).

Observe that we have equality in (12.47) if and only if we have equality in (12.46), which means that the correlation coefficient between $T(X)$ and $J(X, \theta)$ satisfies the condition $|\rho_{T(X),J(X,\theta)}| = 1$. By Theorem 8.5.6 this is equivalent to linear relationship between $T(X)$ and $J(X, \theta)$, that is,

$$T(X) = \gamma_1(\theta)J(X, \theta) + \gamma_2(\theta) \tag{12.48}$$

for some functions $\gamma_1(\theta)$ and $\gamma_2(\theta)$.

An important point here is that T is an estimator of θ, so that the left-hand side does not depend on θ. Consequently, the right-hand side does not depend on θ either; that is, functions γ_1 and γ_2 cancel the dependence of $J(X, \theta)$ on the parameter θ. Let us agree to call the set of conditions that allow all differentiations under the integral sign, combined with the condition of independence of θ of the set of x for which $f(x, \theta) > 0$, the *regularity* conditions. We may now formulate the findings up to now as the statement that *under regularity conditions, any estimator of θ that has finite variance satisfies the Rao–Cramér inequality.*

This leads in a natural way to the following definition.

Definition 12.5.2. Any unbiased estimator T that satisfies the regularity conditions and whose variance attains the minimum equal to the right-hand side of (12.47) is called *efficient*. The ratio $nI(\theta)/\text{Var}_\theta(T)$ is called the *efficiency* of T.

More generally, given two unbiased estimators T_1 and T_2 of θ, the ratio of their variances $\text{Var}_\theta(T_1)/\text{Var}_\theta(T_2)$ is called the *relative efficiency* of T_2 with respect to T_1.

Example 12.5.4. Consider the problem of estimating the mean θ of normal distribution $N(\theta, \sigma^2)$. We know from Example 12.5.1 that $I(\theta) = 1/\sigma^2$. Consider the most obvious estimator for θ, the sample mean

$$\overline{X}_n = \frac{1}{n}(X_1 + \cdots + X_n).$$

Clearly, $\overline{X}_n$ is unbiased, and $\text{Var}(\overline{X}_n) = \sigma^2/n$. The lower bound of variances of all unbiased estimators of θ, based on samples of size n, is $1/(nI(\theta)) = \sigma^2/n$, which shows that $\overline{X}_n$ is efficient.

Example 12.5.5. Continuing Example 12.5.2, suppose that we want to estimate probability of success θ on the basis of n Bernoulli trials. We have here $I(\theta) = 1/[\theta(1-\theta)]$, and therefore the information in the sample of size n is $n/\theta(1-\theta)$. Letting $S = X_1 + \cdots + X_n$ to be the total number of successes, we might try $T = S/n$ as an estimator of θ. Since S is binomial, we have $E(T) = (1/n)E(S) = \theta$, and $\mathrm{Var}(T) = (1/n^2)\mathrm{Var}(S) = \theta(1-\theta)/n$, which shows that T is efficient.

Example 12.5.6. In Section 12.3 we studied five estimators of the range θ in uniform distribution on $[0, \theta]$. For estimators T_1, T_2, and T_3, we found that their variance decreases to zero like $1/n^2$ (and not at the rate $1/n$ as for efficient estimators). Thus, those estimators are, so to say, superefficient. The reason is that $f(x, \theta)$ does not satisfy the regularity conditions: The set of points x at which the density $f(x, \theta)$ is positive is $[0, \theta]$; hence it depends on θ. $\qquad\qquad\square$

Under the regularity conditions, the right-hand side of the Rao–Cramér inequality (12.47) gives the lower bound for variances of unbiased estimators, and therefore if an efficient estimator of θ is found, it is known that better unbiased estimator does not exist. It is necessary to point out, however, that the Rao–Cramér bound is not always attainable. It may happen that the best possible unbiased estimator has a variance larger than the Rao–Cramér bound. For a thorough discussion of these topics and relevant examples, see, for example, Mood et al. (1974).

PROBLEMS

12.5.1 Let $\overline{X}_n$ be the sample mean of n independent observations from a normal distribution with mean θ and known variance σ^2. Find the efficiency of $\overline{X}_{[n/2]}$ (i.e., of the estimator which uses only half of the sample).

12.5.2 Let X, Y be independent, both with normal distribution $\mathrm{N}(\mu, \sigma^2)$. Determine the amounts of information about μ and about σ^2 contained in (i) $X + Y$; (ii) $X - Y$.

12.5.3 Suppose that X has an exponential distribution with density $\lambda e^{-\lambda x}$, $x > 0$. Find the Fisher information $I(\lambda)$.

12.5.4 Let $X_1, \ldots, X_n$ be a random sample from exponential distribution with unknown parameter λ. Construct an efficient estimator of $1/\lambda$ and determine its expectation and variance.

12.5.5 Let $X_1, \ldots, X_n$ be a random sample from Bernoulli distribution with unknown p. Show that the variance of any unbiased estimator of $(1-p)^2$ must be at least $4p(1-p)^3/n$.

12.5.6 Show that an estimator $\overline{X}_n$ of the mean of normal distribution satisfies relation (12.48) and determine functions $\gamma_1(\mu, \sigma^2)$ and $\gamma_2(\mu, \sigma^2)$.

12.5.7 Show that the estimator $T = S/n$, where S is the number of successes in n independent Bernoulli trials, satisfies relation (12.48) and determine functions γ_1 and γ_2.

12.6 CLASSICAL METHODS OF OBTAINING ESTIMATORS

Before introducing further criteria of judging the performance of estimators (besides consistency, unbiasedness, and efficiency), we shall spend some time presenting methods of constructing estimators. In this section, four such methods, oldest and most commonly known, are discussed. Newer methods are sketched in Section 12.7.

Method of Analogy

This method consists simply of building a sample analogy of some quantity that depends on the parameter, compare the two, and solve the resulting equation with respect to the parameter. The formula obtained depends on the sample, hence can be taken as an estimator of the parameter.

Since the most commonly taken quantity is a moment of the random variable, this method of constructing estimators is generally known as *method of moments*. The term *method of analogy*, which describes much better the idea of the method, has been suggested by E. Pleszczyńska. We shall speak of a method-of-moment estimator (or simply a moment estimator), if the quantities used to build the estimate are moments, otherwise we shall speak of a method-of-analogy estimator.

Example 12.6.1. Let $X_1, \ldots, X_n$ be a random sample from the exponential distribution with density $f(x, \theta) = \theta e^{-\theta x}, x > 0$. Then $E(X_i) = 1/\theta$. The sample counterpart of the first moment is the sample mean $\overline{X}_n = (1/n)(X_1 + \cdots + X_n)$. Equating the empirical and theoretical mean we obtain the equation $\overline{X}_n = 1/\hat{\theta}$, which gives the estimator $T_1 = 1/\overline{X}_n$ of the parameter θ.

Example 12.6.2. Continuing Example 12.6.1, suppose that we decided to use the second moment instead of the first. We have here $E(X^2) = \mathrm{Var}(X) + [E(X)]^2 = 2/\theta^2$. The empirical counterpart of the second moment is $\frac{1}{n}\sum_{i=1}^{n} X_i^2$. We therefore obtain the equation

$$\frac{1}{n}\sum_{i=1}^{n} X_i^2 = \frac{2}{\hat{\theta}^2},$$

which, solving for $\hat{\theta}$, gives an estimator

$$T_2 = \sqrt{\frac{2n}{\sum_{i=1}^{n} X_i^2}}.$$ (12.49)

□

Example 12.6.3.   Continuing Example 12.6.2, the median m of exponential distribution is the point at which cdf equals $\frac{1}{2}$, so that we must have

$$F(m) = 1 - e^{-\theta m} = \frac{1}{2},$$

which gives

$$m = \frac{\log 2}{\theta}.$$

The empirical counterpart of median m is the sample median. Assuming for simplicity that the sample size $n = 2k + 1$ is odd, the sample median is the $(k + 1)$st order statistic $X_{k+1:n}$. We therefore have still another estimator of θ, namely

$$T_3 = \frac{\log 2}{X_{k+1:n}}.$$ (12.50)

Example 12.6.4.   Continuing Example 12.6.3, suppose that in the same setup of exponentially distributed observations, we want to estimate the probability $p = P\{X \geq 3\} = e^{-3\theta}$. The method of analogy suggests using the estimator e^{-3T}, where T is any estimator of θ. The three estimators obtained in Examples 12.5.6 to 12.6.3 give now three estimators of p, namely

$$\hat{p}_1 = e^{-3/\overline{X}_n}, \qquad \hat{p}_2 = e^{-3\sqrt{(2n)/\sum X_i^2}}$$

and

$$\hat{p}_3 = \exp\left\{\frac{-3\log 2}{X_{k+1:n}}\right\}.$$

Example 12.6.5.   Let us observe that the estimator $T_5 = (2/n)(X_1 + \cdots + X_n)$ from Example 12.2.4 is also an example of estimator obtained by method of moments: We have here $E(X) = \theta/2$, so doubling the sample mean gives a moment estimator of θ.

□

The method of analogy works also in the case of constructing estimators of several parameters at once. We illustrate this by two simple examples.

Example 12.6.6.   Suppose that we want to estimate both μ and σ^2 in case of observations $X_1, \ldots, X_n$ from some distribution with mean μ and variance σ^2. In this case we have, for instance, the following two expressions for moments of orders 1 and 2:

$$E(X) = \mu, \qquad E(X^2) = \sigma^2 + \mu^2. \tag{12.51}$$

This suggests comparing the empirical first and second moments with μ and $\sigma^2 + \mu^2$, obtaining the equations

$$\frac{1}{n} \sum_{i=1}^{n} X_i = \hat{\mu}$$

$$\frac{1}{n} \sum_{i=1}^{n} X_i^2 = \hat{\sigma}^2 + \hat{\mu}^2.$$

Solving for $\hat{\mu}$ and $\hat{\sigma}^2$ we obtain the pair of estimators for μ and σ^2, namely

$$\hat{\mu} = \overline{X} \quad \text{and} \quad \hat{\sigma}^2 = \frac{1}{n} \sum_{i=1}^{n} X_i^2 - \overline{X}^2 = \frac{1}{n} \sum_{i=1}^{n} (X_i - \overline{X})^2. \tag{12.52}$$

Thus, the sample mean and the sample variance (with divisor n) are method-of-moment estimators of the population mean and variance in the general case.

Example 12.6.7. Suppose that the observations $X_1, \ldots, X_n$ follow a gamma distribution, with the density

$$f(x; \alpha, \lambda) = \frac{\lambda^\alpha}{\Gamma(\alpha)} x^{\alpha-1} e^{-\lambda x}, \qquad x > 0. \tag{12.53}$$

To obtain a method-of-moment estimator of $\theta = (\alpha, \lambda)$, recall that [see (9.51), (9.52), (9.53)]

$$E(X) = \frac{\alpha}{\lambda}, \qquad \mathrm{Var}(X) = \frac{\alpha}{\lambda^2},$$

so that we obtain

$$\lambda = \frac{E(X)}{\mathrm{Var}(X)}, \qquad \alpha = \frac{[E(X)]^2}{\mathrm{Var}(X)}.$$

This gives the method-of-moment estimators

$$\hat{\lambda} = \frac{\sum_{i=1}^{n} X_i}{\sum_{i=1}^{n} (X_i - \overline{X})^2}, \qquad \hat{\alpha} = \frac{(\sum_{i=1}^{n} X_i)^2}{n \sum_{i=1}^{n} (X_i - \overline{X})^2}. \tag{12.54}$$

Example 12.6.8. As the last example of the method-of analogy estimator, consider the problem of "touchy" questions (e.g., concerning the use of drugs, sexual habits, etc.). Generally, the empirical problem concerns some attribute, call it Q, which a person might be reluctant to admit having. The objective is to estimate the frequency θ of persons in a given population whose true reply to the question "Are you a Q-person?" is "yes." The method of collecting the data on such questions consists of using a questionnaire with randomized response (see Example 1.2.11). The respondent activates two random mechanisms, one generating the event A or its complement A^c, the other generating the event B or its complement B^c. These two mechanisms operate independently. The probabilities $P(A) = \alpha$ and $P(B) = \beta$ are known; however, only the respondent knows which of events A and B occurred in a given instance. He is instructed to respond to the question "Are you a Q-person?" if A occurs, and otherwise respond to the question "Did B occur?". The answer "yes" or "not" is recorded by the experimenter, who does not know which question was actually answered.

Conditioning on the occurrence or nonoccurrence of event A, we have here

$$P(\text{"yes"}) = P(\text{"yes"}|A)P(A) + P(\text{"yes"}|A^c)P(A^c)$$
$$= \theta\alpha + \beta(1 - \alpha).$$

If now X respondents out of N tested replied "yes", then X/N is an estimator of $P(\text{"yes"})$ and we have an approximate equality

$$\frac{X}{N} \approx \theta\alpha + \beta(1 - \alpha),$$

which suggests using as an estimator of θ the random variable

$$T = \frac{X/N - \beta(1 - \alpha)}{\alpha}. \qquad \square$$

A problem with the estimators obtained by method of analogy is that they are not unique, in the sense that we do not know which quantities are best to use to provide the theoretical formulas involving the unknown parameter. Consequently, the phrase "a method-of-analogy estimator" does not have a unique meaning. Even the phrase "a method-of-moments estimator" does not have a unique meaning, since we can take moments of various orders. However, as a rule, one should take the moments of lowest orders that depend on θ. To see the type of difficulty here, consider the following example.

Example 12.6.9. Suppose that the observations are known to have distribution uniform on $[-\theta, \theta]$. Then $E(X) = 0$, and the first moment contains no information about θ. One may use here the fact that

$$E(X^2) = \int_{-\theta}^{\theta} x^2 \cdot \frac{1}{2\theta} dx = \frac{\theta^2}{3},$$

which gives the estimator

$$\hat{\theta} = \sqrt{\frac{3}{n}\sum_{i=1}^{n} X_i^2}. \qquad \square$$

Also, as regards consistency, the situation is as follows. The moment estimators are consistent under some very mild conditions. Indeed, the strong law of large numbers for the iid case asserts that $(1/n)(X_1^k + X_2^k + \cdots + X_n^k)$ converges with probability 1 to $E_\theta(X^k)$ if only $E_\theta(|X|^k) < \infty$. Consequently, the empirical moments converge also in probability to the corresponding theoretical moments, and if only the parameter θ is a continuous function of moments (as is usually in case), the consistency of method-of-moment estimators follows.

The situation is not so straightforward for other estimators built on the analogy principle. It is probably simplest to decide each case separately, by establishing whether or not the sample analogue of a given quantity converges in probability (or almost surely) to the corresponding theoretical quantity. However, the main issue with method-of-analogy estimators is that they either coincide with estimators obtained by the maximum likelihood method (discussed below), or else, they are inferior to them.

Maximum Likelihood Estimators

Let us consider again the basic setup, in which we observe a random sample $X_1, \ldots, X_n$ from distribution $f(x, \theta)$. We remind that the symbol $f(x, \theta)$ may stand for the density or for the probability function. If the actual observations are $X_1 = x_1, \ldots, X_n = x_n$, then the probability of this sample (or the joint density at this sample) is

$$f(x_1, \theta) \cdots f(x_n, \theta). \tag{12.55}$$

The product (12.55), regarded as a function of parameter θ, is called the *likelihood function* of the sample, or simply the likelihood function. We shall use the symbol

$$L(\theta) = L(\theta; x_1, \ldots, x_n) = \prod_{i=1}^{n} f(x_i, \theta)$$

or, letting $\mathbf{x} = (x_1, \ldots, x_n)$, we may write

$$L(\theta) = L(\theta, \mathbf{x}) = f_n(\mathbf{x}, \theta)$$

for the likelihood function.

Definition 12.6.1. Given the sample $\mathbf{x} = (x_1, \ldots, x_n)$, the value $\hat{\theta} = \hat{\theta}(\mathbf{x})$ that maximizes the likelihood function is called the *maximum likelihood estimate* (MLE) of θ. □

Let us begin with some examples, which we shall use later as starting points for generalizations.

Example 12.6.10.   In five independent Bernoulli trials with the probability of success θ, we observed three successes and two failures.

We have here $f(x, \theta) = \theta^x(1 - \theta)^{1-x}$, where $x = 0$ or 1, representing failure and success, respectively. Thus, the likelihood of the sample is

$$L(\theta) = \theta^3(1 - \theta)^2, \tag{12.56}$$

where $0 \le \theta \le 1$. It is clear that the information as to which trials led to successes and which to failures does not affect the likelihood, and is irrelevant for the estimation of θ.

To find the maximum of function (12.56), we may differentiate, obtaining the equation

$$
\begin{aligned}
L'(\theta) &= 3\theta^2(1 - \theta)^2 - 2\theta^3(1 - \theta) \\
&= \theta^2(1 - \theta)[3(1 - \theta) - 2\theta] \\
&= \theta^2(1 - \theta)(3 - 5\theta) = 0.
\end{aligned}
$$

The solutions are $\theta = 0$, $\theta = 1$, and $\theta = \frac{3}{5}$. An inspection of $L(\theta)$ shows that this function attains its maximum at the last solution, while $\theta = 0$ and $\theta = 1$ give the minima. Thus, maximum likelihood estimate of θ, in case of our sample, is $\hat{\theta} = \frac{3}{5}$.

Example 12.6.11.   Suppose that we take $n = 3$ observations from Poisson distribution with parameter θ, obtaining the values $x_1 = 2, x_2 = 0, x_3 = 5$. Then the likelihood is

$$
\begin{aligned}
L(\theta) &= \left(\frac{\theta^2}{2!}e^{-\theta}\right)\left(\frac{\theta^0}{0!}e^{-\theta}\right)\left(\frac{\theta^5}{5!}e^{-\theta}\right) \\
&= \frac{\theta^7}{2!5!}e^{-3\theta}.
\end{aligned}
$$

The derivative now is

$$L'(\theta) = \frac{1}{2!5!}[7\theta^6 e^{-3\theta} - 3\theta^7 e^{-3\theta}]$$

and $L'(\theta) = 0$ for $\theta = 0$ and for $\theta = \frac{7}{3}$. Again, an inspection of L shows that the maximum occurs at the second solution, so that the MLE of θ is now $\frac{7}{3}$.

Example 12.6.12.   Suppose that we observe the values $x_1 = 3$ and $x_2 = -2$ from a normal distribution with mean 0 and standard deviation θ. The likelihood now is

$$L(\theta) = \frac{1}{\theta\sqrt{2\pi}}e^{-9/2\theta^2}\frac{1}{\theta\sqrt{2\pi}}e^{-4/2\theta^2} = \frac{1}{\theta^2}\frac{1}{2\pi}e^{-13/2\theta^2}.$$

Since $L(\theta)$ is maximized at the same point at which its logarithm is maximized, we may try to take logarithm first and then differentiate. We have, taking natural logarithms,

$$\log L(\theta) = -2\log\theta - \log(2\pi) - \frac{13}{2\theta^2},$$

hence

$$\frac{d}{d\theta}\log L(\theta) = -\frac{2}{\theta} + \frac{13}{\theta^3},$$

and we obtain the equation

$$-\frac{1}{\theta}\left(2 - \frac{13}{\theta^2}\right) = 0,$$

which gives the solution $\theta = \sqrt{\frac{13}{2}}$. Again, we check easily that the solution gives the point at which the likelihood is maximized, so that MLE of θ is $\sqrt{\frac{13}{2}}$. $\square$

Let us reflect awhile on the examples above. The feature common to all of them was that in each case we attempted to explain the observed data in the best possible way, by finding that particular value of the parameter for which the probability of what we observed was maximal.

Some obvious questions arise here, concerning different aspects of the suggested procedure of finding estimates. First, there are questions of *philosophical* nature concerning the justification of the suggested principle of selecting the estimates. Second, the *statistical* questions concern the properties of the suggested procedure. The point is that the maximum likelihood estimate depends on the sample, so that it is, in effect, a random quantity. Our procedure therefore defines an estimator. What are its properties, such as consistency, bias, and efficiency?

Finally, the *mathematical* questions are: Does MLE always exist? If so, is the maximum of the likelihood function unique? Can it always be obtained by differentiation of likelihood or of its logarithm and solving the resulting equation?

To begin with the last kind of question, the situation is as follows. There are cases when, formally speaking, MLE does not exist; but these cases can often be modified in a natural way so as to remedy the situation.

Example 12.6.13.   Consider the problem of estimating θ on the basis of observations from the uniform distribution on $[0, \theta]$. We have here

$$f(x, \theta) = \begin{cases} 1/\theta & \text{for } 0 < x < \theta \\ 0 & \text{otherwise.} \end{cases}$$

Given the sample $x_1, \ldots, x_n$, the likelihood is

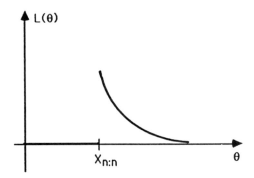

Figure 12.1 Likelihood function for range in uniform distribution.

$$L(\theta) = f(x_1, \theta) \cdots f(x_n, \theta) = \begin{cases} 1/\theta^n & \text{if } 0 < x_i < \theta \\ & \text{for all } i \\ 0 & \text{otherwise.} \end{cases}$$

Thus, the likelihood function is discontinuous: it is a function depicted in Figure 12.1, with discontinuity at the point $t = \max(x_1, \ldots, x_n)$ and decreasing for $\theta > t$. However, for $\theta = t$ we have $L(t) = 0$, so there is no point at which this function attains its maximum and MLE does not exist.

We see here that the cause of the trouble is the choice of the definition of the density. If we defined $f(x, \theta)$ to be $1/\theta$ in the *closed* interval $0 \le x \le \theta$, the likelihood would actually reach its maximum at $\theta = \max(x_1, \ldots, x_n)$.

$\square$

The example above shows a type of situation when MLE does not exist, because of the choice of the density function. Since the density function can be modified at a single point (or at any finite or countable set of points), such a modification may affect the likelihood function. However, we feel intuitively that all such choices (e.g., choosing the value of uniform density on $[0, \theta)$ as $1/\theta$ or 0 at $x = \theta$) are not relevant. The choice is here a matter of description only and is not imposed by the problem. Consequently, we may (and will) always assume that the densities are defined in such a way that the maximum of the likelihood exists.

Example 12.6.14. Assume that we are estimating θ on the basis of a random sample $X_1, \ldots, X_n$, with X_i's being distributed uniformly on the interval $[\theta - \frac{1}{2}, \theta + \frac{1}{2}]$. Thus, we have

$$f(x, \theta) = \begin{cases} 1 & \text{for } \theta - \frac{1}{2} \le x \le \theta + \frac{1}{2} \\ 0 & \text{otherwise.} \end{cases}$$

Consequently, the likelihood for a given sample $X_1, \ldots, X_n$ is given by $L(\theta) = 1$ if $\max(X_1, \ldots, X_n) - \frac{1}{2} \le \theta \le \min(X_1, \ldots, X_n) + \frac{1}{2}$ and 0 otherwise. Thus, all values of $L(\theta)$ in a certain interval equal 1, and the maximum is not unique: All values between $\max(X_1, \ldots, X_n) - \frac{1}{2}$ and $\min(X_1, \ldots, X_n) + \frac{1}{2}$ are MLE's of θ. $\square$

Despite the shortcomings above, the maximum likelihood estimates are reasonable in most cases appearing in practice, and this justifies studying the properties of this method. To illustrate what is involved here, let us continue with the example of an MLE in case of a fixed number of Bernoulli trials.

Example 12.6.15.   Let us consider, as in Example 12.6.10, five Bernoulli trials with probability of success θ. Clearly, the information as to which trials led to success and which led to failures is not essential: What matters is the total number S of successes. The likelihood function for given $S = s$ is

$$L(\theta, s) = \theta^s (1 - \theta)^{5-s}.$$

In the range of interest, $0 \le \theta \le 1$, the function $L(\theta, s)$ is maximized at $s/5$ if $s = 1, 2, 3$, or 4, as can be seen by taking derivatives and solving the equation $L'(\theta) = 0$. For $s = 0$ we have $L(\theta, 0) = (1 - \theta)^5$, and the maximum occurs at $\theta = 0$. Similarly, for $s = 5$ we have $L(\theta, 5) = \theta^5$ and the maximum occurs at $\theta = 1$. Thus, the MLE of θ, for given s, is $s/5$ (although for $s = 0$ and $s = 5$ this fact cannot be established by taking derivatives, since the extremum occurs at the boundary).

Graphically, the six likelihood functions corresponding to various numbers s of successes are presented on Figure 12.2. If we now regard s as a value of random variable S, then the likelihood function becomes random, say $L(\theta, S)$,

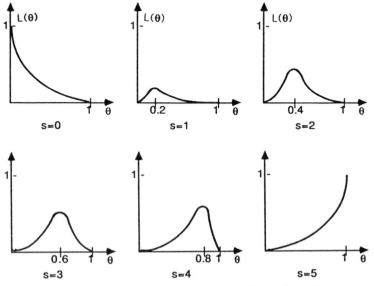

Figure 12.2   Likelihood function for five Bernoulli trials.

and also the corresponding point at which $L(\theta, S)$ attains its maximum, is random, equal to $S/5$. Thus, MLE becomes a random variable, dependent on the sample $(X_1, \ldots, X_n)$ through the statistic $S = X_1 + \cdots + X_5$.   □

In the sequel we regard the likelihood function as a random function of θ, the randomness being induced by the sample $X_1, \ldots, X_n$. The value at which this function is maximized will be denoted by $\hat{\theta}(X_1, \ldots, X_n)$ and will be taken as a maximum likelihood *estimator* of θ. Following tradition, we shall use the same symbol MLE for both the maximum likelihood estimator (random variable) and for a maximum likelihood estimate (its value for a particular sample).

Let us begin with stating some properties of maximum likelihood estimators. The first property, which makes these estimators quite convenient, is known as *invariance*. Consider the likelihood function $L(\theta) = L(\theta, \mathbf{x})$, where $\theta \in \Theta$, and consider first a mapping h of parameter space Θ onto some space Θ'. Suppose that we are interested in estimating the value $h(\theta)$ in the new parameter space Θ'.

Then the maximum likelihood estimators obey the following:

Invariance Principle. *If $\hat{\theta}$ is the MLE of parameter θ, then $h(\hat{\theta})$ is the MLE of parameter $h(\theta)$.*   □

Let us begin with the case when the mapping h is one-to-one. Then there exists the inverse mapping g of Θ' onto Θ, such that $\theta' = h(\theta)$ whenever $\theta = g(\theta')$. If the likelihood $L(\theta, \mathbf{x})$ is maximized at the point $\hat{\theta} = \hat{\theta}(\mathbf{x})$, then the function $L(g(\theta'), \mathbf{x})$ is maximized when $g(\theta') = \hat{\theta}(\mathbf{x})$, hence when $\theta' = h[\hat{\theta}(\mathbf{x})]$. Perhaps the most common application of the invariance principle is the fact that if $\hat{\theta}$ is MLE of the variance σ^2, then $\sqrt{\hat{\theta}}$ is the MLE of the standard deviation σ.

The invariance principle is valid in the case of multidimensional parameters and one-dimensional functions of such parameters. Before proceeding with the proof, it is worthwhile to point out the difference between the present case and the one-dimensional case. Consider then the following example.

Example 12.6.16. Suppose that the sample is taken from a distribution $f(x; \mu, \sigma)$, where μ and σ are the mean and standard deviation (we may consider f as normal, but it is not necessary). Thus, $\theta = (\mu, \sigma)$ is a two-dimensional parameter and assume that the parameter space is $\Theta = \{\mu > 0, \sigma > 0\}$. Suppose that we want to estimate the coefficient of variation $\nu = \sigma/\mu$.

The invariance principle asserts now that if $\hat{\mu}$ and $\hat{\sigma}$ are MLE's of μ and σ, then $\hat{\nu} = \hat{\sigma}/\hat{\mu}$ is the MLE of the coefficient of variation. It should be realized, however, that this conclusion does not follow from previous reasoning, since the function that maps $\theta = (\mu, \sigma)$ into $h(\theta) = \sigma/\mu$ is not one-to-one.   □

The argument showing that invariance principle for MLE's is also valid in the multidimensional case is as follows. Assume that $\theta = (\eta_1, \eta_2, \ldots, \eta_m)$ is an m-dimensional parameter, and let $h(\theta) = h(\eta_1, \eta_2, \ldots, \eta_m)$ be a function of θ to be estimated. We may find $m - 1$ functions $h_2(\theta) = h_2(\eta_1, \eta_2, \ldots, \eta_m), \ldots,$ $h_m(\theta) = h_m(\eta_1, \eta_2, \ldots, \eta_m)$ such that the vector

$$H(\theta) = (h(\theta), h_2(\theta), \ldots, h_m(\theta))$$

is a one-to-one mapping of m-dimensional parameter space Θ into a subset Θ' of m-dimensional space. By the preceding argument, if $\hat{\theta} = (\hat{\eta}_1, \ldots, \hat{\eta}_m)$ is the MLE of θ, then

$$H(\hat{\theta}) = (h(\hat{\eta}_1, \ldots, \hat{\eta}_m), h_2(\hat{\eta}_1, \ldots, \hat{\eta}_m), \ldots, h_m(\hat{\eta}_1, \ldots, \hat{\eta}_m))$$

is the MLE of $H(\theta)$. It follows therefore that $h(\hat{\eta}_1, \ldots, \hat{\eta}_m) = h(\hat{\theta})$ is the MLE of $h(\theta)$. $\qquad\qquad\qquad\qquad\qquad\qquad\qquad\qquad\qquad\qquad\qquad\qquad\qquad\square$

The second important property of maximum likelihood estimators is that they do not depend on the design of the experiment. To explain the issues involved here, we start from formulating the likelihood principle.

Likelihood Principle. *Consider two sets of data,* **x** *and* **y**, *obtained from the same population, although possibly according to different sampling plans. Let their likelihoods be* $L_1(\theta, \mathbf{x})$ *and* $L_2(\theta, \mathbf{y})$.

If $L_1(\theta, \mathbf{y})/L_2(\theta, \mathbf{y})$ *is constant, then both data sets* **x** *and* **y** *should lead to the same conclusion about* θ.

A consequence of this principle is that what matters are the ratios of likelihoods rather that likelihoods themselves. We shall explain by some examples how this principle makes the MLE's independent on the design of experiment.

Example 12.6.17. Assume that we want to estimate the probability of success θ in Bernoulli trials. One experimental design consists simply of fixing the number n of observations and recording the number of successes. If we observe x successes, the likelihood is $L_1(\theta, x) = \binom{n}{x} \theta^x (1 - \theta)^{n-x}$.

Suppose now that one decides to fix x and take observations until one records x successes. In this case, probability that the observations will end on the nth trial is given by negative binomial distribution, so that the likelihood is

$$L_2(\theta, n) = \binom{n-1}{x-1} \theta^x (1 - \theta)^{n-x}.$$

Observe that in the first case x was random and n fixed; in the second it is the other way around. Nevertheless, the ratio of these two likelihoods is $\binom{n}{x} / \binom{n-1}{x-1}$, which does not depend on θ. The MLE of θ is $\hat{\theta} = x/n$

in either case. This means, in particular, that the additional information that in the second case the last experiment led to success does not affect our estimate of θ. □

Example 12.6.18. Consider tennis players A and B, who from time to time play a match one against the other. Let us assume that the probability of A winning a set against B is p, and the results of sets are independent. (This is a somewhat oversimplified assumption; probability p may change with time, and results of sets within a match may be dependent. We shall, however, take this assumption as a starting point for analysis.) We want to estimate the probability p, which reflects relative strengths of players A and B. Assume that A and B are men.[*] Therefore, the probability that A will win in 3 sets is $\alpha_3 = p^3$. Probability that he will win in 4 sets is $\alpha_4 = 3p^3(1-p)$, since he must win the last set (probability p) and two of the first three sets [probability $\binom{3}{2} p^2(1-p)$]. Similarly, the probability of winning in 5 sets is $\alpha_5 = 6p^3(1-p)^2$. The analogous probabilities of winning by B in 3, 4, and 5 sets are β_3, β_4, and β_5, obtained by interchanging the roles of p and $1-p$.

Now, an inspection of the sport section of a newspaper for the last year (say) allows us to collect the data, in the form of six numbers $(a_3, a_4, a_5, b_3, b_4, b_5)$, where a_i is the number of matches won by A (against B) in i sets, and similarly for b_i.

The likelihood of the data is, letting $q = 1 - p$,

$$
\begin{aligned}
L(p; a_3, \ldots, b_5) &= \alpha_3^{a_3} \alpha_4^{a_4} \alpha_5^{a_5} \beta_3^{b_3} \beta_4^{b_4} \beta_5^{b_5} \\
&= (p^3)^{a_3} (3p^3 q)^{a_4} (6p^3 q^2)^{a_5} (q^3)^{b_3} (3q^3 p)^{b_4} (6q^3 p^2)^{b_5} \\
&= C p^{3a_3 + 3a_4 + 3a_5 + b_4 + 2b_5} q^{3b_3 + 3b_4 + 3b_5 + a_4 + 2a_5} \\
&= C p^a q^b,
\end{aligned}
$$

where C is a constant independent of p, while a and b are the total numbers of sets won only by A and by B, respectively. The logarithm of the likelihood is

$$
\log L = \log C + a \log p + b \log(1 - p)
$$

and we easily find the MLE of p:

$$
\hat{p} = \frac{a}{a+b} = \frac{\text{no. of sets won by } A}{\text{no. of sets played}}.
$$

Observe that the number of matches, equal to $a_3 + a_4 + a_5 + b_3 + b_4 + b_5$, does not appear in the estimator of p. The fact that the last set in each match must be won by the winner of the match is not relevant here: The estimator would

---

[*] This assumption is important for subsequent analysis: Men play "best out of five sets" while women play "best out of three sets."

be the same if we fixed in advance the number of sets (not matches!) to be played, and record the number of sets won by A. □

Still another important property of maximum likelihood estimators is that they can be obtained from samples in which some data are only partially observable. To illustrate this property, we shall find the MLE from a censored sample.

Example 12.6.19. Suppose that we want to estimate the parameter λ in exponential distribution with density $f(x, \lambda) = \lambda e^{-\lambda x}, x > 0$. For convenience of terminology, let us interpret the observations as lifetimes of some pieces of equipment. Typical experiment consists here of putting n copies of the equipment in question to test and observing the lifetimes $X_1, X_2, \ldots$. Suppose that the experiment is interrupted after some time T, and some tested items are still working. The data then have the form of a certain number of values $X_1, \ldots, X_m$ of observed lifetimes, while about the remaining $n - m$ items we know only that $X_{m+1} > T, \ldots, X_n > T$.

The likelihood of such a sample is obtained by multiplying the values of density function at points $X_1, \ldots, X_m$ and the probabilities $P\{X_i > T\} = e^{-\lambda T}$ for $i = m + 1, \ldots, n$. The likelihood is therefore

$$L(\lambda) = \lambda e^{-\lambda X_1} \cdots \lambda e^{-\lambda X_m} \left(e^{-\lambda T}\right)^{n-m}$$

$$= \lambda^m e^{-\lambda[X_1 + \ldots X_m + (n-m)T]}.$$

We now have

$$\log L(\lambda) = m \log \lambda - \lambda[X_1 + \cdots + X_m + (n - m)T]$$

and the maximum likelihood estimator is easily found to be

$$\hat{\lambda} = \frac{m}{\sum_{i=1}^{m} X_i + (n - m)T}. \tag{12.57}$$

□

To explain why it is permissible to combine the value of density function and the values of cdf's in the likelihood functions (even though the latter represent probabilities, so that they have to be between 0 and 1, while the former are not restricted in this way), one can recall that one can always multiply likelihood function by some quantity independent of the parameter, without affecting the information about the parameter contained in the likelihood.

One can now interpret the likelihood as the (approximate) probability of the sample, treated as a function of θ: With each data point about which we know its exact value (in the continuous case) we may associate the value $f(x, \theta) \, \Delta x$ as the (approximate) probability that X lies between x and $x + \Delta x$.

The terms involving $\Delta x_1, \Delta x_2, \ldots$ do not depend on θ and can be canceled in the likelihood function.

At the end of this presentation of maximum likelihood estimators, we shall discuss consistency, asymptotic unbiasedness, and asymptotic efficiency of MLE's. We shall not prove, or even formulate, the theorem, as it lies beyond the scope of this book. It is not difficult, however, to provide an informal explanation of why MLE's, under some regularity assumptions, have the properties asserted.

The main idea is as follows. In Section 12.5 we introduced the function $J(X, \theta) = \frac{\partial}{\partial \theta}[\log f(X, \theta)]$, used to define the information $I(\theta)$ and the concept of efficient estimators. If $\mathbf{X}^{(n)} = (X_1, \ldots, X_n)$ is a random vector from distribution $f(x, \theta)$, then $J(\mathbf{X}^{(n)}, \theta) = \sum_{i=1}^{n} J(X_i, \theta)$, with components $J(X_i, \theta)$ being iid random variables. We know from Section 12.5 that $E[J(X_i, \theta)] = 0$ and $\operatorname{Var} J(X_i, \theta) = I(\theta)$.

Consequently, $E[J(\mathbf{X}^{(n)}, \theta)] = 0$ and $\operatorname{Var}[J(\mathbf{X}^{(n)}, \theta)] = nI(\theta)$. By the central limit theorem of Lindeberg–Levy (Theorem 10.5.1) we have

$$\frac{J(\mathbf{X}^{(n)}, \theta)}{\sqrt{nI(\theta)}} \xrightarrow{d} Z, \tag{12.58}$$

where Z is the standard normal random variable.

Let $T = T(\mathbf{X}^{(n)})$ be an unbiased efficient estimator of θ, so that $E_\theta(T) = \theta$ and $\operatorname{Var}_\theta(T) = 1/nI(\theta)$. We know also by (12.48) that

$$T = \gamma_1(\theta) J(\mathbf{X}^{(n)}, \theta) + \gamma_2(\theta),$$

so that

$$E_\theta(T) = \gamma_1(\theta) E_\theta[J(\mathbf{X}^{(n)}, \theta)] + \gamma_2(\theta) = \gamma_2(\theta)$$

and

$$\operatorname{Var}_\theta(T) = \gamma_1^2(\theta) E[J(\mathbf{X}^{(n)}, \theta)]^2 = \gamma_1^2(\theta) \operatorname{Var}[J(\mathbf{X}^{(n)}, \theta)]$$
$$= \gamma_1^2(\theta) nI(\theta).$$

Consequently, we must have

$$\gamma_2(\theta) = \theta, \qquad \gamma_1^2(\theta) = \frac{1}{n^2 I^2(\theta)},$$

which implies that $\gamma_1(\theta) = \pm 1/nI(\theta)$. It follows that

$$T = \pm \frac{1}{nI(\theta)} J(\mathbf{X}^{(n)}, \theta) + \theta,$$

which means that

$$\frac{J(\mathbf{X}^{(n)}, \theta)}{\sqrt{nI(\theta)}} = \pm\sqrt{nI(\theta)}(T - \theta). \tag{12.59}$$

The left-hand side of (12.59) has asymptotic standard normal distribution by (12.58), and the same must be true for the right-hand side (under any choice of sign).

Consequently, we proved the following theorem.

Theorem 12.6.1. *If* $T = T(X_1, \ldots, X_n)$ *is an efficient estimator of* θ *based on a random sample from the distribution* $f(x, \theta)$, *then the random variable*

$$\sqrt{nI(\theta)}(T - \theta)$$

has asymptotically the standard normal distribution.

Actually, the same theorem holds if efficient estimators T are replaced by maximum likelihood estimators $\hat{\theta}_n$ of θ, provided that these estimators are obtained by solving the equation

$$\frac{\partial \log L(\theta, \mathbf{X}^{(n)})}{\partial \theta} = J(\mathbf{X}^{(n)}, \theta) = 0,$$

so that

$$J(\mathbf{X}^{(n)}, \hat{\theta}_n) = 0. \tag{12.60}$$

The proof requires imposing a number of technical assumptions and is quite complicated, but the main idea is simple. We first expand the function $J(\mathbf{X}^{(n)}, \theta)$ about the point $\hat{\theta}_n$ so that

$$J(\mathbf{X}^{(n)}, \theta) = J(\mathbf{X}^{(n)}, \hat{\theta}_n) + (\theta - \hat{\theta}_n) \left. \frac{\partial J}{\partial \theta} \right|_{\theta = \hat{\theta}_n}$$
$$+ \frac{1}{2}(\theta - \hat{\theta}_n)^2 \left. \frac{\partial^2 J}{\partial \theta^2} \right|_{\theta = \hat{\theta}_n} + \cdots.$$

One can show that the second order term converges to zero in probability, so that it can be neglected in the limit, and the same is true for the sum of all higher order terms. Using (12.60), we may then write

$$J(\mathbf{X}^{(n)}, \theta) \approx (\theta - \hat{\theta}_n) \left. \frac{\partial J}{\partial \theta} \right|_{\theta = \hat{\theta}_n}.$$

The left-hand side is now a sum of iid components, as in the proof of Theorem 12.6.1, and it obeys the central limit theorem. Finally, studying the behavior of the derivatives

$$\left. \frac{\partial J(\mathbf{X}^{(n)}, \theta)}{\partial \theta} \right|_{\theta = \hat{\theta}_n}$$

we establish that Theorem 12.6.1 also holds with T replaced by MLE's $\hat{\theta}_n$ (with the same norming constants). This shows that $\hat{\theta}_n$ are asymptotically unbiased, asymptotically efficient, and have asymptotic normal distribution with mean θ and variance $1/nI(\hat{\theta}_n)$. These properties make maximum likelihood estimators a preferred choice by many statisticians.

Bayesian Estimators

In both methods of constructing estimators presented thus far, there was no provision for using any information about θ that the statistician may have prior to taking the sample. This feature may or may not be desirable, depending on a number of factors, especially concerning the kind of information about θ one has, and also concerning some epistemological issues.

Let us begin with describing a possible setup, which is a special case of the setup considered thus far. As before, the observations $X_1, \ldots, X_n$ are assumed to be a random sample from the distribution $f(x, \theta)$, and the objective is to estimate the value of θ. This time, however, assume that θ itself is a value of some random variable Ξ with distribution $\pi(\theta)$. More precisely, we assume that θ is an element from some parameter space Θ, sampled according to probability distribution π. For instance, if Θ is the real line, and Ξ is a continuous random variable with density π, then $P\{\Xi \in A\} = \int_A \pi(\theta)\, d\theta$.

These assumptions describe the situation when the considered statistical problem of estimation is just one of a series of analogous estimation problems which differ by the value of the estimated parameter θ, and the variability of the parameter over similar estimation problems is described by density π, referred to as the *prior* density of θ.

Example 12.6.20.   A grocery store receives shipments of some merchandise (e.g., oranges). The quality of each shipment can be described by a certain parameter θ.

The examples of such parameters may be the percentage of spoiled fruits, the average diameter of a fruit in shipment, the average sugar content in the juice, and so on. Typically, it is not practicable to determine the value of θ for a given shipment exactly, because of cost, or because sampling is destructive, and so on. In such cases, one takes a sample from a given shipment and uses it to estimate θ for this shipment.

If the store has been buying the shipments from the same company for some time, it usually accumulated some data about the variability of values of θ from different shipments. It is natural, therefore, that such experience, in addition to the actual sample from this shipment, should be used in assessment of the parameter θ for a given shipment.

Example 12.6.21.   To some, the example with purchasing oranges may appear not serious enough to justify introducing a new concept (in fact, a new branch of statistics!). Consider, therefore, the case of a physician, who has to diagnose a patient. The data are various symptoms (or their lack), results of tests, and so on. The values of the parameter θ are different possible diagnoses, such as $\theta_1 =$ common cold, $\theta_2 =$ tuberculosis, $\theta_3 =$ AIDS, and so forth. The physician knows the distributions $f(\mathbf{x}, \theta)$ of the results $\mathbf{x} = (x_1, x_2, \ldots, x_n)$ of n specific tests for every θ (here x_1 may be temperature, x_2 and x_3 may be systolic and diastolic blood pressure, x_4 may be glucose level, etc.). At some time, the physician reaches a diagnosis [i.e., reaches the decision (perhaps tentative) about θ and begins treatment]. He may also order further tests if he is not sure about θ or if he still allows for the possibility that $\theta = \theta_2$ or $\theta = \theta_3$ (say). This problem differs from the problem of estimation of θ considered thus far only by the nature of the parameter space and the character of data $x_1, x_2, \ldots, x_n$ which may be qualitative. Observe at this point how the discrete and qualitative nature of parameter space affects the problem in the sense of definition of error: While in estimation the error could be quantified as the difference $T(\mathbf{x}) - \theta$, in medical diagnosis such quantification is not possible (e.g., consider the "error" of treating a patient with TB as having a common cold, or vice versa, treating common cold as TB).

It is clear, however, that the experience and intuition of the diagnosing physician play a crucial role here. In particular, the physician may have some idea as to the prior probabilities $\pi(\theta)$ of various diseases.   □

In the case under consideration, the information from the sample, say $\mathbf{x} = (x_1, \ldots, x_n)$, and prior information about the distribution of θ, in the form of π, can be combined into the conditional density of θ given $\mathbf{x}$, referred to as the *posterior* density. We have here

$$P\{\Xi \in A | \mathbf{X} = \mathbf{x}\} = \int_A \pi(\theta | \mathbf{x}) \, d\theta,$$

where

$$\pi(\theta | \mathbf{x}) = \frac{f_n(\mathbf{x}; \theta) \pi(\theta)}{\int_\Theta f_n(\mathbf{x}; u) \pi(u) \, du}. \tag{12.61}$$

Actually, the integral in the denominator of (12.61) is the reciprocal of the normalizing constant, and we often do not need to determine its value. What really matters is the fact that the posterior density of the parameter given the data is proportional to the likelihood of the data multiplied by the prior density:

$$\pi(\theta | \mathbf{x}) = C f_n(\mathbf{x}, \theta) \pi(\theta) = C L(\theta, \mathbf{x}) \pi(\theta).$$

Example 12.6.22. Suppose that we take observations from independent Bernoulli trials with the same probability of success θ, while θ follows a beta distribution with parameters α and β, that is,

$$\pi(\theta) = C \theta^{\alpha-1} (1 - \theta)^{\beta-1}, \qquad 0 \le \theta \le 1 \tag{12.62}$$

where C is the normalizing constant [given in (9.65)]. We have here, letting $s = \sum x_i$,

$$f_n(\mathbf{x}, \theta) = \prod_{i=1}^{n} \theta^{x_i}(1 - \theta)^{1-x_i} = \theta^s(1 - \theta)^{n-s}.$$

Consequently, the posterior density of θ given $\mathbf{x}$ is, up to the normalizing constant,

$$\pi(\theta|\mathbf{x}) \simeq K\theta^s(1 - \theta)^{n-s}\theta^{\alpha-1}\theta^{\beta-1}$$
$$= K\theta^{\alpha+s-1}(1 - \theta)^{n+\beta-s-1},$$

which we recognize as a beta distribution with parameters $\alpha + s$ and $n + \beta - s$.

Example 12.6.23. Assume that the observations $X_1, X_2, \ldots, X_n$ are independent, each with Poisson distribution with mean θ, while θ has a gamma distribution with parameters a and b, so that

$$\pi(\theta) = C\theta^{a-1}e^{-b\theta}, \qquad \theta > 0, \tag{12.63}$$

where C is the normalizing constant given in (9.51). Now the posterior density of θ is, again letting $s = \sum x_i$,

$$\pi(\theta|\mathbf{x}) = \left[\prod_{i=1}^{n}\left(\frac{\theta^{x_i}}{x_i!}e^{-\theta}\right)\right]C\theta^{a-1}e^{-b\theta}$$
$$= K\theta^{s+a-1}e^{-(n+b)\theta}. \tag{12.64}$$

Here the constant K incorporates also the factorials $x_1! \cdots x_n!$ (in general, we need to keep track only of the terms involving θ). We recognize (12.64) as a gamma distribution with shape parameter $a + s$ and scale parameter $b + n$.

Example 12.6.24. Suppose that $X_1, \ldots, X_n$ is a random sample from normal distribution $N(\theta, \sigma^2)$, with known σ^2. As regards θ, it has a prior normal distribution $N(\mu, \tau^2)$. Let us find the posterior distribution of θ given the sample.

We start with the likelihood function, which can be transformed as follows:

$$f_n(\mathbf{x}, \theta) = \frac{1}{\sigma^n(2\pi)^{n/2}}\exp\left\{-\frac{1}{2\sigma^2}\sum(X_i - \theta)^2\right\}$$
$$\simeq \exp\left\{-\frac{1}{2\sigma^2}\left[\sum(x_i - \bar{x}_n)^2 + n(\theta - \bar{x}_n)^2\right]\right\}$$
$$\simeq \exp\left\{-\frac{n}{2\sigma^2}(\theta - \bar{x}_n)^2\right\}.$$

Here the sign $\simeq$ means proportionality up to constants that do not depend on any unknown parameters [such constants dropped are $\sigma^{-n}(2\pi)^{-n/2}$ in the first proportionality sign, and $\exp\{-(n/2\sigma^2)(x_i - \bar{x}_n)^2\}$ in the second proportionality sign]. The posterior density $\pi(\theta|\mathbf{x})$ is therefore proportional to

$$\exp\left\{-\frac{n}{2\sigma^2}(\theta - \bar{x}_n)^2\right\} \times \exp\left\{-\frac{1}{2\tau^2}(\theta - \mu)^2\right\}.$$

The objective is now to separate the terms involving θ; all other terms will be absorbed in proportionality constant. Expanding the squares, and leaving only terms with θ, we obtain, after some algebra,

$$\pi(\theta|\mathbf{x}) \simeq \exp\left\{-\frac{1}{2q^2}(\theta - m)^2\right\},$$

where

$$m = \frac{\sigma^2}{\sigma^2 + n\tau^2}\mu + \frac{n\tau^2}{\sigma^2 + n\tau^2}\bar{x}_n \qquad (12.65)$$

and

$$q^2 = \frac{\sigma^2\tau^2}{\sigma^2 + n\tau^2}. \qquad (12.66)$$

This means that posterior density of θ is again normal with mean m and variance q^2 given by (12.65) and (12.66), respectively.

Example 12.6.25. Suppose finally that the observations $X_1, \ldots, X_n$ are independent normal $N(0, \theta^2)$, with θ again having a gamma distribution as in Example 12.6.23. We have

$$\pi(\theta|\mathbf{x}) = K\left(\prod_{i=1}^{n}\frac{1}{\theta\sqrt{2\pi}}e^{-\frac{x_i^2}{2\theta^2}}\right)\theta^{a-1}e^{-b\theta}$$

$$= K^*\theta^{a-n-1}e^{-b\theta - \frac{1}{2\theta^2}\sum x_i^2}. \qquad (12.67)$$

This time, the density (12.67) is not a member of any known family of distributions. $\square$

The situation in Examples 12.6.22 and 12.6.23 (as opposed to that in Example 12.6.25) is extremely convenient and therefore deserves special attention. Accordingly, we introduce the following definition:

Definition 12.6.2. A family $\mathcal{F}$ of distributions of parameter θ is said to provide *conjugate* priors for the distribution $f(x, \theta)$ of observations, if whenever the prior distribution of θ is in $\mathcal{F}$, the posterior density $\pi(\theta|\mathbf{x})$ also belongs to $\mathcal{F}$ for any sample $\mathbf{x}$. $\square$

The essence of Examples 12.6.23 and 12.6.25 can be formulated as the following theorems.

Theorem 12.6.2. *Beta densities are conjugate priors for the binomial distribution.*

Theorem 12.6.3. *Gamma densities are conjugate priors for the Poisson distribution.*

Theorem 12.6.4. *Normal densities are conjugate priors for normal distribution.*

We are now in a position to define Bayes estimators. The choice of the estimator depends, naturally, on the loss function. We present general theory and then develop it for the special case of the square loss function.

In general, as in Section 12.4, the penalty for accepting the value of the parameter as θ^*, while in fact it is θ, is expressed by the loss function $\mathcal{L}(\theta^*, \theta)$. In the present case we know the distribution of θ given the observed sample **x**. Thus, we should choose the value θ^* so as to minimize the expected loss

$$E\{\mathcal{L}(\theta^*, \Xi)|\mathbf{x}\} = \int_\Theta \mathcal{L}(\theta^*, \theta)\pi(\theta|\mathbf{x})\,d\theta. \tag{12.68}$$

Note that the left-hand side of (12.68) *does not depend on* θ: It is a function of θ^* and **x**. For each **x** we can therefore try to minimize it, that is, find a value $\theta^*(\mathbf{x})$ such that

$$E\{\mathcal{L}(\theta^*(\mathbf{x}), \Xi)|\mathbf{x}\} \leq E\{\mathcal{L}(\theta^{**}, \Xi)|\mathbf{x}\}$$

for every θ^{**} in Θ. Such a value $\theta^*(\mathbf{x})$ is the best choice of estimate given the sample **x**. When **x** varies according to the distribution $f(\mathbf{x}, \theta)$, we obtain a statistic $\theta^*(\mathbf{X})$. We may now introduce the following definition.

Definition 12.6.3. The statistic $\theta^*(X)$, where $\theta^*(x)$ minimizes the left-hand side of (12.68), is called the *Bayes estimator of θ for the loss function* $\mathcal{L}(\theta^*, \theta)$. □

Minimizing (12.68) in the general case may not be easy. However, when $\mathcal{L}(\theta^*, \theta) = (\theta^* - \theta)^2$, the minimizing value θ^* is well known: It is the mean of the posterior distribution of θ given **x**. As in the case of other estimators, we shall tacitly take the squared error as the loss function (unless we explicitly specify some other loss function). This is the customary choice. It is motivated primarily by the fact that the quadratic loss function allows further development of the theory, through use of well-known results in other branches of mathematics. Accordingly, we shall follow the convention, adopted generally in many statistical texts: *If the loss function is not explicitly specified, then the Bayes estimator of parameter θ is understood as the mean of posterior distribution:*

$$T(\mathbf{X}) = E\{\Xi|\mathbf{X}\} = \int \theta\,\pi(\theta|\mathbf{X})\,d\theta. \tag{12.69}$$

Example 12.6.26. If the observations $\mathbf{X} = (X_1, \ldots, X_n)$ form a random sample from Bernoulli distribution (so that $S = \sum_{i=1}^{n} X_i$ is binomial with parameters n and θ) and θ varies according to beta distribution with parameters α and β, then the Bayes estimator of θ is

$$T(X_1, \ldots, X_n) = \frac{\alpha + S}{\alpha + \beta + n}. \tag{12.70}$$

This follows from the fact, established in Example 12.6.22, that the posterior density of θ is again a beta density with parameters $\alpha + S$ and $\beta + n - S$. We know [see (9.68)] that the mean of a beta distribution with parameters a and b is $a/(a + b)$, which gives (12.70).

Example 12.6.27. Recalling Example 9.11.1, Laplace estimated the probability that the sun will rise tomorrow if it is known to have done so on the past n consecutive days. Thus, the observations form a binomial random variable S, with all trials resulting in success, so $S = n$. The Laplace's estimate $(n + 1)/(n + 2)$ of θ is the Bayes estimate for the uniform prior distribution (i.e., a beta prior distribution with $\alpha = \beta = 1$).

Example 12.6.28. In a similar way we can use Theorem 12.5.1, which asserts that gamma densities are conjugate priors for Poisson distribution. If $X_1, \ldots, X_n$ is a sample from Poisson distribution with parameter θ, and θ has prior density gamma with parameters a and b, then the Bayes estimator of θ is

$$T(X_1, \ldots, X_n) = \frac{a + S}{b + n}. \tag{12.71}$$

This follows from the fact that the mean of a gamma distribution is the ratio of the shape parameter to the scale parameter.

Example 12.6.29. From (12.65) in Example 12.6.24 it follows that the Bayes estimator in case of normal distribution $N(\theta, \sigma^2)$ with known σ^2 and normal prior $N(\mu, \tau^2)$ is

$$T = \frac{\sigma^2}{\sigma^2 + n\tau^2} \mu + \frac{n\tau^2}{\sigma^2 + n\tau^2} \overline{X}_n. \tag{12.72}$$

Thus, T is a weighted average of prior mean μ and sample mean $\overline{X}_n$. Let us analyze the behavior of this estimator under a change of various parameters.

First, with the increase of the sample size n, the estimator puts more and more weight on the sample size $\overline{X}_n$, and in the limit we have $T = \overline{X}_n$. In other words, ultimately the empirical evidence always prevails over prior convictions. Similar limiting conclusions (that $T = \overline{X}_n$) is obtained when $\sigma^2 \to 0$. This means that the more precise the individual information about θ carried by a single observation, the more weight should be attached to empirical evidence. Finally, the same limiting conclusion is obtained if $\tau^2 \to \infty$, that is, when the prior information is more vague (the prior variance τ^2 is interpretable as a measure of uncertainty of the prior information). Going toward the opposite extreme, when $\sigma^2 \to \infty$ or $\tau^2 \to 0$ (observations are subject to large errors, or prior knowledge has high certainty), the Bayes estimator of θ attaches more and more weight to the prior information that the mean of θ is μ. □

The last three examples illustrate the convenience of using conjugate prior distributions. This convenience lies basically in the fact that we have a simple formula for the Bayes estimator, and therefore we can quickly adjust our estimates when new observations became available. To illustrate this point, consider the following example.

Example 12.6.30. Continuing Example 12.6.28, assume that we first take m observations, with the total number of successes $S = X_1 + \cdots + X_m$ and then n observations, with the total number of successes $S' = X'_1 + \cdots + X'_n$. The prior density of θ before taking the first set of observations is beta with parameters α and β.

The situation at the end of the second series of observations may be regarded in two ways:

(a) We have the total of $S + S'$ successes in $m + n$ trials, so that the Bayes estimator is

$$\frac{\alpha + (S + S')}{\alpha + \beta + (m + n)}.$$

(b) We have the total of S' successes in the last n trials when the new prior distribution is posterior distribution after taking the first series of observation; that is, the posterior density is beta with parameters $\alpha_1 = \alpha + S$ and $\beta_1 = \beta + m - S$. The Bayes estimator is now

$$\frac{\alpha_1 + S'}{\alpha_1 + \beta_1 + n},$$

which agrees with that in (a). □

It is essential to realize that what we observed here is an instance of a general theorem on updating evidence (Theorem 4.4.2), which says that if we have two independent (sets of) observations $\mathbf{x}$ and $\mathbf{x}'$, we can use them either "at once," to determine the posterior distribution of θ given $(\mathbf{x}, \mathbf{x}')$, or we can do it in steps: Find the posterior density of θ given one data set (say $\mathbf{x}$) and use it as a new prior to find the posterior density of θ given $\mathbf{x}'$. The results will still be the same. Moreover, the order of choice between $\mathbf{x}$ and $\mathbf{x}'$ is irrelevant and need not coincide with the order in which the data $\mathbf{x}$ and $\mathbf{x}'$ were collected.

All the examples above concerned estimators for the squared error loss function. If the loss function is $\mathcal{L}(\theta^*, \theta) = |\theta^* - \theta|$, then the Bayes estimator is the *median* of the posterior distribution (see Theorem 8.7.3). For the normal case (Example 12.6.29) the mean and median coincide, hence (12.72) is also the Bayes estimator for the absolute error loss. For Examples 12.6.26, 12.6.27, and 12.6.28, the posterior distributions are beta or gamma, and their medians are not expressible by simple formulas in terms of the parameters.

It is clear that the class of *all* distributions on the parameter space Θ is always a set of conjugate priors for any distributions. Such a statement, however, is totally pointless. In fact, a class of conjugate priors is useful only if it leads to a simple formula for the posterior density, allowing explicit formulas for the means, hence for Bayesian estimators against mean square loss. In this perspective, the fact that a given class of distributions is a class of conjugate priors is just a mathematical curiosity, without much significance. Indeed, referring again to Example 12.6.30, we consider there a situation when observations $S = \sum X_i$ are binomial with parameter θ, while θ has a beta distribution. The first assumption is defensible: We often can *make S* have a binomial distribution by devising an appropriate sampling scheme. But the law that governs the variability of θ from case to case is beyond our control, and it is somewhat presumptuous to expect that nature is so kind to a statistician as to "choose" a specific law for θ (e.g., a beta distribution). This criticism, however, is too severe, for in most cases it is not nature but the statistician who chooses the prior distribution. The point is that the class of situations described above, where the parameter θ is a value of a random variable Ξ with distribution π, is rather restricted. Most often, the situation that a statistician faces is "one of a kind," characterized by an unknown value of θ, and it does not make sense to think of a prior distribution π of θ as telling us "how often" we had analogous statistical problems in which the value of the parameter satisfied an inequality of the form $a \le \theta \le b$.

There is a view, accepted by some statisticians, that even in such "one of a kind" situations it makes sense to consider and use the prior distribution of the parameter. Actually, the issue as to whether one allows using prior distributions even if they do not represent frequencies of occurrence of situations characterized by some values of θ is the issue lying in the center of the division of statisticians between Bayesians and non-Bayesians.

The philosophical points of this division are beyond the scope of this book. One may, however, consider the following two competing principles.

1. Statistical conclusions should depend on data only: When two statisticians analyze the same data using the same method, they should reach the same conclusion.
2. Statistical conclusions may depend on the experience, intuition and insight of the statistician who analyzes a given set of data.

Very roughly, statisticians who adhere strictly to principle 1 are non-Bayesians, while those who favor 2 are Bayesians. The latter use the prior distribution π as a means to express their knowledge, intuition, "hunches," and so forth. In this respect, having a class of conjugate priors is usually of great help, not merely in providing formulas for Bayes estimators, but primarily to express one's own prior experience or convictions, or to elicit information about the analyzed problem from the practitioners whom they advise. However, the first of the examples below shows something more fundamental, namely that the concept of prior distribution, reflecting one's personal judgments about a "one-of-a-kind" case, is sometimes unavoidable. Consider the following situation, which without this concept appears paradoxical.

Example 12.6.31. Imagine yourself playing the following game: There are two envelopes, each containing a check issued to your name. The amount on one check is twice as big as the amount on the other. You choose an envelope and inspect the check. At this moment you are offered an option to choose the other envelope. What should you do?

SOLUTION. The standard reasoning is as follows. Let a be the amount of the check that was in the first envelope you selected. The other envelope then contains a check for either $2a$ or $a/2$, each with the same probability. Thus, the expected outcome if you change your decision is

$$\frac{1}{2}(2a) + \frac{1}{2}\left(\frac{a}{2}\right) = \frac{5}{4}a > a, \tag{12.73}$$

which means that *you should always change* the envelope.

This seems paradoxical: Money appears to be created out of nowhere, just by changing the decision. The explanation lies in the fact that calculation (12.73) of the expected value uses probabilities 0.5 that the other envelope contains checks for $2a$ or $a/2$. In fact, one should use here the conditional probabilities, given the observed value a, and those in turn involve prior probabilities.

Indeed, suppose that $\pi(x)$ is the prior probability that the envelopes contain checks for the amounts x and $2x$. Assuming that one has no clairvoyant abilities, and hence always has the chance 0.5 of selecting an envelope with the lesser amount on a check, the unconditional probability of observing the amount a is $\frac{1}{2}\pi(\frac{a}{2}) + \frac{1}{2}\pi(a)$. Given the observed amount a, the probability that the other envelope is for the amount $a/2$ equals

$$\frac{\frac{1}{2}\pi(a/2)}{\frac{1}{2}\pi(a/2) + \frac{1}{2}\pi(a)}.$$

The analogous probability for the amount $2a$ is

$$\frac{\frac{1}{2}\pi(a)}{\frac{1}{2}\pi(a/2) + \frac{1}{2}\pi(a)}.$$

But the condition

$$\frac{\frac{1}{2}\pi(a/2)}{\frac{1}{2}\pi(a/2) + \frac{1}{2}\pi(a)} = \frac{\frac{1}{2}\pi(a)}{\frac{1}{2}\pi(a/2) + \frac{1}{2}\pi(a)}$$

implies $\pi(a/2) = \pi(a)$, and such a condition cannot be satisfied for all a (regardless of whether π represents a density or a discrete probability distribution).

This argument refers to some prior distribution π on the possible amount of the lesser check. Whether this distribution has any frequential interpretation (referring to analogous games played before), or not (if such a game is played only once), is irrelevant here. The only way to escape the paradox is to realize that everyone has some idea as to the probable range of values that may appear on checks in the game. If the check in hand shows a very "small" value, then the other is probably for a higher value. If the check in hand shows a very "high" value, the other is probably smaller. The concepts "very small," "very large," and "probably" are subjective here and refer to the player's idea about the distribution π.

This example provides a rather powerful argument for the need of Bayesian approach to statistical problems.

Example 12.6.32.   A piece of rock (e.g., taken from the moon) is sent to a laboratory to determine its radioactivity level. Assume that the measurement is simply Geiger count N_t, recorded for a certain time t. The role of the number of observations n is now played by observation time t. (If this feature should be confusing to the reader, assume simply that the experiment is run in such a way that Geiger counts $X_1, X_2, \dots$ are recorded, where X_i is the total count in the ith hour of observation. Then $N_t = X_1 + \cdots + X_t$ if t is an integer number of hours.)

We know that N_t has a Poisson distribution with mean θt, where θ is the radiation intensity expressed in the average number of emissions per hour. Suppose that the initial estimate of θ is needed urgently, so that observations can be carried out for only a limited time T. In other words, we have at our disposal a single observation of N_t. The likelihood function is

$$L(\theta) = \frac{(\theta T)^{N_T}}{(N_T)!} e^{-\theta T},$$

so that

$$\log L(\theta) = C + N_T \log \theta - \theta T$$

and the MLE is easily seen to be

$$\hat{\theta} = \frac{N_T}{T}$$

(note that this result concerns the MLE in case of observations running continuously in time; we no longer have sample of size n). To fix the ideas, assume that a total of 50 counts was recorded in $T = 100$ hours of observations, so that the MLE of θ is 0.5.

Suppose now that there are two physicists in the laboratory in question and each has his or her own ideas about what the radioactivity level θ of the specimen tested might be. Dr. Brown favors a certain theory of how the moon was formed, and how and when its rocks became initially radioactive. She thinks that moon rocks of the kind she analyzes should have their level of radioactivity θ about 1, but she is willing to incorporate a fair amount of uncertainty in her judgment, allowing the standard deviation of θ to be as much as 50% of the mean.

On the other hand, Dr. Smith favors a theory which predicts that the moon should have a uniformly low radioactivity, say $\theta = 0.4$ on the average, with standard deviation not exceeding 5% of the mean.

Let us see how these prior convictions will affect the estimates of θ for the specimen in question. We assume that the prior densities belong to the gamma family. If the parameters of a gamma distribution are a (shape) and b (scale), then the mean is a/b and the variance is a/b^2 [see (9.52) and (9.53)]. Consequently, the ratio of the standard deviation to the mean (the coefficient of variation) is

$$CV = \frac{\sqrt{a/b^2}}{a/b} = \frac{1}{\sqrt{a}}.$$

Thus, for Dr. Brown we have

$$CV = \frac{1}{\sqrt{a}} = 0.5, \qquad \frac{a}{b} = 1$$

which gives $a = b = 4$.

For Dr. Smith we have

$$CV = \frac{1}{\sqrt{a}} = 0.05, \qquad \frac{a}{b} = 0.4,$$

so that $a = 400, b = 1000$.

The Bayes estimator is (see Example 12.6.28) $(a + N_T)/(b + T)$, so that we have

$$\hat{\theta} = \frac{4 + 50}{4 + 100} = 0.519$$

for Dr. Brown, and

$$\hat{\theta} = \frac{400 + 50}{1000 + 100} = 0.409$$

for Dr. Smith.

We can see the effects of two factors. One is that MLE (in this case equal 0.5) is being "pulled" toward the mean of the prior distribution. The amount of pull depends on the variance of the prior distribution, reflecting the strength of conviction in the prior distribution (also possibly attributes such as pigheadedness, etc.): Dr. Smith, whose prior has much smaller variance, ends up with an estimate much closer to his prior mean than does Dr. Brown.

Example 12.6.33 (Are Birds Bayesians?). The example below concerns the behavior of certain species of birds in their search for food. The complete theory, including optimization aspects, has been developed by Green (1980). We present here only a fragment concerning assessment (estimation) by a bird.

Assume that the species in question can find food only in "patches," each consisting of a certain number of places where prey can be found. To fix the ideas, we shall think of birds that prey on worms living in pine cones. We assume that a cone has n "holes," each of them containing a prey with probability θ, independently of other holes. Thus, given θ, the number of prey in a cone is binomial with parameters n and θ. The bird has a fixed search pattern, so it does not search the same hole twice, and we assume that the prey cannot hide or escape to another hole during the search. Finally, we assume that θ, the rate of infestation of cones, varies between cones in such a way that θ has beta distribution with parameters α and β.

Let us consider now what could be the best strategy for a bird. First, it is reasonable to assume that the bird try to optimize* the rate of food intake per unit of time. Specifically, a strategy will tell the bird when to leave a cone and start searching the next one. The bird will optimize the rate of food intake, taking into account the average catch at a cone, average time spent on it, and average time of flying to another cone.

Intuitively, if α and β are large, the variability of θ is small [the variance of beta distribution is $\alpha\beta/(\alpha+\beta)^2(\alpha+\beta+1)$]. In such cases, all cones are about the same [variability between the cones is due mainly to variability in binomial distribution with parameters n and $\theta = \alpha/(\alpha+\beta)$]. In such cases, there is very little incentive to fly to the next cone before the current one is searched to the end.

However, if α and β are small, in particular if $\alpha < 1$ and $\beta < 1$, the variability of θ is large. Actually, in the latter case, the distribution of θ has

---

* We are using here a convenient terminology based on analogy with humans. In reality, birds cannot be expected to solve optimization problems, which require computers for humans (see Green, 1980). What we mean here is that in the process of evolution, any mutation toward a better search strategy is likely to become established, and it is possible that birds use strategy that is close to optimal.

density unbounded at $\theta = 0$ and at $\theta = 1$. This means that most cones will be of two categories only: very rich in prey, when θ is close to 1, and very poor in prey, when θ is close to 0. The optimal strategy is then to assess—as quickly as possible—whether θ is close to 1 or to 0, and behave accordingly, leaving the cone in the second case.

Now, if after searching k holes the bird found x worms, its assessment of θ is $(\alpha + x)/(\alpha + \beta + k)$ (i.e., equals the Bayes estimate of θ given x successes in k trials). Without going into detail, the optimal strategy specifies, for each k, the threshold for x, below which the bird ought to leave the cone.

There is an empirical problem to determine whether or not the birds follow the optimal strategy. The experiments involve the use of artificial cones and observation of birds' behavior depending on the findings in the holes searched previously. The preliminary results indicate that birds follow some kind of strategy of breaking or continuing the search of a cone depending on the outcome of the search so far. Whether this is an optimal strategy or not is unclear. But the truly fascinating problem here is: If the birds use a Bayesian strategy, are they capable of changing the prior distribution? In other words, are birds born with knowledge of a search strategy optimal against some fixed α and β characterizing infestation of cones prevalent in the last hundred years (say), or can an individual bird change its search strategy in years of higher infestation of its habitat? $\qquad\qquad\qquad\qquad\qquad\qquad\qquad\qquad\qquad\qquad\square$

Let us investigate briefly the problem of consistency of Bayes estimators.

First, let us observe that in the cases of estimators analyzed (of θ in the binomial case with beta prior, of θ in the Poisson case with gamma prior, and of μ in the normal distribution with normal prior), as the sample size increases, the effect of prior distribution decreases to zero. Indeed, letting S_n denote the binomial random variable with probability of success θ, the Bayes estimator of θ for beta prior satisfies

$$\hat{\theta} = \frac{\alpha + S_n}{\alpha + \beta + n} = \frac{(\alpha/n) + (S_n/n)}{[(\alpha + \beta)/n] + 1} \xrightarrow{P} \theta,$$

since S_n/n converges to θ in probability (and also almost surely).

Similarly, if $X_1, \ldots, X_n$ is a random sample from Poisson distribution with mean θ, and θ has a gamma distribution with parameters a and b, then the Bayes estimator of θ satisfies

$$\hat{\theta} = \frac{a + \sum_{j=1}^{n} X_j}{b + n} = \frac{(a/n) + \overline{X}_n}{(b/n) + 1} \xrightarrow{P} \theta,$$

by the law of large numbers, which asserts that the sample average $\overline{X}_n$ converges to θ in probability (and also almost surely).

Finally, an analogous conclusion for the normal case has already been obtained. We see therefore that Bayes estimators are consistent; in fact, they become increasingly closer to MLE's of the same parameters, regardless of the prior distribution. This property is true for Bayes estimators under some very general conditions, which we shall not state here.

One of the problems that a statistician faces quite often is determination of the sample size: "How big should n be in order that ...?". Various conditions may appear in place of dots; in the case of estimation, these conditions typically state the precision of the estimate in the sense of the probability of errors of a given size. In the case of Bayes estimators the situation is relatively simple. One of the criteria for determining the sample size may be expressed through the posterior distribution. In most typical cases one may wish to have the sample size which ensures that the posterior variance is below a certain minimum (of course, such criteria makes sense only if the estimator used is unbiased or have a small bias).

Example 12.6.34. In sampling from a Bernoulli distribution with unknown p, where p has prior distribution beta with parameters α and β, the posterior distribution of p given k successes in n trials is beta with parameters $\alpha + k$ and $\beta + n - k$. The Bayes estimator of p is $\hat{p} = (\alpha + k)/(\alpha + \beta + n)$, and since $E(k) = np$, we obtain

$$E(\hat{p}) = \frac{\alpha + np}{\alpha + \beta + n} = \frac{(\alpha/n) + p}{[(\alpha + \beta)/n] + 1},$$

which is close to p for large n.

Now, the variance of the posterior distribution is

$$\frac{(\alpha + k)(\beta + n - k)}{(\alpha + \beta + n)^2(\alpha + \beta + n + 1)}.$$

It depends on the value of k to be observed in the sample. However, the numerator $(\alpha + k)(\beta + n - k)$ is bounded by $(\alpha + n)(\beta + n)$, and to determine a sample size n which gives the posterior variance less than some value c, one needs to solve the inequality:

$$\frac{(\alpha + n)(\beta + n)}{(\alpha + \beta + n)^2(\alpha + \beta + n + 1)} \leq c.$$

This is a third-degree inequality in n, which can be solved numerically for each $\alpha, \beta,$ and c. □

Example 12.6.35. In case of sampling from normal distribution $N(\theta, \sigma^2)$, where θ is normal $N(\mu, \tau^2)$, the posterior variance of Bayes estimator (12.66) is $\sigma^2\tau^2/(\sigma^2 + n\tau^2)$. In this case the variance does not depend on observations, so if we want to have posterior variance less than c, we must have $\sigma^2\tau^2/(\sigma^2 + n\tau^2) \leq c$, hence $n \geq \sigma^2(\tau^2 - c)/\tau^2 c$. □

Least Squares Estimators

At the end of this section we present still one more "classical" method of estimation, dating back to Lagrange and Gauss. For an interesting account of discovery of this method, see Stigler (1986). In the present case, the basic setup will be different from the setup analyzed thus far. We shall assume that the data are collected under various conditions, where each of these conditions is characterized by a value of some variable (random or not), say u. We shall refer to u as to the *independent* variable. Next, we assume that for a given value of u one can take observations (one or perhaps more) of some random variable Y_u. Finally, we postulate that

$$Y_u = Q(u) + \epsilon,$$

where $Q(u)$ is a function of u and ϵ is a random variable (generally called *error*), satisfying the conditions

$$E(\epsilon) = 0, \qquad \mathrm{Var}(\epsilon) = \sigma^2 < \infty.$$

It follows that

$$E(Y_u) = Q(u), \qquad \mathrm{Var}(Y_u) = \sigma^2.$$

The function $Q(u)$ is usually called the *regression* of Y on u.

Example 12.6.36. One of the common situations falling under the foregoing scheme arises when we analyze a system that changes in time. Thus, u is the time of taking the observation of some attribute of the system, and $Q(u)$ is the expected value of the observed random variable Y_u, interpreted also as the "true" state of the system.

In some cases we may take only one observation at any given time u; in other cases we may have a number of observations made at the same time, say $Y_u^{(1)}, Y_u^{(2)}, \ldots, Y_u^{(k)}$. In such case, a typical assumption is that $Y_u^{(i)}$ are all independent.

Example 12.6.37. The variable u may also be some aspect that the experimenters can control. For example, a chemist may study the rate of a certain reaction in different temperatures. He then chooses the temperature u and observes one or more times the reaction rate for this u. Then he changes the temperature u and repeats the observations, and so on. The numbers of observations need not be the same for different values u. In general, the choice of distinct values $u_1, \ldots, u_k$ (as well as the choice of k) and the choice of numbers of observations $n_1, n_2, \ldots, n_k$ to be made at selected points $u_1, \ldots, u_k$ belong to the *design* of the experiment. We shall analyze later some problems of optimizing the design of experiments. □

Assume now that we have the experimental data for the design that can be described by the set of pairs

$$(u_i, n_i), \qquad i = 1, \ldots, k, \tag{12.74}$$

where u_i's are distinct values of variable u and n_i's are positive integers.

The data have the form of the array $\{y_{ij}, i = 1, \ldots, k, j = 1, \ldots, n_i, \}$, where y_{ij} is the value of random variable Y_{ij}, representing the jth observation for the value u_i of variable u. We assume that all random variables are independent, with

$$E(Y_{ij}) = Q(u_i), \qquad \text{Var}(Y_{ij}) = \sigma^2. \tag{12.75}$$

In most typical cases, the functional form of function Q (called the regression function) is postulated and assumed to depend on some parameters. For instance, in the case of a linear regression model, we assume that

$$Q(u) = \beta u + \alpha, \tag{12.76}$$

where β and α are the slope and intercept of the regression line. The problem is to estimate one or both parameters α and β.

In more complicated setups, we may postulate a quadratic regression

$$Q(u) = \gamma u^2 + \beta u + \alpha$$

or some other functional form of Q. Generally, we postulate that

$$Q(u) = \varphi(u; \theta_1, \theta_2, \ldots, \theta_r),$$

where φ is a known function and $\theta_1, \ldots, \theta_r$ are some parameters to be estimated. The method of *least squares* is based on the quadratic form

$$U(\theta_1, \ldots, \theta_k) = \sum_{i=1}^{k} \sum_{j=1}^{n_i} [y_{ij} - \varphi(u_i; \theta_1, \ldots, \theta_k)]^2. \tag{12.77}$$

The values $\hat{\theta}_1, \ldots, \hat{\theta}_k$ that minimize U given by (12.77) are called least square (LS) estimates of $\theta_1, \ldots, \theta_k$. As usual, those estimates, regarded as functions of the random variables $\{Y_{ij}\}$, are called *LS-estimators*.

The usual way of finding these estimators is by solving the set of simultaneous equations

$$\frac{\partial U}{\partial \theta_l} = 0, \qquad l = 1, \ldots, k, \tag{12.78}$$

which in the present case take the form

$$\sum_{j=1}^{n_i} [y_{ij} - \varphi(u_i; \theta_1, \ldots, \theta_k)] \sum_{l=1}^{k} \frac{\partial \varphi(u_l; \theta_1, \ldots, \theta_k)}{\partial \theta_l} = 0. \tag{12.79}$$

Example 12.6.38 (Linear Regression).   Suppose that $Q(u) = \alpha u + \beta$. In this case the algebra will simplify somewhat if we order all Y_{ij}'s into a single sequence $y_1, \ldots, y_n$, where $n = \sum_{i=1}^{k} n_i$, and relabel the corresponding values as $u_1, \ldots, u_n$ (so that now we have n values of variable u, but not all of them have to be distinct). We use the same notation without much danger of confusion. The quantity to be minimized becomes

$$U(\alpha, \beta) = \sum_{i=1}^{n} (y_j - \beta u_i - \alpha)^2. \tag{12.80}$$

Differentiating with respect to α and β and setting the derivatives equal to 0, we obtain

$$\beta \sum_{i=1}^{n} u_i^2 + \alpha \sum_{i=1}^{n} u_i = \sum_{i=1}^{n} u_i y_i$$

$$\beta \sum_{i=1}^{n} u_i + n\alpha = \sum_{i=1}^{n} y_i.$$

Letting

$$\frac{1}{n} \sum_{i=1}^{n} y_i = \bar{y}, \qquad \frac{1}{n} \sum_{i=1}^{n} u_i = \bar{u},$$

the solution can be written as

$$\hat{\beta} = \frac{\sum (u_i y_i - \bar{u} \cdot \bar{y})}{\sum (u_i - \bar{u})^2} \tag{12.81}$$

$$\hat{\alpha} = \bar{y} - \hat{\alpha}\bar{u}, \tag{12.82}$$

provided that $\sum (u_i - \bar{u})^2 > 0$. The latter condition is ensured if there are at least two distinct values of u used in the experiment, that is, if not all observations are made for the same value of u.
    An easy check shows that (12.81) does minimize function U.

PROBLEMS

12.6.1   Let $X_1, \ldots, X_n$ be a random sample from the gamma distribution with density $f(x; a, b) = C(a, b) x^{a-1} e^{-bx}, x > 0$, where C is the normalizing constant. Find the method-of-moment estimator of $\theta = (a, b)$, using the first two moments.

12.6.2   Continuing Problem 12.6.1, find method-of-moment estimators of a if b is known, and of b if a is known.

12.6.3 Let $X_1, \ldots, X_n$ be a random sample from the distribution uniform on the union of the two intervals: $[-2, -1]$ and $[0, \theta]$. Find:

  (i) The moment estimator of θ.

  (ii) The MLE of θ.

  (iii) The MLE of θ if positive X_i are recorded exactly, and negative X_i's can only be counted.

  (iv) The MLE of θ if X_i's cannot be observed, and one can only count the numbers of positive and negative ones.

12.6.4 (Bragging Tennis Player) As in Example 12.6.18, consider again the tennis players A and B who from time to time play matches against each other. The probability that A wins a set against B is p.

Assume now that we do not have complete data on all matches between A and B; we learn only of A's victories, so that we know the numbers a_3, a_4, and a_5 of matches won in 3, 4, and 5 sets (we do not even know whether or not he lost any matches with B).

Show that one can find the MLE of p, and find the MLE of the total number of matches and of the number of sets that A lost against B (do not attempt the algebraic solution).

12.6.5 Some phenomena (e.g., headway in traffic) are modeled by a distribution of a sum of a constant and an exponential random variable, so that the density of X has the form

$$f(x;a,b) = \begin{cases} 0 & x < b \\ ae^{-a(x-b)} & x \geq b, \end{cases}$$

where $a > 0$ and $b > 0$ are two parameters. Find:

  (i) The MLE of $\theta = (a, b)$, given a random sample $X_1, \ldots, X_n$.

  (ii) The method-of-moments estimator of θ.

12.6.6 A single observation of a random variable X with geometric distribution gives the result $X = k$. Find the MLE of p if:

  (i) X is the number of failures preceding the first success.

  (ii) X is the number of trials up to and including the first success.

12.6.7 Find the distribution of the MLE of the probability of success p based on two Bernoulli trials.

12.6.8 Let $X_1, \ldots, X_n$ be a random sample from a Poisson distribution with mean λ. Find the MLE of $P(X = 0)$.

12.6.9 Show that the family of gamma distributions provide conjugate priors for the exponential distribution. Determine the parameters of the posterior distribution.

12.6.10 Suppose that there were 15 successes in 24 Bernoulli trials. Find the MLE of the probability p of success if it is known that $p \leq \frac{1}{2}$.

12.6.11 Two independent Bernoulli trials resulted in a failure. What is the MLE of the probability of success p if it is known that p is at most $\frac{1}{4}$? Answer the same question if it is known that p exceeds $\frac{1}{4}$.

12.6.12 Find the MLE of parameter λ in a Poisson distribution given the sample $X_1, \ldots, X_n$:
  (i) Assuming that $X_1 + \cdots + X_n > 0$.
  (ii) When $X_1 + \cdots + X_n = 0$.

12.6.13 Find the moment estimator and MLE of the standard deviation of a Poisson distribution.

12.6.14 Let $X_1, \ldots, X_n$ be a random sample from normal distribution $N(\mu, \sigma^2)$ where μ is known. Find the MLE of standard deviation σ:
  (i) Directly.
  (ii) First finding the MLE of variance σ^2 and then using the invariance rule.

12.6.15 Find the MLE of σ^2 in the case of sample of size $n = 2$ from normal distribution, given only that the difference of the observations taken was 3.

12.6.16 Given the random sample $X_1, \ldots, X_n$ from the uniform distribution on the interval $[\theta_1, \theta_2]$, find the MLE of the mean of the distribution.

12.6.17 Observations are taken from uniform distribution on $[0, \theta]$. Let U_n be the number of observations (out of first n) which are less than 3. Find the MLE of θ.

12.6.18 Suppose that the median of 20 observations taken from a normal distribution with unknown mean and variance is 5, and that only one observation differed from the median by more than 3. Suggest an estimate of the probability that the next two observations will both be between 4 and 5.

12.6.19 Let $X_1, \ldots, X_n$ be a random sample from a log-normal distribution with parameters μ and σ^2 [this means that $\log X_i \sim N(\mu, \sigma^2)$]. Find the MLE of μ and σ^2.

12.6.20 Let $X_1, \ldots, X_m$ be a random sample from $N(\mu_1, \sigma^2)$, and let $Y_1, \ldots, Y_n$ be a random sample from $N(\mu_2, \sigma^2)$, with X_i's being independent from Y_j's. Here σ is common, while μ_1 and μ_2 may be different. Find the MLE of μ_1, μ_2, and σ^2.

12.6.21 Assume that the fraction X of the total time of study devoted to study of statistics in a certain class has the density $f(x; \theta) = (\theta + 1)x^\theta, 0 \le x \le 1$, where $\theta > -1$. Given a random sample $X_1, \ldots, X_n$ find:

(i) MLE of θ.

(ii) Method-of-moments estimator of θ.

(iii) Which classes can expect to have better overall results in a statistics course:

(a) Those with higher θ.

(b) Those with lower θ.

(c) Those with higher $|\theta|$.

(d) θ has nothing to do with grades.

12.7 NEWER METHODS OF OBTAINING ESTIMATORS

In this section we sketch briefly two of the contemporary approaches to the topic of estimation. One of the main underlying ideas is to obtain estimators that would be relatively unaffected by deviations from the assumed model. Such estimators are generally termed *robust*.

To grasp the main issues involved here, consider the problem of estimating the mean μ of a distribution. Then the sample mean $\overline{X}$ is an estimator of μ, and we know that $\overline{X}$ has a number of desirable properties, such as consistency, unbiasedness, and efficiency. These properties, however, are valid under specific assumptions about the underlying distribution. If the actual distribution differs in some respect from the assumed one, $\overline{X}$ may no longer have the asserted properties. In particular, a deviation from the model to which $\overline{X}$ is especially sensitive is called *contamination*. This concept is intended to describe the occurrence of *outliers* or *gross errors*, that is, observations that follow a distribution different from the one assumed in the model, the difference consisting of the mass of outlier distribution being concentrated around much larger (or smaller) values than those typically encountered in the model. An empirical situation of this kind may occur when a decimal point in the data is put at a wrong place. Formally, the model postulates the distribution $f(x, \theta)$, where θ is (say) the mean, while the actual sample is taken from the distribution

$$\varphi(x, \theta) = (1 - \epsilon)f(x, \theta) + \epsilon g(x), \qquad (12.83)$$

where most of the mass of $g(x)$ is far away from the range of "typical" values under $f(x, \theta)$.

It turns out that the sample mean $\overline{X}$ is highly sensitive to deviation from a model of the kind described in (12.83).Various suggestions as to a remedy led to the development of the theory of robust estimation. Here we mention just two of the approaches, L-estimators and M-estimators.

L-Estimators

L-estimators are constructed primarily for estimating the location parameter. Formally, θ will be called a *location parameter* if

$$f(x, \theta) = h(x - \theta)$$

for some probability density (or probability mass function) h.
   Similarly, θ is called a *scale parameter* if

$$f(x, \theta) = \frac{1}{\theta} h\left(\frac{x}{\theta}\right)$$

for some probability density (or probability mass function) h.
   The mean of a distribution may or may not be its location parameter. For instance, it is so in case of the normal distribution. However, in case of exponential distribution, the mean is a scale parameter, while in case of a general gamma distribution the mean is neither a scale nor a location parameter.
   Let $X_1, \ldots, X_n$ be a random sample from the distribution $f(x, \theta)$, and let

$$Y_{1:n} \leq Y_{2:n} \leq \cdots \leq Y_{n:n}$$

denote the order statistics of the sample. By an *L*-estimator we mean a statistic of the form

$$T = \sum_{k=1}^{n} \gamma_{n,k} Y_{k:n},$$

where $\gamma_{n,k}, k = 1, \ldots, n$ is a double array of coefficients. Thus, *L*-estimators are simply linear combinations of order statistics. The class of *L*-estimators contains many well-known estimators: Choosing $\gamma_{n,k} = 1/n$ for $k = 1, \ldots, n$ gives $T = \overline{X}$. The choice $\gamma_{n,1} = 1, \gamma_{n,k} = 0$ for $k \geq 2$, or $\gamma_{n,n} = 1, \gamma_{n,k} = 0$ for $k < n$ gives two extreme order statistics, $Y_{1:n} = \min(X_1, \ldots, X_n)$ and $Y_{n:n} = \max(X_1, \ldots, X_n)$. In a similar way, one can obtain any sample quantile. Choosing $\gamma_{n,[3n/4]+1} = 1, \gamma_{n,[n/4]+1} = -1, \gamma_{n,k} = 0$ for the remaining k, one can obtain the sample interquartile range, and so on.
   Perhaps the most important *L*-estimators are the trimmed and windsorized means, defined as follows.

Definition 12.7.1.   Let $0 < \alpha < \frac{1}{2}$. Then the *α-trimmed mean* is

$$U_n = \frac{1}{n - 2[n\alpha]} \sum_{k=[n\alpha]+1}^{n-[n\alpha]} Y_{k:n}, \tag{12.84}$$

while the *α-windsorized* mean is

$$V_n = \frac{1}{n} \left([n\alpha] Y_{[n\alpha]+1:n} + \sum_{k=[n\alpha]+1}^{n-[n\alpha]} Y_{k:n} + [n\alpha] Y_{n-[n\alpha]:n} \right). \qquad (12.85)$$
$\square$

Thus, α-trimming consists of rejecting from the sample the fraction α of lowest and the fraction α of largest observations, and taking the average of the remaining ones (the middle $1 - 2\alpha$ fraction of observations). On the other hand, α-windsorizing consists of replacing each observation in the lower α-fraction and upper α-fraction of the sample by the sample quantile of order α and $1 - \alpha$, respectively. The windsorized mean is then calculated as the mean of the windsorized sample.

It is clear that the purpose of trimming (or windsorizing) is to eliminate (or decrease) the effect of outliers in the sample.

The main objective of the theory of L-estimates, apart from establishing their asymptotic properties, is to define the notion of optimality and then to determine the optimal level α at which the mean should be trimmed or winsorized.

M-Estimators

Another class of estimators is obtained as follows. Let $h(x, u)$ be a function of two arguments. Given a sample $x_1, x_2, \ldots, x_n$ from the distribution $f(x, \theta)$, one takes as an estimator of θ the solution of the equation

$$\sum_{k=1}^{n} h(x_k, u) = 0. \qquad (12.86)$$

Such estimators are most often obtained by solving an approximate minimization problem. Suppose that we have a "distance" of some sort (not necessarily satisfying any conditions for metric), say $H(x, u)$. As an estimator of θ we choose a point u^* that minimizes the sum

$$\sum_{k=1}^{n} H(x_k, u), \qquad (12.87)$$

interpreted as the sum of distances from u to all sample points. In a sense, u^* is the point closest to the sample, with closeness being expressed by function H. Differentiating (12.87) with respect to u and setting the derivative equal to 0, we obtain equation (12.86) with $h(x, u) = \frac{\partial}{\partial u} H(x, u)$.

This formulation comprises two important classes of estimators, MLE's and least squares estimators. Indeed, if we define the function $H(x, u)$ as $-\log f(x; u)$, then $h(x, u) = -\frac{\partial}{\partial u} \log f(x, u)$ and the M-estimator corresponding to this choice is the maximum likelihood estimator (the minus sign is connected with the fact that the problem is now formulated as a minimization problem).

If we take appropriate functions H and h, we can also obtain different variants of least squares estimators. Similarly, trimmed or windsorized means can be obtained by appropriate choice of the functions H and h. For example, we may take a special form of function $H(x, u)$, namely $H(x, u) = H(x - u)$, for some function H of one argument. The M-estimator then minimizes the sum $\sum_{i=1}^{n} H(X_i - u)$. For $H(x) = x^2$ we have the simplest least squares estimator; if $H(x) = x^2$ for $|x| \leq k$ and $H(x) = k^2$ for $|x| > k$, we obtain a form of windsorized mean.

As with L-estimators, the main direction of research is to study the asymptotic properties of M-estimators under some general assumptions on functions H or h and distributions of X_i.

Resampling Methods

At the end of this section we shall mention briefly some recent methods for improving estimators. These methods require extensive computations, and are applicable now, with the availability of relatively inexpensive and fast calculations.

The first method is called *jackknifing*. We shall illustrate its application to reduce the bias of estimators. Let $X_1, X_2, \ldots, X_n$ be a random sample from a distribution with cdf F. The function F is unknown, and the objective is to estimate some parameter θ, which depends on F, so that $\theta = g(F)$. Some examples here are the mean $\int x \, dF(x)$, the variance $\int [x - E(X)]^2 \, dF(x)$, and so on. Generally, g is a function defined on a class of cdf's (e.g., for all cdf's whose mean exists, etc.), and values of g are real numbers.

We consider the estimators that are the sample counterparts of the corresponding parameters. To be precise, given the sample $x_1, x_2, \ldots, x_n$, we may consider the discrete probability distribution which gives mass $1/n$ to each of the points $x_1, x_2, \ldots, x_n$ (the points may repeat). Let F_n be the cdf of such a distribution; we now take $g(F_n)$ as the estimate of θ. When the points $x_1, \ldots, x_n$ are regarded as values of random variables $X_1, \ldots, X_n$, the values $g(F_n)$ become random, leading to an estimator $\hat{\theta}$, with values $g(F_n)$ for sample $(x_1, \ldots, x_n)$. In some sense, therefore, jackknifing concerns a form of method-of-analogy estimators.

Example 12.7.1. If θ is the mean, then $g(F_n) = \int x \, dF_n(x) = (1/n) \sum_{i=1}^{n} x_i$ is the sample mean $\bar{x}$. In case of estimation of the variance, the estimate is $g(F_n) = \int (x - \bar{x})^2 \, dF_n(x) = (1/n) \sum_{i=1}^{n} (x_i - \bar{x})^2 = s^2$, and so on. □

In general, the estimator $\hat{\theta}$ is biased, with bias

$$E_F(\hat{\theta}) - \theta = E_F[g(F_n)] - g(F). \tag{12.88}$$

We now remove point x_i from the sample and let $F_{n-1}^{(i)}$ be the cdf of the distribution which gives mass $1/(n-1)$ to each of the points of the reduced sample $x_1, \ldots, x_{i-1}, x_{i+1}, \ldots, x_n$. We may calculate the value of an estimator at $F_{n-1}^{(i)}$, obtaining

$$\hat{\theta}^{(i)} = g(F_{n-1}^{(i)}).$$

Let

$$\overline{\theta}^{(\cdot)} = \frac{1}{n} \sum_{i=1}^{n} \hat{\theta}^{(i)}$$

be the average of n estimates $\hat{\theta}^{(1)}, \hat{\theta}^{(2)}, \ldots$ obtained by successively removing elements $x_1, x_2, \ldots$ from the sample.

Now consider the expression

$$\hat{B} = (n-1)(\overline{\theta}^{(\cdot)} - \hat{\theta}). \tag{12.89}$$

It turns out that under assumption (12.93) below, $\hat{B}$ is a good approximation to the bias of $\hat{\theta}$, so that we may improve $\hat{\theta}$ by considering the *jackknife estimator* of θ, defined as

$$\hat{\theta}_J = \hat{\theta} - \hat{B} = \hat{\theta} - (n-1)(\overline{\theta}^{(\cdot)} - \hat{\theta})$$
$$= n\hat{\theta} - (n-1)\overline{\theta}^{(\cdot)}. \tag{12.90}$$

Assume that for each sample size m the expectation of the statistic $g(F_m) = \hat{\theta}(X_1, \ldots, X_m)$ can be expressed as

$$E[g(F_m)] = \theta + \frac{a_1}{m} + \frac{a_2}{m^2} + \cdots, \tag{12.91}$$

where $a_1, a_2, \ldots$ are constants depending on F only (not on m). Thus, the bias of $\hat{\theta}$ is on the order of $1/n$, where n is the sample size. Using the fact that

$$E(\overline{\theta}^{(\cdot)}) = \frac{1}{n} \sum_{i=1}^{n} E[\hat{\theta}^{(i)}] = E[g(F_{n-1})], \tag{12.92}$$

we may write, for the jackknife estimator $\hat{\theta}_J$ given by (12.90),

$$E(\hat{\theta}_J) = nE(\hat{\theta}) - (n-1)E(\overline{\theta}^{(\cdot)})$$
$$= n\left(\theta + \frac{a_1}{n} + \frac{a_2}{n^2} + \cdots\right) - (n-1)\left[\theta + \frac{a_1}{n-1} + \frac{a_2}{(n-1)^2} + \cdots\right]$$
$$= \theta - \frac{a_2}{n(n-1)} - \frac{(2n-1)a_3}{n^2(n-1)^2} - \cdots. \tag{12.93}$$

The cancellation of terms involving a_1 is possible because a_1 is the same for expressions for sample sizes n and $n - 1$. It also explains the use of the factor $n - 1$ in (12.89).We see, therefore, that the bias is now on the order of $1/n^2$, not $1/n$ as for estimator $\hat{\theta}$.

Example 12.7.2.   Let us return once more to the problem of estimating θ in the distribution uniform on $[0, \theta]$, and let us consider the estimator $T_1 = \max(Y_1, \ldots, Y_n) = Y_{n:n}$. The functional g is now of the form

$$g(F) = \inf\{t : F(t) = 1\} ;$$

hence for F uniform on $[0, \theta]$ we have $g(F) = \theta$. If $Y_1, \ldots, Y_n$ is the sample from F, and F_n is the empirical cdf, then $g(F_n) = Y_{n:n}$. We therefore have $E_F[g(F_n)] = E(Y_{n:n}) = [n/(n+1)]\theta$, and expansion (12.91) takes the form

$$E_F[g(F_m)] = \frac{m}{m+1}\theta = \frac{\theta}{1 + \frac{1}{m}}$$

$$= \theta - \frac{\theta}{m} + \frac{\theta}{m^2} - \frac{\theta}{m^3} + \cdots,$$

so that $a_i = (-1)^i - \theta$.

Next, the estimator $\hat{\theta}^{(i)}$, based on the sample with X_i removed, is equal to either $Y_{n:n}$ or $Y_{n-1:n}$, the latter being the case only if the largest element of the sample is removed. Consequently, the average of $\hat{\theta}^{(i)}$ is

$$\overline{\theta}^{(.)} = \frac{n-1}{n}Y_{n:n} + \frac{1}{n}Y_{n-1:n}.$$

The jackknife estimator (12.90) is therefore

$$\hat{\theta}_J = nY_{n:n} - (n-1)\overline{\theta}^{(.)}$$

$$= nY_{n:n} - (n-1)\left(\frac{n-1}{n}Y_{n:n} + \frac{1}{n}Y_{n-1:n}\right)$$

$$= \frac{2n-1}{n}Y_{n:n} - \frac{n-1}{n}Y_{n-1:n}.$$

Since $E(Y_{n:n}) = [n/(n+1)]\theta$ and $E(Y_{n-1:n}) = [(n-1)/(n+1)]\theta$ we obtain for the expectation of $\hat{\theta}_J$,

$$E(\hat{\theta}_J) = \frac{2n-1}{n} \cdot \frac{n}{n+1}\theta - \frac{n-1}{n} \cdot \frac{n-1}{n+1}\theta$$

$$= \frac{n^2+n-1}{n(n+1)}\theta = \theta - \frac{\theta}{n(n+1)}.$$

As we see, the jackknife estimator is biased, but the bias is much smaller than the bias of the estimator $Y_{n:n} = \max(Y_1, \ldots, Y_n)$.   □

The reader may ask here why we bother with the jackknife procedure if we can easily get an unbiased estimator by taking one of the "correction factors," such as in Example 12.2.4, for instance $[(n + 1)/n]Y_{n:n}$, and so on.

The point is that in most situations we do not know the bias: The coefficients $a_1, a_2, \ldots$ in expansion (12.91) are hard or impossible to determine outside of textbook cases. It is also rare that we can find the explicit expression for the jackknife estimator, as in the present case.

The jackknife method, as described above, involves systematic use of all possible samples obtained by removing one element from the original sample and recomputing the estimate. Such computations may be expensive. The situation can be remedied somewhat by replacing the systematic procedure with a random version, where one randomly selects the element to be deleted from the original sample. This procedure has the advantage that it can be stopped at any stage: it is not necessary that the jackknife sample be of the same size as the original sample.

This idea of "randomized jackknifing" leads in a natural way to a sampling scheme called *bootstrap*, introduced by Efron (1982). If $x_1, x_2, \ldots, x_n$ is the original sample, we can consider the bootstrap random variable say $X^{(B)}$, defined as

$$P\{X^{(B)} = x_i\} = \frac{1}{n}, \qquad i = 1, \ldots, n \qquad (12.94)$$

The bootstrap sample is simply a random sample from distribution (12.94) say $X_1^{(B)}, X_2^{(B)}, \ldots, X_m^{(B)}$. This means that $X_i^{(B)}$ are independent, with the same distribution (12.94). In practice, this means that we are sampling the elements of the original sample with replacement. The bootstrap samples are most often used to provide information about variances of estimators.

12.8 SUFFICIENCY

The considerations of this section are motivated by the following observation: Each of the estimators analyzed in this chapter was dependent on the random sample $X_1, \ldots, X_n$ through some statistic, which reduced the data to a single number. The examples of such reductions are statistics $\overline{X} = (1/n)(X_1 + \cdots + X_n)$, $X_{n:n} = \max(X_1, \ldots, X_n)$, and so on.

From a purely formal viewpoint, any such reduction involves some loss of information, simply because while the knowledge of the values $x_1, \ldots, x_n$ allows us to calculate the value of a statistic, the converse is not true: The knowledge of the value of a statistic does not allow us to determine the individual values $x_1, \ldots, x_n$, since different sets of observations may lead to the same value of statistic.

The concept of a sufficient statistic is intended to cover the situations when the loss of information due to reducing the data to the value of a statistic is not relevant for the purpose of estimation of parameter θ. Thus, the definition of sufficiency of a statistic is relative to a given parameter θ.

As before, we consider a random sample $X_1, \ldots, X_n$ from the distribution $f(x, \theta)$, where θ is known to belong to a parameter space Θ. Again, f can be either a density or a probability function of a discrete distribution.

Let $T = T(X_1, \ldots, X_n)$ be a statistic. The intention of the definition below is that the value $T = T(X_1, \ldots, X_n)$ conveys the same information about θ as the whole sample $(X_1, \ldots, X_n)$.

Definition 12.8.1.   The statistic T is said to be *sufficient* for θ if the conditional distribution of the vector $(X_1, \ldots, X_n)$ given $T = t$ does not depend on θ for any value t.   □

Before exploring the consequences of the definition to show that it does indeed correspond to the intended intuition, let us analyze some examples.

Example 12.8.1.   Consider two independent Bernoulli trials X_1, X_2, with probability of success p, and let $T = X_1 + X_2$. The conditional distribution of (X_1, X_2) given T is

$$P\{X_1 = 0, X_2 = 0 | T = 0\} = 1,$$
$$P\{X_1 = 1, X_2 = 1 | T = 2\} = 1.$$

Since $P(T = 1) = P(X_1 = 1, X_2 = 0) + P(X_1 = 0, X_2 = 1) = 2p(1 - p)$, for $T = 1$ we have

$$P\{X_1 = 1, X_2 = 0 | T = 1\} = \frac{P\{X_1 = 1, X_2 = 0\}}{P(T = 1)} = \frac{p(1 - p)}{2p(1 - p)} = \frac{1}{2}.$$

Thus, whatever value of T (0, 1, or 2), the conditional distribution of (X_1, X_2) does not depend on p.

Example 12.8.2.   Let X_t and X_s be the numbers of events in a Poisson process with intensity λ, observed between 0 and t, and between t and $t + s$. We will show that $U = X_t + X_s$ is a sufficient statistic for λ.

Indeed, using the fact that X_t and X_s are independent, we have for $k = 0, 1, \ldots, n$,

$$
\begin{aligned}
P\{X_t = k, X_s = n - k | U = n\} &= \frac{P\{X_t = k\} P\{X_s = n - k\}}{P\{U = n\}} \\[2mm]
&= \frac{\dfrac{(\lambda t)^k}{k!} e^{-\lambda t} \dfrac{(\lambda s)^{n-k}}{(n - k)!} e^{-\lambda s}}{\dfrac{[\lambda(t + s)]^n}{n!} e^{-\lambda(t+s)}} \\[2mm]
&= \binom{n}{k} \left(\frac{t}{t + s}\right)^k \left(\frac{s}{t + s}\right)^{n-k}.
\end{aligned}
$$

Thus, X_t given $U = n$ is binomial with parameters n and $t/(t+s)$, and does not depend on λ.   □

If the conditional distribution of observations $X_1, \ldots, X_n$ given the statistic $T = t$ does not depend on the parameter θ that we want to estimate, then (once we know that $T = t$) the additional knowledge of a particular configuration of $X_1, \ldots, X_n$ observed in the data does not help us in estimating θ. For instance, in Example 12.8.1, $T = 1$ means that we had one success and one failure. The fact that the first trial was a failure and the second was a success is of no additional help in estimating p.

Since the process of determining the conditional distribution of observations given the value of statistic T is sometimes cumbersome, it is desirable to have another method for verifying that a statistic is sufficient. This method is given by the following theorem, due to Neyman.

Theorem 12.8.1 (Factorization Criterion). *Let $X_1, \ldots, X_n$ be a random sample from the distribution $f(x, \theta)$, with $\theta \in \Theta$. A statistic $T = T(X_1, \ldots, X_n)$ is a sufficient statistic for θ if and only if for all $\mathbf{x} = (x_1, \ldots, x_n)$ and all $\theta \in \Theta$, the joint distribution*

$$f_n(\mathbf{x}, \theta) = f(x_1, \theta) \cdots f(x_n, \theta)$$

can be written as

$$f_n(\mathbf{x}, \theta) = u[T(\mathbf{x}), \theta]v(\mathbf{x}), \tag{12.95}$$

where u is a nonnegative function that depends on both θ and $\mathbf{x} = (x_1, \ldots, x_n)$, but dependence on $\mathbf{x}$ is only through the function T, and v is a function of $\mathbf{x} = (x_1, \ldots, x_n)$ that does not depend on θ.

Proof. We shall give the proof in the discrete case; the proof for the continuous case requires rather subtle considerations because densities are defined only up to sets of probability zero. Thus, we now have $f(x, \theta) = P_\theta\{X_i = x\}, i = 1, \ldots, n$.

Let $Q(t)$ be the set of all vectors $\mathbf{x} = (x_1, \ldots, x_n)$ such that $T(\mathbf{x}) = t$, so that

$$P_\theta(T = t) = \sum_{\mathbf{x} \in Q(t)} f_n(\mathbf{x}, \theta).$$

Suppose first that T is a sufficient statistic. Let $T = t$ and let us consider any point $\mathbf{x} \in Q(t)$. The conditional probability $P_\theta\{\mathbf{X} = \mathbf{x} | T = t\}$ does not depend on θ; we may call it $v(\mathbf{x})$, say. Letting $u(t, \theta) = P_\theta\{T = t\}$, we obtain, for any fixed $\mathbf{x} \in Q(t)$,

$$f_n(\mathbf{x}, \theta) = P_\theta\{\mathbf{X} = \mathbf{x}\} = P_\theta\{\mathbf{X} = \mathbf{x} | T = t\}P_\theta\{T = t\}$$
$$= v(\mathbf{x})u(t, \theta),$$

which is the factorization (12.95).

Conversely, suppose that $f_n(\mathbf{x}, \theta)$ satisfies formula (12.97) for some functions u and v. Let us fix t and $\theta \in \Theta$ and compute the conditional probability of a point $\mathbf{x}$ given $T = t$. Clearly, for $\mathbf{x} \notin Q(t)$ we have $P_\theta\{\mathbf{X} = \mathbf{x} | T = t\} = 0$. For $\mathbf{x} \in Q(t)$ we have

$$P_\theta\{\mathbf{X} = \mathbf{x} | T = t\} = \frac{P_\theta\{\mathbf{X} = \mathbf{x}\}}{P_\theta\{T = t\}} = \frac{f_n(\mathbf{x}, \theta)}{\sum_{\mathbf{y} \in Q(t)} f_n(\mathbf{y}, \theta)}$$

$$= \frac{u(t, \theta)v(\mathbf{x})}{\sum_{\mathbf{y} \in Q(t)} u(t, \theta)v(\mathbf{y})} = \frac{v(\mathbf{x})}{\sum_{\mathbf{y} \in Q(t)} v(\mathbf{y})}. \qquad \square$$

We shall now illustrate how Theorem 12.8.1 allows us to find sufficient statistics in many families of distributions.

Example 12.8.3. Consider again Bernoulli trials with probability p of success. We have here, letting $t = \sum x_i$,

$$f_n(x_1, \ldots, x_n, p) = \prod_{i=1}^{n} p^{x_i}(1 - p)^{1-x_i}$$

$$= p^t(1 - p)^{n-t}$$

$$= \left(\frac{p}{1 - p}\right)^t (1 - p)^n.$$

In this case, a sufficient statistic is the total number of successes $T = \sum X_i$, while the function $v(x_1, \ldots, x_n)$ is identically 1.

Example 12.8.4. In Poisson distribution, we have, letting $t = \sum x_i$,

$$f_n(x_1, \ldots, x_n, \lambda) = \prod_{i=1}^{n} \frac{\lambda^{x_i}}{x_i!} e^{-\lambda}$$

$$= \lambda^t e^{-n\lambda} \frac{1}{x_1! \cdots x_n!},$$

which again shows that $T = \sum X_i$ is a sufficient statistic. Here $u(t, \lambda) = \lambda^t e^{-n\lambda}$, while $v(x_1, \ldots, x_n) = (x_1! \cdots x_n!)^{-1}$.

Example 12.8.5. Now let $(x_1, \ldots, x_n)$ be observations from a normal distribution $N(\mu, \sigma^2)$ where σ^2 is known. Then the joint density of the sample can be written as

$$f_n(x_1, \ldots, x_n, \mu) = \frac{1}{\sigma^n(2\pi)^{n/2}} \exp\left\{-\frac{1}{2\sigma^2} \sum(x_i - \mu)^2\right\}$$

$$= \frac{1}{\sigma^n(2\pi)^{n/2}} \exp\left\{-\frac{1}{2\sigma^2}\left(\sum x_i^2 - 2\mu \sum x_i + n\mu^2\right)\right\}$$

$$= \exp\left\{\frac{\mu}{\sigma^2} \sum x_i - \frac{n\mu^2}{2\sigma^2}\right\} v(x_1, \ldots, x_n).$$

Thus, $\sum X_i$ is a sufficient statistic for μ, with

$$v(x_1, \ldots, x_n) = \frac{\exp\{-(1/2\sigma^2)\sum x_i^2\}}{\sigma^n(2\pi)^{n/2}}.$$

Example 12.8.6. Suppose now that μ is known while the unknown parameter is σ^2. We can now take $v(x_1, \ldots, x_n) \equiv 1$, and the sufficient statistic is $\sum(X_i - \mu)^2$.

Example 12.8.7. Consider finally the case of sampling from a uniform distribution on $[0, \theta]$. Here the joint density of the sample is

$$f(x_1, \ldots, x_n, \theta) = \begin{cases} \dfrac{1}{\theta^n} & \text{if } \max(x_1, \ldots, x_n) \le \theta \\ 0 & \text{otherwise.} \end{cases}$$

This function can be written as

$$f(x_1, \ldots, x_n) = \frac{1}{\theta^n} I(\theta, \max(x_1, \ldots, x_n)),$$

where $I(\theta, c) = 1$ if $\theta \ge c$ and 0 if $\theta < c$. This exhibits the function $u(\theta, T)$ [with $v(x_1, \ldots, x_n) \equiv 1$] and shows that $\max(X_1, \ldots, X_n)$ is a sufficient statistic.
□

Let us observe that sufficient statistics are not unique: Every one-to-one transformations of a sufficient statistic is again a sufficient statistic. For instance, in Example 12.8.5, an alternative sufficient statistic to $\sum X_i$ is $\overline{X} = (1/n)\sum X_i$.

The concept of sufficient statistic can be generalized in an obvious way to cover the case when we have a multidimensional sufficient statistic, better known as a set of jointly sufficient statistics. The point is that there are cases where it is not convenient to reduce the information about a parameter to a single number. This situation occurs typically when the parameter Θ is multidimensional. The intuitions motivating the concept are the same as in one-dimensional case.

We shall now formulate the definition using the factorization theorem, that is, using the assertion of the theorem as a basis for the definition.

Definition 12.8.2. The k–dimensional statistic $(T_1, \ldots, T_k)$ is *jointly sufficient* for θ if the density $f_n(\mathbf{x}, \theta)$ satisfies the following condition: For every $\mathbf{x} = (x_1, \ldots, x_n)$ and $\theta \in \Theta$, we have

$$f_n(\mathbf{x}, \theta) = u[T_1(\mathbf{x}), \ldots, T_k(\mathbf{x}), \theta]v(\mathbf{x}),$$

where the nonnegative function $v(\mathbf{x})$ does not depend on θ, while the function u depends on θ, and depends on $\mathbf{x}$ only through the values of statistics $T_1, \ldots, T_k$.
□

Example 12.8.8. Let $(X_1, \ldots, X_k)$ be a random sample from the normal distribution $N(\mu, \sigma^2)$, where now both μ and σ^2 are unknown, so that $\theta = (\mu, \sigma^2)$. We then have

$$f_n(\mathbf{x}, \theta) = \frac{1}{(2\pi\sigma^2)^{n/2}} \exp\left\{ -\frac{1}{2\sigma^2} \sum (x_i - \mu)^2 \right\}$$

$$= \frac{1}{(2\pi)^{n/2}} (\sigma^2)^{-n/2} \exp\left\{ -\frac{\sum x_i^2}{2\sigma^2} + \frac{\mu}{\sigma^2} \sum x_i - \frac{n\mu^2}{2\sigma^2} \right\},$$

and we see that the pair of statistics $T_1 = \sum_{i=1}^n X_i$, $T_2 = \sum_{i=1}^n X_i^2$ is jointly sufficient for θ. $\square$

It is intuitively clear (and can also be shown formally) that if $(T_1, \ldots, T_k)$ are jointly sufficient for θ, and $(T_1^*, \ldots, T_k^*)$ are obtained from $(T_1, \ldots, T_k)$ by a one-to-one transformation, then $T_1^*, \ldots, T_k^*$ are also jointly sufficient for θ. Thus, another pair of jointly sufficient statistics in the case of normal distributions from Example 12.8.8 is $\overline{X} = (1/n)\sum_{i=1}^n X_i$ and $S^2 = (1/n)\sum_{i=1}^n (X_i - \overline{X})^2$. This can also be seen from the representation

$$f_n(\mathbf{x}, \theta) = (2\pi)^{-n/2} (\sigma^2)^{-n/2} \exp\left\{ -\frac{1}{2\sigma^2} \sum (x_i - \mu)^2 \right\}$$

$$= (2\pi)^{-n/2} (\sigma^2)^{-n/2} \exp\left\{ -\frac{1}{2\sigma^2} \sum (x_i - \overline{x})^2 - \frac{n(\overline{x} - \mu)^2}{2\sigma^2} \right\}$$

$$= (2\pi)^{-n/2} (\sigma^2)^{-n/2} \exp\left\{ -\frac{n}{2\sigma^2} [s^2 - (\overline{x} - \mu)^2] \right\}.$$

It is easily noted that the sample $(x_1, \ldots, x_n)$ is always jointly sufficient: It suffices to write formally $T_i(x_1, \ldots, x_n) = x_i$ for $i = 0, 1, \ldots, n$. This is true, but useless. It is also true that the order statistics $(X_{1:n}, X_{2:n}, \ldots, X_{n:n})$ are jointly sufficient. To see this formally, observe that

$$f_n(\mathbf{x}, \theta) = \prod_{j=1}^n f(x_j, \theta) = \prod_{i=1}^n f(x_{i:n}, \theta),$$

since the factors of the last product are a permutation of the factors of the middle product.

Thus, a set of jointly sufficient statistics always exists, and it is natural to look for the maximal reduction of data that retains sufficiency. Here reduction is meant both in the sense of reducing the dimensionality and in the sense of using transformations that are not one-to-one, such as squaring.

The definition of maximal reduction, hence minimality of sufficient statistic, is as follows.

Definition 12.8.3.   A sufficient statistic T is called *minimal* if T is a function of any other sufficient statistic.   □

The definition covers the idea of maximal reduction of the data, leaving sufficiency intact. Indeed, suppose that T is minimal in the sense of Definition 12.8.3 and can still be reduced. This means that there exists a function h which is not one-to-one, such that $U = h(T)$ is a sufficient statistic. Then T is not a function of U, which gives a contradiction.

Example 12.8.9.   Consider one observation from the distribution $N(0, \sigma^2)$. Its density is

$$f(x, \sigma^2) = \frac{1}{\sigma\sqrt{2\pi}} e^{-x^2/2\sigma^2},$$

and it is clear that $T(x) = x$ is a sufficient statistic for σ. However, the statistic $T_1(x) = |x|$ is also sufficient, even though the transformation $|x|$ is not one-to-one. Actually, $|x|$ is a minimal sufficient statistic, and so is $T_2(x) = x^2$, and so on.   □

Definition 12.8.3 extends to the case of minimal jointly sufficient statistics, often called a minimal *set* of jointly sufficient statistics.

Definition 12.8.4.   A set $\{T_1, \ldots, T_k\}$ of jointly sufficient statistics is called *minimal* if it is a function of any other set of jointly sufficient statistics.   □

The intuition behind this definition is as follows. If $(T_1, \ldots, T_m)$ and $(T_1', \ldots, T_k')$ are both jointly sufficient, and the second set is a function of the first, that is, $T_1' = \varphi_1(T_1, \ldots, T_m)$, $T_2' = \varphi_2(T_1, \ldots, T_m), \ldots, T_k' = \varphi_k(T_1, \ldots, T_m)$ for some functions $\varphi_1, \ldots, \varphi_k$, then we may say that

$$(T_1, \ldots, T_m) \succeq (T_1', \ldots, T_k'),$$

where $\succeq$ is an ordering among jointly sufficient sets of statistics. The minimal jointly sufficient statistics are at the "bottom" of this ordering, while the fact that $(Y_{1:n}, \ldots, Y_{n:n})$ is always a set of jointly sufficient statistics means that the set of statistics being ordered by relation $\succeq$ is not empty. Example 12.8.8 shows that the minimal sufficient set of statistics is not unique; however, all these minimal sufficient sets are obtained as one-to-one functions of other minimal sufficient sets.

Example 12.8.10.   Let $X_1, \ldots, X_n$ be random sample from the distribution uniform on $[-\theta, \theta]$, so that

$$f(x; \theta) = \begin{cases} \dfrac{1}{2\theta} & \text{if } -\theta \le x \le \theta \\ 0 & \text{otherwise.} \end{cases}$$

Proceeding as in Example 12.8.7, we easily show that the pair of extreme order statistics $(X_{1:n}, X_{n:n})$ is jointly sufficient for θ. However, this is not a minimal set: The statistic $T(X_{1:n}, X_{n:n}) = \max(|X_{1:n}|, |X_{n:n}|)$ is a single sufficient statistic for θ. $\square$

Typically, the dimensionality of the minimal set of jointly sufficient statistics equals the number of dimensions of the parameter θ. Thus, in case of normal distribution with parameter being the pair (μ, σ^2), minimal sets of jointly sufficient statistics consist of two statistics (e.g., $\sum X_i$ and $\sum X_i^2$). However, there exist distributions that depend on a single parameter, and yet, the only jointly sufficient statistics (hence the minimal sets of jointly sufficient statistics) are the order statistics $(Y_{1:n}, \ldots, Y_{n:n})$. This is true, for instance, for a Cauchy distribution with density

$$f(x; \theta) = \frac{1}{\pi} \frac{1}{1 + (x - \theta)^2}.$$

We omit the proof of this fact.

Consider now the role that sufficient statistics play in estimation. Observe first that the likelihood function is

$$L(\theta, \mathbf{x}) = v(\mathbf{x}) u[T_1(\mathbf{x}), \ldots, T_k(\mathbf{x}), \theta], \tag{12.96}$$

where $(T_1, \ldots, T_k)$ is a set of jointly sufficient statistics (minimal or not). Since the factor $v(\mathbf{x})$ plays no role in the determination of θ at which the likelihood attains its maximum, we may expect that MLE $\hat{\theta}$ of θ will be a function of $T_1, \ldots, T_k$. Since the same kind of representation of the likelihood is valid if $T_1, \ldots, T_k$ form a minimal sufficient set, we may expect that $\hat{\theta}$ will be a function of minimal jointly sufficient statistics.

This argument is valid provided that the MLE is unique (which is usually the case). When the MLE is not unique (for such situations, see Example 12.6.14) one cannot claim that they all are functions of sufficient statistics, but one of them will be.

We can therefore state the following theorem.

Theorem 12.8.2. *If the MLE of θ exists and is unique, then it is a function of a minimal sufficient set of statistics. If the MLE is not unique, then there exists a MLE that is a function of minimal jointly sufficient statistics.*

A very similar situation exists in the case of Bayes estimators. Indeed, if $T_1, \ldots, T_k$ are jointly sufficient for θ, then the posterior density of θ given $\mathbf{x}$ is, for prior density $\pi(\theta)$,

$$\pi(\theta|\mathbf{x}) = \frac{f_n(\mathbf{x}, \theta)\pi(\theta)}{\int f_n(\mathbf{x}, \eta)\pi(\eta)d\eta}$$

$$= \frac{v(\mathbf{x})u(T_1, \ldots, T_k, \theta)\pi(\theta)}{\int v(\mathbf{x})u(T_1, \ldots, T_k, \eta)\pi(\eta)d\eta}$$

$$= \frac{u(T_1, \ldots, T_k, \theta)\pi(\theta)}{H(T_1, \ldots, T_k)}$$

since the factor $v(\mathbf{x})$ cancels; here H is some function of statistics $T_1, \ldots, T_k$. Consequently, the Bayes estimator, equal to the mean of posterior distribution, depends on $\mathbf{x}$ only through statistics $T_1, \ldots, T_k$. This proves the following theorem.

Theorem 12.8.3.   *The Bayes estimator of θ is a function of the minimal jointly sufficient statistics for θ.*

The main importance of sufficient statistics is to a large extent connected with the following theorem. We formulate it for the case of a single sufficient statistic T. However, this theorem can also be formulated when T represents minimal jointly sufficient statistics.

Theorem 12.8.4 (Rao–Blackwell).   *Assume that in a family of distributions $\{f(x, \theta), \theta \in \Theta\}$ a sufficient statistic T exists, and let W be an estimator of $g(\theta)$. Let*

$$W^*(T) = E(W|T). \tag{12.97}$$

Then W^* is an estimator of $g(\theta)$, such that its risk

$$R(W^*, \theta) = E\{[W^* - g(\theta)]^2\}$$

satisfies the condition

$$R(W^*, \theta) \leq R(W, \theta) \tag{12.98}$$

for all $\theta \in \Theta$. Moreover, the inequality in (12.98) is strict for some θ unless W is an unbiased estimator of $g(\theta)$ and W is a function of the sufficient statistic T.   □

The significance of this theorem can best be explained as follows. Suppose that in estimating $g(\theta)$ we use an estimator W that is not a function of a sufficient statistic T. Then there exists an estimator W^* [given by (12.97)] that is a function of the sufficient statistic T, and whose risk is better than the risk of W. In other words, any estimator that is not based on a sufficient statistic can be improved and hence is not admissible (see Definition 12.4.1).

To prove Theorem 12.8.4 we need to show first that formula (12.97) indeed defines an estimator of $g(\theta)$, that is, that the left-hand side does not depend on θ but only on the sample $X_1, \ldots, X_n$. We shall also show that this dependence is through the value of T only, so that the estimator obtained is a function of the sufficient statistic T. The second part of the proof will consist of showing that inequality (12.98) holds for every θ.

To show that formula (12.97) indeed defines an estimator, we shall carry the argument in the discrete case, with $f(x, \theta) = P\{X = x|\theta\}$. Let t be a possible value of the sufficient statistic, and let $Q(t) = \{x = (x_1, \ldots, x_n) : T(\mathbf{x}) = t\}$.

For any fixed t and $\mathbf{x} \in Q(t)$, the conditional distribution of $\mathbf{X}$ given $T = t$ is, using the factorization theorem,

$$P\{\mathbf{X} = x|T = t, \theta\} = \frac{f_n(\mathbf{x}, \theta)}{\sum_{\mathbf{y} \in Q(t)} f_n(\mathbf{y}, \theta)}$$

$$= \frac{v(\mathbf{x})u(t, \theta)}{\sum_{\mathbf{y} \in Q(t)} v(\mathbf{y})u(t, \theta)} = \frac{v(\mathbf{x})}{\sum_{\mathbf{y} \in Q(t)} v(\mathbf{y})},$$

which is independent of θ. On the other hand, $P\{\mathbf{X} = \mathbf{x}|T = t, \theta\} = 0$ if $\mathbf{x} \notin Q(t)$, which is also independent of θ. Consequently, we have

$$E_\theta[W|T = t] = \sum_{\mathbf{x} \in Q(t)} W(\mathbf{x})P\{\mathbf{X} = \mathbf{x}|T = t, \theta\}$$

$$= \frac{\sum_{\mathbf{x} \in Q(t)} W(\mathbf{x})v(\mathbf{x})}{\sum_{\mathbf{y} \in Q(t)} v(\mathbf{y})}, \tag{12.99}$$

which is independent of θ, as asserted.

The estimator $E_\theta[W|T]$ depends on the actual sample observed in the following sense. After observing the sample, say $\mathbf{X} = \mathbf{x}^*$, we compute $T(\mathbf{x}^*) = t$ and take the value $E_\theta[W|t]$ given by (12.99) as the value of our estimator.

Clearly, if $T(\mathbf{x}^*) = T(\mathbf{x}^{**}) = t$, then the value of the estimator $E_\theta(W|T)$ is the same for samples $\mathbf{x}^*$ and $\mathbf{x}^{**}$. This means that this value depends on the sample only through the value of the sufficient statistic T.

It remains now to prove the inequality (12.98). Let us recall the formula for conditional variance: For any random variables X and Y we have

$$\text{Var}(X) = E_Y\{\text{Var}(X|Y)\} + \text{Var}_Y\{E(X|Y)\}. \tag{12.100}$$

Letting $E_\theta(W) = k(\theta)$, we have for the risk of estimator W:

$$\begin{aligned} R(W, \theta) &= E\{[W - g(\theta)]^2\} \\ &= E\{([W - k(\theta)] + [k(\theta) - g(\theta)])^2\} \\ &= E\{[W - k(\theta)]^2\} + [k(\theta) - g(\theta)]^2 \\ &= \text{Var}(W) + [k(\theta) - g(\theta)]^2, \end{aligned} \tag{12.101}$$

since the cross product is easily seen to vanish. The second term is simply the square of the bias [remember that W is used to estimate $g(\theta)$ rather than θ].

We shall now use formula (12.100) taking $X = W$ and $Y = T$. Thus, for every θ we have (omitting for simplicity index θ)

$$\begin{aligned} \text{Var}(W) &= E_T\{\text{Var}(W|T)\} + \text{Var}_T\{E(W|T)\} \\ &\geq \text{Var}_T\{E(W|T)\} = \text{Var}(W^*). \end{aligned} \tag{12.102}$$

Now, using the fact that

$$k(\theta) = E(W) = E_T\{E(W|T)\} = E(W^*), \tag{12.103}$$

we may write, using (12.101), (12.102), and (12.103),

$$\begin{aligned} R(W, \theta) &\geq \text{Var}(W^*) + [k(\theta) - g(\theta)]^2 \\ &= \text{Var}(W^*) + [E(W^*) - g(\theta)]^2 \\ &= E[W^* - g(\theta)]^2 = R(W^*, \theta). \end{aligned}$$

It remains to determine the conditions for equality in (12.98). It is clear that the equality occurs if and only if $k(\theta) \equiv g(\theta)$ and $E_T\{\text{Var}(W|T)\} = 0$. The first condition means that W is an unbiased estimator of $g(\theta)$. The second condition means that $\text{Var}(W|T) \equiv 0$; that is, for every value t of T, the random variable W is constant. But this means that W is a function of T.   □

What the Rao–Blackwell theorem says is that to decrease the risk of an estimator V one should decrease its bias, and if the estimator is not a function of sufficient statistic (or statistics), then replace it by the estimator defined as conditional expectation of V given sufficient statistics.

We now illustrate the second part of this principle. Thus, in the examples below we start from some unbiased estimators and improve them by conditioning with respect to sufficient statistics.

Example 12.8.11.   Let $X_1, \ldots, X_n$ be a random sample from Poisson distribution with parameter λ. Suppose that we want to estimate $P\{X = 0\} = e^{-\lambda}$. We know (see Example 12.8.4) that the sufficient statistic for λ is $\overline{X}$, or equivalently, $T = \sum_{i=1}^n X_i$. We know that T has Poisson distribution with parameter $n\lambda$. Let us use $W = U/n$ as an estimator of $P\{X_i = 0\} = e^{-\lambda} = g(\lambda)$, say, where U is the number of X_i's that are equal to 0. To compare the risks of W and $W^* = E(W|T)$ it is enough to compare their variances, since it will be shown that W and W^* are both unbiased estimators for $g(\lambda)$.

Clearly, U is binomial with parameters n and $g(\lambda)$, so that $E(W) = (1/n)E(U) = g(\lambda)$. Thus, W is unbiased for $g(\lambda)$ and its risk equals

$$R(W, \lambda) = \text{Var}(U/n) = \frac{e^{-\lambda}(1 - e^{-\lambda})}{n}. \tag{12.104}$$

Now let W^* be defined as

$$W^* = E(W|T) = \frac{1}{n}E(U|T). \tag{12.105}$$

Clearly, the number of X_i's that are equal to 0 cannot exceed n. If $T = 1$, then $U = n - 1$, so it remains to consider cases when $T = t > 1$.

Let us start by determining the conditional distribution of one of the observations, say X_1, given $T = t$. We have, for $j \leq t$,

$$P\left\{X_1 = j \,\bigg|\, \sum_{i=1}^{n} X_i = t\right\} = \frac{P\{X_1 = j, \sum_{i=2}^{n} X_i = t - j\}}{P\{\sum_{i=1}^{n} X_i = t\}}$$

$$= \frac{\dfrac{\lambda^j e^{-\lambda}}{j!} \dfrac{[(n-1)\lambda]^{t-j}}{(t-j)!} e^{-(n-1)\lambda}}{\dfrac{(n\lambda)^t}{t!} e^{-n\lambda}}$$

$$= \binom{t}{j} \left(\frac{1}{n}\right)^j \left(1 - \frac{1}{n}\right)^{t-j},$$

which shows that given $\sum X_i = t$, the variable X_1 is binomial with parameters t and $1/n$.

Now let

$$Y_i = \begin{cases} 1 & \text{if } X_i = 0 \\ 0 & \text{if } X_i > 0, \end{cases}$$

so that for each Y_j we have

$$E(Y_j|T = t) = P(X_j = 0|T = t) = P(X_1 = 0|T = t) = \left(1 - \frac{1}{n}\right)^t.$$

Since $U = Y_1 + \cdots + Y_n$, we obtain

$$W^* = \frac{1}{n}E(U|T) = \frac{1}{n}E(Y_1 + \cdots + Y_n|T)$$

$$= \frac{1}{n}\sum_{j=1}^{n} E(Y_j|T) = \left(1 - \frac{1}{n}\right)^T, \tag{12.106}$$

a result that is somewhat unexpected.

Finding the risk of W^* (in order to compare it with the risk of W) requires some work. We know that W^* must be unbiased because $W = U/n$ was unbiased, and therefore we need to compare only variances of W^* and W. Remembering that $T = \sum_{i=1}^{n} X_i$ has Poisson distribution with parameter $n\lambda$, we have generally, for $0 \leq z \leq 1$,

$$Ez^T = \sum_{t=0}^{\infty} z^t \cdot P\{T = t\} = \sum_{t=0}^{\infty} z^t \frac{(n\lambda)^t}{t!} e^{-n\lambda}$$

$$= e^{zn\lambda} \cdot e^{-n\lambda} = e^{n\lambda(z-1)}.$$

On the other hand, $E[z^T]^2 = E[(z^2)]^T = e^{n\lambda(z^2-1)}$. Consequently,

$$\mathrm{Var}(z^T) = E[z^T]^2 - [E(z^T)]^2$$
$$= e^{n\lambda(z^2-1)} - e^{2n\lambda(z-1)}.$$

The estimator W^* is obtained by putting $z = 1 - 1/n$, which gives

$$\mathrm{Var}(W^*) = e^{n\lambda[((n-1)/n)^2-1]} - e^{2n\lambda((n-1)/n-1)}$$
$$= e^{-\lambda(2-1/n)} - e^{-2\lambda} = e^{-2\lambda}(e^{(\lambda/n)} - 1).$$

To compare the risks (variances) of the original estimator $W = U/n$ and the improved estimator W^*, observe that the risk of W, given by (12.104), can be written as

$$\frac{e^{-\lambda}(1 - e^{-\lambda})}{n} = \frac{\lambda e^{-2\lambda}}{n} + \frac{e^{-2\lambda}}{n}(e^{\lambda} - \lambda - 1).$$

On the other hand, for the risk of W^* we have

$$e^{-2\lambda}(e^{\lambda/n} - 1) = e^{-2\lambda}\left(\frac{\lambda}{n} + \frac{\lambda^2}{2n^2} + \frac{\lambda^3}{6n^3} + \cdots\right)$$

$$= e^{-2\lambda}\frac{\lambda}{n} + \frac{e^{-2\lambda}}{n^2}\left(\frac{\lambda^2}{2} + \frac{\lambda^3}{6n} + \cdots\right),$$

which is smaller by a term of order $\lambda^2 e^{-2\lambda}/2n$.

Example 12.8.12.   Consider now the situation of Example 12.3.1, that of estimating θ in uniform distribution on $(0, \theta]$. We know that the statistic $T_1 = \max(X_1, \ldots, X_n)$ is sufficient for θ (see Example 12.8.7) and consider three unbiased estimators of θ that are not functions of T_1, namely

$$T_3 = T_1 + \min(X_1, \ldots, X_n),$$
$$T_4 = 2\overline{X},$$
$$T_5 = (n + 1)\min(X_1, \ldots, X_n).$$

For T_3, the improved estimator is

$$T_3^* = E_\theta(T_3|T_1) = E_\theta\{T_1 + \min(X_1, \dots, X_n)|T_1\}$$
$$= T_1 + E_\theta\{\min(X_1, \dots, X_n)|T_1\}.$$

Given T_1, the distribution of $\min(X_1, \dots, X_n)$ is as follows: We know that one observation equals T_1, while the other $n-1$ are iid, each with a uniform distribution on $[0, T_1]$. Thus, expectation of $\min(X_1, \dots, X_n)$ given T_1 is the same as expectation of a minimum of $X_1', \dots, X_{n-1}'$, each X_i' being uniform on $[0, T_1]$. This conditional expectation equals $(1/n)T_1$. Consequently,

$$T_3^* = T_1 + \frac{1}{n}T_1 = \frac{n+1}{n}T_1,$$

which is equal to the estimator T_2 considered in Example 12.3.1.

Next, for T_4 we have

$$T_4^* = E_\theta\{T_4|T_1\} = E_\theta\left\{\frac{2}{n}(X_1 + \cdots + X_n)|T_1\right\}$$

$$= \frac{2}{n}E(X_1 + \cdots + X_n|T_1).$$

In this case, given T_1, the sum $X_1 + \cdots + X_n$ must have one term equal to T_1, while the others have uniform distribution on $[0, T_1]$, so that again, letting X_i' be uniform on $(0, T_1)$,

$$E(X_1 + \cdots + X_n|T_1) = T_1 + E(X_1' + \cdots + X_{n-1}')$$

$$= T_1 + (n-1)\frac{T_1}{2} = \frac{n+1}{2}T_1,$$

and therefore

$$T_4^* = \frac{2}{n} \cdot \frac{n+1}{2}T_1 = \frac{n+1}{n}T_1.$$

Finally,

$$T_5^* = E_\theta\{(n+1)\min(X_1, \dots, X_n)|T_1\} = (n+1)\frac{1}{n}T_1,$$

where we used the results obtained in determining T_3^*. $\qquad \square$

Observe here that by improving three different estimators (T_3, T_4, and T_5) by conditioning with respect to sufficient statistic T_1 we obtained in each case the same unbiased estimator $T_2 = [(n+1)/n]T_1$. The natural question arises: Was it a coincidence, or a rule? The matter is serious, since if such a result occurs as a rule, we could always select an unbiased estimator for which the conditioning would be especially simple, thus finding the unique unbiased estimator that could not be improved any further, hence with minimal risk. We close this chapter by adding one missing concept, which in some sense answers the question posed above and therefore connects all pieces together.

The situation at present may be summarized as follows. On the one hand, we have the Rao–Cramér inequality, which tells us how much reduction of the variance (hence of the risk, for unbiased estimators) may be possible. This bound may be effectively calculated in textbook cases, but in real-life situations it may be hard or impossible to determine. Also, the bound may be not attainable.

On the other hand, we have the Rao–Blackwell theorem, which tells us that unbiased estimators which are not based on sufficient statistics can be improved; to be more specific, their variance can be reduced by conditioning on sufficient statistics.

Again, the technical difficulties of such conditioning may be formidable. But here the situation can often be simplified: Rather than try to improve an estimator by finding its conditional expectation on a sufficient statistic, one may simply abandon the effort and refrain from using estimators that are not functions of sufficient statistics. Instead, one may try to find an unbiased estimator that is built on sufficient statistics only. This may still be technically difficult but is often easier than determining conditional expectations.

Now, once we find an unbiased estimator that is a function of a sufficient statistic (or a minimal sufficient set of statistics), the Rao–Blackwell theorem tells us that it cannot be improved any further (at least by conditioning). We therefore are tempted to conclude that we found the minimum variance unbiased estimator. Such a conclusion is justified if the estimator in question has variance equal to the right-hand side of Rao–Cramér inequality. Otherwise, the reasoning on which it is based still has a flaw and has to be amended. Indeed, this reasoning works if there is only one unbiased estimator based on a minimal sufficient statistic (or a minimal set of jointly sufficient statistics). But what if there are several such estimators? We know that none of them can be improved by further conditioning, but this does not guarantee that their risks are all minimal.

This observation makes it necessary to single out the cases when such an "anomaly" cannot occur, that is, situations when an unbiased estimator based on a sufficient statistic is unique.

Accordingly, we introduce the following definition:

Definition 12.8.5. A family of distributions $\{f(x; \theta), \theta \in \Theta\}$ is called *complete* if the condition $E_\theta[u(X)] = 0$ for all $\theta \in \Theta$ implies that $u(X) = 0$ with probability 1 for all $\theta \in \Theta$. A sufficient statistic whose distribution is a member of a complete family is said to be a *complete sufficient* statistic.

□

We have here the following theorem:

Theorem 12.8.5. *If the family of distributions $\{f(x; \theta), \theta \in \Theta\}$ is complete, then any unbiased estimator of parameter $g(\theta)$ is unique.* □

Proof. Indeed, suppose that $u_1(X)$ and $u_2(X)$ are unbiased estimators of $g(\theta)$, that is,

$$E_\theta[u_1(X)] = E_\theta[u_2(X)] = g(\theta)$$

for all $\theta \in \Theta$. But then $E_\theta\{u_1(X) - u_2(X)\} = 0$ for all θ, hence $P_\theta\{u_1(X) - u_2(X) = 0\} = 1$ for all θ, which means that the functions u_1 and u_2 coincide.
□

We therefore have the following theorem.

Theorem 12.8.6 (Lehmann and Scheffé). *Let $X_1, \ldots, X_n$ be a random sample from a distribution belonging to the family $\{f(x; \theta), \theta \in \Theta\}$. Let this family be complete, and let $(T_1, \ldots, T_k)$ be minimal jointly sufficient statistics. If the statistic $U^* = U^*(T_1, \ldots, T_k)$ is unbiased for $g(\theta)$, then U^* is the minimal variance unbiased estimator for $g(\theta)$.*

Proof. By completeness, we know that any statistic $U = U(T_1, \ldots, T_k)$ that is unbiased for $g(\theta)$ must satisfy the condition $U = U^*$ with probability 1 (for every θ). On the other hand, if we have any other statistic $W = W(X_1, \ldots, X_n)$ which is unbiased for $g(\theta)$ (not depending on the sufficient statistics $T_1, \ldots, T_k$), then $U = E\{W|T_1, \ldots, T_k\}$ is a statistic depending on $T_1, \ldots, T_k$ which has a variance satisfying $\mathrm{Var}_\theta(U) \leq \mathrm{Var}_\theta(W)$, which shows that the variance of U^* is minimal for each θ.
□

The concept of completeness pertains to a family of distributions. It is usually stated in a phrase appealing to intuition and easy to remember, that a family is complete if "there are no nontrivial unbiased estimators of zero" (a trivial unbiased estimator of zero is a random variable identically equal to zero). We now give some examples of complete families.

Example 12.8.13. Consider random sampling from a Poisson distribution, with $f(x; \theta) = (\lambda^x/x!)e^{-\lambda}, x = 0, 1, 2, \ldots \lambda \in (0, \infty)$. We know that a sufficient statistic is $T = \sum_{i=1}^n X_i$, and the distribution of T is again Poisson with parameter $\theta = n\lambda$. Let $u(T)$ be a statistic, and suppose that $E_\theta[u(T)] = 0$ for all θ. This means that

$$E_\theta[u(T)] = \sum_{n=0}^{\infty} u(n)\frac{\theta^n}{n!}e^{-\theta} = 0.$$

Since $e^{-\theta} \neq 0$, we may cancel it, obtaining the condition $\sum u(n)\theta^n/n! = 0$ for all θ. Letting $g(\theta) = \sum u(n)(\theta^n/n!)$ we have $g(\theta) \equiv 0$. However, the Taylor expansion of $g(\theta)$ is $g(\theta) = \sum [g^{(n)}(0)/n!]\theta^n$, so that $u(n) = g^{(n)}(0)$. But the latter quantity is zero for all n.

Observe here that we used the fact that the family contains *all* Poisson distributions as indexed by θ (or at least sufficiently many of them). If we consider the family of some selected Poisson distributions, say for $\theta_1, \theta_2, \ldots, \theta_r$, we can choose some negative and some positive values of $u(t)$ so as to obtain appropriate cancellations. The simplest case here is to take a family consisting of just one Poisson distribution, with mean $\theta = 5$ (say); then the function $u(T) = T - 5$ is not identically zero, yet its expectation is zero for all members of the family (since there is only one distribution in the family, and the expectation of this distribution is 5).

Example 12.8.14. In case of a uniform distribution on $[0, \theta]$, we know that the sufficient statistic is $T_1 = \max(X_1, \ldots, X_n)$, and its density $f(t, \theta)$ is given by

$$f(t; \theta) = \begin{cases} nt^{n-1}/\theta^n & 0 \le t \le \theta \\ 0 & \text{otherwise.} \end{cases}$$

Suppose that we have a function u which satisfies the condition

$$E_\theta[u(T)] = \int_0^\theta u(t)nt^{n-1}/\theta^n \, dt = 0.$$

This means that

$$\int_0^\theta t^{n-1}u(t) \, dt = 0,$$

and by differentiation with respect to θ we get

$$\theta^{n-1}u(\theta) = 0$$

for all $\theta > 0$. This means that we must have $u(t) \equiv 0$, and therefore the class of all distributions uniform on $[0, \theta]$, $\theta > 0$, is complete.

This explains why in Example 12.8.12, by conditioning the unbiased estimators T_3, T_4, and T_5 on sufficient statistic T_1, we obtain the (unique, in view of completeness) unbiased estimator $[(n+1)/n]T_1$ of θ.   □

Proving completeness of a family starting directly from the definition may not be easy. We therefore present a rather general and easily verifiable sufficient condition for completeness.

Definition 12.8.6. The distribution $f(x, \theta)$ belongs to an *exponential class* if for some set A

$$f(x; \theta) = \begin{cases} c(\theta)b(x) \exp[\sum_{i=1}^m q_i(\theta)k_i(x)] & \text{if } x \in A \\ 0 & \text{otherwise,} \end{cases} \tag{12.107}$$

where $\theta = (\theta_1, \ldots, \theta_m)$ is an m-dimensional parameter and $c(\theta), b(x)$ are strictly positive functions. It is assumed here that the parameter space is a generalized rectangle (possibly infinite in some dimensions) of the form

$$\Theta = \{\theta : a_i \le \theta_i \le b_i, i = 1, \ldots, m\}.$$

Moreover:

(a) The set A where $f(x; \theta)$ is positive does not depend on θ.
(b) The functions $q_1(\theta), \ldots, q_m(\theta)$ are continuous and functionally independent (i.e., none of them is a function of others).
(c) The functions $k_1(x), \ldots, k_m(x)$ are linearly independent (i.e., no linear combination of them is identically zero unless all coefficients are zero), and when $f(x; \theta)$ is a density of a continuous random variable they are differentiable.   □

We have the following theorem:

Theorem 12.8.7. *The exponential family* (12.107) *is complete. Moreover, minimal jointly sufficient statistics T_j have the form*

$$T_j(X_1, \ldots, X_n) = \sum_{i=1}^{n} k_j(X_i) \qquad for\ j = 1, \ldots, m.$$

The joint sufficiency of statistics $T_1, \ldots, T_m$ follows immediately from the form (12.107) The proof is too complex and will be omitted here. For the proof, see Lehmann (1986).

Most common families of distributions are exponential classes (e.g., binomial, Poisson, normal, or gamma). We show it for the Bernoulli and normal distributions, leaving the proofs for the remaining classes as exercises.

Example 12.8.15. For the Bernoulli distribution we have for $x = 0, 1$

$$f(x; p) = p^x(1-p)^{1-x} = (1-p)\left(\frac{p}{1-p}\right)^x$$

$$= (1-p)\exp\left\{x \log\left(\frac{p}{1-p}\right)\right\}.$$

Thus, $A = \{0, 1\}$, $m = 1$, $q_1(p) = \log[p/(1-p)]$, and $k_1(x) = x$. The sufficient statistic is $T = \sum k_1(X_i) = \sum X_i$.

Example 12.8.16. For a normal distribution we have, letting $\theta = (\mu, \sigma)$,

$$f(x, \theta) = \frac{1}{\sigma\sqrt{2\pi}} \exp\left[-\frac{1}{2\sigma^2}(x - \mu)^2\right]$$

$$= \frac{1}{\sigma\sqrt{2\pi}} \exp\left[-\frac{1}{2\sigma^2}(x^2 - 2\mu x + \mu^2)\right]$$

$$= \frac{1}{\sigma\sqrt{2\pi}} \exp\left[\frac{-\mu^2}{2\sigma^2}\right] \exp\left[-\frac{1}{2\sigma^2}x^2 + \frac{\mu}{\sigma^2}x\right],$$

so that we have $c(\theta) = (1/\sigma\sqrt{2\pi})e^{-\mu^2/2\sigma^2}$, $b(x) \equiv 1$, $q_1(\theta) = -1/2\sigma^2$, $q_2(\theta) = \mu/\sigma^2$, $k_1(x) = x^2$, $k_2(x) = x$. Minimal jointly sufficient statistics are therefore $T_1 = \sum_{i=1}^n X_i^2$ and $T_2 = \sum_{i=1}^n X_i$. $\qquad\square$

PROBLEMS

12.8.1 Generalizing Example 12.8.1, let $X_1, \ldots, X_n$ be n independent Bernoulli trials and consider the statistic $T = X_1 + \cdots + X_n$. Show that T is sufficient for probability of success p by finding the joint distribution of $(X_1, \ldots, X_n)$ directly given $T = t$. Find the marginal distribution $P\{X_1 = 1 | T = t\}$.

12.8.2 Find a sufficient statistic for θ if observations are uniformly distributed on the set of integers $0, 1, \ldots, \theta$.

12.8.3 Let $X_1, \ldots, X_n$ be a random sample from a gamma distribution with shape parameter α and scale parameter λ, where α is unknown and λ is known. Determine whether $\overline{X}$ is an admissible estimator of the mean of the distribution (under the squared loss function).

12.8.4 Show that the estimator W defined by (12.105) is strongly consistent. *(Hint:* Use the law of large numbers for $\overline{X}_n$.)

12.8.5 Show that a family of Poisson distributions is an exponential class.

12.8.6 Show that a family of beta distributions is an exponential class.

12.8.7 Show that the family of negative binomial distributions (r known, p unknown) is an exponential class.

12.8.8 Show that the family of gamma distributions is an exponential class, and find minimal jointly sufficient statistics.

12.8.9 Suppose that a random sample is taken from a distribution with density

$$f(x; \theta) = \frac{2x}{\theta^2} \qquad \text{for } 0 \le x \le \theta$$

and $f(x; \theta) = 0$ otherwise. Find the MLE of the median of this distribution, and show that this estimator is a minimal sufficient statistic.

12.9 INTERVAL ESTIMATION

Thus far we dealt with point estimation: An estimator was used to produce a single number, hopefully close to the unknown parameter. The natural question arises: What can be said about the distance between the value produced by the estimator and the true value of the parameter? Here again, the approaches are different, depending on whether or not one can regard the value of θ as a realization of some random variable.

Bayesian Intervals

In the Bayesian scheme the true value θ in a given situation is regarded as the observed value of a random variable Ξ with prior distribution (say, density) π. Given the observation $\mathbf{x}$, the posterior density of Ξ is $\pi(\theta|\mathbf{x})$, and we can assess the probability that Ξ lies between two values a and b as

$$P\{a \le \Xi \le b \,|\mathbf{x}\} = \int_a^b \pi(\theta|\mathbf{x}) \, d\theta. \qquad (12.108)$$

Example 12.9.1. Suppose that we observe $n = 3$ Bernoulli trials with unknown probability of success θ, where θ has the prior distribution beta with parameters $\alpha = \beta = 2$. Assume that we recorded $x = 2$ successes. Then the posterior distribution of the parameter is again beta, with parameters $\alpha + x = 4$ and $n + \beta - x = 3$, so that posterior density is

$$\frac{\Gamma(7)}{\Gamma(4)\Gamma(3)} \theta^3 (1 - \theta)^2 = 60\theta^3 - 120\theta^4 + 60\theta^5, \qquad 0 < \theta < 1.$$

The expected value of the posterior distribution is $\frac{4}{7} = 0.57$, which is the Bayes estimate of θ. The probability that the true value of θ lies below 0.2 equals

$$\int_0^{0.2} (60\theta^3 - 120\theta^4 + 60\theta^5) \, d\theta = 0.017.$$

The probability that the true value of θ lies above 0.8 equals

$$\int_{0.8}^1 (60\theta^3 - 120\theta^4 + 60\theta^5) \, d\theta = 0.099.$$

Consequently, with probability $1 - 0.017 - 0.099 = 0.884$ the true value of θ lies between 0.2 and 0.8

Confidence Intervals

In many cases the value of θ cannot be regarded as a value of a random variable with prior distribution that can be assigned a frequential interpretation. In such cases, non-Bayesian statisticians do not consider it meaningful to speak of "the probability that the value of parameter θ belongs to some set." According to this point of view, all randomness is in the process of selection of the sample. The relevant statements on interval estimation are now expressed through the concept of *confidence intervals*.

Let $\mathbf{X} = \mathbf{X}^{(n)} = (X_1, \ldots, X_n)$ be a random sample from the distribution $f(x, \theta)$. Let α be a fixed number with $0 < \alpha < 1$ (typically, α is a small number, such as $\alpha = 0.05$ or $\alpha = 0.01$).

Definition 12.9.1. A pair of statistics $L = L(\mathbf{X})$ and $U = U(\mathbf{X})$ determine an $(1 - \alpha)$-*level confidence interval* for θ if for all $\theta \in \Theta$,

$$P_\theta \{L(\mathbf{X}) \leq \theta \leq U(\mathbf{X})\} = 1 - \alpha. \qquad (12.109)$$

Similarly, a statistic $L(\mathbf{X})$ is a $(1 - \alpha)$-*level lower confidence bound* if

$$P\{L(\mathbf{X}) \leq \theta\} = 1 - \alpha. \qquad (12.110)$$

$\square$

The upper confidence bound $U(\mathbf{X})$ is defined in a similar way. It is important to observe that the meaning of formula (12.109) is quite different from the meaning of the left-hand side of formula (12.108).

To point out the difference in a most succinct way, observe that the left-hand side of (12.108) refers to posterior probability given $\mathbf{x}$. This means that the sample has already been taken and the probability statement refers to the mechanisms of sampling the particular value of θ. However, according to non-Bayesian philosophy, this sampling does not take place, and therefore probability statements after $\mathbf{x}$ has been observed make no sense.

On the other hand, formula (12.109) refers to randomness in the sample (for a given fixed θ): The interval $[L(\mathbf{X}), U(\mathbf{X})]$ varies from sample to sample and is therefore random. By definition, the probability that it covers the unknown (fixed) value of θ is $1 - \alpha$. This probability, however, refers to the process of taking the sample only: After a specific sample $\mathbf{x}$ has been observed, the observed interval $[L(\mathbf{x}), U(\mathbf{x})]$ either covers or does not cover θ. All randomness is gone, and it makes no sense to speak of "the probability that θ belongs to the interval $[L(\mathbf{x}), U(\mathbf{x})]$" because this probability is now either 0 or 1. Since we do not know which is the case, the term *confidence* is used: We say that the $(1 - \alpha)$-level confidence interval for θ

is $[L(\mathbf{x}), U(\mathbf{x})]$. In frequential terms, the last statements means "the confidence interval $L[(\mathbf{x}), U(\mathbf{x})]$ has been obtained by a method which produces randomly generated intervals, of which, on average, $100(1 - \alpha)\%$ cover the unknown value θ."

It is perhaps regrettable that the terminology here, as it developed historically, is somewhat ambiguous. The term *confidence interval* refers both to interval $L[(\mathbf{X}), U(\mathbf{X})]$ with random endpoints, as well as to the interval $L[(\mathbf{x}), U(\mathbf{x})]$ with endpoints obtained from the sample observed. Sometimes the first interval is called the *probability* interval, and the second is called the *sample* confidence interval. In this book we follow the more common tradition, referring to confidence intervals and sample confidence intervals.

In constructing confidence intervals, we use quantiles of some known distributions, such as standard normal, Student, chi-square, and so on. For typographical reasons, as well as to follow the tradition, we use different notation for distributions symmetric about zero (standard normal and Student). Thus, if Z is a standard normal random variable, then z_α will denote the $(1 - \alpha)$-quantile of Z (or, upper α-percentile), that is,

$$P\{Z \le z_\alpha\} = 1 - \alpha. \tag{12.111}$$

Then $z_{1-\alpha} = -z_\alpha$, and $P\{|Z| \le z_{\alpha/2}\} = 1 - \alpha$. Similarly, the $(1 - \alpha)$-quantile of the Student t distribution with ν degrees of freedom will be denoted by $t_{\alpha,\nu}$.

For asymmetric distributions such as chi-square, the subscript α will be used for upper percentiles of order α, so that $\chi^2_{\alpha,\nu}$ denote the quantile of order $1 - \alpha$ of chi-square distribution with ν degrees of freedom.

Next, we give some examples of confidence intervals.

Example 12.9.2. Consider the random sample $X_1, \ldots, X_n$ from the distribution $N(\theta, \sigma^2)$, where θ is the unknown parameter while σ^2 is known. Then the interval

$$\left(\overline{X} - z_{\alpha/2} \frac{\sigma}{\sqrt{n}}, \overline{X} + z_{\alpha/2} \frac{\sigma}{\sqrt{n}} \right)$$

is a $(1 - \alpha)$-level confidence interval for θ.

Indeed, the inequality

$$\overline{X} - z_{\alpha/2} \frac{\sigma}{\sqrt{n}} < \theta < \overline{X} + z_{\alpha/2} \frac{\sigma}{\sqrt{n}}$$

is equivalent to

$$\theta - z_{\alpha/2} \frac{\sigma}{\sqrt{n}} < \overline{X} < \theta + z_{\alpha/2} \frac{\sigma}{\sqrt{n}},$$

so that both inequalities have the same probabilities. Since $\mathrm{Var}(\overline{X}) = \sigma^2/n$, we have

$$P\left\{\overline{X} - z_{\alpha/2}\frac{\sigma}{\sqrt{n}} < \theta < \overline{X} + z_{\alpha/2}\frac{\sigma}{\sqrt{n}}\right\}$$

$$= P\left\{-z_{\alpha/2} < \frac{\overline{X} - \theta}{\sigma/\sqrt{n}} < z_{\alpha/2}\right\} = 1 - \alpha,$$

which was to be shown.

Example 12.9.3. Let $X_1, \ldots, X_n$ be a random sample of a continuous random variable, with density $f(x, \theta)$ depending on a one-dimensional parameter θ. Furthermore, let $F(x, \theta)$ be the cdf of X, and let $G = 1 - F$. Then each of the random variables $-2 \log F(X, \theta)$ and $-2 \log G(X, \theta)$ has exponential distribution with parameter $\frac{1}{2}$. Indeed,

$$P\{-2 \log F(X, \theta) > t\} = P\{F(X, \theta) < e^{-t/2}\} = e^{-t/2},$$

since $F(X, \theta)$ is uniform on $[0, 1]$. The argument for $G(X, \theta)$ is analogous. Consequently, the sums

$$U = -2\sum_{i=1}^{n} \log F(X_i, \theta) \quad \text{and} \quad W = -2\sum_{i=1}^{n} \log G(X_i, \theta) \qquad (12.112)$$

have gamma distribution (with parameters n and $\frac{1}{2}$), which is the same as the chi-square distribution with $2n$ degrees of freedom. We can therefore write

$$P\{\chi^2_{1-\alpha/2,2n} < -2\sum_{i=1}^{n} \log F(X_i, \theta) < \chi^2_{\alpha/2,2n}\} = 1 - \alpha \qquad (12.113)$$

and

$$P\{\chi^2_{1-\alpha/2,2n} < -2\sum_{i=1}^{n} \log G(X_i, \theta) < \chi^2_{\alpha/2,2n}\} = 1 - \alpha. \qquad (12.114)$$

If the inequality that defines the event in either (12.113) or (12.114) can be solved to the form $L(X_1, \ldots, X_n) < \theta < U(X_1, \ldots, X_n)$, we obtain, by definition, a confidence interval for θ at level $1 - \alpha$.   □

Example 12.9.4. Continuing Example 12.9.3, if $X_1, \ldots, X_n$ have exponential distribution with density $\theta e^{-\theta x}$, then $F(x, \theta) = 1 - e^{-\theta x}$ and $G(x, \theta) = e^{-\theta x}$. Formula (12.114) gives

$$P\{\chi^2_{1-\alpha/2,2n} < 2\theta \sum_{i=1}^{n} X_i < \chi^2_{\alpha/2,2n}\} = 1 - \alpha,$$

and we obtain the $(1 - \alpha)$-level confidence interval:

$$\frac{\chi^2_{1-\alpha/2,2n}}{2\sum_{i=1}^{n} X_i} < \theta < \frac{\chi^2_{\alpha/2,2n}}{2\sum_{i=1}^{n} X_i}.$$

Note that if we used F instead of G, we would obtain the inequality

$$\chi^2_{1-\alpha/2,2n} < -2\sum_{i=1}^{n} \log(1 - e^{-\theta X_i}) < \chi^2_{\alpha/2,2n},$$

which cannot easily be solved for θ. Thus, even when we have an explicit formula for F (and hence for G), use of one of these functions often leads to simpler results than use of the other.   □

In general, how are confidence intervals constructed? For large n, the answer lies in the limit theorem, 12.6.1, about the asymptotic distributions of MLE's obtained from setting the derivative of the likelihood equal to 0 and solving the resulting equation. Indeed, when we know that $\sqrt{nI(\theta)}(\hat{\theta}_n - \theta)$ with $I(\theta)$ defined by (12.34) is asymptotically standard normal (as in Example 12.9.2), we have

$$-z_{\alpha/2} < \sqrt{nI(\theta)}(\hat{\theta}_n - \theta) < z_{\alpha/2}$$

with probability close to $1 - \alpha$. This inequality is equivalent to the inequality

$$\hat{\theta}_n - \frac{z_{\alpha/2}}{\sqrt{nI(\theta)}} < \theta < \hat{\theta}_n + \frac{z_{\alpha/2}}{\sqrt{nI(\theta)}}. \tag{12.115}$$

The latter is not a confidence interval, since the bounds depend on the unknown value of θ. However, for large n we may replace θ by $\hat{\theta}_n$ in $I(\theta)$. Consequently, we can state the following theorem:

Theorem 12.9.1. Let $\hat{\theta}_n$ be the MLE estimator of θ in a problem for which MLE's satisfy the assumption of Theorem 12.6.1. Then for large n,

$$\left(\hat{\theta}_n - \frac{z_{\alpha/2}}{\sqrt{nI(\hat{\theta}_n)}}, \ \hat{\theta}_n + \frac{z_{\alpha/2}}{\sqrt{nI(\hat{\theta}_n)}}\right)$$

is an approximate $(1 - \alpha)$-level confidence interval for parameter θ.

Example 12.9.5. As an illustration, consider the problem of estimating probability p in Bernoulli distribution. The observations $X_1, X_2, \ldots, X_n$ are iid random variables indicating whether consecutive trials lead to success or failure (i.e., $X_i = 1$ or 0 with probabilities p and $1 - p$). The sum $S_n = X_1 + \cdots + X_n$ is the total number of successes in n trials. The MLE of p is $\hat{p} = S_n/n$, and (12.38) gives $I(p) = 1/p(1 - p)$, so that

$$\text{Var}(\hat{p}) = \frac{1}{n^2}\text{Var}(S_n) = \frac{p(1 - p)}{n} = \frac{1}{nI(p)}.$$

By the central limit theorem for binomial distribution, the random variable

$$\frac{S_n - np}{\sqrt{np(1 - p)}} = \sqrt{nI(p)}(\hat{p} - p)$$

is asymptotically standard normal. Consequently, we have the approximate relation

$$P\{-z_{\alpha/2} < \sqrt{nI(p)}(\hat{p} - p) < z_{\alpha/2}\} = 1 - \alpha. \tag{12.116}$$

The inequality that defines the event on the left-hand side of (12.116) is equivalent to

$$|\hat{p} - p| < \frac{z_{\alpha/2}}{\sqrt{n}}\sqrt{p(1 - p)},$$

or

$$\left(1 + \frac{z_{\alpha/2}^2}{n}\right)p^2 - \left(2\hat{p} + \frac{z_{\alpha/2}^2}{n}\right) + \hat{p}^2 < 0. \tag{12.117}$$

The solution of the quadratic inequality (12.117) is

$$\frac{\hat{p} + z_{\alpha/2}^2/2n - z_{\alpha/2}\sqrt{\dfrac{4\hat{p}(1 - \hat{p}) + (z_{\alpha/2}^2)/n}{4n}}}{1 + z_{\alpha/2}^2/n}$$

$$< p < \frac{\hat{p} + z_{\alpha/2}^2/2n + z_{\alpha/2}\sqrt{\dfrac{4\hat{p}(1 - \hat{p}) + z_{\alpha/2}^2/n}{4n}}}{1 + z_{\alpha/2}^2/n}$$

$$\tag{12.118}$$

which gives an approximate $(1 - \alpha)$-level confidence interval for probability p.

If we disregard the terms of order higher than $1/\sqrt{n}$, we obtain a more commonly used further approximation, namely

$$\hat{p} - z_{\alpha/2}\sqrt{\frac{\hat{p}(1 - \hat{p})}{n}} < p < \hat{p} + z_{\alpha/2}\sqrt{\frac{\hat{p}(1 - \hat{p})}{n}}, \tag{12.119}$$

which coincides with the confidence interval given in Theorem 12.9.1. $\square$

A method of construction of confidence intervals for any n is based on the pivotal variables, defined as follows.

Definition 12.9.2. A random variable W is called *pivotal* (for θ) if it depends only on the sample $\mathbf{X}^{(n)}$ and on unknown parameter θ, while its distribution does not depend on any unknown parameters (including θ). $\square$

This means that in principle one should be able to calculate the numerical values of probabilities for W. It is *not enough* to know that W has a normal distribution (say) with one or more parameters unspecified.

If $W = W(\mathbf{X}^{(n)}, \theta)$ is a pivotal random variable, then a $(1 - \alpha)$-level confidence interval may be constructed as follows. First, for given α, one determines the values q_α^* and q_α^{**} such that

$$P\{q_\alpha^* \leq W(\mathbf{X}^{(n)}, \theta) \leq q_\alpha^{**}\} = 1 - \alpha, \qquad (12.120)$$

which is possible, since W is pivotal.

It now remains to convert the inequality $q_\alpha^* \leq W(\mathbf{X}^{(n)}, \theta) \leq q_\alpha^{**}$ into the form

$$L(\mathbf{X}^{(n)}, q_\alpha^*, q_\alpha^{**}) \leq \theta \leq U(\mathbf{X}^{(n)}, q_\alpha^*, q_\alpha^{**}) \qquad (12.121)$$

or, more generally, into the form

$$\theta \in S(\mathbf{X}^{(n)}, q_\alpha^*, q_\alpha^{**}), \qquad (12.122)$$

where S is some set (not necessarily an interval). But as is often the case, if the pivotal quantity W is monotone in θ for every value of $\mathbf{X}^{(n)}$, a solution of the form (12.121) is attainable, and then the statistics L and U provide lower and upper endpoints of a $(1 - \alpha)$-level confidence interval for θ.

Example 12.9.6. The random variables (12.112) from Example 12.9.3 are pivotal, which shows that in the continuous case with a one-dimensional parameter there always exist at least two pivotal variables. $\square$

The next example shows that pivotal variables may also exist in the case of multidimensional parameters.

Example 12.9.7. As in Example 12.9.2, assume that $X_1, \ldots, X_n$ is a random sample from the distribution $N(\theta, \sigma^2)$, except that σ^2 is now unknown. We are still interested in estimating the mean θ. In such cases statisticians use the term *nuisance* parameter for σ^2.

In Example 12.9.2 the random variable $(\overline{X} - \theta)/(\sigma/\sqrt{n})$ was *pivotal* since σ was known. However, if σ is unknown, we have to proceed differently. The idea is to cancel σ in the denominator of $(\overline{X} - \theta)/(\sigma/\sqrt{n})$.

Recall from Theorem 9.12.1 that when $X_i \sim N(\theta, \sigma^2)$ and $S^2 = (1/n) \sum_{i=1}^n (X_i - \overline{X})^2$, the random variable nS^2/σ^2 has a chi-square distribution with $n - 1$ degrees of freedom. Moreover, S^2 is independent of $\overline{X}$. Consequently, the ratio

$$t_{n-1} = \frac{(\overline{X} - \theta)/(\sigma/\sqrt{n})}{\sqrt{nS^2/\sigma^2(n-1)}} = \frac{\overline{X} - \theta}{S} \sqrt{n-1}$$

$$= \frac{\overline{X} - \theta}{\sqrt{(1/n) \sum (X_i - \overline{X})^2}} \sqrt{n-1} \qquad (12.123)$$

has the Student t distribution with $n - 1$ degrees of freedom. Thus, (12.123) is a pivotal random variable (observe that σ is canceled). If $t_{\alpha/2, n-1}$ is the $(1 - \alpha/2)$-quantile of the Student t distribution with $n - 1$ degrees of freedom (which can be found from tables of Student distribution; see Table A3), then we have

$$P\left\{ -t_{\alpha/2, n-1} \leq \frac{\overline{X} - \theta}{S} \sqrt{n-1} \leq t_{\alpha/2, n-1} \right\} = 1 - \alpha,$$

which gives

$$\left(\overline{X} - \frac{t_{\alpha/2, n-1}S}{\sqrt{n-1}}, \overline{X} + \frac{t_{\alpha/2, n-1}S}{\sqrt{n-1}} \right) \qquad (12.124)$$

as a $(1 - \alpha)$-level confidence interval for θ. $\square$

Example 12.9.8 (Confidence Intervals for Variance). Consider again a random sample with normal distribution $N(\mu, \sigma^2)$, this time assuming that the parameter to be estimated is σ^2. We may distinguish two cases: when μ is known, and when μ is unknown. To provide some motivation for considering these cases, consider the problem of assessing the quality of some measuring device. A typical procedure here consists of repeating measurements of the same object. Assuming that the measurement errors are normal, the two cases above correspond to measurements of an object whose true value μ of measured attribute is otherwise known, and measurements of an object whose true value μ of attribute measured is unknown.

In the first case we have the pivotal random variable

$$U = \sum_{i=1}^n (X_i - \mu)^2 / \sigma^2,$$

while in the second case we have the pivotal random variable

$$V = \sum_{i=1}^{n}(X_i - \overline{X})^2/\sigma^2.$$

Since U is a sum of n independent squares of standard normal random variables, it has chi-square distribution with n degrees of freedom. We know also from Theorem 9.12.1 that V has chi-square distribution with $n - 1$ degrees of freedom. Thus, letting $\chi^2_{\alpha/2,\nu}$ and $\chi^2_{1-\alpha/2,\nu}$ denote the $(\alpha/2)$- and $(1 - \alpha/2)$-upper quantiles of chi-square distributions with ν degrees of freedom, we have

$$P\left\{\chi^2_{1-\alpha/2,n} < \frac{\sum_{i=1}^{n}(X_i - \mu)^2}{\sigma^2} < \chi^2_{\alpha/2,n}\right\} = 1 - \alpha$$

and

$$P\left\{\chi^2_{1-\alpha/2,n-1} < \frac{\sum_{i=1}^{n}(X_i - \overline{X})^2}{\sigma^2} < \chi^2_{\alpha/2,n-1}\right\} = 1 - \alpha.$$

We therefore obtain the $(1 - \alpha)$-level confidence intervals

$$\frac{\sum_{i=1}^{n}(X_i - \mu)^2}{\chi^2_{\alpha/2,n}} < \sigma^2 < \frac{\sum_{i=1}^{n}(X_i - \mu)^2}{\chi^2_{1-\alpha/2,n}} \qquad (12.125)$$

and

$$\frac{\sum_{i=1}^{n}(X_i - \overline{X})^2}{\chi^2_{\alpha/2,n-1}} < \sigma^2 < \frac{\sum_{i=1}^{n}(X_i - \overline{X})^2}{\chi^2_{1-\alpha/2,n-1}}. \qquad (12.126)$$

Two obvious questions concerning the intervals (12.125) and (12.126) follow. First, when μ is unknown, one *must* use the interval (12.126). But when μ is known, one may either utilize this knowledge and use interval (12.125) or not utilize it and use interval (12.126). Which is the proper procedure? The answer is that, on average, a shorter interval is obtained if one uses the available information about the expectation μ.

Second, to get the probability $1 - \alpha$ of covering the unknown value of σ^2, it is not necessary to choose values $\chi^2_{1-\alpha/2,n}$ and $\chi^2_{\alpha/2,n}$, which cut equal probabilities at both tails of the chi-square distribution. One could choose two other points a and b such that the probability of a chi-square random variable assuming a value between a and b is $1 - \alpha$. There are infinitely many such pairs (a, b), the pair $(\chi^2_{1-\alpha/2,n}, \chi^2_{\alpha/2,n})$ being just one of them. An obvious criterion would be to choose a pair a, b that minimizes the length of the confidence interval. It is easy to see that this length is proportional to $1/a - 1/b$. We have a similar discussion when μ is unknown.

To solve this optimization problem, let g_m be the density of the chi-square distribution with m degrees of freedom (in our case, we use either $m = n$ or $m = n - 1$). Then the problem can be formulated as

$$\text{minimize } \left(\frac{1}{a} - \frac{1}{b}\right) \text{ subject to } \int_a^b g_m(x)\, dx \geq 1 - \alpha.$$

This problem can be solved by using Lagrange multipliers: Differentiating the function

$$\frac{1}{a} - \frac{1}{b} + \lambda \left[\int_a^b g_m(x)\, dx - (1 - \alpha)\right]$$

with respect to a, b, and λ we obtain three equations:

$$-\frac{1}{a^2} - \lambda g_m(a) = 0, \qquad \frac{1}{b^2} + \lambda g_m(b) = 0, \qquad (12.127)$$

and

$$\int_a^b g_m(x)\, dx = 1 - \alpha. \qquad (12.128)$$

The solution is: Choose a, b satisfying constraint (12.128) and such that $a^2 g_m(a) = b^2 g_m(b)$. These can be solved numerically for each m (see Table A.5). The solution, for large m, is close to the "symmetric" solution, where $a = \chi^2_{\alpha/2,m}, b = \chi^2_{1-\alpha/2,m}$. For small sample sizes it is better to use an exact solution of the optimization problem above, as this gives the shortest possible confidence interval.

Table A.5 gives the optimal left cutoff probability α, and the corresponding upper quantiles $a = \chi^2_{\alpha_1,m}$ and $b = \chi^2_{\alpha_2,m}$, where $\alpha_2 = (1 - \alpha) - \alpha_1$, for $\alpha = 0.1$ and $\alpha = 0.05$ and various numbers of degrees of freedom. The table gives also the relative gain resulting from using the shortest confidence interval (with length proportional to $1/a - 1/b$), as compared to the confidence interval with equal cutoff probabilities, hence with length proportional to $1/\chi^2_{1-\alpha/2,m} - 1/\chi^2_{\alpha/2,m}$.

Example 12.9.9. Suppose that five observations from a normal distribution with unknown μ and σ^2 are such that $\sum(X_2 - \overline{X}) = c$. We want to build a 95% confidence interval for σ^2. The usual procedure, based on cutting off 2.5% at either end of the chi-square distribution with 4 degrees of freedom gives the values $\chi^2_{0.975,4} = 0.484$ and $\chi^2_{0.025,4} = 11.143$ (see Table A.4). This leads to the confidence interval $c/11.143 < \sigma^2 < c/0.484$, or $0.0897\,c < \sigma^2 < 2.0661\,c$, of length $1.9764\,c$.

The shortest 95% confidence interval involves the upper quantiles $a = \chi^2_{0.04969,4} = 0.708$ and $b = \chi^2_{0.90031,4} = 21.047$. The confidence interval is now $0.0475\,c = c/21.047 < \sigma^2 < c/0.708 = 1.4124\,c$. The length is $1.3649\,c$, which is 69% of the length of the "usual" interval.   □

For large n the optimality of the choice of cutoff points a and b does not matter much. Also, in this case, the difference between chi-square distributions with $n - 1$ and n degrees of freedom is negligible. There are cases, however, when an increase in the sample size n does not depend on the experimenter or is very costly. For example, we may wish to test a device that measures some quantity observable only during an earthquake or during a nuclear explosion. In such cases the optimization criteria outlined above are relevant. □

Two natural questions arise: Do pivotal random variables always exist, and if not, can one construct confidence intervals in some other way? As regards the first question, the criteria for existence of pivotal random variables were given by Antle and Bain (1969). In analogy with examples concerning the normal distribution, we have the following theorem.

Theorem 12.9.2

(i) *If θ is a location parameter and $\hat{\theta}$ is the MLE of θ, then $\hat{\theta} - \theta$ is pivotal.*

(ii) *If θ is a scale parameter and $\hat{\theta}$ is the MLE of θ, then $\hat{\theta}/\theta$ is pivotal.*

(iii) *Let θ_1 be a location parameter and θ_2 be a scale parameter, so that*
$$f(x, \theta_1, \theta_2) = \frac{1}{\theta_2} h\left(\frac{x - \theta_1}{\theta_2}\right) \text{ for some density } h. \text{ If } \hat{\theta}_1 \text{ and } \hat{\theta}_2 \text{ are MLE's}$$
of θ_1 and θ_2, then $(\hat{\theta}_1 - \theta_1)/\hat{\theta}_2$ is pivotal.

The answer to the second question is that even if pivotal random variables do not exist, we can construct confidence intervals. We give here examples concerning confidence intervals for quantiles.

Example 12.9.10. Let $X_1, \ldots, X_n$ be a random sample from a distribution with the cdf $F(t)$, and let $Y_{1:n} \leq Y_{2:n} \leq \cdots \leq Y_{n:n}$ be the order statistics of the sample. To simplify the formulation, assume that F is continuous and strictly increasing, and let $\xi_{1/2}$ be the median of the distribution F. Thus, $\xi_{1/2}$ is the unique solution of the equation $F(t) = \frac{1}{2}$. We shall construct confidence intervals for $\xi_{1/2}$. More precisely, we shall consider (random) intervals of the form $[Y_{a:n}, Y_{b:n}]$ for integers a, b satisfying the condition $1 \leq a < b \leq n$ and assess, for each of such intervals, the probability that it covers the parameter $\xi_{1/2}$.

The event $Y_{a:n} \leq \xi_{1/2}$ occurs if and only if at least a elements in the sample are below the median. Similarly, the event $\xi_{1/2} \leq Y_{b:n}$ occurs if and only if at least $n - b$ elements in the sample are above the median. If we call an observation below the median a success and let S denote the number of successes in the sample, then the event $Y_{a:n} \leq \xi_{1/2} \leq Y_{b:n}$ occurs if and only if we have at least a successes and at least $n - b$ failures, so that $a \leq S \leq b$. Since S has distribution $\text{BIN}(n, \frac{1}{2})$, we obtain

$$P\{Y_{a:n} \leq \xi_{1/2} \leq Y_{b:n}\} = \sum_{k=a}^{b} \binom{n}{k} \frac{1}{2^n}.$$

For instance, if $n = 10$, then the interval $[Y_{3:10}, Y_{7:10}]$ between the third and seventh order statistic is a confidence interval for the median with the confidence level

$$\sum_{k=3}^{7} \binom{10}{k} \frac{1}{2^{10}} = 1 - 2 \left[\binom{10}{0} + \binom{10}{1} + \binom{10}{2} \right] \frac{1}{2^{10}} = 0.891.$$

On the other hand, the interval $[Y_{2:10}, Y_{6:10}]$ is a confidence interval for $\xi_{1/2}$ with confidence level $\sum_{k=2}^{6} \binom{10}{k} \frac{1}{2^{10}} = 0.709$.

One-sided confidence bounds for $\xi_{1/2}$ are obtained in the same way: For instance, the order statistic $Y_{a:n}$ is a lower bound for $\xi_{1/2}$ with the confidence level $\sum_{k=a}^{n} \binom{n}{k} \frac{1}{2^n}$.

Example 12.9.11. Continuing Example 12.9.10, let us now fix p and let ξ_p be the pth quantile of the population distribution. Again letting S be the number of successes, where "success" is now an observation below ξ_p, we have S with $\mathrm{BIN}(n, p)$. The interval $[Y_{a:n}, Y_{b:n}]$ is a confidence interval for ξ_p with confidence level

$$\sum_{k=a}^{b} \binom{n}{k} p^k (1 - p)^{n-k}.$$

Again taking $n = 10$, the same interval $[Y_{3:10}, Y_{6:10}]$ can serve as a confidence interval for the first quartile $\xi_{1/4}$, except that now the confidence level will be

$$\sum_{k=3}^{7} \binom{10}{k} \left(\frac{1}{4}\right)^k \left(\frac{3}{4}\right)^{10-k} = 0.526.$$

Clearly, in this case it is better to take confidence intervals based on order statistics where indices are not equidistant from the extremes. □

We complete this chapter with the critique of the concept of confidence intervals. The point is that while a pair of statistics $L = L(\mathbf{X}), U = U(\mathbf{X})$ may satisfy the condition

$$P\{U < \theta < L\} = 1 - \alpha,$$

there are cases when the sample $\mathbf{X}$ may provide additional information, allowing us to claim a higher probability of coverage than $1 - \alpha$. This is illustrated by the following example from De Groot (1986).

Example 12.9.12. Suppose that two independent odservations X_1, X_2 are taken from the distribution uniform on $[\theta - \frac{1}{2}, \theta + \frac{1}{2}]$. Then the pair of order statistics $X_{1:2} = \min(X_1, X_2)$ and $X_{2:2} = \max(X_1, X_2)$ is a 50% confidence interval for θ. Indeed, the interval $[X_{1:2}, X_{2:2}]$ covers θ if and only if one of the observations is below θ and the other is above θ, an event with probability $\frac{1}{2}$. However, if $X_{2:2} - X_{1:2} > \frac{1}{2}$, then the interval $[X_{1:2}, X_{2:2}]$ covers θ with probability 1. Still, strict adherence to the definition of confidence interval requires reporting such an interval as a 50% confidence interval, rather than as a 100% confidence interval.

This example (as well as other examples of this type, e.g., Example 12.6.31) gives some weight to the Bayesian approach to statistical inference.

PROBLEMS

12.9.1 Assume that $[L, U]$ is a confidence interval for parameter θ at confidence level $1 - \alpha$. Show that:

    (i) If g is an increasing function, then $[g(L), g(U)]$ is a $(1 - \alpha)$-confidence interval for $g(\theta)$.

    (ii) If g is a decreasing function, then $[g(U), g(L)]$ is a $(1 - \alpha)$-confidence interval for $g(\theta)$.

12.9.2 Seven measurements of the concentration of some chemical in cans of tomato juice are

$$1.12, 1.18, 1.08, 1.13, 1.14, 1.10, 1.07.$$

Assume that those measurements form a random sample from the distribution $N(\theta, \sigma^2)$.

    (i) Find the shortest 95% and 99% confidence intervals for θ if σ^2 is unknown.

    (ii) Answer part (i) if it is known that $\sigma^2 = 0.0004$.

    (iii) Answer part (i) if σ^2 equals the sample variance s^2.

    (iv) Use the data to obtain 90% confidence intervals for σ^2 and for σ.

12.9.3 A large company wants to estimate the fraction p of its employees who participate in a certain health program.

    (i) It has been decided that if p is below 25%, a special promotion campaign will be launched. In a random sample of 85 employees the number of those who participated in the program was 16. Find a 95% confidence interval for p using formulas (12.118) and (12.119). Should the campaign be launched?

    (ii) Answer the same question if the data are 340 and 64, respectively.

12.9.4 Let $X_1, \ldots, X_n$ be a random sample from a normal distribution with both μ and σ^2 unknown. Let L_α be the length of the shortest confidence interval for μ on confidence level $1 - \alpha$. Find $E(L_\alpha^2)$ as a function of n, σ^2, and α.

12.9.5 Referring to Problem 12.9.4, find the smallest n such that (for given α) we have $E(L_\alpha^2) \le \sigma^2/2$.

12.9.6 Find the probability that the length of a 95% confidence interval for the mean of normal distribution with unknown σ, based on a sample of size $n = 25$, is less than σ.

12.9.7 Let $X_1, \ldots, X_n$ be a random sample from a normal distribution with mean and variance unknown. Construct a $(1 - \alpha)$-level confidence interval for the standard deviation σ of X_1.

12.9.8 Let X_1, X_2 be a random sample of size $n = 2$ from a continuous distribution with median θ. Show that $[X_{1:2}, X_{2:2}] = [\min(X_1, X_2), \max(X_1, X_2)]$ is a 50% confidence interval for θ.

12.9.9 Suppose that the largest observation recorded in a sample of size $n = 35$ from a uniform distribution on $[0, \theta]$ is 5.17. Find a 90% confidence interval for θ.

12.9.10 Use the large-sample distribution of the MLE of mean λ in a Poisson distribution to construct an approximate $(1 - \alpha)$-level confidence interval for λ.

12.9.11 The numbers of new cars of a given make sold per week in 15 consecutive weeks are 5, 5, 6, 3, 5, 8, 1, 4, 7, 7, 5, 4, 3, 0, 9. Assume that these values form a random sample from a Poisson distribution with mean λ. Find a 90% confidence interval for λ. (*Hint:* Use the results of Problem 12.9.10.)

12.9.12 Suppose it is known that the lifetime of a certain kind of device (e.g., a fuel pump in a car) is exponentially distributed with density $f(t; \lambda) = \lambda e^{-\lambda t}, t > 0$ and $f(t; \lambda) = 0$ for $t < 0$. The observed lifetimes of a sample of the devices are $t_1 = 350, t_2 = 727, t_3 = 615, t_4 = 155, t_5 = 962$ (in days). Find 95% confidence intervals for:

(i) λ.

(ii) The expected lifetime of the device.

(iii) The standard deviation of the lifetime of the device.

(iv) The probability that the two copies of the device that you bought will each last more than 2 years.

12.9.13 Let $X_1, \ldots, X_n$ be a random sample from exponential distribution with parameter λ. Construct a confidence interval for λ. (*Hint:* Compare Example 12.9.4.)

12.9.14 Suppose it is known that the arrivals at a checkout counter in a supermarket (i.e., times of arriving at the counter or joining the queue, whichever is earlier) form a Poisson process with arrival rate λ. Counting from noon, the thirteenth customer arrived at 12:18 p.m.

 (i) Find a 90% confidence interval for λ.

 (ii) Construct a 90% confidence interval for the variance of inter-arrival times between consecutive customers.

12.9.15 A sample of 200 trees in a forest have been inspected for the presence of some bugs, out of which 37 trees were found to be infested.

 (i) Assuming a binomial model, give a 90% confidence interval (based on a normal approximation) for the probability p of a tree being infested. Use the exact and approximate formulas.

 (ii) Ordinarily, if a tree is infested, one might expect some of the neighboring trees to be infested. Thus, whether the binomial model is adequate depends on the way the sample was selected. Describe a way of selecting 200 trees which gives a binomial distribution.

REFERENCES

Antle, C. E., and L. J. Bain, 1969. "A Property of Maximum Likelihood Estimators of Location and Scale Parameters," *SIAM Review*, **11**, 251.

De Groot., M., 1986. *Probability and Mathematical Statistics*, Addison–Wesley, Reading, Mass.

Efron, B., 1982. *The Jackknife, the Bootstrap and Other Resampling Plans*, SIAM, Philadelphia.

Gafrikova, V., and M. Niewiadomska-Bugaj, 1992. "Ordering and Comparing Estimators by Means of Gini Separation Index," *Probability and Mathematical Statistics*, **13**, 215–228.

Green, Richard F., "Bayesian Birds: A Simple Example of Oaten's Stochastic Model of Optimal Foraging," *Theoretical Population Biology*, **18.2**, 244–256.

Lehmann, E., 1986. *Testing Statistical Hypotheses*, Wiley, New York.

Mood, A. M., F. A. Graybill, and D. C. Boes, 1974. *Introduction to the Theory of Statistics*, McGraw-Hill, New York.

Stigler, S. M., 1986. *The History of Statistics*, Harvard University Press, Cambridge, Mass.

CHAPTER 13

Testing Statistical Hypotheses

13.1 INTRODUCTION

In this chapter we present the basic concepts and results of the theory of testing statistical hypotheses. To sketch the content of this chapter and provide motivation for various concepts and topics we start from the same empirical setup as for estimation theory.

We will base our decision about some future actions on observations $X_1, X_2, \ldots$ of random variables, with distribution depending on an unknown parameter $\theta \in \Theta$, where Θ is the parameter space.

Suppose that we have some repertoire of actions available, and that if we knew θ, we could select the best among the actions. Thus, with each θ there is associated the action, call it a_θ, which is best for that specific θ. The statistical decision problem is to make an inference about the unknown value of θ on the basis of observations $X_1, X_2, \ldots$, and use this inference to make the choice of an action, this choice being optimal in the sense of some criteria.

The theory of testing statistical hypotheses concerns a special class of such decision problems, characterized by a further simplifying assumption: The set of actions $\{a_\theta, \theta \in \Theta\}$ consists of just two actions, say a and a'. Then the parameter space Θ decomposes into two sets: $\Theta_0 = \{\theta : a_\theta = a\}$, that is, the set where action a is best, and its complement, $\Theta_1 = \{\theta : a_\theta = a'\}$. To identify the best action all one needs to know is whether or not the actual value of θ (under which the observations $X_1, X_2, \ldots$ were taken) is in Θ_0. As opposed to the problem of estimation, where the objective was to approximate θ as closely as possible, here we want to know much less about θ.

The situation is now conceptually very simple: We need a procedure that allows us to decide, on the basis of observation $\mathbf{X} = (X_1, X_2, \ldots, X_n)$, whether or not θ is in Θ_0. Such a decision procedure is called a *test of the hypothesis* that $\theta \in \Theta_0$.

In the case when the sets of possible values of X are disjoint for $\theta \in \Theta_0$ and $\theta \notin \Theta_0$ ($\theta \in \Theta_1$), the problem is trivial. The more realistic and challenging case is when some (or even all) values of $\mathbf{X}$ can occur for $\theta \in \Theta_0$ and for

568

$\theta \notin \Theta_0$; then there can be no procedure that will *always* allow us to reach the correct decision. Indeed, it may happen that θ is in Θ_0, and we will observe a value of **X** that will lead us to the conclusion that $\theta \notin \Theta_0$. It may also happen that θ is not in Θ_0, and we will reach the conclusion that $\theta \in \Theta_0$. These two possibilities are called errors of type I and type II, and the objective is to choose the decision rule that will lead to probabilities of these errors being as small as possible.

We now present several examples of situations that fall under the scheme of hypothesis testing.

Example 13.1.1. A consumer protection agency decides to investigate complaints that some boxes contain less of a product (e.g., cereal) than the amount printed on the box indicates. The boxes are filled and then sealed by a machine. Even with the most sophisticated equipment available, the amounts that the machine puts into boxes will vary. Suppose it was established that these amounts follow a normal distribution with some mean μ (which can be set in the packing plant) and a standard deviation σ. For the sake of concreteness, assume that $\sigma = 1.5$ oz. Assume also that the nominal weight of content of the box is 20 oz.

If the mean μ is set on 20 oz, than about 50% of the boxes will contain less than the nominal amount, and at least for buyers of these boxes, the fact that the other 50% of boxes contain more than the nominal amount may be of no relevance. Consequently, the packing company must set the average μ at a value *above* the nominal weight 20 oz. (Observe that no value μ will guarantee that the content of a box will *always* be above 20 oz.) In this situation a reasonable requirement may be that 99% of boxes must contain at least 20 oz of cereal. If X denotes the content of a randomly chosen box, this means that

$$0.99 = P(X \geq 20) = P\left(Z \geq \frac{20 - \mu}{1.5}\right) = 1 - \Phi\left(\frac{20 - \mu}{1.5}\right),$$

and therefore $\mu = 23.5$ oz. Thus, in order to satisfy the requirement, the company should set the average weight of the content of the box at a value of at least 23.5 oz.

The agency that investigates the customer's complaint has to determine whether or not the average weight content is at least 23.5 oz. Since opening up and weighting all boxes is not feasible, the agency must decide how to do it in a more practical way. For example, one may agree (and such a decision should be made in advance, prior to any sampling) that the total of $n = 16$ boxes of the cereal in question will be bought; the stores will be sampled, and in each store selected the box will be chosen at random from those on the shelf. Then the content of each box will be weighted and the average $\bar{x}$ will be calculated. If $\bar{x} \leq 23$ oz (say), then the hypothesis $\mu \geq 23.5$ will be rejected; otherwise, it will be accepted.

The probabilities of the two types of errors can be calculated as follows. It may happen that $\mu \geq 23.5$, but the sample average $\bar{x}$ falls below 23 oz. We then declare the company to be at fault (of *not* putting the average at least 23.5 oz of cereal to boxes), although in reality the company is not at fault.

It may also happen that we observe $\bar{x} > 23$ (hence we declare the company as not being at fault), whereas, in fact, $\mu < 23.5$. To compute the chance of these two kinds of errors, we will find the probability that the company is declared at fault while the true mean is μ, i.e.,

$$P\{\overline{X} \leq 23|\mu\} = \Phi\left(\frac{23 - \mu}{1.5}\sqrt{16}\right),$$

which is a decreasing function of μ.

Declaring the company at fault is an error if, in fact, $\mu \geq 23.5$ (otherwise, it is a correct decision). The probability of such an error attains the smallest value equal to

$$\Phi\left(\frac{23 - 23.5}{1.5}\sqrt{16}\right) = 0.09.$$

when $\mu = 23.5$. The chances of the other type of error—declaring the company as not at fault, when in fact $\mu < 23.5$—can again be computed for any specific μ as

$$P\{\overline{X} > 23|\mu\} = 1 - \Phi\left(\frac{23 - \mu}{1.5}\sqrt{16}\right).$$

For instance, if $\mu = 22$, then the probability above is less than 0.004.   □

Before going any further a little reflection is in order: What is the structure of a possible test (i.e., of a possible decision rule)? If our decision that $\theta \in \Theta_0$ or $\theta \notin \Theta_0$ should depend on the observation of $\mathbf{X}$ only, then it is enough to specify a set C of points in the space of values of $\mathbf{X}$ with the instruction: If the actual observation $\mathbf{x}$ is in C, then decide that $\theta \notin \Theta_0$; otherwise, decide that $\theta \in \Theta_0$. Such a set C is called a *critical* (or *rejection*) *region* for the hypothesis that $\theta \in \Theta_0$.

In the case analyzed in Example 13.1.1 we have $(X_1, \ldots, X_{16})$ and the suggested critical region for the hypothesis that $\mu > 23.5$ is the set C of all points $(x_1, \ldots, x_{16})$ in 16-dimensional Euclidean space which satisfy the condition $\frac{1}{16}(x_1 + \cdots + x_{16}) \leq 23$. Thus, each test is identified with a critical region C and we are facing an embarrassing richness of potential tests.

We want to find the best critical region C, guided only by the principle of minimizing the probabilities of errors of type I and type II. But this principle is too general and has to be further implemented to secure the existence of the "best" critical region. To appreciate the conceptual difficulties it is enough to realize the following two points:

1. If $\theta \in \Theta_0$, then deciding that $\theta \notin \Theta_0$ is the error of type I. This error occurs whenever the observation $\mathbf{X}$ falls into the critical region C, so that the probability of an error of type I is

$$\alpha(\theta) = P_\theta\{\mathbf{X} \in C\}. \tag{13.1}$$

On the other hand, if $\theta \notin \Theta_0$, then deciding that $\theta \in \Theta_0$ is an error of type II; this occurs whenever $\mathbf{X} \notin C$, so that the probability of an error of type II is

$$\beta(\theta) = P_\theta\{\mathbf{X} \notin C\} = 1 - P_\theta\{\mathbf{X} \in C\}. \tag{13.2}$$

Thus, probabilities of errors are not single numbers but functions of the parameter θ, both expressible in terms of $P_\theta\{\mathbf{X} \in C\}$. This probability is called the *power function* of the test C. The minimization principle requires that the power function be "as small as possible" for $\theta \in \Theta_0$ and "as large as possible" for $\theta \notin \Theta_0$. Obviously, the best test is such that $P_\theta\{\mathbf{X} \in C\} = 0$ if $\theta \in \Theta_0$ and $P_\theta\{\mathbf{X} \in C\} = 1$ if $\theta \notin \Theta_0$. In most cases of practical importance and interest, $P_\theta\{\mathbf{X} \in C\}$ is a continuous function of θ and such a "best" test does not exist.

2. Looking at formulas (13.1) and (13.2) we see that any modification of C has the opposite effects on probabilities of errors of type I and type II: If one of them decreases, then the other increases. This may best be seen if we continue the analysis of box content (Example 13.1.1).

Example 13.1.2. The properties of the decision procedure as to whether the cereal-producing company should be declared at fault depend on the choice of the threshold (23 oz in the last example) and the sample size n. By changing the threshold up or down we change the probabilities of the two types of errors, with the changes always going in opposite directions. The only way to decrease the probabilities of both types of errors at the same time is to increase the sample size n. The "best" test must result from a compromise between the cost of sampling and the consequences of two types of errors. In the case of boxes of cereal, the cost of increasing n from 16 to 100 is of little concern. Of course, raising the sample size in other experiments might be more expensive or not always possible.

The consequences of a decision to declare the company at fault when it is not are again expressible by cost (incurred by the company, which may have to pay a fine, unnecessarily reset the mean to a higher value, etc.). The consequences of declaring the company not at fault when in fact it is are hard to express in terms of cost; they involve many small losses suffered by individual buyers. □

The fact that the probabilities of errors of types I and II are negatively related and that each is a function of θ (defined on sets Θ_0 and Θ_1, respectively) makes formulation of the criterion to be optimized a difficult and challenging task. The conceptual structure of the theory is as follows. First, the problem of optimization is solved for the simplest case, when Θ_0 and Θ_1 each consist of a single point, say θ_0 and θ_1. The solution is given by the celebrated Neyman–Pearson lemma, which specifies the explicit form of the test (i.e., critical region) which has the preassigned probability of error of type I, and a minimal probability of error of type II (or equivalently, a maximum power at value θ_1). Such a test is called *most powerful*. In some sense, the Neyman–Pearson lemma plays a role analogous to that of the Rao–Cramèr inequality in estimation theory: Both set the standard by showing how much can potentially be achieved.

In some classes of testing problems, in which Θ_1 (and often also Θ_0) consist of more than one element, there exists a test C which is most powerful against any $\theta \in \Theta_1$; such tests are called *uniformly most powerful*, (UMP) *tests*. Clearly, a UMP test cannot be improved for any $\theta \in \Theta_1$ and is therefore the best of all possible tests.

There are, however, classes of testing problems in which the UMP test does not exist. In such cases the idea is to reduce (by eliminating tests with some undesirable properties) the class of tests to those that one wants to consider. The problem lies in reducing the class of tests "as little as possible," at the same time eliminating sufficiently many tests so that the UMP test will already exist in the reduced class.

We discuss only one such reduction, reduction to the class of unbiased tests. The UMP unbiased tests may not achieve maximum possible power (as specified by the Neyman–Pearson lemma), but they are uniformly most powerful among the unbiased tests.

To understand the conceptual structure of the theory, especially why a reduction may lead to the appearance of an UMP test, one may use the following metaphorical analogy. Suppose that we want to find the strongest athlete, in the sense of strength of his right arm and the strength of his left arm (regarded separately). It may happen that there will be no strongest athlete in the group: The one with the strongest right arm will have a weak left arm, and vice versa, and no winner will be found. To create an analogy to unbiased tests, one may restrict the competition to those athletes who have equal strength in both arms. Among those, the winner (or winners) can always be found. Such a winner may be called the most powerful "unbiased" athlete.

This analogy immediately suggests another solution: Why not compare athletes with respect to the combined strengths of right and left arms? Or according to the sum of appropriate scores, possibly different for left and right arms? (If you think this idea is farfetched, think of deciding the winner in events such as the decathlon, which is based on the sum of scores for different events).

To continue the athletic metaphor, let us observe that the example was oversimplified to the extent that "most powerful unbiased athlete(s)" can always be found. It may lead the reader to a (wrong) conclusion that UMP unbiased tests always exist. This is not so, and to understand why, suppose that the competition involves comparisons of strengths of right and left arms, and also right and left legs (separately). If no "absolute" winner (strongest on each limb) exists, one may reduce the competition to "unbiased" athletes, those whose strengths of right and left arms are the same, and whose strengths of right and left legs are the same. But now there is no guarantee that "most powerful unbiased athlete" exists, since an athlete with strongest arms need not have strongest legs, and vice versa.

In short, UMP unbiased unbiased tests exist in some classes of problems and do not exist in other classes, so that further reduction may be needed to single out the UMP tests in this reduced class. The best exposition of the theory can be found in a monograph of Lehmann (1983).

Again, this outline of conceptual structure of theory of hypotheses testing does not lead to appreciation of the conceptual difficulties which had to be overcome to build the theory. We shall however present two examples illustrating other points. The first example shows how the reality may force one to use tests which are suboptimal. The second shows the nature of technical difficulties which appear when the term "parameter" is interpreted in a somewhat non-standard sense.

Example 13.1.3. The testing procedure suggested in Example 13.1.1 is that we declare company at fault when $\bar{x} < 23$. Such a test may well be optimal from the point of view of statistical criteria involving probabilities of errors, but it has a disturbing feature: it may happen that we declare the company at fault, when all boxes tested contain actually more than the nominal amount 20 oz. Such a decision may not be defensible in court, in case the company decides to appeal. One should then restrict the consideration to procedures, which have the threshold value set below the nominal value 20 oz., and such procedures may be suboptimal.

Example 13.1.4. Suppose that a pharmaceutical company develops a drug against some disease, hoping it is superior to the drug used thus far. After running the toxicity tests, studying the side effects, and so on (as required by the FDA), the only issue that remains is to compare the merits of the two drugs. In a typical setup, one chooses two groups (samples) of patients. One group is then given the "old" drug, while the other receives the drug just developed. The data consist of two sets of observations, $X_1, \ldots, X_m$ and $Y_1, \ldots, Y_n$, being the results of some medical tests, where m and n represent numbers of people in each group.

Setting the hypotheses in the form discussed above can be accomplished as follows. Let the cdf of observations in the first and second samples be, respectively, $F(t) = P\{X \le t\}$ and $G(t) = P\{Y \le t\}$. The joint cdf of the data (X, Y) may be taken as the product FG, in the sense that

$$P\{X_i \le x_i, i = 1, \ldots, m; Y_j \le y_j, j = 1, \ldots, n\} = \prod_{i=1}^{m} F(x_i) \times \prod_{j=1}^{n} G(y_j).$$

This joint distribution is characterized by the pair $\theta = (F, G)$ of two cdf's. Thus, the parameter space is the class of all pairs of distributions, and the hypothesis tested is that θ belongs to the "diagonal" in this space [i.e., to the set of all pairs of the form (F, F), where F is some cdf]. ☐

Letting $\theta = (F, G)$ goes against the intuition that one may associate with the word *parameter*. In fact, however, there appears to be no agreement among statisticians regarding this term. To see some of the difficulties in choosing a satisfactory definition, observe the following. Moments, as well as some simple functions of moments, such as standard deviation, are parameters according to everybody's intuition. Next, since the median, as well as other quantiles, are also parameters in the common intuition, the definition cannot refer to moments only: This would make it too narrow, since in many classes, the median is not expressible through moments in any simple formula. The same applies to quantities such as $P\{X \le 3\}$, which are parameters, in the sense of being "quantities that one may wish to estimate."

One might then suggest a definition of the form "a function that maps a class of distributions into some space of values." This covers all uses of the word "parameter" but is too wide to be useful. Indeed, suppose that the class of distributions in question consists of two distributions, say of height for males and for females. If we assign letters M and F to these two distributions, then we obtain a parameter that may assume only two values, M and F; some would rather use a term such as "index" here.

Modification of the above, by requesting that the "space of values" be a Euclidean space, is not satisfactory either. Indeed, suppose that we consider the class of all distributions and assign the value zero to distributions with finite mean, and the value 1 to the others. Finally, if (1) quantiles are parameters, and (2) vectors with components, being parameters, are themselves parameters [e.g., (μ, σ) for normal distribution], then one might perhaps allow this principle to be extended infinitely many components. If so, take the set of all quantiles x_p for rational p. This is a countable collection of parameters, hence a parameter. But knowledge of all quantiles for rational p is equivalent to knowledge of the whole cdf. We come therefore to the situation where F itself is a parameter. Strangely enough, the part of the statistics where one deals with problems such as testing the hypothesis that cdf's F and G in two populations are equal is called *nonparametric* statistics (discussed in Chapter 16). Some statisticians say that nonparametric problems are those that involve infinitely many parameters.*

13.2 INTUITIVE BACKGROUND

As already mentioned, we start from the same setup as in Chapter 12: We observe a random sample $X_1, X_2, \ldots, X_n$ from a certain distribution. We know that this distribution belongs to some family $f(x, \theta)$ indexed by a parameter θ, and we know that θ is an element of a parameter space Θ. As before, we do not know the true value of θ that governs the distribution of the sample $X_1, X_2, \ldots, X_n$. The difference between the problem of estimation and the problem of testing is simply that now we are not interested in approximating the true value of θ. What we need to know is only whether θ belongs to a specific subset $\Theta_0 \subset \Theta$.

Example 13.2.1.   Consider a politician who will win or lose the election, depending on whether or not the proportion of voters who will cast their votes for him exceeds 50%. If θ is the fraction of voters who favor the politician in question, then $0 \le \theta \le 1$, so that the parameter space is $\Theta = [0, 1]$, and the set of interest is $\Theta_0 = (\frac{1}{2}, 1]$. A survey may give a sample $X_1, \ldots, X_n$ of random variables with Bernoulli distribution $f(x, \theta) = \theta^x (1 - \theta)^{1-x}, x = 0, 1$, and the question is how to use the observation to determine whether or not $\theta \in \Theta_0$. A typical question may be: Suppose that out of $n = 400$ voters sampled, only 195 will vote for the politician in question and that 205 will vote against him. Should one reject the hypothesis that $\theta > \frac{1}{2}$? The answer depends on many factors, the most crucial being: How likely is it to observe the data so much (195 vs 205) or even more in favor of rejecting the claim $\theta > \frac{1}{2}$ if in fact the true value of θ is above $\frac{1}{2}$?

Example 13.2.2.   Engineers design a new device to measure the cholesterol level of cooking oil. It is claimed that the device will give measurements with a standard error not exceeding 0.1% of the measured content. To test this claim, one takes repeated measurements of the same cooking oil, say $X_1, \ldots, X_n$. Assume that the distribution of X_i is normal, with density

---

* This digression regarding the term *parameter* shows that even in domains so highly formalized as mathematics, it may be prudent, and quite permissible, to leave some terms undefined. An example of such a term outside statistics is *equilibrium*, which has to be defined separately in each context. As regards the term *parameter*, the confusion is minimal and harmless. Some statisticians use the phrase *natural parameters*, which presumably are those in the common core of intuition of all statisticians.

$$f(x; \mu, \sigma) = \frac{1}{\sigma\sqrt{2\pi}} e^{-(x-\mu)^2/2\sigma^2}.$$

The claim is that $\sigma < 0.001\mu$. The actual formulation of the problem, and the procedure to be used, depend on whether or not μ may be assumed known. In the case when μ is known (say, can be controlled by the experimenter), the unknown parameter is σ, and $\Theta = (0, \infty)$. The set Θ_0 is the interval $\Theta_0 = \{\sigma : 0 < \sigma \le 0.001\mu\}$, where μ is known.

In the second case, when μ is not known, we have a bivariate parameter $\theta = (\mu, \sigma)$, and the parameter space is the set $\Theta = \{\theta : 0 \le \mu < \infty, 0 < \sigma < \infty\}$ (clearly, in this case the mean μ cannot be negative). The problem is now to decide whether $\theta \in \Theta_0$, where $\Theta_0 = \{\theta = (\mu, \sigma) : \sigma \le 0.001\mu\}$ (see Figure 13.1).   □

In the generally accepted terminology, the statement that θ belongs to Θ_0 and the statement that θ does not belong to Θ_0 are called *hypotheses*. By some abuse of language, we shall often identify hypotheses with subsets of the parameter space, writing "hypothesis Θ_0" instead of the more precise "hypothesis that $\theta \in \Theta_0$," and so on. We shall frequently use the descriptive notation $H : \theta \in \Theta_0$.

Let $\Theta_1 = \Theta \setminus \Theta_0$, so that out of two hypotheses, $H_0 : \theta \in \Theta_0$ and $H_1 : \theta \in \Theta_1$, exactly one must be true. Logically speaking, denying the truth of one of them is equivalent to accepting the truth of the other. The actually observed sample $\mathbf{x} = (x_1, x_2, \ldots, x_n)$ provides some evidence, typically pointing in favor of one of the two hypotheses. A *decision rule* is a rule that for every sample $\mathbf{x} = (x_1, x_2, \ldots, x_n)$ tells us what to do in case of observing $\mathbf{x}$. These actions may be "accept H_0," "reject H_0 (and therefore accept H_1)," "take another observation," or even "leave the matter unresolved." A decision rule will

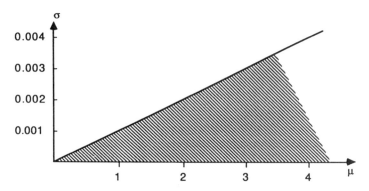

Figure 13.1   Null hypothesis $\sigma \le 0.001\mu$.

be called a *test** (more precisely, a *nonrandomized test*) if the only actions allowed are "accept H_0" and "accept H_1."

Since only two actions are allowed in a test, we must choose one of them regardless of the value of **x**, and therefore any decision rule is equivalent to a partition of the sample space into two sets: the set of those points **x** at which H_0 is to be accepted, and its complement, the set of those points **x** at which H_0 is to be rejected (i.e., H_1 accepted).

Accordingly, we make the following terminological convention. First, we designate one of the hypotheses as the *null hypothesis*, denoted H_0, with the corresponding subset of parameter space denoted Θ_0. The complementary hypothesis is called the *alternative hypothesis*, and is usually denoted by H_1 or H_a. The corresponding subset of parameter space is $\Theta_1 = \Theta \setminus \Theta_0$.

The testing procedure is specified by the subset C of the sample space, called the *critical* region, and the rule

$$\text{reject } H_0 \text{ if } \mathbf{x} \in C.$$

Observe that in this formulation the two hypotheses H_0 and H_1 are treated symmetrically, and designation of one of them as the null hypothesis is arbitrary. Subsequently, this symmetry will be lost, and the null hypothesis will play a different role than the alternative.

Once the problem of testing is formulated as described above, we see that we are facing a problem of choice: The class of all tests coincides with the class of all critical regions, that is, the class of all subsets of the sample space. It becomes necessary therefore to formulate meaningful criteria that will eventually allow us to choose the best test.

To formulate the criteria, we first introduce some auxiliary definitions.

Definition 13.2.1.   A hypothesis $H : \theta \in \Theta_0$ is called *simple* if it specifies completely the distribution of the sample; otherwise, the hypothesis is called *composite*.

Example 13.2.3.   In the voting problem (Example 13.2.1), the hypothesis $\theta \in (\frac{1}{2}, 1]$ is composite. The hypothesis $\theta = \frac{1}{2}$ is simple.

---

* It is important to realize that the phrases "accept H" and "reject H" are to be interpreted as decisions regarding future behavior rather than regarding the truth or falsehood of H. Thus, to "accept H" means (in majority of practical situations) to proceed *as if H were true*." In fact, it would be unrealistic to expect more: Our decision is based on observation of a random sample; hence we are always exposed to the risk of wrong decision, namely rejecting H when in reality H is true, or accepting H when in reality it is false. The best one can do is to control the probabilities of these errors. Instead of stating that "hypothesis H_0 is accepted," the verdict is often phrased in a more cautious way, such as "there is not enough evidence to reject H_0" and the like. The last phrasing will be especially appropriate for the hypotheses analyzed in Chapter 16.

In Example 13.2.2, the hypothesis $\mu = \mu_0, \sigma = \sigma_0$ is simple. On the other hand, the hypothesis $\mu = \mu_0$ is simple or composite, depending on whether σ is known (and hence is not a parameter of the problem) or is unknown.

$\square$

To proceed with formulating the criteria of assessing the performance of a test (critical region C), let us observe that we can make two types of error in our decision:

(a) Reject the null hypothesis if in fact it is true.
(b) Fail to reject null hypothesis if in fact it is false.

These two errors are referred to in statistical literature as type I and type II errors (or errors of the first and second kind). The practical consequences of these two types of errors may be entirely different. It is worthwhile to illustrate such differences by an example.

Example 13.2.4. Assume that medical science has developed some test (or tests) for screening certain disease, say breast cancer, TB, or AIDS. Let **x** stand generally for the result of the test (**x** may be a number, a vector of numerical or qualitative indices, etc.). Let $\theta = 1$ or 0, depending on whether or not the patient has the disease in question (say, TB for the sake of concreteness). Thus, $\Theta = \{0, 1\}$ and $f(\mathbf{x}, \theta)$ are the two distributions of the results of the tests, $f(\mathbf{x}, 0)$ for healthy persons and $f(\mathbf{x}, 1)$ for persons with TB. Let $H_0 : \theta = 1$ be the hypothesis "patient has TB." Then the error of first kind means rejection of the null hypothesis if it is true, that is, declaring a person with TB as healthy. The error of the second kind means failing to reject H_0 when it is false, that is, declaring a healthy person as having TB. The social consequences of these two errors are quite different, the first being much more serious than the second. $\square$

It goes without saying that we would like to minimize the probabilities of both types of errors. The trouble here is twofold: first, the probability of errors depends typically on the value of θ, so that the phrase "probability of rejecting null hypothesis when it is true" does not have a unique meaning if the null hypothesis is composite. In fact, there are many values of θ for which the null hypothesis is true, and the probability in question depends on the particular value of θ. The same remark applies to the alternative hypothesis if it is composite.

The second source of difficulty lies in the fact that under any reasonable definition of probability of errors that may be suggested, the two probabilities are inversely related: If the probability of one type of an error decreases, then the probability of an error of the other type increases. To obtain a convenient tool for expressing the probabilities of errors associated with a test, we introduce the following definition, central to the entire theory of testing statistical hypotheses.

Definition 13.2.2.   If C is the critical region of a test of the null hypothesis H_0: $\theta \in \Theta_0$, the function $\pi_C(\theta)$, defined on the parameter space Θ as

$$\pi_C(\theta) = P_\theta\{\mathbf{X} \in C\}, \qquad\qquad (13.3)$$

is called the *power* (or *power function*) of test C.   □

Thus, power is the probability of rejecting the null hypothesis if the value of parameter is θ. Qualitatively speaking, a good test should have high power when $\theta \notin \Theta_0$, since for such θ the null hypothesis is not true, and rejecting it is a correct decision. On the other hand, for $\theta \in \Theta_0$ the power of a good test should be low, since for such θ the null hypothesis is true, and rejecting it is a wrong decision.

We are now in a position to analyze some simple examples of tests.

Example 13.2.5.   A supermarket buys oranges from a certain company, which claims that the fraction of unacceptable fruit (e.g., rotten, etc.) in each shipment does not exceed 3%. The supermarket is willing to accept this (but not a higher) percentage of bad fruit. The agreed procedure specifies that a random sample of 30 oranges will be selected from each shipment (the method of sampling is typically a part of the agreed protocol). The shipment will be accepted if there is no more than one unacceptable fruit in this sample, and rejected (or bought at a lower price, etc.) otherwise.

Such an acceptance–rejection scheme is an example of a test of a statistical hypothesis. First, assume that we can disregard the finiteness of the population, and assume that the results of sampling constitute independent Bernoulli trials with a probability of "success" (finding a rotten orange) equal θ. The range of θ is (at least theoretically) the entire interval $\Theta = [0, 1]$. The random sample $X_1, \ldots, X_{30}$ is drawn from the distribution $f(x, \theta) = \theta^x(1 - \theta)^{1-x}, x = 0, 1$. The supplier's claim is that $\theta \le 0.03$. Let us take this claim as the null hypothesis, so that $\Theta_0 = [0, 0.03]$, while the alternative is $\Theta_1 = (0.03, 1]$.

The agreed procedure specifies that the supplier's claim (null hypothesis) is rejected if there are two or more bad oranges in the sample, that is, if

$$S_{30} = \sum_{i=1}^{30} X_i \ge 2.$$

This means that the critical region is the set $C = \{2, 3, \ldots, 30\}$. The elements in C are observed values S_{30} which lead to rejection of the supplier's claim that $\theta \in \Theta_0$.

The power function of this test is simply $P\{S_{30} \geq 2\}$, regarded as a function of θ, so that

$$
\begin{aligned}
\pi_C(\theta) &= P_\theta\{S_{30} \geq 2\} = 1 - P_\theta\{S_{30} = 0\} - P_\theta\{S_{30} = 1\} \\
&= 1 - (1-\theta)^{30} - 30\theta(1-\theta)^{29} \\
&= 1 - (1-\theta)^{29}(1+29\theta).
\end{aligned}
$$

In this case $\pi_C(\theta)$ is a function strictly increasing (see Figure 13.2) from $\pi_C(0) = 0$ to $\pi_C(1) = 1$. The highest value of power function on the set Θ_0 (null hypothesis) is attained at $\theta = 0.03$, and equals

$$
\pi_C(0.03) = \max_{\theta \in \Theta_0} \pi_C(\theta) = 0.227.
$$

The supermarket manager would probably be content with such a procedure: If the alternative is true, that is, if the fraction θ of unacceptable fruits exceeds 0.03, then he has at least 22.7% chances of detecting it. In fact, these chances increase rather fast with θ: For instance, for $\theta = 0.05$ we have $\pi_C(\theta) = 0.446$, and for $\theta = 0.1$ we have $\pi_C(\theta) = 0.816$.

On the other hand, the supplier will not be too happy: the chances of having a shipment that meets the standard $\theta \leq 0.03$ rejected can be as high as 22.7% (if θ is very close to 0.03).

Example 13.2.6.   Continuing Example 13.2.5, a possible solution to the problem of accepting orange shipments is to change the procedure: Reject the shipment if three or more (rather than two or more) fruits are unacceptable. The critical region, say C_1 (of rejecting the claim $\theta \leq 0.03$), is now $\{S_{30} \geq 3\}$, so that

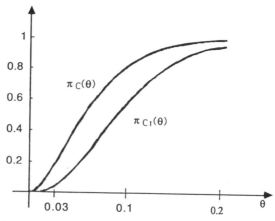

Figure 13.2   Power functions $\pi_C(\theta)$ and $\pi_{C_1}(\theta)$.

$$\pi_{C_1}(\theta) = 1 - (1 - \theta)^{30} - 30\theta(1 - \theta)^{29} - 435\theta^2(1 - \theta)^{28}$$
$$= 1 - (1 - \theta)^{28}(1 + 28\theta + 406\theta^2).$$

Again, this is an increasing function, and (see Figure 13.2)

$$\pi_{C_1}(0.03) = \max_{\theta \in \Theta_0} \pi_{C_1}(\theta) = 0.060.$$

This time the supplier is happier, but the manager is not: He might feel that the chance of accepting a shipment with $\theta = 0.1$ (say), equal to $1 - \pi_{C_1}(0.1) = 0.411$ is too high.   □

These two examples illustrate the fact that an attempt to decrease the probabilities of one type of error (by change of the critical region) leads to an increase in the probabilities of error of the other type.

The only way to decrease both probabilities of error at the same time is to increase the sample size.

Example 13.2.7 (Randomization).   In the case of orange shipments discussed in Examples 13.2.5 and 13.2.6, one could also suggests the following procedure, which does not involve any increase in the sample size, and yet leads to a procedure that may be acceptable to both the manager and the supplier.

The "bone of contention" is the fact that rejection of the shipment if $S_{30} \geq 2$ is too favorable for the manager, while rejection if $S_{30} \geq 3$ is too favorable for the supplier. One can suggest the procedure according to which if S_{30} is 0, or 1, the shipment is accepted by the store; if S_{30} is 3, 4, ..., 30, the shipment is rejected by the store; while if $S_{30} = 2$, a coin is tossed and the shipment is accepted or rejected depending on the outcome of the toss. Actually, it is not necessary that the coin be fair: One can toss a biased coin (or activate some other random mechanism agreed upon by both sides) such that the probability of the shipment being rejected if $S_{30} = 2$ is some fixed number γ.

The power function of such a procedure can still be defined as $\pi(\theta) = P_\theta\{\text{null hypothesis rejected}\}$. In the case under consideration, we have $\pi(\theta) = \gamma P\{S_{30} = 2\} + P\{S_{30} \geq 3\}$. This power function, for $0 < \gamma < 1$, lies between power functions of the two procedures considered in Examples 13.2.5 and 13.2.6, and it is possible that the two sides can negotiate the value of γ that is acceptable for both of them.   □

A randomized procedure described in Example 13.2.7 is somewhat controversial. If testing statistical hypotheses is regarded as a process aimed at establishing the truth or falsehood of some statements about the experimental situation under analysis, then indeed, declaring the truth of one hypothesis on the basis of a flip of a coin may appear appalling and ridiculous. However, in testing statistical hypotheses according to the original intention of Neyman and Pearson (who built the foundations of this theory), the rejection and acceptance of the hypothesis were not statements about the truth and falsehood. Those were intended to be the guidelines for future actions. As Neyman (1950, pp. 259–260) wrote: "The terms "accepting" and "rejecting" a statistical hypothesis are very convenient and are well established. It is important, however, to keep their exact meaning in mind and to discard various additional implications which may be suggested by intuition. Thus, to accept a hypothesis H means only to take an action A rather than B. This does not mean that we necessarily believe that the hypothesis H is true. Also, if the application of the rule of inductive behavior "rejects" H, this means only that the rule prescribes action B and does not mean that we believe that H is false."

In light of this interpretation, randomization of a decision is not such a ridiculous idea as one might think.

Example 13.2.8. Let us consider the situation used repeatedly in our discussion of estimation, namely that of a random sample from the distribution uniform on $[0, \theta]$. We have here $\Theta = (0, \infty)$ and suppose that we want to test the null hypothesis $H_0 : 5 \le \theta \le 6$. The alternative hypothesis is $H_1 : \theta < 5$ or $\theta > 6$. Thus, $\Theta_0 = [5, 6]$ and $\Theta_1 = (-\infty, 5) \cup (6, \infty)$.

We consider several tests and determine their power. First, we know (see Example 12.8.7) that $T_1 = \max(X_1, \ldots, X_n)$ is the sufficient statistic, and we might try to base the testing procedure on it. Clearly, if $T_1 \ge 6$, we should reject the null hypothesis, since it simply cannot be true in this case. Similarly, we should reject the null hypothesis if T_1 is "too small," say $T_1 < 4.6$. Finally, we may argue that the values of T_1 slightly below 6 are also a good indication that H_0 may be false, since we always have $T_1 < 6$. The actual thresholds will depend on n; for the purpose of the present example, let us take the critical region (when H_0 is to be rejected) as

$$C = \{(x_1, \ldots, x_n) : T_1 < 4.6 \text{ or } T_1 > 5.9\}. \tag{13.4}$$

The power of this test is

$$\pi_C(\theta) = P_\theta\{T_1 < 4.6\} + P_\theta\{T_1 > 5.9\}. \tag{13.5}$$

The first term is

$$P_\theta\{T_1 < 4.6\} = \begin{cases} 1 & \text{if } \theta \le 4.6 \\ \left(\dfrac{4.6}{\theta}\right)^n & \text{if } \theta > 4.6. \end{cases}$$

Similarly,

$$P_\theta\{T_1 > 5.9\} = 1 - P_\theta\{T_1 \leq 5.9\}$$

$$= \begin{cases} 0 & \text{if } \theta \leq 5.9 \\ 1 - \left(\dfrac{5.9}{\theta}\right)^n & \text{if } \theta > 5.9. \end{cases}$$

Thus, we obtained

$$\pi_C(\theta) = \begin{cases} 1 & \text{for } \theta \leq 4.6 \\ \left(\dfrac{4.6}{\theta}\right)^n & \text{for } 4.6 < \theta \leq 5.9 \\ \left(\dfrac{4.6}{\theta}\right)^n + 1 - \left(\dfrac{5.9}{\theta}\right)^n & \text{for } \theta > 5.9. \end{cases}$$

The graph of this power function for $n = 15$ is presented in Figure 13.3.

Let us also consider a test based on statistic $\overline{X}$. We know that $E(\overline{X}) = \theta/2$. To provide a comparison with the test (13.4), consider now the test with the critical region

$$C_1 = \{(x_1, \ldots, x_n) : \overline{x} < 2.30 \text{ or } \overline{x} > 2.95\}. \tag{13.6}$$

To determine the exact distribution of $\overline{X}$, while possible in principle, is very cumbersome for large n, so we shall rely on the approximation provided by the central limit theorem, using the fact that $\text{Var}(X_i) = \theta^2/12$. The distribution of $\overline{X}$ is therefore approximately normal $N(\theta/2, \theta^2/12n)$, and we obtain, for the power of the test (13.6),

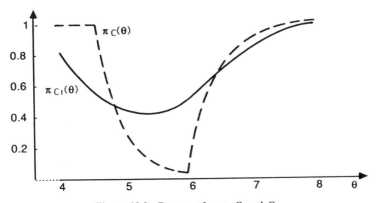

Figure 13.3 Powers of tests C and C_1.

$$\pi_{C_1}(\theta) = P\{\overline{X} < 2.30\} + P\{\overline{X} > 2.95\}$$

$$\approx 1 - P\left\{ \frac{2.30 - \theta/2}{\theta}\sqrt{12n} \le Z \le \frac{2.95 - \theta/2}{\theta}\sqrt{12n} \right\}$$

$$= 1 - \left[\Phi\left(\frac{2.95 - \theta/2}{\theta}\sqrt{12n} \right) - \Phi\left(\frac{2.30 - \theta/2}{\theta}\sqrt{12n} \right) \right].$$

The graph of this function, for $n = 15$, is presented in Figure 13.3.

The qualitative requirement for a "good" test is that it should have the power as low as possible on the null hypothesis, and as high as possible on the alternative hypothesis. A glance at Figure 13.3 shows that the test C, based on the sufficient statistic, is much better than the test C' based on the sample mean $\overline{X}$.

As may be seen, for all θ in the null hypothesis H_0 we have $\pi_C(\theta) < \pi_{C_1}(\theta)$. Actually,

$$\sup_{\theta \in \Theta_0} \pi_C(\theta) = \max(\pi_C(5), \pi_C(6)) = \max(0.2863, 0.2414) = 0.2863.$$

On the other hand, for the function $\pi_{C_1}(\theta)$ on H_0 we have

$$\max(\pi_{C_1}(5), \pi_{C_1}(6)) = \max(0.4094, 0.6033) = 0.6033.$$

For the alternative hypothesis, when we want the power to be as high as possible, we have $\pi_C(\theta) > \pi_{C_1}(\theta)$ except two rather narrow ranges immediately to the left of $\theta = 5$ and to the right of $\theta = 6$. Qualitatively speaking, we may conclude that the test based on critical region C is superior to the test based on critical region C_1.

Example 13.2.9.   Consider finally one of the "classical" examples, testing hypotheses about the mean in normal distribution with known variance. Thus, we assume that $X_1, \ldots, X_n$ is a random sample from the density $f(x, \mu) = (1/\sqrt{2\pi})e^{-(x-\mu)^2/2}$ (we assume here that $\sigma = 1$ which is not a restriction of generality, as it simply amounts to choosing appropriate unit for X_i's). Suppose that we want to test the null hypothesis $H_0 : \mu = \mu_0$ against the alternative $H_1 : \mu \ne \mu_0$. In this case, the null hypothesis is simple, while the alternative is composite.

Common sense and intuition suggest basing the testing procedure on the average $\overline{X} = (X_1 + \cdots + X_n)/n$ of observations. If this average deviates "too much" from μ_0, then null hypothesis should be rejected. This means taking the critical region of the form

$$C = \{(x_1, \ldots, x_n) : |\overline{X} - \mu_0| > k\}, \tag{13.7}$$

where k is some constant.

It is easy to compute the power function of this test. Indeed, we have [remembering that $\text{Var}(\overline{X}) = 1/n$]

$$\begin{aligned}
\pi_C(\mu) &= P_\mu\{|\overline{X} - \mu_0| > k\} \\
&= 1 - P_\mu\{|\overline{X} - \mu_0| \le k\} \\
&= 1 - P_\mu\{\mu_0 - k \le \overline{X} \le \mu_0 + k\} \\
&= 1 - P\{(\mu_0 - \mu - k)\sqrt{n} \le (\overline{X} - \mu)\sqrt{n} \le (\mu_0 - \mu + k)\sqrt{n}\} \\
&= 1 - P\{(\mu_0 - \mu - k)\sqrt{n} \le Z \le (\mu_0 - \mu + k)\sqrt{n}\}.
\end{aligned}$$

If H_0 is true, then $\mu = \mu_0$ and power equals

$$\pi_C(\mu_0) = 1 - P\{|Z| \le k\sqrt{n}\},$$

a quantity that could be chosen at will by the appropriate choice of rejection threshold k. Also, for any fixed $k > 0$ we have $\pi_C(\mu_0) \to 0$ as $n \to \infty$. On the other hand, for fixed $k > 0$ and n, we have $\pi_C(\mu) \to 1$ when $\mu \to \pm\infty$.

This means that for fixed sample size n we can choose the critical region so that the probability of error of the first kind is equal to any preassigned level. The probability of error of second kind approaches 0 as μ moves away from the null hypothesis (see Figure 13.4). How quickly it happens (i.e., how fast the power function approaches 1 as $|\mu - \mu_0|$ increases) depends on n, and therefore the properties of the test can be controlled by choice of the sample size n.

Example 13.2.10. To acquire some geometric intuition, let us consider Example 13.2.9 for $n = 2$. The observation (X_1, X_2) is then a point on the plane, and the rejection region, of the form $|\overline{X} - \mu_0| > k$, is the shaded area in Figure 13.5. The lines parallel to the axis of the nonrejection region are

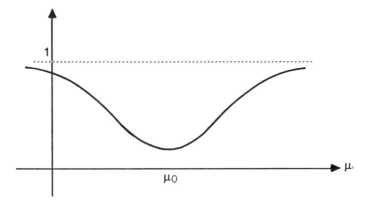

Figure 13.4 Power of the two sided test C.

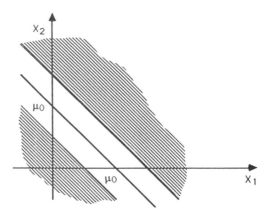

Figure 13.5   Critical region for two-sided test C.

lines when $x_1 + x_2 = c$, hence lines of constant mean $\overline{x}$. The fact that each such line is either completely in the critical region C or completely outside it means that the decision depends on individual observations x_1 and x_2 only through their sum (hence their mean). We say in this case that our test is based on the *test statistic* $\overline{X}$.   □

A moment of reflection on Example 13.2.10 leads to a number of questions. First, is it always the case that we may choose a critical region that can be described in terms of a certain statistic? Clearly, sets of the shape as in Figure 13.5 are convenient for calculating probabilities, since these probabilities concern events expressed through inequalities on $\overline{X}$, and $\overline{X}$ is a random variable whose distribution is well known (in the case of normally distributed components). But perhaps we could get a better test if the critical region had another shape?

Some of the critical regions one might suggest are:

(i) Reject the null hypothesis $H_0 : \mu = \mu_0$ if at least one of the observations X_1, X_2 deviates from μ_0 by more than a specified amount, that is, if $\max(|X_1 - \mu_0|, |X_2 - \mu_0|) \geq c$,

(ii) Reject the null hypothesis if both observations X_1, X_2 deviate from μ_0 by more than a specified amount, that is, if $\min(|X_1 - \mu_0|, |X_2 - \mu_0|) \geq c$.

(iii) Reject the null hypothesis if the point (X_1, X_2) lies outside a circle with center (μ_0, μ_0) and some specified radius, that is, if

$$(X_1 - \mu_0)^2 + (X_2 - \mu_0)^2 \geq r^2.$$

The three types of critical regions are depicted on Figure 13.6. These regions are based on test statistics different than the mean $\overline{X}$, namely on statis-

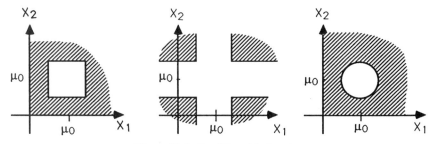

Figure 13.6 Possible critical regions.

tics $T_1 = \max_i(|X_i - \mu_0|)$, $T_2 = \min_i(|X_i - \mu_0|)$, and $T_3 = \sum_i(X_i - \mu_0)^2$. Obviously, one can suggest a number of other statistics to define critical regions.

Moreover, all critical regions C considered thus far had the property that if a point (x_1, x_2) was in C, then the point (x_2, x_1) was also in C. In other words, C was symmetric with respect to the main diagonal $x_1 = x_2$. Such a property reflects our conviction that the order of observations is not relevant: Two data sets that differ only by order of observing their individual data points carry the same information about distribution and should therefore be regarded as equal in their strength of evidence for or against any hypothesis.

Whether or not such a principle of symmetry in treatment of the data points should be used depends on a number of factors. As obvious example when one should not treat the data points symmetrically is when x_1 and x_2 are observations taken by two persons, or in two laboratories, of which one is known to be more reliable (say) than the other. We should then attach more importance to one observation than to the other, because $(3, 7)$ (say) will not carry the same information as $(7, 3)$.

The main issues raised by the examples of this section may be summarized as follows:

(a) A test—being a procedure which ultimately leads to a choice between two hypotheses—is equivalent to specifying a critical region C in the space of observations.

(b) The performance of the test C is described by the power function, defined as $\pi(\theta) = P\{\text{rejecting } H_0 \text{ if } \theta \text{ is the true parameter value}\}$. All criteria of the choice of a test should ultimately be expressed in terms of the power function. Any such criterion should conform to the intuitive requirement that the power of a test should be as high as possible for θ in the alternative hypothesis, and as low as possible for θ in the null hypothesis.

(c) The class of all possible tests (i.e., class of all possible critical regions C) is very rich, and *any* reduction of this class will facilitate a search for the best set (in whichever way the optimality is ultimately defined).

In particular, one such reduction is attained by a restriction exemplified by the symmetry argument above. In the case of n dimensions, this amounts to considering only sets C such that whenever $\mathbf{x} = (x_1, \ldots, x_n)$ is in C and $\mathbf{x}' = (x_{i_1}, \ldots, x_{i_n})$ is obtained by permuting coordinates of $\mathbf{x}$, then $\mathbf{x}'$ is also in C. Another restriction is to consider only tests with critical region C defined in terms of some statistic T. These two (as well as other) restrictions of the class of tests will be discussed in the subsequent parts of this chapter.

PROBLEMS

13.2.1 Consider the following procedure for testing the hypothesis $H_0 : p \geq 0.5$ against the alternative $H_1 : p < 0.5$ in binomial distribution with parameters $n = 10$ and p. We take observation X_1 and reject H_0 if $X_1 = 0$ or accept H_0 if $X_1 \geq 9$; otherwise, we take an observation X_2 (with the same distribution as X_1 and independent of it). Then we accept or reject H_0 depending whether $X_1 + X_2 \geq 5$ or $X_1 + X_2 < 5$. Determine the power function of this procedure.

13.2.2 Consider three tests, C_1, C_2, and C_3, of the same hypothesis, performed independently (e.g., for each of these tests the decision is based on a different sample). Consider now the following three procedures:

Procedure A: Reject H_0 if all three tests reject it; otherwise, accept H_0,

Procedure B: Reject H_0 if at least two tests reject it; otherwise, accept H_0,

Procedure C: Reject H_0 if at least one test rejects it; otherwise, accept H_0.

(i) Express the power functions of procedures A, B, and C through power functions of tests $C_1 - C_3$.

(ii) Assuming the power functions of tests $C_1 - C_3$ being the same, make graphs of power functions of procedures A, B, and C and determine the sets (within H_0 and within H_1) where each of the procedures A, B, and C performs better than the "component" test C_i.

13.2.3 Assume that X is uniformly distributed on $[0, \theta]$, where $\theta > 0$. We want to test the null hypothesis $H_0 : \theta = 3$ against the simple alternative $H_1 : \theta = 2$ (observe that H_0 and H_1 do not exhaust all possibilities). We take the sample of two observations and reject H_0 if $\overline{X} < c$.

(i) Find c such that the probability of a type I error of the test is $\alpha = 0.05$.

(ii) For the test found in (i), find the power function.

13.2.4 Since $\overline{X}$ is not a sufficient statistic for θ, one may expect that the test in Problem 13.2.3 is inferior to a test based on the statistic $\max(X_1, X_2)$.

    (i) Consider a test with critical region $\max(X_1, X_2) \leq c$ and find c which gives the probability of type I error equal to 0.05.

    (ii) Find the power of this test and graph it together with the power of the test found in Problem 13.2.3.

13.2.5 An urn contains five balls, r red and $5 - r$ white. The null hypothesis states that all balls are of the same color (i.e., $H_0 : r = 0$ or $r = 5$). Suppose that we take a sample of size 2 and reject H_0 if the balls are of different colors. Find the power of this test for $r = 0, \ldots, 5$ if the sample is drawn: (i) without replacement; (ii) with replacement. (iii) In each case find the probability of type I error.

13.2.6 For the urn in Problem 13.2.5, suppose that the null hypothesis is $H_0 : r = 1$ or $r = 2$. Draw k balls and let the test be "reject H_0 if all k balls drawn are white." For sampling with and without replacement find the power of the test for all r.

13.3 MOST POWERFUL TESTS

We begin developing the theory from the case of testing a simple null hypothesis against a simple alternative. In the parametric situation of this chapter, it means that we want to test the hypothesis $H_0 : \theta = \theta_0$ against the alternative $H_1 : \theta = \theta_1$, where θ_0 and θ_1 are two parameter values.

To simplify the notation, let us write $f(x, \theta_0) = f_0(x)$ and $f(x, \theta_1) = f_1(x)$. The only assumption about f_0 and f_1 is that they represent different probability distributions. In the discrete case, this means that $P\{X = x | H_0\} = f_0(x) \neq f_1(x) = P\{X = x | H_1\}$ for some x. In the continuous case, when f_0 and f_1 are densities, it is not enough that f_0 and f_1 differ at some isolated point, or even on a countable set of points. Thus, in the continuous case we assume that

$$\int_A f_0(x)\, dx \neq \int_A f_1(x)\, dx$$

for some set A. We present the motivation for various steps of the proof of the reduction principle in the case of a single observations. The extension to the case on n observations will be obvious.

Suppose that we contemplate using some critical region C. Thus, the test is

$$\text{reject } H_0 \text{ if } x \in C. \tag{13.8}$$

Since both hypotheses are simple, we can determine the probabilities of errors of the first and second kind. Thus, the probability of rejecting H_0 if it is true (type I error) equals

$$\alpha = P\{x \in C|H_0\} = \begin{cases} \int_C f_0(x) \, dx \\ \sum_{x \in C} f_0(x) \end{cases} \tag{13.9}$$

depending on whether we deal with a continuous or a discrete case. On the other hand, probability of not rejecting H_0 if H_1 is true (type II error) is

$$\beta = P\{x \notin C|H_1\} = \begin{cases} 1 - \int_C f_1(x) \, dx \\ 1 - \sum_{x \in C} f_1(x) \, dx \end{cases} \tag{13.10}$$

again depending on whether we deal with a continuous or discrete case. Since to decrease β we must increase the integral in (13.10), the problem is to choose C so as to maximize $\int_C f_1(x) \, dx$.

It will be convenient to partition the set of values of x into four sets A_0, A_1, A_2, A_3, depending on which of the two densities is zero and which is positive (see Figure 13.7). Let us first consider the set

$$A_1 = \{x : f_0(x) = 0, f_1(x) > 0\}. \tag{13.11}$$

If any part of this set lies outside C, then we can improve the test by including it into C, that is, take the test with critical region $C^* = C \cup A_1 = C \cup (C^c \cap A_1)$.

Indeed, for the test C^* (in continuous case) we have

$$\alpha^* = \int_{C^*} f_0(x) \, dx = \int_C f_0(x) \, dx + \int_{C^c \cap A_1} f_0(x) \, dx = \alpha$$

since $f_0(x) \equiv 0$ on $C^c \cap A_1$.

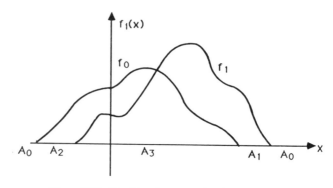

Figure 13.7 Partition into sets A_0, A_1, A_2, and A_3.

Thus, test C^* has the same probability of error of the first kind as test C. However (again, for continuous case), we have

$$
\beta^* = 1 - \int_{C^*} f_1(x)\, dx = 1 - \int_C f_1(x)\, dx - \int_{C^c \cap A_1} f_1(x)\, dx
$$

$$
\leq 1 - \int_C f_1(x)\, dx = \beta,
$$

which means that the probability of type II error decreased. This argument shows that the set A_1 should be totally contained in the critical region C.
   On the other hand, consider the set

$$
A_2 = \{x : f_0(x) > 0, f_1(x) = 0\}. \tag{13.12}
$$

A reasoning analogous to that carried for A_1 shows that the entire set A_2 should lie outside the critical region C. Next, the set

$$
A_0 = \{x : f_0(x) = f_1(x) = 0\}
$$

plays no role in our reasoning: The way in which this set is partitioned between C and its complement has no effect on α and β.
   Now let

$$
A_3 = \{x : f_0(x) > 0 \text{ and } f_1(x) > 0\}. \tag{13.13}
$$

The problem reduces to finding the best way of partitioning A_3 between C and its complement.
   To simplify the argument, consider the discrete case. Suppose that we have already selected some set C as a candidate for a critical set. Suppose that $C \supset A_1$ and $A_2 \subset C^c$, and we contemplate the means of improving C by adding or removing some points from it.
   Let $x^* \notin C$ (with $x^* \in A_3$) and consider the consequences of changing C by including x^* in it. The probability of error of the first kind will change to

$$
\sum_{x \in C} f_0(x) + f_0(x^*) = \alpha + f_0(x^*) > \alpha.
$$

On the other hand, the probability of error of the second kind will change to

$$
1 - \left[\sum_{x \in C} f_1(x) + f_1(x^*) \right] = \beta - f_1(x^*) < \beta.
$$

Thus, α increased by $f_0(x^*)$, while β decreased by $f_1(x^*)$.

In a similar way, suppose that $x^{**} \in C$ and we contemplate removing the point x^{**} from C. Then α will decrease by $f_0(x^{**})$ while β will increase by $f_1(x^{**})$.

Consequently, if $x^* \notin C$ and $x^{**} \in C$, switching the role of these points results in the following change:

$$\begin{aligned} \alpha \quad &\text{changes to} \quad \alpha + (f_0(x^*) - f_0(x^{**})) \\ \beta \quad &\text{changes to} \quad \beta + (f_1(x^{**}) - f_1(x^*)). \end{aligned}$$

Since we want to minimize both α and β, such a change should *always* be carried out if $f_0(x^*) - f_0(x^{**}) \leq 0$ and $f_1(x^{**}) - f_1(x^*) \leq 0$, with at least one inequality being strict.

The inequalities above mean (remembering that x^* and x^{**} are in A_3): $f_1(x^{**}) \leq f_1(x^*)$ and $f_0(x^{**}) \geq f_0(x^*) > 0$, so that (since one inequality is strict)

$$\frac{f_1(x^{**})}{f_0(x^{**})} < \frac{f_1(x^*)}{f_0(x^*)}. \tag{13.14}$$

We therefore obtained the following principle of choosing critical regions C.

Reduction Principle. In choosing the critical regions, one should restrict the considerations to sets based on the *likelihood ratio* $f_1(x)/f_0(x)$, of the form

$$C_k = \left\{ x : \frac{f_1(x)}{f_0(x)} \geq k \right\}. \tag{13.15}$$

$\square$

Indeed, if a critical set is not of the form (13.15), then there exists points x^* and x^{**}, with $x^* \notin C$ and $x^{**} \in C$ and such that (13.14) holds. Then one obtains a better critical region if one exchanges the role of x^* and x^{**}.

It should be observed that we did not solve the question of finding the best test. In fact, we did not even specify the criterion to be optimized. The reduction principle tells us only which tests should not be used, or—to put it differently—it specifies the class of tests from which the choice should be made, provided only that the optimality criterion is compatible with the general motive to decrease *both* probabilities of errors.

Two obvious ways of defining the criterion to be optimized are as follows:

1. Impose an upper bound on one of the probabilities of errors, and minimize the probability of the other kind of error.
2. Minimize a linear combination of the two error probabilities.

These two approaches are closely related, the first being used in the original Neyman–Pearson lemma, dating back to 1933. In formulating the theorem below we shall consider a general situation, of a random sample $\mathbf{X} = (X_1, \ldots, X_n)$ from one of the two distributions, f_0 or f_1. The null hypothesis asserts that the distribution is f_0. For $i = 0, 1$ we shall write $f_i(\mathbf{x}) = f_i(x_1) \cdots f_i(x_n)$. The critical regions under consideration are now subsets of the n-dimensional Euclidean space.

Theorem 13.3.1 (Neyman–Pearson Lemma). *Let C^* be a critical region that has the following properties: There exists a constant $k > 0$ such that:*

   (i) *If $f_1(\mathbf{x})/f_0(\mathbf{x}) > k$, then $\mathbf{x} \in C^*$*
  (ii) *If $f_1(\mathbf{x})/f_0(\mathbf{x}) < k$, then $\mathbf{x} \notin C^*$ [with points at which $f_1(\mathbf{x})/f_0(\mathbf{x}) = k$ partitioned between C^* and its complement in some way].*

Let C be any critical region. Then $\alpha(C) \leq \alpha(C^*)$ implies that $\beta(C) \geq \beta(C^*)$, and if $\alpha(C) < \alpha(C^*)$, then $\beta(C) > \beta(C^*)$.

Proof. Without danger of confusion we may use the symbols $A_0, A_1, A_2,$ and A_3 for the four sets in $\mathbf{R}^n$, depending on which of the joint densities $f_0(\mathbf{x})$ and $f_1(\mathbf{x})$ is zero and which is strictly positive. If we make the convention that $c/0 = \infty$ for any $c > 0$, then the critical region C^* in the lemma is such that

$$A_1 \subset C^*, A_2 \subset (C^*)^c. \tag{13.16}$$

Indeed, we have $f_1(\mathbf{x})/f_0(\mathbf{x}) = \infty > k$ on A_1 and $f_1(\mathbf{x})/f_0(\mathbf{x}) = 0 < k$ on A_2. The theorem follows from the argument for the one-dimensional case preceding this proof: If C does not meet conditions (13.16), then one or both of its error probabilities can be improved, as stated in the assertion of the Neyman–Pearson lemma.

Next, the set A_0 plays no role, and the question remains about points $\mathbf{x}$ in A_3. In the discrete case, the argument given for the one-dimensional case remains valid: If C is not of the form specified in the lemma, then there exist $\mathbf{x}^* \notin C$ and $\mathbf{x}^{**} \in C$ such that (13.15) holds, and switching these points we improve both error probabilities. It remains therefore to prove the theorem in the continuous case. Consider the sets

$$C^* \cap C^c = \left\{ \mathbf{x} : \frac{f_1(\mathbf{x})}{f_0(\mathbf{x})} > k, \mathbf{x} \notin C \right\}$$

$$(C^*)^c \cap C = \left\{ \mathbf{x} : \frac{f_1(\mathbf{x})}{f_0(\mathbf{x})} < k, \mathbf{x} \in C \right\}.$$

We have here

$$\alpha(C^*) - \alpha(C) = \int_{C^*} f_0(\mathbf{x}) \, d\mathbf{x} - \int_C f_0(\mathbf{x}) \, d\mathbf{x}$$

$$= \int_{C^* \cap C^c} f_0(\mathbf{x}) \, d\mathbf{x} - \int_{C \cap (C^*)^c} f_0(\mathbf{x}) \, d\mathbf{x}$$

$$\leq \int_{C^* \cap C^c} \frac{1}{k} f_1(\mathbf{x}) \, d\mathbf{x} - \int_{C \cap (C^*)^c} \frac{1}{k} f_1(\mathbf{x}) \, d\mathbf{x}$$

$$= \frac{1}{k} \int_{C^c} f_1(\mathbf{x}) \, d\mathbf{x} - \frac{1}{k} \int_{(C^*)^c} f_1(\mathbf{x}) \, d\mathbf{x}$$

$$= \frac{1}{k} [\beta(C) - \beta(C^*)].$$

The inequality between the extreme terms completes the proof: If $\alpha(C) \leq \alpha(C^*)$, then the left-hand side is nonnegative and so must be the right-hand side, which means that $\beta(C) \geq \beta(C^*)$. If the left-hand side is strictly positive, so must be the right-hand side. $\qquad\square$

We shall now formulate the analogue of the Neyman–Pearson lemma when the criterion to be minimized is a linear combination of the two error probabilities. The proof, which follows closely the reasoning in the proof of Neyman–Pearson lemma, will be omitted.

Theorem 13.3.2. *Suppose that in testing the simple null hypothesis H_0: $f = f_0$ against the simple alternative $H_1 : f = f_1$, it is desired to find the critical region C^* in n-dimensional space, such that for any other critical region C we have*

$$A\alpha(C^*) + B\beta(C^*) \leq A\alpha(C) + B\beta(C),$$

where $A > 0, B > 0$ are given constants. Then C^* contains all points $\mathbf{x}$ such that $f_1(\mathbf{x})/f_0(\mathbf{x}) > A/B$ and its complement contains all points $\mathbf{x}$ such that $f_1(\mathbf{x})/f_0(\mathbf{x}) < A/B$. The points where $f_1(\mathbf{x})/f_0(\mathbf{x}) = A/B$ can be allocated between C^* and its complement in an arbitrary way.

Since in the case of simple hypotheses $1 - \beta(C)$ is the power of the test with critical region C, the Neyman–Pearson lemma gives us in effect a rule of constructing the *most powerful* test, with the preassigned probability $\alpha(C)$ of type I error (as we shall see, this probability will be determined by the choice of a constant k).

In the sequel we shall use the following definitions. Let C be the critical region for testing the null hypothesis $H_0 : \theta \in \Theta_0$ against the alternative $H_1 : \theta \in \Theta_1$, and let $\pi_C(\theta) = P_\theta\{\mathbf{X} \in C\}$ be the power of the test.

Definition 13.3.1. The *size* of the test C is defined as

$$\overline{\alpha}(C) = \sup_{\theta \in \Theta_0} \pi_c(\theta).$$

$\qquad\square$

Thus, the size of the test of a simple null hypothesis is the probability of type I error, while for composite null hypotheses size is the least upper bound for all probabilities of type I errors.

Definition 13.3.2. Any number α satisfying the inequality $\overline{\alpha}(C) \leq \alpha$ is called the *level* of test C, and a test C satisfying $\overline{\alpha}(C) \leq \alpha$ will be called an α-*level* test.   □

Thus, according to this definition, a test with $\overline{\alpha}(C) = 0.01$ is a 1%-level test, as well as a 5%-level test, and so on. Each test is a 1-level (100%-level) test.

Finally, we define one more concept, that of significance level. This is not so much a property of a test as the constraint imposed by the statistician.

Definition 13.3.3. If in testing the hypothesis H_0 the statistician chooses a number α_0 $(0 < \alpha_0 < 1)$ and decides to use only those tests C whose size satisfies the inequality $\overline{\alpha}(C) \leq \alpha_0$, then α_0 is called the *significance level* chosen by the user.   □

It is often felt that a report that a given hypothesis was rejected at significance level $\alpha = 0.05$ is not informative enough (e.g., perhaps the hypothesis would also be rejected at significance level $\alpha = 0.01$ or even lower?). Thus, it is customary to report the *p-value* of the empirical data, defined as the lowest level at which the null hypothesis would be rejected by the test. More intuitively, the *p*-value of the results actually observed is the probability that—if the experiment were repeated—we would obtain results that give at least as strong evidence for the alternative hypothesis (or equivalently, as strong evidence against the null hypothesis) as the present result. For instance, suppose that we use statistic T for testing, and we reject the null hypothesis if T is large. The result that we observe is $T = t_0$, say. Then the *p*-value of this result is $P\{T \geq t_0 | H_0\}$, so that the smaller is the *p*-value, the stronger is the evidence that suggests rejection of the null hypothesis.*

In case of two-sided tests based on statistic T with a symmetric distribution, the *p*-value of the result t_0 is defined as $2P(T > |t_0|)$. Here the rationale is that the observations more "in favor" of the alternative hypothesis are those above $|t_0|$ and below $-|t_0|$.

In case of tests based on a statistic that does not have a symmetric distribution, the ideas behind the concept of *p*-value become rather fuzzy, and there is no definition on which all statisticians agree. Consequently, one may find various "competing" definitions in different texts. For a review of these definitions, see Gibbons and Pratt (1975). For that reason, some statisticians are against using *p*-values at all. The situation is not helped by the fact that many nonstatistical journals that publish experimental findings (e.g., in medicine,

---

*High *p*-values (close to 1) might serve as indication that the data were manipulated so as to make them "more conforming" to the null hypothesis. The rationale here is as follows: Suppose that the *p*-value of the data is 0.999 (say). This means that the "fit" to H_0 is so perfect that only once in a 1000 times we would observe a better fit. The same principles that lie at the foundation of testing allow us to reach the conclusion that the data were fudged to fit H_0.

biology, education, sociology, psychology, etc.) have a stated policy of not accepting an article unless it shows the *p*-values of the results. Also, many advisors of Ph.D. dissertations require their students to supply the *p*-values in their theses.

Generally speaking, the insistence on using *p*-values is an illustration of the fact that the spread of use of statistical methods is not always accompanied by a parallel increase in the statistical culture.

We shall now illustrate the procedure of test selection by some examples. In four initial examples, we shall also illustrate another aspect: how an empirical hypothesis becomes "translated" into a statistical hypothesis.

Example 13.3.1. Suppose that a psychologist thinks that she found the exact location where memory is stored in the brain. To test whether or not she is right, the following experiment is performed. Ten rats are trained to find food in a maze. In running through the maze, the rat makes three binary choices and as a result ends in one of eight final locations, of which only one contains food. The rats are then subject to surgical destruction of that part of the brain which—according to the researcher's hypothesis—stores the acquired knowledge. After a period of healing, the rats run the maze again, and we observe the number X of rats that reach the arm with food. Without any memory left, each rat has probability $(\frac{1}{2})^3 = 0.125$ of finding the way to the food on the first trial. Suppose that—according to the experimenter—if this probability is as high as 0.3, it would mean that part of the memory must be stored in another (undamaged) region of the brain. If the rats run through the maze independently, and each has the same probability p of choosing the path leading to food, then X has binomial distribution BIN$(10, p)$. We have now two hypotheses: one asserting that $p = 0.125$ (if the researcher is correct in identifying the memory storage region of the brain), and the other asserting that $p = 0.3$ (if the researcher is wrong). Suppose that from the researcher's point of view the error of rejecting hypothesis $p = 0.125$ (i.e., that no memory is left) if in fact it is true is more serious than accepting $p = 0.125$ if in fact $p = 0.3$.

In this case we set the null hypothesis as $H_0 : p = 0.125$ and the alternative is $H_1 : p = 0.3$. Consequently, $f_0(x) = \binom{10}{x}(0.125)^x(0.875)^{10-x}$ and $f_1(x) = \binom{10}{x}(0.3)^x(0.7)^{10-x}$. According to the Neyman–Pearson lemma, we should choose the critical region of the form $\{x : f_1(x)/f_0(x) \ge k\}$ for some k. But

$$f_1(x)/f_0(x) = \frac{(0.3)^x(0.7)^{10-x}}{(0.125)^x(0.875)^{10-x}}$$

$$= \left(\frac{0.7}{0.875}\right)^{10}\left(\frac{0.3 \cdot 0.875}{0.125 \times 0.7}\right)^x = 0.107 \times 3^x.$$

The inequality $\{x : f_1(x)/f_0(x) \geq k\}$ is equivalent to the inequality $x \geq k^*$ for some k^*. This means that the critical region C is formed of the right tail of the distribution of X. Since the possible values of X are 0, 1, ..., 10, we must simply determine the smallest value that belongs to C. Here the choice depends on the significance level α_0 of the test. Suppose that the experimenter decides to use $\alpha_0 = 0.05$. Then he must choose k^* so that $P\{X \geq k^*|p = 0.125\} \leq 0.05$; that is, $P\{X < k^*|p = 0.125\} \geq 0.95$. We have here $P\{X < 4|p = 0.125\} = 0.973$ while $P\{X < 3|p = 0.125\} = 0.881$. It follows that we must reject the null hypothesis $H_0 : p = 0.125$ if $X \geq 4$, that is, if four or more rats find their way to the food on the first try. This test has the probability of type I error equal to $1 - 0.973 = 0.027$, or 2.7%.

The probability of type II error is $P\{X < 4|p = 0.3\} = 0.649$.   □

Some comments about this example appear to be in order. First, we see that type II error has rather high probability, while the probability of type I error is below the level of significance. The test with critical region $C = \{3, 4, ..., 10\}$ will have $\alpha(C) = 0.119$ (i.e., almost 12%), while $\beta(C) = 0.382$. It is clear that among tests satisfying the assertion of the Neyman–Pearson lemma, none has the probability of type I error equal to the desired significance level 0.05. This is due to the discrete nature of the test statistic X.

A procedure with the probability of type I error of $\alpha_0 = 0.05$ exists if we allow randomized procedures. Indeed, suppose that we decide to reject H_0 if $X \geq 4$; accept it if $X \leq 2$; and if $X = 3$, activate some auxiliary random mechanism and reject H_0 with probability γ. Then the probability of rejecting H_0 if it is true equals

$$\gamma P\{X = 3|p = 0.125\} + P\{X \geq 4|p = 0.125\} = 0.092\gamma + 0.027,$$

which equals 0.05 if $\gamma = 0.25$.

A procedure that attains significance level $\alpha_0 = 0.05$ is therefore as follows: If four or more rats reach food on their first trial, then reject H_0. If three rats reach food on the first trial, then toss two coins, and reject H_0 if both show up heads. In all other cases accept H_0.

As already remarked, such randomized procedure may be suggested and agreed upon in the process of acceptance of rejection of merchandise. However, in case of decision regarding scientific hypotheses, it would be disturbing to rely on the toss of a coin.

A way out from the dilemma is to abandon the concept of significance level as a quantity imposed by the experimenter. One can then proceed to minimize a linear combination of errors of type I and type II as in Theorem 13.3.2. A procedure that attains it never requires randomization.

Example 13.3.2. Continuing Example 13.3.1, if we want to find critical region C so as to minimize the linear combination $10\alpha(C) + \beta(C)$, we must include in the critical region all x such that $f_1(x)/f_0(x) > 10$, which means that

$$\frac{(0.3)^x (0.7)^{10-x}}{(0.125)^x (0.875)^{10-x}} > 10$$

or $3^x > 93.46$, hence $x > 4.13$. Thus, if type I error is considered 10 times as serious as type II error, then the null hypothesis should be rejected only when 5 or more rats find their way to the food on the first attempt.

Example 13.3.3.   Suppose that we have a source of drinking water (a well, say) suspected of being contaminated with some bacteria. A fixed volume V of water from the well is sent into the laboratory. Assume that the admissible norm is N bacteria of the kind in question for volume V, with $5N$ bacteria in volume V indicating an undesirable level of contamination. Here $N > 0$, so that some positive level of contamination is acceptable as safe.

The procedure in the laboratory is as follows. The water sent for analysis is thoroughly mixed, and then a sample of volume v of water is drawn from it, where $v \ll V$. A technician inspects the sample under the microscope and reports the number X of bacteria that she sees. It is assumed that the probability that each of the bacteria present in the observed sample is recorded by the technician with probability π, independently of the other bacteria, where the value of π is assumed known.

This process is repeated n times, by the same or perhaps by different technicians, who examine different samples of water, each of volume v. Thus, we have the reports $X_1, X_2, \ldots, X_n$ of the numbers of bacteria observed in different samples. We assume that the probability $1 - \pi$ of overlooking a bacterium is the same for all technicians.

It is desired to test the null hypothesis that the water in the well is safe against the alternative that it contains an undesirable level of bacteria. To solve this problem, it is first necessary to determine the distributions of the observable random variables and translate the empirical hypotheses stated above into statistical hypotheses.

Let $\lambda_0 = N/V$ be the density of bacteria in the well allowed by the safety standards, so that the null hypothesis is $H_0 : \lambda = \lambda_0$, where λ is the actual density of bacteria per unit volume of water in the well. The alternative hypothesis is $H_1 : \lambda = 5N/V = 5\lambda_0 = \lambda_1$, say. We have here $\lambda_1 > \lambda_0$, since $N > 0$ by assumption.

We assume now that in taking the sample of volume v, each of the bacteria has the probability v/V of being included into the sample, independently of other bacteria. The assumption that $v \ll V$ makes the probability v/V small, and from Example 7.3.3 (about "thinning" of Poisson processes) it follows that in each sample of volume v the actual number of bacteria will have Poisson distribution with mean Mv/V, where M is the actual number of bacteria in volume V. Thus, under null hypothesis H_0, in each sample of volume v the number of bacteria will have Poisson distribution with mean $Nv/V = \lambda_0 v$, while under the alternative, the mean will be $\lambda_1 v = 5\lambda_0 v$.

We assume also that the total volume of water inspected by technicians, $v_1 + v_2 + \cdots + v_n$, is still small compared with V, so that the Poisson approximation applies to the total number of bacteria in all samples. Finally, as regards the distribution of the counts $X_1, \ldots, X_n$, from Example 7.3.3 (about "thinning" of Poisson processes) it follows that they also have Poisson distribution, with mean $\lambda_0 \pi$ under hypothesis H_0 and mean $\lambda_1 \pi$ under the alternative hypothesis H_1.

Letting $\mathbf{X} = (X_1, \ldots, X_n)$, we know that the test (whether most powerful, in the sense of the Neyman–Pearson lemma or minimizing the linear combination of error probabilities) is based on the likelihood ratio $f_1(\mathbf{x})/f_0(\mathbf{x})$, with values larger than a threshold leading to rejection of the null hypothesis. We have here for $i = 0, 1$,

$$f_i(\mathbf{x}) = \prod_{j=1}^{n} f_i(X_j) = \prod_{j=1}^{n} \frac{(\pi \lambda_i v)^{X_j}}{X_j!} e^{-\pi \lambda_i v}$$

$$= \frac{(\pi v)^{\sum X_j}}{X_1! \cdots X_n!} e^{-n\pi \lambda_i v} \lambda_i^{\sum X_j}.$$

Consequently, remembering that $\lambda_1 = 5\lambda_0$, we obtain

$$\frac{f_1(\mathbf{x})}{f_0(\mathbf{x})} = e^{-4n\pi\lambda_0 v} 5^{\sum X_j}$$

and we should reject the null hypothesis H_0 if the total count $X_1 + \cdots + X_n$ reported by all technicians exceeds some threshold k.

The actual value of k depends on the numerical values of the parameters. For instance, suppose that we take $n = 50$ samples to be inspected by technicians, and that the probability π of recording a bacteria by each of them is $\pi = 0.1$ (i.e., each technician notices about one bacterium out of each 10 present). Furthermore, assume that the number of bacteria allowed is $10,000$ per liter (10^3 cm^3), with the volume v taken for inspection being $\frac{1}{20}$ of a cubic centimeter. This gives $\lambda_0 = 10^4/10^3 = 10$ with $v = 0.05$ cm^3, so that

$$\pi \lambda_0 v = 0.1 \times 10 \times 0.05 = 0.05.$$

Consequently, the total count $X_1 + \cdots + X_{50}$ in all samples has, under the null hypothesis H_0, Poisson distribution with mean $0.05 \times 50 = 2.5$. Under the alternative, this count is still Poisson, with the mean 12.5.

Suppose that we want to test the null hypothesis on significance level $\alpha = 0.05$. We therefore want k determined from the condition

$$P\{X_1 + \cdots + X_{50} \geq k | \lambda = 2.5\} \leq 0.05,$$

that is, $P\{U \geq k\} \leq 0.05$, where U has Poisson distribution with mean 2.5. We have $P\{U \leq 5\} = 0.9580$, which means that $P\{U \geq 6\} = 0.042$ and we may take $k = 6$. The probability of type II error of this test is $\beta = P(X_1 + \cdots + X_{50} < 6|\lambda = 12.5) = 0.015$. Thus, chances of raising false alarm (in this case type I error) are about 4%, while chances of failing to notice the dangerous level of contamination are only about 1.5%.

Example 13.3.4. Continuing Example 13.3.3, suppose now that the laboratory is under a financial squeeze and decided to save on the cost of the observations. The technicians are instructed to inspect each sample of size v and report only whether or not the bacteria were found. The saving here is that the technician can stop searching the sample as soon as she finds one bacterium (instead of continuing search and counting). Let $Y_1, \ldots, Y_n$ be the data as reported now, with $Y_j = 1$ or 0 depending whether bacteria were found or not in the jth sample. We have, for $i = 0, 1$,

$$P\{Y_j = 1|H_i\} = 1 - P\{Y_j = 0|H_i\} = 1 - e^{-\lambda_i \pi v}. \tag{13.17}$$

Consequently, the total number of samples where bacteria were found is equal $S = Y_1 + \cdots + Y_n$, has binomial distribution, with number of trials n and probability of success given by (13.17), depending on whether H_0 or H_1 is true. Remembering that $\lambda_1 = 5\lambda_0$, the likelihood ratio is now

$$\frac{f_1(s)}{f_0(s)} = \frac{P\{S = s|H_1\}}{P\{S = s|H_0\}} = \frac{\binom{n}{s}(1 - e^{-\pi\lambda_1 v})^s (e^{-\pi\lambda_1 v})^{n-s}}{\binom{n}{s}(1 - e^{-\pi\lambda_0 v})^s (e^{-\pi\lambda_0 v})^{n-s}}$$

$$= \left[\frac{1 - e^{-5\pi\lambda_0 v}}{1 - e^{-\pi\lambda_0 v}} e^{4\pi\lambda_0 v}\right]^s e^{-4\pi\lambda_0 vn}.$$

Since the fraction in brackets exceeds 1, the likelihood increases with s. Again, the critical region is the set $S \geq k$, where k has to be determined from the fact that under null hypothesis S has binomial distribution BIN(n, $1 - e^{-\pi\lambda_0 v}$). In particular, the laboratory may now ask the question: How large should n be so as to ensure the same significance level and the same power as in the more expensive procedure with technicians counting the bacteria? We now have $e^{-\pi\lambda_0 v} = 1 - e^{-0.05} = 0.0488$ and $e^{-\pi\lambda_1 v} = 1 - e^{-0.2} = 0.1813$. We look for n and k such that $P(S \geq k) \leq 0.05$ where $S \sim$ BIN(n, 0.0488), and $P(S' \leq k) \leq 0.015$ where $S' \sim$ BIN(n, 0.1813). An approximate solution can be found from the central limit theorem for binomial distribution (Theorem 10.5.2). We want to have

$$P\left\{Z \geq \frac{k - 0.0488n}{\sqrt{(0.0488)(0.9512)n}}\right\} = 0.05$$

and

$$P\left\{Z \leq \frac{k - 0.1813n}{\sqrt{(0.1813)(0.8187)n}}\right\} = 0.015,$$

so that

$$k = \begin{cases} 0.0488n + (1.645)(0.2154)\sqrt{n} \\ 0.1813n - (2.17)(0.3852)\sqrt{n}. \end{cases}$$

The solution, rounded to integers, is $n = 81, k = 7$. The simplified observation procedure (recording only whether or not the bacteria are present) necessitates an increase of the number of samples from 50 to 81. □

It is clear that in case of continuous random variables we can always find a test (critical region C) such that $\alpha(C)$ equals the desired significance level α_0. We shall present some examples concerning different distributions other than normal, leaving detailed presentation of various tests for normal case to a separate section.

Example 13.3.5. Suppose that the observations $X_1, \ldots, X_n$ form a random sample from exponential distribution with density $f(x, \lambda) = \lambda e^{-\lambda x}, x > 0$. We want to test, on significance level α_0, the null hypothesis $H_0 : \lambda = 5$ against a simple alternative $H_1 : \lambda = 7$. Now the likelihood ratio is

$$\frac{f_1(\mathbf{x})}{f_0(\mathbf{x})} = \frac{\prod_{j=1}^{n}(7e^{-7x_j})}{\prod_{j=1}^{n}(5e^{-5x_j})} = \left(\frac{7}{5}\right)^n e^{-2\sum x_i}$$

and the inequality $f_1(\mathbf{x})/f_0(\mathbf{x}) > k$ is equivalent to $\sum x_i < k^*$ for some k^*.

We remark here that $k^* = -\frac{1}{2}[n \log(5/7) + \log k]$, but the exact relation between k and k^* is not needed to determine the test: What matters most is the direction of inequality. The way we set the solution is that we *reject* H_0 if the likelihood ratio f_1/f_0 (with alternative density on the top) is *large*. Typically, this condition reduces to an inequality for some statistic (such as $x_1 + \cdots + x_n$, or equivalently, $\bar{x}$). In the present case, the rejection region is the left tail (i.e., values of $x_1 + \cdots + x_n$ less than a certain threshold).

To continue, we need now a value k^* such that

$$P\{X_1 + \cdots + X_n < k^* | H_0\} = \alpha_0.$$

Each X_i has exponential distribution with parameter λ_0, hence with the mean $1/\lambda_0$. Consequently, $2\lambda_0 X_i$ has exponential distribution with mean 2, that is, a chi-square distribution with 2 degrees of freedom (see Section 9.9) and therefore

$$\chi_{2n}^2 = 2\lambda_0(X_1 + \cdots + X_n)$$

has chi-square distribution with $2n$ degrees of freedom.

In the case under consideration, we have $\lambda_0 = 5$; hence the critical threshold k^* can be obtained from the tables of chi-square distribution with $2n$ degrees of freedom:

$$P\{X_1 + \cdots + X_n < k^* | H_0\} = P\{10(X_1 + \cdots + X_n) < 10k^* | H_0\}$$
$$= P\{\chi_{2n}^2 < 10k^*\} = \alpha_0.$$

Taking, as an example, $n = 5$ and $\alpha_0 = 0.01$, we obtain $10k^* = 2.558$ from Table A.4 for 10 degrees of freedom, which gives $k^* = 0.2558$.

The probability of type II error is, proceeding as before for $\lambda = 7$,

$$P\{X_1 + \cdots + X_n > 0.2558 | H_1\} = P\{14(X_1 + \cdots + X_n) > 14 \times 0.2558 | H_1\}$$
$$= P\{\chi_{2n}^2 > 3.5812\},$$

which (for $n = 5$) lies between 95% and 97.5% (to get a more exact value, one needs a more precise tables of chi-square distribution or appropriate statistical software).

One may be surprised that the probability of type 2 error is so high. However, we are using here a sample of size $n = 5$ to test the hypothesis that the mean is $\frac{1}{\lambda} = \frac{1}{5} = 0.2$ against the alternative that the mean is $\frac{1}{7} = 0.1428$ (this explains why the rejection region is on the left tail). Since the difference between the two means is small as compared with the standard deviation under the null hypothesis (equal also $\frac{1}{\lambda} = 0.2$), a test with sample size $n = 5$ will have a low power. In addition, we see here the effect of "greed" for high significance: requiring that $\alpha = 0.01$ forces the chances of type II error to be very high.   □

Thus far, we considered testing a simple hypothesis against a simple alternative in a parametric setup, where the distribution belonged to the same family $f(x; \theta)$, and hypotheses were obtained by specifying $H_0 : \theta = \theta_0$ and $H_1 : \theta = \theta_1$. This is, to be sure, the most common case occurring in practice. It is important, however, to realize that the Neyman–Pearson lemma applies to any two simple hypotheses. For instance, suppose that we have a single observation X, which comes from the distribution uniform on $[0, 1]$ if null hypothesis is true, or from the exponential distribution with mean $\frac{1}{3}$ if the alternative is true. Thus the densities are

$$f_0(x) = \begin{cases} 1 & \text{if } 0 \leq x \leq 1 \\ 0 & \text{otherwise} \end{cases}$$

$$f_1(x) = \begin{cases} 3e^{-3x} & \text{if } x \geq 0 \\ 0 & \text{otherwise.} \end{cases}$$

The likelihood ratio equals

$$\frac{f_1(x)}{f_0(x)} = \begin{cases} 3e^{-3x} & \text{if } 0 \le x \le 1 \\ \infty & \text{if } x > 1 \end{cases}$$

and is undefined for $x < 0$. Thus, each critical region should contain the set $\{x : x > 1\}$, and also a set of the form $3e^{-3x} > k$, hence a set of the form $\{x : 0 \le x \le k^*\}$ for some $k^* < 1$. To determine k^* we must have

$$P\{X \in C|H_0\} = \int_0^{k^*} f_0(x)\, dx + \int_1^\infty f_0(x)\, dx = \int_0^{k^*} f_0(x)\, dx = k^*.$$

It follows that if we choose the significance level $\alpha_0 = 0.05$ (say), then $k^* = 0.05$ and we reject H_0 if the observation is either below 0.05 or above 1. The power of this test is

$$1 - \beta = \int_C f_1(x)\, dx = \int_0^{0.05} 3e^{-3x}\, dx + \int_1^\infty 3e^{-3x}\, dx$$
$$= (1 - e^{-0.15}) + e^{-3} = 0.1891.$$

PROBLEMS

13.3.1 A single observation X is taken from a beta distribution with parameters a and b. Find the most powerful test of the null hypothesis H_0: $a = b = 1$ against the alternative H_1: (i) $a = b = 5$, (ii) $a = 2, b = 3$, (iii) $a = b = \frac{1}{2}$. Use significance level $\alpha = 0.05$.

13.3.2 Let X have the negative binomial distribution with parameters r and p. Find the most powerful test of $H_0 : r = 2, p = \frac{1}{2}$ against $H_1 : r = 4, p = \frac{1}{2}$ on significance level $\alpha = 0.05$. Find the probability of type II error. Use a randomized test if necessary. Solve the problem under both interpretations of X.

13.3.3 Let $X_1, \ldots, X_n$ have joint density $f(\mathbf{x}, \theta)$, and let U be a sufficient statistic for θ. Show that the most powerful test of $H_0 : \theta = \theta_0$ against $H_1 : \theta = \theta_1$ can be expressed in terms of U.

13.3.4 Let $X_1, \ldots, X_n$ be a random sample from an exponential distribution with parameter λ. Null hypothesis $H_0 : \lambda = \lambda_0$ is tested against the alternative $H_1 : \lambda = \lambda_1$, where $\lambda_1 > \lambda_0$. Find the most powerful test, and also find the most powerful test based on the statistic $X_{1:n} = \min(X_1, \ldots, X_n)$. Compare the power functions of the two tests if their probabilities of type I error are the same.

13.3.5 Assume that X is normal with mean $\mu = 2$ and unknown σ^2. Find the best critical region in testing $H_0 : \sigma^2 = 2$ against the alternative **(i)** $H_1 : \sigma^2 = 4$, **(ii)** $H_1 : \sigma^2 = 1$.

13.3.6 An urn contains six balls, r red and $6 - r$ blue. Two balls are chosen without replacement. Find the most powerful test of $H_0 : r = 3$ against the alternative $H_1 : r = 5$, with significance level as close to $\alpha = 0.05$ as possible. Find the probability of type II error for all $r \neq 3$.

13.3.7 A multiple-choice exam gives five answers to each of its n questions, only one of them correct. Assume that a student who does not know the answer has probability 0.2 of selecting the correct answer. Let θ be the number of questions to which the student knows the answers, and let X be the student's score on the text (i.e., number of correct responses).

  (i) Determine $f(x; \theta) = P(X = x | \theta)$.

  (ii) Let $n = 50$. Find the most powerful test for testing the null hypothesis $H_0 : \theta = 30$ against the alternative $H_1 : \theta = 40$, at significance level $\alpha = 0.05$.

  (iii) Determine the probability of type II error of the test in (ii) if $\theta = 40$.

13.3.8 The sample space of a test statistic X has five values $\{a, b, c, d, e\}$. We are testing that the distribution of X is f_0 against the alternative f_1, where f_0 and f_1 are given by the table:

X	a	b	c	d	e
f_0	0.2	0.2	0	0.1	0.5
f_1	0.2	0.4	0.3	0	0.1

Find the most powerful test with $\alpha = 0.2$.

13.4 UNIFORMLY MOST POWERFUL TESTS

Looking at the derivation of the most powerful tests in Examples 13.3.3 and 13.3.5, one may note an interesting fact, namely that the final form of the test is, to a large extent, independent on the choice of a specific alternative hypothesis. For instance, in Example 13.3.3, testing water samples for bacteria, we faced the problem of observations $X_1, \ldots, X_n$ coming from one of the two Poisson distributions. Omitting the details referring to size v of water samples and probability π of recording a bacterium if present, we wanted to test a simple hypothesis $H_0 : E(X_i) = \theta_0$ against a simple alternative $H_1 : E(X_i) = \theta_1$, where $\theta_1 > \theta_0$. The likelihood ratio, after canceling the factorials, is

$$\frac{f_1(\mathbf{X})}{f_0(\mathbf{X})} = \left(\frac{\theta_1}{\theta_0}\right)^{\sum X_i} \left(\frac{e^{-\theta_1}}{e^{-\theta_0}}\right)^n.$$

If $\theta_1 > \theta_0$, then the likelihood ratio is an increasing function of $\sum X_i$; hence the rejection region must be of the form

$$C = \{(x_1, \ldots, x_n) : \sum x_i \geq k\}. \tag{13.18}$$

Determination of the value of k involves only the value of θ_0 and level of significance α_0, while the value θ_1 plays no role, provided that $\theta_1 > \theta_0$. Indeed, we know that under a null hypothesis, $U = \sum_{i=1}^{n} X_i$ has a Poisson distribution with mean $n\theta_0$; hence the value k in (13.18) is determined from

$$k = \min \left\{ r : \sum_{j=0}^{r-1} \frac{(n\theta_0)^j}{j!} e^{-n\theta_0} \geq 1 - \alpha_0 \right\} \tag{13.19}$$

where α_0 is the desired significance level.

This means, therefore, that the test with a critical region given by (13.18) and (13.19) is most powerful for *any* alternative hypothesis if only the reasoning leading to (13.18) applies. But the only fact about θ_1 used in the derivation is that $\theta_1 > \theta_0$: This inequality causes the ratio θ_1/θ_0 to exceed 1 and hence the likelihood ratio to increase with the sum $\sum x_i$ (for $\theta_1 < \theta_0$ the likelihood ratio would decrease with the increase of $\sum x_i$ and the critical region would comprise the left tail, i.e., we would reject H_0 if $\sum x_i \leq k$).

The situation when there exists a test that is most powerful against each alternative in a certain class is of sufficient theoretical and practical importance to warrant a separate definition. We shall consider the case of testing a hypothesis $H_0 : \theta \in \Theta_0$ (which may be simple or composite) against composite alternative $H_1 : \theta \in \Theta_1$, on the basis of random sample $\mathbf{X} = (X_1, \ldots, X_n)$.

In the continuous case, we can restrict the considerations to tests based on critical regions. In the discrete case, one can also consider randomized procedures, that is, procedures in which rejection or acceptance of H_0 depends both on observation of $\mathbf{X}$ and also possibly on additional randomization.

Consequently, in the definitions below we consider *procedures* for testing H_0; without much danger of confusion we shall use letter C for a procedure and define its power function as

$$\pi_C(\theta) = P_\theta \{\text{procedure } C \text{ rejects } H_0\}.$$

Let us now fix significance level α_0 $(0 < \alpha_0 < 1)$ and let $K(H_0, \alpha_0)$ be the class of all procedures for testing H_0 whose size is at most α_0, that is, procedures C such that

$$\sup_{\theta \in \Theta_0} \pi_C(\theta) \leq \alpha_0. \tag{13.20}$$

Next, for each $\theta \in \Theta_1$, let

$$g(\theta) = \sup_{C \in K(H_0, \alpha_0)} \pi_C(\theta),$$

so that $g(\theta)$ is the least upper bound for powers of procedures in $K(H_0, \alpha_0)$ against a specific simple alternative θ in H_1.

Definition 13.4.1. A procedure C^* such that

(i) $C^* \in K(H_0, \alpha_0)$
(ii) $\pi_{C^*}(\theta) = g(\theta)$ for every $\theta \in \Theta_1$

will be called *uniformly most powerful* (UMP) procedure for testing H_0 against H_1 on significance level α_0.

In case C^* is nonrandomized (i.e., C^* is the critical region of a test), we shall speak of a UMP test of H_0 against H_1 on the significance level α_0. □

The essence of this definition lies in the fact that *the same* test (procedure) is most powerful against *all* simple hypotheses in H_1. The Neyman–Pearson lemma (or its extension to composite null hypothesis) asserts that the most powerful test exists against any simple alternative, but it may (and often does) happen that these most powerful tests are different for different simple alternatives.

Example 13.4.1. Consider the situation of testing the simple null hypothesis $H_0 : \mu = 0$ against the composite alternative $H_1 : \mu > 0$ in case of normal distribution $N(\mu, \sigma^2)$ with known σ^2. Let us fix $\mu_1 > 0$ and consider the likelihood ratio test of H_0 against $H_1' : \mu = \mu_1$. We have, canceling the common term $1/(\sigma\sqrt{2\pi})^n$ in both densities and letting D stand for a "generic" positive constant (not necessarily the same in each appearance):

$$\frac{f_1(\mathbf{x})}{f_0(\mathbf{x})} = \frac{\exp\{-(1/2\sigma^2)\sum(x_i - \mu_1)^2\}}{\exp\{-(1/2\sigma^2)\sum x_i^2\}} = D e^{(\mu_1/\sigma^2)\sum x_i}.$$

Since $\mu_1 > 0$, this likelihood ratio is an increasing function of $\sum x_i$ (hence also of $\overline{X}$), and the critical region of the most powerful test against $\mu > 0$ is of the form $\{\mathbf{x} = (x_1, \ldots, x_n) : \overline{x} \geq k\}$.

For any given significance level α_0 we determine k from the condition

$$P\{\overline{X} \geq k|H_0\} = P\left\{Z \geq \frac{k}{\sigma}\sqrt{n}\right\} = \alpha_0,$$

so that $k = z_{\alpha_0}\sigma/\sqrt{n}$, where z_{α_0} is the upper α_0-quantile of standard normal distribution.

The only assumption about μ that was used here is $\mu > 0$; consequently, the test with the critical region C,

$$\text{reject } H_0 \text{ if } \overline{X} \geq z_{\alpha_0}\sigma/\sqrt{n} \tag{13.21}$$

is UMP for the alternative $H_1 : \mu > 0$. $\qquad\square$

Example 13.4.2. Continuing Example 13.4.1, observe that the test (13.21) is also UMP for the composite null hypothesis $H_0 : \mu \leq 0$ against the composite alternative $H_1 : \mu > 0$. Indeed, the power of this test is

$$\pi_C(\mu) = P_\mu\left\{\overline{X} \geq z_{\alpha_0}\frac{\sigma}{\sqrt{n}}\right\} = P\left\{\frac{\overline{X} - \mu}{\sigma}\sqrt{n} \geq \frac{z_{\alpha_0}(\sigma/\sqrt{n}) - \mu}{\sigma/\sqrt{n}}\right\}$$

$$= P\left\{Z \geq z_{\alpha_0} - \frac{\mu\sqrt{n}}{\sigma}\right\} \leq P\{Z \geq z_{\alpha_0}\},$$

where the inequality is valid for all $\mu < 0$. Consequently, $\pi_C(\mu) \leq P\{Z \geq z_{\alpha_0}\} = \alpha_0 = \pi_C(0)$; hence the size of the test C is α_0. $\qquad\square$

Example 13.4.3. Example 13.4.2 shows that there exists no UMP test for the hypothesis $H_0 : \mu = 0$ against the alternative $H_1 : \mu \neq 0$. Indeed, (13.21) is the most powerful test against any alternative $\mu > 0$, but it performs very poorly against the alternative $\mu < 0$: Its power on such alternatives is less than α_0. On the other hand, by symmetry, the test with critical region C',

$$\text{reject } H_0 \text{ if } \overline{X} \leq -z_{\alpha_0}\frac{\sigma}{\sqrt{n}} \tag{13.22}$$

is most powerful against any alternative $\mu < 0$ but performs poorly against alternatives $\mu > 0$. The "compromise" test, with the critical region C'',

$$\text{reject } H_0 \text{ if } |\overline{X}| \geq z_{\alpha_0/2}\frac{\sigma}{\sqrt{n}} \tag{13.23}$$

performs quite well for all alternatives $\mu \neq 0$, but its power is below the power of test (13.21) for $\mu > 0$, and below the power of test (13.22) for $\mu < 0$, so that it is not a UMP test. $\qquad\square$

Example 13.4.4. At the end, let us consider the case of a random sample $X_1, \ldots, X_n$ from a gamma distribution with density

$$f(x;\alpha) = \frac{\lambda^\alpha}{\Gamma(\alpha)} x^{\nu-1} e^{-\lambda x}, \qquad x > 0$$

where ν is unknown and λ is known.

Suppose that we want to test the hypothesis $H_0 : \nu = 1$ (so that the distribution is exponential) against the composite alternative $H_1 : 0 < \nu < 1$.

For any fixed ν in the alternative hypothesis, the likelihood ratio is (keeping the convention of letting D to denote a generic constant)

$$\frac{f_1(\mathbf{x})}{f_0(\mathbf{x})} = D \frac{\prod_{j=1}^n x_j^{\nu-1} e^{-\lambda x_j}}{\prod_{j=1}^n e^{-\lambda x_j}} = D \prod_{j=1}^n x_j^{\nu-1}. \tag{13.24}$$

For $\nu < 1$, this likelihood ratio is a decreasing function of the statistic $T = X_1 \cdots X_n$. Consequently, the most powerful test is of the form

$$\text{reject } H_0 \text{ if } T \le k$$

and is also the UMP test against the alternative $H_1 : \nu < 1$. However, the determination of k is not so simple. One needs the distribution of the statistic $T = X_1 X_2 \cdots X_n$ if X_i's are exponentially distributed. For large n one can try using the central limit theorem as follows. We have $X_1 \cdots X_n \le k$ if

$$\log X_1 + \log X_2 + \cdots + \log X_n \le \log k.$$

Clearly, $\log X_i$ are iid random variables, with a cdf given by

$$P\{\log X_i \le t\} = P\{X_i \le e^t\} = 1 - e^{-\lambda e^t},$$

so that the density is

$$f(t) = \lambda e^t e^{-\lambda e^t}, \qquad -\infty < t < \infty.$$

Now the mean and second moment (hence also variance) of $\log X_i$ can be obtained through numerical evaluation of the integrals

$$E(\log X_i) = \lambda \int_{-\infty}^{+\infty} t e^t e^{-\lambda e^t} \, dt,$$

and

$$E(\log X_i)^2 = \lambda \int_{-\infty}^{+\infty} t^2 e^t e^{-\lambda e^t} \, dt.$$

Given numerical values of mean and variance (depending on λ), one can find (for large n) the approximate value of the threshold k for a given significance level. This example shows that the life of a statistician need not be quite easy once one leaves the realm of textbook situations. $\square$

We know from the examples above that UMP tests may not exist. On the other hand, if they do exist, they provide the best available procedures (if one takes into account only the error probabilities, and not extraneous aspects, such as computation costs, etc). It is therefore natural to ask for conditions under which UMP tests exist. The answer is given by the next theorem, which we shall precede by some necessary definitions.

Definition 13.4.2.   We say that the family $\{f(x; \theta), \theta \in \Theta\}$ of distributions has a *monotone likelihood ratio* in statistic T if for any two values $\theta', \theta'' \in \Theta$ with $\theta' < \theta''$, and $\mathbf{x} = (x_1, \ldots, x_n)$, the likelihood ratio

$$f_n(\mathbf{x}; \theta'')/f_n(\mathbf{x}; \theta')$$

depends on $\mathbf{x}$ only through the values $T(\mathbf{x})$, and this ratio is an increasing function of $T(\mathbf{x})$. $\square$

We shall now illustrate the introduced concept by some examples.

Example 13.4.5.   Consider the normal density depending on parameter μ, with σ^2 known, so that

$$f_n(\mathbf{x}, \mu) = \frac{1}{\sigma^n (2\pi)^{n/2}} \exp\left\{-\frac{1}{2\sigma^2}\sum(x_i - \mu)^2\right\}.$$

Let $\mu' < \mu''$, and for typographical reason, let $\mu' = \mu_1, \mu'' = \mu_2$. Then

$$\frac{f_n(\mathbf{x}, \mu_2)}{f_n(\mathbf{x}, \mu_1)} = \exp\left\{-\frac{1}{2\sigma^2}\left[\sum(x_i - \mu_2)^2 - \sum(x_i - \mu_1)^2\right]\right\}$$

$$= \exp\left\{-\frac{1}{2\sigma^2}\left[-2(\mu_2 - \mu_1)\sum x_i + n(\mu_2^2 - \mu_1^2)\right]\right\}$$

$$= De^{[(\mu_2 - \mu_1)/\sigma^2]\sum x_i},$$

where $D > 0$. Consequently, normal distribution for fixed σ^2 have monotone likelihood ratio in $T = \sum X_i$, or equivalently, in $\overline{X}$. $\square$

Example 13.4.6.   Consider now the same normal distribution as in Example 13.4.5, this time for known μ, and varying σ^2. Let $\sigma_1^2 < \sigma_2^2$. Then

$$\frac{f_n(\mathbf{x}, \sigma_2^2)}{f_n(\mathbf{x}, \sigma_1^2)} = D \exp\left\{-\frac{1}{2\sigma_2^2}\sum(x_i - \mu)^2 + \frac{1}{2\sigma_1^2}\sum(x_i - \mu)^2\right\}$$

$$= D \exp\left\{\frac{1}{2}\left(\frac{1}{\sigma_1^2} - \frac{1}{\sigma_2^2}\right)\sum(x_i - \mu)^2\right\}.$$

Since $D > 0$ and $1/\sigma_1^2 - 1/\sigma_2^2 > 0$, we see that the normal family has monotone likelihood ratio in statistic $T = \sum (X_i - \mu)^2$.   □

Example 13.4.7 (Bernoulli Trials). Let X_i $(i = 1, \ldots, n)$ be independent, each with distribution $f(x; p) = p^x (1 - p)^{1-x}, x = 0, 1$, and $0 < p < 1$. Let $0 < p' < p'' < 1$. We then have

$$\frac{f_n(\mathbf{x}, p'')}{f_n(\mathbf{x}, p')} = \left[\frac{p''(1 - p')}{p'(1 - p'')} \right]^{\sum x_i} \left(\frac{1 - p''}{1 - p'} \right)^n.$$

Since $p''(1 - p')/p'(1 - p'') > 1$, the Bernoulli distributions have monotone likelihood ratio in statistic $\sum x_i$ (total number of successes in n trials).   □

It turns out that most of the known families of distributions have monotone likelihood ratio in some statistics; the verification of it is left in the form of various exercises in this section.

The role of families with monotone likelihood ratios for UMP tests is explained by the following theorem. Let $\{f(x; \theta), \theta \in \Theta\}$ be a family indexed by one-dimensional parameter θ. Let $\theta_0 \in \Theta$ and consider the null hypothesis $H_0 : \theta \leq \theta_0$ and the alternative $H_1 : \theta > \theta_0$.

Theorem 13.4.1. *Assume that the family $\{f(x; \theta), \theta \in \Theta\}$ has a monotone likelihood ratio in statistic T. Then for every α_0 with $0 < \alpha_0 < 1$ there exists a test (possibly randomized) that is UMP for testing H_0 against H_1 on significance level α_0. This test satisfies the following two conditions:*

(i) *There exists k such that*

$$\text{if } T(\mathbf{x}) > k, \text{ then } H_0 \text{ is rejected}$$

and

$$\text{if } T(\mathbf{x}) < k, \text{ then } H_0 \text{ is accepted}.$$

(ii) *$P_{\theta_0}\{H_0 \text{ is rejected }\} = \alpha_0$.*

Proof. Observe first that condition (ii) may require randomization. Indeed, in view of (i) and (ii), letting $\gamma = P\{H_0 \text{ is rejected } | T(\mathbf{X}) = k\}$ we must have

$$\alpha_0 = P_{\theta_0}\{T(\mathbf{X}) > k\} + \gamma P_{\theta_0}\{T(\mathbf{X}) = k\}. \tag{13.25}$$

It follows that if $P_{\theta_0}\{T(\mathbf{X}) = k\} = 0$, then we have

$$\alpha_0 = P_{\theta_0}\{T(\mathbf{X}) > k\} = P_{\theta_0}\{T(\mathbf{X}) \geq k\},$$

and the test is nonrandomized, with critical region $C = \{\mathbf{x} : T(\mathbf{x}) \geq k\}$. If $P_{\theta_0}\{T(\mathbf{X}) > k\} < \alpha_0 \leq P_{\theta_0}\{T(\mathbf{X}) \geq k\}$, then by (13.25) we must put

$$\gamma = \frac{\alpha_0 - P_{\theta_0}\{T(\mathbf{X}) > k\}}{P_{\theta_0}\{T(\mathbf{X}) = k\}}.$$

Clearly, we have $0 < \gamma < 1$, and an auxiliary randomization with probability of success equal to γ gives a procedure with significance level α_0.

Let C^* denote the procedure described by the theorem and let $\pi_{C^*}(\theta) = P_{\theta}\{C^*$ leads to rejecting $H_0\}$ be its power. By condition (ii), $\pi_{C^*}(\theta_0) = \alpha_0$.

Without loss of generality, let $\theta_1 > \theta_0$. By the Neyman–Pearson lemma, the most powerful procedure for testing the simple hypothesis $H_0^* : \theta = \theta_0$ against a simple alternative $H_1^* : \theta = \theta_1$ is based on the likelihood ratio, and rejects H_0 if $f_n(\mathbf{x}; \theta_1)/f_n(\mathbf{x}; \theta_0) > k^*$ for some k^*. But by the monotonicity of the likelihood ratio, the last condition means that $T(\mathbf{X}) > k$ for some k. Since the last condition does not depend on the choice of θ_1, the procedure C^* described in the theorem is UMP for testing the simple hypothesis $H_0^* : \theta = \theta_0$ against the composite alternative $H_1 : \theta > \theta_0$.

We shall also show that the power function of procedure C^* is nondecreasing: For any $\theta' < \theta''$ (regardless of their location with respect to θ_0), we have $\pi_{C^*}(\theta') \leq \pi_{C^*}(\theta'')$. This fact will be crucial for completing the proof.

Indeed, consider the class $\mathcal{K}$ of all procedures of testing the simple hypothesis $H_0' : \theta = \theta'$ against the simple alternative $H_1' : \theta = \theta''$, at level of significance $\pi_{C^*}(\theta') = \alpha$, say. We know from the Neyman–Pearson lemma that C^* is the most powerful procedure in class $\mathcal{K}$, that is,

$$\pi_{C^*}(\theta'') = \sup_{C \in \mathcal{K}} \pi_C(\theta''). \tag{13.26}$$

On the other hand, the (randomized) procedure C_0: "reject $H_0^* : \theta = \theta'$ with probability α regardless of observation" satisfies the condition $\pi_{C_0}(\theta) = \alpha$ for all θ. Clearly, $C_0 \in \mathcal{K}$, so that using (13.26) we may write

$$\pi_{C^*}(\theta'') \geq \pi_{C_0}(\theta'') = \alpha = \pi_{C^*}(\theta'),$$

which shows the monotonicity of power of C^*.

To complete the proof of the theorem, we must show that the procedure C^* is UMP not only in the class $\mathcal{B}$ (say) of all procedures C that satisfy the condition $\pi_C(\theta_0) \leq \alpha_0$, but also in the class $\mathcal{B}'$ of all procedures C such that $\sup_{\theta \leq \theta_0} \pi_C(\theta_0) \leq \alpha_0$.

Clearly, $\mathcal{B}' \subset \mathcal{B}$, and the monotonicity of the power function $\pi_{C^*}(\theta)$ shows that C^* is an element of the smaller class $\mathcal{B}'$: Indeed,

$$\sup_{\theta \leq \theta_0} \pi_{C^*}(\theta_0) = \pi_{C^*}(\theta_0) \leq \alpha_0.$$

This, however, completes the proof, since we already know that for every $\theta_1 > \theta_0$, the procedure C^* maximizes the power $\pi_C(\theta_1)$ in the larger class $\mathcal{B}$, hence also maximizes the power in the smaller class $\mathcal{B}'$. $\qquad\square$

PROBLEMS

13.4.1 In each of the families of distributions listed below, check whether they have a monotone likelihood ratio in the parameter specified and find the relevant statistic:

(i) Poisson distribution.

(ii) Exponential distribution.

(iii) Gamma distribution, shape parameter.

(iv) Gamma distribution, scale parameter.

(v) Beta, for each of the parameters separately.

13.4.2 Suppose that the number of defects in magnetic tape of length t (yards) has Poisson distribution with parameter λt.

(i) Assume that 2 defects were found in a piece of tape of length 500 yards. Test the hypothesis $H_0 : \lambda \geq 0.02$ against the alternative $H_1 : \lambda < 0.02$. Use a UMP test at significance level $\alpha \leq 0.01$.

(ii) Find the p-value.

(iii) Find the power of the test at $\lambda = 0.015$.

13.4.3 Let $X_1, \ldots, X_9$ be a random sample from normal distribution $N(\mu, 1)$. We want to test the null hypothesis $H_0 : \mu \leq 0$ against the alternative $H_1 : \mu > 0$. The suggested test is "reject H_0 if $3 \leq \overline{X} \leq 5$." Intuitively, this is a bad test. Find its power function and use it to show that this is indeed a bad test.

13.4.4 Effectivenes of a standard drug in treating specific illness is 60%. A new drug was tested and found to be effective in 48 of 70 cases when it was used. Specify appropriate alternative hypothesis and perform the test at 0.01 level of significance. Find the p-value for the test.

13.4.5 Suppose that $X_1, \ldots, X_n$ is a random sample from the distribution uniform on $[0, \theta]$. Null hypothesis $H_0 : \theta \leq \theta_0$ is to be tested against the alternative $H_1 : \theta > \theta_0$. Argue that the UMP test rejects H_0 if $\max(X_1, \ldots, X_n) > c$. Find c for $\theta_0 = 5, n = 10$, and $\alpha = 0.05$.

13.4.6 For distribution from Problem 13.4.5, suppose that we want to test the null hypothesis $H_0 : \theta \geq \theta_0$ against the alternative $H_1 : \theta < \theta_0$. Show that the UMP test rejects H_0 if $\max(X_1, \ldots, X_n) < c$. Find c, if $\theta_0 = 5, n = 10$, and $\alpha = 0.05$.

13.4.7 In Problem 13.3.7, assume that $n = 50$, and suppose that the instructor wants to fail all students with $\theta \leq 30$, and pass all others. Does there exist a UMP test for the hypothesis $H_0 : \theta \leq 30$? If yes, find the test; if no, justify your answer.

13.4.8 Let $X_1, \ldots, X_n$ be a random sample from a gamma distribution with shape parameter r and scale parameter λ.

    (i) Assume that λ is known, and derive a UMP test for the hypothesis $H_0 : r \leq r_0$ against the alternative $H_1 : r > r_0$.

    (ii) Assume that r is known, and derive a UMP test for the hypothesis $H_0 : \lambda \leq \lambda_0$ against the alternative $H_1 : \lambda > \lambda_0$.

13.4.9 A reaction time to a certain stimulus (e.g., time until solving some problem) is modeled as a time of completion of r processes, running one after another in a specified order. The times $\tau_1, \ldots, \tau_r$ of completion of these processes are assumed to be iid exponential with mean $1/\lambda$.

    Suppose that the observed reaction times (in seconds) are 15.3, 6.1, 8.5, and 9.0. If we know that $r = 3$, test the hypothesis $H_0 : \lambda \geq 0.8$ against the alternative $H_1 : \lambda < 0.8$. Use $\alpha = 0.05$.

13.4.10 Let $X_1, \ldots, X_n$ be a random sample from the distribution $f(x; \theta) = (1/\theta)[\theta/(\theta + 1)]^x$, where $x = 1, 2, \ldots$. Determine the UMP test of the hypothesis $H_0 : \theta = \theta_0$ against the alternative $H_1 : \theta > \theta_0$.

13.5 UNBIASED TESTS

As we already know, there are many practically important situations when the UMP tests do not exist. These situations comprise, among others, those of testing a null hypothesis $H_0 : \theta = \theta_0$ against a two-sided alternative $H_1 : \theta \neq \theta_0$ (i.e., when θ_0 is not on the boundary of the parameter space Θ).

On the other hand, UMP tests are highly desirable, and the idea of searching for such tests appears quite appealing, both theoretically and practically. In this situation, the theoretical efforts became directed toward a reduction of the class of tests. It was hoped that if one rules out some tests, then the reduced class might already contain a UMP test. The objective was to reduce the class of tests by removing "as few tests as possible" and by "ruling out only those tests that have some undesirable properties," yet enough to ensure the existence of UMP tests in the reduced class.

These are qualitative conditions, which can (and have been) implemented in a number of ways. We shall present one of these ways, by requiring unbiasedness. Consider the problem of testing the null hypothesis $H_0 : \theta \in \Theta_0$ against the alternative $H_1 : \theta \in \Theta_1$. Let C be a testing procedure and let $\pi_C(\theta)$ be the power function of C. Here the parameter space Θ may be multidimensional, and one or both of the hypotheses H_0 or H_1 may be composite.

Definition 13.5.1. The test C of H_0 against H_1 is called *unbiased* if

$$\sup_{\theta \in \Theta_0} \pi_C(\theta) \leq \inf_{\theta \in \Theta_1} \pi_C(\theta). \tag{13.27}$$

Since the left-hand side of (13.27) is the size of the test C, we may say that C is unbiased if its power on the alternative hypothesis is never below its size.

In particular, if the null hypothesis is simple, then the power function of an unbiased test reaches its minimum at θ_0. □

It turns out that in some cases when there is no UMP test in the class of all tests, there is a UMP unbiased test.

Example 13.5.1. Consider the case of testing the simple hypothesis H_0 : $\mu = \mu_0$ against the alternative H_1 : $\mu \neq \mu_0$, in case of observations $X_1, \ldots, X_n$ being a random sample from normal distribution $N(\mu, \sigma^2)$ with known σ^2.

Intuition suggests to take $T(\mathbf{X}) = \overline{X} - \mu_0$ as the test statistic, and reject H_0 if either $T(\mathbf{X}) < -k'$ or $T(\mathbf{X}) > k''$ for some suitably chosen positive numbers k' and k''. To have level of significance α_0, we must choose k' and k'' so that

$$1 - \alpha_0 = P_{\mu_0}\{-k' < T(\mathbf{X}) < k''\} = P_{\mu_0}\{-k' < \overline{X} - \mu_0 < k''\}$$
$$= P\left\{-k' \frac{\sqrt{n}}{\sigma} < Z < k'' \frac{\sqrt{n}}{\sigma}\right\},$$

that is,

$$\Phi\left(k'' \frac{\sqrt{n}}{\sigma}\right) - \Phi\left(-k' \frac{\sqrt{n}}{\sigma}\right) = 1 - \alpha_0, \tag{13.28}$$

where Φ is the cdf of standard normal random variable.

Let us investigate the power function of the test above. We have here, letting C be the critical region $\{\overline{X} < \mu_0 - k'\} \cup \{\overline{X} > \mu_0 + k''\}$,

$$\pi_C(\mu) = 1 - P_\mu\{\mu_0 - k' < \overline{X} < \mu_0 + k''\}$$
$$= 1 - P\left\{\frac{\mu_0 - \mu - k'}{\sigma} \sqrt{n} < Z < \frac{\mu_0 - \mu + k''}{\sigma} \sqrt{n}\right\}$$
$$= 1 - \Phi\left(\frac{\mu_0 - \mu + k''}{\sigma} \sqrt{n}\right) + \Phi\left(\frac{\mu_0 - \mu - k'}{\sigma} \sqrt{n}\right). \tag{13.29}$$

Clearly, $\pi_C(\mu)$ is a continuous differentiable function of μ, and C is unbiased if $\left.\dfrac{d}{d\mu} \pi_C(\mu)\right|_{\mu=\mu_0} = 0$. We have, letting φ be the density of standard normal random variable,

$$\frac{d\pi_C(\mu)}{d\mu} = \varphi\left(\frac{\mu_0 - \mu + k''}{\sigma}\sqrt{n}\right)\frac{\sqrt{n}}{\sigma} - \varphi\left(\frac{\mu_0 - \mu - k'}{\sigma}\sqrt{n}\right)\frac{\sqrt{n}}{\sigma}.$$

Substituting μ_0 for μ, we obtain

$$\frac{d\pi_C(\mu)}{d\mu}\bigg|_{\mu=\mu_0} = \left[\varphi\left(\frac{k''}{\sigma}\sqrt{n}\right) - \varphi\left(-\frac{k'}{\sigma}\sqrt{n}\right)\right]\frac{\sqrt{n}}{\sigma},$$

which equals to 0 (remembering that k' and k'' are positive) only if $k' = k''$. Thus, we obtain an unbiased test only if the two parts of the rejection regions are located symmetrically with respect to the null hypothesis value μ_0. Formula (13.28) now gives $k' = k'' = z_{\alpha_0/2}$. □

One can show that the test obtained in Example 13.5.1 is actually the UMP unbiased test for hypothesis $H_0 : \mu = \mu_0$ against the two-sided alternative $H_1 : \mu \neq \mu_0$. It is not, however, UMP test against one-sided alternatives (see Figure 13.8).

There is an argument often advanced against testing a simple null hypothesis of the kind described above. The argument is that a specific hypothesis such as $H_0 : \mu = \mu_0$ simply cannot be true: The chances that the mean is *exactly* equal to μ_0 are zero. This is true if the parameter μ is a random variable, varying from situation to situation according to a continuous prior distribution. It is also true in most cases of Bayesian priors, when μ does not vary, but a statistician's experience can be expressed in terms of a (subjective) probability distribution on μ.

Thus, the argument goes, H_0 should be rejected at once, without testing. As explained at the beginning of this chapter, this conclusion misses the essential intention of the theory of hypotheses testing, which is to serve not as a means of establishing truth of hypotheses, but of establishing good rules of inductive behavior. In this sense, "accepting $H_0 : \theta = \theta_0$" means simply that it is reasonable to act *as if* the parameter value was θ_0, even if in fact θ is only close to θ_0. This suggests testing a composite null hypothesis stating

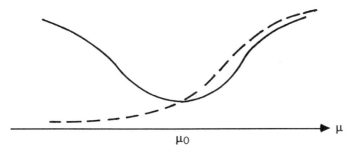

Figure 13.8 Powers of one-sided UMP test (dashed line) and UMP unbiased test (solid line).

that θ lies in some interval, against the alternative that it lies outside it. We shall illustrate this kind of approach using an example of normal distribution.

Example 13.5.2. Let again $X_1, \ldots, X_n$ be a random sample from distribution $N(\mu, \sigma)$, where σ^2 is known. We want to test the null hypothesis $H_0 : \mu_1 \leq \mu \leq \mu_2$ against the alternative $H_1 : \mu < \mu_1$ or $\mu > \mu_2$. To simplify the notation, let

$$\mu^* = \frac{\mu_1 + \mu_2}{2} \tag{13.30}$$

denote the midpoint between the boundaries of the null hypothesis. It appears reasonable to use the test statistic $T(\mathbf{X}) = \overline{X} - \mu^*$, and reject H_0 if $T(\mathbf{X}) > k''$ or $T(\mathbf{X}) < -k'$, where k', k'' are some positive constants. The power of this test (call it C^*) is

$$
\begin{aligned}
\pi_{C^*}(\mu) &= 1 - P_\mu\{-k' < \overline{X} - \mu^* < k''\} \\
&= 1 - P\left\{\frac{\mu^* - \mu - k'}{\sigma}\sqrt{n} < Z < \frac{\mu^* - \mu + k''}{\sigma}\sqrt{n}\right\} \\
&= 1 - \Phi\left(\frac{\mu^* - \mu + k''}{\sigma}\sqrt{n}\right) + \Phi\left(\frac{\mu^* - \mu - k'}{\sigma}\sqrt{n}\right).
\end{aligned}
$$

Again, to have the test unbiased, the power curve must have a minimum at μ^*, which necessitates taking $k' = k''$. Then the size of the test (see Figure 13.9) equals the common value of the power at points μ_1 and μ_2, that is [letting $k = k' = k''$ and $\Delta = (\mu_2 - \mu_1)/2$],

$$
\begin{aligned}
\sup_{\mu_1 \leq \mu \leq \mu_2} \pi_{C^*}(\mu) &= 1 - \Phi\left(\frac{\mu^* - \mu_1 + k}{\sigma}\sqrt{n}\right) + \Phi\left(\frac{\mu^* - \mu_1 - k}{\sigma}\sqrt{n}\right) \\
&= 1 - \Phi\left(\frac{\Delta + k}{\sigma}\sqrt{n}\right) + \Phi\left(\frac{\Delta - k}{\sigma}\sqrt{n}\right).
\end{aligned}
$$

If we now require (for given μ_1, μ_2, and σ, hence given μ^* and Δ) a test with specified significance level α_0 and specified probability β_0 of type II

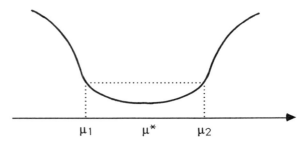

Figure 13.9   Power function of unbiased test.

error at some target value μ_t in the alternative, we can determine threshold k and sample size n from the equations

$$\Phi\left(\frac{\Delta + k}{\sigma}\sqrt{n}\right) - \Phi\left(\frac{\Delta - k}{\sigma}\sqrt{n}\right) = 1 - \alpha_0$$

and

$$\Phi\left(\frac{\mu^* - \mu_t + k}{\sigma}\sqrt{n}\right) - \Phi\left(\frac{\mu^* - \mu_t - k}{\sigma}\sqrt{n}\right) = \beta_0.$$

The solution has to be obtained by a numerical procedure. It can be shown that the resulting test is UMP unbiased.

PROBLEMS

13.5.1 Suppose that $X_1, \ldots, X_n$ is a random sample from the distribution uniform on $[0, \theta]$. We want to test the hypothesis $H_0 : \theta = \theta_0$ against the two-sided alternative $H_1 : \theta \neq \theta_0$. Find an unbiased test that rejects H_0 if $\max(X_1, \ldots, X_n) < c_1$ or $\max(X_1, \ldots, X_n) > c_2$. Find c_1 and c_2 if $\theta_0 = 5, n = 10$, and $\alpha = 0.05$.

13.5.2 Let $X_1, \ldots, X_n$ be a random sample from Poisson distribution with parameter λ. Find the (approximate) UMP unbiased test for the hypothesis $H_0 : \lambda = \lambda_0$ against the two-sided alternative $H_1 : \lambda \neq \lambda_0$, where λ_0 is assumed to be large. [*Hint:* Use the fact that if X has a Poisson distribution with mean λ, then $(X - \lambda)/\sqrt{\lambda}$ converges (as $\lambda \to \infty$) in distribution to a standard normal random variable.]

13.5.3 Assume that the observations $X_1, \ldots, X_n$ form a random sample from $N(0, \sigma^2)$. For testing hypothesis $H_0 : \sigma^2 = \sigma_0^2$ against the alternative $H_1 : \sigma^2 \neq \sigma_0^2$ at level α, find an unbiased test of the form "reject H_0 if $\sum_{i=1}^{n} X_i^2/\sigma_0^2 < C_1$ or $\sum_{i=1}^{n} X_i^2/\sigma_0^2 > C_2$."

13.5.4 Suppose now that the observations of Problem 13.5.3 come from the distribution $N(\mu, \sigma^2)$ with both parameters unknown. The null hypothesis to be tested is $H_0 : \sigma^2 = \sigma_0^2$. Find an unbiased α-level test of the form "reject H_0 if $\sum_{i=1}^{n}(X_i - \overline{X})^2/\sigma_0^2 < C_1$, or $\sum_{i=1}^{n}(X_i - \overline{X})^2/\sigma_0^2 > C_2$."

13.6 TESTING UNDER NUISANCE PARAMETERS

A rather common situation in statistical practice occurs when we are interested in testing hypotheses about a parameter, but the population distribution depends also on some other parameters. Equivalently, we may say that

$\underline{\theta} = (\theta_1, \ldots, \theta_r)$ is a multidimensional parameter, but we are interested in hypotheses involving one component only. Perhaps the most common case is when one wants to test the hypotheses about the mean μ of normal distribution, but variance σ^2 is also unknown.

The parameter or parameters that are not constrained by the null hypothesis are called *nuisance* parameters. The presence of nuisance parameters causes the null and alternative hypotheses to be composite. In this section we show two methods of dealing with nuisance parameters: by building conditional tests and by using the generalized likelihood ratio tests.

Conditional Tests

We begin with some examples and then formulate a general theory. Suppose that the distribution of X depends on a parameter $\theta = (\eta, \tau)$, where η is the parameter tested and τ is the nuisance parameter. Assume that we want to test the null hypothesis that $\eta = \eta_0$ against the alternative that $\eta < \eta_0$. These are in fact composite hypotheses.

$$H_0 : \eta = \eta_0, \tau \text{ arbitrary against } H_1 : \eta < \eta_0, \tau \text{ arbitrary.}$$

It may happen that there exists a sufficient statistic, say T, for τ. In such a case the conditional distribution of X given $T = t$ does not depend on τ, and then it may happen that one can find a test for H_0 on a given level α (for each t separately).

We shall first illustrate such a situation by an example.

Example 13.6.1. Let X and Y be independent binomial random variables: $X \sim \text{BIN}(n_1, p_1)$ and $Y \sim \text{BIN}(n_2, p_2)$. We want to test the hypothesis that $p_1 = p_2$ against one- or two-sided alternative. If the null hypothesis is true, then letting $p = p_1 = p_2$, we have

$$P\{X = x | X + Y = k\} = \frac{P\{X = x, X + Y = k\}}{P\{X + Y = k\}}$$

$$= \frac{P\{X = x\}P\{Y = k - x\}}{P\{X + Y = k\}}$$

$$= \frac{\binom{n_1}{x} p^x q^{n_1 - x} \binom{n_2}{k - x} p^{k - x} q^{n_2 - k + x}}{\binom{n_1 + n_2}{k} p^k q^{n_1 + n_2 - k}}$$

$$= \frac{\binom{n_1}{x} \binom{n_2}{k - x}}{\binom{n_1 + n_2}{k}}, \tag{13.31}$$

which is independent of nuisance parameter p.

The question is which values of X (given $X + Y = k$) constitute the strongest premise against the null hypothesis, and how one should assess the p-value of a result. Here the argument is as follows: We can expect X/n_1 to be close to p_1 and $Y/n_2 = (k - X)/n_2$ to be close to p_2, so that if H_0 is true, one may expect $X/n_1 \approx k/n_2 - X/n_2$. For the two-sided alternative, the "worst" cases are when X is close to 0 or close to k, so that the critical region will comprise two tails of the hypergeometric distribution (13.31).
Letting

$$u(j) = u(j; n_1, n_2, k) = \frac{\binom{n_1}{j}\binom{n_2}{k-j}}{\binom{n_1 + n_2}{k}},$$

we may define the p-value of the result $X = x$ for a two-sided test as

$$2 \min \left\{ \sum_{j \leq x} u(j), \sum_{j \geq x} u(j) \right\}. \tag{13.32}$$

This is *Fisher's exact test*.

Example 13.6.2. Let X and Y be two independent observations of two Poisson random variables, with parameters λ_1 and λ_2, respectively. We want to test the hypothesis $H_0 : \lambda_2 = \delta\lambda_1$ against the alternative $H_1 : \lambda_2 > \delta\lambda_1$, where $\delta > 0$ is some fixed constant. For instance, if $\delta = 1$, we have the hypothesis of equality of parameters in two Poisson distributions.
The joint distribution of (X, Y) is here

$$f(x, y; \lambda_1, \lambda_2) = \frac{\lambda_1^x \lambda_2^y}{x!\, y!} e^{-(\lambda_1 + \lambda_2)}$$

$$= \frac{1}{x!y!} \left(\frac{\lambda_1}{\lambda_1 + \lambda_2} \right)^x \left(\frac{\lambda_2}{\lambda_1 + \lambda_2} \right)^y (\lambda_1 + \lambda_2)^{x+y} e^{-(\lambda_1 + \lambda_2)}.$$

This suggests reparametrization with $\eta = \lambda_2/\lambda_1$ and $\tau = \lambda_1 + \lambda_2$, leading to joint density

$$f(x, y; \eta, \tau) = \binom{x+y}{x} \left(\frac{1}{1 + \eta} \right)^x \left(\frac{\eta}{1 + \eta} \right)^y \frac{\tau^{x+y}}{(x+y)!} e^{-\tau}. \tag{13.33}$$

It is now clear that $T = X + Y$ is a sufficient statistic for τ, and that given $T = t$, the random variable X has the conditional distribution which is binomial with parameters t and $1/(1 + \eta)$. Under null hypothesis, we have $\eta = \delta$, while under the alternative hypothesis, we have $\eta > \delta$. Given $T = t$, we may therefore test the null hypothesis $H_0 : p = 1/(1 + \delta)$ against the alternative $H_1 : p < 1(1 + \delta)$ observing random variable X, with distribution binomial with $x + y = t$ trials and probability of success p.

Example 13.6.3. Suppose traffic engineers suggest that a certain change of traffic light sequence may reduce the number of accidents on some types of intersections. Two intersections, far apart to ensure independence but otherwise identical in all respects (such as traffic intensity, road condition, etc.) are to be tested. The number X of accidents at the intersection with the new traffic light pattern was 7. During the same period, at the intersection with the previous traffic light pattern the number of accidents was $Y = 13$. Does it indicate, at significance level $\alpha = 0.05$, that the new traffic light pattern decreases the probability of an accident?

We want to test the hypothesis that $\lambda_1 = \lambda_2$ against the alternative $\lambda_1 < \lambda_2$ [so that $\delta = 1$ and $1/(1 + \delta) = \frac{1}{2}$]. We have here $t = x + y = 7 + 13 = 20$, and we need to evaluate $P\{X \le 7\}$ when $X \sim \mathrm{BIN}(20, \frac{1}{2})$, so that

$$P\{X \le 7\} = \sum_{j=0}^{7} \binom{20}{j} \left(\frac{1}{2}\right)^{20} = 0.1310.$$

The answer is therefore negative: Even if the new traffic lights pattern did not affect the probability of accident at all, the outcome as obtained (13 against 7), or more extreme, would have more than one chance in 20 of occurring.

□

The construction above is an application of the following theorem.

Theorem 13.6.1. *Suppose that the random sample $\mathbf{X} = (X_1, \ldots, X_n)$ has a joint distribution of the form, for some functions $a, b, u, v_1, \ldots, v_r$*

$$f_n(\mathbf{x}, \eta, \tau_1, \ldots, \tau_r) = a(\eta, \tau_1, \ldots, \tau_r) b(\mathbf{x}) \exp \left\{ \eta u(\mathbf{x}) + \sum_{j=1}^{r} \tau_j v_j(\mathbf{x}) \right\}. \quad (13.34)$$

Let $U = u(\mathbf{X})$, and $\mathbf{V} = (V_1, \ldots, V_r) = (v_1(\mathbf{X}), \ldots, v_r(\mathbf{X}))$. Then $V_1, \ldots, V_r$ are jointly sufficient for $(\tau_1, \ldots, \tau_r)$ for each η, and the conditional distribution of U given $\mathbf{V}(\mathbf{x}) = \mathbf{v}$ depends on η but not on $(\tau_1, \ldots, \tau_r)$. An α-level test of $H_0 : \eta \le \eta_0$ vs. $H_1 : \eta > \eta_0$ is obtained by rejecting H_0 if $u(\mathbf{x}) \ge q(\mathbf{v})$, where $P\{U \ge q(\mathbf{v})|\mathbf{v}\} = \alpha$ if $\eta = \eta_0$ (with the directions of inequality reversed throughout in testing $H_0 : \eta \ge \eta_0$ against $H_1 : \eta < \eta_0$; a two–tailed test is to be used in case of two-sided hypothesis $H_0 : \eta = \eta_0$ against $H_1 : \eta \ne \eta_0$).

Lehmann (1983) shows that under some regularity conditions these tests are UMP unbiased.

Example 13.6.4. To see how the situation of Example 13.6.3 falls under the scheme of Theorem 13.6.1, observe that the distribution of $\mathbf{X} = (X_1, X_2)$

can be reduced to the form (13.33) with $r = 1, u(\mathbf{X}) = X_1, v(\mathbf{X}) = X_1 + X_2$ as follows. Taking $s = \lambda_2/\lambda_1, t = \lambda_1 + \lambda_2$ [hence $\lambda_1 = t/(1 + s), \lambda_2 = st/(1 + s)$], the distribution

$$f(x_1, x_2; \lambda_1, \lambda_2) = \frac{1}{x_1!} \lambda_1^{x_1} e^{-\lambda_1} \times \frac{1}{x_2!} \lambda_2^{x_2} e^{-\lambda_2}$$

reduces, after some algebra, to the form

$$f = \left(\frac{V(\mathbf{x})}{U(\mathbf{x})}\right) \frac{1}{V(\mathbf{x})!} e^{-t} \exp\left\{ U(\mathbf{x}) \ln \frac{1}{s} + V(\mathbf{x}) \left[\ln t + \ln \frac{s}{1+s}\right]\right\}.$$

Introducing new parameters $\eta = \ln(1/s) = \ln(\lambda_1/\lambda_2)$ and $\tau = \ln t + \ln s/(1 + s) = \ln \lambda_2$, and then expressing t in terms of τ and η, we obtain the density in the form (13.33).

Generalized Likelihood Ratio Tests

As shown by the Neyman–Pearson lemma, the analysis of the likelihood ratio is a good method of searching for test statistics. It turns out that an extension of this method [originally called the "lambda principle" by Neyman (1950)] often leads to tests with certain optimality properties. These tests are now called generalized likelihood ratio (GLR) tests.

In what follows we use the following notation. As usual, $\mathbf{X} = (X_1, \ldots, X_n)$ is the random sample from distribution $f(x; \theta)$, where $\theta \in \Theta$. The joint density will be denoted by $f(\mathbf{x}; \theta)$, or $f_n(\mathbf{x}; \theta)$ if it will be necessary to stress the dependence on n. In most typical cases, θ will be a vector-valued parameter, so that the parameter space Θ will be a subset of multidimensional Euclidean space, and we let r stand for the number of dimensions.

Suppose that we want to test the null hypothesis $H_0 : \theta \in \Theta_0$ against the alternative $H_1 : \theta \in \Theta \setminus \Theta_0$. In most typical cases of interest both hypotheses H_0 and H_1 are composite, which corresponds to the sets Θ_0 and Θ_1 containing more than one element. In analogy with most powerful tests in case of a simple null hypothesis and simple alternative (which are based on the likelihood ratio), we might use the ratio

$$v(\mathbf{X}) = \frac{\sup_{\theta \in \Theta_1} f_n(\mathbf{X}, \theta)}{\sup_{\theta \in \Theta_0} f_n(\mathbf{X}, \theta)} \tag{13.35}$$

and reject the null hypothesis if the observation $\mathbf{x}$ gives a high value of $v(\mathbf{x})$. The ratio here is based on the analogy with the likelihood ratio tests given by the Neyman–Pearson lemma: If the numerator in (13.35) greatly exceeds the denominator, then $\mathbf{x}$ is good evidence for the alternative hypothesis. On the other hand, small values of $v(\mathbf{x})$ constitute good evidence for the null hypothesis.

An inconvenience with the use of $v(\mathbf{x})$ is that it may be difficult to compute, especially because typically only one of the two suprema in $v(\mathbf{x})$ is attained. Consequently, it is often easier to use the statistic defined as follows.

Definition 13.6.1.   The ratio

$$\lambda(\mathbf{X}) = \frac{\sup_{\theta \in \Theta_0} f_n(\mathbf{X}, \theta)}{\sup_{\theta \in \Theta} f_n(\mathbf{X}, \theta)}$$

will be called the *generalized likelihood ratio statistic* (or GLR statistic).

□

Under some continuity assumptions, if Θ_0 is a closed set, then the suprema in $\lambda(\mathbf{x})$ are both attained. In particular, if $\hat{\theta}_0 = \hat{\theta}_0(\mathbf{x})$ is the value of the parameter that maximizes $f_n(\mathbf{X}, \theta)$ over the set Θ_0, then the numerator in $\lambda(\mathbf{x})$ becomes $f_n(\mathbf{X}, \hat{\theta}_0(\mathbf{X}))$. Similarly, the value of θ that gives the maximum of the denominator is simply the MLE of θ, denoted $\hat{\theta} = \hat{\theta}(\mathbf{X})$. Thus, a useful computational formula for $\lambda(\mathbf{X})$ is

$$\lambda(\mathbf{X}) = \frac{\max_{\theta \in \Theta_0} f_n(\mathbf{X}, \theta)}{\max_{\theta \in \Theta} f_n(\mathbf{X}, \theta)} = \frac{f_n(\mathbf{X}, \hat{\theta}_0)}{f_n(\mathbf{X}, \hat{\theta})}.$$

Clearly, $\lambda(\mathbf{X}) \leq 1$, since the denominator, being the maximum over a larger set, is at least as large as the numerator. Since $\lambda(\mathbf{X})$ does not depend on any parameter values, it is indeed a statistic.

To see how $v(\mathbf{X})$ and $\lambda(\mathbf{X})$ are related, observe that

$$v(\mathbf{X}) \leq \frac{\max[\sup_{\theta \in \Theta_1} f_n(\mathbf{X}, \theta), \sup_{\theta \in \Theta_0} f_n(\mathbf{X}, \theta)]}{\sup_{\theta \in \Theta_0} f_n(\mathbf{X}, \theta)}$$

$$= \frac{f_n(\mathbf{X}, \hat{\theta})}{\sup_{\theta \in \Theta_0} f_n(\mathbf{X}, \theta)} = \frac{1}{\lambda(\mathbf{X})},$$

hence $v(\mathbf{X})\lambda(\mathbf{X}) \leq 1$. Consequently, large values of $v(\mathbf{X})$ are associated with small values of $\lambda(\mathbf{X})$.

It happens sometimes that the distribution of $\lambda(\mathbf{X})$ under H_0 does not depend on any parameters. In such cases, the α-level critical region is obtained from the condition

$$P\{\lambda(\mathbf{X}) \leq k|H_0\} = \alpha. \tag{13.36}$$

The rationale for this direction of inequality is that small values of $\lambda(\mathbf{X})$ are a premise for the alternative: It means that the maximum of the likelihood over Θ_0 is much smaller than the overall maximum; hence it also means that the maximum over Θ is much higher than the maximum over Θ_0.

It may happen that an exact test of the form (13.36), is not available. However, it can be shown that if a MLE has a asymptotically normal distribution (which is true under very general regularity conditions), then the limiting distribution of $\lambda(\mathbf{X})$ does not involve any parameters. Moreover, if the null hypothesis specifies the values of m among the r components of θ, then asymptotically, $-2 \log \lambda(\mathbf{X})$ has chi-square distribution with $r - m$ degrees of freedom. Thus, the approximate α-level test is

$$\text{reject } H_0 \text{ if } -2 \log \lambda(\mathbf{X}) \geq \chi^2_{\alpha, r-m}.$$

We shall now derive some more important GLR tests as examples.

Example 13.6.5. Let $\mathbf{X} = (X_1, \ldots, X_n)$ be a random sample from normal distribution with unknown μ and σ^2, and consider testing the null hypothesis $H_0 : \mu = \mu_0, \sigma$ arbitrary. Then the parameter space Θ is the upper half-plane, while Θ_0 is the ray $\{(\mu, \sigma^2) : \mu = \mu_0\}$. We now have

$$f(\mathbf{X}; \mu, \sigma^2) = (2\pi\sigma^2)^{-n/2} \exp\left\{ -\frac{1}{2\sigma^2} \sum_{i=1}^{n} (X_i - \mu)^2 \right\} \tag{13.37}$$

The MLE of μ and σ^2 are

$$\hat{\mu} = \overline{X}, \hat{\sigma}^2 = \frac{1}{n} \sum_{i=1}^{n} (X_i - \overline{X})^2. \tag{13.38}$$

Substituting in (13.37), we obtain, after some algebra, the denominator in $\lambda(\mathbf{X})$:

$$\sup_{(\mu, \sigma^2) \in \Theta} f(\mathbf{X}; \mu, \sigma^2) = (2\pi)^{-n/2} (\hat{\sigma}^2)^{-n/2} \exp\left\{ -\frac{1}{2\sigma^2} \sum_{i=1}^{n} (X_i - \hat{\mu})^2 \right\}$$

$$= \left(\frac{n}{2\pi e \sum (X_i - \overline{X})^2} \right)^{n/2}.$$

It remains to find the numerator in $\lambda(\mathbf{X})$, that is,

$$\max_{(\mu, \sigma^2) \in \Theta_0} f(\mathbf{X}; \mu, \sigma^2) = \max_{\sigma^2 > 0} f(\mathbf{X}; \mu_0, \sigma^2)$$

$$= \max_{\sigma^2 > 0} (2\pi\sigma^2)^{-n/2} \exp\left\{ -\frac{1}{2\sigma^2} \sum (X_i - \mu_0)^2 \right\}.$$

Here the maximum is attained at $\hat{\sigma}_0^2 = (1/n) \sum (X_i - \mu_0)^2$, and equals

$$\left(\frac{n}{2\pi e} \right)^{n/2} \left[\sum (X_i - \mu_0)^2 \right]^{-n/2}.$$

Consequently,

$$\lambda(\mathbf{X}) = \left(\frac{\sum(X_i - \mu_0)^2}{\sum(X_i - \overline{X})^2}\right)^{-n/2} = \left(\frac{\sum(X_i - \overline{X})^2 + n(\overline{X} - \mu_0)^2}{\sum(X_i - \overline{X})^2}\right)^{-n/2}$$

$$= \left(1 + \frac{(\overline{X} - \mu_0)^2}{\frac{1}{n}\sum(X_i - \overline{X})^2}\right)^{-n/2} = \left(1 + \frac{1}{n-1}t^2(\mathbf{X})\right)^{-n/2},$$

where

$$t(\mathbf{X}) = \frac{\overline{X} - \mu_0}{\sqrt{(1/n)\sum(X_i - \overline{X})^2}}\sqrt{n-1}$$

is the Student t ratio with $n - 1$ degrees of freedom [see (9.88)]. The inequality $\lambda(\mathbf{X}) \leq k$ is equivalent to the inequality $|t(\mathbf{X})| \geq k^*$.

Thus, the critical region of our test is

$$|t(\mathbf{X})| \geq t_{\alpha/2, n-1}.$$

One can show (see Lehmann, 1983) that this test is UMP unbiased.

Example 13.6.6. More "realistic" examples will be given in Section 13.8, where we present a systematic list of tests for normal distribution. The example below, however, is instructive.

Suppose that we test the null hypothesis that $\mu = 100$ against the alternative that $\mu \neq 100$. The distribution is normal with variance unknown. We have just two observations, $X_1 = 105$ and $X_2 = 105 + a$. The question is: For which a will the null hypothesis be rejected at significance level $\alpha = 0.05$? We have here $\overline{X} = 105 + a/2$, and $\frac{1}{2}[(X_1 - \overline{X})^2 + (X_2 - \overline{X})^2] = a^2/4$. The testing variable is

$$t = \frac{\overline{X} - 100}{\sqrt{a^2/4}}\sqrt{2 - 1} = \frac{5 + a/2}{|a|/2} = \frac{10 + a}{|a|},$$

and we reject the null hypothesis if

$$\frac{|10 + a|}{|a|} > t_{0.025,1} = 12.706.$$

If $a > 0$, we obtain $10 + a > 12.706a$ or $a < 0.85$. If $a < 0$, we obtain similarly $a > -0.73$. Thus, we reject the null hypothesis only for the values of X_2 between 104.27 and 105.85.

This result might appear counterintuitive: It might seem that very large positive a (i.e., very large X_2) should lead to rejecting the hypothesis that $\mu = 100$. This is not so: Since we know nothing about variance, very large a suggests that σ^2 is very large. We then have merely two observations above the hypothetical mean, an event with about 25% chance, not enough to reject H_0.   □

A rather common situation occurring in statistical practice is that of comparison of two samples. We often have two processes (e.g., two kinds of treatment, production process, etc.), typically "old" and "new," and we need to determine whether the "new" process is superior to the "old" one. We then have two samples, $X_1, \ldots, X_m$ and $Y_1, \ldots, Y_n$ which are to be compared. The null hypothesis usually states that these samples come from the same distribution. Formulation of the alternative depends on the actual situation analyzed.

When $X_i \sim N(\mu_1, \sigma_1^2)$ and $Y_j \sim N(\mu_2, \sigma_2^2)$, testing the null hypothesis $H_0 : \mu_1 = \mu_2$ without any assumption about σ_1 and σ_2 is known as the Behrens–Fisher problem. There are different solutions suggested, but thus far there is no test known that would have some of the optimality properties discussed above (e.g., be UMP unbiased). On the other hand, the solutions are known under additional assumptions, in particular if the ratio of variances is known.

Example 13.6.7. Consider the GLR test for the hypotheses $H_0 : \mu_1 = \mu_2$, $\sigma_2^2 = \sigma_1^2$ against the alternative $H_1 : \mu_1 \neq \mu_2$, $\sigma_2^2 = \sigma_1^2$ (a more general case, solved in the same way, is obtained when we assume that $\sigma_2^2 = \gamma \sigma_1^2$, where γ is a known constant).

The likelihood, letting $\sigma_1^2 = \sigma_2^2 = \sigma^2$, has the form

$$f(\mathbf{x}, \mathbf{y}; \mu_1, \mu_2, \sigma^2) = (2\pi\sigma^2)^{-(m+n)/2} e^{-1/2\sigma^2 \left[\sum (x_i - \mu_1)^2 + \sum (y_j - \mu_2)^2\right]}.$$

Maximizing the likelihood over all (μ_1, μ_2, σ^2) and on the subspace $(\mu_1 = \mu_2 = \mu, \sigma^2)$, one can derive the generalized likelihood ratio test. The exact form of the test is given in Section 13.8. We omit the details of calculations, which are similar to those in Example 13.6.5.

Example 13.6.8 (Paired Observations). A situation deceptively similar to that in Example 13.6.7 occurs when we have the data obtained by observing the values of some attribute of different elements of the population, observed "before" and "after." A typical case would be to measure a certain reaction in human subjects before (X_i) and after (Y_i) some treatment. The purpose is to decide whether the treatment has an effect or not. Even though we have here the situation of two samples, $X_1, \ldots, X_n$ (values "before") and $Y_1, \ldots, Y_n$ (values "after"), we cannot apply the method of Example 13.6.7 for $m = n$.

Such a procedure would be incorrect, since in the present case the values X_i, Y_i are not independent (as observations for the same subject). Under some assumptions, however, one can use here one-sample test. The assumptions, which may be satisfied in some cases, are as follows. Imagine that the observations are such that the differences $U_i = Y_i - X_i$ have the same normal distribution with mean μ and variance σ^2. Observe, however, that the values X_i or Y_i separately do not need to have normal distribution.

We may wish to test the null hypothesis $H_0 : \mu = \mu_0$ against a one- or two-sided alternative $H_1 : \mu > \mu_0$ or $H_1 : \mu \neq \mu_0$, σ^2 arbitrary. In most typical applications we take $\mu_0 = 0$ (treatment has no effect). The form of the test is the same as in Example 13.6.7, applied to random variables U_i, and we omit the details.

Example 13.6.9.   To test the efficiency of sleeping pills, a drug company uses a sample of insomniacs. The time (in minutes) until falling asleep is observed for each of them. Few days later, the same persons are given a sleeping pill and the time until falling asleep is measured again. Suppose that the data are

Subject	No Pill (X_i)	With Pill (Y_i)
1	65	45
2	35	5
3	80	61
4	40	31
5	50	20

The proper procedure is to treat the data as paired. The differences $X_i - Y_i = U_i$ are then 20, 20, 19, 9, and 30 and we want to test the hypothesis $E(U_i) = 0$ against the alternative $E(U_i) > 0$. We have here $\overline{U} = 19.6, S_U^2 = \frac{1}{5}\sum(U_i - \overline{U})^2 = 44.24$, so that

$$ t = \frac{\overline{U}}{S_U}\sqrt{5 - 1} = 5.89. $$

Comparing with quantiles of Student t distribution with 4 degrees of freedom, the result has p-value below 0.5%.

However, if we treat the problem as a two-sample problem (which is an *incorrect* procedure), we obtain a different conclusion. The procedure itself is described in Section 13.8), but it is worthwhile to explain here why analysis *of the same numerical values* may lead to two different conclusions, depending on whether these values result from paired or unpaired data.

In essence, in both cases we are comparing two means and trying to determine whether their difference is so small that it may be explained by chance variation (null hypothesis) or that it is large enough to be regarded as "significant." To make such an inference we have to assess the amount of variability, to serve as a base for comparison.

Now, the formulas for an estimate of variance are different in case the data are paired and in case they are not. In general, the first estimate gives a lower value simply because "a person is typically more similar to himself than to another person." Thus, quantitatively speaking, the same difference between the means may turn out to be significant compared with the smaller variance given by formula for paired data and not significant when compared with a higher variance given by the formula for independent samples.

In the present case we have $\overline{X} = 54, S_X = 16.55, \overline{Y} = 32.4, S_Y = 19.41$, so that

$$t = \frac{\overline{X} - \overline{Y}}{\sqrt{5S_X^2 + 5S_Y^2}} \sqrt{\frac{5 + 5 - 2}{\frac{1}{5} + \frac{1}{5}}} = \frac{21.6}{57.04} \sqrt{20} = 1.69.$$

This value, for 8 degrees of freedom, is not significant on the 5% level.
□

In Example 13.6.9 we computed the t-statistic in two ways, treating the data first as paired and then treating them as if they came from two independent samples. This situation is similar to those created by statistical software: When several procedures are available, the statistical package gives a value of test statistic, corresponding p-value, and so on, for each procedure, so conclusions can easily be made. Some of the procedures may be equivalent; however, most of them are usually not. They are appropriate for different ways in which the experiment is designed. The point is that it is *up to the human being to choose correct procedure*, and this choice should never be influenced by final outcomes (i.e., a conclusion about rejection or acceptance of the null hypothesis based on the significance of the value of the test statistic). What is really dangerous is the impression that the statistician always has a choice between two (or even more) procedures and therefore may choose the result which fits more his or her expectations.

In fact, there is *never any choice* in the case under consideration (i.e., between treating data as paired or not): Only one procedure is correct in any given instance (i.e., the actual data are either paired or form two independent samples), and computing the incorrect t-statistic is totally pointless. Statistical software makes life easier for a competent user but obviously confuses incompetent ones.*)

---

* A somewhat metaphorical example illustrating the situation may be as follows. It is easy to build a "universal checkout counter," which would compute the total price and tax for all tax systems existing in the United States. It would print out the result in the form of a tape giving the total sums to pay in New York City, Philadelphia, Miami, and so on. Such a universal counter is an analogue of a statistical package that gives the result of tests when data are paired and when they are coming from independent samples. The customer does not have any choice of the amount to pay: If he does his shopping in Miami, then only one of the amounts is valid for him; others are irrelevant. The same holds for a statistician regarding the case under consideration. In this sense, the common practice in all (or nearly all) statistical textbooks of providing examples of two treatments of the same data (as we did in Example 13.6.9 above) may be in fact "damaging to the brain" by opening the possibility of creating harmful conclusions.

Example 13.6.10. The following data concerning accidents on various types of highways are given by Ohio Department of Transportation, September 1990.

Highway Type	Number of Accidents	Annual Million Vehicle Miles	Accident Rate
Scenic	3,621	1,021	3.55
Other two-lane	36,752	11,452	3.21
Multilane	20,348	6,920	3.23
Interstate	10,460	9,412	1.11

From the table it appears that the accident rate on interstate highways is significantly lower than on other types of highways, and that on the first three types, the accident rates are essentially the same. To test those claims, one can use the likelihood ratio test as follows (see Al-Ghamdi, 1991).

We regard accidents on a given type of highway as events in a Poisson process, when miles of car travel play the role of time (this involves some idealization: neglecting variability of road and weather conditions and also treating each accident as a "single" event involving one car only, etc.).

The contribution to the likelihood arising from the ith type of highway is then

$$\frac{(\lambda_i m_i)^{n_i} e^{-\lambda_i m_i}}{n_i!},$$

where n_i is the number of accidents, m_i is the total number of miles travelled, and λ_i is the (theoretical) accident rate. We want to test the null hypothesis

$$H_0 : \lambda_1 = \lambda_2 = \lambda_3$$

against the alternative

$$H_1 : \lambda_i \neq \lambda_j \text{ for some } i, j = 1, 2, 3.$$

The likelihood is therefore

$$L(\lambda_1, \lambda_2, \lambda_3) = \prod_{i=1}^{3} \frac{(\lambda_i m_i)^{n_i}}{n_i!} e^{-\lambda_i m_i}$$

and is easily seen to be maximized at MLE's $\hat{\lambda}_i = n_i/m_i$. The global maximum is therefore

$$L_1 = \frac{n_1^{n_1} n_2^{n_2} n_3^{n_3} e^{-(n_1+n_2+n_3)}}{n_1! n_2! n_3!}.$$

On the other hand, on the null hypothesis the likelihood equals

$$\prod_{i=1}^{3} \frac{(\lambda m_i)^{n_i}}{n_i!} e^{-\lambda m_i} = \frac{m_1^{n_1} m_2^{n_2} m_3^{n_3} \lambda^{n_1+n_2+n_3} e^{-\lambda(m_1+m_2+m_3)}}{n_1! n_2! n_3!}.$$

The maximum occurs at $\hat{\lambda} = (n_1 + n_2 + n_3)/(m_1 + m_2 + m_3)$ and equals

$$L_0 = \frac{m_1^{n_1} m_2^{n_2} m_3^{n_3}}{n_1! n_2! n_3!} \left(\frac{N}{M}\right)^N e^{-N},$$

where $N = n_1 + n_2 + n_3$, $M = m_1 + m_2 + m_3$. The generalized likelihood ratio therefore equals

$$Q = \frac{L_0}{L_1} = \left(\frac{m_1}{n_1}\right)^{n_1} \left(\frac{m_2}{n_2}\right)^{n_2} \left(\frac{m_3}{n_3}\right)^{n_3} \left(\frac{N}{M}\right)^N.$$

To test the hypothesis we may use the fact that $-2 \log Q$ has asymptotic chi-square distribution with 2 degrees of freedom if H_0 is true. We have here $-2 \log Q = 79.039$, which is highly significant. Thus, the accident rates on the first three types of highways are not the same, contrary to what may seem to be true from a glance at the data. □

We complete this section with a discussion of the problems of reaching the decision before the data collection is complete, or more generally, on basing the decision on a selected part of the data. We shall illustrate the situation by an example that uses the t test, but the solution and the conclusions are valid for any test.

Example 13.6.11. A research lab employs two specialists, Dr. Brown and Dr. Smith. These two generally compete against each other. Dr. Brown claims that he invented a certain method which is superior to the method of Dr. Smith, currently used in the laboratory. The details here make no difference to the problem, and the reader is invited to use his or her imagination and choose the specialties for Drs. Brown and Smith and the details of invention. We assume the following: After some debate, it was decided that Dr. Brown's method would be tested. Five experiments would be run on five consecutive days, starting Monday, and the results $X_1, \ldots, X_5$ recorded.[*] It is known that X_i's form a random sample from a normal distribution with unknown standard deviation. It is also known that for the current method of Dr. Smith, the mean is 10, or perhaps less, so Dr. Brown's method will be declared superior if the mean of X_i's will be higher than 10, that is, if the null hypothesis $H_0; \mu \leq 10$ (asserting that Dr. Brown's method is no better

---

[*] One may feel that important decisions, such as that about the superiority of the scientific method, cannot be decided on the basis of five observations only. Of course, if possible, one should use a larger sample size to get better quality in the inference. This, however, will not change the essence of the problem.

than Dr. Smith's) will be rejected in favor of the alternative $H_1 : \mu > 10$. The significance level, $\alpha = 0.01$, was agreed upon during negotiations.

The observed data for the consecutive five weekdays were: 14.8, 13.6, 13.9, 10.3, and 11.4. We have here $\bar{x} = 12.8$ and $\sqrt{\frac{1}{5}\sum(x_i - \bar{x})^2} = 1.6769$, which gives

$$t = \frac{(12.8 - 10)}{1.6769}\sqrt{4} = 3.339.$$

This value is below the critical value $t_{0.01,4} = 3.747$, so that Dr. Brown's method was not declared superior on the significance level 0.01.

Dr. Brown, however, is not someone who gives up so easily. He noticed that if the test were run on Wednesday (after only 3 observations), we would have $\bar{x} = 14.1$, and $\sqrt{\frac{1}{3}\sum(x_i - \bar{x})^2} = 0.51$, hence $t = 11.369$, which exceeds the critical value $t_{0.01,2} = 6.965$. In fact, if the test were run on Wednesday, the p-value would have been less than 0.005, since $t_{0.005,2} = 9.925$!

The issues involved here are quite serious. Generally, the problem is: When the data are collected sequentially, it may happen that the conclusion reached on the basis of all data values differs from a conclusion that could be reached on the basis of an initial sequence of data points. Is one then justified in reaching the conclusion which is for some reason more convenient (e.g., is in favor of one's own preferred hypothesis, is more likely to get one an extension of a grant, etc.)? In particular, can one discard part of the data?

The answer, of course, is negative, not just because of the moral issues involved. It is equally important to realize that with modified or discarded data it may be hard, or even impossible, to assess the probabilities involved. To take the example of Dr. Brown, to assess the p-value (say) of the result calculated with the use of the three first data points, one would have to assess the conditional probability, given the conclusion of all five data points, something like*

$$P\{(X_1, X_2, X_3) \in C_3 | (X_1, \ldots, X_5) \notin C_5\},$$

where C_3 and C_5 are critical regions for sample sizes $n = 3$ and $n = 5$.

Quite apart from the moral and computational issues involved, there exists a theory of *sequential* testing of hypotheses. At each new data point, one of the three decisions is made: "accept H_0," "accept H_1," or "take another observation." The process stops on making either of the first two decisions. The criteria for making these decisions are chosen in such a way that (1) the probabilities of making wrong decisions (type I and type II errors) are

---

*The analysis here is similar to (but more complicated than) the ballot problem studied in Chapter 3. There we calculated the probability that in the process of counting votes, the losing candidate will lead at least once during counting. A moment of reflection shows that we have here a very similar situation, except that the "votes" (being random variables with values ± 1) are replaced by observations, which may "favor" one or the other hypothesis in varying degree.

bounded by preassigned numbers, and (2) the number of observations taken (being a random variable) has a finite expectation. □

This theory was developed originally by A. Wald. The details can be found in many advanced textbooks on mathematical statistics.

PROBLEMS

13.6.1 John and Peter try to sell the *Encyclopaedia Britannica* in the same neighborhood. John made 36 calls and sold 7 encyclopedias, while Peter made 20 calls and sold only 2 encyclopedias. Does it indicate that John is a beter salesman? Use $\alpha = 0.05$.

13.6.2 Suppose in a group of 10 randomly sampled Democrats only 2 favor a certain issue, while in a sample of 12 Republicans, the same issue is favored by 5 persons. On level $\alpha = 0.05$, does this result indicate that the fractions p_D and p_R of Democrats and Republicans favoring the issue in question are different? Specifically, find the p-value of the data by carrying the two-sided Fisher test.

13.6.3 Company A, which produces batteries, claims that their product is "at least 50% better" than batteries produced by company B. To test the claim, batteries A and B are used one after another in two analogous devices. That is, one device has a battery A installed and is left running until the battery becomes dead. It is then replaced immediately by another battery A, and so on. The second device runs parallel on batteries B.

Assume that lifetimes of batteries are exponential random variables, with densities $\lambda_A e^{-\lambda_A t}$ and $\lambda_B e^{-\lambda_B t}$. Suppose that in some time (say, a week) batteries A had to be replaced 5 times, while batteries B had to be replaced 9 times. Test the advertising claim by determining the p-value of the result. (*Hint:* If interarrival times are exponential, the process is Poisson.)

13.6.4 A dietician introduces a new diet, which she claims will bring a loss of at least 4 lbs in the first week. Each of 10 women followed this diet plan for the same period. The weight losses (in pounds) during the first week were 5.6, 3.8, 7.1, 2.9, 4.6, 4.7, 5.5, 1.3, 6.4, and 4.5. Assuming that these observations form a sample from a normal distribution with unknown mean μ and variance σ^2, test the dietician's claim $H_0 : \mu \geq 4$ against the alternative $H_1 : \mu < 4$. Use $\alpha = 0.05$.

13.6.5 Continuing Problem 13.6.4, test the hypothesis $H_0 : \sigma \leq 0.5$ against the alternative $H_1 : \sigma > 0.5$. Use $\alpha = 0.05$.

13.6.6 The times for the diagnosis and repair of a car with a certain type of problem are assumed to be normally distributed with mean μ and standard deviation $\sigma = 15$ minutes. A mechanic serviced five cars in one day, one after another without any break, and it took him a total of 350 minutes.

    (i) Test, on the level $\alpha = 0.05$, the null hypothesis $H_0 : \mu \leq 60$, against the alternative $H_1 : \mu > 60$.

    (ii) Suppose that you doubt the information that $\sigma = 15$, and decide to test the hypotheses in (i) without assuming anything about σ. If the sum of squares of the five diagnose/repair times is m, how small should m be to reject H_0 on significance level $\alpha = 0.05$?

13.6.7 The function $f(x, 0)$ is given by the following table:

X	a	b	c	d
$\theta = 0$	$1/12$	$1/6$	$1/12$	$2/3$
$0 < \theta < 1$	$\theta/6$	$(1 - \theta)/6$	$1/2$	$1/3$

We want to test the hypothesis $H_0 : \theta = 0$ against the alternative $H_1 : 0 < \theta < 1$.

    (i) Find the generalized likelihood ratio test.

    (ii) Compute the power of the test in (i).

    (iii) Find α and β for the test with rejection region $\{c\}$.

    (iv) Compare tests in (i) and (iii).

13.7 TESTS AND CONFIDENCE INTERVALS

In this section we explain briefly how the theory of testing statistical hypotheses is related to the theory of confidence intervals, discussed in Chapter 12. This connection was noticed by Neyman (who laid the foundations to both theories) as early as 1938.

To simplify the presentation, assume that θ is a one-dimensional parameter. We take a random sample $X_1, \ldots, X_n$ from a distribution $f(x; \theta)$.

A confidence interval (with confidence level $1 - \alpha$) is a random interval $[L, U] = [L(\mathbf{X}), U(\mathbf{X})]$ such that for every $\theta \in \Theta$,

$$P_\theta \{L(\mathbf{X}) \leq \theta \leq U(\mathbf{X})\} = 1 - \alpha. \tag{13.39}$$

Suppose now that we want to test hypothesis $H_0 : \theta = \theta_0$ against the alternative $H_1 : \theta \neq \theta_0$. The equivalence of tests and confidence intervals is based on the fact that if we can construct confidence interval (13.39), then we can also construct an α-level test of H_0 against H_1 and conversely: Given a testing procedure of level α, we can construct a confidence interval.

Indeed, condition (13.39) for $\theta = \theta_0$ allows us to define the set

$$A = \{\mathbf{x} = (x_1, \ldots, x_n) : L(\mathbf{x}) \le \theta_0 \le U(\mathbf{x})\}.$$

Clearly, if we take the set A as the acceptance region of H_0 (equivalently, we let $C = A^c$ to be the critical region for H_0), we obtain a test with level α:

$$P\{H_0 \text{ is rejected} | H_0 \text{ is true }\} = P\{\theta_0 \notin [L(\mathbf{X}), U(\mathbf{X})] | \theta = \theta_0\} = \alpha.$$

Conversely, suppose that for every θ' we can construct an α-level test of hypothesis $H_0 : \theta = \theta_0'$. This means that for every θ' we have a critical region $C_{\theta'}$ such that

$$P\{\mathbf{X} \notin C_{\theta'} | \theta = \theta'\} = 1 - \alpha. \tag{13.40}$$

Now define, for every $\mathbf{x}$,

$$B(\mathbf{x}) = \{\theta' : \mathbf{x} \notin C_{\theta'}\}. \tag{13.41}$$

When $\mathbf{x}$ is the observed value of the random vector $\mathbf{X}$, we obtain a random set $B(\mathbf{X})$. From (13.40) and (13.41) it follows that $P_\theta\{\theta \in B(\mathbf{X})\} = 1 - \alpha$, which means that $B(\mathbf{X})$ is a confidence set for θ with confidence level $1 - \alpha$.

Except for the fact that the second part of the argument provides us with confidence sets (not necessarily intervals), the argument shows that the two theories are essentially equivalent, at least in case of one-dimensional parameters.

The main results of the theory combine optimality properties of tests and confidence intervals. For instance, confidence intervals (sets) associated with UMP tests have the property of being the shortest possible [uniformly most accurate (UMA) confidence intervals]. It is worth remarking here that the theory extends to the case of testing in presence of nuisance parameters. We shall not go into the details here.

13.8 REVIEW OF TESTS FOR NORMAL DISTRIBUTION

We now collect together the major testing procedures for the case of normal distribution. This section is intended as a convenient reference for users rather than as an exposition of new concepts or results. We feel, however, that testing in normal case is important enough to justify the inclusion of such practically oriented section. If derivations were given in other sections, we refer to them. In other cases, we omit the derivations, specifying only the properties of the tests, possibly with indications of the proofs of these properties.

In all tests below, α is the significance level, $1 - \beta$ is the power, and z_p is the upper pth quantile of standard normal random variable Z, so that

$$P\{Z > z_p\} = \Phi(z_p) = p.$$

One-Sample Procedures

The basic setup now is that we observe the random sample $X_1, \ldots, X_n$ from distribution $N(\mu, \sigma^2)$.

Hypotheses About the Mean, Variance Known
Sufficient statistic: $\sum_{i=1}^{n} X_i$, hence also $\overline{X}$.
One-sided alternative:

$$
\begin{array}{lll}
\textbf{(i) } H_0 : \mu = \mu_0 & \text{vs.} & H_1 : \mu > \mu_0 \\
\phantom{\textbf{(i) }} H_0 : \mu \leq \mu_0 & \text{vs.} & H_1 : \mu > \mu_0
\end{array}
$$

$$
\begin{array}{lll}
\textbf{(ii) } H_0 : \mu = \mu_0 & \text{vs.} & H_1 : \mu < \mu_0 \\
\phantom{\textbf{(ii) }} H_0 : \mu \geq \mu_0 & \text{vs.} & H_1 : \mu < \mu_0
\end{array}
$$

Two-sided alternative:

$$
\begin{array}{lll}
\textbf{(iii) } H_0 : \mu = \mu_0 & \text{vs.} & H_1 : \mu \neq \mu_0
\end{array}
$$

$$
\textbf{(iv) } H_0 : \mu_1 \leq \mu \leq \mu_2 \quad \text{vs.} \quad H_1 : \mu < \mu_1 \text{ or } \mu > \mu_2
$$

Other cases:

$$
\textbf{(v) } H_0 : \mu \leq \mu_1 \quad \text{or} \quad \mu \geq \mu_2 \quad \text{vs.} \quad H_1 : \mu_1 < \mu < \mu_2
$$

Test statistics:

$$
\textbf{(i)–(iii)} \quad T_1 = \frac{\overline{X} - \mu_0}{\sigma} \sqrt{n}
$$

$$
\textbf{(iv), (v)} \quad T_2 = \frac{\overline{X} - \mu^*}{\sigma} \sqrt{n} \quad \text{where} \quad \mu^* = \frac{\mu_1 + \mu_2}{2}
$$

Test procedures and their properties:

$$
\begin{array}{l}
\textbf{(i) } \text{reject } H_0 \text{ if } T_1 \geq z_\alpha \\
\textbf{(ii) } \text{reject } H_0 \text{ if } T_1 \leq z_{1-\alpha} \\
\textbf{(iii) } \text{reject } H_0 \text{ if } |T_1| \geq z_{\alpha/2} \\
\textbf{(iv) } \text{reject } H_0 \text{ if } |T_2| \geq k_\alpha \\
\textbf{(v) } \text{reject } H_0 \text{ if } |T_2| \leq v_\alpha,
\end{array}
$$

where k_α and v_α are determined from the equations

$$\Phi\left(\frac{\Delta}{\sigma}\sqrt{n}+k_\alpha\right) - \Phi\left(\frac{\Delta}{\sigma}\sqrt{n}-k_\alpha\right) = 1-\alpha$$

and

$$\Phi\left(\frac{\Delta}{\sigma}\sqrt{n}+v_\alpha\right) - \Phi\left(\frac{\Delta}{\sigma}\sqrt{n}-v_\alpha\right) = \alpha,$$

with $\Delta = (\mu_2 - \mu_1)/2$. Tests (i), (ii), and (v) are UMP tests, (iii) and (iv) are UMP unbiased tests.

Power:

(i) $1 - \beta(\mu) = 1 - \Phi\left(\frac{\mu_0 - \mu}{\sigma}\sqrt{n} + z_\alpha\right)$

(ii) $1 - \beta(\mu) = 1 - \Phi\left(\frac{\mu_0 - \mu}{\sigma}\sqrt{n} - z_\alpha\right)$

(iii) $1 - \beta(\mu) = 1 - \Phi\left(\frac{\mu_0 - \mu}{\sigma}\sqrt{n} + z_{\alpha/2}\right) + \Phi\left(\frac{\mu_0 - \mu}{\sigma}\sqrt{n} - z_{\alpha/2}\right)$

(iv) $1 - \beta(\mu) = 1 - \Phi\left(\frac{\mu^* - \mu}{\sigma}\sqrt{n} + k_\alpha\right) + \Phi\left(\frac{\mu^* - \mu}{\sigma}\sqrt{n} - k_\alpha\right)$

(v) $1 - \beta(\mu) = 1 - 1 - \Phi\left(\frac{\mu^* - \mu}{\sigma}\sqrt{n} + v_\alpha\right) + \Phi\left(\frac{\mu^* - \mu}{\sigma}\sqrt{n} - v_\alpha\right)$

Sample size determination:

    Wanted: sample size n giving power at least $1 - \beta$ at μ

(i), (ii) $n \geq \dfrac{(z_{1-\alpha} + z_\beta)^2}{(\mu - \mu_0)^2}\sigma^2$

    (iii) The equation

$$\Phi\left(\frac{\mu_0 - \mu}{\sigma}\sqrt{n} + z_{\alpha/2}\right) - \Phi\left(\frac{\mu_0 - \mu}{\sigma}\sqrt{n} - z_{\alpha/2}\right) = \beta$$

    must be solved numerically for n.

    (iv) The equation

$$\Phi\left(\frac{\mu^* - \mu}{\sigma}\sqrt{n} + k_\alpha\right) - \Phi\left(\frac{\mu^* - \mu}{\sigma}\sqrt{n} - k_\alpha\right) = \beta$$

    must be solved numerically for n.

    (v) The equation

$$\Phi\left(\frac{\mu^* - \mu}{\sigma}\sqrt{n} + v_\alpha\right) - \Phi\left(\frac{\mu^* - \mu}{\sigma}\sqrt{n} - v_\alpha\right) = \beta$$

    must be solved numerically for n.

We now illustrate the application of some of these tests.

Example 13.8.1 (Generic Problem). All philogaps presently on the market have an average concentration of muzzz of at least 3.7 mg per philogap. A company claims to have discovered a new method of production that will decrease the average muzzz content to level below 3.7 mg. To test this claim, the muzzz content of 15 philogaps of this company were analyzed, and their average muzzz content was found to be 3.52 mg.

It is known that the standard deviation of the muzzz content in a philogap does not depend on the production process, and equals 0.35 mg. It is also known that the muzzz content is normally distributed. On the significance level $\alpha = 0.01$, does this finding indicate that the new production process decreases the concentration of muzzz in philogaps?

Remark. You are probably curious what philogaps and muzzz are. These words, to our best knowledge, mean nothing. If you so wish, substitute "objects" and "attribute A," or "cars" and "miles per gallon," "beer" and "alcohol content," "oranges" and "sugar content in juice," and so on, and change numbers and possibly, inequality directions accordingly.

We set this problem as of type (ii): The null hypothesis is $H_0 : \mu \geq 3.7$ and the alternative is $H_1 : \mu < 3.7$. The value of the test statistic T_1 is $t = [(3.52 - 3.7)/0.35]\sqrt{15} = -1.99$. The 1% quantile of the standard normal distribution is -2.33, so the null hypothesis is not rejected. An average of 3.52 or less in a sample of 15 is more likely than 0.01, even if the mean is, in fact, 3.7. However, the null hypothesis would have been rejected on significance level 0.05 (the 5% quantile is -1.65). One may wish to report the p-value of the result, equal to $P(Z < -1.99) = 0.0233$, or about 2.3%.

Suppose that we want not only a 1% level of significance, but also at least a 95% chance of detecting improvement in the average muzzz content in philogaps by 0.15 mg. In statistical terms, this means that we want the power to be at least 0.95 at $\mu = 3.7 - 0.15 = 3.55$. Then we need to take a sample of at least

$$\frac{(z_{0.01} + z_{0.05})^2}{(3.7 - 3.55)^2}(0.35)^2 = \frac{(2.33 + 1.96)^2}{(0.15)^2}(0.35)^2 = 100.2,$$

that is, at least $n = 101$ elements.

Example 13.8.2. A food-packing company purchased a new machine to fill plastic containers with sour cream. The nominal weight, as listed on the container, is 8 oz. The dial on the machine can be set on average weight $\ldots, 7.98, 8.00, 8.02, 8.04, \ldots$ oz. When it is set on 8.00 oz (say), it puts into successive containers the amounts $X_1, X_2, \ldots$, which are normally distributed with some mean μ and standard deviation (the same for all setting of the dial) $\sigma = 0.005$ oz. This standard deviation reflects the unvoidable container-to-container variability of the amounts of sour cream about their mean μ. Naturally, it is impossible to build a machine that would give the average μ *exactly* equal to the setting on the dial: It may be 7.995, 8.002, and so on.

A consumer protection agency may disregard the instances when the variability will occasionally lead to a container with less than nominal amount of sour cream. However, it might object strongly if the average μ is even slightly less than 8 oz, as this constitutes a systematic theft from the society. On the other hand, if μ exceeds the nominal weight even slightly, it may in time constitute a sizable free gift the company gives to the society.

The company decides that it may take the risk of getting into trouble with the consumer protection agency, or absorb the loss, if the mean μ satisfies the inequality $7.995 \leq \mu \leq 8.015$ but wants to avoid both lower and higher μ. Careful measurements of 50 containers with sour cream gave the average $\bar{x} = 8.017$. Is it a cause for alarm?

We are now in situation (iv), where the null hypothesis states that μ lies between some bounds, and we may proceed as follows. We have here $\mu_1 = 7.995$, $\mu_2 = 8.015$, so that $\mu^* = 8.005$ oz, and $\Delta = 0.01$. Consequently, the observed values of T_2 is $t_2 = [(8.017 - 8.005)/0.005]\sqrt{50} = 16.97$. On the other hand, $(\Delta/\sigma)\sqrt{n} = 14.142$, and the equation

$$\Phi(14.142 + k_\alpha) - \Phi(14.142 - k_\alpha) = 1 - \alpha$$

reduces to $\Phi(14.142 - k_\alpha) = \alpha$. For $\alpha = 0.01$ we must have $14.142 - k_\alpha = -2.33$, hence $k_\alpha = 16.472$. The observed value 16.97 exceeds k_α, and this indicates that the null hypothesis should be rejected on the level 0.01.

It should be stressed here that the conclusion is that $\mu < 7.995$ or $\mu > 8.015$ (despite the fact that the result $\bar{x} = 8.017$ suggests that the second of the two inequalities holds). The point is that the null hypothesis states that $7.995 \leq \mu \leq 8.015$ and its rejection is logically equivalent to the pair of inequalities.

To conclude that $\mu > 8.015$, we should test the null hypothesis $\mu \leq 8.015$. The test statistic would then be

$$t_1 = \frac{8.017 - 8.015}{0.005}\sqrt{50} = 2.82,$$

with corresponding p-value about 0.0025.   □

Hypotheses About the Mean, Variance Unknown
Jointly sufficient statistics: $\sum X_i$ and $\sum X_i^2$, or equivalently, $\overline{X}$ and $\sum(X_i - \overline{X})^2$, and so on.
One-sided alternative:

(i) $H_0 : \mu = \mu_0, \sigma > 0$ vs. $H_1 : \mu > \mu_0, \sigma > 0$
    $H_0 : \mu \leq \mu_0, \sigma > 0$ vs. $H_1 : \mu > \mu_0, \sigma > 0$

(ii) $H_0 : \mu = \mu_0, \sigma > 0$ vs. $H_1 : \mu < \mu_0, \sigma > 0$
    $H_0 : \mu \geq \mu_0, \sigma > 0$ vs. $H_1 : \mu < \mu_0, \sigma > 0$

Two-sided alternative:

(iii) $H_0 : \mu = \mu_0, \sigma > 0$ vs. $H_1 : \mu \neq \mu_0, \sigma > 0$

Test statistics:

$$t = \frac{\overline{X} - \mu_0}{\sqrt{\frac{1}{n} \sum_{i=1}^{n} (X_i - \overline{X})^2}} \sqrt{n - 1}$$

Remark. Many texts use the "simpler" notation, $t = [(\overline{X} - \mu_0)/S]\sqrt{n - 1}$ or $t = [(\overline{X} - \mu_0)/S]\sqrt{n}$. These may be confusing, since one has to bear in mind that in the first case S^2 is the MLE of σ^2; that is, $S^2 = (1/n)\sum(X_i - \overline{X})^2$. In the second case, S^2 is the unbiased estimator of variance: $S^2 = [1/(n-1)]\sum(X_i - \overline{X})^2$. The notation for S^2 is *not* standardized across statistical textbooks and papers.

Test procedures and their properties:

      (i) reject H_0 if $t \geq t_{\alpha,n-1}$

     (ii) reject H_0 if $t \leq -t_{\alpha,n-1} (= -t_{\alpha,n-1})$

    (iii) reject H_0 if $|t| \geq -t_{\alpha,n-1}$

where $t_{p,k}$ is the upper pth quantile of the Student distribution with k degrees of freedom: $P\{X \geq t_{p,n-1}\} = p$.

For each of these tests the power is the function π depending on the two-dimensional variable (μ, σ^2). We have $\pi(\mu_0, \sigma^2) = \alpha$ for all σ^2. However, for $\mu \neq \mu_0$, the values $\pi(\mu_0, \sigma^2) = \alpha$ depend on (unknown) σ^2 and are given by the noncentral Student distribution. Therefore, the sample size determination requires some additional information about σ^2 (a two-stage procedure, etc.). All three tests are UMP unbiased* tests.

Example 13.8.3. A certain make of cars is advertised as attaining gas mileage of at least 32 miles per gallon. Twelve independent tests gave the results 33, 28, 31, 28, 26, 30, 31, 28, 27, 33, 35, 29 miles per gallon. What can one say about the advertisements in light of these data?

Let us make the assumption that the observed mileages have normal distribution. We may then set the problem of evaluation of the advertising claim as that of testing the hypothesis $H_0 : \mu \geq 32$ (claim is true) against the alternative $H_1 : \mu < 32$ (claim is false). We have here $n = 12, \sum X_i = 359, \sum X_i^2 = 10{,}823$. Thus, $\overline{X} = 29.92, \sum(X_i - \overline{X})^2 = 82.92$, so that $t = -2.62$. Since the

---

*As mentioned before, two-sided UMP tests may not exist; however, they may often be found in some restricted classes of tests, as for example in the class of unbiased tests.

critical values for 11 degrees of freedom are $t_{0.025,11} = 2.201, t_{0.01,11} = 2.718$, the result is not significant on the 1% level, but significant on the 2.5% level. In other words, the p-value is between 1% and 2.5%. □

Hypotheses About the Variance, Mean Known

Sufficient statistic: $\sum(X_i - \mu)^2$. Note that the sample size can be 1.
One-sided alternative:

$$\begin{array}{llll}
\textbf{(i)} & H_0 : \sigma^2 = \sigma_0^2 & \text{vs.} & H_1 : \sigma^2 > \sigma_0^2 \\
& H_0 : \sigma^2 \leq \sigma_0^2 & \text{vs.} & H_1 : \sigma^2 > \sigma_0^2
\end{array}$$

$$\begin{array}{llll}
\textbf{(ii)} & H_0 : \sigma^2 = \sigma_0^2 & \text{vs.} & H_1 : \sigma^2 < \sigma_0^2 \\
& H_0 : \sigma^2 \geq \sigma_0^2 & \text{vs.} & H_1 : \sigma^2 < \sigma_0^2
\end{array}$$

Two-sided alternative:

$$\textbf{(iii)} \quad H_0 : \sigma^2 = \sigma_0^2 \quad \text{vs.} \quad H_1 : \sigma^2 \neq \sigma_0^2$$

Test statistic:

$$U = \sum_{i=1}^{n}(X_i - \mu)^2/\sigma_0^2.$$

Test procedures:

(i) reject H_0 if $U > \chi^2_{\alpha,n}$
(ii) reject H_0 if $U < \chi^2_{1-\alpha,n}$
(iii) reject H_0 if $U < \chi^2_{\alpha/2,n}$ or $U > \chi^2_{1-\alpha/2,n}$

where $\chi^2_{p,k}$ is the upper pth quantile of the chi-square distribution with k degrees of freedom: $P\{X > \chi^2_{p,k}\} = p$.

Power function is obtained as follows [calculations are given for case (i), others are similar]:

$$\pi(\sigma^2) = 1 - \beta(\sigma^2) = P_{\sigma^2}\{U > \chi^2_{\alpha,n}\} = P_{\sigma^2}\left\{\frac{\sum(X_i - \mu)^2}{\sigma_0^2} > \chi^2_{\alpha,n}\right\}$$

$$= P_{\sigma^2}\left\{\frac{\sum(X_i - \mu)^2}{\sigma^2} > \sigma_0^2 \frac{\chi^2_{\alpha,n}}{\sigma^2}\right\} = 1 - F_n\left(\sigma_0^2 \frac{\chi^2_{\alpha,n}}{\sigma^2}\right),$$

where F_n is the cdf of the chi-square distribution with n degrees of freedom.
Tests (i) and (ii) are UMP. Test (iii) is asymptotically (for $n \to \infty$) UMP unbiased. A UMP unbiased test in case (iii) is obtained if the thresholds are equal to the endpoints of the shortest $(1 - \alpha)$-level confidence interval for σ^2 (see Section 12.9).

Hypotheses About the Variance, Mean Unknown

Jointly sufficient statistics: $\overline{X}$ and $\sum(X_i - \overline{X})^2$. Tests require that $n \geq 2$.
One-sided alternative:

   (i) $H_0 : \sigma^2 = \sigma_0^2, -\infty < \mu < \infty$ vs. $H_1 : \sigma^2 > \sigma_0^2, -\infty < \mu < \infty$
        $H_0 : \sigma^2 \leq \sigma_0^2, -\infty < \mu < \infty$ vs. $H_1 : \sigma^2 > \sigma_0^2, -\infty < \mu < \infty$

   (ii) $H_0 : \sigma^2 = \sigma_0^2, -\infty < \mu < \infty$ vs. $H_1 : \sigma^2 < \sigma_0^2, -\infty < \mu < \infty$
        $H_0 : \sigma^2 \geq \sigma_0^2, -\infty < \mu < \infty$ vs. $H_1 : \sigma^2 < \sigma_0^2, -\infty < \mu < \infty$

Two-sided alternative:

   (iii) $H_0 : \sigma^2 = \sigma_0^2, -\infty < \mu < \infty$ vs. $H_1 : \sigma^2 \neq \sigma_0^2, -\infty < \mu < \infty$

Test statistic:

$$V = \sum_{i=1}^{n}(X_i - \overline{X})^2/\sigma_0^2.$$

Test procedures:

     (i) reject H_0 if $V > \chi_{\alpha,n-1}^2$

     (ii) reject H_0 if $V < \chi_{1-\alpha,n-1}^2$

     (iii) reject H_0 if $V > \chi_{\alpha/2,n-1}^2$ or $U < \chi_{1-\alpha/2,n-1}^2$

where $\chi_{p,k}^2$ is the upper pth quantile of the chi-square distribution with k degrees of freedom: $P\{X \geq \chi_{k,p}^2\} = p$.

Power functions $\pi(\mu, \sigma^2) = P_{\mu,\sigma^2}\{$ test rejects $H_0\}$ are obtained in the same way as in the case above.

Test (i) is unbiased. Tests (ii) and (iii) are not UMP unbiased.

The following example for the use of these tests was taken from Larsen and Marx (1986).

Example 13.8.4. The A above middle C is the note given to an orchestra, usually by the oboe, for tuning purposes. Its pitch is defined to be the sound of a tuning fork vibrating at 440 hertz (Hz). No tuning fork, of course, will always vibrate at *exactly* 440 Hz; rather, the pitch, Y, is a random variable. Suppose that Y is normally distributed with $\mu = 440$ Hz and variance σ^2 (here the parameter σ^2 is a measure of quality of the tuning fork). With the standard manufacturing process, $\sigma^2 = 1.1$. A new production technique has just been suggested, however, and its proponents claim that it will yield values of σ^2 significantly less than 1.1. To test the claim, six tuning forks are made according to the new procedure. The resulting vibration frequencies are 440.8, 440.3, 439.2, 439.8, 440.6, and 441.1 Hz.

We shall test the hypothesis $H_0 : \sigma^2 = 1.1$ against the alternative $H_1 : \sigma^2 < 1.1$, at the significance level $\alpha = 0.05$. First, we may accept the fact that the new production process indeed gives the mean $\mu = 440$. Then $U = \sum(X_i - 440)^2/1.1 = 1.62$. The value $\chi^2_{0.05,6} = 1.635$, so we conclude that, indeed, the new production process gives a variance of less than 1.1.

Suppose that we have some doubts as to whether or not $\mu = 440$. We can then use the method for unknown μ. Now $V = 2.218$ and $\chi^2_{0.05,5} = 1.145$ and we do not reach the preceding conclusion. $\qquad\qquad\qquad\qquad\qquad\quad$ $\square$

Two-Sample Procedures

The setup now is that we observe two independent random samples $X_1, \ldots, X_m$ and $Y_1, \ldots, Y_m$ from distributions $N(\mu_1, \sigma_1^2)$ and $N(\mu_2, \sigma_2^2)$, respectively.).

Hypotheses About the Means, Variances Known
Jointly sufficient statistics: $\sum_{i=1}^{m} X_i$, $\sum_{i=1}^{n} Y_i$, or $\overline{X}, \overline{Y}$.
One-sided alternative:

$$\textbf{(i)} \ H_0 : \mu_1 = \mu_2 \quad \text{vs.} \quad H_1 : \mu_1 > \mu_2$$

$$H_0 : \mu_1 \leq \mu_2 \quad \text{vs.} \quad H_1 : \mu_1 > \mu_2$$

Two-sided alternative:

$$\textbf{(ii)} \ H_0 : \mu_1 = \mu_2 \quad \text{vs.} \quad H_1 : \mu_1 \neq \mu_2$$

Remark. The opposite inequality in (i) reduces to changing the role of X_i's and Y_j's. The apparently more general hypotheses of the form $\mu_1 = \mu_2 - \Delta$, reduce to the ones above by subtracting constant Δ to all Y_j's.

Test statistic:

$$U = \frac{\overline{X} - \overline{Y}}{\sqrt{\sigma_1^2/m + \sigma_2^2/n}}$$

Test procedures and their properties:

$$\textbf{(i)} \ \text{reject } H_0 \text{ if } U \geq z_\alpha$$
$$\textbf{(ii)} \ \text{reject } H_0 \text{ if } |U| \geq z_{\alpha/2}$$

Power at the point (μ_1, μ_2) with $\mu_2 - \mu_1 = \Delta$

$$\textbf{(i)} \ 1 - \beta(\mu_1, \mu_2) = 1 - \beta(\Delta) = 1 - \Phi\left(z_\alpha + \frac{\Delta}{\sqrt{\sigma_1^2/m + \sigma_2^2/n}}\right)$$

(ii) $1 - \beta(\mu_1, \mu_2) = 1 - \beta(\Delta) = 1 - \Phi\left(z_{\alpha/2} + \dfrac{\Delta}{\sqrt{\sigma_1^2/m + \sigma_2^2/n}}\right)$

$$+ \Phi\left(-z_{\alpha/2} + \dfrac{\Delta}{\sqrt{\sigma_1^2/m + \sigma_2^2/n}}\right).$$

Test (i) is UMP, test (ii) is UMP unbiased.

Hypotheses About the Means, Variances Unknown, but It Is Known That $\sigma_2^2 = \gamma \sigma_1^2$

Jointly sufficient statistics for $(\mu_1, \mu_2, \sigma_1^2)$ are $(\overline{X}, \overline{Y}, \sum(X_i - \overline{X})^2 + \frac{1}{\gamma}\sum(Y_j - \overline{Y})^2)$.

One-sided alternative:

(i) $H_0 : \mu_1 = \mu_2, \sigma_1^2 > 0$ vs. $H_1 : \mu_1 > \mu_2, \sigma_1^2 > 0$
$H_0 : \mu_1 \le \mu_2, \sigma_1^2 > 0$ vs. $H_1 : \mu_1 > \mu_2, \sigma_1^2 > 0$

Two-sided alternative:

(ii) $H_0 : \mu_1 = \mu_2, \sigma_1^2 > 0$ vs. $H_1 : \mu_1 \ne \mu_2, \sigma_1^2 > 0$

Remark: The opposite inequality in (i) reduces to changing the role of X_i's and Y_j's. The apparently more general hypotheses of the form $\mu_1 = \mu_2 + \Delta$, reduce to the ones above by adding constant Δ to all Y_j's.

Test statistics:

$$U = \frac{\overline{X} - \overline{Y}}{\sqrt{\sum(X_i - \overline{X})^2 + \frac{1}{\gamma}\sum(Y_j - \overline{Y})^2}}\sqrt{\frac{m + n - 2}{1/m + \gamma/n}}.$$

Remark: $\gamma = 1$ if $\sigma_1^2 = \sigma_2^2$.

Test procedures:

(i) reject H_0 if $U > t_{\alpha, m+n-2}$
(ii) reject H_0 if $|U| > t_{\alpha/2, m+n-2}$.

Power depends on μ_1, μ_2, σ_1^2 and is expressed through a noncentral Student distribution .

Tests (i) and (ii) are UMP unbiased.

Hypotheses About the Variances; Means Unknown

Jointly sufficient statistics: $\overline{X}, \overline{Y}, \sum(X_i - \overline{X})^2, \sum(Y_j - \overline{Y})^2$.

One-sided alternative:

(i) $H_0 : \sigma_2^2 = \gamma \sigma_1^2$ vs. $H_1 : \sigma_2^2 > \gamma \sigma_1^2$

$H_0 : \sigma_2^2 \le \gamma \sigma_1^2$ vs. $H_1 : \sigma_2^2 > \gamma \sigma_1^2$

Two-sided alternative:

(ii) $H_0 : \sigma_2^2 = \gamma \sigma_1^2, -\infty < \mu_1, \mu_2 < \infty$ vs. $H_1 : \sigma_2^2 \ne \gamma \sigma_1^2, -\infty < \mu_1, \mu_2 < \infty$

Remark: $\gamma = 1$ if hypothesis asserts equality of variances.

Test statistics:

$$F = \frac{\sum (Y_i - \overline{Y})^2 / \gamma (n-1)}{\sum (X_j - \overline{X})^2 / (m-1)}.$$

Test procedures:

(i) reject H_0 if $F \ge F_{\alpha, n-1, m-1}$

(ii) reject H_0 if $F \ge F_{\alpha/2, n-1, m-1}$ or $F \le \dfrac{1}{F_{\alpha/2, m-1, n-1}}$

where $F_{p,k,l}$ is the upper pth quantile of F distribution with (k, l) degrees of freedom.

Tests (i) and (ii) are UMP unbiased.

Hypotheses About the Variances, One or Both Means Known

Jointly sufficient statistics are $\overline{X}, \overline{Y}, \sum (X_i - \overline{X})^2, \sum (Y_j - \overline{Y})^2$.

Whenever a mean is known, it replaces the sample average in ratio F, and the divisor (and number of degrees of freedom) changes into m (respectively, n). For instance, suppose that both μ_1 and μ_2 are known, and we test $H_0 : \sigma_2^2 = \sigma_1^2$ vs $H_1 : \sigma_2^2 > \sigma_1^2$. Then the testing variable is

$$F = \frac{\sum (Y_j - \mu_2)^2 / n}{\sum (X_i - \mu_1)^2 / m}$$

and the null hypothesis is rejected if $F \ge F_{\alpha, n, m}$.

Tests for Binomial Distribution for Large Samples

If X has binomial distribution $\text{BIN}(n, p)$ for n large, one can use either Poisson approximation theorem or the central limit theorem to obtain testing procedures for hypotheses about p. Thus, if n is large but np is small, then asymptotically $X \sim \text{POI}(np)$. The hypothesis $H_0 : p = p_0$ corresponds to $H_0 : \lambda = np_0$ in Poisson distribution, and we can use tests for this distribution.

If neither np nor $n(1 - p)$ is small, then we have $X/n \sim N(p, p(1 - p)/n)$. To test the hypothesis $H_0 : p = p_0$ (against a one- or two-sided alternative) we use the fact that, under H_0, the random variable

$$Z = \frac{X/n - p_0}{\sqrt{p_0(1 - p_0)/n}}$$

is asymptotically standard normal.

For two sample situations, let $X \sim \text{BIN}(n_1, p_1)$, $Y \sim \text{BIN}(n_2, p_2)$ where n_1 and n_2 are both large, and assume that we can use normal approximation. Then we have

$$\frac{X}{n_1} - \frac{Y}{n_2} \sim N\left(p_1 - p_2, \frac{p_1(1 - p_1)}{n_1} + \frac{p_2(1 - p_2)}{n_2}\right).$$

Thus, under the null hypothesis $H_0 : p_1 = p_2$ we have, letting $p_1 = p_2 = p$,

$$\frac{X/n_1 - Y/n_2}{\sqrt{p(1 - p)\,(1/n_1 + 1/n_2)}} \sim N(0, 1).$$

This statistic still involves the nuisance parameter p. However, under H_0 the MLE of p is $(X + Y)/(n_1 + n_2)$, so that we can use the test statistic

$$Z = \frac{\dfrac{X}{n_1} - \dfrac{Y}{n_2}}{\sqrt{\dfrac{X + Y}{n_1 + n_2}(1 - \dfrac{X + Y}{n_1 + n_2})\left(\dfrac{1}{n_1} + \dfrac{1}{n_2}\right)}},$$

which is asymptotically standard normal. Whether we use a one- or two-sided test depends on the alternative hypothesis.

Example 13.8.5.   In primary elections, 28% of Republicans in New Hampshire voted for candidate A. A poll of 180 Republicans in Iowa show that 41 of them will vote for candidate A. Does this result indicate that the Republican support of candidate A is lower in Iowa than in New Hampshire?

SOLUTION.   We have here $p_0 = 0.28$, the known level of support for A in New Hampshire, and we want to test the hypothesis $H_0 : p \geq 0.28$ against the alternative $H_1 : p < 0.28$, where p is the fraction of Republican voters who support A in Iowa. The observed value of the test statistic is

$$Z = \frac{\frac{41}{180} - 0.28}{\sqrt{(0.28)(0.72)/180}} = -1.58$$

which corresponds to the p-value 0.059. Whether such a result justifies taking some special action (e.g., increased campaign appearances, etc.) has to be decided by the candidate and his staff.   ☐

Example 13.8.6. Continuing Example 13.8.5, of 180 persons polled, 110 were women, and 20 of them said they would vote for A (so that there were 70 men, of whom 21 said they would vote for A). Does this result indicate that support for A in Iowa is higher among men, and therefore that the candidate should concentrate his campain on appealing to women voters?

SOLUTION.   We want to test the hypothesis $p_W = p_M$ against the alternative $p_W < p_M$. The value of the test statistic is

$$ z = \frac{\frac{21}{70} - \frac{20}{110}}{\sqrt{\frac{41}{180}\left(1 - \frac{41}{180}\right)\left(\frac{1}{110} + \frac{1}{70}\right)}} = 1.84. $$

The p-value here is about 3.3% and may serve as an indication that indeed, the support for A among Republican women is lower than among Republican men.   ☐

PROBLEMS

13.8.1 (Shoshoni Rectangles)   The following problem is taken from Larsen and Marx (1986). Since antiquity, societies have expressed aesthetic preferences for rectangles having a certain width (w)-to-length (l) ratio. For instance, Plato wrote that rectangles formed of two halves of an equilateral triangle are especially pleasing (for such rectangles $w/l = 1/\sqrt{3}$). Another standard adopted by Greeks is the golden rectangle, defined by the condition that it must be similar to its part remaining after cutting off a square with the side equal to its width (i.e., the shaded area on Figure 13.10 is similar to the whole

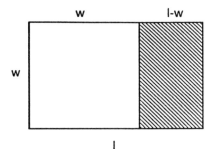

Figure 13.10   Golden rectangles.

rectangle). Thus, we must have $w/l = (l - w)/w$, which gives $w/l = (\sqrt{5} - 1)/2 = 0.618$.

Both Greeks and Egyptians used golden rectangles in their architecture, and even today, the golden rectangle remains an architectural and artistic standard (e.g., items such as drivers' licenses, business cards, or picture frames often have w/l ratios close to 0.618).

The data below show width-to-length ratios of beaded rectangles used by Shoshoni Indians to decorate their leather goods. The question is whether the golden rectangle can be considered an aesthetic standard for Shoshonis.

Width-to-length ratios for 20 rectangles found on Shoshoni handicraft:

0.693 0.749 0.654 0.670 0.662 0.672 0.615 0.606 0.690 0.628
0.668 0.611 0.606 0.609 0.601 0.553 0.570 0.844 0.576 0.933

Use these data to test the hypothesis that Shoshoni Indians use the golden ratio as their aesthetic standard. Choose the appropriate alternative hypothesis, specify the assumptions made, and carry on the test.

13.8.2 A study of pollution was carried out in two lakes, A and B. The level of a specific pollutant was measured using a certain instrument, and the results for lake A were 3.17, 4.22, 2.58, 4.01, and 3.79. In lake B, the measurements were made with the same instrument, but the data reported were the averages of four observations (and the original data are no longer available). The three data points for lake B are 4.04, 4.32, and 4.12. Test the hypothesis that pollution levels are the same in both lakes, against the alternative that the level in lake B is higher.

State all assumptions necessary to perform the test.

13.8.3 Suppose that X_1, X_2 and Y_1, Y_2 are two random samples of size 2 from two normal distributions with unknown means and variances. Let $d_1 = |X_1 - X_2|, d_2 = |Y_1 - Y_2|$ and $r = d_1/d_2$. Use r for testing hypothesis of equality of variances. Specify the critical region at significance level $\alpha = 0.05$ for the one-sided alternative hypothesis (the variance of the first population exceeds that of the second population).

REFERENCES

Al-Ghamdi, Ali Saeed, 1991. "The Application of Mathematical Statistics to Traffic Accidents Data," M.S. Dissertation, Ohio State University, Columbus, Ohio.

Gibbons, J. D., and J. W., Pratt, 1975. "*p*-Values: Interpretation and Methodology," *The American Statistician*, Vol. 29, No. 1, 20–25.

Larsen, R. J., and M. L. Marx, 1986. *An Introduction to Mathematical Statistics and Its Applications*, Prentice Hall, Englewood Cliffs, N.J.

Lehmann, E. L., 1983. *Testing Statistical Hypotheses*. Wiley: New York.

Neyman, J., 1950. *A First Course in Probability and Statistics*. Henry Holt & Co., New York.

CHAPTER 14

Discrimination*

14.1 INTRODUCTION

Discriminant analysis is the branch of scientific inquiry that does not have a unique name. Its problems are known under various names, such as prediction, identification, selection, classification, diagnosis, and screening, to mention just a few. In general, discriminant analysis is associated with identification of distinct classes (or subpopulations) that exist but are unobservable. This identification is often based on some feature data of known origin.

Example 14.1.1. The following example captures the essence of discrimination problems. Suppose we are given only a person's height (Y), and we have to decide whether or not this person is a man. There is a penalty for misclassification: We are charged \$1 for each man misclassified as a woman, and \$2 for each woman misclassified as a man. Since men are, on average, taller than women, a reasonable procedure is to choose a threshold (e.g., 5 feet 6 inches) and then classify every person whose height Y exceeds 5 feet 6 inches as a man, and otherwise classify this person as a woman. Under such a choice of threshold we shall misclassify all men shorter than 5 feet 6 inches and all women taller than 5 feet 6 inches. We can then calculate the expected number of misclassifications of both kinds, the expected penalty, and then possibly change the threshold up or down so as to minimize the expected penalty.

If we choose the statement "The person in question is a man" as the hypothesis to be tested, then two kinds of misclassifications will coincide with errors of type I and type II. □

Problems of discrimination, or identification, can also be regarded as special cases of hypothesis testing, as well as special cases of estimation. Typically, we deal here with a finite number of distributions, and the data **Y** are known to come from one of them. The objective is to determine the actual distribution from which **Y** was sampled.

Formally, this may be regarded as the case when the parameter space

648

Θ is finite, so that we have distributions $f_1, f_2, \ldots, f_r$ corresponding to $\theta = 1, 2, \ldots, r$. In *estimation theory language* the objective is to estimate θ on the basis of the sample. In *testing hypothesis language* the objective is to test the hypothesis that $\theta = \theta_0$, where θ_0 is one of numbers $1, 2, \ldots, r$.

Such an interpretation may be helpful in visualising the place of discrimination theory in statistics, but it contributes little to developement of the discrimination theory itself: In view of the discrete nature of the parameter space, most of the results of estimation theory and testing theory are of limited applicability here. However, there are problems (discussed in Chapter 15) which could properly be classified as discrimination problems, where parameter space is not discrete.

To place the discrimination problems in proper perspective, let us introduce them as follows. In many statistical problems the data consist of pairs (X, Y), where variables X and Y are related in some way. The actual terminology depends on the context, but generally, Y is referred to as a *dependent* variable, and it is assumed to be observable. Various types of situations, and resulting theories, are obtained depending on whether X can be observed or not; if X can be observed, whether its value can be set by the experimenter, and so on.

If X cannot be observed, we typically want to infer something about X on the basis of observation of Y's obtained for to the same value of X. When X is simply an index that identifies the distribution from which Y's are sampled, we have a *discrimination* or *identification* problem. In another setup, we may have the data $(X_i, Y_i), i = 1, \ldots, n$, with only Y_i's being observed, and we want to select those pairs in which X_i exceeds some threshold (e.g., Y may be the performance level observed on an exam while X is the true level of the candidate's acquired knowledge). This is called a *screening* problem. For the presentation of theory of screening and its applications, see e.g. Kowalczyk (1991). If Y can be observed before X, we may wish to *predict* the value of X before its occurs.

Of a variety of discrimination problems, one is discussed in this chapter in some detail. The analysis will concern the simplest case, when the variable X is of binary character, so that $X = 0$ or $X = 1$. We observe Y and have to decide about the value of X; we are, however, allowed to defer the decision.

The theoretical importance of such a decision scheme, despite its simplicity, lies in the fact that one can better grasp the role of loss function and various criteria in determining the solution. From a practical point of view, the analysis, which involves deferring the decision, is a prerequisite for building a theory of sequential procedures, where deferment of the decision regarding X means taking another observation of Y. This scheme of discrimination problem may be implemented in a variety of ways by additional assumption regarding (i) the nature of variable X, (ii) the nature of variable Y, (iii) the form of the data, (iv) the extent to which the conditional distribution of Y given X, or the joint distribution of X and Y, is known, (v) the goal of analysis (i.e., the type of inference about X which is desired). We provide next a series of possible assumptions regarding (i)–(v) and empirical situations that one might wish to analyze.

Variable X

Different types of problems are obtained depending on whether X is deterministic or random, and if so, whether X is a discrete or continuous variable.

Example 14.1.2. The case when X is a binary variable may be illustrated by the problem of determining the sex of a child before its birth on the basis of some observations Y. In this case X assumes two values, M or F, and we know the marginal probabilities $P(X = M)$ and $P(X = F)$. The possibility of inference about X (prior to the birth of the child) depends, qualitatively speaking, on how apart the conditional distributions of Y are, namely $f(y|X = F)$ and $f(y|X = M)$.

Finding the sex of an unborn child is a problem that has fascinated people for a very long time; until recently all methods seemed to give about a 50% success rate, which indicates that whatever observed feature Y one takes, the conditional distribution is almost the same for $X = F$ and $X = M$.

A similar problem, in the context of chickens (rather than humans), is of considerable importance. The point is that for highly automatized modern farms, it is important to separate young chickens by sex as soon as possible after they hatch from eggs, since the feeding and subsequent fate of the male population is different from that of the female population. The classification is done by specialists, called sexers, whose performance is rated by the proportions (evaluated much later) of misclassified male and female chickens.

Example 14.1.3 (Diagnosis). A typical example of a setup with X discrete, but not binary, occurs in medical diagnosis. Here $\mathbf{Y} = (Y_1, \ldots, Y_m)$ may stand for the vector of results of different tests, and/or indicators of some symptoms, while the possible values of X are the diseases.

Example 14.1.4. The problem of estimation of the parameter θ on the basis of sample $\mathbf{Y} = (Y_1, \ldots, Y_m)$ with the Y_i's being iid with distribution $f(y; \theta)$ falls under the present scheme, with the variable X playing the role of the parameter θ. Here the unobservable variable is typically not random, and one cannot speak about the joint distribtuion of θ and Y, except under a Bayesian setup.

A special case of a discrimination problem occurring in regression analysis when the objective is to estimate the unknown value of X is presented in Chapter 15.

Example 14.1.5. In all of the examples above, X was a variable whose value existed at the time Y was observed, but for some reason direct observation of X was impossible. In some cases it may be that X is a value generated by random mechanisms that begin operating only after Y is observed. For instance, Y may represent all relevant data about weather conditions in a given area available up to the present time, while X is the temperature at a specific time and location the next day. In such cases one is usually interested in predicting X.   □

Variable Y

As regards the nature of variable Y, the main types of models are obtained depending on whether or not one is allowed to continue observations. In the latter case, it is assumed that one observes Y and must make the decision about X (e.g., prediction of its value, etc.) on the basis of Y. In this case, the "internal structure" of Y is of little importance: Y can be a single value, or a vector of values $\mathbf{Y} = (\xi_1, \ldots, \xi_n)$, with ξ_i's dependent or not.

If one is allowed to continue observations, it is typically assumed that $\mathbf{Y} = \mathbf{Y}_n = (\xi_1, \ldots, \xi_n)$, where ξ_i's are iid sampled from the same conditional distribution $f(y|X = x, \theta)$ for some unknown x and θ. In this case, it may be reasonable and allowed to postpone the decision about x, take the next observation ξ_{n+1}, and try to make the decision on the basis of $\mathbf{Y}_{n+1}$.

Knowledge of the Joint Distribution of X and Y

Here the typical assumptions are:

(i) The joint distribution is completely known.
(ii) The joint distribution is of some known form, depending on an unknown parameter.
(iii) X is not a random variable, so that one cannot meaningfully speak of joint distribution. In this case, the assumptions concern the conditional distributions $f(y|X = x)$. These may, again, be known completely, or only up to a certain parameter.

Form of the Data

In some problems, the only observations available are the values of Y (one or more), all sampled for the same unknown value of X. In some other cases, in addition to observations of Y for the unknown value of X, we have available also a "learning sample," in form of a set of pairs $(X_1, Y_1), \ldots, (X_n, Y_n)$, where both components X and Y are known. Such data allow us to estimate the parameters needed to make inference about the value X in question.

Example 14.1.6.   Consider a substance that accumulates in human bones with age. The accumulation is linear but subject to fluctuations, so that the concentration Y of the substance in question at death (at age X) satisfies the linear equation $Y = \alpha X + \beta + \xi$, where α and β are some constants (known or not) and ξ is random. The data on (X_i, Y_i) are available. Suppose now that a tomb was discovered of some ancient ruler and it is possible to measure Y. The question then is to estimate X, the age of the ruler at his death.

□

Goal of the Analysis

Finally, as regards the goal of the analysis, in addition to the situations of predicting (or identifying) the value of X, one may have more complicated goals. They are exemplified by problems of screening or selection.

Example 14.1.7 (Screening).   In a typical screening problem, the data have the form (X, Y), with only Y being observable, and the objective is to decide whether or not X exceeds some threshold x_0. This can be formalized as a problem of predicting the variable Z defined as

$$Z = \begin{cases} 1 & \text{if } X \geq x_0 \\ 0 & \text{otherwise.} \end{cases}$$

Alternatively, it can also be formulated as predicting the value of X, with the loss function of prediction X^* defined as

$$\mathcal{L}(X, X^*) = \begin{cases} 0 & \text{if } X \leq x_0 \text{ and } X^* \leq x_0 \\ 0 & \text{if } X \geq x_0 \text{ and } X^* \geq x_0 \\ 1 & \text{otherwise.} \end{cases}$$

A typical case occurs when we have the observations $Y_1, \ldots, Y_n$ for n subjects (e.g., their test scores), while $X_1, \ldots, X_n$ are their (unobservable) levels of acquired knowledge, levels of talent, and so on. The decision rule, based on observed Y_i, classifies the ith subject to the category $X_i \geq x_0$ or to the category $X_i < x_0$.

Example 14.1.8 (Selection).   A selection problem is similar to the screening problem, except for one difference. In screening problems, the decision regarding classification of the ith subject is independent of decisions regarding other subjects. This allows considering each subject separately and developing the theory of optimal decision for the case of a single observation Y and decision about the corresponding X. As opposed to that, in a typical selection problem, the goal is to single out a specified number of subjects with the highest value of X_i, possibly indicating their order. The existence of a constraint regarding the total number of selected subjects makes the choices dependent, and greatly complicates the theory.

Example 14.1.9 (U.S. Olympic Trials).   An appealing decision rule in the case of a problem of selecting the k best is to base the selection on order statistics of Y_i's, and simply choose the subjects with the best observed Y's, that is, subjects with scores $Y_{1:n}, Y_{2:n}, \ldots, Y_{k:n}$ or with scores $Y_{n:n}, Y_{n-1:n}, \ldots, Y_{n-k+1:n}$ (depending on the discipline; obviously, in running, the lowest time is the best, while for instance in the long jump the highest score is the best). The rule of choosing the best Y's is used in the United States for the final selection of athletes to represent the country in a given discipline in Olympics. Here Y_i is the result in the final elimination event, and X_i is the result that the athlete will (or would) achieve later in Olympics.

The advantage of such a rule is mostly its simplicity, allowing the avoidance of bickering. In fact, however, at the time of final elimination for the Olympics, plenty of data are available about each competitor, so that observation of the ith athlete is of the form $\mathbf{Y}_i = (Y_i, Z_{i,1}, Z_{i,2}, \ldots)$, where Y_i is the result in final eliminations, while $Z_{i,1}, Z_{i,2}, \ldots$ are some other results, not necessarily comparable to the results of other athletes. The decision rule adopted is simply to disregard all components except Y_i, and choose the k best Y_i's as the prediction of the k best X_i's.

14.2 PREDICTIVE SUFFICIENCY

In Chapter 12 we introduced the concept of sufficiency of a statistic. In the context of estimation, it was natural to define this concept in terms of the joint density of a random sample $\mathbf{Y} = (Y_1, \ldots, Y_n)$, where Y_i's are iid with distribution $f(y, \theta)$. A statistic $T(Y_1, \ldots, Y_n)$ was sufficient for θ if the conditional distribution of $\mathbf{Y}$ given T was independent on θ. This means that $T(Y_1, \ldots, Y_n)$ was a reduction of data [i.e., reduction of information contained in the sample $\mathbf{Y} = (Y_1, \ldots, Y_n)$], which did not, in fact, reduce information about θ. For the present consideration we may rephrase the definition of sufficiency as follows. First, the role of parameter θ to be estimated is now played by the variable X, so that we may speak of the joint distribution $(X, \mathbf{Y})$. Second, it is not necessary that $\mathbf{Y}$ represents a random sample: In fact, $\mathbf{Y}$ can be an array of observations, perhaps some being of quantitative and some of qualitative character, and not necessarily independent. In the context of estimation we could still say that a statistic $T = T(\mathbf{Y})$ is sufficient for the family of distributions $\mathcal{P}$ if the distribution of $\mathbf{Y}$ given $T = t$ does not depend on index θ (i.e., is the same for every distribution in $\mathcal{P}$). In estimation theory, where θ was a number (or a vector), it was natural to assume that T is an estimator of θ, so values of T were also numbers. In the general setup, the essence of sufficiency is that $T(\mathbf{Y})$ reduces the information contained in $\mathbf{Y}$, but leaves untouched that part of information which allows us to identify the member of the family $\mathcal{P}$ according to which $\mathbf{Y}$ was sampled.

With these ideas in mind, we now introduce the concept of prediction sufficiency.

Let $f(x, \mathbf{y})$ be a bivariate distribution of $(X, \mathbf{Y})$. As usual, $f(x, \mathbf{y})$ stands for the joint density or joint probability function, depending on whether $X, \mathbf{Y}$

are continuous or discrete. If X is discrete and $\mathbf{Y}$ is continuous, then $f(x, \mathbf{y})$ may be specified by the sequence

$$\{x_i, \pi_i, f_i(\mathbf{y}), i = 1, 2, \ldots\},$$

where $\pi_i = P\{X = x_i\}$ and $f_i(\mathbf{y})$ is the conditional density of $\mathbf{Y}$ given $X = x_i$. We now introduce the following definitions.

Definition 14.2.1. The statistic $T = T(\mathbf{Y})$ is called *prediction sufficient* for X if for every $\mathbf{y}, \mathbf{y}'$ such that $T(\mathbf{y}) = T(\mathbf{y}')$, the conditional distributions $f(x|\mathbf{Y} = \mathbf{y})$ and $f(x|\mathbf{Y} = \mathbf{y}')$ coincide.   □

Thus, if T is prediction sufficient, then the conditional distribution of X given $\mathbf{Y} = \mathbf{y}$ depends on $\mathbf{y}$ only through T. In such a case, to predict X we would need to know only the value t of statistic T, not specific $\mathbf{y}$.

Example 14.2.1. Consider the case of a binary variable X. Without loss of generality, assume that $X = 0$ or 1, with $P\{X = 1\} = \pi = 1 - P\{X = 0\}$, and let $f_i(y)$ be the conditional distribution of Y given $X = i$ $(i = 0, 1)$. The joint distribution of (X, Y) is then completely known. The conditional distribution of X given $Y = y$ is given by

$$P\{X = 1 | Y = y\} = \frac{\pi f_1(y)}{(1 - \pi) f_0(y) + \pi f_1(y)} = \frac{\pi [f_1(y)/f_0(y)]}{1 - \pi + \pi [f_1(y)/f_0(y)]}$$

and therefore the likelihood ratio

$$h(Y) = \frac{f_1(Y)}{f_0(Y)}$$

is a prediction-sufficient statistic.   □

In general, situations such as that described above are not frequent. More often, we know that the joint distribution of $(X, \mathbf{Y})$ belongs to a family $\mathcal{P} = \{f(x, \mathbf{y}; \theta), \theta \in \Theta\}$, and θ is unknown. In this case we may still base our prediction of X on a prediction-sufficient statistic $T(\mathbf{Y})$ if we have an auxiliary statistic that is based on $\mathbf{Y}$ only, and is sufficient for θ. Accordingly, we introduce the following definition.

Definition 14.2.2. The pair of statistics $T(\mathbf{Y}), U(\mathbf{Y})$ is *strongly* prediction sufficient in family $\mathcal{P} = \{f(x, \mathbf{y}; \theta), \theta \in \Theta\}$, if for every θ, T is prediction sufficient for X, and moreover, $U(\mathbf{Y})$ is sufficient for θ.   □

In case of strong predictive sufficiency, the actual prediction is based on the distribution $f(x|T(y), \hat{\theta})$, where $\hat{\theta}$ is the unbiased estimator of θ based on sufficient statistic U. When the family of marginal distributions of Y is complete, $\hat{\theta}$ is the minimum variance unbiased estimate of θ, according to Theorem 12.8.6.

Example 14.2.2. Let $\xi_1, \xi_2, \ldots, \xi_N$ be iid Bernoulli random variables with $P\{\xi_i = 1\} = p = 1 - P\{\xi_i = 0\}$, $i = 1, \ldots, N$. Let $\mathbf{Y} = (\xi_1, \ldots, \xi_n)$ for some $n < N$, and let $X = \sum_{i=1}^{N} \xi_i$. Consider the problem of predicting the value of X on the basis of $\mathbf{Y}$; we know the results of the first n trials, and we want to predict the total number of successes in the first N trials.

SOLUTION. Intuitively, the statistics $T(\mathbf{Y}) = U(\mathbf{Y}) = \sum_{i=1}^{n} \xi_j$ are strongly prediction sufficient. To show it formally, observe first that $U(\mathbf{Y})$ is a sufficient statistic for the parameter p (see Example 12.8.3). It remains to show that if $T(\mathbf{Y}) = T(\mathbf{Y}') = t$, then the conditional distribution of X given $T(\mathbf{Y})$ is the same as the conditional distribution of X given $T(\mathbf{Y}')$.

This is obvious, since given $T(\mathbf{Y}) = t$, the distribution of $X - t$ is binomial $\text{BIN}(N - n, p)$, so that

$$P\{X = t + k \mid T(\mathbf{Y}) = t, p\} = \binom{N-n}{k} p^k (1 - p)^{N-n-k}$$

for $k = 0, 1, \ldots, N - n$, and if $T(\mathbf{Y}') = t$, the distribution of X is the same.

The unbiased estimator of p based on U is simply $(1/n)U(\mathbf{Y}) = (1/n)T(\mathbf{Y})$. For the quadratic loss function, the best predictor is the mean of conditional distribution of X given $T(\mathbf{Y})$; hence if $T(\mathbf{Y}) = t$ successes were observed in first n trials, then the predicted total number of successes in N trials is $t + (N - n)\hat{p} = t + (N - n)(t/n) = (N/n)t$. $\square$

Obviously, a prediction-sufficient statistic exists in any family of bivariate distributions, since $T(Y) = Y$ is always prediction sufficient and $U(Y) = Y$ is always sufficient for any parameter θ.

As in the case of ordinary sufficiency, we have the following theorem which characterizes prediction sufficiency.

Theorem 14.2.1. *The statistic $T = T(Y)$ is prediction sufficient in the family of distributions $\mathcal{P} = \{f(x, y, \theta), \theta \in \Theta\}$ if and only if for every $\theta \in \Theta$ there exist nonnegative functions a and b such that*

$$f(x, y, \theta) = a[x, T(y), \theta] b(y, \theta), \tag{14.1}$$

where b does not depend on x, and a depends on y only through values of statistic T.

Proof. As in case of sufficiency, we shall give proof for the case of discrete random variables only: In case of continuous random variables one needs more subtle tools than those we have developed, since the conditional density is defined only up to the sets of probability zero.

Assume now that T is prediction sufficient, so that the conditional distributions of X satisfy the relation

$$P\{X = x | Y = y, \theta\} = P\{X = x | Y = y', \theta\}$$

for every x and $\theta \in \Theta$, whenever $T(y) = T(y')$.

Letting $P\{X = x | Y = y, \theta\} = a(x, t, \theta)$ whenever $T(y) = t$, we may write, for y such that $T(y) = t$,

$$f(x, y, \theta) = a(x, y, \theta) P\{Y = y, \theta\},$$

which is a decomposition of the form (14.1).

Conversely, suppose that formula (14.1) holds. Then

$$P\{X = x | Y = y, \theta\} = \frac{f(x, y, \theta)}{P\{Y = y, \theta\}} = \frac{a[x, T(y), \theta] b(y, \theta)}{\sum_x a[x, T(y), \theta] b(y, \theta)}$$

$$= \frac{a[x, T(y), \theta]}{\sum_x a[x, T(y), \theta]},$$

which is a function depending on y only through the values of statistic T.

□

It ought to be clear that Theorem 14.2.1 is valid when Y is replaced by a multidimensional observation $\mathbf{Y}$.

As in the case of sufficiency of statistics discussed in Section 12.8, we can define the concept of minimal prediction sufficiency.

Definition 14.2.3. The statistic $T = T(\mathbf{Y})$ is called *minimal* prediction-sufficient if T is a function of any other prediction-sufficient statistic.

□

The concept which is intended to be covered by this definition is that of maximal reduction of the data, still leaving prediction sufficiency intact. To see how Definition 14.2.3 covers this intuition, assume that $T(Y)$ is minimal in the sense of definition and can be reduced further without affecting predictive sufficiency. Thus, there exists a function f, not one–to–one, such that the statistic $T' = f(T(\mathbf{Y}))$ is still prediction sufficient. Then T is not a function of predictive-sufficient statistic T', which gives a contradiction.

The marginal distribution of Y may allow the decomposition $b(y, \theta) = b_1(U(y), \theta) b_2(y)$, in which case U is a sufficient statistic for θ based on Y only.

The natural question arises now: If T is minimal prediction sufficient and U is minimal sufficient for θ, do these two statistics coincide (or are they one–to–one functions of one another), as in Example 14.2.2? The answer to this question is negative: Reduction to a statistic sufficient for θ may involve different aspects of Y than reduction to a prediction-sufficient statistic. This is illustrated by the following example.

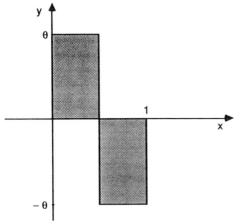

Figure 14.1 Minimal prediction sufficiency with $T \neq U$.

Example 14.2.3. Let (X, Y) have the joint distribution that is uniform on the area presented in Figure 14.1. Here $\theta > 0$ is a parameter of the family, where we assume that $\theta \in \Theta = (0, A)$, with $A < \infty$. Clearly, the marginal distribution of X is uniform on $[0, 1]$. Under a square loss function, the optimal prediction of X (without observation of Y) is $X^* = \frac{1}{2}$, with mean square error equal to $\text{Var}(X) = \frac{1}{12}$.

On the other hand, given $Y = y$, the conditional distribution of X is uniform on $(0, \frac{1}{2})$ if $y > 0$, and uniform on $(\frac{1}{2}, 1)$ if $y < 0$. Thus, the prediction-sufficient statistic is the sign of Y, that is,

$$T(y) = \frac{y}{|y|} = \begin{cases} +1 & \text{if } y > 0 \\ -1 & \text{if } y < 0. \end{cases}$$

The optimal predictors for X^* are then (again, under square loss) $X^* = \frac{1}{4}$ if $y > 0$ and $X^* = \frac{3}{4}$ if $y < 0$.

As regards sufficiency for θ, the minimal sufficient statistic* is $U(Y) = |Y|$. Thus, it is possible that $T \neq U$, in cases when pair (T, U) is strongly prediction sufficient. $\square$

In the preceding example, the conditional distribution of X did not depend on θ. As shown below, it is also possible that $T \neq U$ when the conditional distribution of X depends on θ.

Example 14.2.4. We modify Example 14.2.3 by assuming that for a given $\theta > 0$, the joint distribution of (X, Y) is uniform on two rectangles as shown in Figure 14.2.

---

* Indeed, if $Y_1, Y_2, \ldots, Y_n$ is a random sample from distribution uniform on $[-\theta, \theta]$, then (see Chapter 12) $\max[\max(Y_1, \ldots, Y_n), -\min(Y_1, \ldots, Y_n)]$ is sufficient for θ. In our case $n = 1$, and sufficient statistics reduces to $|Y|$.

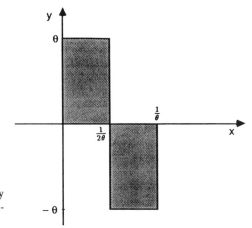

Figure 14.2 Minimal prediction sufficiency and sufficiency when distribution of X depends on θ.

The minimal prediction-sufficient and the minimal sufficient statistic are as before, $T(Y) = \mathrm{sgn}(Y)$ and $U(Y) = |Y|$. This time, however, the conditional distribution of X given Y depends on θ. Given Y, the unbiased estimator of θ based on $U(Y) = |Y|$ is* $\hat{\theta} = 2|Y|$. The prediction X^* (for square error loss) of X is therefore $X^* = 1/(4\hat{\theta}) = 1/(8|Y|)$ for $Y > 0$ and $X^* = 3/(4\hat{\theta}) = 3/(8|Y|)$ for $Y < 0$.

PROBLEMS

14.2.1 Find a prediction-sufficient statistic for X if the joint distribution of (X, Y) is as follows: $X = 1$ or 2 with probability p and $1 - p$, and given X, random variable Y has density $f(y|X = x) = C_x y^x$ for $0 \le y \le 1$, where C_x is the normalizing constant. Find the best (in the sense of mean square error) prediction of X given the random sample $Y_1, \ldots, Y_n$, all observed for the same X.

14.2.2 Find a prediction-sufficient statistic for X if $P(X = 1) = p, P(X = 3) = 1 - p$ and $P(Y = X + 1) = P(Y = X - 1) = \frac{1}{2}$. For every p find the minimal prediction-sufficient statistic.

14.2.3 For Problem 14.2.2 find the best prediction of X for the mean square error.

14.2.4 For Problem 14.2.2 find the best prediction of X in the sense of maximizing the conditional probability of $X = x$ given Y.

---

* Indeed, it was shown in Chapter 12 that $[(n + 1)/n]\max(Y_1, \ldots, Y_n)$ is an unbiased estimator of θ in case of a random sample from distribution uniform on $[0, \theta]$. An extension to case $[-\theta, \theta]$ and then reduction to $n = 1$ gives the result.

14.3 DISCRIMINATION IN THE BINARY CASE

Let us begin with the simplest case, when the variable X assumes only two values, 1 and 2, with $\pi = P\{X = 1\} = 1 - P\{X = 2\}$. The conditional distributions of Y given X are $f_1(y)$ and $f_2(y)$ for $X = 1$ and $X = 2$, respectively. As shown in Example 14.2.1, the prediction-sufficient statistic is the likelihood ratio $h(y) = f_2(y)/f_1(y)$.

Assume that upon observing the value y of the variable Y we have three options: Decide that $X = 1$, decide that $X = 2$, and defer the decision. Moreover, we are also allowed to randomize decisions, that is, let the chance determine which option to take. Generally, any decision rule can be described as a pair of functions $(p_1(y), p_2(y))$, defined on the set Q of possible values of random variable Y and such that $p_i(y) \geq 0, i = 1, 2$, and $p_1(y) + p_2(y) \leq 1$ for all y. The decision rule, $\delta = (p_1, p_2)$, will be interpreted as follows: If y is observed, then decide that $X = 1$ with probability $p_1(y)$, decide that $X = 2$ with probability $p_2(y)$, and defer the decision with probability $1 - p_1(y) - p_2(y)$.

Obviously, any deterministic rule is a special case of the rule δ of the form described above. Indeed, a deterministic rule is a partition of the set Q into three parts, $Q = A_0 \cup A_1 \cup A_2$. We decide that $X = i$ if $y \in A_i (i = 1, 2)$, and defer the decision if $y \in A_0$. This rule is obtained by taking

$$p_1(y) = \begin{cases} 1 & \text{if } y \in A_1 \\ 0 & \text{otherwise} \end{cases} \quad \text{and} \quad p_2(y) = \begin{cases} 1 & \text{if } y \in A_2 \\ 0 & \text{otherwise.} \end{cases}$$

In the sequel, we shall let Δ denote the class of all possible rules $\delta = (p_1, p_2)$, deterministic or randomized.

Let $\delta = (p_1, p_2) \in \Delta$. With any decision rule δ and object of type i $(i = 1, 2)$ we can associate the probabilities of classification $\alpha_{1i}(\delta), \alpha_{2i}(\delta)$ and probabilities $\alpha_{0i}(\delta)$ of deferring the decision. Thus, $\alpha_{21}(\delta)$ will be the probability of misclassifying an object of type 1, that is, classifying it as an object of type 2. Similarly, $\alpha_{12}(\delta)$ will be the probability of misclassifying an object of type 2 (i.e., classifying it as an object of type 1). We have, therefore,

$$\alpha_{21}(\delta) = \int_Q p_2(y)f_1(y) \, dy. \tag{14.2}$$

Similarly,

$$\alpha_{12}(\delta) = \int_Q p_1(y)f_2(y) \, dy, \tag{14.3}$$

and for $i = 1, 2$,

$$\alpha_{0i}(\delta) = \int_Q [1 - p_1(y) - p_2(y)]f_i(y) \, dy. \tag{14.4}$$

In the special case of a deterministic rule δ we have

$$\alpha_{21}(\delta) = \int_{A_2} f_1(y) \, dy, \qquad \alpha_{12}(\delta) = \int_{A_1} f_2(y) \, dy$$

and

$$\alpha_{0i}(\delta) = \int_{A_0} f_i(y) \, dy.$$

We shall now specify various ways of defining optimality of decision rules expressed in terms of probabilities of misclassifications and costs of deferring the decision, and then find an optimal rule for each optimality criterion.

As we shall see, the optimal rules will:

(i) Be based on prediction-sufficient statistic $h(y) = f_2(y)/f_1(y)$.
(ii) Be of special form, involving two thresholds, to be compared with $h(y)$ in order to reach the decision.
(iii) Involve randomization only in a special case when $h(y)$ equals one of the two thresholds.

These facts are closely related to the Neyman–Pearson lemma.

Following the pattern used in Section 12.4, we can introduce the following definitions, in which Δ stands for the class of all decision rules $\delta = (p_1, p_2)$.

Definition 14.3.1. A decision rule $\delta' \in \Delta$ *dominates* the decision rule $\delta \in \Delta$, to be written as $\delta' \succeq \delta$, if

$$\alpha_{ij}(\delta') \le \alpha_{ij}(\delta) \tag{14.5}$$

for $i = 0, 1, 2$ and $j = 1, 2$ with $i \ne j$. If at least one of the inequalities in (14.5) is strict, we shall speak about *strict* domination and write $\delta' \succ \delta$. $\square$

Thus, domination means an improvement of probabilities of misclassification and deferement of decision for elements of either of the two types.

Definition 14.3.2. A decision rule $\delta \in \Delta$ is called *inadmissible* if there exists another decision rule in Δ which dominates it strictly; otherwise, δ will be called *admissible*.

Definition 14.3.3. A class $D \subset \Delta$ of decision rules is called *complete* if for any rule $\delta \in \Delta$ there exists $\delta' \in D$ such that $\delta' \succeq \delta$. $\square$

The intention of these definitions is to eliminate from considerations all rules that need not be taken into account in search for the optimal rule. Indeed, an inadmissible rule is strictly dominated by some other rule, which means that (if only misclassification and deferrement probabilities are taken into account) one should never use an inadmissible rule. On the other hand, if we know a complete class, then the choice can be restricted to this class only. Thus, a special role is played by minimal complete classes of rules.

Clearly, any complete class contains all admissible decision rules. The converse is not true; to understand why, observe that the class of all decision rules is complete. More generally, decision δ' whose existence is postulated in Definition 14.3.3 need not be admissible.

Since a class of admissible decision rules is contained in any complete class, we have the following:

Theorem 14.3.1. *If the class of all admissible decision rules is complete, then it is the smallest complete class.*

The converse is also true; we have

Theorem 14.3.2. *If the smallest complete class exists, then it coincides with the class of all admissible decision rules.*

Observe, however, that the smallest complete class may not exist; this happens when there are no admissible decision rules.

We now introduce a definition that specifies rules of a special form.

Definition 14.3.4. We say that $\delta = (p_1, p_2)$ is a *threshold* rule with respect to a function $g(y)$ if there exist constants k_1, k_2 with $-\infty \le k_1 \le k_2 \le +\infty$ such that

$$p_1(y) = 1 \quad \text{if } g(y) < k_1,$$
$$p_2(y) = 1 \quad \text{if } g(y) > k_2,$$
$$p_1(y) = p_2(y) = 0 \quad \text{if } k_1 < g(y) < k_2.$$

If $g(y) = k_i, i = 1, 2$, then two or even three among the quantities $p_1(y), p_2(y)$, and $1 - p_1(y) - p_2(y)$ may be strictly positive. $\square$

Observe that by taking $k_1 = -\infty$ or $k_2 = +\infty$ (or both) we may obtain rules that always classify an object as type 1, or always classify an object as type 2, or always defer the decision.

We now have the following theorem.

Theorem 14.3.3. *For every rule $\delta \in \Delta$ there exists a threshold rule with respect to the likelihood ratio $h(y) = f_2(y)/f_1(y)$ which dominates δ.*

Here the likelihood ratio $h(y)$ is taken as $+\infty$ if $f_1(y) = 0 < f_2(y)$. If $f_1(y) = f_2(y) = 0$, the value $h(y)$ remains undefined.

The proof is, to a large extent, analogous to proof of the Neyman–Pearson lemma (Theorem 13.3.1), except for the fact that we have here a three-option problem, which is considerably more complicated than a binary decision problem. For a detailed proof, see Bromek and Niewiadomska–Bugaj (1987).

In the sequel we let $\Delta^* \subset \Delta$ denote the class of all threshold rules with respect to likelihood ratio h.

Example 14.3.1. Suppose that when $X = 1$, the conditional distribution of Y is normal $N(\mu_1, \sigma^2)$, while if $X = 2$, then Y is normal $N(\mu_2, \sigma^2)$, where $\mu_1 < \mu_2$. The likelihood ratio is

$$ h(y) = \exp\left\{ -\frac{1}{2\sigma^2}[(y - \mu_2)^2 - (y - \mu_1)^2] \right\} = C \exp\left\{ \frac{\mu_2 - \mu_1}{\sigma^2} y \right\}, $$

where C is a constant. Thus, the likelihood ratio is an increasing function of y and the inequality $h(y) \le k$ is equivalent to the inequality $y \le k'$ for some k'. According to Theorem 14.3.3, every decision rule is dominated by a rule of the form:

> decide that $X = 1$ if $y < \kappa_1$
> decide that $X = 2$ if $y > \kappa_2$
> defer the decision   if $\kappa_1 < y < \kappa_2$.

If $y = \kappa_1$, one can either decide that $X = 1$ or defer the decision, and similarly for $y = \kappa_2$.   □

In the example above the likelihood ratio h was a monotone function of y, and consequently the threshold rules with respect to h were also threshold rules with respect to y. When the likelihood ratio is not monotone, the decision rule need not be a threshold rule with respect to y.

Example 14.3.2. Suppose now that if $X = 1$, then Y has normal distribution $N(0, \sigma_1^2)$, while if $X = 2$, then $Y \sim N(0, \sigma_2^2)$, where $\sigma_1^2 < \sigma_2^2$. The likelihood ratio is

$$ h(y) = \frac{\sigma_1}{\sigma_2} \exp\left\{ \frac{y^2}{2}\left(\frac{1}{\sigma_1^2} - \frac{1}{\sigma_2^2} \right) \right\}, $$

which is an increasing function of y^2. The inequality $h(y) \le k$ is equivalent to $|y| \le k'$. We see that the likelihood ratio is not the minimal prediction-suffficient statistic: the sign of Y is irrelevant, and the minimal prediction-sufficient statistic is $|Y|$. The threshold decision rules are now:

> decide that $X = 1$ if $|y| < \kappa_1$
> decide that $X = 2$ if $|y| > \kappa_2$
> defer the decision   if $\kappa_1 < |y| < \kappa_2$.

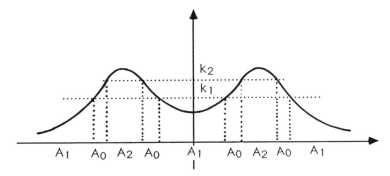

Figure 14.3   Likelihood ratio.

Example 14.3.3.   Consider finally the situation when if $X = 1$, then Y has the double exponential distribution with the density $f_1(y) = \frac{1}{2} \exp\{-|y|\}$, while for $X = 2$ the density is standard normal. Now the likelihood ratio is

$$h(y) = \sqrt{\frac{2}{\pi}} \exp\left\{ -\frac{y^2}{2} + |y| \right\}$$

and again $|Y|$ is a minimal prediction-sufficient statistic [so that $h(Y)$ is not minimal prediction sufficient since $h(y)$ does not determine uniquely $|y|$]. The inequality $h(y) \le k$ is equivalent to $|y| \le k'$ or $|y| \ge k''$ (see Figure 14.3). The threshold rules are given by the sets A_0, A_1, A_2.   □

Knowledge of a complete class of rules (even knowledge of the minimal such class) is not sufficient for determining the optimal rule. The reason is simply that existence of a "preference" relation (in this case dominance $\succeq$) does not imply the existence of a unique "best" rule (i.e., a rule that would dominate all other rules).

If one wants to speak of *the* optimal rule, the description of the problem has to be suitably implemented by additional criteria. The first step here is the specification of loss function. This means expressing the consequences of an application of a discrimination rule on a common scale, with higher values corresponding to less desirable outcomes. In the present case of binary variable X there are six possible outcomes:

True Value	Classification		
of X	Deferred	Type 1	Type 2
1	$\mathcal{L}_{01}$	$\mathcal{L}_{11}$	$\mathcal{L}_{21}$
2	$\mathcal{L}_{02}$	$\mathcal{L}_{12}$	$\mathcal{L}_{22}$

In typical cases there is no penalty for a correct decision (i.e., for classifying object as type i if in fact $X = i$, $i = 1, 2$), which means that $\mathcal{L}_{11} = \mathcal{L}_{22} = 0$. If decision is incorrect or deferred, a penalty may be imposed, so $\mathcal{L}_{ij} \geq 0$ for $j = 1, 2$ and $i = 0, 1, 2$ $(i \neq j)$.

Given the loss function, we formulate various criteria, which can lead to a choice of optimal rule. With every decision rule $\delta \in \Delta$ (threshold or not) one can associate its *risk*, defined as expected loss:

$$R(\delta) = \pi(\alpha_{01}\mathcal{L}_{01} + \alpha_{21}\mathcal{L}_{21}) + (1 - \pi)(\alpha_{02}\mathcal{L}_{02} + \alpha_{12}\mathcal{L}_{12}). \qquad (14.6)$$

Consider now the problem of minimization of risk (14.6) in the class Δ of all rules. By Theorem 14.3.3 we know that the optimum is in the class Δ^* of threshold rules.

From the form of risk function $R(\delta)$ it is clear that if the losses $\mathcal{L}_{01}$ and $\mathcal{L}_{02}$ for deference of the decision are high, it should be best not to defer the decision (this conclusion can be made because all losses $\mathcal{L}_{ij}$ are nonnegative). Specifically, we have here the following theorem, in which we let

$$Q = \frac{\mathcal{L}_{01}}{\mathcal{L}_{21}} + \frac{\mathcal{L}_{02}}{\mathcal{L}_{12}}. \qquad (14.7)$$

Theorem 14.3.4.

(i) *If $Q < 1$ (deferment is cheap), then the risk $R(\delta)$ given by (14.6) attains its minimum for rules $\delta_{\kappa_1, \kappa_2} \in \Delta^*$ with*

$$\kappa_1 = \frac{\pi}{1 - \pi} \frac{\mathcal{L}_{01}}{\mathcal{L}_{12} - \mathcal{L}_{02}}, \qquad \kappa_2 = \frac{\pi}{1 - \pi} \frac{\mathcal{L}_{21} - \mathcal{L}_{01}}{\mathcal{L}_{02}}. \qquad (14.8)$$

If $h(y) = \kappa_1$, then probability $p_1(y)$ may be arbitrary, but $p_2(y) = 0$. If $h(y) = \kappa_2$, then $p_2(y)$ may be arbitrary, but $p_1(y) = 0$.

(ii) *If $Q \geq 1$ (deferment is expensive), then*

$$\kappa_1 = \kappa_2 = \kappa = \frac{\pi}{1 - \pi} \frac{\mathcal{L}_{21}}{\mathcal{L}_{12}}.$$

(iii) *If $Q = 1$, then the probabilities $p_1(y)$ and $p_2(y)$ for y with $h(y) = \kappa$ may be arbitrary, subject only to the constraint $0 \leq p_1(y) + p_2(y) \leq 1$.*

(iv) *Finally, if $Q > 1$, then for y with $h(y) = \kappa$ we may choose arbitrary probabilities $p_1(y)$ and $p_2(y)$, provided that $p_1(y) + p_2(y) = 1$.*

For the proof see Bromek and Niewiadomska–Bugaj (1987). To use Theorem 14.3.4 in practical situations of discrimination between objects of two types on the basis of observation of an attribute Y we need to know:

1. The conditional distributions of Y for objects of type 1 and type 2 [i.e., $f_1(y)$ and $f_2(y)$].

2. The prior probability π that an object is of type 1,
3. The losses $\mathcal{L}_{ij}, i = 0,1,2; j = 1,2 \ (i \neq j)$ resulting from misclassification or deferment of decision for an object of type j.

The knowledge of $f_i(y)$ comes typically from past experience, or from "learning samples," that is, samples of values observed in subpopulations of objects of a given type. The situations vary from "clear-cut" cases of very exact knowledge, to more vague ones, such as "terrorist profile," in the form of a list of features to watch out for by airport security people. Knowledge of the prior probability π may, or may not, come from past experience. In some cases π reflects the experience or intuition of the decision maker, without having a frequential counterpart. The value of π may be elicited from statements such as "it is about twice as likely that the object is of type 1 than of type 2" (which gives $\pi = \frac{2}{3}$), and so on.

Finally, the knowledge of loss function comes from assessment of the seriousness of two misclassification errors, and the relative loss due to deferment of classification (as compared to misclassification). For instance, consider statements such as:

1. It is 10 times as serious to misclassify an object of type 1 than to misclassify an object of type 2.
2. The cost of deferement of decision for objects of type 1 is about $\frac{1}{5}$ of the cost of misclassification; for objects of type 2 the cost of deferment is about half of the cost of misclassification.

These statements give the equations

$$\mathcal{L}_{21} = 10\mathcal{L}_{12}, \quad \mathcal{L}_{01} = \frac{1}{5}\mathcal{L}_{21}, \quad \mathcal{L}_{02} = \frac{1}{2}\mathcal{L}_{12}. \qquad (14.9)$$

Since the units of measurement of loss function can be selected in an arbitrary way, a solution of (14.8) may be

$$\mathcal{L}_{01} = 4, \quad \mathcal{L}_{21} = 20, \quad \mathcal{L}_{02} = 1, \quad \mathcal{L}_{12} = 2. \qquad (14.10)$$

In Examples 14.3.4 to 14.3.6 we shall use the values (14.10) and $\pi = \frac{2}{3}$ to find the decision rules .

First, observe that here we have [see (14.10)]

$$Q = \frac{4}{20} + \frac{1}{2} < 1,$$

hence the appropriate formula is (14.8). We now find that

$$\kappa_1 = \frac{\frac{2}{3}}{\frac{1}{3}} \times \frac{4}{2-1} = 8, \quad \kappa_2 = \frac{\frac{2}{3}}{\frac{1}{3}} \times \frac{20-4}{1} = 32.$$

Example 14.3.4. Suppose that in Example 14.3.1 we have $\mu_1 = 0$, $\mu_2 = 1$, $\sigma^2 = 1$. Then the likelihood ratio is

$$h(y) = \frac{1}{\sqrt{e}} e^y.$$

Consequently, we have $h(y) \leq \kappa$ iff $y \leq 0.5 + \log \kappa$. The optimal threshold rule is therefore

$$\text{decide that } X = 1 \quad \text{if } y \leq 0.5 + \log 8 = 2.58$$
$$\text{decide that } X = 2 \quad \text{if } y \geq 0.5 + \log 32 = 3.97$$
$$\text{defer the decision} \quad \text{if } 2.58 < y < 3.97,$$

where the decisions $X = 1$ and $X = 2$ correspond to distributions $Y \sim N(0, 1)$ and $Y \sim N(1, 1)$ respectively. The choice of the decision at the thresholds is irrelevant in view of the continuity of the distributions and strict monotonicity of the likelihood ratio.

Example 14.3.5. In the situation of Example 14.3.2, suppose that $\sigma_1^2 = 1$, $\sigma_2^2 = 4$. Then the likelihood function is $h(y) = \frac{1}{2} e^{(3/8)y^2}$. The inequality $h(y) \leq k$ is equivalent to $|y| \leq (2\sqrt{2}/\sqrt{3})\sqrt{\log(2\kappa)}$, from which one can compute the thresholds for the classification rule.

Example 14.3.6. Suppose now that we are in the situation of Example 14.3.3, where we have to decide whether Y comes from double exponential distribution or standard normal distribution. Here $h(y) = \sqrt{2/\pi} e^{-(y^2/2)+|y|}$. The function h attains its maximum at $y = \pm 1$, equal to $\sqrt{(2/\pi)} e^{0.5} = 1.315$. It follows that we have $h(y) < \kappa_1 = 8$ for all y, which means that we should *always* decide that $X = 1$ (distribution is double exponential). This is due to the fact that the penalty for misclassifying an observation as normal ($\mathcal{L}_{21} = 20$) is so high that it is best never to venture classifying any observation as coming from a standard normal distribution.

Example 14.3.7. Consider finally the following simplified model of decisions faced by airport security personnel who have to screen passengers to eliminate potential hijackers. Let $X = 1$ (type 1) signify an ordinary passenger and $X = 2$ (type 2) a passenger who intends to hijack the airplane. Based on past knowledge of hijacking the experts compiled a list of features that appear with sufficiently high frequency among hijackers; these concern age, appearance, behavior, and so on. Each of these features alone is not sufficient to arouse suspicion; it is only when sufficiently many of them are observed in one person that there is suspicion, perhaps justifying special action on the part of the security personnel.

To simplify the consideration, let Y be the total number of features observed in a passenger. Assume that it is known (from the data on hijacking as well as from observation of ordinary passengers) that the distribution of Y is Poisson. If $X = 1$ (ordinary passengers), the mean of Y is 5. If $X = 2$ (a hijacker), the mean of Y is 15. The question is: At which observed value of Y should the security personnel start some action: stop the person to ask him or her some questions, try to frustrate the person, and so on (which amounts to deferring the decision that $X = 2$), or detain the person (decide that $X = 2$).

To get the data needed to make optimal decision, one needs some idea about the loss structure. Firstly, $\mathcal{L}_{01}$ and $\mathcal{L}_{21}$ express consequences of deferment of decision and detainment of an innocent passenger, respectively. The first may cause small delays in airplane departures if it occurs often enough. The second could lead to some financial loss if the passenger eventually misses the plane, and so on. Suppose that the ratio of cost of the two consequences is 100, so that we may put $\mathcal{L}_{01} = 1, \mathcal{L}_{21} = 100$. The most serious error, misclassifying a hijacker as an ordinary passenger, yields some very high costs: Even if everything ends happily, the financial costs are considerable. Let us take this cost as 1 million times more than the cost $\mathcal{L}_{21}$, so that $\mathcal{L}_{12} = 10^8$. Finally, assume that the cost of deferring the decision as regards a hijacker are 5% of that of misclassifying him (such could be the case if deferment of a decision leads in 95% of the cases to capturing the hijacker before he boards the plane). We therefore have $\mathcal{L}_{02} = 0.05 \times 10^8 = 5 \times 10^6$.

Assume also that from available data, about one in every million passengers is a hijacker, so that $\pi = 1 - 10^{-6}$. In this case the formula (14.8) gives

$$\kappa_1 = \frac{1 - 10^{-6}}{10^{-6}} \cdot \frac{1}{10^8 - 5 \times 10^6} = 0.01, \quad \kappa_2 = \frac{1 - 10^{-6}}{10^{-6}} \cdot \frac{100 - 1}{5 \cdot 10^6} = 19.80.$$

In the present case the likelihood ratio is

$$h(y) = \frac{(15^y/y!)e^{-15}}{(5^y/y!)e^{-5}} = 3^y e^{-10}.$$

The inequality $h(y) \le \kappa_1$ is equivalent to $y \le 4.96$, while $h(y) \ge \kappa_2$ is equivalent to $y \ge 11.82$. Qualitatively speaking, this is what the security personnel are apparently doing: By making their presence felt (inspection of hand baggage under slightest doubt, ordering passengers to cross the metal detectors again and again, etc.), they create enough frustration to make potential hijackers betray themselves in some way. On the other hand, stopping passengers for a detailed search (i.e., deciding that $X = 2$) occurs only in extreme cases. □

In some situations the conditions of the problem may not allow us to use the solution given in Theorem 14.3.3. In such cases the formulation of the criterion to be optimized needs to be changed. Typical cases are listed below.

For the proofs and detailed solutions, see Bromek and Niewiadomska-Bugaj (1987).

Minimization of Risk When Deferment Is Not Allowed
We have here one threshold only, given by

$$\kappa = \frac{\pi}{1 - \pi} \frac{\mathcal{L}_{21}}{\mathcal{L}_{12}},$$

with arbitrary randomization for y with $h(y) = k$. The resulting minimum of risk coincides with that given by Theorem 14.3.4 for $Q > 1$.

If $\mathcal{L}_{21} = \mathcal{L}_{12}$, then $\kappa = \pi/(1 - \pi)$ and the optimal decision rule is equivalent to: Decide that $X = 1$ if the conditional probability that $X = 1$ given $Y = y$ exceeds 0.5.

Indeed, we have here

$$P\{X = 1 | Y = y\} = \frac{\pi f_1(y)}{\pi f_1(y) + (1 - \pi) f_2(y)}$$

$$= \frac{1}{1 + [(1 - \pi)/\pi] h(y)}$$

and $P\{X = 1 | Y = y\} \geq 0.5$ if and only if $1 + [(1 - \pi)/\pi] h(y) \leq 2$, which is equivalent to $h(y) \leq \pi/(1 - \pi) = \kappa$.

Minimization of Risk Under Constraints on Both Probabilities of Errors
The problem is to minimize the risk in the class of rules δ such that $\alpha_{12} \leq c_2, \alpha_{21} \leq c_1$, where c_1 and c_2 are some constants satisfying the condition $0 \leq c_i \leq 1, i = 1, 2$. A particular case here is when $c_2 = 1$ (no constraints on α_{12}), while $0 < c_1 < 1$, and in addition, no deferment is allowed. This corresponds to the problem of finding the most powerful test of the null hypothesis H_0 : $X = 1$ against the alternative $H_1 : X = 2$ at a significance level at most c_1.

Still another case is obtained by putting the constraints on the ratios of probabilities of making wrong and correct decisions, that is, requiring that

$$\frac{\alpha_{21}}{\alpha_{11}} \leq d_1, \quad \frac{\alpha_{12}}{\alpha_{22}} \leq d_2.$$

All these cases result from a non-Bayesian's reluctance to use the probability π in the criterion, according to the principle that *given the same data and using the same method, two statisticians should arrive at the same conclusion (except when the decision is randomized)*. This principle therefore rules out the use of prior probabilities unless they have frequential interpretation.

Minimax Criteria
One of the optimality criteria that does not depend on prior probability π is the *minimax* criterion, requiring us to minimize larger of the two conditional losses:

$$\max[\mathcal{L}_{21}\alpha_{21}(\delta) + \mathcal{L}_{01}\alpha_{01}(\delta), \mathcal{L}_{12}\alpha_{12}(\delta) + \mathcal{L}_{02}\alpha_{02}(\delta)].$$

If $Q < 1$ (cheap deferment), the solution is given by $\kappa_1 = \rho\kappa_2$, where

$$\rho = \frac{\mathcal{L}_{01}\mathcal{L}_{02}}{(\mathcal{L}_{21} - \mathcal{L}_{01})(\mathcal{L}_{12} - \mathcal{L}_{02})}.$$

If $Q \geq 1$ (expensive deferment), the optimal rule is based on one threshold only (so that there is no deferment of the decision), and the threshold κ and probability p_1 are determined from the equation

$$\mathcal{L}_{21}\left[P\{h(Y) > \kappa | X = 1\} + (1 - p_1)P\{h(Y) = \kappa | X = 1\}\right]$$
$$= \mathcal{L}_{12}\left[P\{h(Y) < \kappa | X = 2\} + p_1 P\{h(Y) = \kappa | X = 2\}\right].$$

PROBLEMS

14.3.1 Suppose that some feature (Y) has geometric distribution for objects of type 1, and Poisson distribution for objects of type 2. Objects of both types are mixed in proportion 1:3. If one object is now sampled at random, how should it be classified, if the ratio of costs of misclassifications of objects types 1 and 2 is 2:3.

14.3.2 For Problem 14.3.1 find the minimax discrimination rule if objects of both types are mixed in an unknown proportion.

REFERENCES

Bromek, T., and M. Niewiadomska-Bugaj, 1987. "Threshold Rules in Two-Class Discrimination Problems," *Probability and Mathematical Statistics*, Vol. 8, 11–16.

Kowalczyk, T., 1991. "Screening Problems," in T. Bromek and E. Pleszczyńska (eds.), *Statistical Inference: Theory and Practice*, PWN, Warszawa and Kluwer, Dordrecht, pp. 86–105.

CHAPTER 15

Linear Models

15.1 INTRODUCTION

This chapter is devoted to statistical problems arising in situations when the observed values (called, depending on the context, "dependent variable," "response," etc.) are influenced by some other variable (referred to as "independent variable," "explanatory variable," "treatment," "factor," etc.). The term "influenced by" can, in principle, be interpreted in a number of ways. In this chapter we consider the case when only expected values are affected, while other characteristics (as well as the type of distribution) remain the same. Thus, an alternative title of the present chapter could be something like "Influencing the Mean."

The theories that we present depend on the assumptions about the independent variable: Can its value be observed or not? If yes, can it be controlled by the experimenter? If no, can it be regarded as random? Depending on the answer to these questions, we have regression analyses or different analyses of variance models (one-factor, two-factor, with or without interaction, with fixed or random effects, etc.).

It is perhaps within the general category of linear models that one has more abundance of specific models than in any other domain of statistics. The numerical procedures for each of these models are available in many statistical packages, so that the results of calculations are literally "at one's fingertips."

However, for the same data set one can typically use various procedures, of which only a few (perhaps just one) are appropriate. Knowledge of which these procedures are comes from understanding the statistical issues involved, and requires communication between the statistician and the experimenter. The fact that other procedures can be carried out (i.e., that the computer can perform the calculations because the data have appropriate dimensions and layout, etc., and that the computer specialist knows how to make the program run) is of no consequence here. A deliberately oversimplified analogy is that a man's right leg can always be amputated if only he still has his right leg. Such a procedure may be appropriate in case of a bone cancer but is typically inappropriate as a treatment of, say, a sprained ankle. To choose the proper procedure the surgeon must not only have technical knowledge of how to amputate a leg (analogous to activating a specific computer program), but he has to know when such a procedure makes sense from the point of view of the ultimate goal of treating the patient.

Many models (theories) are available in the general setup considered in this chapter. In this book we present basic ideas and solutions covering a few of the most representative cases. We hope that this will provide enough information to motivate users of various statistical packages to try to identify conditions when a specific procedure is not the only one available. A thorough and exhaustive presentation of any of these theories may be found in any of numerous books devoted to regression analysis or analysis of variance [see, e.g., Montgomery (1992), Myers (1986), Stapleton (1995) or Myers and Milton (1991)].

In this chapter we outline some common ways to analyze data measured on a scale of at least interval type. We shall discuss the methods of detecting and measuring effects expressed through the mean of the observed random variable. Needless to say, we present only the most common of such methods, regression analysis and analysis of variance.

Example 15.1.1. A simple case of linear regression analysis is as follows. We observe a random variable Y such that $Y = \alpha + \beta X + \xi$, where X is some variable (random or not, possibly controlled by the experimenter) and ξ is the "error" [i.e., a random variable with $E(\xi) = 0$, $\text{Var}(\xi) = \sigma^2 < \infty$]. Both X and Y are assumed to be measured on an interval (or possibly even ratio) scale.

Specific examples may be obtained by taking Y to be time to completion of some chemical reaction, and X the temperature; Y may be some substance that accumulates linearly (up to random fluctuations) in human bones throughout life, X may be the age of the person at death, and so on.

Typically, we have data in the form of pairs (X_i, Y_i), $i = 1, \ldots, n$, possibly with some X_i's repeating. The problems are to estimate α, β, and σ^2 (or test hypotheses about these parameters); predict the value of Y to be observed for some X_0; estimate X corresponding to some observed value of Y. The last problem is a special case of a discrimination problem.

Obvious generalizations involve the model with more than one variable X, for instance when $Y = \alpha + \beta_1 X_1 + \beta_2 X_2 + \xi$, where X_1 and X_2 are some variables. We may have nonlinear models, such as $Y = \alpha + \beta X + \gamma X^2 + \xi$ (quadratic regression), and so on. □

Although not discussed in this book, let us mention here the following important class of problems. Suppose that $Y = \varphi(\mathbf{X}) + \xi$, where φ is some function of $\mathbf{X} = (X_1, \ldots, X_k)$ and ξ is a random variable. We can observe Y, but not $\varphi(\mathbf{X})$ and ξ separately. The function φ is not known. We assume that X_i's can be "set" by the experimenter for making an observation. The problem is to find $\mathbf{X}^* = (X_1^*, \ldots, X_k^*)$ that maximizes $\varphi(\mathbf{X})$.

To be specific, we may have metals $1, \ldots, k$ which are used in an alloy. Here X_i is the amount (or fraction) of the ith metal used, and Y is the observed property of the alloy, say its strength, which we want to optimize. When an alloy is made, its strength Y is observed, but the observations involve some random errors ξ, so for the same proportions X_i $(i = 1, \ldots, k)$ the value of Y may differ from observation to observation. The objective is to find the "best" proportions X_i^*.

The theory of search for such an optimum (in effect: finding the maximum of unknown function φ, under observational errors) is known under various names (e.g., *stochastic search, extremal experiments*, etc.). Recently, this theory is developing quite dynamically. It lies, however, beyond the scope of this book.

Example 15.1.2 (Analysis of Variance—ANOVA). Analysis of variance applies to situations similar to those in regression analysis, but somewhat simplified, where the variables X_i (one, two, or more) cannot be measured. We typically speak of "factors" that operate on some levels.* For instance, we may have the data (measurements of some "response") taken from populations classified according to some criteria: for instance, sex (male, female) and smoking status (never smoked, former smoker, current smoker). We might wish to find out whether any of these factors (sex and smoking status) has an effect on Y (e.g., response to some drugs, etc).

The data take the form of a table:

		1 Never Smoked	2 Former Smoker	3 Current Smoker
1	Male	Y_{11}	Y_{12}	Y_{13}
2	Female	Y_{21}	Y_{22}	Y_{23}

Here Y_{ij} is the response of a subject from group (i, j), where we have one observation in each cell. The model is

$$Y_{ij} = \mu + \alpha_i + \beta_j + \xi_{ij},$$

where we may assume that $\alpha_1 + \alpha_2 = 0, \beta_1 + \beta_2 + \beta_3 = 0$, and ξ_{ij} is the error term, assumed to satisfy $\xi_{ij} \sim N(0, \sigma^2)$. We may wish to test the hypotheses

$$H_0 : \alpha_1 = \alpha_2 = 0$$

and

$$H_0' : \beta_1 = \beta_2 = \beta_3 = 0$$

---

*The use of the word "level" in the context of ANOVA does not imply any specific ordering [e.g., the levels of factors such as "sex" are F (female) and M (male)].

(no sex effect, and no smoking effect) against the alternatives $H_1 : \alpha_1 \neq \alpha_2$ and $H_1' : \beta_k \neq \beta_l$ for some $k, l = 1, 2, 3$. □

This scheme can be modified in a number of ways, of which we will discuss some. The main issue here is that comparing the results within each pair separately (until either a pair of results "significantly different" is found, or until all pairs are checked and found not differing significantly) is *not a correct method*. The reason is that for many pairs, the probability of finding a pair with large difference becomes quite likely because of chance fluctuation, even if (in reality) the null hypothesis is true. The correct method requires testing all pairs *at once*, which is accomplished by the analysis of variance methodology.

15.2 REGRESSION OF THE FIRST AND SECOND KIND

In the sequel we let Y denote the dependent variable (assumed to be one-dimensional), and X (or $\mathbf{X}$) will denote the independent variable. In most cases X will be one-dimensional, but the considerations can usually be extended to m-dimensional vectors $\mathbf{X} = (X_1, \ldots, X_m), m > 1$.

Few comments about the nature of $\mathbf{X}$ and some illustrative examples are in order here. We start with the case when the values of independent variable (or variables) can be observed. Various situations may arise, depending on how the values of $\mathbf{X}$ were chosen. Thus, $\mathbf{X}$ may be random, either one- or multidimensional, nonrandom, or even under the experimenter's control. Typical cases of such situations are exemplified as follows.

Example 15.2.1. Suppose that we investigate some genetic theories, according to which a characteristic Y of offspring, such as height, depends on the characteristics $\mathbf{X} = (X_1, X_2)$ of father and mother (here X_1, X_2 may be heights, but it is also possible to study the effect of some other features on the offspring's height Y).

In this case, $\mathbf{X}$ has some distribution in the population. Whether or not we use this fact in sampling is another matter. For instance, we may sample triplets (Y, X_1, X_2) from the entire population; we may also select some specific pairs $(x_1, x_2), (x_1', x_2'), (x_1'', x_2''), \ldots$ and observe a sample $Y_1, \ldots, Y_n$ from the subpopulation with $(X_1, X_2) = (x_1, x_2)$, another sample $Y_1', \ldots, Y_{n'}'$ from the subpopulation with $(X_1, X_2) = (x_1', x_2')$, and so on.

Example 15.2.2. Sometimes the randomness of Y for a given $X = x$ has to be postulated, since it is not possible to observe more than one Y for the same x. This happens, for instance, if x is the calendar time, and we observe the stock market index Y_x at the end of the day x. Here we treat Y_x as random, since we cannot predict its value with complete precision. We only have the conviction that "if such and such events would have occurred, the value Y_x would be different from actually observed." But for each x only one single value is recorded, and we have no empirical access to the distribution unless we make some assumptions about the nature of randomness of Y_x across different times x.

Example 15.2.3.   It may also happen that **X** is not random. For instance (in the case of one dimension), we may study the relation between some developmental characteristic Y of a child, say the height or size of the vocabulary, as dependent on the age X. The randomness concerns values of Y for the same $X = x$, since in the population of children of age x there is some variability of values of Y.

Example 15.2.4.   Finally, it may happen that the values of X are totally under the experimenter's control. For instance, a chemist may be interested in some characteristic Y of a chemical reaction (say, its duration), depending on temperature X. Then X can be determined arbitrarily by the experimenter, and for a given temperature $X = x$, the randomness of Y may be due to measurement error, or to some other factors.   □

In each of the cases under consideration, the data have the form of a set of pairs (x_i, y_i), $i = 1, 2, \ldots n$, where values of x_i may possibly repeat. Even if x_i's are not random, the formulas are identical with those obtained under the assumption that X has discrete uniform distribution over the set $\{x_1, \ldots, x_n\}$, with probabilities appropriately increased in case of repeated values. Thus, we shall proceed as if X (respectively, **X**) were a random variable. We shall assume throughout this chapter that X and Y have finite variances, hence also finite expectations.

Let us begin with recalling some facts from Chapter 8. Suppose that we know the value of X, say $X = x$, and we want to predict the value of Y. According to Example 8.7.5, the best prediction of Y (in the sense of mean square error) is given by the conditional expectation

$$u(x) = E(Y|X = x). \tag{15.1}$$

In other words, we have

$$\min_{\xi} E[(Y - \xi)^2 | X = x] = E[(Y - u(x))^2 | X = x].$$

Accordingly, we introduce the following definition.

Definition 15.2.1.   The conditional expectation of Y given $X = x$, that is, the function $u(x)$ given by (15.1), will be called the *true regression*, or regression of the *first kind*, of Y on X.   □

In the case of continuous random variable X, it may happen that the regression function $u(x)$ is defined only for almost all x, that is, on a set A such that $P\{X \in A\} = 1$. However, in all cases that are encountered in statistics, there is typically a function $u(x)$ which is "natural" in a given problem. We shall select such a function and comment on the nonuniqueness of a regression function only when such a comment is essential for the problem.

The usefulness of the regression function $u(x)$ is not restricted to the prediction of Y. To mention just one other use of it, suppose that we do not know how X and Y are related; we only formulated the hypothesis that the true regression function is equal to some function u_0. In other words, the null hypothesis is $H_0 : E(Y|X = x) = u_0(x)$ for all x. If we know something about the conditional distribution of Y given X, we may test this hypothesis, rejecting it if the value of Y observed for given $X = x$ is "far" from $u_0(x)$.

Finding the true regression $u(x)$ requires knowledge of the joint distribution of (X, Y) or, more precisely, the conditional distribution of Y given $X = x$. Such knowledge may come only from a sufficiently deep understanding of the stochastic mechanisms that connect X and Y, or from very extensive data on pairs (X_i, Y_i). There are many cases in practice when neither is available. Moreover, even if we do know the joint distribution of (X, Y), determining $u(x)$ may present formidable difficulties.

To cover such types of situations, another type of regression has been introduced. Starting again from problem of prediction, suppose that we want to find the best predictor of Y which is *linear* in X. It will have the form $Y_p = a + bX$, and we will have to find a and b such that Y_p is the best predictor of Y [in the sense of minimizing the mean square error $E(Y - Y_p)^2$]. We have here

$$
\begin{aligned}
E(Y - Y_p)^2 &= E[Y - (a + bX)]^2 \\
&= E(Y^2 + a^2 + b^2 X^2 - 2aY - 2bXY + 2abX) \\
&= E(Y^2) + a^2 + b^2 E(X^2) \\
&\quad - 2aE(Y) - 2bE(XY) + 2abE(X).
\end{aligned}
$$

By differentiating with respect to a and b and setting the derivatives to be 0, we obtain the so-called *normal equations*:

$$
\begin{aligned}
a + bE(X) &= E(Y) \\
aE(X) + bE(X^2) &= E(XY).
\end{aligned}
$$

Multiplying the first equation by $E(X)$ and subtracting it from the second, we obtain

$$
b = \frac{E(XY) - E(X)E(Y)}{\sigma_X^2} = \rho \frac{\sigma_Y}{\sigma_X}, \tag{15.2}
$$

and consequently,

$$
a = E(Y) - \rho \frac{\sigma_Y}{\sigma_X} E(X). \tag{15.3}
$$

Thus, coefficients a and b require only means, variances, and covariance of X, Y. We introduce the following definition.

Definition 15.2.2. The relation

$$Y = a + bX = E(Y) + \rho \frac{\sigma_Y}{\sigma_X}[X - E(X)] \tag{15.4}$$

is called the *linear* regression, or regression of the *second kind*, of Y on X.

□

Before illustrating these concepts with examples, it is necessary to mention that there appears to be no firmly established terminology concerning regression. Some statisticians use this term only to mean the true regression; some others use it to signify a functional relation between X and Y, with randomness arising only from errors in observing Y. The definitions given above, although not universally accepted, seem to cover the "core" of everyday use of the term "regression" by statisticians.

Example 15.2.5. Let us consider the situation when the true regression is the function

$$u(x) = \begin{cases} x & \text{for } 0 \le x \le \frac{1}{2} \\ \frac{1}{2} & \text{for } \frac{1}{2} \le x \le 1. \end{cases}$$

Such regression may occur for various conditional distributions of Y. To fix the ideas, suppose that given x, the random variable Y is distributed uniformly on an interval whose leftmost point is 0. Thus, for $0 \le x \le \frac{1}{2}$, random variable Y_x is uniform on $[0, 2x]$, while for $\frac{1}{2} \le x \le 1$, random variable Y_x is uniform on $[0, 1]$. The situation is presented in Figure 15.1. The shaded area shows all possible points (X, Y). The polygonal line OAB is the graph of the true regression $u(x)$. For use in the next example, note that we have here

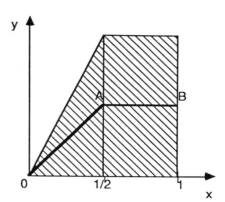

Figure 15.1 True regression.

$$\text{Var}(Y|X = x) = \begin{cases} \frac{x^3}{3} & \text{for } 0 \le x \le \frac{1}{2} \\ \frac{1}{12} & \text{for } \frac{1}{2} \le x \le 1. \end{cases} \tag{15.5}$$

Example 15.2.6. In Example 15.2.5, the distribution of X did not play any role in determining regression. In fact, X could have been nonrandom. On the other hand, to determine the linear regression of Y on X, it is necessary to know the marginal distribution of X. The intuitive justification here is that in linear regression the objective is to find the best approximation of Y by a straight line. Thus, it matters which values of X occur more often and which occur less often. Suppose that the joint distribution of the vector (X, Y) is uniform on the shaded area in Figure 15.2. To determine the linear regression, we need the expectations and variances of X and Y as well as their covariance. We have here, using simple geometry, for the marginals of X and Y,

$$f_X(x) = \begin{cases} cx & \text{for } 0 \le x \le \frac{1}{2} \\ \frac{c}{2} & \text{for } \frac{1}{2} \le x \le 1. \end{cases} \tag{15.6}$$

The value of c is determined from the condition

$$1 = \int_0^1 f_X(x)\, dx = \frac{3}{8}c,$$

hence $c = \frac{8}{3}$. Consequently,

$$E(X) = \int_0^1 x f_x(x)\, dx = \frac{11}{18}.$$

After simple integration we obtain $E(X^2) = \frac{31}{72}$, hence $\text{Var}(X) = \frac{37}{648}$. Similarly, $f_Y(y) = \frac{4}{3}(1 - \frac{y}{2})$ for $0 \le y \le 1$, and consequently,

$$E(Y) = \frac{4}{9}, \qquad \text{Var}(Y) = \frac{13}{162}.$$

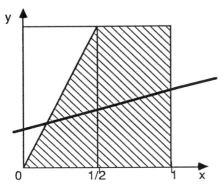

Figure 15.2 Linear regression for uniform distribution of (X, Y).

Finally, the joint density $f(x, y)$ is constant, equal to $\frac{4}{3}$ on the shaded area in Figure 15.2. Thus, $E(XY) = \frac{7}{24}$, and we obtain using (15.4), the equation for linear regression (see Figure 15.2),

$$Y = \frac{13}{37}X + \frac{17}{74}.$$

It is of some interest to compare the average square error of prediction of Y on the basis of X if we use the true and linear regression.

For prediction under true regression, we find the average square error, using (15.5) and (15.6):

$$E_X[\text{Var}(Y|X)] = \int_0^1 \text{Var}(Y|X = x) f_X(x) \, dx$$

$$= \int_0^{1/2} \frac{x^2}{3} \cdot \frac{8}{3} x \, dx + \int_{1/2}^1 \frac{1}{12} \cdot \frac{4}{3} \, dx$$

$$= \frac{5}{72} = 0.0694.$$

On the other hand, for prediction with the use of linear regression we have the error

$$E\left(Y - \frac{13}{37}X - \frac{17}{74}\right)^2 = E(Y^2) + \frac{169}{1369}E(X^2) + \left(\frac{17}{74}\right)^2$$

$$- \frac{26}{37}E(XY) - \frac{17}{37}E(Y) + \frac{442}{2738}E(X) = 0.0732.$$

Thus, in this case the mean square error of linear prediction is about 5.5% higher than the corresponding error of prediction based on true regression. Whether a 5.5% difference is important or negligible depends on the context. The point is that the mean square error of linear prediction is *always* at least as large as the error of prediction based on true regression. Also, one can easily construct examples when the ratio of these two errors is as large as one wishes. In the extreme case, if $Y = g(X)$, where g is a deterministic nonlinear function, g is the true regression and there is no error involved in the prediction of Y using $g(X)$ as predictor. But there is an error involved if a nonlinear function g is approximated by a linear function $a + bX$.

The last example shows that the issue of prediction errors in comparison of two types of regression is only partially statistical: One component is due to the randomness of Y for given $X = x$, and the other is due to replacing the true regression $u(x)$ by a straight line.

One could improve the situation by considering a regression of second order, where the predictor would be of the form $Y_p = a + bX + cX^2$, or a regression of some other special form, and as $Y_p = a + b\sin(cX)$, and so on. The point is that none of these can give the mean square error lower than that based on true regression.

PROBLEMS

15.2.1 Find the true and linear regression of Y on X if (X, Y) have the joint distribution uniform on the parallelogram with vertices at points $(-1, -1), (0, -1), (1, 1)$, and $(0, 1)$. Under the same assumptions, find the true and linear regression of X on Y.

15.2.2 Find an example of random variables (X, Y) such that their true and linear regression of Y on X coincide but (X, Y) do not have a joint normal distribution.

15.2.3 Suppose that the true regression of Y on X is not linear in X. Is it possible that the marginal distribution of X is such that the expected square error of best linear predictor of Y is the same as the expected square error of prediction based on true regression u?

15.2.4 Suppose that $\mathbf{X} = (X_1, X_2)$. Find coefficients a, b_1, b_2 such that $Y_p = a + b_1 X_1 + b_2 X_2$ is the best linear predictor of Y given $\mathbf{X}$.

15.2.5 Find the true regression Y on X if (X, Y) have the joint trinomial distribution

$$P\{X = x, Y = y\} = \frac{n!}{x! y! (n - x - y)!} p_1^x p_2^y (1 - p_1 - p_2)^{n-x-y}$$

for $x, y = 0, 1, \ldots, n, 0 \le x + y \le n$.

15.2.6 Let (x_i, y_j), $i = 1, \ldots, n$, be the data, where at least one value of x_i is not zero. Find an estimate of slope parameter b if you know that we have $a = 0$ [i.e., find the best fit of the model $E(Y|x) = bx$].

15.2.7 An analogue of Problem 15.2 is to find an estimate of intercept a if it is known that $b = 0$ (no calculations are necessary).

15.2.8 The number of eggs in nests of a certain species of birds are 1, 2, or 3, with 2 eggs found in about 80% of nests, and 1 or 3 eggs in about 10% of nests each. In one-egg nests, the egg hatches successfully in 75% of cases. In two-egg nests the probabilities for the number of offspring are 0–20%, 1–30%, 2–50%, while in three-egg nests these probabilities are 0–10%, 1–20%, 2–60%, 3–10%. Find the best linear predictor of Y being the number of offspring based on the observation of X—the number of eggs in the nest.

15.3 DISTRIBUTIONAL ASSUMPTIONS

In addition to the classification of regression models with respect to the two
types of regression, a meaningful and useful classification is obtained when
one considers typical assumptions about the distribution of the response Y
for a given value of X.

The oldest methods that could be included in regression theory appear to
date back to the beginning of the nineteenth century and works of Legendre.
He found a method of best approximation of a set of points by a straight line.
The word "best" was understood in the sense of least squares. In other words,
Legendre found, for a set of data points (x_i, y_i), the coefficients a and b such
that the sum $\sum_i [y_i - (a + bx_i)]^2$ attained its minimum. In this formulation
no assumptions about randomness are needed: The best-fitting line always
exists even if the "best" fit does not mean "good" fit.* The extension to
the best linear fit in two (or k) dimensions (i.e., finding a_1, a_2, b to minimize
the sum $\sum [y_i - (a + b_1 x_i^{(1)} + b_2 x_i^{(2)})]^2)$, is now quite straightforward. Similarly,
the theory extends naturally to other forms of relations (e.g., quadratic or
periodic functions to fit the data are obtained by minimizing the sums $\sum_i [y_i - (a + bx_i + cx_i^2)]^2$ or $\sum_i [y_i - a\cos(bx_i + c)]^2$, etc.).

Viewed in this way, regression theory belongs properly to the domain of
numerical analysis. However, to allow statistical inference, one usually makes
some assumptions about the randomness inherent in the model. One of the
standard assumptions is that for every x we have $Y = \varphi(x) + \epsilon$, when φ is
some deterministic function and ϵ is the "error" random variable, such that:

(i) $E(\epsilon) = 0$.
(ii) $\text{Var}(\epsilon) = \sigma^2 > 0$,
(iii) *The errors $\epsilon, \epsilon', \epsilon'', \ldots$ corresponding to different observations of Y
(for the same, or for distinct values of x) are uncorrelated.*[†]

Under these assumptions, $\varphi(x)$ is the true regression of Y on X. If one
now imposes some parametric model on φ, one can set up the least square
equations for parameters of φ and the parameter σ^2.

If assumptions (i)–(ii) are replaced by

(iv) *The errors $\epsilon, \epsilon', \ldots$ have normal distribution with mean 0 and constant
variance σ^2,*

then the errors are independent.

---

*As a mathematicians' joke says "For any three points, one can find a straight line passing
through these points, provided that the line is sufficiently thick."
†The assumption that variance is independent of x is often called *homoscedasticity*, as opposed
to *heteroscedasticity*.

The independence of errors and the knowledge of distribution have rather profound theoretical consequences: We can write the likelihood of the data and find the maximum likelihood estimators of the parameters.

One can easily see that in this case the MLE's coincide with least square estimators, so it might appear that nothing is gained by replacing (i)–(iii) by (iii) and (iv). In fact, however, under (iv) we know the sampling distribution of estimators, and we are therefore able to use the results from estimation and testing theory to build confidence intervals, tests of various hypotheses, and so on.

Let us mention finally, that if we assume in addition to (i), (ii), and (iv) that

(v) $\varphi(x) = a + bx$
(vi) *X has a normal distribution,*

then the joint distribution of (X, Y) is bivariate normal. The converse is also true, as shown in Theorem 9.10.4. We may summarize these facts as the following:

Theorem 15.3.1.   *In case of bivariate normal distribution (X, Y), the true and linear regression coincide.*

A comment appears necessary here as regards assumptions (iii) and (iv) and their implementation in practical situations. To illustrate potential diffi- culties, suppose that we collect the data on regression, where x is the age of a child, and Y is some response, such as reaction time to a specific stimulus. Suppose that we need two observations for the same age x. Then it is not correct to measure reaction time Y twice for the same child of age x, even if such observations can be regarded as independent. The correct procedure is to take two children of the same age x and observe their reaction times. The reason is that variability of Y has two components: the between-children variability and the within-child variability. Taking two measurements for the same child will involve only the second component, while other observations will involve both components, violating (among others) the assumption of homoscedasticity.

Assumptions (i)–(vi) are by no means the only sensible assumptions one can make in regression analysis. In fact, linearity assumption $E(Y_x) = a + bx$ may often be acceptable only as an approximation, or in some narrow range. A typical case of this kind may be illustrated by the following example, which is of considerable independent interest.

Example 15.3.1 (Logistic Regression).   Assume that the response vari- able Y_x is of binary character (e.g., success and failure). We can then always take the possible values of Y_x as 0 and 1, so that $E(Y_x) = P\{Y_x = 1\} = \pi(x)$. We shall assume that x is a numerical variable (random or not, depending on the situation under study), so that $\pi(x)$ is the true regression of Y on x. The assumption $\pi(x) = a + bx$ can be realistic only in a narrow range of values of x, since we must have $0 \leq \pi(x) \leq 1$ for all x. Still, the inference about the shape of function $\pi(x)$ is of considerable interest, and linearity is a powerful assumption (in the sense of allowing many analytical results). To realistically utilize such an assumption, we can consider odds ratio $\pi(x)/[1 - \pi(x)]$, with range $(0, \infty)$, and its logarithm, with range $(-\infty, \infty)$. Thus, we consider the model

$$\log \frac{\pi(x)}{1 - \pi(x)} = a + bx,$$

or equivalently,

$$\pi(x) = \frac{e^{a+bx}}{1 + e^{a+bx}}. \tag{15.7}$$

The right-hand side of (15.7) is called the *logistic function*, which explains the name *logistic regression*. The problem now is to estimate parameters a and b given the data $(x_i, n_i, N_i), i = 1, \ldots, m$, where x_i's are the values of independent variable at which observations are taken, while n_i and N_i are the number of successes and the number of trials at value x_i. The likelihood of the data is therefore

$$L = \prod_{i=1}^{m} \binom{N_i}{n_i} [\pi(x_i)]^{n_i} [1 - \pi(x_i)]^{N_i - n_i}$$

$$= \left(\prod_{i=1}^{m} \binom{N_i}{n_i} \right) \frac{e^{a \sum n_i + b \sum x_i n_i}}{\prod_{i=1}^{m} (1 + e^{a+bx_i})^{N_i}}.$$

Determining a and b that maximize L requires using numerical iterations. $\square$

PROBLEMS

15.3.1 The random variable Y_x has the density

$$f(y|x) = \begin{cases} 1 & \text{for } a + bx - \frac{1}{2} < y < a + bx + \frac{1}{2} \\ 0 & \text{otherwise.} \end{cases}$$

Find the MLE estimates of a and b given the sample $(x, y) : (-1, 1.3)$, $(0, 1.4)$, $(1, 0.1)$, $(2, -0.4)$.

15.3.2 Let (X, Y) have the distribution uniform on the quadrangle with vertices $(0, 0), (1, 1), (\frac{1}{2}, 1)$, and $(0, 2)$. Find the true regression of Y on X and of X on Y.

15.3.3 For Problem 15.3.2, find $\text{Var}(X|Y)$ and $\text{Var}(Y|X)$.

15.3.4 Assume that the total number X of eggs laid by a bird has Poisson distribution with parameter λ, and that each egg hatches with probability p, independent of other eggs. Find the regression of the number Y of eggs hatched on X. Let X_i, Y_i be the number of eggs laid and hatched by the ith bird. Derive the likelihood of the data for n birds and the equation for estimating λ and p. Find the MLE's of λ and p.

15.3.5 Answer Problem 15.3.4 assuming a more realistic model, where X_i's have the Poisson distribution conditional on the positive value, that is, for $k = 1, 2, \ldots$

$$P\{X_i = k\} = \frac{(\lambda^k/k!)e^{-\lambda}}{1 - e^{-\lambda}}.$$

15.3.6 A still more realistic model is that when all eggs are laid a disaster occurs with probability α, and all eggs are destroyed. If there is no disaster, eggs hatch independently, each with probability p. Find regression $E(Y|X)$, likelihood of the data, and (when possible) estimators of λ, α, and p under the assumptions of Problems 15.3.4 (Poisson distribution of X) and 15.3.5 (conditional Poisson). State carefully the assumptions you make.

15.4 LINEAR REGRESSION IN THE NORMAL CASE

In this section we present the main results in the case when for any given x, the response Y is of the form

$$Y = a + bx + \epsilon,$$

where $\epsilon \sim N(0, \sigma^2)$. Moreover, we assume that the errors ϵ for different observations of Y (for the same as well as for the distinct x) are independent. We shall assume that the data have the form

$$(x_i, y_i), \qquad i = 1, \ldots, n \qquad (15.8)$$

where $y_1, \ldots, y_n$ are the results of independent observations for the corresponding values $x_1, \ldots, x_n$. We shall use the notation

$$\bar{x} = \frac{1}{n}\sum_{i=1}^{n} x_i, \qquad \bar{y} = \frac{1}{n}\sum_{i=1}^{n} y_i$$

and we shall assume that not all x_i's are identical.

The estimates of parameters a, b, and σ^2 are easy to obtain. The likelihood function of the sample (15.8) is

$$L = \prod_{i=1}^{n} \frac{1}{\sigma\sqrt{2\pi}} e^{-(y_i - a - bx_i)^2 / 2\sigma^2}$$

$$= \sigma^{-n} (2\pi)^{-n/2} \exp\left\{ -\frac{1}{2\sigma^2} \sum_{i=1}^{n} (y_i - a - bx_i)^2 \right\},$$

hence

$$\log L = C - \frac{n}{2} \log(\sigma^2) - \frac{1}{2\sigma^2} \sum_{i=1}^{n} (y_i - a - bx_i)^2.$$

Differentiating with respect to a and b, we obtain a pair of equations (in which σ^2 cancels out) which are identical with the equations for least squares estimates of a and b. The solutions can be obtained from formulas (15.2) and (15.3) by treating X as a random variable uniformly distributed on $x_1, \ldots, x_n$. Thus,

$$\hat{b} = \frac{\sum(x_i - \bar{x})(y_i - \bar{y})}{\sum(x_i - \bar{x})^2} \tag{15.9}$$

and

$$\hat{a} = \bar{y} - \hat{b}\bar{x}. \tag{15.10}$$

The denominator in the expression for $\hat{b}$ is not zero in view of the assumption that not all x_i's are equal.

Differentiating $\log L$ with respect to σ^2 and setting the derivative equal to 0, we obtain the MLE of σ^2, namely

$$\hat{\sigma}^2 = \frac{1}{n} \sum_{i=1}^{n} (y_i - \hat{a} - \hat{b}x_i)^2. \tag{15.11}$$

To determine the sampling distribution of estimators (under variability of y_j's only, for fixed x_j's), let us regard y_j as a value of the random variable Y_{x_j}. We have then, letting U, V, and T^2 be the estimators of b, a, and σ^2:

$$U = \frac{\sum(x_i - \bar{x})(Y_{x_i} - \bar{Y})}{\sum(x_i - \bar{x})^2}$$

$$V = \bar{Y} - U\bar{x} \tag{15.12}$$

$$T^2 = \frac{1}{n} \sum_{i=1}^{n} (Y_{x_i} - V - Ux_i)^2.$$

It is important to realize that the randomness of U, V, and T^2 is connected only with the variability of dependent variable* Y_x about its mean $a + bx$. The values x_i may be nonrandom, and even if they arise from sampling, U, V, and T^2 involve only a conditional distribution of Y given $X = x$.

It follows from (15.12) that U, as a linear combination of normally distributed random variables, itself has a normal distribution. To determine the mean, observe that

$$E(\overline{Y}) = \frac{1}{n} \sum_{i=1}^{n} E(Y_{x_i}) = \frac{1}{n} \sum_{i=1}^{n} (a + bx_i) = a + b\overline{x}.$$

Consequently, we have

$$E(U) = \frac{\sum (x_i - \overline{x}) E(Y_{x_i} - \overline{Y})}{\sum (x_i - \overline{x})^2}$$

$$= \frac{\sum (x_i - \overline{x})[a + bx_i - (a + b\overline{x})]}{\sum (x_i - \overline{x})^2}$$

$$= b.$$

To find the variance of U, we may write

$$U = \frac{\sum (x_i - \overline{x}) Y_{x_i} - \overline{Y} \sum (x_i - \overline{x})}{\sum (x_i - \overline{x})^2} = \frac{\sum (x_i - \overline{x}) Y_{x_i}}{\sum (x_i - \overline{x})^2}.$$

Consequently, by assumption of homoscedasticity

$$\sigma_U^2 = \mathrm{Var} \left\{ \frac{\sum (x_i - \overline{x}) Y_{x_i}}{\sum (x_i - \overline{x})^2} \right\}$$

$$= \frac{\sum (x_i - \overline{x})^2 \, \mathrm{Var}(Y_{x_i})}{[\sum (x_i - \overline{x})^2]^2} = \frac{\sigma^2}{\sum (x_i - \overline{x})^2}. \qquad (15.13)$$

Next, the estimator V given by (15.12) is also a linear combination of normal random variables, so it has a normal distribution. Here

$$E(V) = E(\overline{Y}) - E(U)\overline{x} = a + b\overline{x} - b\overline{x} = a,$$

so that U and V are unbiased estimators of b and a.

The derivation of the variance of estimator V is somewhat messy. We have

---

* One has to remember that when we have $x_i = x_j = x$ for some i and j, the random variables Y_{x_i} and Y_{x_j} are independent, even though they may be denoted by the same symbol Y_x.

$$\sigma_V^2 = \text{Var}(\overline{Y}) + \overline{x}^2 \, \text{Var}(U) - 2\overline{x} \, \text{Cov}(U, \overline{Y})$$

$$= \text{Var}\left(\frac{1}{n}\sum Y_{x_i}\right) + \overline{x}^2 \sigma_U^2 - 2\overline{x} \, \text{Cov}\left[\frac{\sum(x_i - \overline{x})Y_{x_i}}{\sum(x_i - \overline{x})^2}, \frac{1}{n}\sum Y_{x_j}\right]$$

$$= \frac{\sigma^2}{n} + \overline{x}^2 \frac{\sigma^2}{\sum(x_i - \overline{x})^2} - \frac{2\overline{x}}{n\sum(x_i - \overline{x})^2} \, \text{Cov}\left[\sum(x_i - \overline{x})Y_{x_i}, \sum Y_{x_j}\right].$$

In the last term, all covariances corresponding to $i \neq j$ vanish, while all others are equal σ^2. Consequently,

$$\text{Cov}\left[\sum(x_i - \overline{x})Y_{x_i}, \sum Y_{x_j}\right] = \sum(x_i - \overline{x})\sigma^2 = 0,$$

and after some algebra, we obtain

$$\sigma_V^2 = \sigma^2 \frac{\sum x_i^2}{\sum(x_i - \overline{x})^2}. \tag{15.14}$$

In a similar manner, one can show that

$$\text{Cov}(U, V) = -\sigma^2 \frac{\overline{x}}{\sum(x_i - \overline{x})^2} \tag{15.15}$$

(we leave the proof as an exercise).

These results suggest that (U, V) has a bivariate normal distribution. In fact, the following theorem holds.

Theorem 15.4.1. *The estimators (V, U, T^2) are jointly sufficient for the parameter $\theta = (a, b, \sigma^2)$. Moreover, (U, V) have bivariate normal distribution with means, variances, and covariance given by (15.13)–(15.15). Finally, T^2 is independent of (U, V) and nT^2/σ^2 has chi-square distribution with $n - 2$ degrees of freedom.*

Proof. One can consider the joint moment generating function

$$m(t_1, t_2, t_3) = E(e^{t_1 U + t_2 V + t_3 T^2}).$$

Substituting the expression for U, V, and T^2, and integrating with respect to the joint density of $(Y_1, \ldots, Y_n)$, one can show (after considerable algebra) that $m(t_1, t_2, t_3) = m_1(t_1, t_2) \cdot m_2(t_3)$, which proves the independence of (U, V) and T^2. The form of functions m_1 and m_2 will then show that claims about the distributions of (U, V) and T^2 are also valid. We omit the details. $\square$

From formulas (15.13)–(15.15) one can also derive some conclusions as to the design of experiments aimed at estimating the regression parameters. To minimize the variance σ_U^2 of the estimator of slope, one should attempt to maximize $\sum(x_i - \bar{x})^2$. If possible, this means taking half of the observations at the minimal possible value of x, and the other half at the maximal possible value of x. Such a recommendation, however, presupposes total faith in linearity of regression. In practice, it is usually prudent to take some additional observations at intermediate values of x.

To attain higher precision of the estimate of intercept a, one should attempt to have $\bar{x}$ as close to zero as possible. The condition $\bar{x} = 0$ implies also that estimators U and V are independent.

We now show some examples, which will illustrate the application of Theorem 15.4.1 to the construction of confidence intervals and tests for parameters of the regression model.

Example 15.4.1 (Is It Good to Be a Royal Prince?). Poland had altogether 13 kings whose fathers were also Polish kings (starting in the fifteenth century, Polish kings were elected, and election of the late king's son, although often likely, was by no means automatic). In one pair the son died young in a battle; deaths in the remaining 12 pairs were from natural causes. The ages at death of the fathers (x_i) and sons (y_i) are listed below.

Father	x_i	Son	y_i
Mieszko I	62	Bolesław the Brave	59
Bolesław the Brave	59	Mieszko II	44
Casimir I the Restorer	42	Bolesław the Bold	42
Władysław I the Short	73	Casimir III the Great	60
Władysław II Jagiełło	83	Casimir IV Jagiellonian	65
Casimir IV Jagiellonian	65	John I Albert	42
Casimir IV Jagiellonian	65	Alexander Jagiellonian	45
Casimir IV Jagiellonian	65	Sigismund the Old	81
Sigismund I the Old	81	Sigismund II Augustus	52
Sigismund III Vasa	66	Władysław IV Vasa	53
Sigismund III Vasa	66	John II Casimir Vasa	63
Augustus II the Strong	63	Augustus III	67

As may be seen, only in 2 of 12 pairs did the son live longer than his father. Can this be attributed to chance, or does it indicate some systematic trend (e.g., being an heir to the throne one is "pampered" in childhood, growing less resistant to stress and additional duties brought later by ruling, etc.). We have here

$$\bar{x} = 65.83, \qquad \bar{y} = 56.08,$$

which means that on average, sons lived nearly 10 years less than fathers. For the regression coefficients, and their variances and covariances, we obtain

$$\hat{a} = 29.214, \qquad \hat{b} = 0.408$$

and also $\sum x_i^2 = 53,224, \sum(x_i - \bar{x})^2 = 1215.67$. Finally,

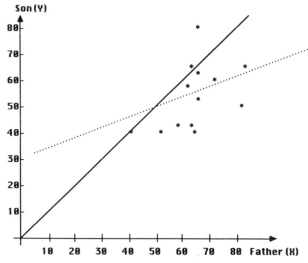

Figure 15.3 Ages at death of Polish kings and their heirs.

$$T^2 = 115.034.$$

The individual points, as well as the regression line

$$y = 29.214 + 0.408x$$

are presented in Figure 15.3. We shall return to the analysis in Example 15.4.2.   □

We begin by constructing confidence intervals for the regression intercept and slope a and b. We know that $(U - b)/\sigma_U$ has a standard normal distribution. This quantity, however, involves the nuisance parameter σ. Since nT^2/σ^2 is independent of U and has chi-square distribution with $n - 2$ degrees of freedom, the ratio

$$t = \frac{(U - b)/\sigma_U}{\sqrt{nT^2/[\sigma^2(n - 2)]}} = \frac{(U - b)\sqrt{(n - 2)\sum(x_i - \bar{x})^2}}{\sqrt{\sum(Y_{x_i} - Ux_i - V)^2}} \tag{15.16}$$

has Student t distribution with $n - 2$ degrees of freedom. Consequently, (15.16) is a pivotal quantity for b (see Definition 12.9.2). Letting $\hat{a}$ and $\hat{b}$ denote the observed values of estimators V and U of intercept and slope, the $(1 - \gamma)$-level confidence inteval for slope b is

$$\hat{b} \pm \frac{t_{\gamma/2, n-2}}{\sqrt{(n - 2)\sum(x_i - \bar{x})^2}} \sqrt{\sum(y_j - \hat{a} - \hat{b}x_j)^2}.$$

In a similar way, we may derive the $(1-\gamma)$-level confidence interval for the intercept a in regression line, namely

$$\hat{a} \pm t_{\gamma/2,n-2}\sqrt{\frac{\sum x_i^2}{(n-2)\sum(x_i-\bar{x})^2}}\sqrt{\sum(y_i-\hat{a}-\hat{b}x_i)^2}.$$

Example 15.4.2.   To continue Example 15.4.1, the 95% confidence intervals for regression slope and intercept are now, respectively,

$$[0.408-0.751; 0.408+0.751] = [-0.343, 1.159]$$

and

$$29.214-50.00; 29.214+50.00] = [-20.786; 79.214].$$

Since the confidence interval for regression slope b covers the value 0, we cannot exclude the possibility that the true regression is a constant a. This means that the age of son at death does not depend on the father's age at death. Thus, the evidence is not conclusive as regards the effects on life duration of being born to a royal family.   □

Actually, we can construct a simultaneous confidence set for both regression parameters using the F distribution. One can show (we omit the proof) that the random variable

$$Q = \frac{1}{\sigma^2}[n(U-b)+2n\bar{x}(U-b)(V-a)+\sum x_i^2(V-a)^2]$$

has the chi-square distribution with 2 degrees of freedom. Consequently, the random variable

$$\frac{Q/2}{nT^2/(n-2)} = \frac{n-2}{2} \times \frac{[n(U-b)+2n\bar{x}(U-b)(V-a)+\sum x_i^2(V-a)^2]}{\sum(Y_{x_i}-Ux_i-V)^2}$$

has the F distribution with $(2,n-2)$ degrees of freedom. Thus, the ellipsoid in the (a,b)-plane

$$n(\hat{b}-b)^2+2n\bar{x}(\hat{b}-b)(\hat{a}-a)+\sum x_i^2(\hat{a}-a)^2 \le \frac{2}{n-2}F_{\gamma,2,n-2}\sum(y_i-\hat{a}-\hat{b}x_i)^2$$

$$(15.17)$$

is an $(1-\gamma)$-level confidence set for regression parameters.

Let us also observe that one can use the results obtained thus far to build estimators and construct tests of hypotheses about linear combinations of regression coefficients a and b. We illustrate the situation with an example.

690   LINEAR MODELS

Example 15.4.3. Suppose that we need to estimate the parameter $\theta = Aa + Bb$, where A and B are given constants. A special case is obtained here if $A = 1, B = x_0$, so that the objective is to estimate the mean response at $X = x_0$ [i.e., $E(Y_{x_0}) = a + bx_0$].

Clearly, an unbiased estimator of θ is $W = AV + BU$, whose value for the sample is $\hat{\theta} = A\hat{a} + B\hat{b}$. The distribution of W is normal, since W is a sum of two normally distributed random variables. We have

$$\sigma_W^2 = A^2\sigma_V^2 + B^2\sigma_U^2 + 2AB\,\mathrm{Cov}(U,V)$$

$$= \frac{\sigma^2}{\sum(x_i - \bar{x})^2}[A^2\sum x_i^2 - 2AB\bar{x} + B^2].$$

Proceeding as before, we show that the random variable

$$\frac{[A(V-a) + B(U-b)]\sqrt{(n-2)\sum(x_i - \bar{x})^2}}{\sqrt{[A^2\sum x_i^2 - 2AB\bar{x} + B^2][\sum(Y_{x_i} - Ux_i - V)^2]}}$$

has Student t distribution with $n-2$ degrees of freedom. This gives the $(1 - \gamma)$-level confidence interval for θ as

$$\hat{\theta} \pm t_{\gamma/2,n-2}\sqrt{\frac{A^2\sum x_i^2 - 2AB\bar{x} + B^2}{(n-2)\sum(x_i - \bar{x})^2}}\sqrt{\sum(y_i - \hat{a} - \hat{b}x_i)^2}. \qquad \square$$

The results given above can also be used to test hypotheses about the regression coefficients a and b. The testing procedures (likelihood ratio tests) use the same Student-type ratios as the confidence intervals above. We shall give the results for the slope coefficient b, leaving the derivation of the tests for intercept a as exercises.

Suppose that we want to test the null hypothesis

$$H_0 : b = b_0$$

against a one- or two-sided alternative. Then, given H_0, the random variable $(U - b_0)/\sigma_U$ has standard normal distribution, and consequently, the test statistic obtained upon division by $\sqrt{nT^2/[\sigma^2(n-2)]}$, that is,

$$t = \frac{U - b_0}{\sqrt{\sum(Y_{x_i} - Ux_i - V)^2}}\sqrt{(n-2)\sum(x_i - \bar{x})^2} \qquad (15.18)$$

has t distribution with $n-2$ degrees of freedom. A one- or two-sided rejection region should be used, depending on the alternative hypothesis.

PROBLEMS

15.4.1 The data on the score on the entrance exam (x) and GPA on graduation (y) for 12 randomly selected students of a certain university
are

x	355	361	402	365	375	404
y	3.66	3.49	3.86	3.24	3.55	3.92
x	349	380	420	395	309	375
y	3.11	3.19	3.76	3.75	3.12	3.48

Assume normality and homoscedasticity.
   (i) Compute the MLE's of a, b, and σ^2.
   (ii) Test the hypothesis that there is no relation between the grade
on the entrance exam and the GPA, against the alternative that
higher scores on the entrance exam tend to be associated with
higher GPA's.
   (iii) Find the shortest 95% confidence intervals for a and for b.
   (iv) Find the joint confidence set for (a, b) with confidence level 0.95,
sketch it, and compare with answers to part (iii).

15.4.2 Suppose that the observations are taken only at two values, x_1 and
x_2, of independent variable. Let $\bar{y}_1$ and $\bar{y}_2$ be the average observed
responses for $x = x_1$ and for $x = x_2$, respectively. Show that the estimated regression line passes through points $(x_1, \bar{y}_1)$ and $(x_2, \bar{y}_2)$.

15.4.3 Derive the test for the null hypothesis $H_0 : a = a_0$ against the one-
or two-sided alternative.

15.4.4 Using the ideas explained in deriving the confidence set for (a, b),
derive the testing procedure for the null hypothesis $H_0 : a = a_0, b = b_0$ against the general alternative $H_1 : H_0$ is false. Consider two cases:
(i) σ known, and **(ii)** σ unknown.

15.4.5 Suppose it is known that the true regression (assuming a normal case)
is linear of the form $E(Y_x) = bx$. Derive the MLE's for b and for σ^2.

15.5 TESTING LINEARITY

In this section we make the following two additional assumptions regarding
the design of the experiment for collecting the data on regression parameters:

   (i) There are at least three distinct values among $x_1, x_2, \ldots, x_n$.
   (ii) There exist at least two repeated values among $x_1, \ldots, x_n$.

We shall show that under (i) and (ii) it is possible to construct a test for linearity of regression, that is, a test of the null hypothesis

$$H_0 : E(Y_x) = a + bx \text{ for some } a, b$$

against the alternative

$$H_1 : E(Y_x) = u(x), \text{ where } u(x) \text{ is not a linear function of } x.$$

The test will be based on the construction of two independent unbiased estimators of σ^2. One of them will estimate σ^2 regardless of whether or not H_0 is true, and the other will be an unbiased estimator of σ^2 only under the null hypothesis.

For the considerations of the present section only, it will be convenient to change the labeling of the sample (x_i, y_i), $i = 1, \ldots, n$ as follows. Let $x'_1, x'_2, \ldots, x'_r$ be all those x_i at which multiple observations were made, and let $n_1, \ldots, n_r$ be the numbers of observations made for those values. Furthermore, let the observations made for x'_j be $y'_{j,1}, \ldots, y'_{j,n_j}$, regarded as values of iid random variables $Y'_{j,1}, \ldots, Y'_{j,n_j}$. Finally, the remaining values of independent variable will be denoted by $x'_{r+1}, \ldots, x'_m$, with the corresponding observations $y'_{r+1}, \ldots, y'_m$ being the values of random variables $Y'_{r+1}, \ldots, Y'_m$ (if there are no such values, then $m = r$).

We have, therefore,

$$n_1 + \cdots + n_r + (m - r) = n \tag{15.19}$$

with $n_i \geq 2, i = 1, \ldots, r$.

Consider now the following decomposition of the sum of the squared deviations:

$$
\begin{aligned}
s^2 &= \sum_{i=1}^{n} (y_i - \hat{a} - \hat{b}x_i)^2 \\
&= \sum_{j=1}^{r} \sum_{t=1}^{n_j} (y'_{j,t} - \hat{a} - \hat{b}x'_j)^2 + \sum_{k=r+1}^{m} (y'_k - \hat{a} - \hat{b}x'_k)^2 \\
&= \sum_{j=1}^{r} \sum_{t=1}^{n_j} (y'_{j,t} - \bar{y}'_j + \bar{y}'_j - \hat{a} - \hat{b}x'_j)^2 + \sum_{k=r+1}^{m} (y'_k - \hat{a} - \hat{b}x'_k)^2 \\
&= \sum_{j=1}^{r} \sum_{t=1}^{n_j} (y'_{j,t} - \bar{y}'_j)^2 + \sum_{j=1}^{r} n_j (\bar{y}'_j - \hat{a} - \hat{b}x'_j)^2 + \sum_{k=r+1}^{m} (y'_k - \hat{a} - \hat{b}x'_k)^2 \\
&= s_1^2 + s_2^2 + s_3^2. \tag{15.20}
\end{aligned}
$$

In passing from the third to the fourth expressions, the cross-products were omitted. One can check that, indeed, all cross-products equal zero.

Now, the sums s_1^2, s_2^2, and s_3^2 are observed values of random variables

$$S_1^2 = \sum_{j=1}^{r} \sum_{t=1}^{n_j} (Y'_{j,t} - \overline{Y}'_j)^2,$$

$$S_2^2 = \sum_{j=1}^{r} n_j (\overline{Y}'_j - V - Ux'_j)^2,$$

$$S_3^2 = \sum_{k=r+1}^{m} (Y'_k - V - Ux'_k)^2.$$

Under the assumption of normality of distributions of Y_x, the random variables S_1^2 and S_2^2 are independent; random variables S_3^2 and S_1^2 are independent as well. The proof of these facts is similar to the proof of Theorem 9.12.1 and we omit it.

Finally, as regards the distributions, we know that S_1^2/σ^2 has a chi-square distribution with the number of degrees of freedom equal to

$$\sum_{j=1}^{r} (n_j - 1) = \sum_{j=1}^{r} n_j - r = n - m$$

in view of (15.19). Since $r \geq 1$ and $n_j \geq 2$ for all j, we have $n - m > 0$. It is important to realize that this statement about the distribution of S_1^2 holds *regardless of whether or not the null hypothesis H_0 about linearity of regression is true*.

On the other hand, *if the null hypothesis H_0 is true*, then we also have

$$S_2^2/\sigma^2 + S_3^2/\sigma^2 \sim \chi_{m-2}^2.$$

Again, the number of degrees of freedom is positive, in view of the assumption that m, the number of distinct values of x_i, is at least 3.

To construct the testing procedure, observe finally that any violations of the null hypothesis will tend to increase the expected value of the sum $S_2^2 + S_3^2$, since $E(Y - \xi)^2$ is minimized for $\xi = E(Y)$.

It follows from the above that the random variable

$$F = \frac{(S_2^2 + S_3^2)/(m-2)}{S_1^2/(n-m)}$$

has the F distribution with $(m-2, n-m)$ degrees of freedom, provided that the null hypothesis H_0 is true, while any lack of fit to the linear model will tend to inflate the value of F. Thus, the testing procedure with significance level α is:

Reject the hypothesis of linearity of regression if

$$\frac{[\sum_{j=1}^{r} n_j(\bar{y}_j' - \hat{a} - \hat{b}x_j')^2 + \sum_{k=r+1}^{m}(y_k' - \hat{a} - \hat{b}x_k')^2]/(m-2)}{\sum_{j=1}^{r}\sum_{t=1}^{n_j}(y_{j,t}' - \bar{y}_j')^2/(n-m)} > F_{\alpha,m-2,n-m}.$$

Example 15.5.1. To develop some sort of intuition concerning the linearity test, we shall analyze the situation in a deliberately oversimplified case:* two observations for $x = 0$ are d and $-d$, an observation for $x = 1$ is c, and an observation for $x = 2$ is 0. We shall find the range of values c (for fixed d) and the range of values d (for given c) when the linearity hypothesis should be rejected, on a level of significance $\alpha = 0.05$, say.

Intuitively, for fixed d, linearity will be rejected if the middle point deviates too far from the x-axis in any direction, and (for fixed $c \neq 0$) when d is close to 0.

We have here

$$\bar{x} = \frac{0+0+1+2}{4} = \frac{3}{4}$$

and $\sum(x_i - \bar{x})^2 = 11/4$. Moreover, $\bar{y} = (d - d + c + 0)/4 = c/4$. This gives

$$\hat{b} = \frac{c}{11}, \qquad \hat{a} = \frac{2c}{11}.$$

The estimated regression line is therefore $\hat{a} + \hat{b}x = (c/11)(2 + x)$. To compute the F ratio, we have

$$s_1^2 = 2d^2$$

with $n - m = 1$. As regards the numerator, we find that

$$s_2^2 + s_3^2 = 2\left[0 - \frac{c}{11}(2+0)\right]^2 + \left[c - \frac{c}{11}(2+1)\right]^2 + \left[0 - \frac{c}{11}(2+2)\right]^2$$

$$= \frac{88}{121}c^2$$

with $m - 2 = 1$ degree of freedom. For $\alpha = 0.05$ the 95% quantile of distribution of $F_{1,1}$ is 161.45. Thus, the hypothesis of linearity should be rejected if

$$\frac{\frac{88}{121}c^2}{2d^2} > 161.45 \quad \text{or} \quad \frac{c^2}{d^2} > 443.99.$$

Since $d > 0$, this is equivalent to the inequality

$$|c| > 21.07d. \qquad \qquad \square$$

---

* For more realistic examples, see the Problems for this section.

PROBLEMS

15.5.1 Suppose that we have six data points in addition to those in Problem 15.4.1:

x	355	402	402	309	375	375
y	3.44	3.91	3.95	3.24	3.52	3.31

Test the hypothesis (using all 18 data values) that regression of the GPA on the score on the entrance exam is linear.

15.5.2 The output of a certain device is suspected to decrease linearly with the temperature. Two observations were taken for each temperature, and the data (in appropriate units) are as follows:

Temp.	55	65	75	85	95	105
Output	2.01	2.01	2.02	1.48	1.93	1.90
	2.03	2.02	2.00	1.48	1.95	1.94

On the significance level 0.05, test the hypothesis that the output is a linear function of temperature.

15.5.3 Check the identity (15.20).

15.5.4 Derive, if possible, a test of linearity of regression under the assumptions of this section, the only difference being that for values $x'_{r+1}, \ldots, x'_m$ we have the data on averages $y'_{r+1}, \ldots, y'_m$ and the corresponding multiciplicities $n'_{r+1}, \ldots, n'_m$, but we do not have access to individual observations.

15.6 PREDICTION

Consider now the problem of prediction. As before, the data have the form of a set of pairs $(x_i, y_i), i = 1, \ldots, n$, where y_i is the observed value of a random variable Y_{x_i} assumed to be normal with mean $a + bx_i$ and standard deviation σ. The random variables $Y_{x_1}, \ldots, Y_{x_n}$ are independent, and at least two among $x_1, \ldots, x_n$ are distinct. The problem is to predict Y_{x_0} as precisely as possible.

More generally, we may want to predict the average of k independent observations of Y_{x_0}. By prediction we mean here providing an interval, as short as possible, such that the value of the predicted random variable will fall into this interval with preassigned probability, say $1 - \gamma$. Let $\overline{Y}_{x_0}$ denote the average of k observations to be taken at the value x_0 of the independent variable. We are looking for an interval $[c_1, c_2]$ such that

$$P\{c_1 \le \overline{Y}_{x_0} \le c_2\} = 1 - \gamma.$$

The solution is obtained as follows. Note first that the average $\overline{Y}_{x_0}$ has distribution $N(a + bx_0, \sigma^2/k)$. Consequently,

$$Z = \frac{\overline{Y}_{x_0} - a - bx_0}{\sigma}\sqrt{k}$$

is a standard normal random variable, and we have

$$\{-z_{\gamma/2} \le Z \le z_{\gamma/2}\} = 1 - \gamma.$$

A simple argument based on symmetry about zero of normal density shows that $(-z_{\gamma/2}, z_{\gamma/2})$ is the shortest prediction interval for Z. Thus, the corresponding shortest prediction interval for $\overline{Y}_{x_0}$ (given that the regression parameters a, b, and σ^2 are known) is

$$a + bx_0 \pm z_{\gamma/2}\frac{\sigma}{\sqrt{k}},$$

and its length is $2(\sigma/\sqrt{k})z_{\gamma/2}$. The actual prediction interval has to take into account the fact that the regression parameters are estimated. The construction is based on an analogue to the pivotal quantity. Thus, the random variable

$$L = \overline{Y}_{x_0} - Ux_0 - V$$

has normal distribution (being a linear combination of normal random variables), and $E(L) = 0$ in view of the fact that U and V are unbiased estimators of b and a. We have, using the fact that $\overline{Y}_{x_0}$ is independent of (U, V),

$$\sigma_L^2 = \text{Var}(L) = \text{Var}(\overline{Y}_{x_0}) + x_0^2 \text{Var}(U) + \text{Var}(V) + 2x_0 \text{Cov}(U, V)$$

$$= \frac{\sigma^2}{k} + x_0^2\sigma_U^2 + \sigma_V^2 + 2x_0 \text{Cov}(U, V)$$

$$= \frac{\sigma^2}{k} + x_0^2\frac{\sigma^2}{\sum(x_i - \bar{x})^2} + \frac{\sigma^2 \sum x_i^2}{n\sum(x_i - \bar{x})^2} - 2x_0\frac{\sigma^2\bar{x}}{\sum(x_i - \bar{x})^2}$$

$$= \sigma^2\left[\frac{1}{k} + \frac{x_0^2 + \frac{1}{n}\sum x_i^2 - 2x_0\bar{x}}{\sum(x_i - \bar{x})^2}\right]$$

$$= \sigma^2\left[\frac{1}{k} + \frac{1}{n} + \frac{(x_0 - \bar{x})^2}{\sum(x_i - \bar{x})^2}\right].$$

Consequently, the random variable

$$\frac{L}{\sigma_L} = \frac{\overline{Y}_{x_0} - Ux_0 - V}{\sigma\sqrt{1/k + 1/n + (x_0 - \bar{x})^2/\sum(x_i - \bar{x})^2}}$$

has standard normal distribution. Dividing by $\sqrt{nT^2/[\sigma^2(n-2)]}$ we obtain the random variable

$$t = \frac{(\overline{Y}_{x_0} - Ux_0 - V)\sqrt{n-2}}{\sqrt{1/k + 1/n + (x_0 - \overline{x})^2/\sum(x_i - \overline{x})^2}\sqrt{\sum(\overline{Y}_{x_0} - Ux_i - V)^2}},$$

which has t distribution with $n-2$ degrees of freedom. Substituting the observed values $\hat{a}$ and $\hat{b}$ of V and U, the prediction interval for $\overline{Y}_{x_0}$ with prediction probability $1 - \gamma$ becomes

$$\hat{a} + \hat{b}x_0 \pm \frac{t_{\gamma/2, n-2}}{\sqrt{n-2}}\sqrt{\left[\frac{1}{k} + \frac{1}{n} + \frac{(x_0 - \overline{x})^2}{\sum(x_i - \overline{x})^2}\right]\left[\sum(y_{x_i} - \hat{a} - \hat{b}x_i)^2\right]}. \quad (15.21)$$

Let us remark here that the prediction interval (for fixed n and k) is shortest if $x_0 = \overline{x}$ (i.e., it is "easier" to predict values of dependent variable for x_0 close to $\overline{x}$).

Observe also that as $k \to \infty$, the length for prediction interval for *known* a, b, and σ tends to 0. In the present case, an increase of k has much less effect, and as $k \to \infty$ the length of the prediction interval tends to a positive quantity, depending on n and on the location of observations $x_1, x_2, \ldots, x_n$. This is consistent with our intuition, according to which in the present case uncertainty of prediction has two sources: randomness of dependent variable about its estimated mean, and uncertainty as to the exact location of the true mean.

PROBLEMS

15.6.1 Using all data points from Problems 15.4.1 and 15.5.1, find the 95% prediction interval for the GPA of a student who scored 400 on an entrance exam.

15.6.2 Find the prediction interval with probability $1 - \gamma$ of coverage for an observation to be taken at the value x_0 by independent variable, given the data $(x_i, y_i), i = 1, \ldots, n$, and assuming a normal model of the form $Y_x = bx + \epsilon$ with $\epsilon \sim N(0, \sigma^2)$.

15.6.3 Under the conditions of Problem 15.6.2 find the prediction interval for the mean of k observations taken for a value x_0 of the independent variable.

15.6.4 Assume that the weight at harvest of a certain fruit grown in a greenhouse has normal distribution $N(a + bt, \sigma^2)$, where t is the average temperature. A sample of five fruits for $t = 80°F$ gave weights 1.02, 1.03, 0.98, 1.05, 1.02, while a sample of seven fruits from a greenhouse with $t = 86°F$ (other conditions being equal) gave 1.03, 1.03, 1.09, 1.07, 1.04, 1.02, 1.08. Give a 95% prediction interval for the average weight of four fruits grown:

 (i) In the first greenhouse.

 (ii) In the second greenhouse.

 (iii) In the greenhouse with $t = 84°F$.

15.7 INVERSE REGRESSION

Inverse regression, sometimes called *discrimination*, is the problem of inference about the unknown value x_0 of the independent variable, on the basis of a number of observations of the response for this value.

Thus, we consider the following setup. The data consist of two groups of observations: One group is, as before, the sample

$$(x_i, y_i), \qquad i = 1, \ldots, n,$$

where y_i is the observed value of the random variable $Y_i \sim N(a + bx_i, \sigma^2)$, with the usual assumption of independence. The second group is the sample

$$(y_1', \ldots, y_m')$$

of observations assumed to be values of iid random variables, with distribution $N(a + bx_0, \sigma^2)$, where x_0 is unknown. The objective is to estimate x_0, with a, b, σ^2 being unknown.

The likelihood of the data is

$$L = \prod_{i=1}^{n} \frac{1}{\sigma\sqrt{2\pi}} e^{-(1/2\sigma^2)(y_i - a - bx_i)^2} \prod_{j=1}^{m} \frac{1}{\sigma\sqrt{2\pi}} e^{-(1/2\sigma^2)(y_j' - a - bx_0)^2}$$

$$= C(\sigma^2)^{-(m+n)/2} \exp\left\{ -\frac{1}{2\sigma^2} \left[\sum_{i=1}^{n}(y_i - a - bx_i)^2 + \sum_{j=1}^{m}(y_j' - a - bx_0)^2 \right] \right\}.$$

Taking logarithms and differentiating with respect to a, b, σ^2, and x_0, we obtain, after some cancellations, the equations

$$\sum_{i=1}^{n}(y_i - a - bx_i) + \sum_{j=1}^{m}(y_j' - a - bx_0) = 0$$

$$\sum_{i=1}^{n}(y_i - a - bx_i)x_i + x_0 \sum_{j=1}^{m}(y_j' - a - bx_0) = 0$$

$$(m+n)\sigma^2 - \sum_{i=1}^{n}(y_i - a - bx_i)^2 - \sum_{j=1}^{m}(y'_j - a - bx_0)^2 = 0$$

$$\sum_{j=1}^{m}(y'_j - a - bx_0) = 0.$$

Letting $\bar{y}' = (1/m)\sum_{j=1}^{m} y'_j$, the last equation gives the estimate

$$\hat{x}_0 = \frac{\bar{y}' - \hat{a}}{\hat{b}},\qquad\qquad (15.22)$$

assuming, of course, that $\hat{b} \neq 0$. The third equation gives

$$\hat{\sigma}^2 = \frac{1}{m+n}\left[\sum_{i=1}^{n}(y_i - \hat{a} - \hat{b}x_i)^2 + \sum_{j=1}^{m}(y'_j - \bar{y}')^2\right].$$

Using (15.22) in the first two equations, one can easily check that in each case the sum involving y'_j equals zero, which means that the expressions for $\hat{a}$ and $\hat{b}$ are given by (15.9) and (15.10).

These results are consistent with our intuition. Indeed, since we do not know x_0, the observations $y'_1, \dots, y'_m$ cannot provide any information about the slope and intercept of the regression line. On the other hand, if only $m > 1$, the values $y'_1, \dots, y'_m$ provide additional information about σ^2.

To set a confidence interval for $\hat{x}_0$ we may proceed as in the case of prediction. The random variable

$$W = \bar{Y}' - Ux_0 - V$$

has normal distribution with mean zero and variance

$$\sigma_W^2 = \sigma^2\left(\frac{1}{m} + \frac{1}{n} + \frac{(x_0 - \bar{x})^2}{\sum(x_i - \bar{x})^2}\right).$$

Consequently, W/σ_W has a standard normal distribution. To eliminate σ, we note that the random variable

$$\frac{(m+n)\hat{\sigma}^2}{\sigma^2} = \frac{1}{\sigma^2}\left[\sum_{i=1}^{m}(Y_{x_i} - Ux_i - V)^2 + \sum_{j=1}^{m}(Y'_{x_{0,j}} - \bar{Y}')^2\right]$$

has chi-square distribution with $m + n - 3$ degrees of freedom. Indeed, the two sums are independent, and their numbers of degrees of freedom are $n - 2$ and $m - 1$. Consequently, the random variable

$$\frac{W/\sigma_W}{\sqrt{(m+n)\hat{\sigma}^2/[\sigma^2(m+n-3)]}} \tag{15.23}$$

has the Student t distribution with $m+n-3$ degrees of freedom. The observed value of random variable given by (15.23) is

$$t = \frac{(\bar{y}' - \hat{a} - \hat{b}x_0)\sqrt{m+n-3}}{\sqrt{\frac{1}{m} + \frac{1}{n} + \frac{(x_0 - \bar{x})^2}{\sum (x_i - \bar{x})^2}} \times \sqrt{\sum_{i=1}^{n}(y_i - \hat{a} - \hat{b}x_0)^2 + \sum_{j=1}^{m}(y_j' - \bar{y}')^2}}$$

and the confidence interval is obtained by converting the inequality

$$-t_{\gamma/2,m+n-3} < t < t_{\gamma/2,m+n-3} \tag{15.24}$$

into an inequality for x_0. Observe, however, that (15.24) is now a *quadratic* inequality.

Example 15.7.1. The amounts of a chemical compound that dissolve in a given amount of water at different temperatures are given in the following table:

Temperature x (°C)	Amount y (grams)		
5	3	4	2
10	7	7	6
15	10	13	11
20	15	18	17
25	21	18	19

Two measurements for an unknown temperature x_0 were 14 and 16. What can one say about x_0?

SOLUTION. We have here $n = 15, m = 2$, and the relevant quantities are

$$\hat{a} = -1.4, \ \hat{b} = 0.853, \ \bar{x} = 15, \ \sum (x_i - \bar{x})^2 = 750,$$

$$\sum (y_i - \hat{a} - \hat{b}x_i)^2 = 21.46, \ \bar{y}' = 15, \ \sum (y_i' - \bar{y}')^2 = 2.$$

Thus, the point estimate of x_0 is $\hat{x}_0 = 19.23$. For $\gamma = 0.05$, we find from tables of the Student distribution for $n + m - 3 = 14$ degrees of freedom the quantile $t_{0.025,14} = 2.145$. To obtain a 95% confidence interval for x_0 we must solve the inequality (15.24), which in our case takes the form

$$\frac{|15 - 0.853x_0 + 1.4|}{\sqrt{\frac{1}{2} + \frac{1}{15} + (x_0 - 15)^2/750}} \times \frac{\sqrt{2 + 15 - 3}}{\sqrt{21.46 + 2}} < 2.145$$

or

$$\frac{|346.95 - 18.05x_0|}{\sqrt{x_0^2 - 30x_0 + 650}} < 2.145.$$

We obtain, after some algebra,

$$x_0^2 - 38.56 + 365.45 < 0$$

which gives the confidence interval

$$16.76 < x_0 < 21.80. \qquad \square$$

PROBLEMS

15.7.1 Using data points from Problems 15.4.1 and 15.5.1 estimate the score on an entrance exam of a student who graduated with an GPA of 3.95.

15.7.2 Five measurements of Y at $x = 10$ are 10.5, 10.6, 9.7, 11.1, and 12.3. Six measurements of Y at $x = 20$ are 3.1, 3.6, 3.1, 4.0, 5.2, and 2.9. Assuming that the regression of Y on X is linear, estimate the value of x if two observations made at this value are 6.3 and 7.1.

15.8 BLUEs

Most of the results presented thus far rely on the assumptions that the random variable Y_x is normally distributed with constant variance and that the observations are independent. These assumptions allow us to use the likelihood and provide access to the distributions of MLE's. The natural question one may ask is what to do if the assumptions above are not satisfied.

First, as regards homoscedasticity, there exist numerous variance-stabilizing transformations. These transformations have been suggested by statisticians as an *ad hoc* remedy against heteroscedasticity: Instead of data of the form (x_i, y_i) one uses the data (x_i, y_i^*), with $y_i^* = g(y_i)$, where g is a suitably selected function.

One may also use transformation of y's which depend on x's, that is, replacing the pair (x_i, y_i) by $(x_i, g_{x_i}(y_i))$, where g_x is some function [e.g., replacing (x_i, y_i) by $(x_i, y_i/x_i)$, etc.]. Which transformation should be used in a given situation is a problem that may be hard to resolve and need not have a unique answer, especially when little is known about the distribution of Y_x. It is, rather, common sense, combined with statistical intuition and experience, that should serve as a guide.

A question interesting from both a theoretical and a practical point of view is what to do if the distribution of Y_x is not normal. Suppose, for instance, that the model analyzed is that of linear regression $Y_x = a + bx + \epsilon_x$, where one assumes that the errors ϵ_x satisfy the conditions

    (i) $E(\epsilon_x) = 0$ (unbiasedness).
    (ii) $E(\epsilon_x^2) = \sigma^2$ (homoscedasticity).
    (iii) $E(\epsilon_x \epsilon_{x'}) = 0$ (orthogonality).

The least squares estimators of regression coefficients a and b (which are also the maximum likelihood estimators, in the normal case) are

$$\hat{b} = \frac{\sum (x_i - \bar{x})(y_i - \bar{y})}{\sum (x_i - \bar{x})^2} = \frac{\sum (x_i - \bar{x}) y_i}{\sum (x_i - \bar{x})^2}$$

and

$$\hat{a} = \bar{y} - \hat{b}\bar{x}.$$

As regards σ^2, one may use the estimate based on residuals, namely

$$\hat{\sigma}^2 = \frac{1}{n-2} \sum_{j=1}^{n} (y_j - \hat{a} - \hat{b}x_j)^2,$$

being the observed value of the estimator

$$S^2 = \frac{nT^2}{n-2} = \frac{1}{n-2} \sum_{j=1}^{n} (Y_j - V - U_{x_j})^2$$

with U, V, and T given by (15.12). Clearly, U, V, and S^2 are unbiased estimators of b, a, and σ^2, with variances and covariances obtained as before. This is true because the calculations of moments did not rely on the assumption of normality.

It is possible to show that under an assumption of normality U and V are also minimum variance estimators of b and a (i.e., that their variances coincide with the bounds given by the Rao–Cramèr inequality).

The question is: Is there any reason to use U and V as estimators of b and a when normality assumption does not hold? The mere availability of the computational formulas in statistical packages is hardly a justification. Unbiasedness of U and V is a desirable property only if it could be related to MSE: As we have seen, there are situations when the use of biased estimators is recommended, as leading to smaller mean square error.

The answer is positive and is given by the following theorem (which we shall state here without proof); it is generally referred to as the Gauss–Markov theorem.

Theorem 15.8.1. *Consider the observations* $(x_i, Y_j), i = 1, 2, \ldots, n$, *with* $Y_i = a + bx_i + \epsilon_i$, *where the errors* ϵ_i *satisfy conditions* (i)-(iii) *of unbiasedness, homoscedasticity, and orthogonality. Let* $\mathcal{L}$ *be the class of all linear combinations of observations* Y_i, *that is, of statistics of the form*

$$r_1 Y_1 + r_2 Y_2 + \cdots + r_n Y_n,$$

where r_i *are constants* [*depending possibly on the vector* $(x_1, \ldots, x_n)$]. *Furthermore, let* $\mathcal{L}_a \subset \mathcal{L}, \mathcal{L}_b \subset \mathcal{L}$ *be the subsets of* $\mathcal{L}$ *consisting of statistics that are unbiased estimators of a and of b. Then the statistics V and U given by* (15.12) *have minimal variances in classes* $\mathcal{L}_a$ *and* $\mathcal{L}_b$.

The acronym used here is BLUE, which stands for "best linear unbiased estimator." Thus, V is BLUE for a and U is BLUE for b.

Note that, in view of unbiasedness, "best" estimators mean those with minimum mean square error. Recall that in the normal case U and V are best (in the sense of MSE) estimators of regression parameters in the class of *all* estimators, linear or not. In the present case, under a weaker assumption, the conclusion is also weaker: U and V are best estimators in a more restricted class of estimators, namely those which depend linearly on the observations.

PROBLEMS

15.8.1 Suppose it is known that the number of errors in a text (e.g., a computer program) of length x is a Poisson random variable with unknown mean λ. If we observe n texts of lengths $x_1, x_2, \ldots, x_n$, then the number of errors they contain, $Y_1, \ldots, Y_n$, satisfy

$$E(Y_j) = \text{Var}(Y_j) = \lambda x_j.$$

Find the BLUE of λ.

15.8.2 Carry out the calculations in the following direct proof of the Gauss–Markov theorem showing that LS estimators of a and b are BLUE. For a one has to determine the constants $a_i, i = 1, \ldots, n$, such that the statistic $T = \sum_{i=1}^{n} a_i Y_i$ satisfies the conditions

(1) $E(T) = a$.

(2) The variance of T is the smallest among all linear estimators for which condition (1) holds.

Show first that condition (1) means that

$$\sum a_j = 1, \qquad \sum a_j x_j = 0. \qquad (15.25)$$

Next, using the fact that Y_j's are uncorrelated and homoscedastic, show that $\mathrm{Var}(T) = \sigma^2 \sum a_j^2$, so that we have to minimize $\sum a_j^2$ subject to constraints (15.25). Using Lagrange multipliers, this means that one must minimize

$$\sum a_j^2 - \lambda_1 \left(\sum a_j - 1 \right) - \lambda_2 \sum a_j x_j.$$

Take derivatives with respect to $a_1, \ldots, a_n$, λ_1, and λ_2, solve the resulting $n + 2$ equations, and check that the solution a_j gives the LS estimator of a.

15.8.3 Provide the same argument as in Problem 15.8.2 for the LS estimator of b.

15.8.4 Given the data $(x_i, y_i), i = 1, \ldots, n$, find the LS estimators of the quadratic regression $a + bx + cx^2$.

15.9 REGRESSION TOWARD THE MEAN

We start with some anecdotes, which may at first appear puzzling, and then investigate the nature of regression. We show, in particular, how regression relates to the construction of a highly successful theory in the domain of social sciences.

Example 15.9.1. A time-honored dilemma faced by parents and educators is whether it is better to punish bad behavior or reward good behavior. After centuries of rather strict adherence to the first option, the current psychological theories suggest that children (as well as adults, dogs, etc.) should be rewarded for good behavior (success, good performance, etc.), while punishment should be avoided.

An army sergeant claims that such psychologists' ideas are plain rubbish, and he has highly convincing evidence that psychologists are wrong: In fact, punishment works, while reward does not. Indeed, suppose that he trains his men in performing some task (jointly or individually), such as making up a bed, shooting at a moving target, or identifying an enemy plane. His training principles are such that whenever his men perform poorly he punishes them— by verbal abuse, additional training, and so on. When they perform well he either says nothing (which is a form of a praise) or rewards them somehow.

He then notices that punishment "works"—in most cases, after being punished, his men perform better. On the other hand, reward typically does not work, and most often has negative effects: After his men are praised for good performance, they typically do worse next time.

To see what really happens here, let us assume that neither sergeant nor psychologists are correct, and in fact, reward or punishment has no effect on the level of performance. Let us disregard the initial phase of training, when typically the average level of performance quickly increases, and consider the later phases, when the increase in skills is rather slow. As a first approximation, assume that the performance levels $X_1, X_2, \ldots$ at successive trials are iid random variables, oscillating about their common mean $\mu = E(X_i)$. Intuitively, if $X_n = x_n > \mu$ (above-average performance on the nth trial), then the event $X_{n+1} < x_n$ (decline in performance on the next trial) is more likely than the event $X_{n+1} > x_n$. The opposite inequality holds if $x_n < \mu$.

These inequalities are true for all values x_n if X_i's have normal distribution, or generally, have a symmetric distribution (then the mean and median coincide). For asymmetric distributions this property need not be true for all x_n, but is true for all x_n sufficiently large or sufficiently small.

This reasoning shows that the sergeant's criticism of psychological theories is not well grounded: One can expect to observe a decline in performance following a good performance, and improvement following a bad performance, regardless of punishment or reward. By the same token, however, this argument shows that current psychological theories, suggesting that any performance, good or bad, be praised, require more empirical evidence of validity other than mere pointing out that after praising poor results, the performance tends to improve.   □

The phenomenon described in the example above, known as *regression toward the mean*, is not restricted to the case of independent identically distributed observations. In was discovered in the nineteenth century by Galton, who studied various hereditary traits. He noticed that (using height as an example) tall fathers tend to have tall sons, but their sons tend to be closer to the average than the fathers. Similarly, short fathers tend to have short sons, but their sons tend to be closer to the average than their fathers. Galton called it a "tendency toward mediocrity." Galton's choice of the word "mediocrity" may also explain why he chose the term "regression," a word with somewhat derogatory connotations.

It should be realized that regression, understood as "affecting the mean of dependent variable Y by independent variable X," need not imply any causal relationship between the values of X and Y. One of the more common types of relation between X and Y which leads to regression phenomenon, yet does not involve any causal effects of X on Y, is exemplified by the following situation. Imagine a person who measures some attribute of objects, say their length. In each case he takes two observations, which typically differ somewhat bacause of the measurement error. Suppose that he measures objects of different lengths, and calls the first and second measurements of the ith object x_i and y_i.

Imagine now that points (x_i, y_i) are plotted, resulting in a scatter diagram. If the measurement errors are small, and/or the objects measured differ in their true lengths, we shall observe that the points (x_i, y_i) have a strong linear relationship with a slope close to 1.

Such a relationship will appear stronger when the variability of lengths of measured objects is higher. This effect is utilized in the construction of some psychological questionnaires.

Example 15.9.2 (Psychological Test Scores). In "softer" areas of psychology, such as those dealing with personality or motivation (as opposed to areas such as memory studies, with more quantifiable experiments) a researcher introduces typically some *construct* (e.g., "neuroticism," "self-esteem," etc.). Those constructs are then used to explain and/or predict some behavior. The explanation has the form of specific hypotheses, such as "persons with low self-esteem are more likely to be aggressive." In addition to theoretical justification of such hypotheses, there arises a problem of testing them empirically. Clearly, one needs here a tool for measuring the level of the construct (e.g., a tool to measure the level of neuroticism, self-esteem, etc.).

A typical tool has the form of a questionnaire and a scoring rule. (For some questions, it is the answer "yes" that contributes to the total score, for other questions it is the answer "no"; this is done to eliminate bias arising from a possible tendency toward some types of answers.)

When the questionnaire is applied to a subject s, one obtains the score X_s. Upon repetition, the score could be different, say X'_s. One of the central assumptions of the theory of psychological tests is that the expected scores X_s and X'_s are equal:

$$E(X_s) = E(X'_s) = T_s;$$

moreover, the deviations $\epsilon_s = X_s - T_s$ and $\epsilon'_s = X'_s - T_s$ satisfy the conditions

$$\text{Var}(\epsilon_s) = \text{Var}(\epsilon'_s) = \sigma_s^2, \qquad E(\epsilon_s \epsilon'_s) = 0.$$

For fixed s, the value T_s (called the *true value* of the measured construct for person s) is a constant, while ϵ_s is a random variable (reflecting intraperson variability of response to questionnaire, upon hypothetical repetitions of measurement).

Assume now that person s is sampled from the population according to some probability distribution.* Using Theorem 8.7.4, we now have

$$\sigma_X^2 = \text{Var}(X) = E_s\{\text{Var}(X|T)\} + \text{Var}\{E_s(X|T)\}$$
$$= E(\sigma_s^2) + \text{Var}(T_s) = \sigma_\epsilon^2 + \sigma_T^2,$$

and clearly, $\sigma_{X'}^2 = \sigma_\epsilon^2 + \sigma_T^2$. Similarly,

---

* It is not necessary that this distribution be uniform on the population of subjects; in most cases this distribution is uniform only on a subset of the population. For instance, if the questionnaire is designed, say, specifically for management executives or chess grandmasters, then the distribution of s assigns nonzero probabilities only to persons from the subpopulations in question.

$$\begin{aligned}
\text{Cov}(X, X') &= E(XX') - E(X)E(X') \\
&= E\{(T + \epsilon)(T + \epsilon')\} - [E(T)]^2 \\
&= E(T^2) - [E(T)]^2 = \sigma_T^2,
\end{aligned}$$

since $E(\epsilon T) = E_s\{E(\epsilon T)\} = E_s\{T_s E(\epsilon)\} = E(0) = 0$, and similarly for the other products. Thus,

$$\rho(X, X') = \frac{\sigma_T^2}{\sigma_T^2 + \sigma_\epsilon^2}. \tag{15.26}$$

The last ratio is called the *reliability* of the test, and formula (15.26) shows that reliability equals the test–retest correlation. This correlation approaches 1 with the increase of σ_T^2, that is, with the increase of variance of true scores of the test in the population under study.

The reliability of a psychological test is (as distinct from instruments for physical measurements) not intrinsic for the test only, but depends also on how diverse the population is to which it is applied. For instance, a questionnaire measuring, say, neuroticism might have a low reliability if applied to highly selected groups, say cosmonauts, monks, or brothers of Nobel Prize winners, but have high reliability when applied to a group such as all employees of a company.

15.10 ANALYSIS OF VARIANCE

In the remainder of this chapter we deal with testing for the existence of effects of the independent variable X on the response Y in cases when the values of X cannot be observed. What is available for observation is a classification of experimental situations into groups within which one can assume the same (unspecified) value of X.

Example 15.10.1. Suppose that we have a measurement tool (e.g., a psychological questionnaire) to measure some attribute, say level of aggressiveness. One might be interested in checking whether or not levels of aggressiveness among teenagers (Y) are the same in, say, large towns, small towns, and rural communities, and whether or not they are the same for boys and girls. The data will then consist of the results of measurements of Y in six samples of teenagers, corresponding to division according to gender and community size.   □

The standard terminology of the analysis of variance (ANOVA) is that of *levels of factors*. In Example 15.10.1 we had two factors: gender, with two levels (M and F), and size of community, with three levels: large town, small town, rural. These factors cross each other, in the sense that every level of one factor can be combined with every level of the other factor.

The central assumptions of analysis of variance models are very much the same as in regression models, namely:

(i) *The response variable Y has, for each level of factors, a normal distribution with the same (unknown) variance σ^2 (homoscedasticity).*

(ii) *The factors may affect only the mean of the response.*

(iii) *Distinct observations for the same or different levels of factors are independent.*

In the sequel we show the tests for the hypothesis that a given factor has no effect on the response variable, against the alternative that it has some effect.

The main issue here is that these tests can be carried out for various factors *on the same data.* In fact, ANOVA models originated from questions arising in agriculture, where one is interested in the response variable (e.g., size of harvest Y), as dependent on combinations of various factors, such as type of soil, time of planting, time of harvesting, type of cultivation, use of various fertilizers, and so on. Since a typical experiment lasts for one season, it is imperative to find a design that would allow us to study the effects of various factors using the same data. We begin with the simplest case of one factor only.

15.11 ONE-WAY CLASSIFICATION

Consider the situation when the data are partitioned into groups, each corresponding to one level of the factor. Alternatively, the same setup may be described as "independent samples from different populations."

We let n_i denote the number of observations from the ith group, where $i = 1, \ldots, I$. We have here $I \geq 2$ and $n_i \geq 2$ for at least one i.

Let

$$n_1 + n_2 + \cdots + n_I = N$$

be the total number of observations, and let $y_{i1}, y_{i2}, \ldots, y_{in_i}$ be the observations corresponding to the ith level of the factor. These observations are regarded as the recorded values of the random variables $Y_{i1}, Y_{i2}, \ldots, Y_{in_i}$.

According to the assumptions stated at the beginning of this section, all random variables Y_{ij} are independent, normally distributed, with $\text{Var}(Y_{ij}) = \sigma^2$. Moreover, since the effect of a factor is expressed only through the mean, we must have

$$E(Y_{ij}) = \mu_i, \qquad j = 1, \ldots, n_i.$$

The objective is to test the null hypothesis

$$H_0 : \mu_1 = \mu_2 = \cdots = \mu_I$$

against the alternative

$$H_1 : \mu_i \neq \mu_{i'} \text{ for some } i, i'.$$

In the sequel, it will be convenient to let

$$\mu_i = \mu + \alpha_i, \qquad i = 1, \ldots, I,$$

where

$$\mu = \frac{\mu_1 + \cdots + \mu_I}{I}. \tag{15.27}$$

We have, therefore,

$$\sum_{i=1}^{I} \alpha_i = 0,$$

and the hypothesis tested can be formulated as

$$H_0 : \alpha_1 = \alpha_2 = \cdots = \alpha_I = 0$$

(no effect of the factor), against the alternative

$$H_1 : \alpha_i \neq 0 \text{ for at least one } i.$$

Before we develop the testing procedure, let us observe that if $I = 2$, we have the problem of comparing the means of two normal populations with the same (unknown) variance σ^2. This problem was solved in Chapter 13 (Section 13.8), and the testing procedure used a Student ratio with $n_1 + n_2 - 2$ degrees of freedom. The critical region consisted of two tails of the Student distribution.

It would seem that if $I > 2$, we can use this result for the present case, by comparing pairs of levels of the factor until either we find a pair where the difference is significant (and then we reject H_0), or all pairs are tested with no significant difference found (in which case we accept H_0). The reason that such a procedure is unacceptable lies in the fact that it is impossible to determine its significance level, because (i) the procedures for overlapping pairs of factor levels are not independent, and (ii) even for nonoverlapping pairs, if the null hypothesis is true, the chances of at least one incorrect rejection of null hypothesis increase quickly with the number of tested pairs.

Consequently, the objective is to find a procedure that could test the null hypothesis of no effect of the factor with preassigned level of significance. The construction here will be based on a partition of the sum of squared deviations from the mean, very similar to the technique used for testing linearity of regression in Section 15.6.

In the derivation below, the subscript + will stand for averaging over the values of the index replaced by +. Thus,

$$\overline{Y}_{i+} = \frac{1}{n_i} \sum_{j=1}^{n_i} Y_{ij}, \quad \overline{y}_{i+} = \frac{1}{n_i} \sum_{j=1}^{n_i} y_{ij}$$

and

$$\overline{Y}_{++} = \frac{1}{N} \sum_{i=1}^{I} \sum_{j=1}^{n_i} Y_{ij} = \frac{1}{N} \sum_{i=1}^{I} n_i \overline{Y}_{i+}$$

and similarly for $\overline{y}_{++}$.

In the identities below we omit the cross-products. We encourage the reader to verify that all the cross-products are indeed zero. Using the fact that by (15.27) we have $E(Y_{ij}) = \mu + \alpha_i$, we decompose the sum of squared deviations of the variables from their means as follows:

$$S^2 = \sum_{i=1}^{I} \sum_{j=1}^{n_i} (Y_{ij} - \mu - \alpha_i)^2$$

$$= \sum_{i=1}^{I} \sum_{j=1}^{n_i} (Y_{ij} - \overline{Y}_{i+} + \overline{Y}_{i+} - \mu - \alpha_i)^2$$

$$= \sum_{i=1}^{I} \sum_{j=1}^{n_i} (Y_{ij} - \overline{Y}_{i+})^2 + \sum_{i=1}^{I} n_i (\overline{Y}_{i+} - \mu - \alpha_i)^2 \tag{15.28}$$

$$= \sum_{i=1}^{I} \sum_{j=1}^{n_i} (Y_{ij} - \overline{Y}_{i+})^2 + \sum_{i=1}^{I} n_i (\overline{Y}_{i+} - \overline{Y}_{++} + \overline{Y}_{++} - \mu - \alpha_i)^2$$

$$= \sum_{i=1}^{I} \sum_{j=1}^{n_i} (Y_{ij} - \overline{Y}_{i+})^2 + \sum_{i=1}^{I} n_i (\overline{Y}_{i+} - \overline{Y}_{++} - \alpha_i)^2 + N(\overline{Y}_{++} - \mu)^2.$$

Under the assumption that the Y_{ij} are normally distributed, the three terms in the last row are independent random variables. Moreover, the first term, upon division by σ^2, has the chi-square distribution with number of degrees of freedom equal to $(n_1 - 1) + \cdots + (n_I - 1) = N - I$ *regardless whether or not the null hypothesis is true*. The second sum, again upon division by σ^2, has the chi-square distribution with $I - 1$ degrees of freedom, provided that $E(\overline{Y}_{i+} - \overline{Y}_{++}) = \alpha_i$ for every i. Thus, under H_0, the sum

$$\frac{1}{\sigma^2} \sum_{i=1}^{I} n_i (\overline{Y}_{i+} - \overline{Y}_{++})^2 \tag{15.29}$$

has chi-square distribution with $I - 1$ degrees of freedom. Moreover, any violation of H_0 will increase the expectation of the sum (15.29). This suggests the use of an appropriate F ratio to test the hypothesis H_0.

Finally, if $\mu = 0$, then $N(\overline{Y}_{++})^2/\sigma^2$ has the chi-square distribution with one degree of freedom, which allows us to test another null hypothesis H_0' : $\mu = 0$ against the alternative H_1' : $\mu \neq 0$. In terms of observations, the testing procedure is most often displayed in the form of the following ANOVA table.

Source of Variation	Degrees of Freedom	Sum of Squares	Mean Sum of Squares	F Ratio
Mean	1	SSM	$MSM = \dfrac{SSM}{1}$	$F_M = \dfrac{MSM}{MSR}$
Factor	$I-1$	SSA	$MSA = \dfrac{SSA}{I-1}$	$F_A = \dfrac{MSA}{MSR}$
Residual	$N-I$	SSR	$MSR = \dfrac{SSR}{N-I}$	
Total	N	SST		

where

$$SSM = N(\bar{y}_{++})^2$$

$$SSA = \sum_{i=1}^{I} n_i(\bar{y}_{i+} - \bar{y}_{++})^2$$

$$SSR = \sum_{i=1}^{I} \sum_{j=1}^{n_i} n_i(y_{ij} - \bar{y}_{i+})^2$$

$$SST = \sum_{i=1}^{I} \sum_{j=1}^{n_i} y_{ij}^2.$$

Using F ratios as marked in the table we reject the hypothesis $H_0 : \alpha_1 = \cdots = \alpha_I = 0$ (on significance level γ) if the ratio F_A exceeds the $(1-\gamma)$-quantile $F_{\gamma,I-1,N-I}$ of the F distribution with numbers of degrees of freedom $(I-1, N-I)$. It is important to remember that in ANOVA *one always uses a one-sided critical region* (since any violation of H_0 tends to increase the numerator without affecting the denominator).

If for some reason one is interested in the hypothesis $H_0' : \mu = 0$, then one should reject it in favor of $H_1' : \mu \neq 0$ if the ratio F_M exceeds the quantile $F_{\gamma,1,N-I}$. Such a test may be useful if observations can be positive, as well as negative (e.g., deviations from the required norm in a technological process).

PROBLEMS

15.11.1 To test the mileage achieved by cars produced by different companies, but of comparable price, size, and so on, one make of cars was selected from each of the three major American companies and from two foreign companies. For each of the makes selected a number of new cars were chosen and their mileages recorded. The data (in miles per gallon) are:

$n_1 = 5$	$n_2 = 4$	$n_3 = 5$	$n_4 = 3$	$n_5 = 6$
25.1	27.1	29.9	25.4	29.2
26.2	26.4	21.4	28.2	29.3
24.9	26.8	22.2	27.1	30.4
25.3	27.2	22.5		28.5
23.9		20.8		28.9
				29.2

On significance level 0.05, test the hypothesis that the average mileage is the same for all makes of cars tested.

15.11.2 Assume that we take a random sample of size n from a normal distribution $N(\mu, \sigma^2)$. Next, the observations are divided into k groups of sizes $n_1, \ldots, n_k$, where $n_i \geq 2$ for $i = 1, \ldots, k$ and $n_1 + \cdots + n_k = n$. Let S_i^2 be the sum of squared deviations of observations in the ith group from the mean of this group.
   (i) Find the distribution of $S_1^2 + \cdots + S_k^2$.
   (ii) Find the distributions of S_k^2 / S_1^2.
   (iii) Find x such that

$$P\left\{ \frac{S_1^2 + S_2^2}{S_1^2 + S_2^2 + S_3^2 + S_4^2 + S_5^2} < x \right\} = 0.99.$$

15.11.3 Verify that the cross-products of the partition (15.28) are indeed zero.

15.11.4 Compare the test developed in this chapter with Student t test for the case when the factor operates at two levels only. Which procedure (if either) is better?

15.12   TWO-WAY CLASSIFICATION

Assume now that we have data concerning possible effects of two factors, say A and B. Factor A appears on I levels, while factor B appears on J levels, each level of A combined with each level of B. Let y_{ij} be the observation on the ith level of A and the jth level of B. Such data can therefore be arranged into a matrix $[y_{ij}]$ with I rows and J columns. For the moment we assume that we have one observation in each of the IJ cells formed by crossing factors A and B. As before, we regard y_{ij} as the recorded value of random variable Y_{ij}, assumed to have normal distribution

$$Y_{ij} \sim N(\mu + \alpha_i + \beta_j, \sigma^2).$$

Here α_i and β_j represent the effects of factors A and B on the ith and jth levels, respectively. Without loss of generality we may assume that

$$\sum_{i=1}^{I} \alpha_i = \sum_{j=1}^{J} \beta_j = 0. \tag{15.30}$$

We shall construct a test for the null hypothesis

$$H_0^{(A)} : \alpha_1 = \cdots = \alpha_I = 0$$

against the alternative

$$H_1^{(A)} : \alpha_i \neq 0 \text{ for some } i,$$

as well as a test for

$$H_0^{(B)} : \beta_1 = \cdots = \beta_J = 0$$

against the alternative

$$H_1^{(B)} : \beta_j \neq 0 \text{ for some } j.$$

The tests are built on a partition of sum of squares, as in the case of one factor. Omitting again the cross-products (which are zero), we have

$$S^2 = \sum_{i=1}^{I} \sum_{j=1}^{J} (Y_{ij} - \mu - \alpha_i - \beta_j)^2$$

$$= \sum_{i=1}^{I} \sum_{j=1}^{J} (Y_{ij} - \overline{Y}_{++} + \overline{Y}_{++} - \mu - \alpha_i - \beta_j)^2$$

$$= \sum_{i=1}^{I} \sum_{j=1}^{J} (Y_{ij} - \overline{Y}_{++} - \alpha_i - \beta_j)^2 + IJ(\overline{Y}_{++} - \mu)^2$$

$$= \sum_{i=1}^{I} \sum_{j=1}^{J} (Y_{ij} - \overline{Y}_{i+} - \overline{Y}_{+j} + \overline{Y}_{++})^2 + \sum_{i=1}^{I} J(\overline{Y}_{i+} - \overline{Y}_{++} - \alpha_i)^2$$

$$+ \sum_{j=1}^{J} I(\overline{Y}_{+j} - \overline{Y}_{++} - \beta_j)^2 + IJ(\overline{Y}_{++} - \mu)^2.$$

Substituting the observed data for random variables and letting $\mu = 0$, $\alpha_i = \beta_j = 0$ for all i, j, we obtain the following sums of squares:

$$SSM = IJ(\bar{y}_{++})^2,$$

$$SSA = \sum_{i=1}^{I} J(\bar{y}_{i+} - \bar{y}_{++})^2,$$

$$SSB = \sum_{i=1}^{J} I(\bar{y}_{+j} - \bar{y}_{++})^2,$$

$$SSR = \sum_{i=1}^{I}\sum_{j=1}^{J}(y_{ij} - \bar{y}_{i+} - \bar{y}_{+j} + \bar{y}_{++})^2,$$

which add up to the total sum of squares

$$SST = \sum_{i=1}^{I}\sum_{j=1}^{J} y_{ij}^2.$$

The four terms in the last decomposition of S^2 are independent random variables. The last (SSR) does not involve any parameters, and (upon division by σ^2) has a chi-square distribution with number of degrees of freedom equal to

$$IJ - (I-1) - (J-1) - 1 = (I-1)(J-1),$$

regardless of the truth or falsehood of any of the hypotheses. The second sum has a chi-square distribution if $\alpha_1 = \cdots = \alpha_I = 0$, and a similar statement holds for the third sum if $\beta_1 = \cdots = \beta_J = 0$. The numbers of degrees of freedom are $I - 1$ and $J - 1$, respectively. Any deviation from the null hypothesis tends to increase the corresponding sum of squares.

The ANOVA table is now:

Source of Variation	Degrees of freedom	Sum of Squares	Mean Sum of Squares	F Ratio
Mean	1	SSM	$MSM = \dfrac{SSM}{1}$	$F_M = \dfrac{MSM}{MSR}$
A	$I-1$	SSA	$MSA = \dfrac{SSA}{I-1}$	$F_A = \dfrac{MSA}{MSR}$
B	$J-1$	SSB	$MSB = \dfrac{SSB}{J-1}$	$F_B = \dfrac{MSB}{MSR}$
Residual	$(I-1)(J-1)$	SSR	$MSR = \dfrac{SSR}{(I-1)(J-1)}$	
Total	$N = IJ$	SST		

We reject the hypothesis $H_0^{(A)}$ (that factor A has no effect) if

$$F_A > F_{\gamma, I-1, (I-1)(J-1)}$$

and we reject the hypothesis $H_0^{(B)}$ if

$$F_B > F_{\gamma, J-1, (I-1)(J-1)}$$

where γ is the desired level of significance and $F_{\gamma, m, n}$ stands for the $(1 - \gamma)$-quantile of F distribution with (m, n) degrees of freedom.

PROBLEMS

15.12.1 The nutritional value of a certain vegetable is measured on 18 specimens grown in two varieties in three geographical regions. The data are as follows:

Variety	Geographical Region		
	A	B	C
1	6.3	9.2	6.8
	11.5	5.1	7.2
	9.2	8.1	5.5
2	11.0	5.4	7.1
	7.3	5.0	7.8
	8.2	6.1	8.4

Study the effect of variety and geographical location on the nutritional value, taking the average of three observations in each cell as a response. Use the significance level $\alpha = 0.05$.

15.12.2 Twelve overweight subjects participated in a study to compare three weight-reducing diets. Subjects were grouped according to initial weight, and each of three subjects from each initial weight group was randomly assigned to a diet. The weight losses (in pounds) at the end of experimental period are:

Initial Weight	Diet		
	A	B	C
150–174	10	23	24
175–199	12	21	26
200–224	12	31	21
Over 224	20	28	33

(i) Do these data provide sufficient evidence that (after eliminating the effect of initial weight) the diets are different in their effectiveness? Use the significance level $\alpha = 0.01$.

(ii) Does the initial weight affect the loss of weight due to dieting?

15.12.3 Assume that there are three factors, so that the model is $E(Y_{ijk}) = \mu + \alpha_i + \beta_j + \gamma_k$, where $\sum \alpha_i = \sum \beta_j = \sum \gamma_k = 0$. Moreover, assume Y_{ijk} is normally distributed with variance σ^2. Derive the tests for the hypotheses analogous to those in case of two factors.

15.13 ANOVA MODELS WITH INTERACTION

The model considered in Section 15.12 was of the form $E(Y_{ij}) = \mu + \alpha_i + \beta_j$, which assumes that the effects of the factors A and B are additive. The hypotheses $H_0^{(A)}$ and $H_0^{(B)}$ are tested within this model.

In general, the effects of A and B need not be additive, and one might wish to consider the case when $E(Y_{ij})$ is an entirely arbitrary function of i and j. Such a function can always be represented in the form

$$E(Y_{ij}) = \mu + \alpha_i + \beta_j + \gamma_{ij},$$

$\sum_i \alpha_i = \sum_j \beta_j = \sum_i \gamma_{ij} = \sum_j \gamma_{ij} = 0$ for all i,j (we leave the proof of this as an exercise). The constants γ_{ij} are referred to as an *interaction* (between factors A and B). Testing for the presence of an interaction is based on an idea similar to that used in testing linearity of regression. It requires a model-independent estimate of σ^2, which in turn can be achieved if we have more than one observation for each combination of levels of factors A and B. We assume therefore that the data now have the form of a three-dimensional array $y_{ijk}, i = 1,\ldots,I; j = 1,\ldots J; k = 1,\ldots,K$, where $I \geq 2, J \geq 2, K \geq 2$. Leaving the details of the derivation as an exercise, the decomposition of the sum of squared deviations of the random variables Y_{ijk} from their means is

$$
\begin{aligned}
S^2 &= \sum_{i=1}^{I}\sum_{j=1}^{J}\sum_{k=1}^{K}(Y_{ijk} - \mu - \alpha_i - \beta_j - \gamma_{ij})^2 \\
&= IJK(\overline{Y}_{+++} - \mu)^2 \\
&\quad + \sum_{i=1}^{I}JK(\overline{Y}_{i++} - \overline{Y}_{+++} - \alpha_i)^2 + \sum_{j=1}^{J}IK(\overline{Y}_{+j+} - \overline{Y}_{+++} - \beta_j)^2 \\
&\quad + \sum_{i=1}^{I}\sum_{j=1}^{J}K(\overline{Y}_{ij+} - \overline{Y}_{i++} - \overline{Y}_{+j+} + \overline{Y}_{+++} - \gamma_{ij})^2 \\
&\quad + \sum_{i=1}^{I}\sum_{j=1}^{J}\sum_{k=1}^{K}(Y_{ijk} - \overline{Y}_{ij+})^2.
\end{aligned}
\tag{15.31}
$$

As before, the last sum (upon division by σ^2) has chi-square distribution with $IJ(K-1)$ degrees of freedom, given only normality and homoscedasticity (regardless of any other hypotheses tested). Thus it may serve as a denominator in all F ratios used for testing, while the two single sums and the double sum can be used to test the hypotheses about the effects of A, of B, and of their interaction. To put it differently, each of the sums above, divided by its number of degrees of freedom, is an unbiased estimator of σ^2 if the appropriate hypothesis is true (except the last sum, for which this holds regardless of any hypothesis). Letting

$$SSM = IJK(\bar{y}_{+++})^2,$$

$$SSA = \sum_{i=1}^{I} JK(\bar{y}_{i++} - \bar{y}_{+++})^2,$$

$$SSB = \sum_{j=1}^{J} IK(\bar{y}_{+j+} - \bar{y}_{+++})^2,$$

$$SSAB = \sum_{i=1}^{I}\sum_{j=1}^{J} K(\bar{y}_{ij+} - \bar{y}_{i++} - \bar{y}_{+j+} + \bar{y}_{+++})^2,$$

$$SSR = \sum_{i=1}^{I}\sum_{j=1}^{J}\sum_{k=1}^{K} (y_{ijk} - \bar{y}_{ij+})^2,$$

$$SST = \sum_{i=1}^{I}\sum_{j=1}^{J}\sum_{k=1}^{K} y_{ijk}^2,$$

the corresponding ANOVA table is as follows:

Source of Variation	Degrees of freedom	Sum of Squares	Mean Sum of Squares	F Ratio
Mean	1	SSM	$MSM = \dfrac{SSM}{1}$	$F_M = \dfrac{MSM}{MSR}$
A	$I-1$	SSA	$MSA = \dfrac{SSA}{I-1}$	$F_A = \dfrac{MSA}{MSR}$
B	$J-1$	SSB	$MSB = \dfrac{SSB}{J-1}$	$F_B = \dfrac{MSB}{MSR}$
AB	$(I-1)(J-1)$	$SSAB$	$MSAB = \dfrac{SSI}{(I-1)(J-1)}$	$F_{AB} = \dfrac{MSI}{MSR}$
Residual	$IJ(K-1)$	SSR	$MSR = \dfrac{SSR}{IJ(K-1)}$	
Total	$N = IJK$	SST		

The tests reject the hypothesis of lack of effects of a given factor or lack of interaction if the corresponding F ratio exceeds the critical value obtained from the right tail of the F distribution with a given number of degrees of freedom.

Example 15.13.1. A researcher studies the effects of gender and type of stimulus ("soothing" or "exciting") on the aggressive behavior of parrots. Six male and six female birds of a given species are placed each in a separate cage, isolated from other cages. The six birds of a given gender are randomly divided into two groups. The cages are covered for the night. Before uncovering the cage in the morning, the birds hear a tape. One is a "soothing" tape, with the voice of the experimenter talking quietly to the birds. The other tape contains the angry voice of the experimenter. The observed value Y is the number of times the bird attacks the experimenter's hand when she uncovers the cage and puts food into the plate. The data are as follows (the three numbers in each cell representing the data for three birds):

	Soothing (S)	Angry (A)
M	8, 6, 13	22, 28, 33
F	5, 10, 6	12, 14, 9

The within-cell averages $\bar{y}_{ij+}$ and row and column averages y_{i+}, y_{+j}, as well as y_{++}, are

	S	A	
M	9	27.67	18.33
F	7	11.67	9.33
	8	19.67	13.83

The sum SSR of squared deviations from the cell means equals $SSR = 113.33$. We therefore have the following table:

	df	SS	MS	F
Mean	1	2296.33	2296.33	162.10
Sex	1	243.00	243.00	17.15
Stimulus	1	408.33	408.33	28.82
Interaction	1	147.00	147.00	10.38
Residuals	8	113.33	14.17	

At significance level 0.05, the value $F_{0.05,1,8} = 5.32$, so that we conclude that both the gender of the parrot and the type of stimulus have an effect, and there is interaction between gender and type of stimulus. The mean response is significantly different from zero. □

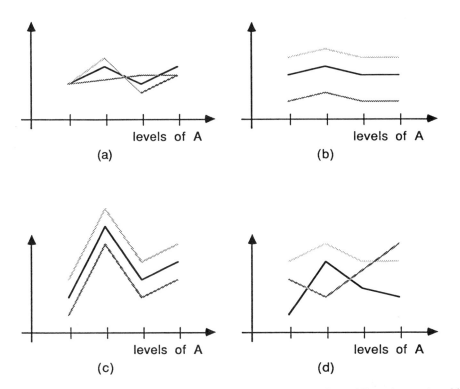

Figure 15.4 (a) No effects of A, B, or AB; (b) no effect of A, effect of B, no interaction; (c) effect of A, no effect of B, no interaction; (d) effect of A, effect of B, effect of interaction.

To grasp better the meaning of interaction, one may use the following graphical representation. Let us arrange the categories of one factor (say A) along the horizontal axis (there may be no numerical values attached, and the categories need not have any "natural" order). Then along the horizontal axis we can plot average responses $\bar{y}_{ij+}$ for various j. The effects of A, B and of interaction can now be interpreted as follows (see Figure 15.4). The effect of A means that at least one of the curves differs significantly from the horizontal line (last two figures). The effect of B, but not of interaction, means that curves for various levels of B are parallel to one another, but significantly different from one another (Figures 15.4b and c). Finally, interaction reveals itself by lack of parallelism of the curves (Figure 15.4d).

PROBLEMS

15.13.1 Show that if $E(Y_{ij}) = c(i, j)$, where $i = 1, \ldots, I, j = 1, \ldots, J$, then one can always find constants μ, α_i, β_j, and γ_{ij} such that

$$\sum_{i=1}^{I} \alpha_i = \sum_{j=1}^{J} \beta_j = \sum_{i=1}^{I} \gamma_{ij} = \sum_{j=1}^{J} \gamma_{ij} = 0,$$

and $c(i, j) = \mu + \alpha_i + \beta_j + \gamma_{ij}$.

15.13.2 The scores on level of emotional maturity of young adult males classified by age and extent of use of marijuana are given below.

	Marijuana Use		
Age	Never	Occasionally	Daily
15–19	25	18	17
	28	23	24
	22	19	19
20–24	28	16	18
	32	24	22
	30	20	20
25–29	25	14	10
	35	16	8
	30	15	12

Perform an analysis of variance for these data, using a significance level $\alpha = 0.05$, testing for effect of age, extent of use, and interaction.

15.13.3 Derive formula (15.31).

15.13.4 Use the data given in Problem 15.12.1 to test for the existence of an interaction between variety and geographical location.

15.13.5 Derive the test for the case of three-way ANOVA with interactions. The model now is

$$E(Y_{ijkl}) = \mu + \alpha_i + \beta_j + \gamma_k + \delta_{ij} + \delta'_{ik} + \delta''_{jk} + \epsilon_{ijk},$$

where $\sum_i \alpha_i = \sum_j \beta_j = \sum_k \gamma_k = \sum_i \delta_{ij} = \sum_j \delta_{ij} = \sum_i \delta'_{ik} = \sum_k \delta'_{ik} = \sum_j \delta''_{jk} = \sum_k \delta''_{jk} = \sum_i \epsilon_{ijk} = \sum_j \epsilon_{ijk} = \sum_k \epsilon_{ijk} = 0$, with index l referring to the lth observation in cell (i, j, k).

15.14 FURTHER EXTENSIONS

The ANOVA models presented in the last three sections can easily be extended to the case of more than two factors, with or without interaction, provided that we consider the completely balanced designs: Each level of a factor can be combined with all combinations of levels of other factors, and in each cell we have the same number of observations. Testing for interactions is possible only if the latter number (of observations per cell) is at least 2. The trouble with such extensions of the theory is that they are seldom useful. On the one hand, the number of observations needed soon becomes unattainable practically. For instance, with three factors, each on five levels, we need $5^3 = 125$ observations, and twice that if we want to test for interactions.

This situation led to research in several major directions. The most important was to invent experimental schemes that would allow testing for the presence of effects (as well as estimation of those effects) with as small a number of experiments as possible. If one resigns from the stringent requirement that every level of a factor has to appear with every combination of levels of other factors, there are many possibilities of experiments (e.g., forming so-called Latin or Greco-Latin squares). To use a simple example, imagine that we have three factors, each appearing on five levels. Representing one factor as a row, the other as a column, and the third as a letter (with levels a, b, c, d, e), one may arrange the experiment as follows:

a	b	c	d	e
b	c	d	e	a
c	d	e	a	b
d	e	a	b	c
e	a	b	c	d

Each level of the first factor (row) combines exactly once with each level of the second factor (column), and exactly once with the third factor (letter). The same is true for the other two factors. However, of $5^3 = 125$ possible combinations, only 25 actually appear.

It is possible to include here the fourth factor (say Greek letters, $\alpha, \beta, \gamma, \delta, \epsilon$) as follows:

$a\alpha$	$b\beta$	$c\gamma$	$d\delta$	$e\epsilon$
$b\epsilon$	$c\alpha$	$d\beta$	$e\gamma$	$a\delta$
$c\delta$	$d\epsilon$	$e\alpha$	$a\beta$	$b\gamma$
$d\gamma$	$e\delta$	$a\epsilon$	$b\alpha$	$c\beta$
$e\beta$	$a\gamma$	$b\delta$	$c\epsilon$	$d\alpha$

Now each row and each column has exactly one of the Roman letters and exactly one of the Greek letters, and each Roman letter is combined exactly once with each Greek letter.

One can therefore plan an agricultural experiment (say) in which in each row are plants of one of five varieties of seed, in each column one uses one of the five varieties of fertilizers, each Roman letter corresponds to one of the five different amounts of watering, and each Greek letter corresponds to one of the five different times of planting.

There appears to be a seemingly countless variety of experimental designs to cover all contingencies that may occur in practice, each design with its own testing or estimation procedures. By introducing appropriate criteria, one can search for designs that optimize these criteria.

REFERENCES

Myers, R. H. (1986). *Classical and Modern Regression with Applications*, Duxbury Press, Boston, Mass.

Montgomery, D. C.(1992). *Introduction to Linear Regression Analysis*, 2nd ed. Wiley, New York.

Myers, R. H.and J. S. Milton (1991). *A First Course in the Theory of Linear Statistical Models*. PWS–Kent.

Stapleton, J. (1995). *Linear Statistical Models*. Wiley, New York.

CHAPTER 16

Rank Methods

16.1 INTRODUCTION

This chapter is devoted to methods of handling data measured on ordinal scales only (more precisely: when only ordinal relations are taken into account). These methods* fall under the general heading "nonparametric statistics" and involve techniques based on ranks. There are several advantages of these methods. They require the assumption that the data are sampled from a continuous distribution, but the type of this distribution is irrelevant. Yet, nonparametric methods are almost as powerful as normal theory methods. They can, therefore, be applied when the population distribution is not normal (e.g. is skewed), or if the sample size is not large enough to test normality. Another advantage of methods based on ranks is that they are computationally very simple.

Example 16.1.1. Suppose that we have four observations from one population, say (after ordering) $x_1 = 3, x_2 = 7, x_3 = 8$, and $x_4 = 15$, and five observations from another population, say $y_1 = 0, y_2 = 1, y_3 = 4, y_4 = 5$, and $y_5 = 6$. If the data are indeed expressed on the ordinal (but not any higher) scale, then these data contain the same information as data such as $x_1 = 30, x_2 = 32, x_3 = 1000, x_4 = 1001, y_1 = -5, y_2 = 0, y_3 = 31, y_4 = 35, y_5 = 36$. In effect, all information that *can* be used is contained in the sequence

$$Y\,Y\,X\,Y\,Y\,Y\,Y\,X\,X\,X$$

One can use only the fact that in nine observations, those from the first sample occupy places 3, 7, 8, and 9. □

---

*For an exhaustive presentation of nonparametric methods, see, for example, Hollander and Wolfe (1973).

Methods based on ranks *should* be used in all cases when the data are measured on an ordinal scale. They *may* be used in any case when the data are measured on an interval or ratio scale. In other words, one can always use methods for a weaker scale. In reality, however, when in doubt, many persons use stronger methods. This is customary—as already mentioned—for example in education, when students' grades are averaged, normalized with the use of standard deviation, and so on. In fact, it is far from clear what the measurement status of student grades is, so prudence would suggest refraining from treating them as if they were expressed on an interval scale. Tradition, however, requires giving average scores on exams, and it is perhaps true that people have developed common intuition that may be used as a "relational system" that permits treating the grades as if they were expressed on an interval (or even ratio) scale.

It is mostly the assumption of continuity that distinguishes the analysis of this chapter from the analysis of those categorical data that happen to be of ordinal character (analyzed in Chapter 17). Such data are exemplified by the statistical analysis of, say, questionnaire responses such as "strongly agree," "agree," "neutral," "disagree," "strongly disagree." The essential difference is that under the assumption of population distribution being continuous we can disregard ties in our theoretical considerations. Practically, it means that ties in the data are rather exceptional.

We begin with a study of the behavior of empirical cdf's by showing that the empirical cdf converges almost surely and uniformly to the cdf of the underlying random variable. This is a general fact, true regardless of whether or not the random variable is continuous. If we restrict the analysis to the continuous case, we can construct a test for the hypothesis that the random sample $X_1, \ldots, X_n$ comes from a population with specific continuous cdf.

Next, we consider the two-sample problem, namely of testing the hypothesis that the random samples $(X_1, \ldots, X_m)$ and $(Y_1, \ldots, Y_n)$ come from the same continuous distribution. We also introduce tests for randomness. Finally, we present procedures for testing hypotheses about population medians in the case of one, two, and more than two samples.

16.2 GLIVENKO–CANTELLI THEOREM

We begin again with the most common setting in statistics and probability, that of a random sample

$$\mathbf{X}_n = (X_1, \ldots, X_n) \qquad (16.1)$$

from some population. Specifically, this means that $X_1, \ldots, X_n$ is the initial fragment of a sequence $\mathbf{X}_\infty = \{X_i, i = 1, 2, \ldots\}$ of iid random variables. We let F denote the common cdf of X_i's, so that $F(t) = P\{X_i \leq t\}$. At this moment we make no assumptions about the nature of F, so that F is an arbitrary nondecreasing right-continuous function, satisfying

$$\lim_{t \to \infty} F(t) = 1, \qquad \lim_{t \to -\infty} F(t) = 0$$

(see Chapter 6). For any n, we define the empirical cdf of the sample (16.1), letting, for any t,

$$F_n(t) = \frac{\text{number of data values that do not exceed } t}{n}.$$

Thus, $F_n(t)$ is a step function that increases by $1/n$ (or by a multiple of $1/n$) at each point of the sample. It is important to keep in mind that $F_n(t)$ is a random function, depending formally on the sequence $\mathbf{X}_\infty$, but in fact depending only on the first n observations.

If we fix the value t, then $nF_n(t)$ is a random variable, equal to the number of X_i's among the first n observations, satisfying the condition $X_i \leq t$. Thus, $nF_n(t)$ has binomial distribution with parameters n and $p = P\{X_i \leq t\} = F(t)$. Consequently, by the strong law of large numbers for binomial random variables, we have for every t

$$F_n(t) = \frac{nF_n(t)}{n} \to F(t) \tag{16.2}$$

with probability 1.

To appreciate the meaning of this result, and the meaning of its extension below, let us write explicitly $F_n(t, \mathbf{X})$, where $\mathbf{X} = \mathbf{X}_\infty$. Since $\mathbf{X}$ is the element of the sample space corresponding to sampling an infinite sequence of values of iid random variables, each with cdf F, the phrase "with probability 1" or "almost certainly" means "for all sequences $\mathbf{X}$, except sequences in a set $\mathcal{N}$ with $P(\mathcal{N}) = 0$."

Specifically, (16.2) means that for every t there exists a set $\mathcal{N}_t$ of sequences $\mathbf{X}$ such that $P(\mathcal{N}_t) = 0$ and

$$\lim_{n \to \infty} F_n(t, \mathbf{X}) = F(t) \tag{16.3}$$

if $\mathbf{X} \notin \mathcal{N}_t$.

We shall now prove an extension of this result to the convergence for all t at once, and uniform. We have the following theorem.

Theorem 16.2.1 (Glivenko–Cantelli). *As $n \to \infty$, we have*

$$\sup_t |F_n(t) - F(t)| \to 0$$

with probability 1.

This means that we can find one "exceptional" set $\mathcal{N}$, that is, a set $\mathcal{N}$ of sequences $\mathbf{X}$ with $P(\mathcal{N}) = 0$ and such that

$$\sup_t |F_n(t, \mathbf{X}) - F(t)| \to 0$$

as $n \to \infty$ for all $\mathbf{X} \notin \mathcal{N}$.

Proof. We show first that such a common set $\mathcal{N}$ exists for convergence $F_n(t)$ to $F(t)$ for all t (convergence being not necessarily uniform). Let A be the set of all points at which F is discontinuous. Since F is nondecreasing and continuous on the right, the condition $t \in A$ means that $F(t) - F(t-) > 0$, where

$$F(t-) = \lim_{\tau \to t^-} F(\tau) = \sup_{\tau < t} F(\tau).$$

If $t \in A$, we define the random variables

$$U_n(t) = U_n(t, \mathbf{X}) = \begin{cases} 1 & \text{if } X_n = t \\ 0 & \text{if } X_n \neq t, \end{cases} \tag{16.4}$$

so that $E[U_n(t)] = P\{U_n(t) = 1\} = F(t) - F(t-)$. Clearly, the random variables $U_1(t), U_2(t), \dots$ are iid, and by the strong law of large numbers, we have

$$\frac{1}{n} \sum_{j=1}^{n} U_j(t, \mathbf{X}) \to F(t) - F(t-)$$

for all $\mathbf{X} \notin \mathcal{N}_t^*$, where $P(\mathcal{N}_t^*) = 0$. Here $\mathcal{N}_t^*$ is the "exceptional" set chosen for specific t.

Clearly, the set A is at most countable. Indeed, if $A_k \subset A$ is the set of points $t \in A$ with $F(t) - F(t-) \geq 1/k$, then A_k has at most k elements because of monotonicity of F and the condition $F(\infty) - F(-\infty) = 1$. Since $A = \bigcup_k A_k$, the set A is at most countable, and the condition (16.4) holds for all $\mathbf{X} \notin \bigcup_{t \in A} \mathcal{N}_t^*$, with

$$P(\bigcup_{t \in A} \mathcal{N}_t^*) \leq \sum_{t \in A} P(\mathcal{N}_t^*) = 0.$$

Now let Q be the set of all rational t (or any other countable dense subset of real line). We may assert that $F_n(t) \to F(t)$ for all $t \in Q$ with probability

1, since by (16.3) we have $F_n(t, \mathbf{X}) \to F(t)$ for all t if $\mathbf{X} \notin \bigcup_{t \in Q} \mathcal{N}_t$, and again $P(\bigcup_{t \in Q} \mathcal{N}_t) \leq \sum_{t \in Q} P(\mathcal{N}_t) = 0$.

Thus, we showed that with probability 1 we have $F_n(t) \to F(t)$ for all $t \in Q$ and $F_n(t) - F_n(t-) \to F(t) - F(t-)$ for all $t \in A$. We shall now show that with probability 1 we also have $F_n(t) \to F(t)$ for all $t \notin A$, and then we show that the convergence is uniform. These last statements do not involve any probability considerations: They are true for any $\mathbf{X} \notin \left(\bigcup_{t \in A} \mathcal{N}_t^*\right) \cup \left(\bigcup_{t \in Q} \mathcal{N}_t\right)$, so that we may from now on suppress the dependence on $\mathbf{X}$ and consider a fixed sequence of cdf's $F_n(\cdot)$.

Thus, let t be a continuity point of F, and let $t \notin Q$. We have to show that $F_n(t) \to F(t)$. Let $\epsilon > 0$ and let $t_1, t_2 \in Q$ satisfy the relations $t_1 < t < t_2$ and

$$F(t_2) - F(t_1) < \frac{\epsilon}{2} \tag{16.5}$$

(which is possible because t is a continuity point of F). Next choose N such that for $n \geq N$ we have

$$|F_n(t_1) - F(t_1)| < \frac{\epsilon}{2} \quad \text{and} \quad |F_n(t_2) - F(t_2)| < \frac{\epsilon}{2}. \tag{16.6}$$

This is possible because $F_n(t_i) \to F(t_i)$ as $n \to \infty$ for $i = 1, 2$. By monotonicity of F we have $F(t_1) \leq F(t) \leq F(t_2)$; hence by (16.5)

$$F(t_2) - \frac{\epsilon}{2} < F(t) < F(t_1) + \frac{\epsilon}{2}.$$

Using (16.5), (16.6), and the monotonicity of $F_n(\cdot)$ we can now write, for $n \geq N$, that

$$F(t) \leq F(t_2) \leq F(t_1) + \frac{\epsilon}{2} \leq \left(F_n(t_1) + \frac{\epsilon}{2}\right) + \frac{\epsilon}{2} \leq F_n(t) + \epsilon$$

and also

$$F(t) \geq F(t_2) - \frac{\epsilon}{2} \geq \left(F_n(t_2) - \frac{\epsilon}{2}\right) - \frac{\epsilon}{2} \geq F_n(t) - \epsilon,$$

which gives

$$|F_n(t) - F(t)| \leq \epsilon.$$

It remains to prove now that the convergence $F_n(t) \to F(t)$ is uniform, that is, $\sup_t |F_n(t) - F(t)| \to 0$ as $n \to \infty$. We shall give the proof after Chung (1971).

Assume the contrary. Then there exists $\epsilon_0 > 0$ and sequences n_k and t_k such that $|F_{n_k}(t_k) - F(t_k)| > \epsilon_0$ for all t. Since $F_n(\cdot)$ and $F(\cdot)$ are cdf's, we cannot have $t_k \to +\infty$ or $t_k \to -\infty$; hence the sequence $\{t_k\}$ is bounded. Without loss of generality we may assume that $t_k \to t^*$.

Now there exists either a subsequence of $\{t_k\}$ which converges to t^* monotonically from below or monotonically from above. Similarly, since $|F_{n_k}(t_k) - F(t_k)| > \epsilon_0$, there exists either a subsequence at which $F_{n_k}(t_k) > F(t_k) + \epsilon_0$ or a subsequence at which $F_{n_k}(t_k) < F(t_k) - \epsilon_0$. Restricting the analysis to these subsequences, we distinguish four cases:

(a) $t_k \uparrow t^*, t_k < t^*, F_{n_k}(t_k) > F(t_k) + \epsilon_0$.
(b) $t_k \uparrow t^*, t_k < t^*, F_{n_k}(t_k) < F(t_k) - \epsilon_0$.
(c) $t_k \downarrow t^*, F_{n_k}(t_k) > F(t_k) + \epsilon_0$.
(d) $t_k \downarrow t^*, F_{n_k}(t_k) < F(t_k) - \epsilon_0$.

We now select $t', t'' \in Q$ such that $t' < t^* < t''$.

In case (a) we may write, for all k sufficiently large, using monotonicity of F_{n_k} and F,

$$\epsilon_0 < F_{n_k}(t_k) - F(t_k) \le F_{n_k}(t^*-) - F(t')$$
$$\le F_{n_k}(t^*-) - F_{n_k}(t^*) + F_{n_k}(t'') - F(t'') + F(t'') - F(t').$$

If we now let $k \to \infty$, the difference $F_{n_k}(t^*-) - F_{n_k}(t^*)$ converges to $-(F(t^*) - F(t^*-))$, and the difference $F_{n_k}(t'') - F(t'')$ converges to 0. Letting $t'' \downarrow t^*, t' \uparrow t^*$ along values in Q, the last difference converges to $F(t^*) - F(t^*-)$. Thus, the right-hand side can be made as small as possible, which gives a contradiction.

In case (b) we may write

$$\epsilon_0 < F(t_k) - F_{n_k}(t_k) < F(t^*-) - F_{n_k}(t')$$
$$\le F(t^*-) - F(t') + F(t') - F_{n_k}(t').$$

Letting $k \to \infty$ we obtain $F(t') - F_{n_k}(t') \to 0$, and letting next $t' \uparrow t^*$ we obtain $F(t^*-) - F(t') \to 0$, which leads to a contradiction.

In case (c) we have

$$\epsilon_0 < F_{n_k}(t_k) - F(t_k) < F_{n_k}(t'') - F(t^*)$$
$$\le F_{n_k}(t'') - F_{n_k}(t') + F_{n_k}(t') - F(t') + F(t') - F(t^*).$$

Now $F_{n_k}(t'') - F_{n_k}(t') \to F(t'') - F(t')$, while $F_{n_k}(t') - F(t') \to 0$. Letting $t'' \downarrow t^*$ and using continuity of F on the right, the term $F(t'') - F(t')$ converges to $F(t^*) - F(t')$, and we again obtain a contradiction.

Finally, in case (d) we may write

$$\epsilon_0 \le F(t_k) - F_{n_k}(t_k) \le F(t'') - F_{n_k}(t^*)$$
$$\le F(t'') - F(t') + F(t') - F_{n_k}(t') + F_{n_k}(t^*-) - F_{n_k}(t^*).$$

As $k \to \infty$ we have $F(t') - F_{n_k}(t') \to 0$, while $F_{n_k}(t^*-) - F_{n_k}(t^*) \to F(t^*-) - F(t^*)$. Now letting $t' \uparrow t^*$ and $t'' \downarrow t^*$, we have $F(t'') - F(t') \to F(t^*) - F(t^*-)$, which also leads to a contradiction.

PROBLEMS

16.2.1   Let X be a random variable such that $P\{X = a\} = \frac{1}{2}, P\{X = b\} = \frac{1}{3}$, and $P\{X = c\} = \frac{1}{6}$. Let $X_1, X_2, \ldots, X_{200}$ be a random sample from the distribution of X. Suppose that among the first 100 observations of X_i, 55 were equal a and 38 were equal b. Among the next 100 observations, 51 were equal to a, and 30 were equal to b. Determine $\sup_t |F_{200}(t) - F(t)|$ and $\sup_t |F_{100}(t) - F(t)|$ in all six cases $a < b < c$, $a < c < b$, $b < a < c$, and so on, as well as all cases such as $b = c < a, a = b < c, \ldots$, and $a = b = c$.

16.2.2   Let $F(t) = 0$ for $t < 0$, $F(t) = p$ for $0 \le t < 1$, and $F(t) = 1$ otherwise. Use the central limit theorem to evaluate directly the distribution of $\sup_t |F_n(t) - F(t)|$ and show that $F_n(t)$ tends to $F(t)$ almost surely and uniformly in t.

16.3   ONE-SAMPLE KOLMOGOROV–SMIRNOV TEST

Let

$$D_n = \sup_t |F_n(t) - F(t)|. \tag{16.7}$$

The fact that the distance D_n converges a.s. to zero if F_n's are empirical cdf's of random samples drawn from distribution F suggests studying the rate of this convergence, that is, finding constants $c_n \to \infty$ such that the sequence of random variables $\{c_n D_n, n \ge 1\}$ has a limiting distribution. Research in this direction led to a remarkable discovery, due to Kolmogorov and Smirnov, that if F is continuous, then this limiting distribution exists for $c_n = \sqrt{n}$, and moreover, it does not depend on F.

The following theorem specifies this limiting distribution and serves as a foundation for a test of the hypothesis that a random sample comes from a specific distribution. We omit the proof here.

Theorem 16.3.1 (Kolmogorov and Smirnov).   *Let F be a continuous cdf, and let $X_1, X_2, \ldots$ be a sequence of iid random variables, with $P\{X_i \le t\} = F(t)$. Then for every $z > 0$,*

$$\lim_{n \to \infty} P\{\sqrt{n} D_n \le z\} = Q(z),$$

where

$$Q(z) = 1 - 2\sum_{k=1}^{\infty}(-1)^{k-1}e^{-2k^2z^2}. \tag{16.8}$$

The function $Q(z)$ is a cdf of a continuous distribution, called *Kolmogorov distribution*. The values of this cdf are given in Table A7.

Suppose now that we observe a random sample $(X_1, \ldots, X_n)$ from a continuous cdf F and we want to test the hypothesis

$$H_0 : F = F_0$$

against the alternative

$$H_1 : F \neq F_0,$$

where F_0 is a continuous cdf.

Under the null hypothesis, for large n, the distribution of $\sqrt{n}D_n$ is given by $Q(x)$. If the true distribution of X_i's is F^*, we may write

$$\sup_t |F_n(t) - F_0(t)| \leq \sup_t \{|F_n(t) - F^*(t)| + |F^*(t) - F_0(t)|\}$$
$$\leq \sup_t |F_n(t) - F^*(t)| + \sup_t |F^*(t) - F_0(t)|$$
$$= D_n + \sup_t |F^*(t) - F_0(t)|.$$

Upon multiplication by $\sqrt{n}$, the first term, $\sqrt{n}D_n$, has the limiting distribution (16.8), while the second term, $\sqrt{n}\sup_t |F^*(t) - F_0(t)|$, tends to infinity if $F^*(t) \not\equiv F_0(t)$.

The last property means that the test should reject H_0 if the observed value of the statistic $\sqrt{n}D_n$ exceeds the critical value determined from the right tail of the distribution (16.8). This test has—in the limit—power 1 against *any* alternative.

Example 16.3.1. A small town had 30 fires last year: on 5 and 18 January, 3, 4, 21, and 26 February, 5, 10, and 13 March, 6 April, 16 and 25 May, 19 June, 10 and 21 July, 12 and 15 August, 1, 8, and 21 September, 2, 6, 7, 19, and 29 November, 3, 9, 12, 17, and 24 December. Are this data consistent, on the 0.05 level, with the hypothesis that the occurrences of fires follow a Poisson process?

SOLUTION. One of the solutions may be based on the fact that if the fires form a Poisson process, then their occurrences are distributed throughout the year according to the uniform distribution. Here the analysis of the data is illustrated in Table 16.1. Since $F_n(\cdot)$ is a step function, it suffices to inspect only the differences $|F(x_i) - F_n(x_i)|$ and $|F(x_i) - F_n(x_{i-1})|$. In Table 16.1 we have

$$F(x_{21}) - F_{30}(x_{20}) = 0.838 - 0.667 = 0.171,$$

Table 16.1

Fire i	Day x_i	$F_n(x_i)$ $=i/30$	$F(x_i)$ $=x_i/365$	Fire i	Day x_i	$F_n(x_i)$ $=i/30$	$F(x_i)$ $=x_i/365$
1	5	0.033	0.014	16	224	0.533	0.614
2	18	0.067	0.049	17	227	0.567	0.622
3	34	0.100	0.093	18	244	0.600	0.668
4	35	0.133	0.096	19	251	0.633	0.688
5	52	0.167	0.142	20	264	0.667	0.723
6	57	0.200	0.156	21	306	0.700	0.838
7	64	0.233	0.175	22	310	0.733	0.849
8	69	0.267	0.189	23	311	0.767	0.852
9	72	0.300	0.197	24	323	0.800	0.885
10	96	0.333	0.263	25	333	0.833	0.912
11	136	0.367	0.373	26	337	0.867	0.923
12	145	0.400	0.397	27	343	0.900	0.940
13	170	0.433	0.466	28	346	0.933	0.948
14	191	0.467	0.523	29	351	0.967	0.962
15	202	0.500	0.553	30	363	1.000	0.995

and an inspection of the data shows that this difference is maximal. Thus, $\sqrt{n}D_n = \sqrt{30} \times 0.171 = 0.937$. Since (see Table A7) we have $P\{\sqrt{n}D_n > 0.94\} = 0.3399$, there is no reason to reject the null hypothesis that the fires form a Poisson process. $\qquad\square$

It is important to realize that the Kolmogorov–Smirnov test applies only when the null hypothesis specifies the distribution F completely. The test cannot be used in cases when, say, the null hypothesis specifies only that F is a normal distribution. One may be tempted to estimate the parameters from the sample, leading to a specific normal distribution F^*, and then apply the test using the statistic $\sqrt{n}D_n^* = \sqrt{n}\sup_x |F_n(x) - F^*(x)|$. The point is that $F^*(x)$ is then a *random* cdf, depending on the same sample as that used to determine F_n, and the limiting distribution of $\sqrt{n}D_n^*$ is *not* given by formula (16.8).

Example 16.3.2. The following example of an application of the Kolmogorov–Smirnov test is of some independent interest. Suppose that we need samples of points which are distributed uniformly on a circle. In practice, a need of this kind may arise when we must evaluate numerically the integral of a (complicated) function defined on the circle in question and must resort to Monte Carlo techniques, estimating the integral as the expected value of the function at a randomly selected point. A more simple example may be that the "function" here is the thickness of a layer of coal, so that "estimating the integral over a circle" is really estimating the total amount of coal under a circular surface.

Anyway, a quality of estimation depends crucially on how good the method is that is used for generating random points in the circle. Without loss of generality, we may assume that the circle is $x^2 + y^2 \leq 1$. Letting ξ_1, ξ_2 be a pair of independent random variables distributed uniformly on $[0, 1]$, we let $x = 2\xi_1 - 1, y = 2\xi_2 - 1$ and then check whether $x^2 + y^2 \leq 1$. If yes, the pair (x, y) is accepted. If not, we sample a new pair (ξ_1, ξ_2), and proceed in this way until we obtain an accepted pair (x, y). One can continue this process to generate as many accepted pairs as needed. The distribution of (x, y) is easily seen to be uniform on a circle with radius 1 centered at the origin as long as ξ_1, ξ_2 are iid uniform on $[0, 1]$. Thus it is only the quality of the random number generator in the software that determines the quality of the resulting random points.

However, some sampling programs attempt to "save computer time": When $x^2 + y^2 > 1$, rather than rejecting both x and y, the value of x is kept, and the computer generates a new ξ_2, transforms it to a new $y = 2\xi_2 - 1$, and tests if we have $x^2 + y^2 \leq 1$ (i.e., once x is sampled, it is retained, and only y is added to it).

For an estimation of savings in computer time and for the distribution of the resulting points (x, y), see Problem 16.3.3. To test the uniformity of distribution, $n = 100$ points were generated on the circle according to the "timesaving" scheme. Two tests were performed, both reducing the problem to a one-dimensional Kolmogorov–Smirnov test. First, measuring the angles from an arbitrary direction, the angles between this direction and the line connecting point (x, y) with the origin should be distributed uniformly on $[0, 2\pi]$. Taking the positive x-axis as the direction chosen, and counting counterclockwise, the values $\varphi_i = \arctan(x_i/y_i) + (\pi/2)[1 - \text{sign}(y_i)]$ should have a uniform distribution on $[0, 2\pi]$. The ordered values φ_i are shown in the first three columns of Table 16.2.

The empirical cdf of these values increases by 0.01 at each $\varphi_{i:100}$, while $F(t) = t/2\pi$. For our data, $\sqrt{n}D_n = \sqrt{n} \sup |F_n(t) - F(t)|$ equals 1.136, which corresponds to the p-value between 0.16 and 0.18 (see Table A7).

On the other hand, if the points have uniform distribution on a circle with radius 1, then the distance R of a random point from the center has cdf $F_R(t) = P\{R \leq t\} = \pi t^2/\pi = t^2$ for $0 \leq t \leq 1$. Thus, the theoretical cdf of distance R is the parabola t^2. The three last columns in Table 16.2 give the ordered distances. We can then compute the values of the Kolmogorov–Smirnov statistic $\sqrt{n}D_n = \sqrt{n} \sup |F_n(t) - t^2|$, which equals 1.737; the corresponding p-value (see Table A7) is less than 0.005.

The question now is: What is the p-value of the observed result? We have performed two tests, one giving a nonsignificant result corresponding to a p-value of 0.15 and the other giving a highly significant result with a p-value of 0.005. One may be tempted to take the lower p-value and on this basis reject the hypothesis of the uniformity of distribution on the circle. Such a procedure, however, amounts in effect to "trying someone twice for the same crime."

Table 16.2

$\varphi_{i:100}$			$R_{i:100}$		
0.0537	1.7784	3.7217	0.0432	0.6757	0.8811
0.0570	1.8854	3.8933	0.1238	0.6969	0.8827
0.0571	1.8967	3.9110	0.1411	0.7045	0.8834
0.1170	1.9798	3.9215	0.1872	0.7145	0.8846
0.1242	2.0339	4.0757	0.2365	0.7168	0.9027
0.1525	2.0721	4.0778	0.2774	0.7268	0.9104
0.3353	2.1705	4.1081	0.3865	0.7312	0.9134
0.4496	2.1833	4.1494	0.4014	0.7326	0.9169
0.4866	2.2170	4.1922	0.4036	0.7365	0.9200
0.5360	2.3099	4.3407	0.4125	0.7462	0.9223
0.5655	2.3482	4.5311	0.4262	0.7547	0.9254
0.5957	2.4588	4.5437	0.4457	0.7557	0.9278
0.6237	2.4618	4.5571	0.4905	0.7833	0.9333
0.6343	2.5279	4.7786	0.4966	0.7836	0.9349
0.7253	2.5874	4.8246	0.5178	0.7924	0.9421
0.8785	2.5840	5.0184	0.5298	0.7967	0.9443
0.8945	2.8752	5.2162	0.5351	0.8175	0.9460
0.9055	2.9165	5.2770	0.5371	0.8268	0.9468
0.9839	2.9489	5.3605	0.5701	0.8290	0.9494
1.0182	2.9534	5.3661	0.5729	0.8311	0.9496
1.1167	2.9606	5.4251	0.5734	0.8337	0.9503
1.1671	2.9669	5.6105	0.5788	0.8412	0.9536
1.2113	2.9774	5.7050	0.5865	0.8453	0.9566
1.3437	3.0322	5.8654	0.5888	0.8472	0.9591
1.4048	3.0663	5.8959	0.6185	0.8476	0.9616
1.4319	3.1596	6.0020	0.6232	0.8485	0.9621
1.4435	3.2035	6.0355	0.6251	0.8491	0.9687
1.4962	3.3165	6.1418	0.6261	0.8521	0.9709
1.5223	3.3226	6.1496	0.6279	0.8632	0.9795
1.5231	3.4023	6.1781	0.6355	0.8656	0.9815
1.5716	3.4204	6.2117	0.6509	0.8669	0.9880
1.5774	3.4407	6.2705	0.6556	0.8692	0.9988
1.6020	3.4981		0.6578	0.8713	
1.7543	3.5595		0.6717	0.8724	

To see what is involved here, suppose that instead of the p-value we just carry on two tests on significance levels α_1 and α_2, respectively. The decision procedure is

$$\text{"reject } H_0 \text{ if either test 1 or test 2 rejects it"} \qquad (16.9)$$

The probability of a type 1 error therefore equals

$$\alpha = P\{\text{test 1 rejects or test 2 rejects}|H_0\}$$
$$= P\{\text{test 1 rejects}|H_0\} + P\{\text{test 2 rejects}|H_0\}$$
$$-P\{\text{both tests reject}|H_0\}$$
$$= \alpha_1 + \alpha_2 - P\{\text{both tests reject}|H_0\}.$$

If test 1 and test 2 are performed on two independent samples, then the attained significance level of procedure (16.9) is

$$\alpha = \alpha_1 + \alpha_2 - \alpha_1 \alpha_2. \tag{16.10}$$

In the present case, testing was performed on the same data set. Normally, the determination of significance level in such case is very difficult, since it requires knowledge of the joint distribution of two statistics used in test 1 and test 2.

In the case under consideration, however, the situation is simple, since φ and R are independent (under null hypothesis), and therefore formula (16.10) applies giving the result 0.154.

PROBLEMS

16.3.1 It is known that the observed values $x_1, \ldots, x_n$ satisfy the condition $\frac{1}{3} \le x_i \le \frac{2}{3}$ for all i, and the null hypothesis that the distribution is uniform on $[0, 1]$ is not rejected by the Kolmogorov–Smirnov test on significance level $\alpha = 0.05$. What can one say about n?

16.3.2 Suppose that the data are as in Example 16.3.1, except that there were only 25 fires, none of them in November. Test that the fires occur according to a Poisson process.

16.3.3 Find the joint density of (X, Y) resulting from the "time saving scheme" of Example 16.3.2. Also find the expected number of random variables ξ_i necessary to sample in order to obtain one pair (X, Y) under both schemes.

16.4 TWO-SAMPLE KOLMOGOROV–SMIRNOV TEST

One of the commonly occurring problem in statistics is to determine whether or not two samples come from a population with the same distribution. This is the two-sample problem; we encountered its special cases on some occasions (e.g., in considering the Student test for equality of two normal distributions with the same variance). We now consider this problem in its generality, under the only assumption that the samples are drawn from continuous distributions. Thus, let $X_1, \ldots, X_m$ and $Y_1, \ldots, Y_n$ denote two independent random samples, with $P\{X_i \leq t\} = F(t)$ and $P\{Y_j \leq t\} = G(t)$, where F and G are two continuous cdf's. The objective is to test the hypothesis

$$H_0 : F = G$$

against

$$H_1 : F \neq G.$$

As opposed to the one-sample case of Section 16.3, the null hypothesis is now composite. Nevertheless, it is still possible to find a statistic that can serve as a basis of the test for any simple hypothesis contained in H_0. We have the following theorem, also due to Kolmogorov and Smirnov.

Theorem 16.4.1. *Let $\mathbf{X}_m = (X_1, \ldots, X_m)$ and $\mathbf{Y}_n = (Y_1, \ldots, Y_n)$ be iid random variables with a continuous cdf, and let $F_m(t)$ and $G_n(t)$ be empirical cdf's of $\mathbf{X}_m$ and $\mathbf{Y}_n$, respectively. Furthermore, let*

$$D_{n,m} = \sup_t |F_m(t) - G_n(t)|.$$

Then we have

$$\lim_{m,n \to \infty} P\left\{ \sqrt{\frac{mn}{m+n}} D_{m,n} \leq t \right\} = Q(t),$$

with $Q(t)$ given by (16.8).

We omit the proof. Let us comment only about the use of Theorem 16.4.1 for testing H_0 against H_1. We know that if H_0 is true, then the statistic $\sqrt{[mn/(m+n)]}D_{m,n}$ has a limiting distribution given by cdf $Q(t)$, regardless of which particular cdf governs the sampling of X_i's and Y_j's. Suppose now that H_0 is not true. As $m, n \to \infty$, we have $F_m(t) \to F(t)$ and $G_n(t) \to G(t)$ almost surely and uniformly in t, by the Glivenko–Cantelli theorem. We then have

$$D_{m,n} = \sup_t |F_m(t) - G_n(t)|$$

$$= \sup_t |F_m(t) - F(t) + F(t) - G(t) + G(t) - G_n(t)|$$

$$\leq \sup_t |F_m(t) - F(t)| + \sup_t |F(t) - G(t)| + \sup_t |G_n(t) - G(t)|.$$

We can then write

$$\sqrt{\frac{mn}{m+n}}D_{m,n} \le \sqrt{\frac{m}{1+m/n}}\sup_t |F_m(t) - F(t)|$$

$$+\sqrt{\frac{mn}{m+n}}\sup_t |F(t) - G(t)|$$

$$+\sqrt{\frac{n}{1+n/m}}\sup_t |G_n(t) - G(t)|$$

$$\le \sqrt{m}\sup_t |F_m(t) - F(t)|$$

$$+\sqrt{\frac{mn}{m+n}}\sup_t |F(t) - G(t)|$$

$$+\sqrt{n}\sup_t |G_n(t) - G(t)|.$$

The two extreme terms in the last expressions have limiting distributions, while the middle term tends to infinity for any single hypothesis contained in the alternative. Thus, again, the test rejects the null hypothesis when the value of the statistic $\sqrt{[mn/(m+n)]}D_{m,n}$ is large enough.

Example 16.4.1. Is a Poisson process observed at every other event also a Poisson process? We know, of course, that the answer is negative: If we observe every other event in a Poisson process, then the interevent times are sums of two exponential random variables, hence are not exponential. Let us verify this fact empirically. Table 16.3 gives $m = 12$ interarrival times in a Poisson process with mean 1 (where every event is observed) and $n = 16$ interarrival times for every other event in another Poisson process with mean $\frac{1}{2}$ (so the mean interarrival times are the same). The left column gives the observed interarrival times in both samples, jointly ordered, while the next

Table 16.3

T_i	$F_m(\cdot)$	$G_n(\cdot)$	T_i	$F_m(\cdot)$	$G_n(\cdot)$
0.049	0.083	0.000	0.942	0.750	0.375
0.198	0.166	0.000	0.969	0.750	0.437
0.237	0.250	0.000	1.033	0.833	0.437
0.259	0.333	0.000	1.094	0.833	0.500
0.310	0.416	0.000	1.375	0.833	0.562
0.352	0.500	0.000	1.392	0.833	0.625
0.381	0.500	0.062	1.555	0.833	0.687
0.546	0.500	0.125	1.625	0.833	0.750
0.547	0.500	0.187	1.697	0.833	0.812
0.569	0.583	0.187	2.019	0.916	0.812
0.801	0.583	0.250	2.065	1.000	0.812
0.803	0.666	0.250	2.114	1.000	0.875
0.878	0.750	0.250	2.244	1.000	0.937
0.895	0.750	0.312	2.534	1.000	1.000

two columns give the values of $F_m(t)$ and $G_n(t)$ at the observed points [$F_m(t)$ increases at points from the first sample, while $G_n(t)$ increases at points from the second sample]. It may be seen, therefore, that the six shortest interarrival times are all in a Poisson process observed at every event. Here the value $D_{m,n}$ is 0.5 and the statistic $\sqrt{[mn/(m+n)]}D_{m,n}$ equals 1.309, corresponding to a p-value of about 0.065.

PROBLEMS

16.4.1 What is the minimal possible value of the statistic $D_{m,n}$ if k values of X_i precede the third in magnitude value Y_j? If $m = 100, k = 30$, and $n = 200$, is there enough evidence to reject (on the level $\alpha = 0.05$) the null hypothesis that the distributions of X's and Y's are the same?

16.4.2 Suppose that of 30 fires in Example 16.3.1 those on 5 and 18 January, 3 and 21 February, 10 March, 6 April, 25 May, 19 June, and 3 December were caused by arson, while in the remaining cases arson was excluded. Use the Kolmogorov–Smirnov statistic to test the hypothesis that the occurrence of "arson" and "nonarson" fires within a year follow the same distribution.

16.4.3 Suppose that one sample contains $2m$ data points, while the other contains $2m + k$ data points. The first $2m$ and the last $2m$ data points in the joint sample alternate between samples. Thus, the ordered data have the form

$$\underbrace{YX \cdots YX}_{2m} \underbrace{YY \cdots Y}_{k} \underbrace{XY \cdots XY}_{2m}. \qquad (16.11)$$

For given m find $k = k(m)$ for which the Kolmogorov–Smirnov test will reject the hypothesis that both samples are drawn from the same population (use $\alpha = 0.05$).

16.4.4 Solve Problem 16.4.3 if the string of k consecutive Y's occurs at the beginning of the joint ordering.

16.4.5 Assume that each of two samples contains $2m + k$ elements with the following ordering:

$$\underbrace{YX \cdots YX}_{2m} \underbrace{YY \cdots Y}_{k} \underbrace{XY \cdots XY}_{2m} \underbrace{XX \cdots X}_{k}. \qquad (16.12)$$

For given m find $k = k(m)$ for which the Kolmogorov–Smirnov test will reject the hypothesis that both samples are drawn from the same population (use $\alpha = 0.05$).

16.5 SOME ONE-SAMPLE RANK TESTS

We begin this section by presenting the Wilcoxon signed rank test. This test, used for testing hypotheses about a location parameter in symmetric distributions (median), is an excellent example of the simplicity and versatility of nonparametric methods, often combined with amazing elegance.

Assume that we have a random sample $(X_1, \ldots, X_n)$ from a continuous distribution. Assume also that this distribution is symmetric; that is, there exists a point θ such that for all $x > 0$ we have

$$F(\theta - x) = 1 - F(\theta + x),$$

where F is a cdf. Equivalently, we may say that the density f of X_i's satisfies the condition

$$f(\theta + x) = f(\theta - x)$$

for every x.

We shall present a test of the null hypothesis $H_0 : \theta = \theta_0$ against either the one-sided alternative $H_1 : \theta > \theta_0$ or the two–sided alternative $H_1 : \theta \neq \theta_0$. We can use this test also for $H_0 : \theta \leq \theta_0$ against $H_1 : \theta > \theta_0$. The case of null and alternative hypotheses involving opposite inequalities can be obtained by an obvious change of signs. To define the test statistic, consider the absolute differences $V_1 = |X_1 - \theta_0|, V_2 = |X_2 - \theta_0|, \ldots, V_n = |X_n - \theta_0|$. Since the underlying distribution is continuous, we may assume that all V_i's are distinct and that none equals 0.

Let us arrange V_i's in increasing order and assign ranks $R_1, R_2, \ldots, R_n$ to them, with rank 1 assigned to the smallest V_i. Furthermore, let

$$\eta_i = \begin{cases} +1 & \text{if } X_i > \theta_0 \\ -1 & \text{if } X_i < \theta_0. \end{cases}$$

The *Wilcoxon signed rank statistic* is defined as

$$S_n = \sum_{i=1}^{n} \eta_i R_i.$$

To construct a test for any of the hypotheses mentioned above, we need to:

1. Find the distribution (or at least, limiting distribution as n becomes large) of the statistic S_n under the null hypothesis.
2. Study the effect of values of θ in the alternative hypothesis on the values of S_n.

Suppose therefore that the true value of θ is θ_0. In this case the signs η_i are equally likely to be positive or negative:

$$P\{\eta_i = 1\} = P\{\eta_i = -1\} = \frac{1}{2}.$$

Moreover, the random variables $\eta_1, \ldots, \eta_n$ are independent (since each is determined by a different X_i), and also η_i is independent of R_i, by symmetry of the distribution of X_i about θ_0. The values $R_1, \ldots, R_n$ form a permutation of numbers $1, \ldots, n$ so that we can write

$$S_n = \sum_{i=1}^{n} i \eta_i.$$

Since $E(\eta_i) = 0$, $\text{Var}(\eta_i) = E(\eta_i^2) = 1$, we also have

$$E(S_n) = 0$$

and

$$\text{Var}(S_n) = \sum_{i=1}^{n} i^2 \, \text{Var}(\eta_i) = \sum_{i=1}^{n} i^2 = \frac{n(n+1)(2n+1)}{6}.$$

It is possible, though tedious, to determine the distribution of S_n for small values of n; the exact distribution of S_n can be found in almost any sufficiently large collection of statistical tables. For large n one can prove that $S_n / \sqrt{\text{Var}(S_n)}$ has a limiting standard normal distribution.

Indeed, S_n is the sum of n independent random variables $\eta_1 + 2\eta_2 + \cdots + n\eta_n$, where $i\eta_i = \pm i$ with probability $\frac{1}{2}$ each. To apply the Lindeberg–Feller theorem (10.5.5), we have $\text{Var}(i\eta_i) = E[(i\eta_i)^2] = i^2$, so that $s_n^2 = \text{Var}(S_n) = \sum_{i=1}^{n} i^2 \sim n^3$. Since $E(i\eta_i) = 0$ for $i = 1, 2, \ldots$, we have to show that

$$\frac{1}{s_n^2} \sum_{i=1}^{n} \int_{|x| \geq \epsilon s_n} x^2 \, dF_i(x) \to 0$$

for every $\epsilon > 0$, where F_i is the cdf of $i\eta_i$. Since $s_n \sim n^{3/2}$, we shall have $\epsilon s_n > n$ for n large enough, and each integral will be equal to 0, since $i\eta_i$ is either i or $-i$, hence $|i\eta_i| \leq n$. This shows that the Lindeberg–Feller condition is satisfied and the proof is complete.

To determine now whether to use the right tail, left tail, or both tails of distribution of S_n (limiting or exact), observe that if the true value of θ exceeds θ_0, then

$$P\{\eta_i = +1\} = P\{X_i > \theta_0\} > P\{X_i > \theta\} = \frac{1}{2}.$$

Consequently, positive signs are more likely than negative ones, and this will tend to increase the value of S_n. Thus, in testing $H_0 : \theta = \theta_0$ or $H_0 : \theta \leq \theta_0$ against $H_1 : \theta > \theta_0$, large values of S_n are evidence for the alternative, and we should use the right tail as the critical region.

Table 16.4

i	$|Y_i - 350|$	η_i	rank	i	$|Y_i - 350|$	η_i	rank
1	35	−1	5	11	68	+1	12
2	143	+1	16	12	190	+1	20
3	16	+1	4	13	65	−1	11
4	59	−1	9	14	10	+1	2
5	151	+1	17	15	76	+1	13
6	153	+1	18	16	125	+1	15
7	38	+1	6	17	14	−1	3
8	176	+1	19	18	105	+1	14
9	42	−1	7	19	49	−1	8
10	60	+1	10	20	9	+1	1

Example 16.5.1. Asume that f is a density symmetric about 0, that is, satisfies the condition $f(x) = f(-x)$ for all x. Let $X_1, \ldots, X_n$ be a random sample from distribution with density f, and suppose that we observe the values $Y_i = \theta + X_i, i = 1, \ldots, n$. Thus, θ is the median of Y_i (and also its mean, if $E|X_i| < \infty$). Let the observed values of $n = 20$ observations of Y_i be 315, 493, 366, 291, 501, 503, 388, 526, 308, 410, 418, 540, 285, 360, 426, 475, 336, 455, 301, 359. We want to test the hypothesis that the median θ satisfies the inequality $\theta \le 350$, against the alternative $\theta > 350$. The consecutive values $|Y_i - 350|$, the signs η_i of the differences $Y_i - 350$, and ranks R_i are listed in Table 16.4.

The value of the statistic S_{20} equals 124. The asymptotic variance of S_{20} is $20 \cdot 21 \cdot 41/6 = 2870$, and therefore the observed value of statistic $S_{20}/\sqrt{\text{Var}(S_{20})}$ is 2.31. The corresponding p-value is 0.0104, and we may conclude that the median of the population exceeds 350. ☐

Wilcoxon one-sample test can also be applied for comparison of two related populations. This setup was introduced already in Chapter 13, when the test for the mean difference in paired data was developed.

Example 16.5.2 (Paired Data). When comparing the effects of two treatments A and B, part of variation in the data is caused by other factors (such as patient's age, gender, state of health, etc., in medical experiments). To eliminate, or at least significantly reduce that variability, the experimenter should apply treatments A and B to the same, or very similar ("almost the same") subjects. Therefore the data will concern here pairs, either formed naturally (e.g. twins), or carefully matched as closely as possible (as regards age, gender, etc.). It is important to realize that the pairing process is not random: In fact, the strength of the final conclusion depends largely on how well the members of each pair are matched on as many factors as possible.

Next, within each pair, one allocates one member to treatment A and the other to treatment B, this allocation being random. Now let X_i and Y_i denote the results of treatment A and B, respectively, for members of the ith pair of subjects, and let $Z_i = X_i - Y_i$. If the treatments do not differ, then Z_i is as likely to be positive as negative (i.e., its median is zero) and will have a symmetric distribution, because members of the pair are matched and treatments are allocated at random to members of the pair. Thus, the hypothesis "treatments do not differ" and "treatment A is superior" are now expressed as $H_0 : \theta = 0$ and $H_1 : \theta > 0$. The Wilcoxon signed rank procedure can be used to test these hypotheses. □

We present here one more procedure, called the runs test. In case of a single sample, it may be used to test the hypothesis that the sample elements were selected randomly. In case of two samples, it may be used to test the equality of the two underlying distributions. To illustrate the type of questions we shall be analyzing in one sample case, suppose that we took a sample of size $n = 10$ and observed that the first 5 observations were all negative while the last 5 observations were all positive. Can this be taken as an indication that the process of taking the sample was not random?

The general idea of run test is as follows. Let us partition the observations $X_1, \ldots, X_n$ into two classes, say A and $B = A^c$, in such a way that the partition is induced by the values of X_i only, not by their order (e.g., A may be the set of observations that are positive, observations that exceed the hypothesized median, etc.). Formally, this means that if the vector $(X_1, \ldots, X_n)$ leads to a choice of observations with indices $i_1, \ldots, i_k$ to form the set A and $\pi(1), \ldots, \pi(n)$ is a permutation of indices $(1, \ldots, n)$, then the choice of set A from the permuted vector of observations $(X_{\pi(1)}, \ldots, X_{\pi(n)})$ is the set with indices $\pi(i_1), \ldots \pi(i_n)$. The randomness of the sample, combined with the fact that with probability 1 that there are no ties among the sample values, implies that each set of indices of appropriate size is equally likely to be the set A of the partition.

In a special case, we may have two samples, $X_1, \ldots, X_m$ from a distribution F and $Y_1, \ldots, Y_n$ from a distribution G, and we may wish to test the null hypothesis that $F \equiv G$ (i.e., samples come from the same distribution). We then consider the joint sample of X's and Y's and the partition corresponds to the two constituent samples. In this case, if the null hypothesis is true, each arrangement of X's and Y's is equally likely.

The test in now based on the intuitive idea that if the sampling is really random (or, in the two-sample case, if samples come from the same distribution), then the partition into sets A and $B = A^c$ is "random." One of the possible measures of deviations from randomness is to observe the number of runs. To fix the ideas, imagine that we have m symbols A and n symbols B arranged in some order, such as

$$BB\underline{A}B\underline{A}BAAA\underline{BB}A\underline{BB}$$

(so that we have $m = 5$ and $n = 7$). A run is a string of elements of one kind, bordered by either elements of the other kind or by the end of the string. For instance, the string above has four runs of elements B, as underlined, and three runs of elements A (not underlined). The test is based on the intuitive expectation that too small number of runs, such as

$$AAAAABBBBBBB$$

or too large number of runs, such as

$$BABABABABABB,$$

indicate lack of randomness.

To develop the test, one needs the distribution of the number of runs under the null hypothesis. Let R be the total number of runs. In the sequel we shall assume that $m \le n$. Clearly, the smallest value of R is 2, while the largest possible value of R is obtained for alternating runs of A's of length 1. If $m < n$, then the maximal value of R is therefore $2m + 1$, while if $m = n$ it is $2m$.

It remains to determine the probabilities $P\{R = r\}$ for all possible values of r. Consider first the case when r is even, say $r = 2k$. The sequence must then contain k runs of each kind, which alternate, starting either with a run of A's or with a run of B's. Let us imagine m elements A arranged in a string. Dividing it into k runs means choosing $k - 1$ out of $m - 1$ places separating consecutive A's. This can be done in $\binom{m-1}{k-1}$ ways. In a similar way, the string of B's of length n can be divided into k runs in $\binom{n-1}{k-1}$ distinct ways.

A joint string with $2k$ runs is now formed by dividing A's and B's into k runs each, as described above, and joining them by taking alternating runs. For instance, suppose that $k = 3, m = 5$, and $n = 7$. The string of five A's can be partitioned into three runs in $\binom{4}{2} = 6$ ways, namely

$$\begin{array}{ccc}
AAA & A & A \\
AA & AA & A \\
AA & A & AA \\
A & AA & AA \\
A & AAA & A \\
A & A & AAA.
\end{array}$$

In a similar way, a string of seven B's can be divided into three strings in $\binom{6}{2} = 15$ ways. Taking one such partition for A's and one for B's, we obtain two arrangements giving $r = 2 \cdot 3 = 6$ runs. For example, taking the partition $(AAA|A|A)$ and $(B|BBBB|BB)$, one obtains two arrangments:

$$AAABABBBBABB \quad \text{and} \quad BAAABBBBABBA.$$

Consequently,

$$P(R = 2k) = \frac{2\binom{m-1}{k-1}\binom{n-1}{k-1}}{\binom{m+n}{m}}, \tag{16.13}$$

where the denominator gives the total number of arrangements of m objects A and n objects B. For $r = 2k + 1$ we must have either k runs of A's and $k + 1$ runs of B's, or vice versa. Reasoning analogous to that used in obtaining (16.13) leads to

$$P(R = 2k + 1) = \frac{\binom{m-1}{k-1}\binom{n-1}{k} + \binom{m-1}{k}\binom{n-1}{k-1}}{\binom{m+n}{m}}. \tag{16.14}$$

Formulas (16.13) and (16.14) therefore give the distribution of the number of runs R.

Example 16.5.3. A machine produces items, whose nominal diameter is c. Because of inherent variability, the diameters of items produced are random, sometimes above c and sometimes below it. The machine was designed in such a way that the diameter of the each item produced has no effect on the diameter of the next one. The diameters of 12 consecutively produced items were recorded and classified as "above c" (A) or "below c" (B). The resulting sequence was $BAAAAABBBBBB$, so that $m = 5, n = 7$, and $R = 3$. The small number of runs led to the suspicion that there may be some systematic low-frequency oscillation in the operation of the machine which tends to produce long runs of items with dimensions above c, followed by long runs of items with dimensions below c.

To test the hypothesis on "total randomness" of dimensions of items against the alternative of "low-frequency oscillation" we must choose the left tail of the distribution of R (low number of runs) as the critical region, or equivalently, determine the p-value of the result observed, that is, $P(R \le 3)$. In this case the calculations are quite straightforward: Using formulas (16.13) and (16.14) we obtain, taking $m = 5, n = 7$, and $k = 1$,

$$P(R \le 3) = P(R = 2) + P(R = 3)$$
$$= \frac{1}{\binom{m+n}{m}} \left\{ 2\binom{m-1}{k-1}\binom{n-1}{k-1} \right.$$
$$\left. + \left[\binom{m-1}{k-1}\binom{n-1}{k} + \binom{m-1}{k}\binom{n-1}{k-1} \right] \right\}$$

$$= \frac{1}{\binom{12}{5}} \left\{ 2\binom{4}{0}\binom{6}{0} + \left[\binom{4}{0}\binom{6}{1} + \binom{4}{1}\binom{6}{0}\right] \right\}$$

$$= \frac{2+6+4}{\binom{12}{5}} = \frac{12}{792} = 0.0152.$$

Thus, observing only 3 runs in this situation is a rather strong indication of a low-frequency oscillation effect. □

Calculations such as these in Example 16.5.3 are cumbersome for large m and n. Fortunately, one can use the normal approximation here. We shall state the results without proof.

First, one can show that

$$E(R) = \frac{2mn}{m+n} + 1$$

$$\text{Var}(R) = \frac{2mn(2mn - m - n)}{(m+n)^2(m+n-1)}.$$

We also have the following theorem.

Theorem 16.5.1. *If $m \to \infty, n \to \infty$ in such a way that $m/n \to \eta$ with $0 < \eta < \infty$, then the random variables*

$$\frac{R - 2m/(1+\eta)}{\sqrt{4\eta m/(1+\eta)^3}}$$

tend in distribution to $N(0,1)$.

Thus, replacing η by m/n, we may expect that the random variable

$$\frac{R - 2mn/(m+n)}{2mn} \sqrt{(m+n)^3} \tag{16.15}$$

has approximate standard normal distribution, provided that m and n are large.

One can show that the approximation is very good for $m, n \geq 20$, and quite acceptable when $m, n \geq 10$.

Example 16.5.4. Returning to Example 16.5.3, we have $m = 5, n = 7$, and using (16.15) we have an approximation

$$P\{R \le 3\} = P\left\{\frac{R - 2 \cdot 5 \cdot 7/(5+7)}{2 \cdot 5 \cdot 7}\sqrt{(5+7)^3} \le \frac{3 - 2 \cdot 5 \cdot 7/(5+7)}{2 \cdot 5 \cdot 7}\sqrt{(5+7)^3}\right\}$$

$$\approx P\left\{Z \le \frac{3 - 5.83}{70} \times 41.57\right\}$$

$$= P\{Z \le -1.68\} = 0.0465$$

As compared with the exact p-value 0.0152, this approximation is not good. This shows that for small sample sizes one should try to determine the exact p-value (by direct evaluation, use of special tables, or an appropriate statistical package).

PROBLEMS

16.5.1 Prove the asymptotic normality of the Wilcoxon signed rank statistic S_n using the Liapunov theorem.

16.5.2 Of 15 data points, one is between 0 and 1, two between -2 and -1, three between 2 and 3, four between -4 and -3, and five between 4 and 5. Use the Wilcoxon signed rank statistic to test the hypothesis that (i) the median is 0; (ii) the median is 1.

16.5.3 Some texts define the Wilcoxon signed rank statistic as $\tilde{S}_n = \sum \tilde{\eta}_i R_i$ where $\tilde{\eta}_i = 1$ if $X_i > \theta_0$ and 0 otherwise. Determine the mean and variance of $\tilde{S}_n$ and show that tests based on S_n and on $\tilde{S}_n$ are equivalent in the following sense: Under a null hypothesis, $S_n = \tilde{S}_n - \tilde{S}_n^*$, where $\tilde{S}_n$ and $\tilde{S}^*$ have the same distribution.

16.5.4 Twelve pairs of subjects, matched within each pair as to age, gender, general state of health, and initial weight, were put on two types of diets. The data on loss of weight (in pounds) after five weeks are as follows:

Pair	1	2	3	4	5	6	7	8	9	10	11	12
Diet A	15	33	21	17	14	25	25	31	18	5	46	11
Diet B	18	17	10	10	32	11	8	26	-3	19	5	8

Use the Wilcoxon signed ranked test to test the hypothesis that both diets have the same effect, against the alternative that diet B is less efficient than diet A. Use $\alpha = 0.01$.

16.5.5 A machine is set to produce items, each with diameter of 1 inch. The diameters of 15 consecutive items produces are 1.11, 1.15, 0.98, 1.11, 1.08, 1.06, 0.97, 0.97, 1.05, 1.02, 0.98, 0.99, 0.96, 1.03, 1.01. Use a runs test, taking 1 inch as a threshold to test the hypothesis that the measurements represent random deviations from the required standard.

16.5.6 Solve Problems 16.4 to 16.4 using a runs test. Compare the results obtained by different methods (use the same $\alpha = 0.05$). Explain the differences if they exist for $m = 10$ and $k = 5$.

16.6 TWO-SAMPLE RANK TESTS

The situation analyzed in this section is the usual setup of two independent random samples, $\mathbf{X} = (X_1, \ldots, X_m)$ and $\mathbf{Y} = (Y_1, \ldots, Y_n)$, with X_i's being iid with a continuous cdf F and Y_i's being iid with continuous cdf G. We want to test the null hypothesis

$$H_0 : F \equiv G$$

against some alternatives, whose form we discuss later.

We present one of the most important of the rank tests, the Wilcoxon–Mann–Whitney two-sample test. The underlying idea, which led to the suggestion of the test statistic and allowed us to derive its distribution under the null hypothesis, is as follows.

Let us combine both samples and then arrange all observations in increasing order:

$$U_1 \leq U_2 \leq \cdots \leq U_{m+n},$$

where each U_i belongs to one of the samples. Because of the assumed continuity of F and G, we may disregard the possibility of ties in the joint sample, so we can assume that all inequalities among U_i's are strict.

Let us assign ranks from 1 to $m + n$ to consecutive elements U_i. Now, if the null hypothesis is true, then the m ranks of elements of the first sample and the n ranks of elements of the second sample are mixed "randomly," in the sense that each of the $\binom{m+n}{m}$ allocations of the m ranks of elements of the first sample has the same probability $1 / \binom{m+n}{m}$.

Wilcoxon suggested using the statistic W_X, defined as the sum of ranks of elements of the sample $(X_1, \ldots, X_m)$ in the joint ordering of both samples. Formally, we may write

$$W_X = \sum_{i=1}^{m+n} i I_X(U_i),$$

where, $I_X(U_i) = 1$ if U_i comes from sample $\mathbf{X}$, and 0 otherwise.

Equivalently, we may use the statistic W_Y, being the sum of ranks of elements of the second sample in the joint ordering. The statistics W_X and W_Y carry the same information. Indeed, we have

$$W_X + W_Y = \sum_{i=1}^{m+n} iI_X(U_i) + \sum_{i=1}^{m+n} iI_Y(U_i) = \sum_{i=1}^{m+n} i(I_X(U_i) + I_Y(U_i))$$

$$= \sum_{i=1}^{m+n} i = \frac{(m+n)(m+n+1)}{2},$$

so that

$$W_Y = \frac{(m+n)(m+n+1)}{2} - W_X.$$

Let us begin with finding the expectation and variance of W_X. We have

$$E(W_X) = E\left(\sum_{i=1}^{m+n} iI_X(U_i)\right) = \sum_{i=1}^{m+n} iE[I_X(U_i)]$$

$$= \sum_{i=1}^{m+n} iP\{I_X(U_i) = 1\}.$$

Since all allocations of the m ranks of elements of the first sample among $m+n$ elements of both samples are equally likely, the probability that the ith ranking element comes from the first sample is $P(I_X(U_i) = 1) = m/(m+n)$. We have, therefore,

$$E(W_X) = \sum_{i=1}^{m+n} i\frac{m}{n+m} = \frac{m(m+n+1)}{2}.$$

The calculation of variance is somewhat more tedious (we leave it as exercise). The result is

$$\text{Var}(W_X) = \frac{mn(m+n+1)}{12}. \tag{16.16}$$

One can also show (we omit the proof) that as $m \to \infty, n \to \infty$, the random variable

$$Z_{m,n} = \frac{W_X - E(W_X)}{\sqrt{\text{Var}\, W_X}} = \frac{W_X - m(m+n+1)/2}{\sqrt{mn(m+n+1)/12}}$$

converges in distribution to the standard normal random variable.

To design a testing procedure, it is now necessary to specify the alternative hypothesis. In other words, we have to determine the class of hypotheses such that if (F, G) belongs to this class, then the values of $Z_{m,n}$ will tend to be larger (or smaller, or larger in absolute value) than the values of standard normal random variable Z.

One of such classes of alternative hypotheses is obtained by taking

$$H_1 : G(t) = F(t - \theta) \text{ for all } t \tag{16.17}$$

for some θ. Taking the derivative, the hypothesis (16.17) can be expressed as

$$H_1 : g(t) = f(t - \theta) \text{ for all } x.$$

To grasp the meaning of this hypothesis and its consequence for W_X, let us consider the case $\theta > 0$. Then $G(t) = F(t - \theta) \le F(t)$, which means that $P\{Y \le t\} \le P\{X \le t\}$; hence the values of X tend to be smaller than values of Y (since whatever the value t, random variable X is more likely to be below t than random variable Y). Consequently, the observations $X_1, \ldots, X_m$ will tend to be located closer to the left end, hence have smaller ranks. Thus, small values of W_X support the alternative hypothesis $H_1: \theta > 0$. The case $\theta < 0$ is analogous, while for the alternative $H_1 : G(t) = F(t - \theta), \theta \ne 0$ one should take the two-sided test.

Actually, the class of alternatives against which the Wilcoxon statistic W_X could be used is larger. Let us agree to say that random variable X is *stochastically larger* than Y if $P\{X \le t\} \le P\{Y \le t\}$ for all t. In the present notation, X is stochastically larger than Y if $F(t) \le G(t)$ for all t. Thus, stochastic dominance of Y by X (or X by Y) is an alternative that will tend to inflate (or decrease) the statistic W_X.

To get a somewhat closer insight into the power properties of the Wilcoxon test, let us mention that at about the same time as Wilcoxon introduced his statistic W_X, Mann and Whitney introduced another statistic, pertaining to the same two-sample problem, namely

$$R = \text{ number of pairs } (X_i, Y_j) \text{ with } X_i > Y_j.$$

The statistic W_X can be written as

$$W_X = \sum_{j=1}^{m} R_j,$$

where R_j is the rank of $X_{j:m}$. This means that R_j equals the number of elements in the combined sample that do not exceed $X_{j:m}$.

By definition, there are j elements in the sample $X_1, \ldots, X_m$ which are less or equal to $X_{j:m}$, so that

$$R_j = j + \text{ number of } Y_i \text{ with } Y_i < X_{j:m}.$$

Consequently,

$$W_X = \sum_{j=1}^{m} [j + \text{number of } i\text{'s with } Y_i < X_{j:m}]$$

$$= \frac{m(m+1)}{2} + R.$$

It follows that

$$E(R) = E(W_X) - \frac{m(m+1)}{2}$$

$$= \frac{m(m+n+1)}{2} - \frac{m(m+1)}{2} = \frac{mn}{2}$$

and

$$\text{Var}(R) = \text{Var}(W_X) = \frac{mn(m+n+1)}{12}.$$

It also follows from asymptotic normality of W_X that as $m, n \to \infty$, the statistic

$$Z_{m,n} = \frac{R - mn/2}{\sqrt{mn(m+n+1)/12}} \tag{16.18}$$

converges in distribution to standard normal. Since the tests based on W_X and on R are equivalent, they became known as Wilcoxon–Mann–Whitney tests.

Observe now that the total number of all possible pairs with elements of the pair coming from different samples is mn. Consequently, R/mn is a consistent estimator of the probability $P\{Y < X\}$, so that as $m \to \infty, n \to \infty$, we have

$$\frac{R}{mn} \xrightarrow{P} P\{Y < X\} = \int_{-\infty}^{+\infty} G(t)f(t) \, dt$$

$$= \int_{-\infty}^{+\infty} [1 - F(t)]g(t) \, dt.$$

Under the null hypothesis $F \equiv G$, we have $P\{Y < X\} = \frac{1}{2}$. If $P\{Y < X\} = \xi \neq \frac{1}{2}$, then we may write, using (16.18),

$$Z_{m,n} = \sqrt{\frac{2mn}{m+n+1}} \left(\frac{R}{mn} - \frac{1}{2} \right)$$

$$= \sqrt{\frac{2mn}{m+n+1}} \left(\frac{R}{mn} - \xi \right) + \sqrt{\frac{2mn}{m+n+1}} \left(\xi - \frac{1}{2} \right).$$

The first term converges in distribution to a standard normal random variable, while the second diverges to $+\infty$ or $-\infty$ if only $\xi \neq \frac{1}{2}$. This shows that in the limit as $m, n \to \infty$, the Wilcoxon–Mann–Whitney test (based on W_X or R) has asymptotic power 1 for all alternatives (F, G) with $\xi \neq \frac{1}{2}$.

Example 16.6.1. Consider two athletes, A and B; one of them is to be selected to represent the country in some competition. Assume that both athletes attained some stable level of proficiency in their discipline, so that their results (in competitions, or their best daily training results, etc.) may be taken as random samples of some random variables, say $X^{(A)}$ and $X^{(B)}$. Furthermore, assume that the discipline is such that tied results are unlikely [e.g., A and B are discus throwers (rather than, say, athletes specializing in the 100-meter dash)]. Suppose we know $n = 15$ results of A and $n = 20$ results of B, none of the results repeating. After arranging these results jointly from worst to the best, we obtain the sequence

A	A	A	B	A	B	B	A	A	A	B	B
1	2	3	4	5	6	7	8	9	10	11	12
B	B	A	B	A	B	B	A	B	B	A	A
13	14	15	16	17	18	19	20	21	22	23	24
B	B	A	B	B	B	A	B	B	B	A	
25	26	27	28	29	30	31	32	33	34	35.	

This means that the three lowest results are of athlete A, fourth is of athlete B, and so on. The best result also belongs to A.

Let us define the "better" athlete (of a pair) as the one who has a better than even chance of defeating the other one. Thus, A is better than B if

$$\xi = P(A \text{ beats } B) > \frac{1}{2}.$$

The ideas that lie at the foundations of the U.S. system of selecting Olympic representation is that in the case under consideration, A is better than B because the best result of A is better than the best result of B. Let us therefore test the hypothesis that

$$\xi = P(A \text{ beats } B) = P(X > Y) \le \frac{1}{2}$$

(A is equal to B, or inferior to B) against the alternative that $\xi > \frac{1}{2}$. The Wilcoxon statistic (sum of ranks of A) is

$$W_X = 1 + 2 + 3 + 5 + 8 + 9 + 10 + 15 + 17 + 20 + 23 + 24 + 27 + 31 + 35 = 230.$$

For $m = 15, n = 20$ we have $E(W_X) = 270$, $\text{Var}(W_X) = 900$, so that the p-value is $P\{Z > (230 - 270)/30\} = P\{Z > -1.33\} = 0.9082$. There is therefore no reason to reject the null hypothesis that A is no better than B, despite the fact that the best result is attained by athlete A. □

PROBLEMS

16.6.1 Prove formula (16.16). [*Hint:* Show first that

$$\mathrm{Var}(I_X(U_i)) = \frac{mn}{(m+n)^2} \quad \text{and} \quad \mathrm{Cov}(I_X(U_i), I_X(U_j)) = -\frac{mn}{(m+n)^2(m+n-1)}$$

for $i \neq j$.]

16.6.2 Samples of sizes $m = 20$ and $n = 10$ from two populations are selected. The numbers $r_k, k = 1, 2, \ldots, 10$, of elements of the first sample that precede the kth element of the second sample are: $1, 1, 2, 4, 4, 6, 8, 9, 11, d$.

    (i) Find the values of the Wilcoxon and Mann–Whitney statistics as a function of d.

    (ii) Find the value of the Kolmogorov–Smirnov statistic as a function of d.

    (iii) Suggest the appropriate alternative hypothesis and determine d for which hypothesis about the same median may be rejected based on the Mann–Whitney or Wilcoxon test.

    (iv) Base the answer for part (iii) on the Kolmogorov–Smirnov test.

16.6.3 Use a runs test for the data of Problem 16.6.2.

16.6.4 Assume that n is odd and that all m elements of the first sample are below the median of the second sample. Find the range of the test statistic and the rejection region for the (i) Wilcoxon test, (ii) Mann–Whitney test, (iii) run test, and (iv) Kolomogorov–Smirnov test for the appropriate alternative hypothesis. Use $\alpha = 0.05, n = 51, m = 20$.

16.7 ONE-WAY ANOVA

Finally, we present one more rank procedure, the Kruskal–Wallis test, generalizing the one-way analysis of variance. Thus, we consider I random samples, of sizes $n_1, n_2, \ldots n_I$, where $I \geq 2$. We let X_{ij} be the jth element ($j = 1, \ldots, n_i$) in the ith sample, and we assume that $X_{i,1}, \ldots, X_{i,n_i}$ are iid random variables with cdf given by

$$F_i(x) = G(x - \theta_i)$$

for all x. Here G is assumed to be a cdf of some continuous random variable. We want to test the null hypothesis

$$H_0 : \theta_1 = \theta_2 = \cdots = \theta_I$$

against the alternative

$$H_1 : \text{ not all } \theta_i \text{ are equal.}$$

If G is a cdf symmetric about 0 [i.e., satisfying $G(-x) = 1 - G(x)$ for all $x > 0$], then the median and the mean (if it exists) of F_i are easily seen to be θ_i, and null hypothesis asserts that the medians (or means) of all populations are the same. The variance need not exist, but the fact that all populations have the same distribution up to a location parameter corresponds to the assumption of homoscedasticity of Chapter 15.

Let us order all observations X_{ij} from smallest to largest, and let R_{ij} be the rank of observation X_{ij} in the joint ordering. Furthermore, let $R_{i+} = \sum_{j=1}^{n_i} R_{ij}$ be the sum of ranks corresponding to elements of the ith sample (clearly, if we have two populations, then R_1 and R_2 correspond to Wilcoxon statistics W_X and W_Y from the preceding section).

The test of null hypothesis H_0 is based on the Kruskal–Wallis statistic

$$B = \frac{12}{n(n+1)} \sum_{i=1}^{I} \frac{R_{i+}^2}{n_i} - 3(n+1), \tag{16.19}$$

where $n = n_1 + \cdots + n_I$ is the total sample size.

We have the following theorem.

Theorem 16.7.1. *If H_0 is true, and all sample sizes $n_1, n_2, \ldots, n_I$ increase to infinity in such a way that $n_i/n \to p_i > 0$ for all i, then the distribution of statistic B converges to the chi-square distribution with $I - 1$ degrees of freedom.*

We shall not give a full proof, but outline the argument, at the same time explaining why the large values of the statistic B are compatible more with the alternative than with the null hypothesis.

First, under the null hypothesis, we have

$$E\left(\frac{R_{i+}}{n_i}\right) = \frac{n+1}{2} \tag{16.20}$$

and

$$\text{Var}\left(\frac{R_{i+}}{n_i}\right) = \frac{(n+1)(n-n_i)}{12n_i}. \tag{16.21}$$

We leave the proof as an exercise. Now letting

$$V_i = \frac{R_{i+}/n - E(R_{i+}/n)}{\sqrt{\text{Var}(R_{i+}/n)}}$$

we may write

$$\sum_{i=1}^{I}\left(1-\frac{n_i}{n}\right)V_i^2 = \sum_{i=1}^{I}\left(1-\frac{n_i}{n}\right)\left[\frac{R_{i+}/n_i - E(R_{i+}/n_i)}{\sqrt{\mathrm{Var}(R_{i+}/n_i)}}\right]^2$$

$$= \sum_{i=1}^{I}\frac{n-n_i}{n}\frac{(R_{i+}/n_i - (n+1)/2)^2}{(n+1)(n-n_i)/12n_i}$$

$$= \frac{12}{n(n+1)}\sum_{i=1}^{I}\frac{R_{i+}^2}{n_i} - 3(n+1). \tag{16.22}$$

We leave the last equation as an exercise.

First, the expression (16.22) shows that any deviation from the null hypothesis will tend to increase B. Thus, the critical region is always the right tail of the appropriate chi-square distribution, as in ANOVA tests under normal assumptions.

Second, one can show that V_i converges in distribution to a standard normal variable, or equivalently, the distribution of R_{i+} is asymptotically normal. Once this fact is established, the proof of the theorem can be completed first by observing that the random variables $R_{1+}, \ldots, R_{I+}$ are constrained by the condition $\sum_{i=1}^{I} R_{i+} = n(n+1)/2$, which reduces the number of degrees of freedom to $I-1$. It then remains to check, for instance, that the asymptotic mean and variance of B agree with those of the chi-square distribution with $I-1$ degrees of freedom.

The proof of the asymptotic normality of R_{i+} lies beyond the scope of this book; it relies on one of the central limit theorems for the sum of exchangeable random variables.

PROBLEMS

16.7.1 Prove relations (16.20) and (16.21).

16.7.2 Show that the statistics B given by (16.19) and (16.22) are equal.

16.7.3 Statistics 102 is a continuation of Statistics 101, taken only by some students who passed Stat 101. In the class of 15 students of Stat 102, five took Stat 101 from instructor X, four took it from instructor Y, and the rest took the course from instructor Z. Ordered according to their performance, the students of the three instructors are

$$Z\,X\,Z\,Z\,Z\,Y\,X\,Z\,Y\,X\,Y\,X\,Z\,X\,Y$$

(this means that the best student was taught by instructor Z, second by instructor X, and so on). On the significance level $\alpha = 0.05$, test the hypothesis that the performance of students in Stat 102 does not depend on who taught them Stat 101.

REFERENCES

Chung, K. L., 1971. *A Course in Probability Theory*, Academic Press, New York.

Hollander, M., and D. Wolfe, 1973. *Nonparametric Statistical Methosa*, John Wiley & Sons, New York.

CHAPTER 17

Analysis of Categorical Data

17.1 INTRODUCTION

The term *categorical data* refers to outcomes of statistical observations made either on a nominal scale, or on a discrete ordinal scale (so that ties are expected to occur often). Typical examples of the nominal scale occur when data represent frequencies of categories of some qualitative attribute (e.g., responses in a questionnaire about state of residence or religious affiliation). The discrete ordinal scale consists of a set of naturally ordered categories. For example, opinion on a specific issue may be classified as "favorable" or "unfavorable"; education achieved may be classified as "high school," "junior college," "four-year college or university," "graduate school"; and so on. In both cases, there is a natural order of categories.

A special case is played by data of binary character: for instance, male vs. female, smoker vs. nonsmoker, and so on. There is no "natural" ordering here, but one can always assign values 0 and 1 (or any other two values) to the categories.

We start from procedures for testing the hypothesis that the data were sampled according to a distribution of some type. An example here may be as follows.

Example 17.1.1. A typical case of data on a nominal scale occurs in genetic experiments. Suppose that we have a gene with two forms, A and a, so that each individual (plant, say) belongs to one of the three categories AA, Aa, or aa. If none of the forms is dominant, the three genotypes can be identified (they coincides with phenotypes). According to genetic theory, the probabilities of three genotypes are $\theta^2, 2\theta(1-\theta)$, and $(1-\theta)^2$, where θ is the unknown frequency of allele A. Suppose that out of 200 plants we have 100 of type AA, 89 of type Aa, and 11 of type aa. Are those data in agreement with genetic theory?

Example 17.1.2. As another example, suppose that we have data on a continuous random variable, which we group into classes (e.g., "below 10," "at least 10 but less than 20," etc.). We want to test the null hypothesis that these data follow a normal distribution.

In Chapter 16 the hypotheses about the distribution were tested by the Kolmogorov–Smirnov test. The present situation, however, differs in two respects. First, instead of individual values we have only the class counts. Second, only the family of distributions (normal) is specified.   □

In the case above, the null hypothesis is composite, that is, asserts that the data follow one of the distributions from a certain family. A similar situation occurs when we are interested is studying relationship between two (or more) variables according to which the data are classified. In such problems the null hypothesis states only that the variables are independent. This is again a composite hypothesis, if the two marginal distributions are not completely specified.

Example 17.1.3. Suppose that we have joint data on nationality and incidence of various types of cancer. In case of any dependence between these two attributes, the next problem is to search for a genetic, dietary, or other reason for lack of independence.   □

In Example 17.1.3 both variables (nationality and type of cancer) were of a nominal character. When both variables are measured on an ordinal scale, one may additionally be interested in the strength of association between variables. Such questions are discussed in the second part of this chapter, where we present selected methods of analysis of ordinal aspects of categorical data. Examples of problems of this kind are very common.

Example 17.1.4. For data measured originally on an ordinal scale and later grouped into classes, the methods of Chapter 16 are not applicable. Suppose that we study a relationship between level of education (classified only as "I, no college"; "II, some college"; and "III, at least four-year college/university") and the frequency of changing jobs (again classified as high or low). The data have the form

Frequency of Job Change	Education		
	I	II	III
Low	n_{11}	n_{12}	n_{13}
High	n_{21}	n_{22}	n_{23}

Here n_{ij} is the frequency of occurrence of a given category in a random sample of persons. We want to test the hypothesis that there exists an association (positive? negative?) between frequency of changes of the job and level of education. The question is to define the concept of "positive association" and to develop methods of testing it.   □

The associations between ordinal variables as exemplified above are often too important to be disregarded by using only nominal methods but too crude or too far from linear to be studied by correlation coefficient.

The theory of categorical data is a domain with a long tradition, the "chi-square test" being one of the oldest examples of statistical procedures that use only frequencies. At the same time, this domain has been developing rapidly during the last two decades, and we shall also try to provide some information about the newer methods in categorical data analysis.*

17.2 CHI-SQUARE TESTS

In this section we present one of the oldest and best known statistical tests. We begin with the case of a discrete distribution concentrated on a finite number of values. Thus, let X be a random variable with possible values $x_1, \ldots, x_r$ and the corresponding probabilities $p_i = P\{X = x_i\}, i = 1, \ldots, r$. Let $X_1, \ldots, X_n$ be a random sample of X, and let N_i be the frequency of value x_i in the sample.

The vector $(N_1, \ldots, N_r)$ will be called the *count vector* of the sample. Since $N_1 + \cdots + N_r = n$, one of the coordinates of the count vector is redundant, and we shall often refer to $(N_1, \ldots, N_{r-1})$ as to the count vector. It is intuitively clear that the count vector carries all necessary information about the distribution of X. Formally, the likelihood of the data is

$$L = Cp_1^{N_1} p_2^{N_2} \cdots p_r^{N_r} = Cp_1^{N_1} p_2^{N_2} \cdots p_{r-1}^{N_{r-1}} (1 - \sum_{i=1}^{r-1} p_i)^{n - \sum_{i=1}^{r-1} N_i},$$

where $C = n!/(N_1! N_2! \cdots N_r!)$. Consequently, the count vector is a jointly sufficient statistic for the vector $(p_1, p_2, \ldots, p_{r-1})$.

One important remark here is that procedures allowing us to make an inference about the distribution $(p_1, p_2, \ldots, p_{r-1})$ of X on the basis of the count $(N_1, \ldots, N_{r-1})$ do not depend on $x_1, \ldots, x_r$. It explains why x_i's need not even be numbers: they may represent qualitative categories, cells in a cross-classification, and so on.

Suppose that we want to test the hypothesis

$$H_0 : p_i = p_i^0, \qquad i = 1, \ldots, r \tag{17.1}$$

against the general alternative

$$H_a : \text{hypothesis } H_0 \text{ is false.}$$

---

* An exhaustive presentation of the field may be found in Agresti (1990).

Here $(p_1^0, \ldots, p_r^0)$ is some fixed probability distribution, where we assume that $p_i^0 > 0$ for $i = 1, \ldots, r$.

Under the null hypothesis, for each i, the count N_i has binomial distribution $\text{BIN}(n, p_i^0)$. The test will be based on the following theorem.

Theorem 17.2.1. *Let $(N_1, \ldots, N_r)$ be the count vector of random sample of size n from multinomial distribution with probabilities $(p_1^0, \cdots, p_r^0)$, so that $p_1^0 + \cdots + p_r^0 = 1$ and $N_1 + \cdots + N_r = n$. Then the statistic*

$$Q^2 = \sum_{j=1}^{r} \frac{(N_j - np_j^0)^2}{np_j^0} \tag{17.2}$$

has the limiting (as $n \to \infty$) χ_{r-1}^2 distribution. The statistic Q^2 is often referred to as Pearson's chi-square.

Proof. We present a simple argument showing that the theorem is true for $r = 2$. In this case we have, remembering that $p_2^0 = 1 - p_1^0$ and $N_2 = n - N_1$,

$$\begin{aligned}
Q^2 &= \frac{(N_1 - np_1^0)^2}{np_1^0} + \frac{(N_2 - np_2^0)^2}{np_2^0} \\
&= \frac{(N_1 - np_1^0)^2}{np_1^0} + \frac{(n - N_1 - n(1 - p_1^0))^2}{n(1 - p_1^0)} \\
&= \frac{(N_1 - np_1^0)^2}{np_1^0} + \frac{(N_1 - np_1^0)^2}{n(1 - p_1^0)} \\
&= \left[\frac{N_1 - np_1^0}{\sqrt{np_1^0(1 - p_1^0)}} \right]^2 \sim Z^2 = \chi_1^2.
\end{aligned}$$

For arbitrary r the algebra is more complicated: The statistic Q^2 is represented as a sum of $r - 1$ squares of random variables, each converging in distribution to a standard normal random variable, and such that their coefficients of correlation tend to zero as n increases. We omit the details.
□

It will be helpful to use the following descriptive notation for the chi-square statistic Q^2:

$$Q^2 = \sum \frac{(\text{observed} - \text{expected})^2}{\text{expected}}.$$

The "observed" stands simply for the counts N_j, while the "expected" in the present case of a simple null hypothesis are np_j^0.

As a practical rule, one obtains a reasonable approximation of the distribution of Q^2 if expected counts are at least 5, and the approximation is good if the expected counts exceed 10. When there are many cells, the approximation is good enough even if few expected frequencies are as small as 1.

To test the null hypothesis $H_0 : p_j = p_j^0, j = 1, \ldots, r$ against the alternative H_a : hypothesis H_0 is false, we need to determine the critical region. Since any violations of the null hypothesis in the chi-square test will tend to increase the value of statistic Q^2, we should take the right tail as the critical region, with $\chi^2_{\alpha, r-1}$ as a critical value.

Example 17.2.1. According to genetic theory, the seeds collected from a field of pink pea should produce plants with white, pink, and red flowers in the proportion 1:2:1. Of 400 plants grown from such seeds, 93 were white, 211 were pink, and 96 were red. Does this result contradict genetic theory?

SOLUTION.   We have here 400 observations of a three-valued random variable; according to the null hypothesis we have $p_1^0 = \frac{1}{4}, p_2^0 = \frac{1}{2}$, and $p_3^0 = \frac{1}{4}$, so that the expected counts np_i^0 are 100, 200, and 100. The observed value of the test statistic Q^2 is

$$Q^2 = \frac{(93 - 100)^2}{100} + \frac{(211 - 200)^2}{200} + \frac{(96 - 100)^2}{100} = 1.255.$$

Since we have here 2 degrees of freedom, even the 10% critical value $\chi^2_{0.1,2} = 4.605$ is not exceeded, and consequently, the data do not provide enough evidence against null hypothesis.   □

As already mentioned, the chi-square test is often used for testing the hypothesis that the data of a continuous type follow a specific distribution. In this case the count vector $(N_1, \ldots, N_r)$ is obtained as follows. First, we divide the range of values of the observed random variable X into r sets $C_1, C_2, \ldots, C_r$ that form a partition (i.e., they are disjoint and cover all possibilities). Typically, the sets C_j are intervals, but this is not necessary; the sets C_j need not be connected, and may consist of a number of noncontiguous intervals. The count N_j, given the sample of n observations of random variable X, is defined as a frequency of observations in set C_j. Clearly, $N_1 + \cdots + N_r = n$. If f is the density of X specified by the null hypothesis, then we have

$$p_j^0 = \int_{C_j} f(x) \, dx, \qquad j = 1, \cdots, r.$$

The test statistic Q^2 depends not only on the sample, but also on the choice of the partition into sets C_j. Sometimes the choice of the partition is "natural," while in other cases the partition is chosen in a rather arbitrary way.

Example 17.2.2. Fox and James (1987) give the following data about the birth signs of 851 prominent chess players:

Capricorn	63	(Lasker, Keres, Chiburdanidze)
Aquarius	79	(Spassky, Bronstein)
Pisces	101	(Fischer, Tarrasch, Geller, Larsen)
Aries	76	(Smyslov, Kasparov, Korchnoi, Portisch)
Taurus	77	(Miles, Nunn, Steinitz)
Gemini	67	(Petrosian, Karpov, Short, Euwe)
Cancer	54	(Morphy, Anderssen)
Leo	67	(Botvinnik)
Virgo	63	(Philidor)
Libra	69	(Rubinstein, Fine)
Scorpio	71	(Capablanca, Alekhine, Nimzowitsch, Tal)
Sagitarius	64	(Reshevsky, Pillsbury).

It seems that Pisces have a significantly higher number of prominent chess players.

We may wish to test the null hypothesis that the birthday is not related to chess talent. In this case, one may expect that the birthdays of $n = 851$ prominent chess players form a random sample from the uniform distribution on the year (which we may conveniently regard as a continuous distribution). The expected count for each sign will be $851 \cdot \frac{1}{12} = 70.917$, and the value of statistic Q^2 will be

$$\frac{(63 - 70.917)^2}{70.917} + \cdots + \frac{(64 - 70.917)^2}{70.917} = 21.53.$$

For 11 degrees of freedom we have $\chi^2_{0.05,11} = 19.675$ and $\chi^2_{0.025,11} = 21.920$, which means that the p-value of the observed result is less than 5% but more than 2.5%. Someone with a firm belief that there is some truth in astrology may take this result as an argument in his favor. A sceptic who is convinced that the configuration of stars and planets at the time of one's birth cannot affect this person's talents will regard the observed result as an example of an error of the first kind. In the present case, the p-value is 0.0283, which means that such a result will occur, on average, about once in 35 times if the null hypothesis is true. More precisely, suppose that we collect data on large sets of people with some talent (e.g., composers, painters, writers, etc.). Even if configuration of stars has no effect on *any* talent, on the average in one out of 35 such studies we may expect to observe the result as being as "significant" as the one for chess players.

Example 17.2.3. The chi-square test can also be used by taking the left tail (indicating a good fit) for detecting whether the data were tampered with. The point is that the p-value gives the probability that in a repetition of the experiment, one would observe worse (in the discrete case, no better) fit than the one actually observed. Thus, if the p-value is close to 1, say 0.99, it means that on the average only once in 100 repetitions may one expect to observe a better fit. That strongly suggests that the data have been "doctored" to make them conform better to the null hypothesis. An interesting example of such a type of inference is provided by Fisher's analysis of data on heredity by G. Mendel. For a detailed explanation, see Freedman et al. (1992). Here we sketch briefly the ideas and results.

Mendel studied laws of inheritance of various characteristics, eventually introducing the concept of gene. A typical experiment of Mendel looks as follows. Plants of genotype AA are crossed with plants of genotype aa. All seeds are then hybrids Aa. A number of such seeds are grown and the plants are cross-pollinated. According to genetic theory, the ratio of plants of genotypes AA, Aa, and aa are 1:2:1. If AA and Aa cannot be distinguished, then the ratio of genotype aa to all others is 1:3, and so on. For example, of 800 plants, about 200 can be expected to be aa. In all Mendel experiments, the observed numbers differ suspiciously little from the expected. For example, suppose that in the last case Mendel would report 205 plants aa out of 800. The chi-square fit of such a result is

$$\frac{(205 - 200)^2}{200} + \frac{(595 - 600)^2}{600} = 0.167,$$

which corresponds to a p-value of about 90%.

A single result with so high a p-value would not be unusual, but when Fisher combined all Mendel's data using the combined chi-square test, the p-value was 0.99996 (i.e., only 4 times out of 100,000 could one expect a better fit). Thus, either Mendel had extraordinary luck, or his data were "beautified" to conform better to his theory. □

In the cases considered above, the null hypothesis was simple; that is, the hypothetical distribution was specified completely. More often, the null hypothesis comprises a class of distributions. We now consider cases of "parametric" character, when the hypothetical distribution of a random variable X depends on some parameter θ (possibly vector valued) in a specified family of distributions. Thus, we assume that X is a discrete random variable, with r possible values $x_1, \ldots, x_r$, and such that

$$P\{X = x_j\} = p_j(\theta),$$

where $\theta = (\theta_1, \ldots, \theta_k) \in H_k$ is a point in a subset H_k of the k-dimensional Euclidean space. Moreover, we assume that the number of classes r satisfies the inequality

$$r \geq k + 2 \tag{17.3}$$

and that $p_j(\theta) > 0$ for all $j = 1, \ldots, r$ and all $\theta \in H_k$.

As before, we assume that we have a random sample of values of X, leading to the count vector $(N_1, \ldots, N_r)$, where $N_1 + \cdots + N_r = n$. This time, we cannot use the test statistic (17.2) because the expected class frequencies $np_j(\theta)$ depend now on the unknown parameter θ. According to Theorem 17.2.2, one may use instead estimated expected frequencies obtained as functions of MLE of θ. Given the count $(N_1, N_2, \ldots, N_r)$ the likelihood of the data is

$$L(\theta; N_1, \ldots, N_r) = [p_1(\theta)]^{N_1} \cdots [p_r(\theta)]^{N_r}. \qquad (17.4)$$

Let $\hat{\theta} = (\hat{\theta}_1, \ldots, \hat{\theta}_k)$ denote the value of the parameter θ which maximizes the likelihood (17.4). We then have the following theorem:

Theorem 17.2.2.   *The statistic*

$$Q^2 = \sum_{j=1}^{r} \frac{[N_j - np_j(\hat{\theta})]^2}{np_j(\hat{\theta})} \qquad (17.5)$$

has, as $n \to \infty$, the limiting chi-square distribution with $r - 1 - k$ degrees of freedom.

We omit the proof which can be found in Cramér (1946).

Example 17.2.4.   Consider a gene with two alleles, A and a. Let the frequency of gene A in the population be θ. Under random mating, the frequencies of individuals of genotypes AA, Aa, and aa are $\theta^2, 2\theta(1 - \theta)$, and $(1 - \theta)^2$. To test the theory, n individuals are randomly selected, and the count of the three genotypes is N_1, N_2, N_3. Then the likelihood of the data is

$$L = [\theta^2]^{N_1}[2\theta(1 - \theta)]^{N_2}[(1 - \theta)^2]^{N_3} = 2^{N_2}\theta^{2N_1+N_2}(1 - \theta)^{N_2+2N_3}.$$

Differentiating $\log L$, we obtain easily the MLE of θ, namely

$$\hat{\theta} = \frac{2N_1 + N_2}{2n}. \qquad (17.6)$$

For a numerical example, suppose that $n = 200, N_1 = 25, N_2 = 10$, and $N_3 = 165$. We then have $\hat{\theta} = \frac{60}{400} = 0.15$. Consequently, using (17.5), we get

$$Q^2 = \frac{(25 - 30)^2}{30} + \frac{(10 - 51)^2}{51} + \frac{(165 - 144.5)^2}{144.5} = 36.70.$$

Since this result exceeds $\chi^2_{0.005,1} = 7.879$, the p-value is less than 0.005, so the evidence against the null hypothesis of random mating provided by such data would be very strong. $\square$

In the case of data of a continuous type and counts resulting from grouping, the MLE's of parameters are typically very hard to obtain. To illustrate the situation, suppose that X has exponential distribution with density $\theta e^{-\theta x}, x > 0$. Given the data $X_1, \ldots, X_n$, the MLE of θ is easy to obtain as $\hat{\theta} = n / \sum X_i = 1/\overline{X}$. To use Theorem 17.2.2, one must, however, partition the positive part of the real axis into sets $C_1 = [0, t_1), C_2 = [t_1, t_2), \ldots, C_{r-1} = [t_{r-2}, t_{r-1})$, and $C_r = [t_{r-1}, \infty)$, where $0 < t_1 < t_2 < \cdots < t_{r-1}$. Letting $t_0 = 0$ and $t_r = \infty$, we have

$$p_j(\theta) = \int_{t_{j-1}}^{t_j} \theta e^{-\theta x}\, dx = e^{-\theta t_{j-1}} - e^{-\theta t_j}, \qquad j = 1, \ldots, r.$$

The likelihood now becomes

$$L = \prod_{j=1}^{r} \left(e^{-\theta t_{j-1}} - e^{-\theta t_j} \right)^{N_j}$$

and $\hat{\theta}$ is the solution of the equation $\dfrac{d}{d\theta} \log L = 0$.

Similarly, in case of the normal distribution with $\theta = (\mu, \sigma^2)$, the MLE of θ given full data is

$$\hat{\mu} = \overline{x}, \qquad \hat{\sigma}^2 = \frac{1}{n} \sum_{i=1}^{n} (x_i - \overline{x})^2.$$

In the case when the observations are grouped and N_j is the count of observations falling in to the interval $[t_{j-1}, t_j)$, the MLE is the solution of the system of equations

$$\frac{\partial \log L}{\partial \mu} = 0, \qquad \frac{\partial \log L}{\partial (\sigma^2)} = 0$$

where

$$L = \prod_{j=1}^{r} \left(\frac{1}{\sigma \sqrt{2\pi}} \int_{t_{j-1}}^{t_j} e^{-\frac{(x-\mu)^2}{2\sigma^2}}\, dx \right)^{N_j}.$$

Again, the solution has to be obtained through a numerical procedure.

In situations such as that above, one is tempted to use Theorem 17.2.2, substituting for $\hat{\theta}$ the MLE's obtained from the complete data, and not the grouped data. The limiting distribution of the statistic Q^2 in such a case is unknown; however the following theorem is true (see Chernoff and Lehmann, 1954).

Theorem 17.2.3. *Assume that the conditions of Theorem 17.2.2 hold, and let $\hat{\theta}^{*2} = (\hat{\theta}_1^*, \ldots, \hat{\theta}_k^*)$ be the MLE of parameter θ based on the observations $X_1, \ldots, X_n$ [and not on the counts $(N_1, \ldots, N_r)$]. Then the statistic*

$$Q^{*2} = \sum_{j=1}^{r} \frac{[N_j - np_j(\hat{\theta}^*)]^2}{np_j(\hat{\theta}^*)} \tag{17.7}$$

satisfies, as $n \to \infty$, the condition

$$P\{\chi^2_{r-1-k} \geq t\} \geq \lim_{n \to \infty} P\{Q^{*2} \geq t\} \geq P\{\chi^2_{r-1} \geq t\}$$

for every $t > 0$. □

We omit the proof. The following comments, however, explain the meaning of this theorem and provide some intuitive justification.

Theorem 17.2.1 asserts that if the null hypothesis is simple (specifies completely the distribution), then the limiting distribution of Q^2 (under the null hypothesis) is χ^2_{r-1}. On the other hand, if the null hypothesis is composite and we have to estimate the expected counts by finding MLE's of the parameters $\theta_1, \ldots, \theta_k$ given the counts, we lose k degrees of freedom.

Now, if we use estimated parameter $(\theta_1, \ldots, \theta_k)$, we do better than when we use frequencies. Consequently, in this case the p-value of the result lies between the p-value under full information about the distribution and the p-value under the "poorest" information, namely when we have to use the counts.

Example 17.2.5. Assume that we have raw data (not given here) for the numbers $x_{1905}, x_{1906}, \ldots, x_{1991}$ of cloudless nights in the last $n = 87$ years at some prospective telescope site. Suppose that it is known that $x_{1905} + x_{1906} + \cdots + x_{1991} = 21,163$ and $x^2_{1905} + \cdots + x^2_{1991} = 5,226,819$. However, the actual data are not available, and instead we have the following counts N_i of years with given numbers of cloudless nights:

Interval	N_i
160 or below	1
161 to 180	3
181 to 200	7
201 to 220	17
221 to 240	18
241 to 260	26
261 to 280	9
281 to 300	4
301 or above	2

We want to test the hypothesis that the number of cloudless nights at the site in question is normally distributed. MLE's of μ and σ^2 computed from original data are

$$\bar{x} = \frac{21,163}{87} = 243.25 \quad \text{and} \quad \hat{\sigma}^2 = \frac{5,226,819}{87} - (243.25)^2 = 907.82;$$

hence $\hat{\sigma} = 30.13$.

The estimated class probabilities are, letting X denote the number of cloudless nights,

$$p_1 = P\{X \le 160\} = \Phi\left(\frac{160 - 243.25}{30.13}\right) = \Phi(-2.76) = 0.0029$$

$$p_2 = P\{160 < X \le 180\} = \Phi(-2.10) - \Phi(-2.76) = 0.015$$

and similarly for subsequent intervals.

The actual and expected counts for the consecutive classes are therefore

Interval	$87p_i$	N_i
160 or below	0.252	1
161 to 180	1.305	3
181 to 200	4.959	7
201 to 220	12.676	17
221 to 240	19.192	18
241 to 260	22.281	26
261 to 280	15.356	9
281 to 300	7.056	4
301 or above	2.612	2

Since the expected class sizes are too small for the first two classes and also for the last class, we combine the first three classes together as well as the last two classes, obtaining following results:

Interval	$87p_i$	N_i
200 or below	6.516	11
201 to 220	12.676	17
221 to 240	19.192	18
241 to 260	22.281	26
261 to 280	15.356	9
281 or above	9.668	6

The observed value of the statistic Q^2 is now 9.278, which gives the p-value between 0.026 (3 df) and 0.098 (5 df). $\qquad\qquad\qquad\qquad\qquad\qquad$ □

We shall now apply the chi-square for testing some hypotheses which can be formulated in the case of contingency tables. The latter give the counts for the data that can be classified into groups according to two or more classification systems. We always assume tacitly that the observations are independent and—unless stated otherwise—identically distributed. If n observations are taken, then we let N_{ij} be the number of observations that belong to the ith class in the first classification system, and to the jth class in the second classification system.

In general, the number of classification variables may be larger, but we consider only the case of two classification systems. The counts can be arranged into a matrix $[N_{ij}]$, where we have $\sum_{ij} N_{ij} = n$. Such a count matrix is called a *two-way contingency table*. Examples of such pairs of classification systems may be gender vs. educational level, smoking status vs. cause of death, and so on.

Let p_{ij} be the probability that a single observation will fall to the cell (i, j); that is, it will belong to the ith category in the first classification, and the jth category in the second classification. The marginal probabilities are then

$$p_{i+} = \sum_j p_{ij}, \quad p_{+j} = \sum_i p_{ij}$$

and the most obvious null hypothesis here is that of independence of the two classifications: $H_0 : p_{ij} = p_{i+}p_{+j}$ for all i, j versus the alternative $H_a : H_0$ is false. Now let $\hat{p}_{ij}$ denote the MLE of the probability p_{ij}, based on the count matrix $[N_{ij}]$. Then the test statistic (17.7) takes on the form

$$Q^2 = \sum_{i=1}^{r} \sum_{j=1}^{c} \frac{(N_{ij} - n\hat{p}_{ij})^2}{n\hat{p}_{ij}}. \tag{17.8}$$

This statistic, according to the Theorem 17.2.2, has the limiting (as $n \to \infty$) chi-square distribution with the number of degrees of freedom $rc - 1 - k$, where k is the number of estimated parameters.

To determine k, and also to find the estimators $\hat{p}_{ij}$, observe that under the null hypothesis H_0 we have (by the invariance property of MLE's) $\hat{p}_{ij} = \hat{p}_{i+}\hat{p}_{+j}$. Now, there are r values of the marginal probabilities p_{i+}, and c values of marginal probabilities p_{+j}, but in each of these marginal distributions one value is a function of others, for instance $p_{r+} = 1 - \sum_{i=1}^{r-1} p_{i+}$ and similarly, $p_{+c} = 1 - \sum_{j=1}^{c-1} p_{+j}$. Thus, the number of estimated parameters is $k = (r - 1) + (c - 1)$, and consequently, the number of degrees of freedom of the limiting distribution of (17.8) is

$$rs - 1 - (r - 1) - (c - 1) = (r - 1)(c - 1). \tag{17.9}$$

Finally, we know that the MLE of a probability in multinomial distribution is the relative frequency, so that

$$\hat{p}_{i+} = \frac{N_{i+}}{n}, \qquad \hat{p}_{+j} = \frac{N_{+j}}{n},$$

where

$$N_{i+} = \sum_j N_{ij}, \quad N_{+j} = \sum_i N_{ij}. \tag{17.10}$$

Thus, $\hat{p}_{ij} = (N_{i+}N_{+j})/n^2$, and substitution in (17.8) gives the following theorem.

Theorem 17.2.4.   *Assume that n independent observations are taken (with replacement) from a population, and each observation is classified according to two attributes, with the number of possible categories of the two attributes being, respectively, $r > 1$ and $c > 1$. Let $[N_{ij}]$ be the count matrix. For testing the hypothesis*

H_0 : *classifications according to the two attributes are independent*

against the general alternative

$$H_a : \text{hypothesis } H_0 \text{ is false,}$$

one can use the statistic

$$Q^2 = \sum_{i=1}^{r} \sum_{j=1}^{c} \frac{(N_{ij} - N_{i+}N_{+j}/n)^2}{N_{i+}N_{+j}/n}, \tag{17.11}$$

which (under H_0) has limiting chi-square distribution with $(r-1)(c-1)$ degrees of freedom.

Any deviations from H_0 tend to increase Q^2, so that the critical region contains values of Q^2 exceeding $\chi^2_{\alpha,(r-1)(c-1)}$.

Naturally, the provision that sampling must be with replacement (which guarantees that observations are identically distributed) can be dropped for sufficiently large populations, since then the difference between sampling with and without replacement becomes negligible.

Example 17.2.6.   The Special Election Issue of *Newsweek* (Nov./Dec. 1992) gives the exit poll results for the 1992 presidential election. One of the tables is the following:

		Clinton	Bush	Perot
White	87%	41%	38%	21%
Black	9%	82%	11%	7%

Although the conclusion seems quite clear,* let us try to analyze this table and test the hypothesis that the preferences for the three candidates are independent of race. The total sample size given is $n = 15,241$ voters, and the margin of error is given as 1.1 percentage points.

Observe first that the marginal percentages of whites and blacks do not

---

* Giving percentages in this case is justified, since the total size of the sample is suficiently large. In many cases, notably in advertising, the percentages can be deceptive: "Sixty percent of doctors prefer Bufferin to aspirin" is quite likely to mean that six doctors out of the *total of 10 asked* stated such a preference, a fact of rather negligible statistical significance, obscured by using percentages.

add up to 100%, which means that the data for 4% of the voters (Hispanics, etc.) were not taken into account. Thus, we may estimate the sample size for our test to be about $0.96 \times 15{,}241 = 14{,}631$. Now the marginal totals for whites and blacks can be estimated as

$$N_{W+} = 14{,}631 \times \frac{0.87}{0.87 + 0.09} = 13{,}259,$$

and therefore $N_{B+} = 1{,}372$.

Next, the percentages in each row add to 100%; this allows us to estimate the counts $N_{W,Clinton}, N_{W,Bush}$, and so on. The entire contingency table takes on the form

	Clinton	Bush	Perot	
White	5,436	5,039	2,784	13,259
Black	1,125	151	96	1,372
	6,561	5,190	2,880	14,631

We can now compute the observed value of the statistic Q^2 given by the formula (17.11). We obtain here $Q^2 = 845.2$, which exceeds the critical value for chi-square distribution with 2 degrees of freedom chosen for any reasonable level of significance α. In this case the results were so obvious that no statistical test was needed: The race and preference are definitely dependent.

Continuing this analysis we may ask what confidence level the pollsters use in announcing their "margin of error." In this case, it is given as ± 1.1 percentage points. Now, we are estimating here the true proportion p on the basis of sample of size $n = 14{,}631$. If X is the number of observations of a given category, then X has a binomial distribution with parameters n and p; hence X is approximately normal $N(np, np(1-p))$. Consequently, the estimated percentage, $100\hat{p} = 100X/n$ has approximate normal distribution with mean $E(100\hat{p}) = E(100\frac{X}{n}) = 100p$ and

$$\text{Var}(100\hat{p}) = \text{Var}\left(\frac{100}{n}X\right) = \frac{100^2}{n^2}\text{Var}(X) = \frac{100^2 p(1-p)}{n}.$$

The $(1-\alpha)$ confidence interval for the mean $100p$ is $100\hat{p} \pm z_{\alpha/2}\sqrt{\text{Var }100\hat{p}}$.

The fact that the error given is the same for all data (instead of depending on the observed perecentage) suggests that an upper bound $p(1-p) \le \frac{1}{4}$ is used. We therefore have the inequality

$$z_{\alpha/2}\sqrt{\frac{100^2 p(1-p)}{n}} \le z_{\alpha/2}\frac{100}{\sqrt{4 \times 14{,}631}}.$$

The right hand side equals 1.1 for $z_{\alpha/2} = 2.575$, which suggests that the pollsters used a 99% confidence level.

PROBLEMS

17.2.1 Show that in case of a 2×2 contingency table, the statistic Q^2 given by (17.11) is proportional to $(N_{11}N_{22} - N_{21}N_{12})^2$, and find the proportionality constant.

17.2.2 Mrs. Smith, who teaches the third grade in an elementary school, classified each student according to whether the grade on the first midterm was below or above the median for this midterm, and then did the same for the second midterm. The results obtained are:

	Second Midterm	
First Midterm	Below	Above
Below	30	5
Above	5	30

She then computed Q^2 and found the p-value significant. Find the p-value. What legitimate conclusion can be made on the basis of this p-value?

17.2.3 A certain type of toy is sold with three batteries included. The number of defective batteries (X) in a random sample of 200 toys were as follows:

X	0	1	2	3
Count	51	92	40	17

Test the hypothesis that the number of defective batteries in a toy has binomial distribution.

17.2.4 Assume that the genders of children in the family are independent. In a human population, the probability that a child is a male is very close to $\frac{1}{2}$. The data of numbers of boys (X) in the random sample of 100 families with four children are as follows:

X	0	1	2	3	4
Count	7	21	40	27	5

Test the hypothesis that the distribution of the number of boys in a family of four children is indeed BIN$(4, \frac{1}{2})$.

17.2.5 Suppose that counts of female offsprings in the animal families of a certain species with four offspring are 3, 8, 28, 40, 21. Test the hypothesis that the corresponding distribution is binomial.

17.2.6 Given the data on numbers of hits of various areas of London by V2 rockets (see Example 9.8.9), test the hypothesis that the numbers of hits have Poisson distribution.

17.3 HOMOGENEITY AND INDEPENDENCE

The chi-square test for independence described in Theorem 17.2.4 concerns the case when the counts N_{ij} arise from cross-classification of independent and identically distributed observations. In many experiments one of classification variables may not be random but controlled by the experimenter. Therefore, such data cannot be treated as a sample from a bivariate distribution.

Example 17.3.1 (Prospective and Retrospective Studies). In most cases of contingency tables, the objective is to analyze the hypothesized relationship between *cause* and *effect* (referred to also as *stimulus* and *response*; *explanatory* and *response* variable; *independent* and *dependent* variable, etc.). Let X and Y denote these variables, with X having r categories (also referred to as *levels* or *treatments*, depending on the context) and Y having c categories.

In prospective studies one selects groups of subjects corresponding to various levels of X, and then classifies each group separately according to levels of Y. For instance, in the social sciences one may be interested in productivity (Y) in its dependence on stress level (X). The data may result from selecting r groups of subjects, exposing the ith group to the ith stress level, and then observing levels of productivity. Consequently, the totals N_{i+} (sizes of groups exposed to levels of stress) are not random. Similarly, in medical research, one may select two groups of patients and then administer the analyzed treatment to one group and placebo to the other group. The response Y can be observed in both groups. Again, the sizes of the groups are not random but are under the control of the experimenter.

In retrospective studies the situation is similar, except that now the marginal counts N_{+j} of response categories are controlled by the experimenter. An example might be provided by typical data on smoking habits and lung cancer. A sample of subjects who died from lung cancer is selected and then compared with a sample (possibly matched with respect to various attributes, such as sex, age, etc.) of subjects who died from other causes. The sizes of these two samples are chosen largely at will. The two samples are then classified according to categories related to smoking. $\square$

For the analysis of contingency tables in which one of the marginal frequency vectors is fixed, let us first introduce the appropriate notation and then formulate the hypothesis to be tested.

The data form, as before, a count matrix $[N_{ij}]$, with marginal counts $N_{i+} =$

$\sum_j N_{ij}$ and $N_{+j} = \sum_i N_{ij}$. The total number of observations is $n = \sum_{ij} N_{ij}$. For the sake of argument, assume that the counts $N_{i+}, i = 1, \ldots, r$, are not random. For each i, the vector $(N_{i1}, N_{i2}, \ldots, N_{ic})$ is assumed to represent the counts from N_{i+} iid observations, sampled from the distribution corresponding to the ith level of the first attribute. The probabilities in this distribution will be denoted by $(p_{1|i}, p_{2|i}, \ldots, p_{c|i})$, where $p_{1|i} + p_{2|i} + \cdots + p_{c|i} = 1$. Here $p_{j|i}$ stands for the probability that the observation will fall into the jth class of the second attribute if the sample is taken from the population of objects with the ith class on the first attribute. Note that $p_{j|i}$ is not a conditional probability as long as i is not random. We use, however, the symbols appropriate for conditional probabilities, since in the special case when i is random, we have the obvious relation

$$p_{ij} = p_{j|i} p_{i+}. \tag{17.12}$$

The hypothesis of interest here is the hypothesis of homogeneity, which may be stated as follows: H_0: The distributions $(p_{1|i}, \ldots, p_{c|i})$ do not depend on i; that is, for each $j = 1, \ldots, c$ we have $p_{j|1} = \cdots = p_{j|r} = p_j$.

As before, we shall obtain a test for H_0 against the general alternative H_a: The hypothesis H_0 is false. Despite the differences between the independence hypothesis from the preceding section and the homogeneity hypothesis above, they are tested by the same statistic. Indeed, for any fixed i, the component of the chi-square sum is

$$Q_i^2 = \sum_{j=1}^c \frac{(N_{ij} - N_{i+}\hat{p}_{j|i})^2}{N_{i+}\hat{p}_{j|i}},$$

where $\hat{p}_{j|i}$ is the MLE of the probability $p_{j|i}$. Clearly, under the null hypothesis of homogeneity, we have

$$Q_i^2 = \sum_{j=1}^c \frac{(N_{ij} - N_{i+}\hat{p}_j)^2}{N_{i+}\hat{p}_j}, \tag{17.13}$$

and $\hat{p}_j = N_{+j}/n$ (i.e., the MLE of p_j is the overall frequency of the jth category in all populations corresponding to various levels i). Adding over i, we obtain the test statistic

$$Q^2 = \sum_{i=1}^r \sum_{j=1}^c \frac{(N_{ij} - N_{i+}N_{+j}/n)^2}{N_{i+}N_{+j}/n}. \tag{17.14}$$

The number of degrees of freedom equals $r(c-1) - k$, where k is the number of estimated parameters. Indeed, each Q_i^2 would have $c - 1$ degrees of freedom if the value p_j were known. The parameters estimated are $p_1, \ldots, p_{c-1}$, so that $k = c - 1$, and we obtain the following theorem.

Theorem 17.3.1. *If the null hypothesis H_0 is true, and if $N_{i+} \to \infty$ for $i = 1, \ldots, r$, then the statistic Q^2 given by (17.14) has the limiting chi-square distribution with $r(c-1) - (c-1) = (r-1)(c-1)$ degrees of freedom.*

Any violation of the null hypothesis will tend to increase the value of Q^2, so that again, the null hypothesis will be rejected if the observed value of the statistic Q^2 exceeds $\chi^2_{\alpha,(r-1)(c-1)}$.

PROBLEMS

17.3.1 Show that the statistic Q^2 for testing independence can be written as

$$Q^2 = \sum_{i=1}^{r} \sum_{j=1}^{c} \frac{N_{ij}^2}{E_{ij}} - n,$$

where $E_{ij} = N_{i+} N_{+j}/n$.

17.3.2 For the 3×4 contingency table

$$
\begin{array}{cccc}
k & k & k & k \\
k & k & 0 & k \\
k & k & k & k
\end{array}
$$

find k such that the hypothesis about independence of two classifications will be rejected at the 0.05 significance level.

17.3.3 For the 3×4 contingency table

$$
\begin{array}{cccc}
5 & 5 & 5 & 5 \\
5 & 5 & k & 5 \\
5 & 5 & 5 & 5
\end{array}
$$

find k such that the hypothesis about independence of two classifications will be rejected at the 0.05 significance level.

17.3.4 For a $2 \times c$ contingency table the homogeneity hypothesis states that the two rows represent samples from the same discrete distribution. Derive the statistic Q^2 for this case.

17.3.5 Professionals from various disciplines were asked about their job-related stress. A random sample of size 100 was selected from each group of professionals (engineers, doctors, and lawyers) and each person was asked to evaluate the level of job-related stress as low, moderate, or high. The results of a study are given below.

	L	M	H
Doctors	5	25	70
Engineers	25	25	50
Lawyers	10	30	60

Specify the hypothesis to be tested, perform the test, and make the appropriate conclusions.

17.3.6 A random sample of 29 university students was selected and each student was classified according to two variables: high school GPA and college GPA. Each variable had the same two categories: "I, below 3.0," and "II, at least 3.0." Formulate a hypothesis to be tested. Perform the test for the data below, and make appropriate conclusions.

High School	College GPA	
GPA	I	II
I	5	3
II	12	9

17.4 CONSISTENCY AND POWER

The chi-square tests of either independence or homogeneity serve as tests against a general alternative asserting simply that the null hypothesis is false. In fact, if the independence hypothesis is not true, then (as the sample size increases) the probability of rejection of the null hypothesis tends to 1. We may rephrase this property by stating that the power of the chi-square test for independence tends to 1 for any simple hypothesis contained in the alternative as the sample size increases. This property of the test is called *consistency*.

The situation is similar in the case of tests for homogeneity, except that now the sample sizes refer to rows (levels of explanatory variable) and are determined not by chance but by the experimenter. Again, if the null hypothesis is not true, then at least two of the rows of the matrix $[p_{i|j}]$ are different. The power of the chi-square test will tend to 1 on a simple alternative in which the rows labeled i_0 and i_0' are different if the sample sizes N_{i_0+} and $N_{i_0'+}$ both tend to infinity.

In general, analysis of the power of chi-square tests for independence or homogeneity appears quite difficult, since it involves determining the exact or limiting distribution of the statistic under specific distributions of the underlying population.

However, if one looks at the main motivation of analysis of power of a test, namely the answer to the question "Which hypothesis should be accepted if one rejects null hypothesis?", then one can suggest the following approach.

When the null hypothesis is not valid, it typically is due to the fact that there is a strong association between some specific values of the two attributes analyzed, while for other values the null hypothesis may be at least approximately satisfied.

In symbols, the null hypothesis (of independence) asserts that all absolute differences

$$|p_{ij} - p_{i+}p_{+j}| \tag{17.15}$$

are zero. If the null hypothesis is not valid, then there exist absolute differences that are strictly positive. What typically happens in such cases is that few of those differences are quite high, while others may be zero or close to zero (rather than all of these differences being small). In these cases one would like to identify the cells for which the differences (17.15) are high.

To present the solution, let us first derive an alternative to the chi-square test, namely the generalized likelihood ratio (GLR) test (see Definition 13.4.2). Consider the case of testing the hypothesis of independence, which states that $p_{ij} = p_{i+}p_{+j}$ for all i, j. The union of H_0 and H_a allows the probabilities p_{ij} to be arbitrary (subject only to the constraint that $\sum_{i,j} p_{ij} = 1$). The data form the contingency table $[N_{ij}]$.

The likelihood of the data is (up to a multiplicative constant) equal to $L = \prod_i \prod_j (p_{ij})^{N_{ij}}$. It is maximized at $\hat{p}_{ij} = N_{ij}/n$, where $n = \sum_{ij} N_{ij}$ is the total sample size. Consequently, the denominator in the GLR, equal to the maximum over all parameter space, is

$$\max_{H_0 \cup H_a} L = \prod_i \prod_j \frac{(N_{ij})^{N_{ij}}}{n^{N_{ij}}} = n^{-n} \prod_i \prod_j (N_{ij})^{N_{ij}}.$$

On the other hand, the likelihood over the null hypothesis equals (up to the same multiplicative constant)

$$L = \prod_i \prod_j (p_{i+})^{N_{ij}} (p_{+j})^{N_{ij}}.$$

This is maximized at $\hat{p}_{i+} = N_{i+}/n$ and $\hat{p}_{+j} = N_{+j}/n$, and we obtain

$$\max_{H_0} L = \prod_i \left(\frac{N_{i+}}{n}\right)^{N_{i+}} \prod_j \left(\frac{N_{+j}}{n}\right)^{N_{+j}}$$

$$= n^{-2n} \prod_i \prod_j (N_{i+})^{N_{i+}} (N_{+j})^{N_{+j}}.$$

The generalized likelihood ratio therefore equals

$$\lambda = \frac{\prod_i (N_{i+})^{N_{i+}} \prod_j (N_{+j})^{N_{+j}}}{n^n \prod_i \prod_j (N_{ij})^{N_{ij}}} = \prod_i \prod_j \left(\frac{N_{i+}N_{+j}}{nN_{ij}}\right)^{2N_{ij}}. \tag{17.16}$$

Under the null hypothesis H_0, the statistic

$$G^2 = -2\log \lambda = 2 \sum_{i=1}^r \sum_{j=1}^c N_{ij} \log \frac{N_{ij}}{(N_{i+}N_{+j})/n} \tag{17.17}$$

has the limiting (as $n \to \infty$) chi-square distribution with the number of degrees of freedom equal to the difference in the number of estimated parameters in the denominator and in the numerator. Thus, the number of degrees of freedom is

$$(rc - 1) - [(r - 1) + (c - 1)] = (r - 1)(c - 1).$$

Symbolically, we may write $Q^2 = \sum_{i,j} (N_{ij} - E(N_{ij}))^2/E(N_{ij})$, where N_{ij} and $E(N_{ij})$ stand for observed and expected (under null hypothesis) counts. On the other hand, using the fact that

$$\log(1 + x) = x + \frac{x^2}{2} + \cdots$$

we may write

$$G^2 = 2 \sum_{i,j} N_{ij} \left\{ \log \frac{N_{ij}}{E(N_{ij})} \right\} = 2 \sum_{i,j} N_{ij} \log \left(1 + \frac{N_{ij} - E(N_{ij})}{E(N_{ij})}\right)$$

$$\approx 2 \sum_{i,j} N_{ij} \frac{N_{ij} - E(N_{ij})}{E(N_{ij})} + \sum_{i,j} \frac{(N_{ij} - E(N_{ij}))^2}{E(N_{ij})} \times \frac{N_{ij}}{E(N_{ij})}.$$

Now, as $n \to \infty$ we have $N_{ij}/E(N_{ij}) \to 1$ in probability, and also $E\left[\sum(N_{ij} - E(N_{ij}))\right] = 0$. This suggests that the first sum is close to zero and the second sum is close to Q^2.

Although the argument above falls short of being a proof, it suggests that Q^2 and G^2 are close to one another for large samples. In fact, one can show that $Q^2 - G^2 \to 0$ in probability as $n \to \infty$. The idea of using the statistic G^2 to investigate the power lies in the additivity property of chi-square distribution, according to which if X has χ_k^2 distribution, and $k = k_1 + k_2 + \cdots + k_s$ is a sum of positive integers, then there exist independent random variables $Y_1, \ldots, Y_s$ such that $X = Y_1 + \cdots + Y_s$ and $Y_i \sim \chi_{k_i}^2$.

In the case under consideration, we have the random variable G^2 given by (17.17), which (under null hypothesis) has the limiting chi-square distribution with $(r - 1)(c - 1)$ degrees of freedom. It may be shown that G^2 can be represented as a sum of independent random variables $G_1^2, \ldots, G_s^2$, each corresponding to a subtable of the original contingency table.

The subtables are obtained by taking a part of the original table and then collapsing some of the categories. The necessary conditions under which the components G_i^2 in the sum

$$G^2 = G_1^2 + G_2^2 + \cdots + G_s^2, \qquad (17.18)$$

evaluated for the subtables are independent are as follows:

1. The degrees of freedom for the components $G_1^2, \ldots, G_s^2$ must sum to the number $(r - 1)(c - 1)$ of the degrees of freedom of G^2.
2. Each cell count N_{ij} of the original table must appear in exactly one subtable.
3. Each of the marginal counts N_{i+} and N_{+j} of the original table must appear as marginal count in exactly one subtable.

If conditions 1–3 are satisfied then the values of statistic G^2 computed for the entire table and for the subtables satisfy (17.18), and the components are independent.

Example 17.4.1.   Consider a 3×3 contingency table

	Y_1	Y_2	Y_3	Marginals
X_1	N_{11}	N_{12}	N_{13}	N_{1+}
X_2	N_{21}	N_{22}	N_{23}	N_{2+}
X_3	N_{31}	N_{32}	N_{33}	N_{3+}
Marginals	N_{+1}	N_{+2}	N_{+3}	

An example of a decomposition satisfying conditions 1–3 of independence may be as follows:

Subtable 1:

	Y_1	Y_2	
X_1	$\mathbf{N_{11}}$	$\mathbf{N_{12}}$	$N_{11} + N_{12}$
X_2	$\mathbf{N_{21}}$	$\mathbf{N_{22}}$	$N_{21} + N_{22}$
	N_{11} $+N_{21}$	N_{12} $+N_{22}$	

Subtable 2:

	Y_1 or Y_2	Y_3	
X_1	$N_{11} + N_{12}$	$\mathbf{N_{13}}$	$\mathbf{N_{1+}}$
X_2	$N_{21} + N_{22}$	$\mathbf{N_{23}}$	$\mathbf{N_{2+}}$
	$N_{11} + N_{12}$ $+N_{21} + N_{22}$	N_{13} $+N_{23}$	

Subtable 3:

	Y_1	Y_2	
X_1 or X_2	$N_{11} + N_{21}$	$N_{12} + N_{22}$	$N_{11} + N_{12}$ $+N_{21} + N_{22}$
X_3	$\mathbf{N_{31}}$	$\mathbf{N_{32}}$	$N_{31} + N_{32}$
	$\mathbf{N_{+1}}$	$\mathbf{N_{+2}}$	

Subtable 4:

	Y_1 or Y_2	Y_3	
X_1 or X_2	$N_{11} + N_{12}$ $+N_{21} + N_{22}$	N_{13} $+N_{23}$	N_{1+} $+N_{2+}$
X_3	$N_{31} + N_{32}$	$\mathbf{N_{33}}$	$\mathbf{N_{3+}}$
	$N_{+1} + N_{+2}$	$\mathbf{N_{+3}}$	

The number of degrees of freedom of the original table is 4, and for each of the subtables it is 1, so the first condition is met.

As regards conditions 2 and 3, one can check that they are met by simple inspection: In the four subtables, all entries and marginals from the original table are identified by boldface.

The algebraic verification that the condition (17.18) holds is somewhat cumbersome, and we leave it to the reader. Here we observe only that the partition of this example is not the only one possible for a 3×3 table: There are eight other partitions, obtained by taking a possible 2×2 subtable as subtable 1 and building the remaining three subtables by suitable grouping of categories. The partition specified above would be chosen if there are reasons to believe that the membership in the first two rows is strongly associated with membership in the first two columns.

Example 17.4.2. In a study of marijuana use in colleges, 445 students were sampled and classified according to the response variable (use of marijuana or other drugs) into three categories: "never," "occasionally," "regularly." As a possible explanatory variable the experimenters selected the number of parents ("neither one," "exactly one," "both") who were alcohol or drug users. The data that follow are those of Devore (1991):

Parents' Drug Use	Student's Drug Use			
	Never (N)	Occasionally (O)	Regularly (R)	
Neither	141	54	40	235
One	68	44	51	163
Both	17	11	19	47
	226	109	110	445

For 3×3 tables, the statistic G^2 has the limiting chi-square distribution with 4 degrees of freedom. The value for the table above is $G^2 = 22.254$, which is highly significant (p-value equals 0.00018). Incidentally, for this table we have $Q^2 = 22.373$, which illustrates the closeness of G^2 and Q^2 for large samples. Thus, we may conclude that there is a relationship between the parental and student use of drugs.

One of the possible questions one may ask here is whether a positive or a negative example is stronger. In other words, taking for granted that the frequency of marijuana use tends to increase with the number of parents who use alcohol or drugs, the question is whether the effect of a bad example of one parent tends to outweigh the good example of the other parent. To get an insight into this question, we consider two decompositions of G^2. The first decomposition corresponds to the following four subtables:

Subtable 1:

	N	O
Neither	141	54
One	68	44

Subtable 2:

	$N + O$	R
Neither	195	40
One	112	51

Subtable 3:

	N	O
Neither or one	209	98
Both	17	11

Subtable 4:

	$N + O$	R
Neither or one	307	91
Both	28	19

The values of G^2 for these for tables are, respectively, $G_1^2 = 4.344$, $G_2^2 = 10.957$, $G_3^2 = 0.616$, and $G_4^2 = 6.336$, and we check that $G^2 = G_1^2 + G_2^2 + G_3^2 + G_4^2$. The corresponding p-values for 1 degree of freedom are, respectively, 0.037, 0.0009, 0.432, and 0.012.

An alternative decomposition is as follows:

Subtable 1*:

	O	R
One	44	51
Both	11	19

Subtable 2*:

	N	$O + R$
One	68	95
Both	17	30

Subtable 3*:

	O	R
Neither	54	40
One or both	55	70

Subtable 4*:

	N	$O + R$
Neither	141	94
One or both	85	125

Now the values are $G_{1*}^2 = 0.871$, $G_{2*}^2 = 0.470$, $G_{3*}^2 = 3.893$, and $G_{4*}^2 = 17.019$. The corresponding p-values are 0.351, 0.493, 0.048, and 0.00004.

These results appear to indicate that a bad example, even of one parent, prevails. Indeed, subtables 1* and 2* show that if at least one parent uses drugs or alcohol, then it practically does not matter whether it is only one of the parents or both of them. On the other hand, subtables 3* and 4* show that there is a significant effect if neither parent uses drugs or alcohol. The effect consists of increasing greatly the likelihood that a student will never use the marijuana, and—in case he uses it—of decreasing the frequency of use.

The conclusions are strengthened if one analyzes also the first decomposition. Here either the category "one" is compared with "neither" or the category "neither or one" is compared with the category "both." As one can see, subtables 1, 2, and 4 show a significant effect of a parent using alcohol or drugs. Subtable 3 shows that the positive effect of the other parent is limited. Indeed, "neither or one" is the same as "one or two" positive models, while "both" is the same as "neither" positive models. Subtable 3 shows that the presence or absence of a positive model does not have any significant effect on frequency of use of marijuana as long as this frequency is low (or zero). Subtable 4 suggests that the presence of a positive parent role model has the effect of lowering the probability of regular use of marijuana during the college years.

PROBLEMS

17.4.1 Find a decomposition of G^2 into independent components by decomposing:

(i) A $2 \times k$ table.

(ii) A 3×4 table.

17.4.2   The data on incidence of a certain disease, classified by age and gender, are as follows:

	Below 20	20-39	40-59	60 and above
Men	10	20	30	40
Women	20	40	60	300

Find the values of the statistics Q^2 and G^2. Find the decomposition of G^2 and verify that the "source" for lack of independence is the very high incidence of this disease among women over 60.

17.4.3   At the beginning of the semester a random sample of 52 students was selected out of students in all introductory statistics classes. Students were then classified according to their GPA (I, "below 3.0"; II, "between 3.0 and 3.5"; and III, "above 3.5") and their attitude toward statistics course [(i) "I hate to take this class but I have to"; (ii) "I do not mind taking this class but it does not seem to be my favorite one"; (iii) "I look forward to taking this class"]. The results of classifications are given below.

	(i)	(ii)	(iii)
I	7	4	1
II	5	7	6
III	3	8	11

Test the hypotheses of independence of both classification variables. In the case of lack of independence, try to identify the cause of dependence using decomposition.

17.5 ORDINAL RELATIONS

In this section we assume that the categories are ordered according to some relation which is relevant for the phenomenon studied. This relation (in case of each attribute) will have the formal properties of the relation "less than." As examples one can take response categories in some psychological or sociological questionnaires. Thus, a given activity may be classified according to remembered or perceived frequency [e.g., "never," "sometimes," "often," "very often"]; a statement may be categorized into one of the classes labeled "strongly disagree," "disagree," "neutral," "agree," "strongly agree," and so on. In each case, there is a well-defined order of the classes, and this order (for at least one of the attributes) cannot be expressed on a scale of type higher than ordinal. This means that while one may assign numerical values to categories, these numbers can only be used for comparison, with higher values representing categories more to the right. It is meaningless to add these numbers or to compare differences between them.*

Accordingly, we shall consider first the situation of 2×2 tables, and then $r \times c$ tables, with $r \geq 2, c \geq 2$. The rows will be labeled $1, \ldots, r$ and the columns will be labeled $1, \ldots, c$, and these indices will be assumed to reflect the increasing direction of the order on the two attributes in question. As before, N_{ij} denotes the count in the (i, j)th cell of the contingency table, while N_{i+}, N_{+j}, and n stand for the sum of the ith row, jth column, and the total number of observations.

We shall always represent problems in such a way that the first variable (whose categories correspond to rows) will be the explanatory variable, and the second variable (columns) will be the response. The population probabilities will be denoted by $p_{j|i}$, defined as the probability of the jth response level if the explanatory variable is on the ith level. Whenever the values of the explanatory variable result from random sampling from a bivariate population, we shall use the symbols p_{ij} and p_{i+}, and we then have the relation

$$p_{ij} = p_{i+}p_{j|i}.$$

The null hypothesis will assert that there is no relationship between the explanatory and response variables. This can also be expressed as the homogeneity hypothesis

$$H_0 : p_{j|1} = p_{j|2} = \cdots = p_{j|r}$$

for every j [i.e., probability distribution $(p_{1|i}, p_{2|i}, \ldots, p_{c|i})$ does not depend on i].

$$H_0 : p_{ij} = p_{i+}p_{+j}$$

for all i, j.

So far the formulation of the problem is exactly the same as in the case

---

* It is necessary to mention here that such rigorous restraint is rather an exception than a rule. In the social sciences one tends to assign numbers, such as $-2, -1, 0, 1$, and 2 to categories "strongly disagree," "disagree," "neutral," "agree," and "strongly agree" and then proceed to calculate the "average level" of agreement in a group of persons, variance, standard deviation, and so on. The same situation occurs in education, where one assigns numbers to grades A, $A-$, $B+, \ldots$ and then calculates average grades, to be used in comparing students, schools, and even educational levels of countries. The common justification seems to be that "everybody is doing it" and "with large data sets things average out somehow." But assigning numbers to grades A, $A-$, $B+, \ldots$ and then calculating GPA (grade point average) means that (say) $B+$ in a test in mathematics counts the same as $B+$ for an essay in English, or $B+$ in a test in music appreciation, regardless of who is teaching the subject and who is grading the exams, an assumption that is hardly acceptable.

To be sure, even if someone came up with a better method, the cost of its implementation would be prohibitive, and opposition would be very strong. But the fact remains that the conclusion derived from such an analysis is quite similar to the following: If Mary likes (on a scale from 1 to 10) her boyfriend Jack as 8, and the same Mary likes kiwi fruit as 9 (again, on the scale from 1 to 10), then Mary likes kiwi fruit 12.5% more than she likes Jack.

of chi-square tests discussed in preceding sections. The difference lies in the specification of the alternative hypotheses. The chi-square tests are "omnibus" tests: Their power tends to 1 (as the sample sizes increases to infinity) under *any* simple hypothesis outside the null hypothesis H_0. On the other hand, in this section we consider tests of H_0 against alternatives which assert that there is some *monotone* relation between explanatory and response variables. Thus, H_0 and H_a do not exhaust all possibilities. By requiring that the tests have power tending to 1 only on some simple hypotheses outside H_0, one may (in general) obtain tests with power (on selected alternatives) higher than the power of chi-square tests.

The monotone relation mentioned above means that increasing values of one variable tend to accompany increasing values of the other variable (this is positive association: negative association is explicated in a similar way). This concept of association can be formalized in a number of different ways [for details see Agresti (1984)]. We present here one of the possible definitions.

Before formulating the alternative hypotheses, let us introduce some definitions. Let X and Y be two random variables with cdf's $F(t)$ and $G(t)$. The random variables X and Y may be defined on the same sample space and be dependent or independent; they may also be defined on different sample spaces.

We say that X is stochastically larger than Y (to be written as $X \geq_{st} Y$) if $F(t) \leq G(t)$ for all t. The strict domination $X >_{st} Y$ is defined similarly, with the requirement that $X \geq_{st} Y$, and in addition $F(t_0) < G(t_0)$ for some t_0. In the case of categorical variables with ordered categories, we do not have the concept of the cdf directly available. We may, however, define for each $i = 1, \ldots, r$ the function

$$F^{(i)}(j) = \sum_{m \leq j} p_{m|i}, \qquad j = 1, \ldots, c. \qquad (17.19)$$

The function $F^{(i)}$ has many of the essential features of a cdf and may be used to define the relation $>_{st}$ for random variables with values on an ordinal scale only. We can say that a random variable with distribution $(p_{1|i}, p_{2|i}, \ldots, p_{c|i})$ is stochastically larger than a random variable with distribution $(p_{1|i'}, p_{2|i'}, \ldots, p_{c|i'})$ if for all j we have $F^{(i)}(j) \leq F^{(i')}(j)$, and $F^{(i)}(j_0) < F^{(i')}(j_0)$ for some j_0.

We can now formulate the class of alternative hypotheses of interest in analyzing contingency tables with attributes measured on a discrete ordinal scale. Thus, the null hypothesis H_0 of independence or homogeneity may be tested against the alternative

$$H_a^+ (\text{positive association}) : F^{(i)}(j) \geq F^{(i')}(j)$$

for $i \leq i'$ and all j, or the alternative

$$H_a^- (\text{negative association}) : F^{(i)}(j) \leq F^{(i')}(j)$$

for $i \leq i'$ and all j.

PROBLEMS

17.5.1 Consider a $2 \times c$ contingency table, and express the positive (negative) association in terms of probabilities in the two rows.

17.5.2 Solve Problem 17.5.1 for a $r \times 2$ table.

17.6 2 × 2 CONTINGENCY TABLES

We begin with the simplest case, when both variables have only two categories (i.e., are treated as binary variables). We let those levels be denoted by 1 and 2; we then have two distributions of the response, corresponding to the values of the explanatory variable, namely

$$(p_{1|1}, p_{2|1}) \quad \text{and} \quad (p_{1|2}, p_{2|2})$$

where $p_{2|1} = 1 - p_{1|1}$ and $p_{2|2} = 1 - p_{1|2}$.

The null hypothesis of homogeneity asserts the equality of two conditional distributions and therefore reduces to

$$H_0 : \frac{p_{1|1}}{p_{1|2}} = 1.$$

When response 1 was in some sense undesirable (e.g., death, relapse of disease, etc.) the ratio $p_{1|1}/p_{1|2}$ was often called the *relative risk*. This term is now used generally regardless of the context (similarly to the use of "success" and "failure" in binomial distribution).

The alternative hypothesis of positive association asserts that higher value of explanatory variable gives a higher probability of a higher value of the response variable, that is, $p_{2|1} < p_{2|2}$. The latter inequality is equivalent to $1 - p_{1|1} < 1 - p_{1|2}$ hence $p_{1|1}/p_{1|2} > 1$. Similarly, negative association means that the relative risk is less than 1.

An alternative formulation, which also suggests a testing procedure, is as follows. Consider a binary distribution $(x, 1 - x)$, where x is the probability of some event A. Then the ratio

$$\eta = \frac{x}{1 - x} \tag{17.20}$$

is called the *odds* (for the event A). It is clear that η determines x uniquely, namely $x = \eta/(1 + \eta)$. As x increases from 0 to 1, the odds η increase from zero to infinity. The odds for the complement of the event A are $1/\eta$. In the case of the distributions $(p_{1|1}, p_{2|1})$ and $(p_{1|2}, p_{2|2})$, the odds (for the response 1) are

$$\eta_1 = \frac{p_{1|1}}{p_{2|1}} \quad \text{and} \quad \eta_2 = \frac{p_{1|2}}{p_{2|2}}.$$

To formulate the null and alternative hypotheses, it appears natural to consider the *odds ratio*

$$\theta = \frac{\eta_1}{\eta_2}. \tag{17.21}$$

We have here

$$\theta = \frac{p_{1|1}p_{2|2}}{p_{1|2}p_{2|1}} = \frac{p_{11}p_{22}}{p_{12}p_{21}},$$

the latter expression being meaningful when the explanatory variable is random.

The null hypothesis, of either homogeneity or independence, takes on the form

$$H_0 : \theta = 1, \tag{17.22}$$

while the alternative H_a^+ of positive and H_a^- of negative association are, respectively,

$$H_a^+ : \theta > 1 \quad \text{and} \quad H_a^- : \theta < 1. \tag{17.23}$$

To fix one's attention, suppose that the null hypothesis $H_0 : \theta = 1$ is to be tested against the alternative of positive association $H_a^+ : \theta > 1$. The data have the form of a 2×2 count matrix (contingency table)

$$\begin{bmatrix} N_{11} & N_{12} \\ N_{21} & N_{22} \end{bmatrix}.$$

The question is: how to determine the *p*-value of the observed contingency table, that is, the *probability* (calculated under the assumption that the null hypothesis is true) *of observing—if the experiment were to be repeated—a contingency table that would be at least as much in favor of the alternative as the contingency table actually observed.*

To implement this central idea of determining the *p*-value of the observed result, one needs to:

1. Explicate the notion of ordering contingency tables according to the relation of "being more in favor of the alternative hypothesis."
2. Specify the probabilities of occurrence of various contingency tables under the null hypothesis. Note that the null hypothesis of independence is composite, hence does not lead directly to numerical values of probabilities.

As to result 1, one can construct a statistic that mimics the odds ratio θ. Clearly, such a statistic results from replacing the probabilities p_{ij} by the corresponding frequencies N_{ij}/n, where n is the total sample size. This leads to the statistic, say U, defined as

$$U = \frac{N_{11}N_{22}}{N_{12}N_{21}}$$

(and $U = \infty$ if N_{12} or N_{21} is zero).

Thus, if the observed contingency table is $[n_{ij}]$, then the p-value of this result is defined formally as

$$P_0 \left\{ [N_{ij}] : \frac{N_{11}N_{22}}{N_{12}N_{21}} \geq \frac{n_{11}n_{22}}{n_{12}n_{21}} \right\}. \tag{17.24}$$

It remains now to specify the probability distribution P_0 on a suitably selected class of 2×2 contingency tables. The main requirement is that P_0 should not depend on any parameters, so that its numerical value could be determined for every contingency table for which P_0 is defined.

Now, if we take a random sample of size n from a population whose elements are classified according to two dichotomous classifications, then the probability of a particular contingency table with sum of entries equal n is given by the multinomial distribution

$$P\{N_{ij} = n_{ij}, i, j = 1, 2\} = \frac{n!}{n_{11}!n_{12}!n_{21}!n_{22}!} p_{11}^{n_{11}} p_{12}^{n_{12}} p_{21}^{n_{21}} p_{22}^{n_{22}}. \tag{17.25}$$

If the null hypothesis H_0 is true, then $p_{ij} = p_{i+}p_{+j}$, and substitution to (17.25) gives

$$P_{H_0}\{N_{ij} = n_{ij}, i, j = 1, 2\} = \frac{n!}{n_{11}!n_{12}!n_{21}!n_{22}!} p_{1+}^{n_{1+}} p_{2+}^{n_{2+}} p_{+1}^{n_{+1}} p_{+2}^{n_{+2}}, \tag{17.26}$$

where $n_{i+} = n_{i1} + n_{i2}$ and $n_{+j} = n_{1j} + n_{2j}$ for $i, j = 1, 2$.

It is now clear that to obtain probabilities of contingency tables that do not depend on the parameter, it is necessary to restrict the definition of P_0 to the class of tables with given marginals n_{i+} and n_{+j} ($i, j = 1, 2$). This suggests taking as P_0 the conditional probabilities given both marginals, since the product $p_{1+}^{n_{1+}} p_{2+}^{n_{2+}} p_{+1}^{n_{+1}} p_{+2}^{n_{+2}}$ will then cancel. Intuitively, this means that we shall compare the original table to other tables with the same marginal totals.

We have, for the probability of marginals being (n_{1+}, n_{2+}) and (n_{+1}, n_{+2}), the product of two binomial probabilities:

$$P\{N_{1+} = n_{1+}\} = \binom{n}{n_{1+}} p_{1+}^{n_{1+}} (1 - p_{1+})^{n - n_{1+}}$$

and

$$P\{N_{+1} = n_{+1}\} = \binom{n}{n_{+1}} p_{+1}^{n_{+1}} (1 - p_{+1})^{n - n_{+1}}.$$

Consequently,

$$P\{N_{1+} = n_{1+}, N_{+1} = n_{+1}\} = P\{N_{1+} = n_{1+}\}P\{N_{+1} = n_{+1}\}$$

$$= \binom{n}{n_{1+}}\binom{n}{n_{+1}} p_{1+}^{n_{1+}} p_{2+}^{n_{2+}} p_{+1}^{n_{+1}} p_{+2}^{n_{+2}}. \quad (17.27)$$

Dividing (17.26) by (17.27), we obtain, after rearranging the multinomial coefficients, the following theorem in which $N_{ij}(x)$ is the 2×2 table with given marginals and $N_{11} = x$.

Theorem 17.6.1. *Under the null hypothesis* $H_0 : \theta = 1$, *for any integers* n, n_{1+}, n_{+1} *satisfying the conditions* $n > 0, 0 \le n_{1+} \le n, 0 \le n_{+1} \le n$,

$$P\{[N_{ij}] = [N_{ij}(x)] | N_{1+} = n_{1+}, N_{+1} = n_{+1}\}$$

$$= \frac{\binom{n_{+1}}{x}\binom{n - n_{+1}}{n_{1+} - x}}{\binom{n}{n_{1+}}} = \frac{\binom{n_{1+}}{x}\binom{n - n_{1+}}{n_{+1} - x}}{\binom{n}{n_{+1}}}. \quad (17.28)$$

Observe now that we have

$$\frac{n_{11}n_{22}}{n_{12}n_{21}} = \frac{x(n - n_{1+} - n_{+1} + x)}{(n_{1+} - x)(n_{+1} - x)},$$

which is an increasing function of x. This means that ordering of the tables specified in (i) coincides with ordering with respect to the element $x = n_{11}$. Consequently, the p-value defined by (17.24) becomes, in view of Theorem 17.6.1 and (17.28), the sum

$$\sum_{x \ge n_{11}} \frac{\binom{n_{1+}}{x}\binom{n - n_{1+}}{n_{+1} - x}}{\binom{n}{n_{+1}}}.$$

Example 17.6.1. To study whether or not there exists a positive association between musical and mathematical abilities, a group of fourth graders was classified according to their scores (high or low). The results were as follows:

Music	Mathematics		
	High	Low	
High	4	1	5
Low	2	5	7
	6	6	12

Do the observed data lead to rejecting the null hypothesis of independence of musical and mathematical abilities in favor of the alternative of a positive association?

SOLUTION.   There is only one contingency table with the same marginals as the original one, which is more in favor of the alternative, namely

$$
\begin{array}{cc|c}
5 & 0 & 5 \\
1 & 6 & 7 \\
\hline
6 & 6 & 12
\end{array}
$$

The p-value of the observed result is therefore the sum of probabilities of both tables; hence it equals

$$
\frac{\binom{5}{4}\binom{7}{2}}{\binom{12}{6}} + \frac{\binom{5}{5}\binom{7}{1}}{\binom{12}{6}} = 0.121.
$$

Thus, on the level 0.1 (and hence also on any lower level) the null hypothesis could not be rejected: there is about a 12% chance of observing a result at least as much in favor of the alternative hypothesis due only to random fluctuations (i.e., if in fact the null hypothesis is true).   □

Let us mention here that the test as described above is applicable also to the cases of prospective or retrospective studies, when one of the marginals is not random. Indeed, if the marginals n_{1+} and n_{2+} are selected by the experimenter, then the contingency table is

$$
\begin{array}{cc|c}
n_{11} & n_{12} & a \\
n_{21} & n_{22} & b \\
\hline
n_{+1} & n_{+2} & n
\end{array}
$$

(where we let a and b denote the nonrandom marginals, to make them graphically distinct from the random marginals). The corresponding table of conditional probabilities, under the null hypothesis of homogeneity, is

$$
\begin{bmatrix} p_{1|1} & p_{2|1} \\ p_{1|2} & p_{2|2} \end{bmatrix} = \begin{bmatrix} \gamma & 1-\gamma \\ \gamma & 1-\gamma \end{bmatrix}.
$$

The likelihood of the data, under the null hypothesis, is therefore a product of two binomial probabilities, namely

$$\binom{a}{n_{11}} \gamma^{n_{11}} (1 - \gamma)^{n_{12}} \binom{b}{n_{21}} \gamma^{n_{21}} (1 - \gamma)^{n_{22}}$$

$$= \binom{a}{n_{11}} \binom{n - a}{n_{+1} - n_{11}} \gamma^{n_{+1}} (1 - \gamma)^{n_{+2}}.$$

On the other hand, the probability of the observed column marginals is, again under H_0,

$$\binom{n}{n_{+1}} \gamma^{n_{+1}} (1 - \gamma)^{n_{+2}}.$$

Thus, the conditional probability, given the column marginal, is free of γ, namely it equals

$$\frac{\binom{a}{n_{11}} \binom{n - a}{n_{+1} - n_{11}}}{\binom{n}{n_{+1}}},$$

which agrees with (17.28).

Example 17.6.2. Returning to Example 17.6.1, the p-value of the observed contingency table would be the same as calculated there (i.e., about 0.12) if we selected 6 students with high and 6 with low math scores, and then classified them according to their music scores.

Example 17.6.3. Twenty occasional headache sufferers participated in testing a newly developed headache remedy. They were given the drug and later asked whether or not it was significantly better than the drug they usually took. In fact, however, every fourth subject tested received not a drug but a placebo. The data are as follows:

	Significant Improvement	No Significant Improvement	
Drug	11	4	15
Placebo	3	2	5
	14	6	20

Do these data indicate that the new drug is better than the drugs usually taken?

SOLUTION. We are testing here the hypothesis of homogeneity against the alternative of a positive association between taking the new drug and beneficial effects for patients. To determine the p-values observe that the count n_{11}, given the marginals, may be only 11, 12, 13, or 14 if the table is to be at least as much in favor of the alternative as the observed one. Thus the tables in question are:

$$\begin{bmatrix} 11 & 4 \\ 3 & 2 \end{bmatrix}, \begin{bmatrix} 12 & 3 \\ 2 & 3 \end{bmatrix}, \begin{bmatrix} 13 & 2 \\ 1 & 4 \end{bmatrix}, \begin{bmatrix} 14 & 1 \\ 0 & 5 \end{bmatrix}.$$

Their probabilities are respectively

$$\frac{\binom{15}{11}\binom{5}{3}}{\binom{20}{14}} = 0.352, \quad \frac{\binom{15}{12}\binom{5}{2}}{\binom{20}{14}} = 0.117$$

$$\frac{\binom{15}{13}\binom{5}{1}}{\binom{20}{14}} = 0.014, \quad \frac{\binom{15}{14}\binom{5}{0}}{\binom{20}{14}} = 0.0004.$$

Hence the p-value is about 0.48, and there in no evidence to claim that the new drug is superior to the drugs used so far.   □

The testing procedure described above is known as the Fisher exact test, and it works well against one-sided alternatives. In testing the null hypothesis $H_0 : \theta = 1$ against a two-sided alternative $H_0 : \theta \neq 1$, one encounters the usual problem of defining p-values in the case of two-sided alternatives. There seems to be no agreement among statisticians as to what the proper procedure is in such situations. In case of continuous and symmetric distributions of the test statistic (e.g., Student), the p-value for the result observed is usually taken as a doubled p-value for the one-sided alternative. In the case of asymmetric distribution, there is little justification for doubling the one-sided p-value. Additionally, in the case of discrete distributions, another source of difficulty is that according to such a procedure one can obtain a value exceeding 1. Some authors suggest taking as p-value the sum of all probabilities of tables that are at most as likely as the observed one [see Freeman and Halton (1951)].

The "extremist" view here is that p-values should not be used in the case of two-sided alternatives. This view is perhaps too extreme, since there is a general feeling that observed values of testing statistics differ in their "strength" as a premise against the null hypothesis, depending on their location in the critical region.

PROBLEMS

17.6.1 Out of 10 men and 10 women, 2 and 4, respectively, were found to be allergic to a specific drug.

(i) Does these data indicate that there is a difference between men and women in their propensity to develop allergic reaction to the drug in question?

(ii) Does these data indicate that men are less likely than women to develop allergic reaction to the drug in question?

Find corresponding p-values for parts (i) and (ii) and explain why are they different.

17.6.2 Use the data from Problem 17.3.6 to test if highschool GPA is positively related to college GPA.

17.7 $r \times c$ CONTINGENCY TABLES

Finally, we extend the results for the 2×2 table to the general case of $r \times c$ tables, in which the categories corresponding to rows and those corresponding to columns are ordered. The assumptions of Theorem 17.6.1 carry over to the present case. By conditioning on both marginals, we obtain (under the null hypothesis of independence or homogeneity) a probability distribution defined on the class of all $r \times c$ tables with given marginals, which does not involve any unknown parameters.

For the case when both marginals are random, the situation is as follows. Suppose that we take n independent observations, each with the same distribution, and classify them according to two systems, with r and c categories, respectively. The cell counts are N_{ij}, and the marginal counts are N_{i+} and N_{+j}, where $i = 1, \ldots, r$ and $j = 1, \ldots, c$. We then have the following theorem.

Theorem 17.7.1. *Under the assumption of independence, for any contingency table* $[n_{ij}], i = 1, \ldots, r, j = 1, \ldots, c,$ *with* $\sum_{i,j} n_{ij} = n$ *and with marginal counts* n_{i+} *and* n_{+j} *we have*

$$P\{[N_{ij}] = [n_{ij}]|N_{i+} = n_{i+}, N_{+j} = n_{+j}\} = \frac{\prod_{i=1}^{r}(n_{i+})! \prod_{j=1}^{c}(n_{+j})!}{n! \prod_{i=1}^{r} \prod_{i=1}^{c}(n_{ij})!}. \quad (17.29)$$
□

To implement Definition 17.24 for the p-value of an observed contingency table, say A_0, it remains now to order all $r \times c$ tables with marginals the same as those of A_0, according to their "strength of support" for the alternative hypothesis of (say) positive association. In presenting the solution we use an approach different from that used for 2×2 tables. Instead of expressing the null hypothesis in terms of a single parameter and then finding its empirical counterpart (as we did for 2×2 tables), we introduce several indices (statistics, i.e., functions of the observed counts N_{ij}). Each of these indices will provide, on an intuitive ground, an ordering of contingency tables according to the strength of support for the alternative.

Symbolically, let A_0 be the observed contingency table, and let A be any contingency table with the same marginals as A_0. Furthermore, let $t = t(A)$ be a real–valued function defined on the considered class of contingency tables. Then the p-value of the observed table is defined as

$$\sum_{A:t(A)\geq t(A_0)} P_0(A). \tag{17.30}$$

The only problem remaining is the choice of the statistic t so that its values can be taken as reflecting the order of "strength of support" of the alternative hypothesis. The choices here are based on the following idea. Consider a pair of observations. Since each observation is classified as belonging to one of the rows and one of the columns, such a pair determines two pairs of coordinates, say (r', c') and (r'', c'').

Definition 17.7.1. The pair of observations is called *concordant* if

$$(r'' - r')(c'' - c') > 0;$$

it is called *discordant* if

$$(r'' - r')(c'' - c') < 0;$$

and it is called *tied* if $(r'' - r')(c'' - c') = 0$. $\qquad\qquad\square$

It appears obvious that every concordant pair provides support in favor of the alternative of positive association: The differences $r'' - r'$ and $c'' - c'$ are both nonzero and of the same sign. This means that higher evaluation on one variable is accompanied by higher evaluation on the other variable. For the same reason, every discordant pair provides support for the alternative of a negative association: A higher classification on one variable is accompanied by a lower evaluation on the other variable.

In the sequel, we let

$$C = \text{number of concordant pairs}$$

and

$$D = \text{number of discordant pairs.}$$

If one may assume that any concordant (or discordant) pair equally supports the alternative, then the overall support for the alternative of (say) positive association is a function of the difference $C - D$. Standardizing this difference, we let

$$\hat{\gamma} = \frac{C - D}{C + D}. \tag{17.31}$$

The choice of the symbol stresses the fact that $\hat{\gamma}$ (estimator of some parameter γ) depends on the observed contingency table.

One can now define the *p*-value of an observed contingency table (in testing the alternative of positive association) as the sum of probabilities of all the tables with the same marginals as the one observed, and with the value of the index $\hat{\gamma}$ at least as high as for the original table.

Example 17.7.1. For 2×2 tables, the numbers of concordant and discordant pairs of observations are

$$C = n_{11}n_{22}, \quad D = n_{12}n_{21},$$

hence

$$\hat{\gamma} = \frac{n_{11}n_{22} - n_{12}n_{21}}{n_{11}n_{22} + n_{12}n_{21}} = \frac{\theta - 1}{\theta + 1} = 1 - \frac{2}{\theta + 1},$$

where

$$\theta = \frac{n_{11}n_{22}}{n_{12}n_{21}}.$$

Thus, the ordering according to the values of $\hat{\gamma}$ coincides with the ordering according to the values of θ.

Example 17.7.2. For the case of marijuana use in colleges (Example 17.4.2), the categories represented by the rows and those represented by the columns are ordered by a natural way: by number of parents who use drugs or alcohol, and by frequency of marijuana use by the student. The table is:

Parents' Drug Use	Student's Drug Use			
	N Never	*O* Occasionally	*R* Regularly	
Neither	141	54	40	235
One	68	44	51	163
Both	17	11	19	47
Marginals	226	109	110	445

A concordant pair is formed by observations such that one of them is below and to the right of the other. Thus

$$C = 141 \times (44 + 51 + 11 + 19) + 54 \times (51 + 19)$$
$$+ 68 \times (11 + 19) + 44 \times 19 = 24{,}281.$$

Similarly,

$$D = 40 \times (68 + 44 + 17 + 11) + 54 \times (68 + 17)$$
$$+ 51 \times (17 + 11) + 44 \times 17 = 12{,}366.$$

Thus, $\hat{\gamma} = 0.325$. □

To evaluate the p-value of an observed contingency table, one should use a statistical package such as SAS. The case below should give the reader an idea of situations where the p-value is accessible without a computer.

Example 17.7.3. Consider the (hypothetical) case of testing a method of predicting the time of occurrence of an earthquake. The method was used in a total of 13 earthquakes, classified as major or minor. The columns represent the precision of prediction (e.g., "good," "fair," and "poor"). Suppose the contingency table is as follows:

	Good	Fair	Poor	
Major	4	2	1	7
Minor	1	4	1	6
	5	6	2	13

We have here $C = 4 \times (4 + 1) + 2 \times 1 = 22$ and $D = 1 \times (1 + 4) + 2 \times 1 = 7$, hence $\hat{\gamma} = 15/29 = 0.517$.

The null hypothesis is that of independence (or homogeneity); the alternative is of positive dependence between row and column variable. In other words, the null hypothesis is that the method performs equally well (or equally poorly) for major and for minor earthquakes. The alternative is that predictions tend to be better for major earthquakes.

There are only three 2×3 tables with the same marginals as the original table and a higher value of $\hat{\gamma}$, namely

$$\begin{bmatrix} 4 & 3 & 0 \\ 1 & 3 & 2 \end{bmatrix}, \quad C = 26, D = 3, \hat{\gamma} = 0.793$$

$$\begin{bmatrix} 5 & 1 & 1 \\ 0 & 5 & 1 \end{bmatrix}, \quad C = 31, D = 5, \hat{\gamma} = 0.838$$

$$\begin{bmatrix} 5 & 2 & 0 \\ 0 & 4 & 2 \end{bmatrix}, \quad C = 34, D = 0, \hat{\gamma} = 1.$$

The p-value therefore equals the sum of probabilities of the original table and the three tables above. Using (17.29), factoring out the factorials of the marginals and the total count, the p-values equals

$$\frac{(7!6!)(5!6!2!)}{13!}\left[\frac{1}{4!2!1!1!4!1!}+\frac{1}{4!3!0!1!3!2!}+\frac{1}{5!1!1!0!5!1!}+\frac{1}{5!2!0!0!4!2!}\right]$$

$$= 0.0874 + 0.0583 + 0.0070 + 0.0087 = 0.1614.$$

Thus, there is not enough justification to reject the null hypothesis in favor of the alternative.   □

The index $\hat{\gamma}$ utilizes only the information contained in the numbers C and D of concordant and discordant pairs. The numbers of tied pairs are not used, since a pair of observations tied on one or both variables provide evidence neither in favor of the null hypothesis nor in favor of the alternative. However, one could argue that a large number of tied pairs (as compared with $C + D$) is an argument that the difference $C - D$ might be insignificant. This led to the introduction of two indices that take tied pairs into account, Kendall's tau-b and Somers' d.

The total number of all possible pairs of observations (disregarding the order) is

$$\binom{n}{2} = \frac{n(n-1)}{2}.$$

Therefore, the total number of tied observations is $n(n-1)/2 - C - D$. We let

T_X = number of pairs of observations tied on the first coordinate (row),

T_Y = number of pairs of observations tied on the second coordinate (column),

T_{XY} = number of pairs of observations tied on both coordinates (falling to the same cell).

Observe that the categories of ties are not disjoint: any pair tied on both coordinates is counted twice, in T_X and in T_Y.

Definition 17.7.2. The statistic

$$\tau_b = \frac{C - D}{\sqrt{[n(n-1)/2 - T_X]\,[n(n-1)/2 - T_Y]}} \tag{17.32}$$

is called *Kendall's tau-b*.   □

Definition 17.7.3. The statistic

$$d = \frac{C - D}{n(n-1)/2 - T_X} \tag{17.33}$$

is called *Somers' d*.   □

In the formulas above we have

$$T_X = \sum_{i=1}^{r} \frac{n_{i+}(n_{i+} - 1)}{2} \quad \text{and} \quad T_Y = \sum_{j=1}^{c} \frac{n_{+j}(n_{+j} - 1)}{2}. \tag{17.34}$$

Example 17.7.4. Returning to the marijuana data (Examples 17.4.2 and 17.7.2), we have $T_X = 41{,}779$ and $T_Y = 37{,}306$. The total number of all possible pairs is $445 \times 444/2 = 98{,}790$. We have $\tau_b = 0.201$ and $d = 0.209$ (while $\hat{\gamma} = 0.325$). $\qquad\qquad\qquad\qquad\qquad\qquad\qquad\qquad\qquad\qquad\qquad\qquad\quad$ $\square$

Calculation of the p-value based on τ_b or d for the alternative of positive association again uses the sum of all probabilities of tables with the same marginals as the one observed, and the index τ_b, or d, at least as high as that for the observed table.

We finish this chapter with some comments concerning the use of the three indices, $\hat{\gamma}$, τ_b and d. First, observe that the denominators in formulas (17.32) and (17.33) depend only on marginal distribution, hence remain constant for all contingency tables whose probabilities are added in calculating the p-value. The terms of the sums involve exactly the same tables, those with the difference $C - D$ at least as large as that for the observed contingency table. This proves the following theorem.

Theorem 17.7.2. *The p-values of a given contingency table calculated for the orderings induced by Kendall's tau-b and by Somers' d coincide.*

Second, the p-values computed for ordering given by $\hat{\gamma}$ and Kendall's tau-b (or Somers' d) may differ.

Proof. It is enough to show two contingency tables with the same marginals, whose order according to $\hat{\gamma}$ does not agree with the order according to $C - D$. Such tables are, for instance,

$$A = \begin{bmatrix} 4 & 2 & 0 \\ 1 & 4 & 2 \end{bmatrix}, \qquad B = \begin{bmatrix} 5 & 0 & 1 \\ 0 & 6 & 1 \end{bmatrix}.$$

For table A we have $C = 28, D = 2$, hence $C - D = 26$, $\hat{\gamma} = 0.867$. For table B we have $C = 35, D = 6$, hence $C - D = 29 > 26$ and $\hat{\gamma} = 0.707 < 0.867$. $\qquad$ $\square$

Finally, it ought to be mentioned that the very fact of introduction of an index is—to a certain extent—an attempt to reduce the problem to one dimension. That enables one to compare objects (in this case, contingency tables) and select the best. This may be, however, a deceptive comfort, since not all things are comparable in such a simple way. In the present case, just thinking in terms of an index (say, $\hat{\gamma}$) leads to danger of attaching significance to certain values of that index. For instance, one may tend to take $\hat{\gamma} = 1$ first as an indication of a very strong positive association (which it is) and then tend to attach to it a fixed significance level. To see this, observe that the tables

$$\begin{bmatrix} 2 & 2 & 0 \\ 0 & 2 & 2 \end{bmatrix} \quad \text{and} \quad \begin{bmatrix} 3 & 3 & 0 \\ 0 & 2 & 2 \end{bmatrix}$$

both have $D = 0$, hence $\hat{\gamma} = 1$. The p-value for the first table equals

$$\frac{(4!4!)(2!4!2!)}{8!2!2!2!2!} = 0.086$$

while the p-value for the second table is about half of that for the first table:

$$\frac{(6!4!)(3!5!2!)}{10!3!2!3!2!} = 0.0476.$$

This is simply the effect of regarding as equivalent two situations characterized by the same value of $\hat{\gamma}$. This example shows that values of $\hat{\gamma}$ are not comparable for experiments with different marginal counts.

Computations of the p-value become cumbersome especially when the number of row and/or column categories increases. When cell counts are sufficiently large, normal approximation of sampling distribution may be used (see, e.g., Agresti, 1990).

PROBLEMS

17.7.1 Compute the values of $\hat{\gamma}$, τ_b and d for each of the following tables:

(i) $\begin{bmatrix} 4 & 10 & 15 \\ 6 & 20 & 44 \\ 7 & 22 & 50 \end{bmatrix}$

(ii) $\begin{bmatrix} 12 & 8 & 6 & 3 \\ 10 & 6 & 4 & 2 \\ 5 & 3 & 1 & 0 \end{bmatrix}$

(iii) $\begin{bmatrix} 150 & 130 & 120 & 110 \\ 25 & 15 & 10 & 5 \end{bmatrix}$

(iv) $\begin{bmatrix} 150 & 130 & 120 & 110 \\ 5 & 10 & 15 & 25 \end{bmatrix}$

17.7.2 A study reported in *Science* magazine investigated the relationship between gender, handedness (right- or left-handed), and relative foot size (left foot bigger than right foot, left foot within one-half shoe size of right foot, or right foot bigger). A random sample of 150 adults gave the following data:

Relative foot size	Right-handed		Left-handed	
	Male	Female	Male	Female
$L > R$	2	55	6	0
$L \approx R$	10	18	6	2
$L < R$	28	14	0	9

Test the association between gender and handedness considering three groups of people: these whose left foot is bigger than the right foot, these for whom both feet are almost the same, and these whose right foot is bigger. Compare the results obtained.

17.7.3 Measure the association of GPA and attitude toward statistics courses based on the data from Problem 17.4.3 using $\hat{\gamma}$, τ_b, and d coefficients.

17.7.4 Hypothetical data on incidence of a certain disease, classified by age and sex, are:

	below 20	20–39	40–59	60 and over
Men	5	10	15	20
Women	10	20	30	40

Calculate the value of statistics Q^2, G^2, $\hat{\gamma}$, τ_b and d. Next, add the value 100 to one entry and calculate the values of statistics Q^2, G^2, $\hat{\gamma}$, τ_b and d for each of resulting 8 tables, such as

$$\begin{bmatrix} 105 & 10 & 15 & 20 \\ 10 & 20 & 30 & 40 \end{bmatrix},$$

$$\begin{bmatrix} 5 & 110 & 15 & 20 \\ 10 & 20 & 30 & 40 \end{bmatrix}, \dots$$

Comment on the different ways in which such an addition changes the values of particular statistics.

REFERENCES

Agresti, A., 1984. *Analysis of Ordinal Categorical Data*, John Wiley & Sons, Inc., New York.

Agresti, A., 1990. *Categorical Data Analysis*, John Wiley & Sons, New York.

Chernoff, H. and E. L. Lehmann, 1954. "The Use of the Maximum Likelihood Estimates in χ^2 Tests for Goodness of Fit". *Annals of Mathematical Statistics*, **25**, pp. 579–586.

Cramér, H., 1946. *Mathematical Methods of Statistics*. Princeton Univ. Press: Princeton, N.J.

Devore, J.L., 1991. *Probability and Statistics for Engineering and the Sciences*, 3rd ed., Brooks & Cole, Pacific Grove, Calif.

Fox, M. and R. James, 1987. *The Complete Chess Addict*, Faber and Faber, London.

Freedman, D., R. Pisani, R. Purves, and A. Adhikari, 1992. *Statistics*, 2nd ed. W.W. Norton, New York.

Freeman, G. H. and J. H. Halton, 1951. Note on an Exact Treatment of Contingency, Goodness-of-Fit and Other Problems of Significance. Biometrika 38: 141–149.

APPENDIX A

Statistical Tables

Table A1. Cumulative binomial probabilities $\sum_{k=0}^{x} b(k;n,p) = \sum_{k=0}^{x} \binom{n}{k} p^k (1-p)^{n-k}$.

							p					
n	x	0.05	0.10	0.15	0.20	0.25	0.30	0.33	0.35	0.40	0.45	0.50
2	0	0.9025	0.8100	0.7225	0.6400	0.5625	0.4900	0.4489	0.4225	0.3600	0.3025	0.25
	1	0.9975	0.9900	0.9775	0.9600	0.9375	0.9100	0.8911	0.8775	0.8400	0.7975	0.75
3	0	0.8574	0.7290	0.6141	0.5120	0.4219	0.3430	0.3008	0.2746	0.2160	0.1664	0.12
	1	0.9927	0.9720	0.9392	0.8560	0.8437	0.7840	0.7452	0.7182	0.6480	0.5747	0.50
	2	0.9999	0.9990	0.9966	0.9920	0.9844	0.9730	0.9641	0.9571	0.9360	0.9089	0.87
4	0	0.8145	0.6561	0.5220	0.4096	0.3164	0.2401	0.2015	0.1785	0.1296	0.0915	0.06
	1	0.9860	0.9477	0.8905	0.8192	0.7383	0.6517	0.5985	0.5630	0.4752	0.3916	0.31
	2	0.9995	0.9963	0.9880	0.9728	0.9492	0.9163	0.8918	0.8735	0.8208	0.7585	0.68
	3	1.0000	0.9999	0.9995	0.9984	0.9961	0.9919	0.9881	0.9850	0.9744	0.9590	0.93
5	0	0.7738	0.5905	0.4437	0.3277	0.2373	0.1681	0.1350	0.1160	0.0778	0.0503	0.03
	1	0.9774	0.9185	0.8352	0.7373	0.6328	0.5282	0.4675	0.4284	0.3370	0.2562	0.18
	2	0.9988	0.9914	0.9734	0.9421	0.8965	0.8369	0.7950	0.7648	0.6826	0.5931	0.50
	3	1.0000	0.9995	0.9978	0.9933	0.9844	0.9692	0.9564	0.9460	0.9130	0.8688	0.81
	4	1.0000	1.0000	0.9999	0.9997	0.9990	0.9976	0.9961	0.9947	0.9898	0.9815	0.96
6	0	0.7351	0.5314	0.3771	0.2621	0.1780	0.1176	0.0905	0.0754	0.0467	0.0277	0.01
	1	0.9672	0.8857	0.7765	0.6554	0.5339	0.4202	0.3578	0.3191	0.2333	0.1636	0.10
	2	0.9978	0.9841	0.9527	0.9011	0.8306	0.7443	0.6870	0.6471	0.5443	0.4415	0.34
	3	0.9999	0.9987	0.9941	0.9830	0.9624	0.9295	0.9031	0.8826	0.8208	0.7447	0.65
	4	1.0000	0.9999	0.9996	0.9984	0.9954	0.9891	0.9830	0.9777	0.9590	0.9308	0.89
	5	1.0000	1.0000	1.0000	0.9999	0.9998	0.9993	0.9987	0.9982	0.9959	0.9917	0.98
7	0	0.6983	0.4783	0.3206	0.2097	0.1335	0.0824	0.0606	0.0490	0.0280	0.0152	0.00
	1	0.9556	0.8503	0.7166	0.5767	0.4449	0.3294	0.2696	0.2338	0.1586	0.1024	0.06
	2	0.9962	0.9743	0.9262	0.8520	0.7564	0.6471	0.5783	0.5323	0.4199	0.3164	0.22
	3	0.9998	0.9973	0.9879	0.9667	0.9294	0.8740	0.8318	0.8002	0.7102	0.6083	0.50
	4	1.0000	0.9998	0.9988	0.9953	0.9871	0.9712	0.9566	0.9444	0.9037	0.8471	0.77
	5	1.0000	1.0000	0.9999	0.9996	0.9987	0.9962	0.9935	0.9910	0.9812	0.9643	0.93
	6	1.0000	1.0000	1.0000	1.0000	0.9999	0.9998	0.9996	0.9994	0.9984	0.9963	0.99
8	0	0.6634	0.4305	0.2725	0.1678	0.1001	0.0576	0.0406	0.0319	0.0168	0.0084	0.00
	1	0.9428	0.8131	0.6572	0.5033	0.3671	0.2553	0.2006	0.1691	0.1064	0.0632	0.03
	2	0.9942	0.9619	0.8948	0.7969	0.6785	0.5518	0.4764	0.4278	0.3154	0.2201	0.14
	3	0.9996	0.9950	0.9786	0.9437	0.8862	0.8059	0.7481	0.7064	0.5941	0.4770	0.36
	4	1.0000	0.9996	0.9971	0.9896	0.9727	0.9420	0.9154	0.8939	0.8263	0.7396	0.63
	5	1.0000	1.0000	0.9998	0.9988	0.9958	0.9887	0.9813	0.9747	0.9502	0.9115	0.85
	6	1.0000	1.0000	1.0000	0.9999	0.9996	0.9987	0.9976	0.9964	0.9915	0.9819	0.96
	7	1.0000	1.0000	1.0000	1.0000	1.0000	0.9999	0.9999	0.9998	0.9993	0.9983	0.99
9	0	0.6302	0.3874	0.2316	0.1342	0.0751	0.0404	0.0272	0.0207	0.0101	0.0046	0.00
	1	0.9288	0.7748	0.5995	0.4362	0.3003	0.1960	0.1478	0.1211	0.0705	0.0385	0.01
	2	0.9916	0.9470	0.8591	0.7382	0.6007	0.4628	0.3854	0.3373	0.2318	0.1495	0.08
	3	0.9994	0.9917	0.9661	0.9144	0.8343	0.7297	0.6585	0.6089	0.4826	0.3614	0.25
	4	1.0000	0.9991	0.9944	0.9804	0.9511	0.9012	0.8602	0.8283	0.7334	0.6214	0.50
	5	1.0000	0.9999	0.9994	0.9969	0.9900	0.9747	0.9596	0.9464	0.9006	0.8342	0.74
	6	1.0000	1.0000	1.0000	0.9997	0.9987	0.9957	0.9922	0.9888	0.9750	0.9502	0.91
	7	1.0000	1.0000	1.0000	1.0000	0.9999	0.9996	0.9991	0.9986	0.9962	0.9909	0.98
	8	1.0000	1.0000	1.0000	1.0000	1.0000	1.0000	1.0000	0.9999	0.9997	0.9992	0.99

able A1. *(Continued)*

	x	0.05	0.10	0.15	0.20	0.25	0.30	0.33	0.35	0.40	0.45	0.50
							p					
	0	0.5987	0.3487	0.1969	0.1074	0.0563	0.0282	0.0182	0.0135	0.0060	0.0025	0.0010
	1	0.9139	0.7361	0.5443	0.3758	0.2440	0.1493	0.1080	0.0860	0.0464	0.0233	0.0107
	2	0.9885	0.9298	0.8202	0.6778	0.5256	0.3828	0.3070	0.2616	0.1673	0.0996	0.0547
	3	0.9990	0.9872	0.9500	0.8791	0.7759	0.6496	0.5684	0.5138	0.3823	0.2660	0.1719
	4	0.9999	0.9984	0.9901	0.9672	0.9219	0.8497	0.7936	0.7515	0.6331	0.5044	0.3770
	5	1.0000	0.9999	0.9986	0.9936	0.9803	0.9527	0.9268	0.9051	0.8338	0.7384	0.6230
	6	1.0000	1.0000	0.9999	0.9991	0.9965	0.9894	0.9815	0.9740	0.9452	0.8980	0.8281
	7	1.0000	1.0000	1.0000	0.9999	0.9996	0.9984	0.9968	0.9952	0.9877	0.9726	0.9453
	8	1.0000	1.0000	1.0000	1.0000	1.0000	0.9999	0.9997	0.9995	0.9983	0.9955	0.9893
	9	1.0000	1.0000	1.0000	1.0000	1.0000	1.0000	1.0000	1.0000	0.9999	0.9997	0.9990
	0	0.4632	0.2059	0.0874	0.0352	0.0134	0.0047	0.0025	0.0016	0.0005	0.0001	0.0000
	1	0.8290	0.5490	0.3186	0.1671	0.0802	0.0353	0.0206	0.0142	0.0052	0.0017	0.0005
	2	0.9638	0.8159	0.6042	0.3980	0.2361	0.1268	0.0833	0.0617	0.0271	0.0107	0.0037
	3	0.9945	0.9444	0.8227	0.6482	0.4613	0.2969	0.2171	0.1727	0.0905	0.0424	0.0176
	4	0.9994	0.9873	0.9383	0.8358	0.6865	0.5155	0.4148	0.3519	0.2173	0.1204	0.0592
	5	0.9999	0.9978	0.9832	0.9389	0.8516	0.7216	0.6291	0.5643	0.4032	0.2608	0.1509
	6	1.0000	0.9997	0.9964	0.9819	0.9434	0.8689	0.8049	0.7548	0.6098	0.4522	0.3036
	7	1.0000	1.0000	0.9994	0.9958	0.9827	0.9500	0.9163	0.8868	0.7869	0.6535	0.5000
	8	1.0000	1.0000	0.9999	0.9992	0.9958	0.9848	0.9711	0.9578	0.9050	0.8182	0.6964
	9	1.0000	1.0000	1.0000	0.9999	0.9992	0.9963	0.9921	0.9876	0.9662	0.9231	0.8491
	10	1.0000	1.0000	1.0000	1.0000	0.9999	0.9993	0.9984	0.9972	0.9907	0.9745	0.9408
	11	1.0000	1.0000	1.0000	1.0000	1.0000	0.9999	0.9997	0.9995	0.9981	0.9937	0.9824
	12	1.0000	1.0000	1.0000	1.0000	1.0000	1.0000	1.0000	0.9999	0.9997	0.9989	0.9963
	13	1.0000	1.0000	1.0000	1.0000	1.0000	1.0000	1.0000	1.0000	1.0000	0.9999	0.9995
	0	0.3585	0.1216	0.0388	0.0115	0.0032	0.0008	0.0003	0.0002	0.0000	0.0000	0.0000
	1	0.7358	0.3917	0.1756	0.0692	0.0243	0.0076	0.0036	0.0021	0.0005	0.0001	0.0000
	2	0.9245	0.6769	0.4049	0.2061	0.0913	0.0355	0.0189	0.0121	0.0036	0.0009	0.0002
	3	0.9841	0.8670	0.6477	0.4114	0.2252	0.1071	0.0642	0.0444	0.0160	0.0049	0.0013
	4	0.9974	0.9568	0.8298	0.6296	0.4148	0.2375	0.1589	0.1182	0.0510	0.0189	0.0059
	5	0.9997	0.9887	0.9327	0.8042	0.6172	0.4164	0.3083	0.2454	0.1256	0.0553	0.0207
	6	1.0000	0.9976	0.9781	0.9133	0.7858	0.6080	0.4921	0.4166	0.2500	0.1299	0.0577
	7	1.0000	0.9996	0.9941	0.9679	0.8982	0.7723	0.6732	0.6010	0.4159	0.2520	0.1316
	8	1.0000	0.9999	0.9987	0.9900	0.9591	0.8867	0.8182	0.7624	0.5956	0.4143	0.2517
	9	1.0000	1.0000	0.9998	0.9974	0.9861	0.9520	0.9134	0.8782	0.7553	0.5914	0.4119
	10	1.0000	1.0000	1.0000	0.9994	0.9961	0.9829	0.9650	0.9468	0.8725	0.7507	0.5881
	11	1.0000	1.0000	1.0000	0.9999	0.9991	0.9949	0.9881	0.9804	0.9435	0.8692	0.7483
	12	1.0000	1.0000	1.0000	1.0000	0.9998	0.9987	0.9966	0.9940	0.9790	0.9420	0.8684
	13	1.0000	1.0000	1.0000	1.0000	1.0000	0.9997	0.9992	0.9985	0.9935	0.9786	0.9423
	14	1.0000	1.0000	1.0000	1.0000	1.0000	1.0000	0.9999	0.9997	0.9984	0.9936	0.9793
	15	1.0000	1.0000	1.0000	1.0000	1.0000	1.0000	1.0000	1.0000	0.9997	0.9985	0.9941
	16	1.0000	1.0000	1.0000	1.0000	1.0000	1.0000	1.0000	1.0000	1.0000	0.9997	0.9987
	17	1.0000	1.0000	1.0000	1.0000	1.0000	1.0000	1.0000	1.0000	1.0000	1.0000	0.9998

Table A2. Standard normal cumulative distribution function $\Phi(z) = \int_{-\infty}^{z} \frac{1}{\sqrt{2\pi}} e^{-t^2/2} dt$

z	0.00	0.01	0.02	0.03	0.04	0.05	0.06	0.07	0.08	0.09
0.0	0.5000	0.5040	0.5080	0.5120	0.5160	0.5199	0.5239	0.5279	0.5319	0.535
0.1	0.5398	0.5438	0.5478	0.5517	0.5557	0.5596	0.5636	0.5675	0.5714	0.575
0.2	0.5793	0.5832	0.5871	0.5910	0.5948	0.5987	0.6026	0.6064	0.6103	0.614
0.3	0.6179	0.6217	0.6255	0.6293	0.6331	0.6368	0.6406	0.6443	0.6480	0.651
0.4	0.6554	0.6591	0.6628	0.6664	0.6700	0.6736	0.6772	0.6808	0.6844	0.687
0.5	0.6915	0.6950	0.6985	0.7019	0.7054	0.7088	0.7123	0.7157	0.7190	0.722
0.6	0.7257	0.7291	0.7324	0.7357	0.7389	0.7422	0.7454	0.7486	0.7517	0.754
0.7	0.7580	0.7611	0.7642	0.7673	0.7704	0.7734	0.7764	0.7794	0.7823	0.785
0.8	0.7881	0.7910	0.7939	0.7967	0.7995	0.8023	0.8051	0.8078	0.8106	0.813
0.9	0.8159	0.8186	0.8212	0.8238	0.8264	0.8289	0.8314	0.8340	0.8365	0.838
1.0	0.8413	0.8438	0.8461	0.8485	0.8508	0.8531	0.8554	0.8577	0.8599	0.862
1.1	0.8643	0.8665	0.8686	0.8708	0.8729	0.8749	0.8770	0.8790	0.8810	0.883
1.2	0.8849	0.8869	0.8888	0.8907	0.8925	0.8944	0.8962	0.8980	0.8997	0.901
1.3	0.9032	0.9049	0.9066	0.9082	0.9099	0.9115	0.9131	0.9147	0.9162	0.917
1.4	0.9192	0.9207	0.9222	0.9236	0.9251	0.9265	0.9279	0.9292	0.9306	0.931
1.5	0.9332	0.9345	0.9357	0.9370	0.9382	0.9394	0.9406	0.9418	0.9429	0.944
1.6	0.9452	0.9463	0.9474	0.9484	0.9495	0.9505	0.9515	0.9525	0.9535	0.954
1.7	0.9554	0.9564	0.9573	0.9582	0.9591	0.9599	0.9608	0.9616	0.9625	0.963
1.8	0.9641	0.9649	0.9656	0.9664	0.9671	0.9678	0.9686	0.9693	0.9699	0.970
1.9	0.9713	0.9719	0.9726	0.9732	0.9738	0.9744	0.9750	0.9756	0.9761	0.976
2.0	0.9772	0.9778	0.9783	0.9788	0.9793	0.9798	0.9803	0.9808	0.9812	0.981
2.1	0.9821	0.9826	0.9830	0.9834	0.9838	0.9842	0.9846	0.9850	0.9854	0.985
2.2	0.9861	0.9864	0.9868	0.9871	0.9875	0.9878	0.9881	0.9884	0.9887	0.989
2.3	0.9893	0.9896	0.9898	0.9901	0.9904	0.9906	0.9909	0.9911	0.9913	0.991
2.4	0.9918	0.9920	0.9922	0.9925	0.9927	0.9929	0.9931	0.9932	0.9934	0.993
2.5	0.9938	0.9940	0.9941	0.9943	0.9945	0.9946	0.9948	0.9949	0.9951	0.995
2.6	0.9953	0.9955	0.9956	0.9957	0.9959	0.9960	0.9961	0.9962	0.9963	0.996
2.7	0.9965	0.9966	0.9967	0.9968	0.9969	0.9970	0.9971	0.9972	0.9973	0.997
2.8	0.9974	0.9975	0.9976	0.9977	0.9977	0.9978	0.9979	0.9979	0.9980	0.998
2.9	0.9981	0.9982	0.9982	0.9983	0.9984	0.9984	0.9985	0.9985	0.9986	0.998
3.0	0.9987	0.9987	0.9987	0.9988	0.9988	0.9989	0.9989	0.9989	0.9990	0.999

Table A3. Upper quantiles $t_{\alpha,\nu}$ of Student t distribution with ν degrees of freedom.

ν	0.400	0.300	0.200	0.100	0.050	α 0.025	0.010	0.005	0.001	0.0005
1	0.325	0.727	1.376	3.078	6.314	12.706	31.821	63.657	318. 309	636.619
2	0.289	0.614	1.061	1.886	2.920	4.303	6.965	9.925	22.327	31.599
3	0.277	0.584	0.978	1.638	2.353	3.182	4.541	5.841	10.215	12.924
4	0.271	0.569	0.941	1.533	2.132	2.776	3.747	4.604	7.173	8 .610
5	0.267	0.559	0.920	1.476	2.015	2.571	3.365	4.032	5.893	6 .869
6	0.265	0.553	0.906	1.440	1.943	2.447	3.143	3.707	5.208	5 .959
7	0.263	0.549	0.896	1.415	1.895	2.365	2.998	3.499	4.785	5 .408
8	0.262	0.546	0.889	1.397	1.860	2.306	2.896	3.355	4.501	5 .041
9	0.261	0.543	0.883	1.383	1.833	2.262	2.821	3.250	4.297	4 .781
10	0.260	0.542	0.879	1.372	1.812	2.228	2.764	3.169	4.144	4.587
11	0.260	0.540	0.876	1.363	1.796	2.201	2.718	3.106	4.025	4.437
12	0.259	0.539	0.873	1.356	1.782	2.179	2.681	3.055	3.930	4.318
13	0.259	0.538	0.870	1.350	1.771	2.160	2.650	3.012	3.852	4.221
14	0.258	0.537	0.868	1.345	1.761	2.145	2.624	2.977	3.787	4.140
15	0.258	0.536	0.866	1.341	1.753	2.131	2.602	2.947	3.733	4.073
16	0.258	0.535	0.865	1.337	1.746	2.120	2.583	2.921	3.686	4.015
17	0.257	0.334	0.863	1.333	1.740	2.110	2.567	2.898	3.646	3.965
18	0.257	0.534	0.862	1.330	1.734	2.101	2.552	2.878	3.610	3.922
19	0.257	0.533	0.861	1.328	1.729	2.093	2.539	2.861	3.579	3.883
20	0.257	0.533	0.860	1.325	1.725	2.086	2.528	2.845	3.552	3.850
21	0.257	0.532	0.859	1.323	1.721	2.080	2.518	2.831	3.527	3.819
22	0.256	0.532	0.858	1.321	1.717	2.074	2.508	2.819	3.505	3.792
23	0.256	0.532	0.858	1.319	1.714	2.069	2.500	2.807	3.485	3.768
24	0.256	0.531	0.857	1.318	1.711	2.064	2.492	2.797	3.467	3.745
25	0.256	0.531	0.856	1.316	1.708	2.060	2.485	2.787	3.450	3.725
26	0.256	0.531	0.856	1.315	1.706	2.056	2.479	2.779	3.435	3.707
27	0.256	0.531	0.855	1.314	1.703	2.052	2.473	2.771	3.421	3.690
28	0.256	0.530	0.855	1.313	1.701	2.048	2.467	2.763	3.408	3.674
29	0.256	0.530	0.854	1.311	1.699	2.045	2.462	2.756	3.396	3.659
30	0.256	0.530	0.854	1.310	1.697	2.042	2.457	2.750	3.385	3.646
40	0.255	0.529	0.851	1.303	1.684	2.021	2.423	2.704	3.307	3.551
60	0.254	0.527	0.848	1.296	1.671	2.000	2.390	2.660	3.232	3.460
00	0.254	0.526	0.845	1.290	1.660	1.984	2.364	2.626	3.174	3.390
∞	0.253	0.524	0.842	1.282	1.645	1.960	2.326	2.576	3.090	3.291

Table A4. Upper quantiles $\chi^2_{\alpha,\nu}$ of the chi-square distribution with ν degrees of freedom.

ν	α									
	0.995	0.990	0.975	0.950	0.900	0.100	0.050	0.025	0.010	0.0
1	0.000	0.000	0.000	0.004	0.016	2.706	3.841	5.024	6.635	7.8
2	0.010	0.020	0.051	0.103	0.211	4.605	5.991	7.378	9.210	10.5
3	0.072	0.115	0.216	0.352	0.584	6.251	7.815	9.348	11.345	12.8
4	0.207	0.297	0.484	0.711	1.064	7.779	9.488	11.143	13.277	14.8
5	0.412	0.554	0.831	1.145	1.610	9.236	11.070	12.833	15.086	16.7
6	0.676	0.872	1.237	1.635	2.204	10.645	12.592	14.449	16.812	18.5
7	0.989	1.239	1.690	2.167	2.833	12.017	14.067	16.013	18.475	20.2
8	1.344	1.646	2.180	2.733	3.490	13.362	15.507	17.535	20.090	21.9
9	1.735	2.088	2.700	3.325	4.168	14.684	16.919	19.023	21.666	23.5
10	2.156	2.558	3.247	3.940	4.865	15.987	18.307	20.483	23.209	25.1
11	2.603	3.053	3.816	4.575	5.578	17.275	19.675	21.920	24.725	26.7
12	3.074	3.571	4.404	5.226	6.304	18.549	21.026	23.337	26.217	28.3
13	3.565	4.107	5.009	5.892	7.042	19.812	22.362	24.736	27.688	29.8
14	4.075	4.660	5.629	6.571	7.790	21.064	23.685	26.119	29.141	31.3
15	4.601	5.229	6.262	7.261	8.547	22.307	24.996	27.488	30.578	32.8
16	5.142	5.812	6.908	7.962	9.312	23.542	26.296	28.845	32.000	34.2
17	5.697	6.408	7.564	8.672	10.085	24.769	27.587	30.191	33.409	35.7
18	6.265	7.015	8.231	9.390	10.865	25.989	28.869	31.526	34.805	37.1
19	6.844	7.633	8.907	10.117	11.651	27.204	30.144	32.852	36.191	38.5
20	7.434	8.260	9.591	10.851	12.443	28.412	31.410	34.170	37.566	39.9
21	8.034	8.897	10.283	11.591	13.240	29.615	32.671	35.479	38.932	41.4
22	8.643	9.542	10.982	12.338	14.041	30.813	33.924	36.781	40.289	42.7
23	9.260	10.196	11.689	13.091	14.848	32.007	35.172	38.076	41.638	44.1
24	9.886	10.856	12.401	13.848	15.659	33.196	36.415	39.364	42.980	45.5
25	10.520	11.524	13.120	14.611	16.473	34.382	37.652	40.646	44.314	46.9
26	11.160	12.198	13.844	15.379	17.292	35.563	38.885	41.923	45.642	48.2
27	11.808	12.879	14.573	16.151	18.114	36.741	40.113	43.195	46.963	49.6
28	12.461	13.565	15.308	16.928	18.939	37.916	41.337	44.461	48.278	50.9
29	13.121	14.256	16.047	17.708	19.768	39.087	42.557	45.722	49.588	52.3
30	13.787	14.953	16.791	18.493	20.599	40.256	43.773	46.979	50.892	53.6
40	20.707	22.164	24.433	26.509	29.051	51.805	55.758	59.342	63.691	66.7
50	27.991	29.707	32.357	34.764	37.689	63.167	67.505	71.420	76.154	79.4
60	35.534	37.485	40.482	43.188	46.459	74.397	79.082	83.298	88.379	91.9
120	83.852	86.923	91.573	95.705	100.624	140.233	146.567	152.211	158.950	163.6

Table A5. Quantiles of chi-square distribution for determining the shortest confidence interval for σ.

	$\alpha = 0.10$				$\alpha = 0.05$			
ν	α_1	$\chi^2_{\alpha_1,\nu}$	$\chi^2_{\alpha_2,\nu}$	%	α_1	$\chi^2_{\alpha_1,\nu}$	$\chi^2_{\alpha_2,\nu}$	%
1	0.09998	0.016	18.189	24.92	0.04999	0.004	19.511	24.99
2	0.09988	0.211	18.056	49.02	0.04998	0.103	21.640	49.48
3	0.09948	0.582	17.647	61.21	0.04988	0.351	20.726	61.82
4	0.09882	1.056	18.100	68.50	0.04969	0.708	21.047	69.10
5	0.09800	1.594	18.907	73.41	0.04943	1.139	21.806	73.94
6	0.09708	2.175	19.871	76.95	0.04911	1.623	22.736	77.41
7	0.09612	2.788	20.927	79.65	0.04876	2.147	23.791	80.04
8	0.09516	3.426	22.041	81.77	0.04839	2.703	24.910	82.11
9	0.09420	4.084	23.182	83.48	0.04802	3.284	26.083	83.78
10	0.09328	4.758	24.352	84.90	0.04764	3.886	27.270	85.16
11	0.09238	5.447	25.530	86.09	0.04726	4.505	28.472	86.32
12	0.09152	6.147	26.719	87.11	0.04689	5.141	29.689	87.31
13	0.09070	6.858	27.915	87.98	0.04653	5.790	30.915	88.16
14	0.08990	7.579	29.109	88.75	0.04619	6.451	32.153	88.91
15	0.08916	8.308	30.313	89.42	0.04585	7.123	33.386	89.56
16	0.08844	9.045	31.514	90.02	0.04552	7.804	34.619	90.14
17	0.08774	9.788	32.711	90.55	0.04521	8.495	35.859	90.67
18	0.08710	10.539	33.916	91.03	0.04490	9.193	37.091	91.13
19	0.08648	11.295	35.117	91.46	0.04461	9.899	38.328	91.56
20	0.08588	12.056	36.315	91.85	0.04433	10.612	39.563	91.94
21	0.08530	12.823	37.510	92.21	0.04405	11.331	40.791	92.29
22	0.08476	13.595	38.708	92.54	0.04379	12.056	42.024	92.61
23	0.08424	14.371	39.903	92.84	0.04354	12.787	43.254	92.91
24	0.08374	15.151	41.095	93.12	0.04329	13.523	44.478	93.18
25	0.08326	15.935	42.286	93.37	0.04306	14.264	45.706	93.43
26	0.08278	16.723	43.471	93.61	0.04283	15.009	46.928	93.67
27	0.08234	17.514	44.659	93.83	0.04261	15.759	48.149	93.89
28	0.08192	18.310	45.846	94.04	0.04240	16.513	49.369	94.09
29	0.08150	19.108	47.027	94.23	0.04219	17.271	50.583	94.28
30	0.08110	19.909	48.208	94.41	0.04199	18.032	51.797	94.46
40	0.07780	28.063	59.927	95.75	0.04031	25.823	63.834	95.78
50	0.07536	36.418	71.499	96.57	0.03903	33.855	75.694	96.59
60	0.07346	44.918	82.949	97.13	0.03803	42.064	87.418	97.14
80	0.07066	62.232	105.571	97.83	0.03653	58.859	110.535	97.84
100	0.06866	79.847	127.915	98.25	0.03544	76.015	133.321	98.26

Table A6(a). Upper quantiles $F_{0.05,\nu_1,\nu_2}$ of the F distribution.

ν_2	ν_1									
	1	2	3	4	5	6	7	8	9	10
1	161.45	199.50	215.71	224.58	230.16	233.99	236.77	238.88	240.54	241.8
2	18.51	19.00	19.16	19.25	19.30	19.33	19.35	19.37	19.38	19.4
3	10.13	9.55	9.28	9.12	9.01	8.94	8.89	8.85	8.81	8.7
4	7.71	6.94	6.59	6.39	6.26	6.16	6.09	6.04	6.00	5.9
5	6.61	5.79	5.41	5.19	5.05	4.95	4.88	4.82	4.77	4.7
6	5.99	5.14	4.76	4.53	4.39	4.28	4.21	4.15	4.10	4.0
7	5.59	4.74	4.35	4.12	3.97	3.87	3.79	3.73	3.68	3.6
8	5.32	4.46	4.07	3.84	3.69	3.58	3.50	3.44	3.39	3.3
9	5.12	4.26	3.86	3.63	3.48	3.37	3.29	3.23	3.18	3.1
10	4.96	4.10	3.71	3.48	3.33	3.22	3.14	3.07	3.02	2.9
11	4.84	3.98	3.59	3.36	3.20	3.09	3.01	2.95	2.90	2.8
12	4.75	3.89	3.49	3.26	3.11	3.00	2.91	2.85	2.80	2.7
13	4.67	3.81	3.41	3.18	3.03	2.92	2.83	2.77	2.71	2.6
14	4.60	3.74	3.34	3.11	2.96	2.85	2.76	2.70	2.65	2.6
15	4.54	3.68	3.29	3.06	2.90	2.79	2.71	2.64	2.59	2.5
16	4.49	3.63	3.24	3.01	2.85	2.74	2.66	2.59	2.54	2.4
17	4.45	3.59	3.20	2.96	2.81	2.70	2.61	2.55	2.49	2.4
18	4.41	3.55	3.16	2.93	2.77	2.66	2.58	2.51	2.46	2.4
19	4.38	3.52	3.13	2.90	2.74	2.63	2.54	2.48	2.42	2.3
20	4.35	3.49	3.10	2.87	2.71	2.60	2.51	2.45	2.39	2.3
21	4.32	3.47	3.07	2.84	2.68	2.57	2.49	2.42	2.37	2.3
22	4.30	3.44	3.05	2.82	2.66	2.55	2.46	2.40	2.34	2.3
23	4.28	3.42	3.03	2.80	2.64	2.53	2.44	2.37	2.32	2.2
24	4.26	3.40	3.01	2.78	2.62	2.51	2.42	2.36	2.30	2.2
25	4.24	3.39	2.99	2.76	2.60	2.49	2.40	2.34	2.28	2.2
26	4.23	3.37	2.98	2.74	2.59	2.47	2.39	2.32	2.27	2.2
27	4.21	3.35	2.96	2.73	2.57	2.46	2.37	2.31	2.25	2.2
28	4.20	3.34	2.95	2.71	2.56	2.45	2.36	2.29	2.24	2.1
29	4.18	3.33	2.93	2.70	2.55	2.43	2.35	2.28	2.22	2.1
30	4.17	3.32	2.92	2.69	2.53	2.42	2.33	2.27	2.21	2.1
40	4.08	3.23	2.84	2.61	2.45	2.34	2.25	2.18	2.12	2.0
60	4.00	3.15	2.76	2.53	2.37	2.25	2.17	2.10	2.04	1.9
120	3.92	3.07	2.68	2.45	2.29	2.18	2.09	2.02	1.96	1.9
∞	3.84	3.00	2.60	2.37	2.21	2.10	2.01	1.94	1.88	1.8

Table A6(a). *(Continued)*

ν_2	ν_1								
	12	15	20	24	30	40	60	120	∞
1	243.91	245.95	248.01	249.05	250.10	251.14	252.20	253.23	254.30
2	19.41	19.43	19.45	19.45	19.46	19.47	19.48	19.49	19.50
3	8.74	8.70	8.66	8.64	8.62	8.59	8.57	8.55	8.53
4	5.91	5.86	5.80	5.77	5.75	5.72	5.69	5.66	5.63
5	4.68	4.62	4.56	4.53	4.50	4.46	4.43	4.40	4.36
6	4.00	3.94	3.87	3.84	3.81	3.77	3.74	3.70	3.67
7	3.57	3.51	3.44	3.41	3.38	3.34	3.30	3.27	3.23
8	3.28	3.22	3.15	3.12	3.08	3.04	3.01	2.97	2.93
9	3.07	3.01	2.94	2.90	2.86	2.83	2.79	2.75	2.71
10	2.91	2.85	2.77	2.74	2.70	2.66	2.62	2.58	2.54
11	2.79	2.72	2.65	2.61	2.57	2.53	2.49	2.45	2.40
12	2.69	2.62	2.54	2.51	2.47	2.43	2.38	2.34	2.30
13	2.60	2.53	2.46	2.42	2.38	2.34	2.30	2.25	2.21
14	2.53	2.46	2.39	2.35	2.31	2.27	2.22	2.18	2.13
15	2.48	2.40	2.33	2.29	2.25	2.20	2.16	2.11	2.07
16	2.42	2.35	2.28	2.24	2.19	2.15	2.11	2.06	2.01
17	2.38	2.31	2.23	2.19	2.15	2.10	2.06	2.01	1.96
18	2.34	2.27	2.19	2.15	2.11	2.06	2.02	1.97	1.92
19	2.31	2.23	2.16	2.11	2.07	2.03	1.98	1.93	1.88
20	2.28	2.20	2.12	2.08	2.04	1.99	1.95	1.90	1.84
21	2.25	2.18	2.10	2.05	2.01	1.96	1.92	1.87	1.81
22	2.23	2.15	2.07	2.03	1.98	1.94	1.89	1.84	1.78
23	2.20	2.13	2.05	2.01	1.96	1.91	1.86	1.81	1.76
24	2.18	2.11	2.03	1.98	1.94	1.89	1.84	1.79	1.73
25	2.16	2.09	2.01	1.96	1.92	1.87	1.82	1.77	1.71
26	2.15	2.07	1.99	1.95	1.90	1.85	1.80	1.75	1.69
27	2.13	2.06	1.97	1.93	1.88	1.84	1.79	1.73	1.67
28	2.12	2.04	1.96	1.91	1.87	1.82	1.77	1.71	1.65
29	2.10	2.03	1.94	1.90	1.85	1.81	1.75	1.70	1.64
30	2.09	2.01	1.93	1.89	1.84	1.79	1.74	1.68	1.62
40	2.00	1.92	1.84	1.79	1.74	1.69	1.64	1.58	1.51
60	1.92	1.84	1.75	1.70	1.65	1.59	1.53	1.47	1.39
120	1.83	1.75	1.66	1.61	1.55	1.50	1.43	1.35	1.25
∞	1.75	1.67	1.57	1.52	1.46	1.39	1.32	1.22	1.00

Table A6(b). Upper quantiles $F_{0.01,\nu_1,\nu_2}$ of the F distribution.

ν_2	1	2	3	4	5	6	7	8	9	10
					ν_1					
1	4052.2	4999.5	5403.4	5624.6	5763.7	5859.0	5928.4	5981.1	6022.5	6055.9
2	98.50	99.00	99.17	99.25	99.30	99.33	99.36	99.37	99.39	99.40
3	34.12	30.82	29.46	28.71	28.24	27.91	27.67	27.49	27.35	27.23
4	21.20	18.00	16.69	15.98	15.52	15.21	14.98	14.80	14.66	14.55
5	16.26	13.27	12.06	11.39	10.97	10.67	10.46	10.29	10.16	10.05
6	13.75	10.92	9.78	9.15	8.75	8.47	8.26	8.10	7.98	7.87
7	12.25	9.55	8.45	7.85	7.46	7.19	6.99	6.84	6.72	6.62
8	11.26	8.65	7.59	7.01	6.63	6.37	6.18	6.03	5.91	5.81
9	10.56	8.02	6.99	6.42	6.06	5.80	5.61	5.47	5.35	5.26
10	10.04	7.56	6.55	5.99	5.64	5.39	5.20	5.06	4.94	4.85
11	9.65	7.21	6.22	5.67	5.32	5.07	4.89	4.47	4.63	4.54
12	9.33	6.93	5.95	5.41	5.06	4.82	4.64	4.50	4.39	4.30
13	9.07	6.70	5.74	5.21	4.86	4.62	4.44	4.30	4.19	4.10
14	8.86	6.51	5.56	5.04	4.69	4.46	4.28	4.14	4.03	3.94
15	8.68	6.36	5.42	4.89	4.56	4.32	4.14	4.00	3.89	3.80
16	8.53	6.23	5.29	4.77	4.44	4.20	4.03	3.89	3.78	3.69
17	8.40	6.11	5.18	4.67	4.34	4.10	3.93	3.79	3.68	3.59
18	8.29	6.01	5.09	4.58	4.25	4.01	3.84	3.71	3.60	3.51
19	8.18	5.93	5.01	4.50	4.17	3.94	3.77	3.63	3.52	3.43
20	8.10	5.85	4.94	4.43	4.10	3.87	3.70	3.56	3.46	3.37
21	8.02	5.78	4.87	4.37	4.04	3.81	3.64	3.51	3.40	3.31
22	7.95	5.72	4.82	4.31	3.99	3.76	3.59	3.45	3.35	3.26
23	7.88	5.66	4.76	4.26	3.94	3.71	3.54	3.41	3.30	3.21
24	7.82	5.61	4.72	4.22	3.90	3.67	3.50	3.36	3.26	3.17
25	7.77	5.57	4.68	4.18	3.85	3.63	3.46	3.32	3.22	3.13
26	7.72	5.53	4.64	4.14	3.82	3.59	3.42	3.29	3.18	3.09
27	7.68	5.49	4.60	4.11	3.78	3.56	3.39	3.26	3.15	3.06
28	7.64	5.45	4.57	4.07	3.75	3.53	3.36	3.23	3.12	3.03
29	7.60	5.42	4.54	4.04	3.73	3.50	3.33	3.20	3.09	3.00
30	7.56	5.39	4.51	4.02	3.70	3.47	3.30	3.17	3.07	2.98
40	7.31	5.18	4.31	3.83	3.51	3.29	3.12	2.99	2.89	2.80
60	7.08	4.98	4.13	3.65	3.34	3.12	2.95	2.82	2.72	2.63
120	6.85	4.79	3.95	3.48	3.17	2.96	2.79	2.66	2.56	2.47
∞	6.63	4.61	3.78	3.32	3.02	2.80	2.64	2.51	2.41	2.32

ble A6(b). *(Continued)*

	12	15	20	24	30	40	60	120	∞
					ν_1				
1	6106.3	6157.3	6208.7	6234.6	6260.7	6286.8	6313.1	6339.4	6366.0
2	99.42	99.43	99.45	99.46	99.47	99.47	99.48	99.49	99.50
3	27.05	26.87	26.69	26.60	26.50	26.41	26.32	26.22	26.13
4	14.37	14.20	14.02	13.93	13.84	13.75	13.65	13.56	13.46
5	9.89	9.72	9.55	9.47	9.38	9.29	9.20	9.11	9.02
6	7.72	7.56	7.40	7.31	7.23	7.14	7.06	6.97	6.88
7	6.47	6.31	6.16	6.07	5.99	5.91	5.82	5.74	5.65
8	5.67	5.52	5.36	5.28	5.20	5.12	5.03	4.95	4.86
9	5.11	4.96	4.81	4.73	4.65	4.57	4.48	4.40	4.31
0	4.71	4.56	4.41	4.33	4.25	4.17	4.08	4.00	3.91
1	4.40	4.25	4.10	4.02	3.94	3.86	3.78	3.69	3.60
2	4.16	4.01	3.86	3.78	3.70	3.62	3.54	3.45	3.36
3	3.96	3.82	3.66	3.59	3.51	3.43	3.34	3.25	3.17
4	3.80	3.66	3.51	3.43	3.35	3.27	3.18	3.09	3.00
5	3.67	3.52	3.37	3.29	3.21	3.13	3.05	2.96	2.87
6	3.55	3.41	3.26	3.18	3.10	3.02	2.93	2.84	2.75
7	3.46	3.31	3.16	3.08	3.00	2.92	2.83	2.75	2.65
8	3.37	3.23	3.08	3.00	2.92	2.84	2.75	2.66	2.57
9	3.30	3.15	3.00	2.92	2.84	2.76	2.67	2.58	2.49
0	3.23	3.09	2.94	2.86	2.78	2.69	2.61	2.52	2.42
1	3.17	3.03	2.88	2.80	2.72	2.64	2.55	2.46	2.36
2	3.12	2.98	2.83	2.75	2.67	2.58	2.50	2.40	2.31
3	3.07	2.93	2.78	2.70	2.62	2.54	2.45	2.35	2.26
4	3.03	2.89	2.74	2.66	2.58	2.49	2.40	2.31	2.21
5	2.99	2.85	2.70	2.62	2.54	2.45	2.36	2.27	2.17
6	2.96	2.81	2.66	2.58	2.50	2.42	2.33	2.23	2.13
7	2.93	2.78	2.63	2.55	2.47	2.38	2.29	2.20	2.10
8	2.90	2.75	2.60	2.52	2.44	2.35	2.26	2.17	2.06
9	2.87	2.73	2.57	2.49	2.41	2.33	2.23	2.14	2.03
0	2.84	2.70	2.55	2.47	2.39	2.30	2.21	2.11	2.01
0	2.66	2.52	2.37	2.29	2.20	2.11	2.02	1.92	1.80
0	2.50	2.35	2.20	2.12	2.03	1.94	1.84	1.73	1.60
0	2.34	2.19	2.03	1.95	1.86	1.76	1.66	1.53	1.38
∞	2.18	2.04	1.88	1.79	1.70	1.59	1.47	1.32	1.00

Table A7. Tail probabilities $1 - Q(z)$ of the Kolmogorov distribution.

z	0.00	0.01	0.02	0.03	0.04	0.05	0.06	0.07	0.08	0.09
0.3	1.0000	1.0000	0.9999	0.9999	0.9998	0.9997	0.9995	0.9992	0.9987	0.99
0.4	0.9972	0.9960	0.9945	0.9926	0.9903	0.9874	0.9840	0.9800	0.9753	0.97
0.5	0.9639	0.9572	0.9497	0.9415	0.9325	0.9228	0.9124	0.9013	0.8896	0.87
0.6	0.8643	0.8508	0.8367	0.8222	0.8073	0.7920	0.7764	0.7604	0.7442	0.72
0.7	0.7112	0.6945	0.6777	0.6609	0.6440	0.6272	0.6104	0.5936	0.5770	0.56
0.8	0.5441	0.5280	0.5120	0.4962	0.4306	0.4653	0.4503	0.4355	0.4209	0.40
0.9	0.3927	0.3791	0.3657	0.3527	0.3399	0.3275	0.3154	0.3036	0.2921	0.28
1.0	0.2700	0.2594	0.2492	0.2392	0.2296	0.2202	0.2111	0.2024	0.1939	0.18
1.1	0.1777	0.1700	0.1626	0.1555	0.1486	0.1420	0.1356	0.1294	0.1235	0.11
1.2	0.1122	0.1070	0.1019	0.0970	0.0924	0.0879	0.0834	0.0794	0.0755	0.07
1.3	0.0681	0.0646	0.0613	0.0582	0.0551	0.0522	0.0495	0.0469	0.0443	0.04
1.4	0.0397	0.0375	0.0354	0.0335	0.0316	0.0298	0.0282	0.0266	0.0250	0.02
1.5	0.0222	0.0209	0.0197	0.0185	0.0174	0.0164	0.0154	0.0145	0.0136	0.01
1.6	0.0120	0.0112	0.0105	0.0098	0.0092	0.0086	0.0081	0.0076	0.0071	0.00
1.7	0.0062	0.0058	0.0054	0.0050	0.0047	0.0044	0.0041	0.0038	0.0035	0.00
1.8	0.0031	0.0029	0.0027	0.0025	0.0023	0.0021	0.0020	0.0018	0.0017	0.00
1.9	0.0015	0.0014	0.0013	0.0012	0.0011	0.0010	0.0009	0.0009	0.9998	0.00
2.0	0.0007	0.0006	0.0006	0.0005	0.0005	0.0004	0.0004	0.0004	0.0003	0.00
2.1	0.0003	0.0003	0.0002	0.0002	0.0002	0.0002	0.0002	0.0002	0.0001	0.00
2.2	0.0001	0.0001	0.0001	0.0001	0.0001	0.0001	0.0001	0.0001	0.0001	0.00
2.3	0.0001	0.0000	0.0000	0.0000	0.0000	0.0000	0.0000	0.0000	0.0000	0.00

Answers to Selected Even-Numbered Problems

CHAPTER 1

1.2.2 TT, HTT, THT, THHT, HTHT, HHTT, HHHH, THHH, HTHH, HHTH, HHHT.

1.2.4 134, 234, 135, 235.

1.2.6 (i) 10 pairs (MS, MG, MP, MJ, SG, SP, SJ, GP, GJ, PJ); (ii) 8 pairs (MS, MG, MP, MJ, SM, SG, SP, SJ); (iii) 6 (MG+SP, MG+SJ, MP+SJ, MP+SG, MJ+SG, MJ+SP).

1.2.8 $b^2 > 4ac$, which gives a region in 3-dimensional space.

1.2.10 (i) THT; (ii) A_2, A_4, A_6; (iii) HHH, TTT, HTT, THH; (iv) B_2, B_5.

1.2.12 Yes. If the answer is "yes" and the interviewer manages to find out that the respondent was not born in April, the privacy is not mantained.

1.3.2 False.

1.3.4 True.

1.3.6 True.

1.3.10 $X = B \div A$.

1.3.12 $D_1 = E_1$, $D_2 = E_2$, $D_3 = E_3$, $D_4 = E_9$, $D_5 = E_{11}$, $D_6 = E_{10}$, $D_7 = E_6$, $D_8 = E_5 = E_4$, $D_9 = E_7$, $D_{10} = E_8$.

1.3.14 (i) $x = 0, y = 4$; (ii) either $x = 0$ or $y \geq 3$; (iii) $x = 0$, no inference about y possible; (iv) $x \leq 4$, no inference about y possible.

1.4.2 $\lim A_n = \varnothing$.

1.4.4 (i) 2^{2^n}; (ii) $2^{3 \times 2^{n-2}}$; (iii) 2^{n+1}; (iv) 2; (v) Answers are the same as for the field.

CHAPTER 2

2.3.2 True: (i), (ii), (v); False: (iii), (iv), (vi), (vii), (viii).

812 ANSWERS TO SELECTED EVEN-NUMBERED PROBLEMS

2.3.4 (i) 13/24
2.4.2 5/18.
2.4.4 5/6.
2.4.6 (i) 0.6; (ii) 0.4; (iii) 0.5.
2.4.8 (i) 0.1; (ii) 0.3; (iii) 1; (iv) 0.6.
2.4.10 0.1.
2.4.12 (i) 0.56; (ii) 0.88; (iii) 0.32.
2.4.14 4/9.
2.5.2 0.25.
2.5.4 25.
2.7.2 Yes, Tom's utility of the ticket satisfies $u(T) < 150$.

CHAPTER 3

3.2.2 $x = 12$.
3.2.4 $(2n)!/(2^n n!)$.
3.2.6 (i) 20; (ii) 13; (iii) 14.
3.2.8 1,024,000,000.
3.2.10 (i) 153; (ii) a hat; (iii) 138.
3.2.14 (i) $2/n$; (ii) $1/n$; (iii) 0.5; (iv) $2(n-3)/[n(n-1)]$.
3.2.16 (i) 28,800; (ii) 86,400.
3.2.18 $k \geq n$.
3.2.20 (i) $v(A) = 0.6$; (ii) $v(A) = 0.4, v(B) = 0.25, v(x) = 7/60$ for other members.
3.2.22 0.016 for each non-permanent member, 0.168 for each permanent member.
3.3.2 (i) $\binom{2}{2}\binom{98}{8}/\binom{100}{10}$; (ii) $\binom{2}{1}\binom{98}{9}/\binom{100}{10}$; (iii) $\binom{2}{0}\binom{98}{10}/\binom{100}{10}$.
3.3.4 $2^{50}/\binom{100}{50}$.
3.3.6 4!49!
3.3.8 (i) 4; (ii) 36; (iii) 5108; (iv) 624; (v) 3744; (vi) 109,824.
3.3.10 $p = 0.0222$. If Queen is replaced by Jack, then $p = 0.0204$.
3.3.12 (i) 0.8964; (ii) 0.04255; (iii) 0.0611.
3.3.14 $p = A(n, k)/k^n$.
3.3.16 $p = (n + 1)/\binom{2n}{n}$.
3.3.18 (i) 252; (ii) 0; (iii) 5.
3.3.20 $2\binom{a+b}{b+r+k}/\binom{a+b}{a}$.
3.4.6 $\binom{-2}{k} = (-1)^k(k + 1)$.
3.5.2 $\binom{n}{d}\binom{d}{c}\binom{c}{b}\binom{b}{a}$ or $\binom{n}{a}\binom{n-a}{b-a}\binom{n-b}{c-b}\binom{n-c}{d-c}$.

3.5.4   107,775,360.

3.6.2 1.11×10^{-14}.

CHAPTER 4

4.2.2   (i) 1;   (ii) $P(B)/[P(A) + P(B)]$.

4.2.4 $a/(1 + a)$.

4.2.6   1/3.

4.2.8   1/4.

4.2.10   (i) 0.492;   (ii) 0.123;   (iii) 1;   (iv) 1/4;   (v) 0.5, 0.105, 1, 0.210;
(vi) 0.25 for n odd, less than 0.25 for n even;   (vii) P(sum odd) $\rightarrow$ 0.5,
P(product odd) $\rightarrow$ 0.125, P(product odd|sum odd) $\rightarrow$ 0.25.

4.2.12 $3/4^n$.

4.2.14   (i) 0.145;   (ii) 0.5;   (iii) 0.885.

4.3.2   (i) 32/189;   (ii) 46/189;   (iii) 111/189.

4.3.4   Both formulas are false.

4.3.6 $1 - p$.

4.3.8   1.

4.4.2   (i) 23/45;   (ii) 13/45;   (iii) 14/23.

4.4.4   15/29.

4.4.6   15/22.

4.4.8   1/3.

4.5.2 $1 - \sqrt{2}/2$.

4.5.4   0 for $k \geq 1$ and $1 - k$ for $k < 1$.

4.5.6 A and B are not independent–all other pairs are.

4.5.8 $0 < P(A) < 1$.

4.5.10   (i) a;   (ii) a.

4.5.12   6/11.

4.5.14   6/11.

4.5.16   0.006.

4.5.18   Plan (a) is better.

4.5.20   (i) No;   (ii) No.

4.6.2   (i) If B is the event consisting of the occurrence of all events in the set
K of size t and non–occurrence of all remaining events, then $P(B) = p_t/\binom{N}{t}$;
(ii) $\sum_{t=0}^{N-1} p_t(1 - t/N)$ and $\sum_{t=2}^{N} p_t\binom{N-2}{t-2}/\binom{N}{t}$

CHAPTER 5

5.1.2 $P(X_{n+1} = j | X_n = i) = p_j$ for $i = 0, \ldots, m$. If $i > m$ then $P(X_{n+1} = j | X_n = i) = p_{j-i+m}$ for $j = 0, 1, \ldots$

5.2.2 The formulas remain valid if we replace p by $p/(1-r)$ and q by $q/(1-r)$.

5.2.4 Rows of the matrix are: $(0, 1/2, 1/2), (1/4, 0, 3/4), (1/8, 7/8, 0)$.

5.2.6 $p_{i,0} = r_i/(r_i + r_{i+1} + \ldots), p_{i,i+1} = 1 - r_i/(r_i + r_{i+1} + \ldots)$

5.3.2 $p_{i,i-2}^{(2)} = i(i-1)/N^2, p_{i,i}^{(2)} = [i(N-i+1) + (N-i)(i+1)]/N^2, p_{i,i+2}^{(2)} = (N-i)(N-i-1)/N^2, p_{i,j}^{(2)} = 0$ for all other j.

5.3.4 $p_{up}^{(2)} = r^2[(100-C)A + (100-C)(100-B) + C(100-C)]/20000$.

5.4.4 (i) Boris Borisovich is most frequent; (ii) Ivan Borisovich is less frequent than Boris Ivanovich; (iii) 1/3.

5.4.6 $\binom{N}{j}\binom{N}{N-j}/\binom{2N}{N}$.

5.5.2 $p_{30-30} = \sum_{n=0}^{\infty}(2pq)^n p^2 = p^2/(1 - 2pq)$.

CHAPTER 6

6.2.4 0 for $t < 1$, $k(k+1)/42$ for $k \le t < k+1, k = 1, \ldots, 5$, and 1 for $t \ge 6$.

6.2.6 $P\{W \le t\} = [F_X(t)]^2$.

6.2.8 0 for $t < 0$, $\pi(t/a)^2$ for $0 \le t < a/2$, $\sqrt{(2t/a)^2 - 1} + 2(t/a)^2 - [\pi/2 - 2\arctan\sqrt{(2t/a)^2 - 1}]$ for $a/2 \le t < a\sqrt{2}/2$, and 1 for $t \ge a\sqrt{2}/2$.

6.2.10 (i) 0.7; (ii) 0.0608; (iii) 0.7786.

6.2.12 1 if $d/D \le 0.2929$, 2 if $0.2929 < d/D \le 0.7071$, 3 if $0.7071 \le d/D < 0.7979$, and 4 if $d/D \ge 0.7979$. Median is not unique for $d/D = 0.7071$.

6.3.2 (i) $(1 - e^{-\lambda/2})/(1 - e^{-\lambda})$; (iii) $(e^{-\lambda/2} - e^{-\lambda})/(1 - e^{-\lambda})$.

6.3.6 (i) 1/12; (ii) 0.75.

6.3.8 (i) $P\{X_4 = -4\} = P\{X_4 = 4\} = 1/16, P\{X_4 = -2\} = P\{X_4 = 2\} = 4/16, P\{X_4 = 0\} = 6/16$; (ii) 5/16; (iii) $P\{X_4 > 0|X_4 \ge 4\} = 5/11$ for $n = 4$, and $P\{X_5 > 0|X_5 \ge 0\} = 1$ for $n = 5$.

6.3.10 (i) $a = 5, b = 22$; (ii) 0.3195; (iii) 0.2065.

6.3.12 (i) 5/6; (ii) 0.1150; (iii) 0.5.

6.4.2 $\lambda^n e^{-\lambda}/n!, n = 0, 1, 2, \ldots$

6.4.4 $g(y) = 1/\sqrt{1 + 4y}$ for $0 \le y \le 2$ and 0 otherwise.

6.4.6 (i) $X = \sqrt[5]{33U - 32}$; (ii) 32/33.

6.4.8 (i) uniform on $[0, 1]$; (ii) uniform on $[-1, 1]$.

6.4.10 (i) $g(y) = 2/\sqrt{1 - 4y}$ for $0 \le y < 0.25$, and 0 otherwise; (ii) $h(w) = 2$ for $0.5 \le w \le 1$ and 0 otherwise.

6.4.12 $g(u) = 2u\lambda e^{-\lambda u^2}$ for $u \ge 0$ and 0 for $u < 0$.

6.4.14 $g_E(u) = m^{-3/2}k\sqrt{2u}e^{-2bu/m}$ for $u > 0$.

6.5.2 (i) $(\alpha/2)e^\alpha$; (ii) $h(t) = 1/(2 - t)$ for $0 < t < 1$ and $h(t) = \alpha$ for $t \ge 1$.

6.5.4 0.1876.

CHAPTER 7

7.1.2 $f(x,y) = x/15 + y/15$ for $0 \le x \le 2, 0 \le y \le 3$.

7.1.4 (ii) 22/36, 0, 1/36;   (iii) 2/3, 2/3, 10/26, 3/16.

7.1.6 $k = 2$.

7.1.8 (i) $c = 6$;   (ii) 3/20;   (iii) 0 if $u < 0, v < 0$, u^2v^3 if $0 \le u, v \le 1$, v^3 if $u > 1, 0 \le v \le 1$, u^2 if $v > 1, 0 \le u \le 1$, and 1 if $u > 1, v > 1$.

7.2.2 (i) $e^{-2\lambda}$;   (ii) $(\lambda^x/x!)e^{-\lambda}$;   (iv) $e^{-2\lambda}\sum_{y=1}^{\infty}\sum_{x=0}^{y-1}\lambda^{x+y}/(x!y!)$;   (v) $0.5 - 0.5e^{-2\lambda}\sum_{x=0}^{\infty}\lambda^{2x}/(x!)^2$.

7.2.6 (4) for X, (7) for Y.

7.2.8 (i) 1/14;   (ii) 0;   (iii) 4/7;   (iv) 11/14;   (v) No.

7.2.10 (i) e^{-5} (ii) $1/(1+y)^2$.

7.2.12 (i) $G_0(u) = [b/(a+b)]e^{-(a+b)u}, G_1(u) = [a/(a+b)]e^{-(a+b)u}$;   (ii) U has EXP$(a+b)$ distribution.

7.2.14 $\int_0^{\infty} h_1(x)\exp\{-\int_0^x[h_1(u) + h_2(u)]du\}dx$.

7.3.2 (i) $\alpha > -1$;   (ii) $A = (\alpha+1)(\alpha+2)$;   (iii) $f_X(x) = (\alpha+2)(1-x)^{\alpha+1}, f_Y(y) = (\alpha+2)y^{\alpha+1}$;   (iv) $g(x|y) = (\alpha+1)(1-x/y)^{\alpha}/y, g(y|x) = (\alpha+1)[1/(1-x)][(y-x)/(1-x)]^{\alpha}$ for $0 < x < y < 1$.

7.3.4 $f_D(z) = (z/\sigma^2)\exp\{-z^2/(2\sigma^2)\}, z > 0$.

7.3.6 (i) $f(x,y) = 1/(2\pi R\sqrt{x^2+y^2})$ for $x^2+y^2 \le R^2$;   (ii) $f_X(x) = [1/(\pi R)]\log\left[R/x + \sqrt{(R/x)^2 - 1}\right]$;   (iii) $\varphi(r) = 2r/R^2, 0 \le r \le R$.

7.3.8 (i) $k = 1$;   (ii) $f_X(x) = 2x^2 + 1/3, f_Y(y) = y^2 + 2/3$;   (iii) 0.3011.

7.4.2 $P(Y - X = j) = (1/11)(5/6)^j$ for $j \ge 1$ and $P(X = Y) = 1/11$ for $j = 0$.

7.4.4 $f_{Z_1} = n[1 - F(z)]^{n-1}f(z), f_{Z_n} = nF^{n-1}(z)f(z)$.

7.4.6 $f_U(u) = e^{-u/2}/\sqrt{2\pi u}$.

7.4.8 $f_W(w) = b(1 - \sqrt{w})^2$ for $0 \le w \le 1$.

7.4.10 $g(u,v) = f[(u-b)/a, (v-d)/c]/|ac|$.

7.4.12 (i) $k = 1/[4(a+b)]$;   (ii) $f_Z(z) = [(a-b)\sqrt{z} + b]/[4(a+b)z^2]$.

7.4.14 $1 - e^{-c^2/2}$.

7.4.16 (i) For $U = aX + bY, V = X, f_U(u) = \int f(v, (u - av)/b)/|b|dv$;   (ii) For $U = XY, V = X, f_U(u) = \int f(v, u/v)|u|/v^2 dv$;   (iii) For $U = X/Y, V = X$, $\int f(v, v/u)|v|/u^2 dv$.

7.4.18 $f_{X/Y}(t) = 1/(t+1)^2$.

7.5.2 (i) $q = 2/(a+b+c)$;   (ii) $f_{X_1,X_2}(x_1,x_2) = (2ax_1 + 2bx_2 + c)/(a+b+c)$ for $0 \le x_1, x_2 \le 1$;   (iii) $g_{(X_1,X_2)|X_3}(x_1,x_2|x_3) = 2(ax_1 + bx_2 + cx_3)/(a+b+2cx_3)$;   (iv) $g_{X_3|(X_1,X_2)}(x_3|x_1,x_2) = 2(ax_1 + bx_2 + cx_3)/(2ax_1 + 2bx_2 + c)$;   (v) 0.5.

7.5.4 (i) $k = 24$;   (ii) 5/16;   (iii) 1/16;   (iv) 5/16;   (v) $f_Z(z) = 4(1-z)^3$ for $0 < z < 1$, and 0 otherwise.

CHAPTER 8

8.2.2 $\sigma\sqrt{2/\pi}$.

8.2.4 $\sqrt{\pi/2}$.

8.2.6   3.36.

8.2.10 $E(X) = (b^2 - a^2 + d^2 - c^2)/[2(b - a + d - c)]$.

8.4.2   (i) $a = 23/66, b = 5/66$;   (ii) $a = 5/24, b = 7/48$.

8.4.4   (i) 0;   (ii) 1/3;   (iii) 31/3;   (iv) 2.

8.4.6 $1/(n + 1)$.

8.4.8   15.

8.5.2 $m_X(t) = (e^{nt} - 1)/[n(e^t - 1)]$ for $t \neq 1$, and $m_X(t) = 1$ for $t = 0$.

8.5.4 λ^4.

8.5.8 $(1 - 2p)/\sqrt{p(1 - p)}$.

8.5.10   Skewness is $\lambda^{-1/2}$, kurtosis is $3 + 1/\lambda$.

8.6.2 $\varphi_X(t) = \exp\{i\mu t - \sigma^2 t^2/2\}$.

8.6.4 $\varphi_{X-X'}(t) = |\varphi(t)|^2$.

8.7.2   0.7977.

8.7.4   (i) $164 - 30c$;   (ii) $(c - 2)/(2\sqrt{3})$;   (iii) $-1.464 \leq c \leq 5.464$.

8.7.6   (i) -1.375;   (ii) -27.50;   (iii) $Var(X) = 3124.61, Var(Y) = 62343.75$ (iv) $\rho_{XY} = -0.00056$.

8.7.8 $0, 1 - \rho^2, 0$.

8.7.10 $E(X) = a, E(Y) = 0.5, Var(X) = a^2/6, Var(Y) = 1/12, \rho = \sqrt{2}/2$.

CHAPTER 9

9.2.2 r/n.

9.3.2   0.5.

9.3.4 $p < 0.5$.

9.3.6 $\rho(X_1, X_1 + X_2) = \sqrt{n_1(n_1 + n_2)}$.

9.4.2 $\gamma = 1 - 100M/[C(M + 1)], \pi = M/(1 + M); \alpha_2 = 100M^2/[C(M + 1)^2]$.

9.4.4 $P(X \leq x) = 1 - q^{[x]+1}$.

9.5.2 $p_{k,r} = \sum_{j=0}^{3-k} \binom{j+3-r}{j} p^{4-r} q^j$.

9.6.2 $E(X) = 6/5, Var(X) = 9/25$.

9.7.2 $P(X$ is even$) = 0.5 + 0.5e^{-2\lambda}$.

9.7.6   (i) 0.4164;   (ii) 0.1677.

9.7.8   (i) 0.1094;   (ii) 0.0042.

9.7.12   (i) 0.0588;   (ii) 0.1567.

9.8.2 $7n/30$.

9.8.4 $11T/30$.

9.9.2 $f_{X_1-X_2}(t) = 0.5\lambda e^{-\lambda|t|}$ for $-\infty < t < \infty$.

9.9.4 $F_T(t) = 1 - e^{-t\sum_{i=1}^{5}\lambda_i}, f_T(t) = (\sum_{i=1}^{5}\lambda_i)e^{-t\sum_{i=1}^{5}\lambda_i}, t > 0$.

9.9.6 $F_T(t) = \{1 - [1 - (1 - e^{-\lambda_3 t})(1 - e^{-\lambda_4 t})] \times e^{-\lambda_2 t}\} \times (1 - e^{-\lambda_1 t}e^{-\lambda_5 t})$.

9.9.8 $E(T) = (r+1)^{1/(r+1)}\Gamma[1 + 1/(r+1)], r > -1$.

9.10.2 (i) 0.583; (ii) -1.035; (iii) 1.96; (iv) 1.549.

9.10.4 $E(X) = 10.72$.

9.10.6 (i) 0.3520; (ii) 0.0793.

9.10.8 About 48 feet above the average level.

9.11.2 $\Gamma(\alpha + \beta)\Gamma(k + \alpha)\Gamma(m + \beta)/[\Gamma(\alpha)\Gamma(\beta)\Gamma(k + m + \alpha + \beta)]$.

9.11.4 $\alpha = \beta = (1/k^2 - 1)/2$, provided $|k| < 1$.

CHAPTER 10

10.2.4 0.3679, 0.6262.

10.5.2 $a = 11.15, b = 1.24$.

10.5.4 $n \geq 10$.

10.5.6 $n \geq 8$.

10.5.8 0.9875.

10.5.10 $\chi^2_{0.05,47} = 62.95, \chi^2_{0.01,47} = 69.59, \chi^2_{0.05,113} = 137.73, \chi^2_{0.01,113} = 148.03$.

CHAPTER 12

12.2.2 $T = \sqrt{(3/n)\sum_{i=1}^{n}X_i^2}$.

12.3.4 No.

12.4.4 $b = 1 - a, a = \sigma_2^2/(\sigma_1^2 + \sigma_2^2)$.

12.4.6 $U_n = \{\# \text{ of } X_i's \text{ among } X_1,\ldots,X_n \text{ such that } X_i \leq x\}/n$, $Var(U_n) = \theta(1 - \theta)/n$.

12.4.8 (ii) $MSE_\theta(\overline{X}) = 1/(3n), MSE_\theta[(U + V)/2] = 2/[(n + 1)(n + 2)]$.

12.4.10 $k = 1/(n + 1)$.

12.5.2 (i) $I_{X+Y}(\mu) = 2/\sigma^2, I_{X+Y}(\sigma^2) = 1/(2\sigma^4)$; (ii) $I_{X-Y}(\mu) = 0, I_{X-Y}(\sigma^2) = 1/(2\sigma^4)$

12.5.4 $\overline{X}$.

12.5.6 $\gamma_1(\mu, \sigma^2) = \sigma^2/n, \gamma_2(\mu, \sigma^2) = \mu$.

12.6.2 $T_a = b\overline{X}, T_b = a/\overline{X}$.

12.6.4 MLE of the total number of matches is $(a_3 + a_4 + a_5)/Q(\hat{p})$, MLE of the number of matches lost by A is $\hat{l} = (a_3 + a_4 + a_5)[1 - Q(\hat{p})]/Q(\hat{p})$, where $Q(p) = p^3(10 - 15p + 6p^2)$.

12.6.6 (i) $k/(k + 1)$ (ii) $(k - 1)/k$.

12.6.8 $e^{-\overline{X}}$.

12.6.10 0.5.

12.6.12 (i) $\overline{X}$ (ii) MLE does not exist.

12.6.14 $\sqrt{(1/n)\sum(X_i - \mu)^2}$.

12.6.16 $[\min(X_1,\dots,X_n) + \max(X_1,\dots,X_n)]/2$.

12.6.18 0.0587.

12.6.20 $\hat{\mu}_1 = \overline{X}, \hat{\mu}_2 = \overline{Y}, \hat{\sigma}^2 = [\sum_{i=1}^{m}(X_i - \overline{X})^2 + \sum_{i=1}^{n}(Y_i - \overline{Y})^2]/(m + n)$.

12.8.2 $\max(X_1,\dots,X_n)$.

12.8.6 $f(x;\alpha,\beta) = \{\Gamma(\alpha+\beta)/[\Gamma(\alpha)\Gamma(\beta)]\}\exp\{(\alpha-1)\log x + (\beta-1)\log(1-x)\}$.

12.8.8 $f(x;\alpha,\lambda) = [\lambda^\alpha/\Gamma(\alpha)]\exp\{(\alpha-1)\log x - \lambda x\}$. $T_1 = \prod X_i$, and $T_2 = \sum X_i$ are minimal jointly sufficient.

12.9.2 (i) 1.117 ± 0.035 and 1.117 ± 0.053 (ii) 1.117 ± 0.016 and 1.117 ± 0.020 (iii) 1.117 ± 0.026 and 1.117 ± 0.034 (iv) $[0.00068, 0.00520]$ and $[0.026, 0.072]$.

12.9.4 $E(L_\alpha^2) = 4\sigma^2[t_{\alpha/2,n-1}]^2/n$.

12.9.6 $P\{\chi_{24}^2 < 35.179\}$.

12.9.10 $\overline{x} \pm z_{\alpha/2}\sqrt{\overline{x}/n}$.

12.9.12 (i) $[0.00058, 0.00365]$; (ii) $[274.05, 1728.62]$; (iii) the same as in (i); (iv) $[0.0049, 0.4294]$.

12.9.14 (i) $[0.428, 1.081]$ (ii) $[0.856, 5.459]$.

CHAPTER 13

13.2.2 (i) $\pi_a(\theta) = \pi_1(\theta)\pi_2(\theta)\pi_3(\theta)$, $\pi_b(\theta) = \pi_1(\theta)\pi_2(\theta)\pi_3(\theta) + \pi_1(\theta)\pi_2(\theta)[1 - \pi_3(\theta)] + \pi_1(\theta)[1 - \pi_2(\theta)]\pi_3(\theta) + [1 - \pi_1(\theta)]\pi_2(\theta)\pi_3(\theta)$, $\pi_c(\theta) = 1 - [1 - \pi_1(\theta)][1 - \pi_2(\theta)][1 - \pi_3(\theta)]$; (ii) A is always better than any test C_i, B is better than any test C_i if $0 < \pi(\theta) < 0.5$, C is always worse.

13.2.4 (i) 0.671; (ii) $\pi(\theta) = 1$ for $0 \le \theta \le 0.671$, $\pi(\theta) = 0.45/\theta^2$ for $\theta > 0.671$.

13.2.6 (i) $\pi(r) = \binom{5-r}{k}/\binom{5}{k}$ for $r \le 5 - k$; (ii) $\pi(r) = (1 - r/5)^k$.

13.3.2 X- the number of failures preceding r-th success. Reject H_0 if $X \ge 7$, accept H_0 if $X \le 5$, while for $X = 6$ reject H_0 with probability 0.5413. $\beta = 0.7837$.

13.3.4 Reject H_0 if $2\lambda_0 \sum X_i \le \chi_{0.05,2n}^2$, $\pi(\lambda) = P\{\chi_{2n}^2 \le (\lambda/\lambda_0)\chi_{0.05,2n}^2$. Reject H_0 if $X_{1:n} < -(\log 0.95)/(\lambda_0 n)$, $\pi_{X_{1:n}} = 1(0.95)^{\lambda/\lambda_0}$.

13.3.6 $1 - r(r - 1)/30$.

13.3.8 Reject H_0 if c or b observed.

13.4.2 (i) Reject H_0; (ii) 0.0028; (iii) 0.059.

13.4.4 $H_1 : p > 0.6$, p-value $= 0.072$.

13.4.6 $c = 3.71$.

13.4.8 (i) $T_1(\mathbf{X}) = \prod X_i$, reject H_0 on large values of T_1; (ii) $T_2(\mathbf{X}) = \sum X_i$, reject H_0 on small values of T_2.

13.4.10 The UMP test rejects H_0 if $\sum X_i \geq r$, where r is the smallest integer such that $\sum_{x=1}^{r} C_0[\theta_0/(\theta_0 + 1)]^x \geq \alpha$.

13.5.2 Reject H_0 if $|X - \lambda_0|/\sqrt{\lambda_0} > z_{\alpha/2}$.

13.6.2 0.536.

13.6.4 Do not reject H_0.

13.6.6 (i) Not significant at 0.05 level; (ii) $m < 24,940$.

13.8.2 $t = -1.987$, not enough evidence at $\alpha = 0.05$ that the polution level in lake B is higher.

CHAPTER 14

14.2.2 $T(Y_1, \ldots, Y_n) = (T_0, T_4)$, where $T_0 = \prod_{i=1}^{n} I_{Y_i \neq 0}, T_4 = \prod_{i=1}^{n} I_{Y_i \neq 4}$.

14.2.4 If $T_0 = 0$ predict $X = 1$, if $T_4 = 0$ predict $X = 3$. If $Y_1 = \cdots = Y_n = 2$, predict $X = 1$ if $p > 0.5$ and $X = 3$ if $p < 0.5$. If $p = 0.5$ any choice (1 or 3) is equally good.

CHAPTER 15

15.2.4 $a = E(Y) - b_1 E(X_1) - b_2 E(X_2), b_1 = (\sigma_Y/\sigma_X)(\rho_{X_1,Y} - \rho_{X_1,X_2} \times \rho_{X_2,Y})/(1 - \rho_{X_1,X_2}^2), b_2 = (\sigma_Y/\sigma_X)(\rho_{X_2,Y} - \rho_{X_1,X_2} \times \rho_{X_1,Y})/(1 - \rho_{X_1,X_2}^2)$.

15.2.6 $\hat{b} = \sum_{i=1}^{n} x_i y_i/(\sum_{i=1}^{n} x_i^2)$.

15.2.8 $0.335 + 0.475X$.

15.3.4 $\hat{\lambda} = \sum X_i/n, \hat{p} = \sum Y_i/\sum X_i$.

15.5.2 Reject H_0, $F = 451.14 > F_{0.05,4,6} = 4.53$.

15.6.2 $\hat{b}x_0 \pm t_{\gamma/2,n-1}\sqrt{1 + x_0^2/\sum x_i^2}\sqrt{\sum(y_i - \hat{b}x_i)^2}/\sqrt{n - 1}$.

15.6.4 (i) $[0.980, 1.061]$; (ii) $[1.014, 1.089]$; (iii) $[1.006, 1.076]$.

15.7.2 $13.646 < x_0 < 17.883$.

15.8.4 $\hat{a} = (1/n)(\sum y_i - b \sum x_i - c \sum x_i^2), \hat{b} = [S_{XY}S_{X^2X^2} - S_{XX^2}S_{X^2Y}]/[(S_{XX^2})^2 - S_{XX}S_{X^2X^2}], \hat{c} = [S_{XY}S_{XX^2} - S_{XX}S_{X^2Y}]/[(S_{XX^2})^2 - S_{XX}S_{X^2X^2}]$.

15.11.2 (i) $(S_1^2 + \ldots + S_k^2)/\sigma^2$ has chi-square distribution with $= n - k$ degrees of freedom; (ii) $[S_2^2/(n_k - 1)]/[S_1^2/(n_1 - 1)]$ has $F(n_k - 1, n_1 - 1)$ distribution; (iii) $x = 1/[1+m_1/(m_2 F_{0.01,m_2,m_1})]$, where $m_1 = n_3+n_4+n_5-3$ and $m_2 = n_1+n_2-2$.

15.11.4 The tests are equivalent.

15.12.2 $F_W = 2.94 < F_{0.01,3,6} = 10.72, F_D = 15.16 > F_{0.01,2,6} = 9.76$. There is no effect of initial weight, but the type of diet affects final results.

15.13.2 $F_A = 6.325 > F_{0.05,2,18} = 3.55, F_B = 38.855 > F_{0.05,2,18} = 3.55, F_{AB} = 4.970 > F_{0.05,4,18} = 2.93$—significant effect of the age, marijuana use and their interaction on the level of emotional maturity.

15.13.4 $F_{AB} = 1.428 < F_{0.05,2,12} = 3.89$—no interaction between the variety and geographical location.

CHAPTER 16

16.3.2 $\sqrt{n}D_n = 0.815$, p-value $= 0.52$, do not reject H_0.

16.4.2 Reject H_0 at 0.05 level.

16.4.4 $k > m \times (5.55 + 1.36\sqrt{1.85 + 16m})/(2m - 1.85)$.

16.5.2 (i) Do not reject H_0 at 0.05 level; (ii) Do not reject H_0 at 0.05 level.

16.5.4 $Z = 1.647$, significant at 0.1 level.

16.5.6 $Z = 5.668$ for Problems 16.4.3 and 16.4.4 , $Z = 4.808$ for Problem 16.4.5. Reject H_0 in all cases.

16.6.2 (i) $W_X = 364 + d$, $R = 309 + d$; (ii) $D_{20,10} = 0.35$ if $d \geq 13$, 0.40 if $d = 12$, and 0.45 if $d = 11$; (iii) H_1: median in population 1 exceeds that of population 2, $d \leq 42$ for $\alpha = 0.05$ (iv) The largest possible value of test statistic, 1.16, is not significant at 0.1 level.

16.6.4 (i) Range: $210 \leq W_X \leq 710$, critical region: $W_X \leq 590$; (ii) Range: $0 \leq R \leq 500$, critical region: $R \leq 381$; (iii) Range: $2 \leq R \leq 42$, critical region: $R \leq 23$; (iv) Range: $D_{20,51} = 26/51$, critical region: $\sqrt{20 \times 51/71}D_{20,51} > 1.36$.

CHAPTER 17

17.2.2 $Q^2 = 35.714$, p-value below 0.005.

17.2.4 $Q^2 = 1.307 < \chi^2_{0.1,4} = 7.779$. Do not reject H_0.

17.2.6 $Q^2 = 1.019 < \chi^2_{0.05,3} = 7.815$. Do not reject H_0.

17.3.2 $k \geq 14$.

17.3.6 $Q^2 = 0.069 < \chi^2_{0.1,1} = 2.706$.

17.4.2 $Q^2 = 33.686$, $G^2 = 35.249$.

17.6.2 p-value equals 0.5675.

17.7.2 For "gender-M/F" and "handedness-R/L" variables values of $\hat{\gamma}$ are: $-1, -0.6875, 1$. p-values for testing for negative association in the first two tables are 0 and 0.0583, p-value for testing for positive association in the third table is 0.

17.7.4 Values of statistics $Q^2, G^2, \hat{\gamma}, \tau_b$ and d are all 0 for original table. For the tables modified by addition we have respectively: 86.957, 96.696, 0.72, 0.50, 0.6; 68.38, 72.12, 0.52, 0.35, 0.4; 53.64, 54.99, 0.10, 0.06, 0.07; 41.67, 41.99, $-0.63, -0.38, -0.4$; 32.61, 37.21, $-0.56, -0.31, -0.45$; 20.12, 20.08, -0.07, $-0.04, -0.05$; 15.63, 13.06, 0.46, 0.23, 0.30.

Index

WILEY SERIES IN PROBABILITY AND STATISTICS

*Now available in a lower priced paperback edition in the Wiley Classics Library.

*Now available in a lower priced paperback edition in the Wiley Classics Library.

*Now available in a lower priced paperback edition in the Wiley Classics Library.

*Now available in a lower priced paperback edition in the Wiley Classics Library.